**CA 1964**

- DESTRUCT LINEAR-SHA...
- RACEWAY CAP
- SAFE & ARM (DESTRUCT)
- DESTRUCT LINEAR-SHAPED CHARGE
- SPIN FIN
- UMBILICAL CONNECTION POINTS
- INTERSTAGE II AND III
- SEPARATION JOINT
- SEPARATION JOINT
- GUIDANCE AND CONTROL SECTION
- SEPARATION JOINT
- INSTRUMENT WAFER (R & D ONLY)
- STAGE III MOTOR
- STAGE II MOTOR
- BASE HEAT DEFLECTOR
- SAFE & ARM (STAGE II)
- IGNITER (STAGE II)
- IGNITER (STAGE III)
- SAFE & ARM (IGNITER)
- INTERSTAGE I & II
- SAFE & ARM (DESTRUCT)
- NOZZLE CONTROL UNIT (STAGE III)
- DESTRUCT JET PENETRATOR
- EXTERNAL INSULATION

# MINUTEMAN

# MINUTEMAN

*A Technical History of the Missile
That Defined American Nuclear Warfare*

DAVID K. STUMPF

Foreword by
Lieutenant General Jay W. Kelley, USAF (ret.)

The University of Arkansas Press
Fayetteville
2020

Copyright © 2020 by The University of Arkansas Press. All rights reserved. No part of this book should be used or reproduced in any manner without prior permission in writing from the University of Arkansas Press or as expressly permitted by law.

ISBN: 978-1-68226-154-5
eISBN: 978-1-61075-735-5

Manufactured in the United States of America

25  24  23  22  21     5  4  3  2  1

Designed by Liz Lester

♾ The paper used in this publication meets the minimum requirements of the American National Standard for Permanence of Paper for Printed Library Materials Z39.48–1984.

Library of Congress Cataloging-in-Publication Data

Names: Stumpf, David K., 1953– author.
Title: Minuteman: a technical history of the missile that defined American nuclear warfare / David K. Stumpf; foreword by Lieutenant General Jay W. Kelley, USAF (ret.).
Other titles: Technical history of the missile that defined American nuclear warfare
Description: Fayetteville: The University of Arkansas Press, [2020] | Includes bibliographical references and index. | Summary: "Minuteman describes the technological breakthroughs necessary to field a weapon system that has served as a powerful component of the strategic nuclear triad for more than half a century."—Provided by publisher.
Identifiers: LCCN 2020019520 (print) | LCCN 2020019521 (ebook) | ISBN 9781682261545 (cloth) | ISBN 9781610757355 (ebook)
Subjects: LCSH: Minuteman (Missile)—History. | Intercontinental ballistic missile bases—United States—History.
Classification: LCC UG1312.I2 S775 2020 (print) | LCC UG1312.I2 (ebook) | DDC 358.1/754820973—dc23
LC record available at https://lccn.loc.gov/2020019520
LC ebook record available at https://lccn.loc.gov/2020019521

*To my parents, Paul K. and Ruth R. Stumpf,
who taught me to persevere.*

*To my in-laws, Dean S. and Nellie M. Pocock,
who inspired me to do things right the first time.*

# CONTENTS

| | | |
|---|---|---|
| | List of Figures | ix |
| | List of Tables | xv |
| | Air Force Acronyms | xix |
| | Foreword | xxv |
| | Preface | xxix |
| | Acknowledgments | xxxi |
| | Introduction | 3 |
| 1 | The Air Force and Strategic Missiles | 5 |
| 2 | Evolution of the Minuteman Force Levels | 17 |
| 3 | Minuteman in Context | 33 |
| 4 | Solid Propellant Comes of Age | 41 |
| 5 | From Polaris Came Minuteman | 55 |
| 6 | Siting and Facility Design | 83 |
| 7 | Construction and Activation | 125 |
| 8 | Motors and Airframe | 157 |
| 9 | Mark 5 and 11 Series Reentry Vehicles | 183 |
| 10 | Guidance and Control | 205 |
| 11 | Targeting Minuteman | 237 |
| 12 | Research and Development Flight Programs | 249 |
| 13 | Operational Flight and Evaluation Programs | 271 |
| 14 | Operational Base Missile Test Programs | 315 |
| 15 | Aspects of Command and Control | 337 |
| 16 | Keeping Pace: Modernization and Upgrades | 379 |
| 17 | Force Reduction | 399 |
| | Epilogue | 409 |

| | |
|---|---:|
| Appendix A: Construction, Acceptance, and Activation Summaries | 411 |
| Appendix B: Flight Test Programs | 421 |
| Appendix C: Operational Flight Test and Evaluation Programs | 429 |
| Appendix D: Airborne Launch Control Center Panels | 437 |
| Notes | 445 |
| Bibliography | 497 |
| Index | 519 |

# LIST OF FIGURES

| | | |
|---|---|---|
| 1.1 | ICBM base distribution from 1961 to 2019. | 12 |
| 1.2 | An Atlas F raised to the surface and ready to launch during a test exercise by 556 SMS, Plattsburg AFB, New York. | 13 |
| 1.3 | A "battleship" Titan I rises to the surface at the Operational Suitability Test Facility at Vandenberg AFB, California. | 14 |
| 1.4 | Titan II N-7 (61-2730) lifts off from Launch Complex 395-C, Vandenberg AFB, on 16 February 1963. | 15 |
| 2.1 | Minuteman force mix from 1962 to 2019. | 31 |
| 3.1 | Soviet missile base locations circa 1980. | 38 |
| 3.2 | Approximate ranges for Minuteman IA from Malmstrom AFB, Minuteman IB from Ellsworth AFB, and Minuteman III from Minot AFB. | 39 |
| 4.1 | Ercoupe NC286655 becomes airborne on 12 August 1941. | 45 |
| 4.2 | The Thunderbird test rocket at Inyokern, California, 25 November 1947. | 49 |
| 4.3 | Illustration of the burn pattern of a five-point-star grain. | 50 |
| 4.4 | The effect on specific impulse of the addition of powdered aluminum or magnesium to a polyvinyl chloride propellant with ammonium perchlorate as the oxidizer. | 53 |
| 5.1 | The only existing photograph of the Big Stoop flight test vehicle. | 58 |
| 5.2 | The progression, left to right, from the liquid-propellant Jupiter to the solid-propellant Jupiter-S and finally to the operational solid-propellant Polaris A-1. | 58 |
| 5.3 | A reproduction of the original Minuteman General Operational Requirement 171 illustration describing the operating requirements for the Minuteman missile. | 77 |
| 6.1 | Original prospective deployment areas for Minuteman outlined as Areas 1 through 4. | 85 |
| 6.2 | The location and distribution of the Minuteman missile flights at each wing. | 89 |
| 6.3 | Typical layout of a Wing II through V LF showing the security fence, the early design intrusion detection system antennas, and the service area. | 97 |
| 6.4 | Plot plans of the LFs of each of the wings. | 97 |
| 6.5 | Typical LF underground layout showing the two-level LER and launch tube. | 98 |
| 6.6 | The three variations of missile support or suspension systems. | 100 |
| 6.7 | LER upper level. | 104 |
| 6.8 | LER lower level. | 105 |
| 6.9 | PAS cross-section. | 106 |

| | | |
|---|---|---|
| 6.10 | Launcher closure door tracks. | 107 |
| 6.11 | Launcher closure door revetment at LF-02, Vandenberg AFB. | 109 |
| 6.12 | LSB, Wings I and II. | 110 |
| 6.13 | LSB, Wing VI and 564 SMS. | 111 |
| 6.14 | Aerial view of LCF Alpha-01, 10 SMW, 341 SMW, Malmstrom AFB. | 113 |
| 6.15 | LCC and LCEB positions relative to the LCF structure. | 113 |
| 6.16 | Typical site plan for Wings I and II LCF showing location of the LCC relative to the aboveground structures. | 114 |
| 6.17 | Typical site plan for Wings III through V LCF. | 114 |
| 6.18 | Typical site plan for Wing VI and 564 SMS LCF. | 115 |
| 6.19 | Floor plan of LCC Delta-01, Ellsworth AFB, circa 1991. | 116 |
| 6.20 | Sectional views of LCC Delta-01, Ellsworth AFB, circa 1991. | 117 |
| 6.21 | Details of a WS-133B LCC. | 119 |
| 6.22 | Cutaway drawing of LCEB Oscar-01, Whiteman AFB, circa 1991. | 120 |
| 6.23 | Floor plan and sectional view of LCEB Oscar-01, Whiteman AFB, circa 1991. | 121 |
| 6.24 | Cutaway drawing of typical WS-133B LCEB. | 123 |
| 7.1 | The lower launch tube was 52 feet long and 12 feet in diameter; the cylinder was fabricated from 0.25-inch plate steel. | 133 |
| 7.2 | The lower launch tube insertion is nearly complete at a Delta Flight LF. | 133 |
| 7.3 | The LER foundation is complete, and the interior 0.25-inch steel plate liner sections for the LER are being placed at a Delta Flight LF. | 135 |
| 7.4 | Rebar hoops are being installed on the exterior of the wall of the LER. | 136 |
| 7.5 | Beginning of the concrete pour for the launcher closure door and surround. | 137 |
| 7.6 | Workers prepare the sealing surface for the PAS hatch. | 138 |
| 7.7 | LCC capsule upper liner under construction at Grand Forks AFB LCC Oscar-0. | 140 |
| 7.8 | Wing VI LCC Golf-0 blast door installation. | 141 |
| 7.9 | LCC Delta-01 capsule concrete forms being placed, Ellsworth AFB. | 142 |
| 7.10 | LCC concrete pour is complete, showing the access shaft and intake and exhaust air delay lines. | 143 |
| 7.11 | Wings III through V LCEB rebar placement. | 145 |
| 7.12 | Wing V geology permitted the use of large augers to excavate most of the LF shafts. | 148 |
| 7.13 | 15-foot-diameter auger in operation. The uniformly smooth walls of the launch tube excavation, which greatly simplified construction at F. E. Warren AFB. | 149 |
| 8.1 | Examples of the jetevator and axial nozzle design used in the Navy's Polaris A-1 (both stages) and A-2 (second stage) fleet ballistic missiles, respectively. | 159 |
| 8.2 | The offset nozzle simulator used an Aerojet 5KS4500 designed to approximate the exhaust flow typical in one quadrant of the four-nozzle Minuteman Stage I motor. | 162 |

| | | |
|---|---|---|
| 8.3 | Design detail of Minuteman Stage I nozzle. | 163 |
| 8.4 | Cutaway illustration of Minuteman I Stage I motor. | 163 |
| 8.5 | All three stages of the Minuteman IA and IB used four steerable nozzles for thrust vector control through differential gimballing of pairs of nozzles. | 164 |
| 8.6 | Cutaway illustration of Minuteman II Stage II. | 167 |
| 8.7 | Minuteman II and III Stage II as well as Minuteman III Stage III LITVC. | 169 |
| 8.8 | Filament winding pattern similar to that used for the fabrication of Minuteman Stage III motor casings. | 171 |
| 8.9 | Minuteman I and II Stage III. | 171 |
| 8.10 | Aft view of post-boost system rocket engine and identification of components. | 177 |
| 8.11 | Minuteman variant comparison (best dimensions available). | 180 |
| 9.1 | Wind tunnel Schlieren photographs illustrating the detached bow shock wave generated by a blunt reentry vehicle body, compared to the attached shock wave with a pointed reentry body. | 184 |
| 9.2 | Short-, medium-, and long-range ballistic missile trajectory characteristics. | 185 |
| 9.3 | The General Electric Mark 2 reentry body components and the Avco Mark 2 version, which never flew but was a competing contract to General Electric. | 187 |
| 9.4 | The RVX-1 components. | 193 |
| 9.5 | The recovered Avco RVX-1–5 on display at Avco facilities. | 194 |
| 9.6 | Comparison of the Avco Mark 4 Mod 1, General Electric Mark 3 Mod IX, and Mark 3 Mod IIB. | 197 |
| 9.7 | Comparison of the size and shape of the Avco Mark 5 and the Mark 4. | 199 |
| 9.8 | Major components of the Mark 5 flight test vehicle. | 199 |
| 9.9 | The nose tip of the Mark 5 reentry vehicle with the Avcoat coating removed showing the cells of Avcoite ceramic material that took the brunt of the reentry heating. | 200 |
| 9.10 | Effect of reentry on heat shield material of a recovered Mark 5 reentry vehicle. | 200 |
| 9.11 | Comparison of Minuteman IA and IB reentry vehicles. | 203 |
| 9.12 | The deployment history of Minuteman Mark 5, Mark 11 series, and Mark 12 reentry vehicles, 1962 to 1975. | 203 |
| 10.1 | The exploded-view diagram of the NS10 stable platform identifying system components. | 210 |
| 10.2 | Minuteman I NS10Q series stable platform. | 210 |
| 10.3 | The major components of the Minuteman I NS10 guidance system. | 211 |
| 10.4 | Diagram of an early model of the Autonetics G6B4 two-axis, free-rotor, gas-bearing gyroscope showing major components. | 212 |
| 10.5 | Cross-section of the operational G6B4 gyroscope. | 212 |

| | | |
|---|---|---|
| 10.6 | Stable platform for the Minuteman II NS17 and Minuteman III NS20 missile guidance systems. | 222 |
| 10.7 | Comparison of the size of the Minuteman II and III D37 and Minuteman I D17B computers. | 224 |
| 10.8 | The Minuteman II missile guidance system was configured to be housed in the same missile section as that of Minuteman I, which had a height of 31.5 inches. | 228 |
| 10.9 | Minuteman III NS20 missile guidance system. | 233 |
| 11.1 | Members of a CTT measuring the angle (Angle A) between two external alignment monuments as part of the RMAV process. | 240 |
| 11.2 | Targeting and alignment procedures diagram for establishing reference mirror azimuths using external reference monuments. | 241 |
| 11.3 | Typical mechanical relationship between the installed missile and optical alignment equipment on LER Level 1 for aligning Minuteman I and II guidance sets. | 242 |
| 11.4 | Two members of the Targeting and Alignment Team, A1C Duane Bowser and A1C Ronald Sammons, verify the secondary reference mirror azimuth alignment prior to positioning the collimator. | 243 |
| 12.1 | Aerial view of the Edwards AFB Minuteman silo launch test facilities. | 250 |
| 12.2 | Horizontal in-silo launch test structure at Edwards AFB, September 1959. | 251 |
| 12.3 | One-third-scale model used in the horizontal test structure. STM 101 being lowered into the silo launch test facility at Edwards AFB. | 252 |
| 12.4 | Interior configuration of in-ground test silo facility at Edwards AFB. | 253 |
| 12.5 | The first tethered launch on 15 September 1959 was highly successful. | 254 |
| 12.6 | Configuration of pad LFs at Patrick AFB, Florida, circa 1961. | 256 |
| 12.7 | Minuteman IA FTM 401 at the instant of Stage I ignition on 1 February 1961, LC-31A. | 258 |
| 12.8 | Unsuccessful launch of FTM 404, the first full-flight test of in-silo launches at LC-32B on 30 August 1961. | 259 |
| 12.9 | Impact plot for the first 9 successful Minuteman IA launches out of 15 attempts. | 261 |
| 12.10 | The first Minuteman II flight test missile, FTM 449, undergoing final preparation. | 265 |
| 13.1 | Map of the locations of the Minuteman LFs and LCFs at Vandenberg AFB, California. | 273 |
| 13.2 | Kwajalein Atoll. | 275 |
| 13.3 | Eniwetok Atoll. | 276 |
| 13.4 | Impact plot for the Minuteman IB Follow-on Operational Test Program. | 291 |
| 13.5 | Modifications made to Vandenberg LF-06 and LF-03 in support of the Reentry System Launch Program. | 294 |

| | | |
|---|---|---|
| 13.6 | The Casmalia Express, Minuteman IB TDT-1, shortly after destruction by the range safety officer on 15 June 1993. | 295 |
| 13.7 | The configuration of the Phoenix Islands target area. | 304 |
| 13.8 | FTM 202 receives its reentry vehicle shroud on 16 October 1968 at the Cape. SSgt. Stephen Kravitsky inspects a Minuteman III at the 321 SMW in 1989. | 307 |
| 13.9 | Minuteman III Dust-Hardening Program design change detail. | 308 |
| 13.10 | Cloud sampling pattern by WB-57-F aircraft minutes prior to reentry vehicle impact for flights PVM-3 and PVM-4. | 311 |
| 13.11 | Salvo launch of Minuteman III Glory Trip 40GM and 68GM from LF-09 and LF-08, respectively, on 10 July 1979. | 313 |
| 14.1 | Programmed trajectory profile for Long Life I. | 317 |
| 14.2 | Primary launch crew for Long Life I. | 318 |
| 14.3 | Profile of the OBLSS. | 323 |
| 14.4 | Proposed trajectory and target area for the first launch of GIANT PATRIOT Operational Base Launch Program. Flight profile showing probable impact distance from the LF for the Stage I motor casing and interstage panels. | 327 |
| 14.5 | Profile of a Minuteman II missile modified for a GIANT PACE SELM test. | 332 |
| 14.6 | The MOMS or SELM test could include the opening of the launcher closure using the ballistic gas generators as in an operational launch. | 335 |
| 15.1 | 341 SMW LCC India-01, May 1963, WS-133A. MCCC's Control Console. | 344 |
| 15.2 | 341 SMW LCC India-01, May 1963, WS-133A. DMCCC's missile combat crew Communications Console. | 345 |
| 15.3 | Minuteman II MCCC's Launch Control Console, circa 1991, WS-133A-M, Whiteman AFB. | 348 |
| 15.4 | Minuteman II DMCCC's Communication Console, circa 1991, WS-133A-M, Whiteman AFB. | 349 |
| 15.5 | 564 SMS MCCC's Command Console, WS-133B. | 350 |
| 15.6 | 564 SMS DMCCC's Status Console, WS-133B. | 351 |
| 15.7 | REACT-A Console at LCC November-01, F. E. Warren AFB. | 355 |
| 15.8 | The 13 May 1963 launch of MER 203 from Naval Missile Facility Point Arguello Launch Complex A. | 357 |
| 15.9 | ERCS equipment additions to Minuteman facilities. | 359 |
| 15.10 | ERCS flight profile. | 362 |
| 15.11 | ERCS signal coverage. | 364 |
| 15.12 | ALCS, EC-135 aircraft. | 365 |
| 15.13 | ALCC MCCC-A Launch Control Panel illustration. | 367 |
| 15.14 | ALCC DMCCC-A Launch Monitor Panel illustration. | 368 |
| 16.1 | Minuteman LF scale model used in HEST program at Kirtland AFB, New Mexico. | 381 |

| | | |
|---|---|---|
| 16.2 | Illustration of the sequential detonation of explosives to create the moving shock wave. | 381 |
| 16.3 | Layout of the first HEST at F. E. Warren AFB. | 382 |
| 16.4 | Preparation of layers of Primacord explosives for HEST II showing specific angle on wooden frames. | 383 |
| 16.5 | HEST III took place on 22 September 1966 at Grand Forks AFB. | 384 |
| 16.6 | Launcher closure upgrade installation details. | 389 |
| 16.7 | The current missile suspension system being lowered into an LF during the Rivet MILE program. | 390 |
| 16.8 | The major modifications to the shock-isolated platform in the upper level of the LER. | 391 |
| 17.1 | Typical LF site dismantlement cross-section. | 403 |
| 17.2 | The lower launch tube ready for concrete cap placement. | 404 |
| 17.3 | The launcher closure door is pulled from the top of LF Romeo-29 on 25 February 2014 and buried in a nine-foot-deep hole. | 407 |

# LIST OF TABLES

| | | |
|---|---|---|
| Table 1.1 | Minuteman Program Correlation Chart | 16 |
| Table 2.1 | Air Force Planned Minuteman Force Levels, 1960 | 19 |
| Table 2.2 | Department of Defense and Air Force Minuteman Force Levels, 23 September 1961 | 20 |
| Table 2.3 | Department of Defense and Air Force Minuteman Programmed Force Levels, 21 November 1962, FY 1963 to 1968 | 21 |
| Table 2.4 | Department of Defense and Air Force Minuteman Programmed Force Levels, 6 December 1963 | 21 |
| Table 2.5 | Minuteman Force Levels, Soviet Fatalities, and Industrial Destruction | 22 |
| Table 2.6 | Department of Defense and Air Force Minuteman Programmed Force Levels, 3 December 1964, FY 1966 to 1970 | 22 |
| Table 3.1 | Minuteman Payload Summary | 37 |
| Table 3.2 | Minuteman Range, Accuracy, and Reliability Estimates, 1970 | 37 |
| Table 5.1 | Status of Solid Propulsion Systems, February 1958 | 68 |
| Table 5.2 | Minuteman General Characteristics, February 1958 | 69 |
| Table 5.3 | Minuteman R&D Cost Estimates (in Millions), February 1958 | 69 |
| Table 5.4 | ICBM Force Objectives and Estimated Dollar Requirements | 70 |
| Table 6.1 | US Programmed Forces and Estimated Soviet Threat, 1961 and 1965 | 93 |
| Table 6.2 | Characteristics of Basing Systems with Recommended Separation Distances | 93 |
| Table 6.3 | Weapon Effects for a 2-, 5-, and 30-Megaton Air or Surface Burst | 94 |
| Table 6.4 | Minuteman Facility Hardness and Survivability Criteria, 1964 | 95 |
| Table 7.1 | Minuteman Wing Construction Timeline | 126 |
| Table 7.2 | Minuteman Launch Facility and Launch Control Facility Nomenclature | 129 |
| Table 7.3 | Minuteman Wing Geographical Statistics | 131 |
| Table 7.4 | CEBMCO Start and Completion Dates | 131 |
| Table 7.5 | Summary of Construction Material Requirements for Operational Facilities for Wing VI, Grand Forks AFB | 144 |
| Table 7.6 | Construction Contracts WS-133A and WS-133B Operational Facilities | 150 |
| Table 8.1 | Summary of Minuteman Motor Designations, 1968 | 164 |
| Table 8.2 | Minuteman I through III Stage I M55A1 Motor Specifications | 165 |
| Table 8.3 | Minuteman Stage II Motor Specifications | 169 |
| Table 8.4 | Minuteman Stage III Motor Specifications | 173 |
| Table 8.5 | Minuteman Missile Ablative Coating Configurations | 178 |
| Table 8.6 | Minuteman Airframe Production | 179 |

| | | |
|---|---|---|
| Table 9.1 | Air Force Reentry Vehicle Designators through Minuteman III | 184 |
| Table 10.1 | NS10Q2 Operation during Flight, Minuteman IB Trajectory | 217 |
| Table 13.1 | Minuteman Launch Facilities at Vandenberg AFB: Total Launches | 272 |
| Table 13.2 | Major Air Force Western Test Range Target Areas, 1980 | 274 |
| Table 13.3 | Typical Minuteman IA Flight Sequence | 280 |
| Table 16.1 | Summary of Minuteman Force Composition by Wing Resulting from the Force Modernization Program | 387 |
| Table 16.2 | Integrated Improvement Program Dates | 393 |
| Table 16.3 | Selected Minuteman Modification Completion Dates | 398 |
| Table 17.1 | Minuteman Deployment and Deactivation Summary | 407 |
| Table A.1 | 341 SMW, Malmstrom AFB, Construction and Acceptance Milestone Dates | 411 |
| Table A.2 | 44 SMW, Ellsworth AFB, Construction and Acceptance Milestone Dates | 412 |
| Table A.3 | 455 SMW, Minot AFB, Construction and Acceptance Milestone Dates | 413 |
| Table A.4 | 351 SMW, Whiteman AFB, Construction and Acceptance Milestone Dates | 414 |
| Table A.5 | 90 SMW, F. E. Warren AFB, Construction and Acceptance Milestone Dates | 415 |
| Table A.6 | 321 SMW, Grand Forks AFB, Construction and Acceptance Milestone Dates | 416 |
| Table A.7 | 564 SMS, Malmstrom AFB, Construction and Acceptance Milestone Dates | 416 |
| Table A.8 | Activation Summary for the 341 SMW and 564 SMS, Malmstrom AFB | 417 |
| Table A.9 | Activation Summary for the 44 SMW, Ellsworth AFB, and 455 SMW, Minot AFB | 418 |
| Table A.10 | Activation Summary for the 351 SMW, Whiteman AFB, and 90 SMW, F. E. Warren AFB | 419 |
| Table A.11 | Activation Summary for the 321 SMW, Grand Forks AFB | 420 |
| Table B.1 | Minuteman IA Flight Test Program, Atlantic Missile Range, 1961–1963 | 421 |
| Table B.2 | Minuteman IB Flight Test Program, Atlantic Missile Range, 1962–1964 | 423 |
| Table B.3 | Minuteman II Flight Test Program, Atlantic Missile Range, 1964–1968 | 426 |
| Table B.4 | Minuteman III Flight Test Program, Air Force Eastern Test Range, 1968–1970 | 428 |
| Table C.1 | Minuteman IA Follow-on Operational Test Program, Vandenberg AFB, 1965–1966 | 429 |
| Table C.2 | Minuteman IB Demonstration and Shakedown Operation Program, Vandenberg AFB | 430 |
| Table C.3 | Summary of Successful Minuteman IB OT Mixed Marble II Exercises | 431 |

| | | |
|---|---|---|
| Table C.4 | Minuteman IB Operational Test Program, Mark 5 and Mark 11 Reentry Vehicle Recovery Flights to the Eniwetok Lagoon | 431 |
| Table C.5 | Minuteman IB Follow-on Operational Test Program Impact Results | 432 |
| Table C.6 | Reentry System Launch Program, Minuteman IB Flight List | 432 |
| Table C.7 | Wing VI Minuteman II Research and Development Flight Tests | 433 |
| Table C.8 | Wing VI Minuteman II Demonstration and Shakedown Operation Flight Tests | 434 |
| Table C.9 | Wings I, II, IV Minuteman II Force Modernization Flight Tests | 435 |
| Table C.10 | Minuteman II Operational Test and Special Test Program Summary | 436 |

# AIR FORCE ACRONYMS

| | |
|---|---|
| A&E | architect–engineer |
| A&T | assembly and test |
| ABL | Allegany Ballistics Laboratory |
| ABMA | Army Ballistic Missile Agency |
| ABM | antiballistic missile |
| ABNCP | Airborne National Command Post |
| ABRES | Advanced Ballistic Reentry System |
| ACCS | Airborne Command Control Squadron |
| ACIC | Aeronautical Chart and Information Center |
| AEC | Atomic Energy Commission |
| AEDC | Arnold Engineering Development Center |
| AFB | Air Force Base |
| AFBMC | Air Force Ballistic Missile Committee |
| AFBMD | Air Force Ballistic Missile Division |
| AFFTC | Air Force Flight Test Center |
| AFMTC | Air Force Missile Test Center |
| AFSC | Air Force Systems Command |
| AIRS | Advanced Inertial Reference Sphere |
| ALCC | Airborne Launch Control Center |
| ALCS | Airborne Launch Control System |
| AMC | Air Materiel Command |
| AMR | Atlantic Missile Range |
| AMSA | Advanced Manned Strategic Aircraft |
| AR | aircraft rocket |
| ARC | Atlantic Research Corporation |
| ARDC | Air Research and Development Command |
| ARSIP | Accuracy, Reliability, Supportability, Improvement Program |
| ARTV | advanced reentry test vehicle |
| BMC | Ballistic Missile Committee |
| CDB | command data buffer |
| CE | circular error |
| CEBMCO | Corps of Engineers Ballistic Missile Construction Office |
| CEP | circular error probable |
| CQAP | Component Quality Assurance Program |
| CSD-M | Command Signal Decoder-Missile |
| CTT | Combat Targeting Team |
| DDA | digital differential analyzer |
| DDRS | declassified documents reference system |
| DEFCON | defense (readiness) condition |
| DIABLO | Determination of Impact of Autonetics Boundary on Logistics and Operation |
| DMCCC | deputy missile combat crew commander |

| | |
|---|---|
| DMCCC-A | deputy missile combat crew commander-airborne |
| DoD | Department of Defense |
| DOT&E | director of testing and evaluation |
| DTU | data transfer unit |
| DX | Defense Extraordinary Priority |
| EAM | Emergency Action Message |
| ECC | emergency combat capability |
| EEP | expanded execution plans |
| EIS | environmental impact statement |
| ELC | emergency launch capability |
| EMP | electromagnetic pulse |
| EO | executive order |
| ERCS | Emergency Rocket Communications System |
| ESA | electrical surge arrester |
| ETR | Eastern Test Range |
| EWO | emergency war order |
| FDE | Flight Development Evaluation |
| FY | fiscal year |
| G&G | geodetic and geophysical |
| GALCIT | Guggenheim Aeronautical Laboratory, California Institute of Technology |
| GBSD | ground-based strategic deterrent |
| GCA | gyrocompass assembly |
| GIP | Guidance Improvement Program |
| GMG | guided missile group |
| GMR | ground maintenance response |
| GOR | general operating requirement |
| GRAB | Galactic Radiation and Background |
| GRP | Guidance Replacement Program |
| GUP | Guidance Upgrade Program |
| HAC/RMPE | Higher Authority Communication/Rapid Message Processing Element |
| HEFP | Hybrid Explicit Flight Program |
| HEST | high-explosive simulation technique |
| HETF | Hill Engineering Test Facility |
| HF | high frequency |
| HRS | Hard Rock Silo |
| HST | heading sensitivity test |
| IAW | in accordance with |
| IBMS | intercontinental ballistic missile |
| ICBM | intercontinental ballistic missile |
| IDF | integrated demonstration flight |
| ILCS | Improved Launch Control System |
| IMPSS | Improved Minuteman Physical Security System |
| IOC | initial operational capability |
| IRBM | intermediate-range ballistic missile |
| JATO | jet-assisted takeoff |

| | |
|---|---|
| JCS | Joint Chiefs of Staff |
| JPL | Jet Propulsion Laboratory |
| LCC | launch control center |
| LCEB | launch control equipment building |
| LCF | launch control facility |
| LCSB | launch control support building |
| LECG | Launch Enable Control Group |
| LEPS | launcher environmental protection system |
| LER | launcher equipment room |
| LES | launch enable switch |
| LF | launch facility |
| LITVC | liquid-injection thrust vector control |
| LSB | launcher support building |
| MCCC | missile combat crew commander |
| MCCC-A | missile combat crew commander-airborne |
| MCU | mechanical code unit |
| MESP | Minuteman Extended Survivable Power |
| MG | missile group |
| MGS | missile guidance set |
| MICCS | Minuteman Command Control System |
| MILE | Minuteman Integrated Life Extension |
| MILS | Missile Impact Locator System |
| MIMS | missile maintenance squadron |
| MIRV | multiple independently targetable reentry vehicles |
| MIT | Massachusetts Institute of Technology |
| MMT | missile maintenance team |
| MOMS | modified operational missiles |
| MPT | Missile Procedures Trainer |
| MRSS | Mobile Range Safety System |
| MRT | minimum reaction time |
| MRV | multiple reentry vehicles |
| MTBF | mean time between failure |
| MTS | missile training squadron |
| NACA | National Advisory Committee for Aeronautics |
| NCU | nozzle control unit |
| NOTS | Naval Ordnance Test Station |
| NRDC | National Research and Defense Council |
| NSC | National Security Council |
| OBL | operational base launch |
| OBLSS | Operational Base Launch Safety System |
| OGP | Operations Ground Program |
| OP | operations plan |
| ORDCIT | Ordnance, California Institute of Technology |
| ORT | operational readiness training |
| OSD | Office of the Secretary of Defense |

| | |
|---|---|
| OTE | operational test and evaluation |
| PACCS | Post-Attack Command-and-Control System |
| PAS | Personnel Access System |
| PBAA | polybutadiene acrylic acid |
| PBAN | polybutadiene acrylic acid–acrylonitrile polymer |
| PBCS | post-boost control system |
| PBPS | post-boost propulsion system |
| PBV | post-boost vehicle |
| PDM | programmed depot maintenance |
| PDS | Precision Deployment System |
| PIGA | pendulous integrating gyro-accelerometer |
| PRP | Propellant Replacement Program, Personnel Reliability Program |
| PSAT | perturbation self-alignment technique |
| PSRE | propulsion system rocket engine |
| QUAINT | quantized integrating torquer |
| RATO | rocket-assisted takeoff |
| REACT | Rapid Execution and Combat Targeting |
| ROC | required operational capability |
| RSLP | Reentry System Launch Program |
| RTV | reentry test vehicle |
| SAC | Strategic Air Command |
| SAMSO | Space and Missile Systems Organization |
| SATAF | Site Activation Task Force |
| SATCAL | self-alignment technique calibration |
| SCLC | simulated combat launch capability |
| SCRSAS | Self-Contained Range Safety Abort System |
| SELM | simulated electronic launch Minuteman |
| SERV | safety-enhanced reentry vehicle |
| SERV/W | safety-enhanced reentry vehicle/warhead |
| SETD | systems engineering and technical development |
| SIEGE | simulated electromagnetic ground environment |
| SIN | Support Information Network |
| SLBM | submarine-launched ballistic missile |
| SLEP | Service Life Extension Program |
| SMS | strategic missile squadron |
| SMW | strategic missile wing |
| SRV | single reentry vehicle |
| SSAS | software status authentication system |
| START | Strategic Arms Reduction Treaty |
| STL | Space Technology Laboratories |
| TBM | tactical ballistic missile |
| TORUS | transient omnidirectional radiating unidistant and static simulator |
| TRW | Thompson-Ramo-Wooldridge |
| UHF | ultra high frequency |
| USSTRATCOM | United States Strategic Command |

| | |
|---|---|
| VERDAN | versatile digital analyzer |
| VHF | very high frequency |
| VRSA | voice reporting signal assembly |
| VSA | vibrating string accelerometer |
| WDD | Western Development Division |
| WGS | World Geodetic System |
| WSA | weapon storage area |
| WSCE | weapon system controller equipment |
| WSEG | Weapons System Evaluation Group |
| WTR | Western Test Range |
| XSPV | experimental solid-propellant vehicle |

# FOREWORD

Minuteman! Just hearing or reading the word inspires thoughts of readiness, action, and commitment placed in our hearts and minds ages ago by our early founders and defenders. David Stumpf has presented us an awesome document about Minuteman, the weapon system, which proudly took the name and essence of those first to respond and act in defense of the few to enable a nation of the very many. As some would say, always leave things better than ya found 'em. Minuteman has done just that, and more. It has added a critical new dimension to the term "Minuteman" for America: strategic nuclear deterrence.

From Whiteman in Missouri to north of I-80 and way north of the Mason–Dixon Line. Only the best go north, only the chosen are frozen, and freezin's the reason! Remember that? Of course you do—it's true! Minuteman is in the heartland. On a farm, a ranch—yeah, in the outback, I suppose. You and I have heard and seen those farmers, ranchers, and townfolk talk about "their" missile site with pride! And they shed a tear too when their missile site was blown up and destroyed in the interest of arms control. As a former senior military advisor to the Arms Control and Disarmament Agency, I could tell you many stories in that regard.

We had MOBs (main operating bases), an LF (launch facility), an LCC (launch control center), and an LCF (launch control facility). At least it was an LCF until someone renamed it a MAF (missile alert facility). We drove out and back, many miles and hours. Remember we're talking about north of I-80 and what that means—snow! Years later, some of the distant sites were supported by helicopter. Great idea, except bad weather is bad for road vehicles *and* helicopters. Made no difference whether you were ops, maintenance, or cops, we all endured the same challenges. And for many years, we wore the same badge—the pocket rocket. Bad day, in my opinion, when we split up the career field with two different badges! We shared a common bond: ensuring Minuteman was good to go—responsive and effective.

Of course, we had our tests and challenges: the standboards, the ORI, the NSI, and the dreaded SMES (3901st). We were tested and evaluated continually, and that was good because we were damn good, and so was the weapon system! Minuteman and its ops and maintenance were evaluated at home in the LCC and LF and at Vandenberg AFB for an operational test launch. Pull the missile, send it to Vandenberg, modify for test, and launch on command. Could be from an LCC, or what's this? An airplane! A strength of the Minuteman system was the ability to launch from either an LCC or ALCS (Airborne Launch Control System). I recall quite well turning the key for an ALCS launch of an MM II. But there was so much more. And remember, in all these tests, we could launch one, or a ripple of two or more, or a salvo of several—and we did! Do you remember the effort to try an OT launch from Malmstrom AFB (Wing I)? Never happened because of where the first and second stages would land, but there was serious effort in working the details! And how about the seven-second launch from Ellsworth (Wing II)? This actually happened! We tested the LF and LCC through SELM (Simulated Electronic Launch Minuteman) and actually blew the door off in MOM (modified operational missile). It goes on and on. Some of us remember Combat Targeting Teams, code change, and rev change. And how about this—as a young captain, I was the SAC project officer for the MM III Mk 12 reentry system from R&D to first on alert. Can you imagine trying to explain to SAC flying generals—or anyone else for that matter—just how the MM III Mk 12 MIRV worked?

Most of us know all about campering a site—how long did it take to raise a B-plug? We sometimes forget another important military capability we developed: distant coordination. From LCC (the capsule, we called it), to security, to an LF many miles away for maintenance or a security alarm, all weather. All done with tight security and without visual contact among any of the participants. Awesome and aggravating all at the same time!

Some remember Curtain Raiser, many remember Olympic Arena, and most remember Guardian Challenge! But we all remember the awesome responsibility. At first intimidating, humbling, just the thought of what these weapons would do if ever launched. But soon, in the daily grind, distant from the LF and the weapon, we transitioned into a different mode of man and machine: just keep it operating. Green and white were good—any other color, not so good.

Our unit heritage goes way back to the bomb groups of WW II. Some LCCs had blast doors with nose art from B-17s and B-24s. Of course, others had more contemporary décor, but with relevance to mission, such as "Pizza Hut—we deliver hot in 30 minutes!"—with reference to warhead and missile flight time.

You remember it, I remember it: upon opening a blast door, the sounds and smells in the LF and LCF, and occasionally, the no-notice surprise of an evaluator. And we remember the roads. Rolling a TE (transporter-erector) or a PT (payload transporter) on those roads north of I-80 could be so very hazardous and dangerous. Remember the US Marshals leapfrogging from intersection to intersection? Unless you were there and had operated such equipment, it's hard to understand. I sat in a TE accident briefing to CINCSAC. The CINC asked how many in the room had ever driven an 18-wheeler? The only person holding up their hand was the CINC! And he had, as a young man back in the day, driven 18-wheelers. He believed the young airman did his best under the circumstances. It was a good day!

We remember two-man crews, 24-hour alerts, three-man crews, 72-hour alerts, and we remember when women came on the crew force—same-gender crews! Remember all the discussion? And we grew up! How many different uniforms did you wear on crew duty? White suit (coveralls), two-piece blues, green bag, blue bag, ABUs, and more!

The following is by my friend Col. USAF (ret.) Paul Murphy. Paul was a crew member, also a missile maintenance officer and later director of missile maintenance at HQ SAC. Read it and think big—you were part of this!

> ALERT
> (Col. Paul Murphy)
>
> Oh I have slipped beneath the surface dirt,
> Downward I've crept and not felt hurt.
> White-suited I've entered my capsule womb,
> And sat for hours in a concrete tomb.
> Chances are you've never seen my place,

> Nor done my job sitting in inner space.
> But all your flights soaring through the air,
> Would not be, were I not there.
> Minutes creep by in a slow, unending parade,
> Lights flick on, glow brightly, then fade.
> A missile sits in deadly rest;
> A key remains secure;
> Above me a world goes on because of
> Boredom I endure.

If you were on an ops crew, surely you recall boredom. Reckon it all depends on how you handled it. Could have been study for a postgrad degree, read a good book, accomplish some recurring training, or perhaps you worked to be the best crew MCCC or DMCCC. And you also recall how that boredom could change almost instantly to very meaningful and purposeful reaction.

Lest we think David's work an epitaph for Minuteman, think again! Minuteman is alive and well, thank you very much! Fewer in number, for sure, but then, they are far better than the early MM Is some were part of. The ops, maint, and cops that sustain her 24/7 are as good as or better than we were back in the day. So all you pocket rocket mafia stand tall, head high, and flag flying! No duty was ever more important to America's security. We just have to deal with the fact that not a lot of America even knows we were, and you are, there for them!

*Jay W. Kelley*
*Lt. Gen., USAF (ret.)*
*November 2019*

# PREFACE

I had originally intended to write my next book on the Advanced Ballistic Reentry System (ABRES) program. In March 2015 I called the Air Force Global Strike Command History Office to find out if there was any interest for such a book and if there was unclassified archival material available. Yancy Mailes, the historian at the time, said there was little, if any, unclassified material in their archive concerning ABRES. He then asked if I was the author of the Titan II history released in 2000, which he had enjoyed reading. If so, while there was no funding available from his office or the Department of Defense Legacy Program that had funded the Titan II book, a similar book on the Minuteman program was needed.

This piqued my interest as I recalled that during my research for declassified details on various launch programs for Titan II, I had come across detailed declassified information on Minuteman IA and IB operational flight programs. Further research located a declassified copy of Robert F. Piper's *The Development of the SM-80 Minuteman,* published in April 1962. Piper's history covered the Minuteman program from inception up to initial construction decisions but was, for the most part, not technical in nature.

There were three major aspects to the Minuteman story I was particularly interested in describing in detail: the development of solid propellants to the point where Minuteman was feasible, the development of inertial guidance systems light enough to be carried by the Minuteman missile, and the advances necessary for a lightweight reentry vehicle. Additionally, decisions on siting, design, and construction, as well as flight test programs, would be thoroughly discussed. Further research broadened the subject matter considerably. Would I find enough information without a clearance? The accumulation of over 250 GB of data testifies to the material available during four and a half years of research with open source material. Coverage of Minuteman III, which is still deployed, did suffer from limited declassified material.

J. D. "Dill" Hunley's *Preludes to US Space Launch Vehicle Technology: Goddard Rockets to Minuteman III* was an excellent starting point for the story of solid-propellant development. Dill graciously reviewed the early manuscript and helped me tighten up the text considerably, as well as suggesting additional content. With Dill's help I was able to locate and interview Charles B. Henderson, a key player in the metallized propellant breakthrough that led to the high-energy Minuteman and Polaris solid propellants.

On a visit to the Vandenberg AFB History Office, I was also fortunate to locate a copy of *A Brief History of Minuteman Guidance* by Robert Nease and Daniel Hendrickson, which provided a wealth of information specific to Minuteman guidance system development. My interviews with Robert Nease and Robert Knox were memorable, to say the least. Knox's critique of the guidance chapter and his patience with my questions are greatly appreciated. Marshall McMurran's *Achieving Accuracy* provided detailed information on Minuteman computers. Donald MacKenzie generously shared with me transcripts of interviews with key inertial guidance system engineers from his book *Inventing Accuracy.*

Researching early reentry vehicle technology development proved fruitful, but details about Minuteman were scarce. Pursuing Avco Corporation, the manufacturers of the Minuteman Mark 5 and Mark 11 series reentry vehicles, I called Textron Systems, which had purchased Avco, in Wilmington, Massachusetts, and was put in contact with Philip Fote. Phil was instrumental in the development of the reentry vehicles for Atlas, Titan I, and—most important to me—Minuteman at Avco. I was fortunate enough to have two long interviews with Phil at Textron Systems. He also provided me with Avco historical documents and some excellent photographs of recovered reentry vehicles. Unfortunately, Phil passed away before the book was complete, but he did get a chance to review the reentry vehicle chapter in detail and provide critical comments, which I greatly appreciate.

While David N. Spires's *On Alert: An Operational History of the United States Air Force Intercontinental Ballistic Missile Program, 1945–2011* covered all the ICBM programs, there was a wealth of information on Minuteman operations. I encourage readers to get a copy of this book, as it covers several areas that I did not.

Chapter 5 was previously published in an extended form in *Quest: The History of Spaceflight Quarterly* 26, no. 3 (2019). Chapter 9 was previously published in extended form in the Fall 2017 issue of *Air Power History*.

It must be reiterated that all of the research material used in this book came from open sources. The manuscript was reviewed in full by the Air Force Global Strike Command and released for publication with no redactions.

The Minuteman program has over 60 years of history, which would take several volumes to cover in its entirety. I have told a large portion of the story of Minuteman, and hopefully, this book will encourage research on other aspects to come to light. Digital appendices covering additional aspects of the program are available on the University of Arkansas Press website at www.uapress.com/minuteman/.

*David K. Stumpf*
*July 2020*

# ACKNOWLEDGMENTS

Acknowledgments are as hard to write as the book itself. Undoubtedly there will be somebody left out, so I apologize in advance.

The start of my research took place in Seattle. A $2,000 grant from the Association of Air Force Missileers (AAFM) funded my first research road trip to the Boeing Aerospace Company Archive in Seattle. Charlie Simpson, Col., USAF (ret.), at the time executive director of the AAFM, presented my request to the AAFM Board of Directors and answered questions on Minuteman reliability factors, which I greatly appreciate.

With the patient help of Tom Lubbesmeyer, an archivist at Boeing, I was able to scan over 3,000 pages of Minuteman Historical Summary documents during my three-day visit. Access to these documents proved to be the foundation for the first half of the book, as they covered 1958 to 1970 in amazing detail.

This book would not have been possible without the incredibly patient assistance of Mitch Cannon, CMSgt., USAF (ret.), a legendary Minuteman resource. All the illustrations were remastered by Mitch, except where noted. He also provided constant on-call assistance with difficult parts of the manuscript as well as reading the entire rough draft for content. His comments are greatly appreciated. Mitch deserves substantial credit for any success this book achieves.

Greg Ogletree, Maj., USAF (ret.), provided a wealth of information on the Airborne Launch Control System, as well as access to his expansive collection of technical orders, slides, and photographs. Greg did an exhaustive critical review of the book in its most massive form, which is truly appreciated. He found the time to critically review captions as well. Greg also deserves substantial credit for any success this book achieves. More than once, Greg—who had planned to tackle this project himself "someday"—said he was glad I was writing the book, and now I know what he meant.

Dr. Rick Sturdevant, deputy director of history, Headquarters Air Force Space Command, generously agreed to help me pare down the original manuscript when I reached the 220,000-word mark and still had two chapters to go. Rick also provided timely encouragement when the project seemed overwhelming. His admonition to "just write the damn thing" is what got me started with the actual manuscript.

Monte Watts, a former captain and Minuteman launch officer, introduced me to Mitch Cannon and provided enthusiastic support throughout the project. Monte arranged for photography of a Minuteman I guidance system artifact at the F. E. Warren ICBM and Heritage Museum. He also provided me with photographs of the NS10 guidance system stable platform on display at the office of the 576th Flight Test Squadron, Vandenberg AFB. A visit to actually see the last remaining NS10 stable platform is now on my bucket list!

Chuck Penson provided invaluable research assistance with the Minuteman documents in the Titan Missile Museum Archive and elsewhere, as well as constant encouragement.

Michael Byrd, at the time the 20 AF Historian, provided summary documents on wing construction dates as well as the deployment dates of the Mark 5, 11 series, and Mark 12 reentry vehicles. Jeremy Prichard, who replaced Michael, also provided enthusiastic assistance in the latter stages of my research.

Dave Fields, even crazier than I am about missiles, provided me with copies of multitudes of construction photographs culled from the Library of Congress and elsewhere, which saved considerable research time.

Keith Baylor volunteered to help me in any way he could and was the lucky person who put together the chronologies that appear in the digital appendices. Keith was a trajectory officer for Minuteman at Strategic Air Command headquarters and took on the task of constructing the simulation of the trajectory found in the digital appendices. Thank you, Keith, and thank your family for me. I know that both the chronologies and the simulation were a lot of work.

Carla Pampe, chief, Civic Outreach, Air Force Global Strike Command Public Affairs, was my point of contact and more than once helped with introductions to various history offices. Carla shepherded the manuscript through review at Global Strike Command in record time. DeAngela White and Jeremy Foster, also at Global Strike Command, provided assistance with document research and declassification. Rex Ellis, chief, Treaty Compliance Office, kindly provided me with information on treaty history.

Dr. Mary (Dixie) Dysart, Archie DiFante, Cathy Cox, and Tammy Horton at the Air Force Historical Research Agency, Maxwell AFB, Alabama, provided requested documents quickly and enthusiastically while at the same time making sure I realized there were other people needing documents just as urgently. Dr. Dysart, thank you for your early and enthusiastic support. Archie, thank you for your patience with my seemingly endless declassification requests and often stern admonition when I was treading on thin ice with my requests. Cathy and Tammy, your enthusiasm frequently made my day.

Randy Ross, David Fort, Aaron Arthur, John Wilson, Patrick Kerwin, and Michael S. Binder helped me navigate the National Archives and Research Administration's Samuel C. Phillips Collection.

Adriana Ercolana was an efficient and enthusiastic research assistant for me at the National Archives as well as the Army Corps of Engineers History Office. Her patient copying of many documents from the Phillips Collection, as well as entire Minuteman Corps of Engineers Ballistic Missile Construction Office histories, was worth every penny.

Garrett Moore, CWO, USA (ret.), provided not only research assistance at the National Archives but patiently worked with me on the targeting chapter as well. My stay with Garrett and his wife in Washington, DC, was a welcome relief from a long line of hotels during my first research trip.

Gordon Barnes, Lt. Col., USAF (ret.), and Dr. Louis Decker encouraged me to visit them in St. Louis for a fascinating discussion on the targeting process. Both were a great help in making sure the targeting chapter was technically correct.

Laurie Austin and Katie Rice provided cheerful and timely assistance in locating documents relating to Secretary of Defense McNamara at the John F. Kennedy Presidential Library. My efforts to correctly cite my reference documents stem from difficulties finding the McNamara material I needed due to poor citations.

Kevin M. Bailey and Michelle Kopfer at the Dwight D. Eisenhower Presidential Library guided me during a visit of several days and provided long-distance assistance for document citations I neglected to record.

Julie Gibson at Textron Systems was a cheerful intermediary between me and Philip Fote, chief missile systems engineer. Phil was pretty busy when I contacted him, and Julie made sure nothing slipped between the cracks. It was a real pleasure to meet Phil and spend time discussing the reentry vehicle technology, and Julie made it all happen. D. Scott MacBridge gave me a memorable tour of the unclassified parts of Textron Systems, which was amazing.

Craig Brunetti, National Air and Space Museum (NASM), gave me a tour of Minuteman and early reentry vehicle artifacts in the NASM collection and the chance to see in person much of what I was writing about. Craig also helped with locating documents and photographs for me.

William True and Sharon Newey Moorehead at Aerojet Rocketdyne helped me uncover the existence of a copy of a Large Solid Rocket Feasibility Program progress report. A copy of the final report exists but remains classified. William pursued getting part of the progress report declassified for me. William and I grew up across the street from each other in Davis, California—strange how things work out.

Eric Smith, archivist at the US Space and Rocket Center, Huntsville, Alabama, spent an afternoon with me in the storage facility at the center, where I got to see many of the original ablation research artifacts.

Ward Hemenway, Col., USAF (ret.), was a hands-on resource for stories about the early reentry vehicle deployment and also, much to my pleasant surprise, gave me his Avco manufacturer's model of the Mark 5 reentry vehicle after our interview.

Tim Pavek provided me with detailed information about the Minuteman II deactivation process at Ellsworth as well as access to his collection of slides. Thank you, Tim, for all the email assistance after my visit.

Bob Kelchner, CMSgt., USAF (ret.), Andy Doll, SMSgt., USAF (ret.), and John Mills, TSgt., USAF (ret.), answered endless questions and read multiple versions of the targeting and alignment section of the book. Thank you all.

Leigh Tange, CMSgt., USAF (ret.), and Rick Johnson, CMSgt., USAF (ret.), helped arrange an after-hours comprehensive tour of the launch facility trainer at Ellsworth and answered numerous questions about reentry vehicles and guidance systems over the years. Rick pointed out Leigh at the 44th Strategic Missile Wing reunion as a critical source of information, much of which can be found in the digital appendices.

Ken LaRock and Melissa Shaw from the National Museum of the United States Air Force provided photographs and documentation of the Minuteman 5 reentry vehicles on display.

Sandy Fye at the National Museum of Nuclear Science and History, formerly known as the National Atomic Museum, enthusiastically supported my research at the museum, including a tour of the storage facility, where I got to see a Mark 11 reentry vehicle, as well as escorting me as I measured several of the artifacts.

Jennifer Cuddeback and John Wilson painstakingly researched requests for poorly referenced documents at the Lyndon B. Johnson Presidential Library.

Jonathan McDowell graciously hosted me for three days as I mined his incredible archive of missile and space history. Jonathan's launch database proved to be the key for the flight test history detail in the book, and he graciously allowed me to include the Minuteman portion in the digital appendices.

Jay Bogess provided several Minuteman resources from his collection. I don't recall how Jay found me, but I'm glad he did, as the documents were unique and provided confirmation of several points in the Minuteman II story.

Martha Davis provided a high school English teacher's viewpoint on the text, which was critical to the early organization of the book. She also patiently explained the intricacies of English grammar to me yet again.

Joseph T. Page II provided several key documents from his collection and helped me navigate the unclassified material at the Vandenberg AFB History Office early in the project. Shawn Riem, former Vandenberg historian, and Scott Bailey, current Vandenberg historian, responded promptly to my requests for information. Thank you both.

John Turner, Public Affairs, 341st Missile Wing, provided information on the deactivation process for the 564th Missile Squadron, which had proved hard to find.

Craig Allen, Lt. Col., USAF (ret.), graciously provided me with a slide scanner early in the research effort, which was immediately put to use copying his unique archive of reentry vehicle photographs. Craig also gave me his contractor model of a Minuteman III. During the same visit to the Ogden area, Jim Meyers, Col., USAF (ret.), provided insight on the differences between Minuteman IB and III since he was present at the Minot upgrade. His slides of the airborne test, which is detailed in the digital appendices, added to the story as well.

During the October 2018 AAFM meeting in Cheyenne, Wyoming, I had the opportunity to interview Robert Parker, Maj. Gen., USAF (ret.), Ronald Gray, Brig. Gen., USAF (ret.), and Gary L. Curtin, Maj. Gen., USAF (ret.), about the early days of the ALCS operations. Thank you all for spending time with me during the meeting.

Thank you, Jay Kelley, Lt. Gen., USAF (ret.), for your support throughout this project and for agreeing to write the foreword for me. Our discussion of the Mark 12 reentry vehicle program was a highlight in my research.

David Scott Cunningham, editor, University of Arkansas Press, enthusiastically supported the idea for the book when I needed a publisher in order to request an endorsement from the Office of the Secretary of the Air Force. Jenny Vos, my point of contact when I sent in the manuscript, patiently answered my questions. Janet Foxman and copyeditor Denise Logsdon deserve medals for their efforts to educate me on the ins and outs of copyediting as the book progressed. Liz Lester, designer, crafted a beautiful book. Thank you all.

In 2017 my Parkinson's disease tremor got to the point where it impaired my ability to type. Through the generosity of Connie O'Brien and John Brady, who not only recommended Dragon Naturally Speaking software to me but, much to my surprise, gave me a copy, I was able to press on with writing the book.

This book would not have been possible in any way, shape, or form, without the complete understanding of my lovely wife of 43 years, Sarahni. She endured the passion I have for this project, which those who know me know can truly be overwhelming. More than once, we agreed that my ability to find key individuals in the history of the program was all the more reason to write the book. She never wavered in her support, even throughout my various medical problems, encouraging me to continue when it would have been far easier to encourage me to pass the project on to others.

Thank you, everyone, for your invaluable assistance in helping me make a contribution to the history of the Minuteman ICBM program and its role in assuring peace through the many turbulent years of the Cold War.

*David K. Stumpf*
*July 2020*

# MINUTEMAN

# INTRODUCTION

On 11 December 1962, Alpha Flight of the 10th Strategic Missile Squadron (10 SMS), 341st Strategic Missile Wing (341 SMW), Malmstrom Air Force Base (AFB), Montana, was declared combat ready and took its place in the nation's nuclear deterrent force inventory. Construction had begun just 22 months earlier on 16 March 1961, with the start of the excavation for the Alpha Flight launch control center (LCC).

On 25 February 1958, the Minuteman program had been briefed to Secretary of Defense Neil McElroy by Col. Edward N. Hall, USAF. The seemingly impossible task of bringing this new system, with its advanced solid-propellant capability of nearly instant response, into operational status in under five years was a tribute to the coordinated efforts of Space Technology Laboratories (STL), Boeing, and its associate contractors: Aerojet, Thiokol, Hercules Powder Company, North American Aviation's Autonetics Division, and Fuller-Webb, the prime contractor for the construction of the 341 SMW facilities.

Conceived by Hall as the ultimate intercontinental ballistic missile (ICBM) weapon system, Minuteman represented a dramatic change in deployment strategy. As originally imagined by Hall, there was little need for any maintenance presence, as missiles needing repair would be removed, replaced, and returned to a depot for repair. The result was the radical decision for the deployment of Minuteman in unmanned launch facilities (LFs). This was in sharp contrast to the significantly more complicated manned silo/launch control complexes for the liquid-fueled Atlas D, E, and F and Titan I and II.

The genius of Hall's concept is exemplified by the fact that the Minuteman program has been a major part of the nation's nuclear deterrent for over half a century. There have been numerous upgrades to the LCCs and LFs, along with significant improvements to the missile and payload through the deployment of four variations: Minuteman IA (LGM-30A), IB (LGM-30B), II (LGM-30F), and III (LGM-30G).

Originally deployed as part of the strategy of massive retaliation, the Minuteman command-and-control systems were flexible enough to be adapted to the strategies of counterforce, mutually assured destruction, and sufficiency.

What was the breakthrough that enabled solid propellant to be powerful enough for use in ICBMs? How was a lightweight inertial guidance system achieved? What were the advances in reentry vehicle technology that facilitated a relatively small warhead on Minuteman when compared to the liquid-fueled missile warheads? What modifications were made that allowed Minuteman to remain current among all the changes in technology and the treaties governing strategic nuclear weapons?

The answers to these questions and more can be found in this history, with one caveat. Since 400 Minuteman IIIs are still deployed, classification issues have exempted many details on Minuteman III from consideration. Additionally, with 57 years of deployment history, some selection of events and programs was necessary. This being said, there is still a rich history revealed, as well as one yet to be told.

transcontinental range.⁷ Characterizing Spaatz as no engineer, Senator Brien McMahon (D-CT), the committee chairman, asked for Bush's opinion of the article. Bush replied:

> If you were talking about 400 or 500 miles, I would say by all means. That is what the Germans did with their V-2 . . . But 3,000 miles? That is not just a single step beyond, it is a vastly different thing, gentlemen. I think we can leave that out of our thinking. I wish the American people would leave it out of their thinking.⁸

In February 1946, in a *National Geographic* article, Arnold reviewed the wide-ranging Army Air Forces' accomplishments during World War II, emphasizing how World War I biplane technology metamorphosed into the B-29 Superfortress. He described how jet-propelled fighters and bombers soon would supersede propeller-driven aircraft. Much to Bush's bemusement, Arnold claimed:

> It is now entirely possible, with the engineering information attainable, to build a ground-to-ground missile capable of traveling more than a thousand miles, and it is probable that in the not-distant future it will be possible to send remote-controlled missiles to any spot on the earth's surface.⁹

In *Modern Arms and Free Men*, published in 1949, Bush discussed the role of science in preserving democracy and repeated his lack of concern about intercontinental missiles:

> We may need to fear intercontinental bombing by manned aircraft at high altitudes . . . But there need be little fear of the intercontinental missile in the form of a pilotless aircraft, for it is not so effective from the standpoint of cost or performance as the airplane with a crew. Can such a missile be made to hit anything at the end of its flight? The V-2 could be made to hit with reasonable frequency within 15 miles of a point of aiming at a range of 200 miles. A similar missile flying 2,000 miles could be depended to hit within 150 miles of its target with reasonable frequency. This probability could certainly be improved. Its costs would be astronomical. For the near future, the really important and significant field of guided missiles lies in much shorter ranges, above those readily handled by guns but not so large as to run up size and costs to prohibitive heights.¹⁰

## Exploring Solutions

### Truman Administration 1945 to 1953

The Truman administration was the first to confront decision-making about long-range missile development. Spurred into action by the Soviet Union's detonation of an atomic bomb on 29 August 1949 and outbreak of the Korean War on 25 June 1950, President Truman asked Kaufman T. Keller, Chrysler Corporation board chairman, to be the Defense Department's director of the Office of Guided Missiles. Appointed on 30 October 1950, Keller and a small staff reviewed the status of major US missile programs. He obtained approval and funds for establishment of production lines of several high-priority missile programs—Nike, Terrier, and Sparrow—a major expansion of flight test programs. The

Atlas ICBM program, after being funded in April 1946, only to be canceled in July 1947, regained favor as MX-1593 on 16 January 1951. By January 1952 his evaluation of 22 programs had resulted in their approval for special development by the secretary of defense.

## Eisenhower Administration 1953 to 1961

Keller's final report, released on 17 September 1953, described Atlas as "a highly complex and long-term project still in the study stage." On 12 November 1953, the individual service secretaries received authority to approve guided missile programs within their departments.[11]

The ICBM concept and concomitant funding for development spawned numerous congressional and military committees. Eisenhower became the first president faced with a decision to acquire and deploy intermediate-range ballistic missiles (IRBMs) and ICBMs. He pondered several questions before deciding. Was the Soviet Union building and deploying similar missile forces? What was an appropriate response, sheltering the population from attack through developing early warning systems or a strong retaliatory response? Could the United States afford both?

Eisenhower began his first term by reorganizing the National Security Council (NSC) to provide "an integration of views which would be the product of continuous association between skilled representatives of all elements germane to national security."[12] He appointed a special assistant for national security affairs, expanded NSC membership, and made it clear the organization should frequently use outside civilian experts with fresh viewpoints.[13]

Concerned about the Truman administration's militaristic approach to the Soviet Union and its consequences for the US economy, on 9 May 1953, Eisenhower directed the NSC to create three panels to review national security policy alternatives. Code-named Project Solarium, the panels spent six weeks investigating current policies and proposing future directions. Task Force A, led by George Kennan, studied containment policy, essentially a continuation of Truman administration ideas. Task Force B, led by Air Force Gen. James McCormack, studied explicitly stated areas the United States would automatically defend against Soviet attack. Task Force C, led by Adm. Richard Conolly, studied "rolling back" Communism. On 16 July 1953, each task force presented its final report to the NSC. Several more months of analysis by the NSC staff led, on 30 October 1953, to NSC Report 162/2, which introduced a massive retaliation policy that relied on strategic nuclear weapons.[14]

In October 1953 Trevor Gardner, assistant secretary of the Air Force for research and development, established the Strategic Missiles Evaluation Group, nicknamed the Teapot Committee, to focus solely on strategic missiles then being developed: Snark, Navaho, and Atlas. The committee included John von Neumann, chairman (Princeton); Clark B. Millikan, Charles C. Lauritsen, Louis G. Dunn (all from the California Institute of Technology); Hendrik W. Bode (Bell Telephone Laboratories); Alan E. Puckett (Hughes Aircraft); George B. Kistiakowsky (Harvard); Jerome B. Wiesner (MIT); and Lawrence A. Hyland (Bendix Aviation). Col. Bernard A. Schriever, assistant for development planning under the Air Force deputy chief of staff for development, served as the committee's military representative.

Its report, released on 10 February 1954, recommended accelerating the Atlas program, noting "a radical reorganization of the project considerably transcending the Convair framework is required if a military useful vehicle is to be had within a reasonable span of time." Only after that reorganization should spending increase. Finding the performance specifications obsolete, the Teapot Committee concluded:

> a. The military requirement on CEP (circular error probable) should be relaxed from the present 1,500 feet to at least two, probably three nautical miles.
> b. The warhead weight might be reduced as far as 1,500 pounds, the precise figure to be determined after the CASTLE tests and by missile system optimization. Warhead diameter should also be considered a variable, somewhat flexible parameter.
> c. The reentry problem should be reinvestigated with special attention to "detachable drag-skirt" considerations. This will involve a study of the interplay of reentry heating, choice of trajectory, resulting aiming precision, and terminal vulnerability considerations. Without this, no impact Mach number should be rigidly specified, and in particular the present specification (M-6) should be discarded.
> d. The guidance problem should be re-examined in the light of the radically relaxed CEP. In particular, more serious consideration should now be given to an entirely missile-contained inertial guidance.
> e. The present concept of launching base system and supporting facilities for the IBMS leads to too vulnerable an operation. The design of the missile and the nature and layout of the supporting facilities should be adjusted to provide an optimum combination of low vulnerability, high firepower, and short starting time.[15]

Above all, the members reiterated:

> The most urgent and immediate need in the IBMS program is the setting up of the above-mentioned new IBMS development-management agency for the entire program, including the Convair effort. This program can then be subsequently extended and accelerated in some optimal manner to be determined by the studies of this new group. The setting up of various parallel projects as required will then also follow. The nature of the task for this new agency requires that overall technical direction be in the hands of an unusually competent group of scientists and engineers capable of making systems analyses, supervising the research phases, and completely controlling the experimental and hardware phases of the program, present ones as well as subsequent ones, that will have to be initiated. The type of directorial team needed is that of the caliber and strength that may require the creation of the special group by a "drafting" operation performed by the highest-level government executives on university, industry, and government organizations.[16]

The result was a total reorganization of the Air Force strategic missile program, with the only unaltered goal being earliest possible deployment of a credible system. Having each of the Atlas subsystems supported by at least one alternative contractor as insurance against design failure gave birth to the Titan I program.

On 14 May 1954, Gen. Thomas D. White, Air Force chief of staff, gave Atlas the service's highest development priority. At the same time, he assigned overall management to a newly created Air Research and Development Command (ARDC) West Coast field office. Two months later, on 1 July 1954, Gen. Thomas S. Power, ARDC commander, designated the new field office as the Western Development Division (WDD).[17] A month later, Brig. Gen. Schriever became WDD commander and Atlas program overseer. With Atlas Scientific Advisory Committee input, the WDD staff considered three approaches to a new organizational framework. The most favored had the Air Force continuing system responsibility but employing Ramo-Wooldridge Corporation for systems engineering and technical advice. Convair would still handle all Atlas missile fabrication, but Ramo-Wooldridge would provide technical direction on the overall program. With the acceptance of the new direction on 8 September 1954, the Atlas program—indeed the strategic missile program as we now know it—came into being.[18]

Earlier in 1954, President Eisenhower had asked Lee Dubridge, chairman of the Science Advisory Committee of the Office of Defense Mobilization, to investigate whether technological means existed to obtain early warning of a Soviet surprise attack and defend against such an attack. On 26 July 1954, Eisenhower asked Dr. James Killian, president of MIT and a Science Advisory Committee member, to chair a committee to study these problems. The Technological Capabilities Panel, also known as the Killian Committee, met from September 1954 to February 1955 and submitted its report to the NSC on 14 February 1955.[19]

With respect to the strategic missile program, the Killian Committee report recommended:

1. The development of an intercontinental ballistic missile (with about 5,500 nautical mile range and megaton warhead) continue to receive the very substantial support necessary to complete it at the earliest possible date.
2. There be developed a ballistic missile (with about 1,500 nautical mile range and megaton warhead) for strategic bombardment; both land-basing and ship-basing should be considered.[20]

The Killian report strongly influenced Eisenhower's views on ICBMs, leading to even greater prioritization of strategic missile development.[21]

Debates had been raging throughout his administration on the role of civil defense in protecting the American population from Soviet attack. From January to June 1956, Representative Chet Holifield (D-CA), who believed offensive striking power had received too much emphasis at the expense of civil defense, had chaired House Military Operations Subcommittee hearings on civil defense planning. As a result of the hearings, in January 1957, the Civil Defense Administration recommended the United States build a $32-billion shelter system to protect the entire population from radioactive fallout. Not wanting to spend such a vast sum on questionable protection, Eisenhower asked his Science Advisory Committee to have experts study the issue. The resulting Gaither Committee, named after Horace Rowan Gaither Jr., its first director, set out to recommend active and passive defensive measures for protecting the civilian population but wound up recommending offensive retaliatory capabilities to present a greater deterrent to the Soviet Union. In its report, submitted on 7 November 1957, shortly after the Sputnik shock, the Gaither Committee specifically recommended:

> 1. To increase Strategic Air Command's strategic offensive power (to match Russia's expected early ICBM capability):
>    a. Increase the initial operational capability (IOC) of our IRBMs (Thor and/or Jupiter) from 60 to 240.
>    b. Increase the IOC of our ICBMs (Atlas and Titan) from 80 to 600.[22]

In the decade from 1949 to 1959, significant advances in nuclear warhead design and guided missile technology enabled the guided missile concept to expand rapidly from aircraft armament and tactical weapons to the IRBM and ICBM. The paperback edition of *Modern Arms and Free Men* (1959) included in its introduction Bush's contention that he had been right all along:

> When I wrote this book in 1948–49, I held that intercontinental ballistic missiles were then possible but not practical. I did so because officers of the Air Force were proclaiming loudly that such missiles were just around the corner. . . . That they must be controlled by the Air Force which would proceed to build them. This announcement was part of a rather disgraceful competition between the services for appropriation and control, from which we have now rather well recovered. I said that if the ICBM appeared at all, it would be after many years and I was dead right. . . . I recite all this because I have often been accused of opposing the whole program and I dislike being taken to task when I was, for once, right.[23]

The 1950s saw development of a wide variety of guided missiles, ranging from turbojet cruise missiles to IRBMs and ICBMs. Of necessity, the initial focus for the longer-range missiles involved liquid propulsion. Numerous authors have covered this aspect of ICBM development already.[24]

★ ★ ★

At the end of the 1950s, the second generation of ICBMs, Titan II and Minuteman, were well into development. Titan II used storable liquid propellants, which eliminated loading liquid oxygen at the last minute, thereby decreasing the time between receipt of a valid launch command and launch from 15 minutes with Atlas and Titan I to 58 seconds. While the Titan II carried the largest thermonuclear warhead in the US ICBM program, even storable liquid propellants limited its deployment to only 54 missiles in three strategic missile wings.[25]

While Titan II represented the ultimate liquid-propellant ICBM, the solid-propellant Minuteman embodied the combination of a more rapid response—capable of launch in under 35 seconds—and safety in numbers, with a final deployment of 1,000 missiles, compared to 103 for Atlas, 54 for Titan I, and 54 for Titan II (see Figures 1.1–1.4).[26] A correlation chart for the Minuteman variants and wings is found in Table 1.1.

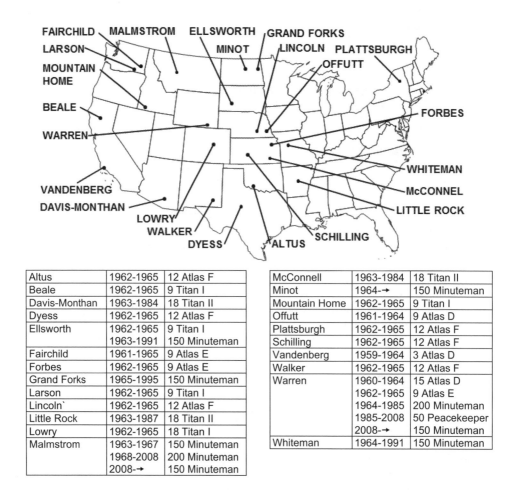

FIGURE 1.1  ICBM base distribution from 1961 to 2019. *Adapted from "Missileers' Heritage."*

FIGURE 1.2 An Atlas F raised to the surface and ready to launch during a test exercise by 556 SMS, Plattsburg AFB, New York. Atlas F was deployed in the silo-lift, 1x12 configuration (12 launch complexes with one missile at each complex). *Official USAF photograph, courtesy of K. McCleary.*

FIGURE 1.3  A "battleship" Titan I rises to the surface at the Operational Suitability Test Facility at Vandenberg AFB, California. This missile was a nonflight test article used to validate facility systems and procedures. Note the size of the man standing at the right side of the silo. *Courtesy Lockheed Martin Vandenberg Launch Operations.*

FIGURE 1.4 Titan II N-7 (61-2730) lifts off from Launch Complex 395-C, Vandenberg AFB, on 16 February 1963. This was the first in-silo launch in the Titan II program. The "belly-band" reinforcing bands characteristic of missiles N-1 through N-11 are clearly evident on the Stage I tanks. Observers immediately noted that the missile was unfortunately spinning clockwise at this point. *Inset:* white arrow indicates guidance system wiring harnesses that were pulled out of the missile by the failure of the 2B1E umbilical release mechanism, causing the rotation; black arrow points to the telemetry antenna windows in the early developmental Mark 6 reentry vehicles. (Inset created by author.) *Official USAF photograph, courtesy of Andrew Hall.*

TABLE 1.1  Minuteman Program Correlation Chart[a]

|  |  | MINUTEMAN | | | WEAPON SYSTEM | | | MISSILE | | | |
|---|---|---|---|---|---|---|---|---|---|---|---|
|  |  | MM I | MM II | MM III | WS-133A | WS-133B | WS-133A-M | LGM-30A | LGM-30B | LGM-30F | LGM-30G |
| MINUTEMAN | MM I |  |  |  | X |  |  | X | X |  |  |
|  | MM II |  |  |  |  | X | X |  |  | X |  |
|  | MM III |  |  |  |  | X | X |  |  |  | X |
| WEAPON SYSTEM | WS-133A | X |  |  |  |  |  | X | X |  |  |
|  | WS-133B |  | X | X |  |  |  |  |  | X | X |
|  | WS-133A-M |  | X | X |  |  |  |  |  | X | X |
| WING | I | X | X | X | X |  | X | X | X | X | X |
|  | II | X | X |  | X |  | X |  | X | X |  |
|  | III | X |  | X | X |  | X |  | X |  | X |
|  | IV | X | X |  | X |  | X |  | X | X |  |
|  | V | X |  | X | X |  | X |  | X |  | X |
|  | VI |  | X | X |  | X |  |  |  | X | X |
|  | I X 564 SMS |  | X | X |  | X |  |  |  | X | X |

a) Adapted from "Minuteman Correlation Chart," *Minuteman Service News*, September–October 1970; "Minuteman Correlation Chart," *Minuteman Service News*, March–April 1973.

# 2

# EVOLUTION OF THE MINUTEMAN FORCE LEVELS

## Eisenhower Administration 1953 to 1961

Col. Edward Hall originally conceived Minuteman as literally hundreds, possibly thousands, of simple, unmanned missile launchers spaced three miles apart to prevent elimination of more than one LF by a single five-megaton warhead. He based the force size on the number needed to destroy Soviet targets enumerated in current and potential war plans, while factoring in reliability. One calculation used a reliability of 50 percent by 1964, which meant deploying twice as many missiles as if they were 100 percent reliable. Hall thought it would be cheaper and faster to deploy less reliable missiles rather than focusing development efforts on higher reliability and, consequently, delaying operational deployment. In the late 1950s, Air Force intelligence estimated the Soviet Union would have 3,000 missiles deployed by 1965. Given all these considerations, Hall estimated that an initial force structure of 4,000 Minuteman missiles by 1965 would meet survivability, reliability, and capability for attacking the Soviet Union.[1]

The first formal Minuteman development plan, issued on 21 February 1958, called for a 400-missile operational force by the end of Fiscal Year (FY) 1964. William Holaday, the defense secretary's director of guided missiles, approved this estimate for research and development purposes on 27 February 1958.[2]

Over a year passed before a significant change occurred in the proposed Minuteman force level. On 27 October 1959, the Air Force Directorate of Research and Development issued a Minuteman program summary stating that 805 missiles would be deployed by the end of FY 1964, with 150 being rail-mobile and 655 silo-based.[3]

On 1 April 1960, President Eisenhower approved deployment of 150 Minuteman missiles as Wing I. The FY 1961 budget included funding for Wing I's third squadron of 50 missiles, signifying approval for both funding and force level. Malmstrom AFB, Montana, would be the home of Wing I.[4]

On 3 April 1960, Lt. Gen. Bernard A. Schriever asked Dr. Clark B. Millikan, chairman of the Office of Secretary of Defense Ballistic Missile Committee (OSD-BMC), to review technical and operational aspects of the entire ICBM program, which was facing critical schedule and budgetary problems.[5] The Air Force Ballistic Missile Division (AFBMD) and STL were reviewing the entire program to reaffirm the present mix of missile programs or recommend changes, but Schriever also wanted an independent review. Millikan organized the Lauritsen Committee, named after its chairman, Dr. Charles A. Lauritsen. Its members included Dr. Lawrence A. Hyland, Trevor Gardner, Dr. Hendrik W. Bode, and Dr. Jerome B.

Wiesner. All had participated in the earlier von Neumann study (see Chapter 1) and were cognizant of missile development issues.[6]

The Lauritsen Committee's report, which went to Schriever on 31 May 1960, said development problems with Thiokol's Minuteman Stage I motor and the Autonetics guidance system would delay deployment 6 to 12 months. An earlier decision, on 13 May 1959, to accelerate the schedule by one year, meant loss of a year of testing and development.[7] Since meeting the schedule for deploying the first wing of 150 missiles was a national priority, the committee recommended the Air Force accept reduced reliability, accuracy, and range for the first wing. Schriever had first addressed this issue in a message to ARDC in February 1959.[8] On 2 August 1960, the Air Force Ballistic Missile Committee (AFBMC) agreed with the Lauritsen Committee's recommendations, including Improved Minuteman (Minuteman II) development to address reliability, accuracy, and range issues for future wings.[9]

## Kennedy Administration 1961 to 1963

Soon after taking office, President John F. Kennedy, prompted by the Weapons System Evaluation Group (WSEG) Report 50, Evaluation of Strategic Offensive Weapons Systems, published on 27 December 1960, requested reappraisal of the Defense Department budget for FY 1962.[10] The WSEG report was in response to a Joint Chiefs of Staff (JCS) request in September 1959 for an independent, apolitical evaluation of the optimum mix of strategic weapon systems for use in a general war. The JCS specifically asked WSEG for "an evaluation of offensive weapons systems that might be used in a strategic role, particularly during the 1964–1967 period," with specific attention to each system's "reliability, reaction time, command and control, ability to penetrate anticipated defenses, accuracy, ability to withstand enemy attack, as well as acquisition and maintenance costs and probable useful life." WSEG Report 50 did not recommend an optimum mix of strategic offensive weapons but instead evaluated the proposals of the individual services and remained ambivalent on the choice between hardened and dispersed versus mobile Minuteman deployment.[11]

Shortly after his confirmation, Secretary of Defense Robert S. McNamara spent an entire day studying the report and questioning the project's leader and staff. Strategic weapon system reliability and command-and-control issues dominated the discussion.[12] McNamara learned the Air Force, as described in WSEG Report 50, planned the Minuteman force shown in Table 2.1.[13]

The Kennedy administration submitted its FY 1962 Military Construction Bill to Congress in early March 1961. The administration initially requested six additional 50-missile Minuteman squadrons for the hardened-and-dispersed force (300 missiles added to the 150 already slated for the first Minuteman wing at Malmstrom AFB) and three 30-missile rail-mobile squadrons, for a total of 540 missiles. On 28 March 1961, McNamara revised the request to a total of nine hardened-and-dispersed squadrons and no rail-mobile squadrons. Cost analysis had shown that a 50-missile hardened-and-dispersed squadron cost nearly the same as a 30-missile rail-mobile squadron. The revision resulted in a net gain of 60 missiles, making a revised force total of 600.[14]

On 23 September 1961, McNamara issued a draft presidential memorandum, "Recommended Long-Range Nuclear Delivery Forces 1963–1967." For the Minuteman

TABLE 2.1  Air Force Planned Minuteman Force Levels, 1960[a]

| | NUMBER OF DEPLOYED MISSILES AT THE END OF YEAR | | |
|---|---|---|---|
| FISCAL YEAR | FIXED | MOBILE | TOTAL |
| 1961 | – | – | – |
| 1962 | – | – | – |
| 1963 | 120 | 30 | 150 |
| 1964 | 649 | 156 | 805 |
| 1965 | 1,225 | 300 | 1,525 |
| 1966 | 2,000 | 300 | 2,300 |
| 1967 | 2,000 | 300 | 2,300 |

a) WSEG Report 50: Evaluation of Strategic Offensive Weapons Systems, 27 December 1960, Appendix B to Enclosure F.

component, McNamara recommended purchasing 100 hardened-and-dispersed missiles and 50 rail-mobile missiles. As explained in his recommendation, he remained unconvinced about the necessity of the rail-mobile Minuteman but was willing to keep them for budgeting purposes. The JCS had recommended 300 and 50, respectively.

McNamara summarized the major factors that influenced his decision-making about force levels through FY 1967 (Table 2.2). He sought to avoid either a "minimum deterrent posture," which he defined as "one in which, after a Soviet attack, we would have a capability to retaliate ... and be able to destroy most of Soviet urban society, but in which we would not have a capability to counterattack against Soviet military forces," or "full first-strike capability." Full first-strike capability "would be achieved if our forces were so large and so effective, in relation to those of the Soviet Union, that we would be able to attack and reduce Soviet retaliatory power to the point at which it could not cause severe damage to US population and industry." McNamara rejected both alternatives and instead offered forces that "will provide major improvements in our strategic posture: in its survivability, its flexibility, and its ability to be used in a controlled and deliberate way under a range of contingencies."[15] Several histories report that the Air Force wanted up to 10,000 Minuteman missiles, pointing to this number as an example of excessive arms procurement. The only documented evidence, however, is a memo from Carl Kaysen to President Kennedy that mentions informal discussions of needing 8,000 to 10,000 missiles.[16]

To further support his force level recommendation, McNamara noted the 1,200 to 1,700 high-priority targets for FY 1965 were expected to grow to 1,350 to 2,200 by the end of FY 1967. His recommended FY 1965 force level for all delivery systems was 3,440 nuclear weapons, totaling 4,789 megatons, compared to the services' proposals of 4,052 and 5,600, respectively. For FY 1967 his recommendations grew to 4,180 and 5,890, while the services proposed 5,450 and 7,620, respectively.[17]

McNamara's recommendations sparked extensive discussion within the NSC that raised the point that achieving a 600-missile Minuteman force by the end of FY 1964 meant

TABLE 2.2 Department of Defense and Air Force Minuteman Force Levels, 23 September 1961[a]

| | NUMBER OF DEPLOYED MISSILES AT THE END OF YEAR | | | | | |
| | DEPARTMENT OF DEFENSE | | | AIR FORCE | | |
| FISCAL YEAR | FIXED | MOBILE | TOTAL | FIXED | MOBILE | TOTAL |
| --- | --- | --- | --- | --- | --- | --- |
| 1961 | – | – | – | – | – | – |
| 1962 | – | – | – | – | – | – |
| 1963 | 150 | – | 150 | – | – | – |
| 1964 | 600 | – | 600 | – | – | – |
| 1965 | 700 | 50 | 750 | 1,200 | 50 | 1,250 |
| 1966 | 800 | 100 | 900 | 1,700 | 300 | 2,000 |
| 1967 | 900 | 100 | 1,000 | 2,300 | 300 | 2,600 |

a) Recommended Long-Range Nuclear Delivery Forces 1963–1967, 23 September 1961, National Security Policy, Part 2, Draft Memorandum from Secretary of Defense McNamara to President Kennedy, "Foreign Relations of the United States, 1961–1963," Vol. VIII, Document 46.

continuing the program on a crash basis. Air Force briefings had indicated a slowdown in deployment pending significant range, accuracy, and payload improvements before Minuteman deployed with the second wing.[18]

On 21 November 1962, McNamara issued a draft presidential memorandum for FY 1963 to 1968 with changes in strategic forces composition (Table 2.3). The Air Force sought to deploy Minuteman IB to a level of 900 missiles, then replace up to 150 with the Improved Minuteman (Minuteman II) while simultaneously deploying 1,200 Minuteman IIs by 1968. McNamara recommended a ceiling of 800 Minuteman Is with an additional 500 Minuteman IIs by 1968.[19]

## Johnson Administration 1963 to 1969

On 6 December 1963, McNamara proposed a lower total force with a greater number of Minuteman IIs, reasoning that Minuteman II offered 30 to 40 percent greater effectiveness than Minuteman IB. The Air Force had also decreased its total force level recommendation, emphasizing additional Minuteman IIs (Table 2.4). McNamara now planned for 1,200 Minuteman missiles, while the Air Force wanted 1,400.[20] He recommended an increase of only 50 LFs instead of the 200 previously proposed in the FY 1965 budget. McNamara argued that the overall effectiveness of 400 Minuteman Is and 800 Minuteman IIs was greater than what the previously approved force level of 800 Minuteman Is and 500 Minuteman IIs could achieve. He recommended retrofitting 400 of the 800 Minuteman Is in Wings I through V with Minuteman IIs plus continuing conceptual development of an advanced ICBM, presumably Minuteman III (Table 2.4).[21]

McNamara explained his recommendations in relation to the policy of "assured

TABLE 2.3  Department of Defense and Air Force Minuteman Programmed Force Levels, 21 November 1962, FY 1963 to 1968[a]

| FISCAL YEAR | DEPARTMENT OF DEFENSE | | | AIR FORCE | | |
|---|---|---|---|---|---|---|
| | MM I | MM II | TOTAL | MM I | MM II | TOTAL |
| 1963 | 150 | – | 150 | 150 | – | 150 |
| 1964 | 600 | – | 600 | 600 | – | 600 |
| 1965 | 800 | – | 800 | 900 | – | 900 |
| 1966 | 800 | 150 | 950 | 900 | 300 | 1,200 |
| 1967 | 800 | 350 | 1,150 | 850 | 800 | 1,650 |
| 1968 | 800 | 500 | 1,300 | 750 | 1,200 | 1,950 |

a) Memorandum for the President, Draft, Subject: Recommended FY 1963–1968 Strategic Retaliatory Forces, 21 November 1962.

TABLE 2.4  Department of Defense and Air Force Minuteman Programmed Force Levels, 6 December 1963[a]

| | FISCAL YEAR | | | | | | |
|---|---|---|---|---|---|---|---|
| | 1963 | 1964 | 1965 | 1966 | 1967 | 1968 | 1969 |
| DEPARTMENT OF DEFENSE | | | | | | | |
| MINUTEMAN I | 150 | 600 | 800 | 750 | 610 | 480 | 400 |
| MINUTEMAN II | – | – | – | 200 | 390 | 620 | 800 |
| TOTAL | 150 | 600 | 800 | 950 | 1,000 | 1,100 | 1,200 |
| AIR FORCE | | | | | | | |
| MINUTEMAN I | 150 | 600 | 800 | 780 | 700 | 620 | 540 |
| MINUTEMAN II | – | – | – | 170 | 550 | 780 | 860 |
| TOTAL | 150 | 600 | 800 | 950 | 1,250 | 1,400 | 1,400 |

a) Memorandum for the President, Draft, Subject: Recommended FY 1965–1969 Strategic Retaliatory Forces.

destruction." He used both "expected estimates" and "pessimistic estimates" reflecting the most likely and worst-case outcomes, respectively. The "expected estimate" assumed 85 percent reliability for the Minuteman force; 50 percent was the "pessimistic estimate." The value at the zero-Minuteman force level represented the results of an attack by the rest of the strategic forces proposed for FY 1969. Clearly the already authorized force of 950 Minuteman missiles would be adequate for "assured destruction" (Table 2.5).[22]

McNamara's draft memorandum also discussed retrofitting Wings I through V with Minuteman IIs, beginning in FY 1966 with 400 and ending in FY 1969 with 800 (Force

TABLE 2.5  Minuteman Force Levels, Soviet Fatalities, and Industrial Destruction[a]

|  | NUMBER OF MINUTEMAN | | | | |
|---|---|---|---|---|---|
|  | 0 | 950 | 1000 | 1200 | 1400 |
| EXPECTED ESTIMATES | | | | | |
| PERCENT FATALITIES | 50 | 69 | 70 | 71 | 72 |
| PERCENT INDUSTRIAL CAPACITY DESTROYED | 57 | 82 | 87 | 89 | 90 |
| PESSIMISTIC ESTIMATES | | | | | |
| PERCENT FATALITIES | 17 | 29 | 30 | 32 | 33 |
| PERCENT INDUSTRIAL CAPACITY DESTROYED | 30 | 50 | 51 | 53 | 54 |

a) Memorandum for the President, Draft, Subject: Recommended FY 1965–1969 Strategic Retaliatory Forces.

Modernization; see Chapter 16). Minuteman II had eight targets stored in the missile guidance system, enabling rapid retargeting compared to Minuteman I, which only had two (see Chapter 10). McNamara planned to collocate 250 additional Minuteman IIs across Wings I through V and retrofit 400 more through FY 1969.[23] While the JCS chairman and other service chiefs agreed with McNamara's force level recommendation, Air Force Chief of Staff Gen. LeMay's recommendation for 1,950 Minuteman missiles by 1969 was not included in McNamara's recommendation (Table 2.6).[24]

On 20 November 1964, McNamara set the final Minuteman force level at 1,000 missiles. Two weeks later, on 3 December 1964, in his draft presidential memorandum for FY 1966 to 1970 strategic offensive forces, McNamara explained that 400 one-megaton weapons would destroy 40 percent of the urban population in Soviet cities and 30 percent

TABLE 2.6  Department of Defense and Air Force Minuteman Programmed Force Levels, 3 December 1964, FY 1966 to 1970[a]

| FISCAL YEAR | DEPARTMENT OF DEFENSE | | | AIR FORCE | | |
|---|---|---|---|---|---|---|
|  | MM I | MM II | TOTAL | MM I | MM II | TOTAL |
| 1964 | 600 | – | 600 | 150 | – | 150 |
| 1965 | 800 | – | 800 | 600 | – | 600 |
| 1966 | 800 | 80 | 880 | 750 | 200 | 950 |
| 1967 | 700 | 300 | 1,000 | 610 | 390 | 1,000 |
| 1968 | 550 | 450 | 1,000 | 480 | 620 | 1,100 |
| 1969 | 400 | 600 | 1,000 | 400 | 800 | 1,200 |
| 1970 | 250 | 750 | 1,000 | 300 | 900 | 1,200 |

a) Memorandum for the President, Draft, Subject: Recommended FY 1966–1970 Strategic Retaliatory Forces.

of the entire population. Doubling the number to 800 would only marginally increase the destruction, and further increases would produce smaller and smaller returns due to attacking smaller and smaller cities. Therefore, a Minuteman force of 1,000 missiles was more than sufficient.[25]

McNamara and the Air Force argued about the need for an advanced manned strategic aircraft (AMSA). Further analysis comparing an AMSA bomber, Polaris, Minuteman, and the B-52 armed with a short-range attack missile showed a future Minuteman III was much more cost-effective than the proposed AMSA.[26]

On 5 February 1965, Deputy Secretary of Defense Cyrus Vance testified before the House Armed Services Committee on the proposed FY 1966 budget and projected program needs through FY 1970. Vance read from McNamara's prepared statement:

> Last year we had tentatively planned to fund another 200 Minuteman silos in each Fiscal Year 1966–67 (for a total of 1,200 missiles).
>
> On the basis of our analysis of the general nuclear war problem in the early 1970s, I am convinced that another 200 Minuteman silos are not required at this time. We now believe that we can markedly increase the "kill" capabilities of the Minuteman force through a number of qualitative improvements which now appear feasible. The Minuteman force presently planned for Fiscal Year 1970 will have a total destruction capability of at least 30 to 40 percent greater than a force of the same size consisting of only Minuteman I. This is equivalent to adding 300 to 400 missiles to a force of 1,000 Minuteman. With the additional improvements which now appear possible, the destruction capabilities of the Minuteman force could be further increased in the future, if that appears desirable, by a factor of two compared to a force of the same size consisting of only Minuteman I.[27]

On 1 November 1965, McNamara made his recommendations for strategic offensive and defensive forces for FY 1967 through 1971. Confronting the issue of funding for range, payload, and accuracy improvements to Minuteman, he recommended proceeding with Minuteman III development and anticipated deploying 50 as Minuteman I replacements by the end of FY 1969. His force composition recommendation would change, but his force level recommendation did not, because replacing Minuteman Is with Minuteman IIIs resulted in 700 Minuteman IIs and 300 Minuteman IIIs by 1974.[28]

On 23 February 1966, McNamara testified before the Senate Armed Forces Committee concerning the proposed budget for FY 1967. Asked about the differences among Minuteman I, II, and III, he characterized the difference between Minuteman II and Minuteman I as like that between the B-52 and B-47 bomber, respectively, with Minuteman III being even more advanced than Minuteman II. When asked if the secretary of the Air Force and the Air Force chief of staff concurred on force levels, McNamara admitted earlier disagreement but said everyone now agreed on a 1,000-missile force.[29]

On 22 September 1966, McNamara recommended maintaining the Minuteman force level at 1,000 missiles, with 600 Minuteman IIs and 400 Minuteman IIIs by the end of FY 1972.[30] At this time the Force Modernization Program had begun, and the Minuteman force consisted of 918 missiles: 137 Minuteman IAs (Mark 5), 490 Minuteman IBs (Mark 11), 171 Minuteman IBs (Mark 11A), and 120 Minuteman IIs (Mark 11B).[31]

Achievement of the full 1,000-Minuteman force level occurred on 21 April 1967, with a combat readiness declaration for the 564th Strategic Missile Squadron (564 SMS, also known as collocated Squadron 20, or the "Odd Squad") at Malmstrom AFB. At this point the Minuteman force consisted of 137 Minuteman IAs (Mark 5), 410 Minuteman IBs (Mark 11), 173 Minuteman IBs (Mark 11A), and 280 Minuteman IIs (Mark 11B).[32]

## Nixon Administration 1969 to 1974

Early in 1969, Minuteman IAs left Strategic Air Command's (SAC) inventory. Although the last missiles of the 490 SMW Lima Flight were taken off strategic alert on 15 January 1969, five Minuteman IAs remained on emergency combat capability (ECC) status in Kilo Flight. On 12 February 1969, crews pulled the last two missiles from LFs Kilo-02 and Kilo-10, and Boeing accepted those facilities for conversion to Minuteman II.[33]

The Air Force never achieved McNamara's recommended force level of 700 Minuteman IIs but instead peaked at 500; the Force Modernization Program's Minuteman II portion was completed at Malmstrom AFB with the 341 SMW's full complement of 150 Minuteman IIs on 27 May 1969. Another 50 Minuteman IIs already were deployed at the 564 SMS at Malmstrom, with the remaining 300 evenly divided between Wing V, Whiteman AFB, and Wing VI, Grand Forks AFB.[34]

On 15 March 1971, Secretary of Defense Melvin Laird presented the Nixon administration's FY 1972 defense budget to Congress. He said the Minuteman III program had a total force objective of 550 missiles, the remainder being Minuteman II. On 1 June 1973, among other aspects of the Force Modernization Program, Headquarters Air Force cut FY 1975 Minuteman procurement funding from $356 million to $227 million to reach the congressionally approved austere budget but still meet Laird's 550-missile goal. Earlier Air Force plans, however, had included deploying at least 700 Minuteman IIIs and replacing the remaining Minuteman IIs with the proposed Minuteman IIIA.[35]

## Ford Administration 1974 to 1976

The Air Force plans were dashed for a variety of reasons, the foremost being the Strategic Arms Limitation Talks II framework agreement reached by President Gerald R. Ford and General Secretary Leonid I. Brezhnev on 24 November 1974. It limited the number of strategic delivery vehicles for both nations to 2,400 and stipulated a ceiling of 1,320 missiles equipped with multiple independently targetable reentry vehicles (MIRVs)—Minuteman III and Poseidon.[36]

On 3 September 1974, the last operational Minuteman IB LF, the 320 SMS LF Juliet-03 at F. E. Warren AFB, went off strategic alert and was turned over to Boeing for Minuteman III modification.[37]

On 3 January 1975, Secretary of Defense James R. Schlesinger announced the decision to limit Minuteman III deployment to 550, thereby preserving the option to arm Poseidon and planned Trident submarine-launched ballistic missiles (SLBMs) with MIRVs. Consequently, the Minuteman III production line at Plant 77, Hill AFB, in Utah would close.[38]

The Minuteman Force Modernization Program's latter phase involved replacing 350 Minuteman IBs—150 at Wing III, Minot AFB, and 200 at Wing V, F. E. Warren AFB—with Minuteman IIIs. Minuteman III also replaced Minuteman II missiles at Wing VI, Grand Forks AFB, and the 564 SMS at Malmstrom AFB—150 and 50, respectively. The 150 Minuteman IIs at Grand Forks went to Wing II, Ellsworth AFB, replacing Minuteman IBs. After the exchange at the 564 SMS's LF Tango-49 on 12 July 1975, the Minuteman force reached the goal set by Laird and upheld by Schlesinger. Minuteman force levels remained at 550 Minuteman IIIs and 450 Minuteman IIs, with various airframe, guidance, and facility upgrades (see Chapters 8, 10, and 16) for the next 13.5 years.[39]

## Reagan Administration 1981 to 1989

On 2 October 1981, President Ronald Reagan announced his decision to scrap the multiple protective shelter basing mode for the new MX ICBMs (renamed Peacekeeper a year later) and place them in existing Minuteman silos. After considerable discussion, on 2 November 1981, a Senate resolution expressed that body's lack of enthusiasm for Reagan's decision. On 23 October 1981, perhaps in response to the Senate's disapproval, Headquarters Air Force announced plans for Peacekeeper interim deployment in "super hardened" LFs pending studies on a permanent, large-scale basing mode. Two months later, on 30 January 1982, Headquarters Air Force announced modification of the Peacekeeper program plan, putting 50 Peacekeepers in existing Minuteman LFs.[40]

On 8 September 1982, Congress passed Public Law 97-252, "Department of Defense Authorization Act, Fiscal Year 1983." Clearly unhappy with Peacekeeper decisions, Congress included a provision that most Peacekeeper funding could not be spent until the president completed his review of long-term basing modes. Congress, the president, and the Air Force continued an often acrimonious debate about Peacekeeper basing, proposing a multitude of innovative possibilities, such as multiple protective shelters, closely spaced basing, and more. Nobody seemed to want the new missiles in their state. On 22 November 1982, President Reagan informed Congress he had decided to base 100 Peacekeeper missiles in 100 super-hard capsules in a closely based spacing mode at F. E. Warren AFB.[41]

On 7 January 1983, the Ballistic Missile Organization's Detachment 1 activated at F. E. Warren AFB to coordinate conversion of Minuteman III LFs to accept Peacekeeper launch canisters. Due to its larger diameter, Peacekeeper would use cold-launch ejection by pressurized steam from the launch tube, with subsequent Stage I ignition once clear of the launcher. On 31 October 1983, Detachment 1 completed the logistics plan for integrating Peacekeeper missiles into existing Minuteman LFs. Also, in October, a Draft Final Environmental Impact Statement on Peacekeeper was issued. This marked the first formal announcement of plans to deploy 100 Peacekeeper missiles in 319th and 400th Strategic Missile Squadrons' (319 and 400 SMS) Minuteman III LFs at F. E. Warren AFB.[42]

On 23 May 1985, the Senate approved the Nunn–Warner Amendment to the Department of Defense Authorization Act of 1986, limiting Peacekeeper deployment to 50 Minuteman III LFs. Four months later, on 18 September 1985, the Senate and House Conference Committee approved the amendment.[43] Peacekeeper LF conversion began on 3 January 1986 with Minuteman III removal from the 400 SMS's LF Quebec-02 at F. E.

Warren AFB. Peacekeeper became fully operational on 30 December 1988 with final installation at LF Quebec-10, establishing Minuteman force levels at 500 Minuteman IIIs and 450 Minuteman IIs.[44]

## Bush Administration 1989 to 1993

On 4 February 1991, Secretary of the Air Force Donald Rice and Air Force Chief of Staff Gen. Merrill McPeak announced probable strategic cutbacks due to budget constraints and expectation of new limitations from ongoing treaty negotiations. The Air Force intended to begin retiring Minuteman IIs in FY 1992, most likely starting with Wing II at Ellsworth, because facilities there were the oldest. In April 1991 the Air Force announced its decision to consolidate Minuteman III forces to three bases.[45]

On 31 July 1991, President George H. W. Bush and Soviet President Mikhail Gorbachev signed the "Treaty between the United States of America and the Union of Soviet Socialist Republics on Further Reduction and Limitation of Strategic Offensive Arms," originally called the START Treaty but renamed START I after the signing of START II. START I limited ICBM and SLBM forces to no more than 1,600 launchers and 4,900 warheads and only 500 land-based ICBMs with no more than 1,200 warheads.[46]

On 27 September 1991, President Bush announced his decision to remove all Minuteman IIs from strategic alert immediately, with deactivation beginning soon thereafter (see Chapter 17).[47] Since START I permitted dismantlement over seven years, the decision to accelerate the process unilaterally caused considerable consternation in the Senate. Several days earlier, SAC had issued Program Plan 91-7, "Minuteman II Deactivation/Minuteman III Conversion, 341st Missile Wing, 23 September 1991," for Malmstrom AFB. Rivet Dome II became the code name for the deactivation process, with Rivet Add denoting the conversion process.[48]

On 5 December 1991, Congress passed the National Defense Authorization Act for FY 1992 and 1993. It contained no funding for Minuteman III redeployment between bases and stipulated no funds from proceeding years could be used for such transfer. Furthermore, no Minuteman IIIs in storage could be placed in Minuteman II LFs, pending the defense secretary's submission of a plan for restructuring strategic forces consistent with START I. While this did not affect Minuteman II deactivation, facility conversion to support Minuteman III at the 341 SMW was delayed.[49] The secretary of defense met the FY 1992 to 1993 Department of Defense Appropriations Bill requirements, which enabled emplacement of the first Minuteman III, previously stored at Hill AFB, in 12 SMS's LF Juliet-09 on 17 November 1992. Twenty-nine more would be emplaced before the remaining 120 came from another, still unnamed, base.[50]

On 3 January 1993, President Bush and Russian President Boris Yeltsin signed the "Treaty between the United States of America and the Russian Federation on the Further Reduction and Limitation of Strategic Offensive Arms" (START II, also known as the De-MIRVing Treaty). It limited the United States to 3,500 deployed strategic nuclear warheads and no MIRVed ICBMs—MIRVs were still allowed on SLBMs. The US land-based ICBM force could have no more than 500 launchers and 500 warheads.[51]

## Clinton Administration 1993 to 2001

By September 1994 initial conversion of 30 of the 341 SMW's LFs to house Minuteman III missiles was finished. Since the 150 Minuteman IIIs at Grand Forks remained on alert, the Minuteman III force level increased to 530, remaining there until the eventual redeployment of 120 Wing VI Minuteman III missiles to the 341 SMW at Malmstrom (see Chapter 17).[52]

On 22 September 1994, Secretary of Defense William J. Perry released the first post–Cold War Nuclear Policy Review (NPR) at a press conference. He described a new policy based on mutually assured safety rather than mutually assured destruction.[53] To satisfy START II de-MIRVing stipulations, plans originally called for each Minuteman III wing to convert one squadron per year, beginning in late 1994, to a new single-reentry vehicle carrier platform, or bulkhead, yet to be designed and manufactured. A later decision to use the retired Peacekeeper Mark 21 reentry vehicle/W87 warhead combination, however, delayed de-MIRVing pending a Minuteman III guidance system upgrade to accommodate the Mark 21.

The source of the remaining 120 Minuteman III missiles for Malmstrom became evident on 1 October 1995, when the Base Realignment and Closure Commissions recommended inactivating the 321 SMW at Grand Forks AFB and transferring its missiles to the 341 SMW at Malmstrom. A survey of three of the four remaining Minuteman fields (F. E. Warren was excluded due to 50 Peacekeeper missiles deployed at the 400 SMS) indicated the high water table at Grand Forks reduced survivability and increased on-site support. The missile alert rate at Grand Forks had also been consistently lower than at Minot and Malmstrom.[54]

On 5 October 1995, a crew removed the first of 120 Minuteman III missiles designated for shipment to Malmstrom from the 446 SMS LF Alpha-04. The remaining Minuteman IIIs from Grand Forks went to temporary storage at Hill AFB or the Camp Navajo Depot in Arizona. On 4 June 1998, the last of the 150 Minuteman IIIs from Grand Forks was pulled from 447 SMS's LF Golf-15, and the Minuteman III force level returned to 500, remaining there until 2007.[55]

On 26 January 1996, the Senate ratified START II. De-MIRVing the Minuteman IIIs at F. E. Warren AFB commenced on 23 November 1998, when bulkheads finally became available. Instead of removing two of the three warheads on each Minuteman III, however, the Air Force opted to remove two of the three reentry vehicles from each of the 150 Minuteman IIIs at F. E. Warren to meet the limit of 1,200 required by December 2001 under START I—three warheads on each of the 200 Minuteman IIIs at Malmstrom and the 150 at Minot, for 1,050, plus the 150 at F. E. Warren.[56]

## Bush Administration 2001 to 2009

Conversion of 90 SMW Minuteman IIIs at F. E. Warren AFB to a single-reentry vehicle payload was completed on 6 August 2001. In accordance with the START I protocols, crews destroyed the bulkheads capable of carrying multiple warheads.[57]

The Bush administration released the second Nuclear Posture Review on 31 December 2001. It sought reduction of operationally deployed strategic nuclear warheads to between 1,700 and 2,200 within a decade. Like the Clinton administration, the Bush administration found it unnecessary to predicate US nuclear planning on countering the "Russian threat" after the Cold War's end. Maintenance of the current force structure and warheads taken off alert would constitute a "responsive" force. "Downloading," removal of one or more reentry vehicles from MIRVed ICBMs and SLBMs, would reduce total warheads to 3,800 by FY 2007. This was not based on START I, since the United States had achieved that goal in December 2001.[58]

On 24 May 2002, President Bush and Russia's President Vladimir Putin signed the Strategic Offensive Reductions Treaty (formally "Treaty on Strategic Offensive Reductions" and often abbreviated SORT, also known as the Moscow Treaty), which stipulated further reduction in strategic nuclear weapons to between 1,700 and 2,200 warheads by 31 December 2012.

The Defense Department's "Report of the 2006 Quadrennial Defense Review," dated 6 February 2006, announced the decision to remove hundreds of Minuteman III warheads to reach the 500-warhead level. To achieve that limit, each Minuteman III would carry one warhead, later changed to 50 missiles carrying two warheads and 400 carrying one warhead due to pending retirement of 50 missiles from the force.[59] Gen. James E. Cartwright, USMC, commander, US Strategic Command (USSTRATCOM), provided details at a hearing before the Senate Armed Services Committee on 29 March 2006. He said the primary reason for the reduction to 450 deployed missiles was a need for test assets to permit the test and evaluation launch program at Vandenberg to continue past the year 2020. For a 500-missile force, there were 539 propulsion system rocket engines (PSREs) and 552 boosters available. Assuming continued use of approximately four PSREs and booster assets per year for flight testing, both components would reach depletion by 2012, leaving no Minuteman III components for use in the Rocket System Launch Program or Missile Defense Agency targets. Conversely, with a 450-missile force, PSREs would last until 2024 and boosters until 2023.[60]

The 50 missiles carrying two warheads was not a violation of the START II Treaty because the Russians, responding to US withdrawal from the Anti-Ballistic Missile (ABM) Treaty, had withdrawn from START II in June 2002. All remaining Minuteman IIIs retained the capability of carrying one to three warheads. This complied with Moscow Treaty limits of 1,700 to 2,200 total strategic missile warheads. Gen. Cartwright did not specifically name where the 50-missile reduction would occur, but he mentioned the logic of three 150-missile wings.[61]

Montana's congressional delegation immediately began campaigning to delay the reduction in force since removal of the 564 SMS's 50 Minuteman IIIs seemed likely. The design of the squadron's internal communication system was identical to that of the 321 SMW at Grand Forks (WS-133B) and significantly different than the other nine Minuteman III (WS-133A-M) squadrons. This incurred excessive costs due to separate training and maintenance personnel at Malmstrom. Montana's congressional delegation succeeded in delaying Air Force action with an amendment to the National Defense Authorization Act for FY 2007 prohibiting withdrawal of any Minuteman IIIs from the active force until the secretary of defense provided a detailed strategic justification for the reduction.[62]

Congress received the justification in March 2007. On 26 April 2007, an Air Force news release acknowledged plans to inactivate the 564 SMS. Following the usual release of an environmental impact statement (EIS) and the public comment period, the Air Force directed removal of the squadron's missiles on 29 June 2007. Actual removal commenced with LF Sierra-38 on 12 July and continued until 28 July 2007, when the 50th missile left—LF Tango-41.[63] With inactivation of the 564 SMS, the Minuteman force level was now 450, deployed in three wings of 150 missiles each: 341 SMW, Malmstrom AFB, Montana; 91 SMW, Minot AFB, North Dakota; and 90 SMW, F. E. Warren AFB, Wyoming.

On 4 November 2009, a report commissioned by the Senate ICBM Coalition, chaired by Senators Kent Conrad (D-ND) and Michael Enzi (R-WY), urged retention of the entire 450 Minuteman III force level with single warheads. The coalition wanted to be clear that the ICBM force remained crucial to national security.[64]

## Obama Administration 2009 to 2017

The 2010 Nuclear Posture Review Report, released on 6 April 2010, announced the de-MIRVing of all Minuteman IIIs. The report stated, "This step will enhance the stability of the nuclear balance by reducing the incentive for either side to strike first." The report also reaffirmed the Clinton administration practice of day-to-day (i.e., peacetime) "open-ocean targeting" for all strategic missiles.[65]

Two days later, on 8 April 2010, President Barack Obama and Russian President Dmitry Medvedev signed "the Measures for the Further Reduction and Limitation of Strategic Offensive Arms Treaty" (the New START Treaty) in Prague, Czech Republic. It limited both countries to 700 deployed ICBMs, SLBMs, and heavy bombers. Furthermore, the treaty limited the number of warheads deployed on ICBMs, SLBMs, and heavy bombers to 1,550 deployed and 800 nondeployed ICBM launchers, SLBM launchers, and heavy bombers. Within those limits either country could structure its strategic forces as it saw fit. Both countries had seven years from 5 February 2011, the day the treaty entered into force, to meet those limits.[66]

In congressional testimony on 2 March 2011, Gen. C. Robert Kehler, USSTRATCOM commander, described the strategic force status under the new treaty:

> The treaty's limit of 700 deployed ICBMs, deployed SLBMs, and deployed heavy bombers, supports strategic stability by allowing the United States to retain a robust triad of nuclear delivery systems while downloading all Minuteman III ICBMs to a single warhead each.
>
> 800 deployed and nondeployed launchers of ICBMs, launchers of SLBMs and nuclear-capable heavy bombers allow the retention of up to 100 ICBM and SLBMs launchers and nuclear-capable bombers in a nondeployed status.

Kehler also announced plans to retain up to 420 of the currently deployed 450 Minuteman IIIs, plus 240 Trident II D5 SLBMs and 60 nuclear-capable heavy bombers, rendering a combined force of 720 deployed ICBMs, SLBMs, and heavy bombers. That was 20 over the treaty limits, which the United States had another seven years to meet. This decision gave the Air Force sufficient time to refine further reductions. The 50 Peacekeeper LFs, empty since 2005, and the 50 LFs at the 564 SMS that held Minuteman IIIs until 2008 could now be filled with gravel and removed from the deployed launcher list.[67]

At a House Subcommittee on Strategic Forces hearing on 2 November 2011, Representative Austin Scott (R-GA) asked Kehler about rumors of a significant cost-saving Minuteman III force reduction. Scott wondered whether meeting the deployed strategic launcher limit necessitated inactivating an entire Minuteman III wing or just a squadron. He also asked when de-MIRVing would begin. Kehler said the New START Treaty gave both sides considerable flexibility in managing deployed forces and STRATCOM was formulating the most cost-efficient force structure. He reassured subcommittee members that strategic implications were first and foremost in any decision to reduce Minuteman. STRATCOM and the Air Force planned to begin de-MIRVing the remaining 300 MIRVed Minuteman IIIs beginning in FY 2012.[68]

On 14 March 2014, Secretary of the Air Force Deborah Lee James testified before the House Armed Services Committee regarding the National Defense Authorization Act for FY 2015. In response to questioning by Representative Robert Bishop (R-UT), she alluded to a "warm" basing concept for Minuteman III but deflected further questioning.[69]

On 8 April 2014, the Obama administration sent Congress a report describing implementation of the New START Treaty, as required by Section 1042 of the National Defense Authorization Act for FY 2012. The report contained a budget for eliminating nonoperational ICBM launchers—the 50 Peacekeeper facilities at F. E. Warren AFB and 50 Minuteman III facilities at Malmstrom AFB. It said de-MIRVing of the remaining 300 Minuteman IIIs at Minot and Malmstrom AFBs would end in FY 2014. To many analysts' surprise, the report also budgeted for converting 54 Minuteman III LFs to nondeployed status—50 from among the three wings and 4 used for test launches at Vandenberg AFB.[70] Empty LFs at the operational wings would remain in "warm" status, meaning fully operational and ready to accept missiles, and the number of empty LFs would rotate among the three wings depending on maintenance issues, which allowed the full complement of missiles to be on alert during maintenance.[71]

To prevent any further force level reductions, the National Defense Authorization Act for FY 2015, Section 1644, required that each LF containing a deployed missile as of the legislation's enactment date, 19 December 2014, be preserved at "warm" status.[72] Removal began in May 2015 at all three bases and ended at Malmstrom in April 2017, Minot in May 2017, and F. E. Warren in June 2017 (Figure 2.1).[73]

As of November 2019, the Minuteman III force level stood at 400 deployed missiles, with actual numbers at the three bases fluctuating due to maintenance requirements. The Warm Silo Agreement resulted in an initial distribution of 17 empty LFs at Malmstrom, 16 at Minot, and 17 at F. E. Warren. The wings played leapfrog with the warm silos as emptied launchers underwent maintenance, corrosion repair, lubrication, and other activities in a process called programmed depot maintenance (PDM, previously known as Rivet MILE; see Chapter 16). Before those launchers could accept a re-emplaced missile, another

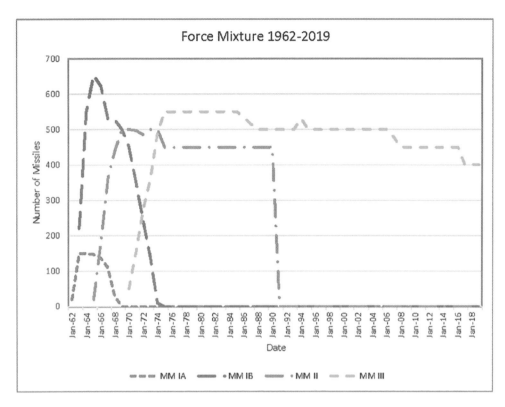

FIGURE 2.1 Minuteman force mix from 1962 to 2019. The figure shows the number of missiles deployed at the end of the calendar year. As of December 2019, there are 450 Minuteman III launchers available for missile emplacement, with a maximum of 400 Minuteman III deployed.

launcher had to be emptied to meet treaty requirements. This process coincided with a different PDM task concerning missile stages, with missiles periodically coming and going from each base.[74]

★ ★ ★

Just how did the Minuteman fit into US strategic forces over six decades? From the era of massive retaliation in the 1950s to counterforce and mutually assured destruction in the 1960s and beyond, details of the Minuteman's role remain classified. Using available published documents on the Single Integrated Operational Plan for strategic nuclear forces, however, Minuteman's role can be placed in context.

# 3

# MINUTEMAN IN CONTEXT

A quantum leap in nuclear warfare came with the deployment of ICBMs in the early 1960s. Flight times of 30 to 35 minutes gave much less warning than bombers that took hours to reach their targets. While IRBMs were deployed first, their flight time was much less than 30 minutes, but forward basing made them much more vulnerable to Soviet attack.

While ICBM accuracy differed from bombers, first-generation Atlas and Titan I missiles proved satisfactory against extensive air defense installations, bomber-filled air bases, and large-area military facilities, such as army depots and naval bases. The second-generation Titan II carried the single largest US ICBM warhead, approximately 10 megatons, for use against buried command-and-control facilities or large-area targets. As originally conceived, what Minuteman lacked in accuracy or warhead yield—a circular error probable (CEP) of 0.5 to 1.0 nautical mile (nm) and 0.5 to 1.5 megatons, respectively—it would make up in quantity.[1]

## Minuteman Era Begins

Minuteman entered the strategic nuclear forces arena on 27 October 1962, during the Cuban Missile Crisis. The 10 Minuteman IA missiles of Alpha Flight, 10 SMS, 341 SMW, Malmstrom AFB, achieved ECC status in response to Soviet IRBM and medium-range bomber deployment in Cuba (see Chapter 15). An additional five Minuteman ICBMs went on strategic alert status at Vandenberg AFB during the crisis. They joined 113 Atlas and 54 Titan I missiles on strategic alert across the country.[2]

On 11 December 1962, Alpha and Bravo Flights, 10 SMS, 341 SMW, were placed on strategic alert.[3] This began over a half century of continuous strategic alert for the Minuteman force. On 3 May 1967, only 4.5 years later, the land-based ICBM force reached its maximum size, with 54 Titan IIs and 1,000 Minuteman ICBMs on strategic alert. Atlas and Titan I missiles had been deactivated in 1964 and 1965.[4]

The decision to keep Minuteman deployment on schedule rather than waiting for resolution of range and reliability issues left coverage of Soviet and Chinese targets incomplete because the 150 Minuteman IAs at Malmstrom AFB had a reduced range of 4,780 to 4,950 nm. Deployment of Minuteman IB, with a 5,500-nm range, soon rectified this.[5] Targets in most of the Soviet Union and the northeastern part of China were within reach when all six operational wings, the full 1,000-missile force, achieved strategic alert in April 1967 (see Chapter 2).

## Eisenhower Administration 1953 to 1961

The ICBM's role in US war plans was unclear at the onset of its development. Prior to December 1960, the commanders-in-chief of nuclear-capable US military organizations prepared their own plans for using nuclear weapons within their areas of operation. This resulted in considerable targeting overlap. In August 1960 Secretary of Defense Thomas Gates initiated a new targeting approach for nuclear weapons, the Joint Strategic Target Planning Staff (JSTPS). The JSTPS generated a coordinated nuclear war plan, the Single Integrated Operational Plan (SIOP), using all US nuclear strike capabilities and evaluating two conditions for targeting: first, eliminate Chinese-Soviet strategic nuclear delivery capability, including military and governmental command-and-control centers, and second, attack major industrialized areas in China and Soviet-bloc countries. A national strategic target list contained more than 80,000 entries, which were screened to 3,729 critical facilities. A total of 1,060 designated ground zeros or aiming points meant that many targets were collocated and could be destroyed by one nuclear weapon of sufficient yield.[6]

The resulting plan, Single Integrated Operational Plan 62 for FY 1962 (SIOP-62), was completed during the Eisenhower administration. Adm. Arleigh Burke, chief of naval operations, staunchly criticized it. The Navy viewed SIOP-62 "not as an objective plan but a capabilities plan, aimed at utilizing all available forces to achieve maximum destruction." SAC had damage criteria that required the nuclear equivalent of 300 to 500 kilotons to achieve the damage caused by the 13-kiloton Hiroshima bomb. Navy strategic planners claimed fallout alone from these massive overtargeting policies would endanger areas far removed from the Soviet Union or China, saying, "The fallout already exceeds JCS limits for points such as Helsinki, Berlin, Budapest, Northern Japan, and Seoul." Professor George B. Kistiakowsky, President Eisenhower's science advisor, reviewed the preliminary SIOP-62 and agreed with the initial attack premise but felt follow-on portions resulted in the ability "to kill 4 to 5 times over somebody who is already dead." The Eisenhower administration did not implement SIOP-62, preferring to leave that task to the incoming Kennedy administration.[7]

## Kennedy Administration 1961 to 1963

SIOP-62 went into effect on 15 April 1961, and President John F. Kennedy was briefed on 13 September 1961. Of its 1,060 aim points, 800 were primarily military targets, the remainder being urban or industrial. The SIOP listed 14 launch options. Option 1, the Alert Option, would generate 1,004 bombers and missiles with 1,685 nuclear weapons, which could launch within 15 minutes of receiving the command. The final option, Option 14, was all 2,244 delivery systems, with a total of 3,267 nuclear weapons. Although 14 options gave a sense of flexible response, all amounted, in reality, to a massive attack, because the other options built upon Option 1. Each option differed by preparation time for launching an attack. The longer the preparation time, the greater the forces generated and launched. Getting nuclear forces launched before a Soviet attack destroyed them became a major concern. With enough time decisions about withholding attacks on Soviet-bloc countries were possible, but the 14 options did not include them. As target numbers continued to increase, the Minuteman force had an important role in SIOP planning and execution.[8]

Lack of flexibility in SIOP-62, a remnant of Eisenhower's massive retaliation policy, dissatisfied Kennedy. In the surreal world of nuclear strategy, these essentially all-or-none options would have made it difficult for the Soviet leadership to distinguish between an attack oriented solely toward military facilities and one on urban or industrial centers, making their likely response total retaliation. Gen. Lyman L. Lemnitzer, chairman of the JCS, countered that even if they made the distinction, the Soviet response would be equally devastating. He emphasized the benefit of using our weapons rather than losing them in the retaliatory attack. A simplified plan, such as SIOP-62, made retaliation against a surprise attack much more feasible if communications were already disrupted.[9]

Secretary of Defense McNamara also urged greater flexibility, briefing the new SIOP-63 to President Kennedy on 14 September 1962. Its "optimum mix" contained three parts or tasks—nuclear targets, other military targets, and urban or industrial targets—and five primary attack options: Soviet strategic nuclear forces; military facilities located away from cities, such as air defenses that covered bomber routes; Soviet conventional forces in close proximity to urban areas; Soviet command-and-control centers; and total attack, including all of the above. Within each of the five major options were suboptions: air- versus ground-burst weapons, clean versus dirty bombs, larger versus smaller warheads, and civil defense versus evacuation. Ironically, while these options and suboptions gave the appearance of greater flexibility, all employed thousands of nuclear warheads.[10]

## *Johnson Administration 1963 to 1969*

In October 1964, when President Lyndon Johnson requested an update on the status of US nuclear forces, the "alert" force had 1,798 megatons of nuclear warheads. That included 150 Minuteman IAs, each with a 1-megaton W-59 warhead, and 510 Minuteman IBs carrying the 1.2-megaton W-56. This represented 42 percent of US nuclear-capable alert forces.[11]

## *Nixon Administration 1969 to 1974*

With some modifications SIOP-63 remained in effect until the Nixon administration. The time period of 1963 to 1974 encompassed the era of "mutually assured destruction." The intent of the American response to a nuclear attack had been, according to McNamara, "to inflict an unacceptable degree of damage upon any single aggressor, or combination of aggressors, even after absorbing a surprise first strike."[12]

In late January 1969, Dr. Henry Kissinger, President Nixon's national security advisor, began a comprehensive review of SIOP-63. It sparked a multiyear process that culminated on 1 January 1976 in SIOP-5, which contained a series of limited-attack scenarios to prevent escalation to all-out nuclear war.[13] The new plan aimed for 70 percent destruction of the economic and industrial base the Soviet Union needed for economic recovery after a large-scale strategic nuclear exchange, a goal stated in "Policy Guidance for the Employment of Nuclear Weapons, 3 April 1974."[14] As John B. Walsh, deputy director, Strategic and Space

Systems, Defense Research and Engineering, explained when questioned by Representative Jack Edwards (R-AL), this included targeting Eastern Europe to prevent its use by the Soviet Union as a base for economic recovery, as had happened after World War II. Target classes remained all-encompassing but were now segregated into four main categories, each further delineated by a subcategory that governed choice of attack option: Soviet strategic nuclear forces, conventional forces and bases, military and civilian command-and-control centers, and economic- or industrial-base facilities.[15] Significantly increased delivery capabilities with Minuteman III ICBMs, plus Poseidon and Trident SLBMs, providing a 0.1– to 0.5-nautical-mile CEP, made possible the refined number of attack options (Tables 3.1 and 3.2; Figures 3.1 and 3.2).[16]

## Into the Future

With Titan II deactivation, announced on 24 April 1981 and completed on 14 July 1987, Minuteman became the sole land-based ICBM strategic deterrent until the deployment of 50 Peacekeeper missiles beginning on 3 January 1986.[17] In February 1995 Secretary of Defense William Perry summarized the land-based strategic deterrent's future. Assuming START II became effective, Peacekeeper deactivation would begin in 2003, leaving a force of 500 Minuteman IIIs as the sole land-based deterrent through at least 2010, with no current plans for replacement.[18] In December 2001 Secretary of Defense William Cohen's annual report included plans for the modernization of the Minuteman III force to ensure capability through 2020.[19] The Defense Department's Nuclear Posture Review Report of April 2010 explained the continuation of the Minuteman III Life Extension Program to keep the system operational through 2030, as mandated by Congress.[20]

At the end of 2019, Minuteman remained a potent strategic deterrent, 57 years after initial deployment. Like the ageless, repeatedly modified B-52, today's Minuteman III only externally resembles the Minuteman III first deployed in the 1970s, testimony to the versatility of its original design more than a half century ago.

★ ★ ★

Just what breakthrough made Minuteman even feasible? While the Air Force claimed credit for critical developments in solid-propellant performance, further analysis reveals the Navy as a major player that beat the Air Force to deployment of a large-diameter, solid-propellant missile, Polaris A-1, aboard the USS *George Washington* (SSBN-598) on 15 November 1960, fully two years before Minuteman IA deployment.[21]

TABLE 3.1  Minuteman Payload Summary

| MISSILE | RV | NO. OF RVS | PEN. AIDS | DEPLOYMENT DATES | WARHEAD (MT) |
|---|---|---|---|---|---|
| MM IA | Mark 5 | 1 | -- | 1962–1969[a] | W59 (1)[b] |
| MM IB | Mark 11, 11A | 1 | -- | 1963–1975[a] | W56 (1.2)[b] |
| MM II | Mark 11B, 11C | 1 | Mark 1, 1A | 1963–1991 | W56 |
| MM II | ERCS | 1 | -- | 1967–1991 | -- |
| MM III | Mark 12 | 1–3 | chaff, decoys | 1970–2010[c] | W62 (0.170)[c] |
| MM III | Mark 12A | 1–3 | chaff, decoys | 1979[c] | W78 (0.330)[c] |
| MM III | Mark 21 | 1–2 | -- | 2007[c] | W87 (0.300)[c] |

a) "History of Strategic Air Command, FY 1970," Historical Study Number 117, Vol. II, Narrative;
b) Hansen, *The Swords of Armageddon, US Nuclear Weapons Development Since 1945*, Vol. III;
c) Norris and Kristensen, "US Nuclear Forces, 2010," *Bulletin of Atomic Scientists*, May–June 2010;
"US Nuclear Forces, End of 1991," *Bulletin of Atomic Scientists*, January–February 1992.

TABLE 3.2  Minuteman Range, Accuracy, and Reliability Estimates, 1970

| MISSILE | RANGE (NM) | CEP (NM) | RELIABILITY |
|---|---|---|---|
| MINUTEMAN IA (Mark 5) | 4,780–4,910[a,f] | 1.00[d] | 0.60[d] |
| MINUTEMAN IB (Mark 11) | 5,500[a] | 0.41[e] | 0.80[d] |
| MINUTEMAN IB (Mark 11A) | 5,500[a] | 0.65[e] | 0.80[d] |
| MINUTEMAN II (Mark 11B, C) | 5,300–6,300[c] | 0.50[b] | 0.60[d] |
| MINUTEMAN III (Mark 12) REENTRY VEHICLE 1 | 5,000[c] | 0.25[b] | -- |
| REENTRY VEHICLE 2 | 5,650[c] | 0.25 | -- |
| REENTRY VEHICLE 3 | 6,270[c] | 0.25 | -- |

a) "USAF Ballistic Missile Programs, 1962–1967"; b) "USAF Ballistic Missile Programs, 1969–1970";
c) "History of the Strategic Air Command, FY 1975"; d) "History of the Strategic Air Command, FY 1969"; e) "History of the Strategic Air Command, FY 1970"; f) Lauritsen Committee Report.

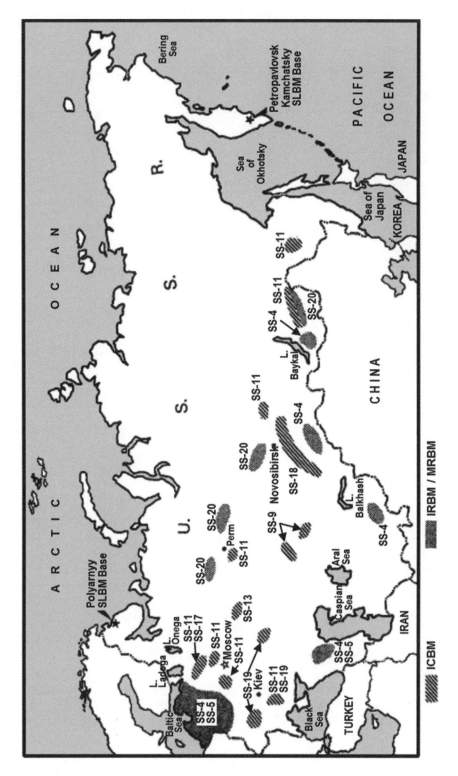

FIGURE 3.1 Soviet missile base locations circa 1980. While Minuteman was most likely not targeted on specific missile complexes, most of the Soviet sites were within range, something that the Soviets had to consider. The missile bases were frequently collocated with the "core areas," defined as "concentrations of military, political, or economic activity vital to the welfare of the Soviet Union" that were conceivably targets. *Adapted from "US-Soviet Military Balance: 1960–1980."*

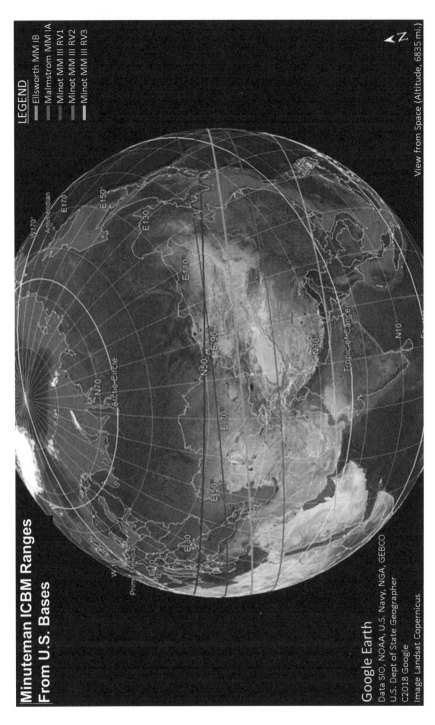

FIGURE 3.2 Approximate ranges for Minuteman IA from Malmstrom AFB, Minuteman IB from Ellsworth AFB, and Minuteman III from Minot AFB. Minuteman II Mark 11B and 11C had similar ranges to the Minuteman III first and third reentry vehicles, respectively (Table 3.2). *Adapted from "History of Strategic Air Command, 1976."*

# 4

# SOLID PROPELLANT COMES OF AGE

While development of lightweight nuclear weapons and reentry vehicles and inertial guidance systems was critical to the feasibility of the Minuteman ICBM program, development of more energetic solid propellants that could replace the cumbersome liquid-propellant systems was the first link.[1]

## In the Beginning

The history of solid-propellant rockets is vague before the extensive work on bombardment rockets by two British experimentalists, William Congreve and William Hale, in the early and mid-nineteenth century. Congreve and Hale brought a terror weapon used by the armies in India into the European and American arsenals. The capabilities of their rockets were limited by the available black powder propellant with a specific impulse, $I_{sp}$, of 80 to 100 seconds.[2] Lack of a combustion chamber per se, since both the Congreve and Hale rockets used a conical internal-burning grain design (cross-sections of several of the Congreve rockets appear to also show a cylindrical grain), further limited their performance.[3] Congreve made two major contributions to the nascent field of rocketry. First, he used metal cases instead of cardboard, which provided for better storage. Second, he mounted the guide stick to the longitudinal (the center of the rocket instead of the side) axis of the rocket, thereby eliminating the tendency for the rockets to yaw and lose accuracy. Hale went one step further by eliminating the guide stick completely by introducing spin stabilization, first with small tangential holes in the rocket case and later with curved vanes at the rocket's base, which slightly redirected the exhaust to provide a spinning motion.[4]

In the early 1900s, Robert H. Goddard conducted the first concerted solid-propellant rocket research in the United States. Better known for his research and development of liquid-propellant sounding rockets from 1920 to 1940, Goddard actually began his research at Princeton University in 1912 with theoretical work on increasing the efficiency of solid propellants. He carried out experiments for boosting an instrument payload to high altitudes to study the upper atmosphere at Clark University during 1915 and 1916.[5]

Goddard picked up where Congreve and Hale had left off, realizing that he had to greatly improve his calculated 2 percent efficiency of their designs if he hoped to reach the upper atmosphere with a sounding rocket. He defined efficiency as "the ratio of the velocity of the ejected gas to the heat energy of the powder." Goddard systematically improved methods for determining thrust and efficiency, which culminated in detailed theoretical work on a "reloading, or multiple-charge" rocket at Worcester Polytechnic Institute in 1917.[6]

Goddard used these new techniques to demonstrate that his application of the de Laval nozzle design (a converging, then diverging shape) improved thrust to nearly 63 percent.[7] His work also demonstrated that thrust in a vacuum was the same as in the atmosphere, a key to the future use of rockets in space.[8] Using his newfound knowledge on propellant efficiencies, Goddard decided to move from the limitations of solid propellants to a liquid-propellant system using gasoline and liquid oxygen, which delivered the increased thrust needed for reaching higher altitudes.[9]

## Guggenheim Aeronautical Laboratory, California Institute of Technology

The genesis of the Minuteman program has most often been equated with the work of Col. Edward Hall at the Air Force's WDD during 1955 and 1956. This is only partially true, as the drive to determine the feasibility of scaling up the size of solid-propellant rockets began much earlier. The first research into solid-propellant rockets larger than two inches in diameter for rocket-assisted takeoff (RATO) of aircraft occurred in the mid-1930s at the Guggenheim Aeronautical Laboratory, California Institute of Technology (GALCIT, Caltech), under the tutelage of Dr. Theodore von Kármán. Over the next two decades, the Caltech team, along with Allegany Ballistics Laboratory (ABL) and others, including Hall, evolved their research to the point where large-diameter, solid-propellant rockets for weapon systems appeared feasible.

Experiments had begun at Caltech in 1936 with the work of Frank J. Malina, John W. Parsons, and Edward S. Forman. Apollo M. O. Smith and Hsue-Shen Tsien soon joined them. The regular explosions, as they refined techniques, earned this group the sobriquet "the Suicide Club."[10]

In 1938 Malina authored two papers, one with Smith and another with Tsien, on the theoretical flight analysis of a sounding rocket, expanding on Goddard's earlier work and refining design parameters. In the first paper, Malina and Smith showed that if a high-efficiency rocket motor could be built as a "step" rocket with stages of 600, 200, and 100 pounds, it could reach an altitude of nearly 1,000 miles and a maximum speed of 11,000 miles per hour. They concluded, "This analysis definitely shows that if a rocket motor of high efficiency can be constructed, far greater altitudes can be reached than is possible by any other known means."[11]

Six months later, publication of a further refinement by Tsien and Malina indicated a solid rocket, propelled by successive impulses from currently available motors, could reach above 20 miles in altitude. In addition, the authors compared the efficiencies of the solid-propellant, constant-volume motor to a constant-pressure, liquid-propellant engine, explaining that abandonment of solid-propellant motors resulted from their lack of sustainable thrust. With successive impulses, using solid-propellant cartridges, it seemed the simplicity of such a design over that of a liquid-propellant engine warranted further experimentation.[12]

All these ideas placed Caltech at the forefront of rocket propulsion research in the United States. Von Kármán had a long-time association with Gen. Arnold, and when Arnold took command of the Army Air Corps in September 1938, he informed von Kármán about the military's interest in rocket propulsion. Arnold gathered scientists to make presentations in Washington, DC, at a December 1938 meeting of the National Academy of Sciences Committee on Army Air Corps Research. That meeting focused on the most pressing problems facing the new B-17 heavy bomber: radar, aircraft windshield deicing, and RATO. Caltech President Robert Millikan and von Kármán were members of the committee and chose Malina to represent GALCIT, while Vannevar Bush and Jerome Hunsaker represented the MIT. Bush and Hunsaker took on the deicing problem, while dismissing RATO work as a "Buck Rogers" undertaking.[13] Bush commented to Millikan and von Kármán that he didn't understand how "a serious engineer or scientist could play around with rockets."[14] Since the term "rocket" was somewhat controversial at the time, often subjecting its users to ridicule, scientists began referring to RATO as JATO, and the name stuck.[15]

In January 1939 the committee asked von Kármán and the GALCIT group to study the JATO problem and prepare a formal research proposal. On 1 July 1939, GALCIT received a $10,000 grant (GALCIT Project No. 1) from the National Academy of Science to study the JATO concept using both solid- and liquid-propellant systems.[16]

The next month, Parsons and Forman published a paper demonstrating experimental feasibility of rocket propulsion by successive impulses (actual construction of such a rocket proved impractical). Hailed by the *Astronautics* journal editor as "without a doubt the most painstaking investigation into dry fuel motors since the early Goddard days," this article confirmed and extended Goddard's earlier work and that of their coworkers. However, the sounding rocket program soon took a backseat to the fully funded JATO study.[17]

By early 1940 von Kármán had tired of the repeated explosions from the solid-propellant tests. Malina recalled von Kármán spent an evening writing four differential equations, which he then asked Malina to solve:

> He said to me: "Let us work out the implications of these equations; if they show that the process of a restricted burning-powder rocket is unstable, we will give up, but if they show that the process is stable, then we will tell Parsons to keep trying."[18]

The equations indicated that for solid-propellant rockets to provide stable thrust for use in JATO units, the ratio of area of the throat of the exit nozzle to the burning area of the propellant had to remain constant. End-burning charges, which burned like a cigarette, should perform along those lines in theory since the surface area of the burning propellant in an end burner remains constant because the burned area of the propellant is equal to the area of the circle formed by the motor case.[19] The internal grain designs of Congreve and Hale did not have this constant thrust capability because the conical or cylindrical surface area increased as propellant was consumed. That worked for skyrockets but not for the thrust needed to assist in takeoff of bomb-laden aircraft.

By July 1940 progress was reported on both liquid- and solid-propellant systems. Red fuming nitric acid replaced liquid oxygen as a more readily storable oxidizer. Solid propellants with burn times of 10 seconds or more became routine, a considerable improvement over the previous 1-second duration. The Army Air Corps assumed direct sponsorship of the work after learning about program progress. In addition, a close relationship between

Caltech and the Air Materiel Command (AMC) at Wright Field, Dayton, Ohio, began as Aircraft Laboratory Project MX-121.[20]

One year later, in early 1941, with a considerably expanded staff and facilities, the decision was made to conduct a flight test with the solid-propellant JATO. Theoretical analysis was made on the effect JATO thrust would have on a small aircraft, and AMC chose the Ercoupe, a low-wing twin-tailed monoplane, for the tests. The first successful JATO of an aircraft took place at March Field, Riverside, California, on 12 August 1941, with Captain Homer A. Boushey Jr., Army Air Corps, at the controls. Three solid-propellant JATO bottles were attached under each wing and designed so that both the exhaust nozzles and combustion chambers would fly free of the aircraft in case of failure. The distance required for takeoff was shortened from 580 to 300 feet; with an overload of 285 pounds, the distance was shortened from 905 to 438 feet. During the tests 152 JATO units were used without a failure (Figure 4.1).[21] Unfortunately for the program, when JATO units containing the GALCIT 27 propellant used in the tests were taken from accelerated storage conditions, they tended to explode upon ignition.

Current manufacturing techniques for propellant systems took dry powder of various compositions and compressed it under 40,000 pounds per square inch (psi) in a sequential series of 22 layers into the rocket casing.[22] It already was known that prolonged storage led to cracking and unstable burning. Continued work by Parsons, Mark M. Mills, and Fred S. Miller led, in June 1942, to an asphalt–potassium perchlorate castable propellant, designated GALCIT 53, which was the solution to this problem. Potassium perchlorate was mixed with roofing asphalt at 350 degrees Fahrenheit (F). The interior of the motor casing was lined with a mixture of asphalt and oil that served, when cooled, to inhibit the propellant from burning along the casing wall. The propellant mixture was then cooled slightly and poured into the casing and set aside for final solidification. Rocket motors using GALCIT 61-C, a derivative of GALCIT 53, had a storage temperature range of minus 9 to 120 degrees F and were used by the Navy in JATO units until the end of World War II.[23]

Given the Navy's interest in JATO systems, the need for commercial manufacture of JATO units became apparent. The researchers looked into forming a business arrangement with several aircraft companies. The Caltech administration had no objections since Caltech was not in the production business, but the researchers could not find an interested aircraft company. They decided to form their own company, the Aerojet Engineering Corporation, to produce both liquid- and solid-propellant JATO systems. Aerojet was formally incorporated on 19 March 1942, with von Kármán as president and Malina as treasurer.[24]

The JATO development work continued, using both liquid- and solid-propellant units, with the GALCIT Project No. 1 contract limiting the work to design of rockets for use with aircraft only. Then, in 1943, research took an abrupt turn when British intelligence asked von Kármán to identify structures in aerial photographs of new German facilities on the French coast. Joined by Malina and Tsien, von Kármán reported the structures looked like missile launch platforms and missile assembly buildings. The British had come to the same conclusion. The equipment turned out to be V-1 assembly and launch facilities.[25]

The Army Air Corps Materiel Command liaison officer for Caltech, Col. W. H. Joiner, saw the need for Army Air Forces development of similar capabilities. Joiner asked the

FIGURE 4.1
Ercoupe NC286655 becomes airborne on 12 August 1941. The Porterfield aircraft in the foreground had started its takeoff roll simultaneously with the Ercoupe. The JATO cut the takeoff roll from 580 to 300 feet and reduced takeoff time from 13.1 to 7.5 seconds. *Photograph courtesy of Jet Propulsion Laboratory.*

GALCIT group to study the feasibility of adapting the JATO work for use in long-range rockets. Von Kármán, Malina, and Tsien completed the feasibility study in November 1943. Their analysis covered liquid- and solid-propellant rockets, plus ramjet and compressor jet propulsion. The state of the art in the United States at this time led them to conclude that long-range rockets using solid or liquid propellants were immediately feasible, but the range was approximately 10 to 11 miles with a warhead weighing 50 to 200 pounds. Using reasonable assumptions, however, they concluded that a rocket weighing 10,000 pounds with a range of 100 miles could be built.[26] Based on the report, von Kármán proposed to the Army Air Force Materiel Center an expansion of current research to develop ramjet engines and warhead-carrying rockets. The term "Jet Propulsion Laboratory" first appeared in this report.[27]

## Jet Propulsion Laboratory, California Institute of Technology

Army Air Forces showed no interest in pursuing the proposal, but the Army Ordnance liaison officer for Caltech, Capt. R. B. Staver, forwarded a copy to his superiors. On 15 January 1944, von Kármán received a request for an expanded program. This resulted in a one-year, $1.6-million contract in June 1944, known as Ordnance/California Institute of Technology 1 (ORDCIT-1). On 1 July 1944, work began on guided missile development. The initial focus included four objectives: develop a missile carrying a minimum of 100 pounds of high explosive, with a range of up to 150 miles, a target dispersion of no more than 2 percent of the range, and a speed greater than a contemporary fighter aircraft.[28]

The first solid-propellant missile developed on the new contract was Private A (XF10S1000-A; X for experimental, F for fin-stabilized, 10 for 10-inch-diameter, 1,000 for pounds of thrust, and A for first in series).[29] Private A was an R&D project, not destined to be a weapon system. The missile design was as simple as possible: 8 feet long, 10 inches in diameter, carrying a 60-pound payload, and powered by a 192-pound charge of the asphalt-based GALCIT 61-C restricted end-burning propellant, providing 1,000 pounds of thrust for 34 seconds. The choice of a solid-propellant motor stemmed from JPL's JATO development experience. Four tail fins provided the only guidance, with the missile being boosted along a 36-foot launch rail by a cluster of four armament rockets. JPL conducted 24 flights in December 1944, with an average range of 18,000 yards, a maximum altitude of 14,500 feet, and an estimated maximum speed of 1,300 feet per second (886 miles per hour). Solid-propellant rocket motors now had a place in military weapon systems development.[30]

Meanwhile, research by JPL's solid-propellant group continued to improve on the storage and operational temperature limitations of the asphalt-based propellants. If solid-propellant rockets were to be useful, not only as JATO units but as aerial weapons, the tendency for the asphalt-based propellant to crack below minus 9 degrees F or turn to liquid above 120 degrees F had to be rectified.

In February 1944 work began on finding a suitable replacement propellant, starting with a newly developed butyl rubber compound, Buna-S. Guided by Charles Bartley, formulation work was begun and test firing carried out by Lawrence Settlemire in mid-1945. Buna-S had all the temperature properties asphalt lacked, but it had one significant drawback: poor binding properties with the insulation at the casing surface. Neoprene formed an excellent bond with the insulating liner but was not conducive to large-scale manufacturing.[31] During a conversation with Shell Development Company's Frank M. McMillan at the annual American Chemical Society meeting, Bartley learned about LP-2, a new liquid polysulfide polymer Thiokol Chemical Corporation had developed that could be cured to form rubber.[32] The solution had been found. In liquid form LP-2 could be readily mixed with the oxidizer and, when a catalyst was added, solidified and easily bonded to the casing insulation. The new propellant mixture had significantly lower chamber pressures compared to the earlier propellants. Therefore, chamber walls could be thinner and the motor lighter for better performance. Over the next two years, Bartley's team at JPL worked to refine formulation of the new propellant.[33]

Beyond propellant formulation, the team faced another major problem. Inherent to end-burning rocket motors was the transitory rise in thrust at ignition to a stable thrust higher than calculations predicted. In industry terminology, the thrust was "progressive" and then "neutral." These results implied an increase in the surface area of the burning propellant, but crack development, the logical source of surface area increase, and delamination of the propellant–casing bond, were no longer a problem. The original cigarette-like burning surface apparently changed into a cone-shaped surface that stabilized with a larger surface area. Terminating thrust shortly after ignition confirmed this. While the actual cause of the phenomenon proved difficult to determine, the practical solution became casting the propellant with the cone shape in place. Their papers made no mention of Congreve's or Hale's much earlier and similar internal grain designs.

Two new employees, John I. Shafer and Henry L. Thackwell, were among many brought in to investigate the problem. By February 1946 their efforts indicated the polysulfide-based propellant had enormous possibilities for much larger motors, provided resolution of the coning problem could be achieved.[34]

Investigation of the coning phenomenon expanded into developing new internal grain shapes to achieve higher thrust with available propellants through increased surface-burning area. British scientists had carried out work to this effect in 1935, in the push to develop anti-aircraft rockets for London's defense against bomber attacks. A British scientist, Harold J. Poole, postulated that using an internal-burning, star-shaped grain, available thin-walled tubing could be used because the unburnt propellant would provide sufficient insulation. Use of a combustion inhibitor at the case–propellant interface also served as a binder to the case wall. While straightforward in theory, the British dropped the star-configuration concept after successful flight tests because problems arose with the case bonding seal being damaged by nitroglycerin leaking from the cordite propellant during storage. To make matters worse, slippage of the propellant charge under acceleration caused motor failure. In late 1937 a loose charge propellant, meaning not case-bonded, replaced the star-grain design, with satisfactory results.[35]

Poole discussed the British research with Caltech and other laboratories during his US visit in 1943.[36] Two engineers, R. J. Thompson and R. R. Newton at ABL in West Virginia, used the concept in their theoretical design of the Vicar high-speed rocket, which developed into a flight test model called Curate, whose 1945 flight likely marked the first successful US flight test of an internal-burning, star-grain design.[37]

Having learned about the British work in late 1946 and unclear as to why it took three years, Bartley asked Shafer to review it along with ABL's published report. By early 1947 the stage was set for the next evolutionary step in solid-propellant rocketry: large-diameter rockets with ranges equal to or surpassing the 150 to 200 nm of the V-2. The new star-grain design seemed to promise use of thinner casing walls and facilitate longer burning times and higher speeds. The problems the British had with insulation would not be an issue because the polysulfide propellant bonded directly to the casing wall, thereby eliminating the need for a restrictive layer of insulation.[38]

Interestingly, an external cruciform-grain solid-propellant rocket, the 3.5-inch-diameter aircraft rocket (AR), was used in the latter stages of World War II. A second Caltech group, funded by the Office of Scientific Research and Development and working under National Research and Defense Council (NRDC) auspices, developed the AR for the Navy. In 1946 Edward W. Price, a physicist at the Naval Ordnance Test Station (NOTS), China Lake, California, where the NRDC group had relocated from its Eaton Canyon facilities near Pasadena, independently developed an improved AR version, the high-velocity AR, using an internal-burning, eight-point-star grain, double-base powder mixture of nitroglycerin and nitrocellulose. Price experienced the same problem the British had encountered with the cellulose acetate plastic casing liner. A nationwide search found a suitable plastic liner, ethylcellulose, which solved the problem.[39]

In October 1947 Martin Summerfield, Shafer, Thackwell, and Bartley proposed a sounding rocket, the experimental solid-propellant vehicle (XSPV)—nearly 17 feet tall, 12 inches in diameter, and theoretically capable of reaching an altitude of 810,000 feet.[40] A new polysulfide-based propellant, JPL-121, with a 10-point-star grain, became key to the ambitious proposal. Army Ordnance and the Navy heard about JPL's progress with polysulfide propellant formulations and requested an evaluation of the potential for large, solid-propellant motors. Since only static tests had been done with the new propellant, confirmation of its ability to withstand high acceleration required actual flight testing. A six-inch-diameter flight test vehicle, nicknamed Thunderbird, with an internal-burning, 10-point-star grain, was fabricated with JPL-121 polysulfide propellant, using ammonium perchlorate as the oxidizer instead of potassium perchlorate. Successful flight tests in late 1947 and early 1948 prompted the engineers to revisit their earlier calculations (Figures 4.2 and 4.3). The Thunderbird program had validated Shafer's internal star-grain design and demonstrated structural integrity under a heretofore unattained acceleration of 120 times gravity. In April 1948, using the data from these tests, Summerfield, Shafer, Thackwell, and Bartley made a more audacious proposal to build a much more powerful solid-propellant rocket than the V-2, one capable of carrying a nine-ton payload to the same range as the V-2's one-ton payload.[41]

The Army authorized the Sergeant sounding rocket, a test vehicle based on the XSPV, in the spring of 1948. A 15-inch diameter made Sergeant a bold state-of-the-art advancement for JPL. To reach the desired altitude of 700,000 feet, a steel case only 0.065 inch thick was specified, since the internal star-grain propellant design would insulate the case from the combustion gases. Preliminary static test firings with a heavy steel case in February 1949 proved the versatile polysulfide propellant easily enabled an increase in propellant diameter from the 6-inch-diameter Thunderbird to 15 inches. Moving to the much thinner, weight-saving flight test vehicle casing turned out to be a bad idea, with all 12 static firings ending explosively. Those failures spelled the demise of Sergeant's development. JPL Director Louis Dunn, not waiting for researchers to determine the cause of the cracking phenomenon, canceled all further work on the project and limited the solid motor group to fundamental research. Salvaged pieces of the exploded casing and propellant revealed that cracks had developed from the sharp star points and migrated to the casing surface, burning and causing excessive pressure, with subsequent casing failure.[42]

The thin-walled steel casing explained the cracking problem, because the pressure of combustion gases caused more casing expansion than predicted, which allowed the propellant to expand and resulted in the observed cracks. Three solutions to the problem were developed: use of a thicker steel case, reformulation of the propellant for somewhat more elasticity, and modification of the star pattern by rounding the points. Since Dunn's edict prevented further testing, Thackwell abruptly left JPL to join Thiokol Chemical Corporation, where he continued his research on polysulfide propellants. JPL turned its attention to liquid-propellant research and development of the Corporal guided missile for the Army.

After he arrived at Thiokol's Redstone Division, Redstone Arsenal, Alabama, Thackwell's enthusiasm for development of a truly large-diameter, solid-propellant rocket using the polysulfide propellant garnered the attention of Richard W. Porter, manager of General Electric's Project Hermes, which was experimenting with captured V-2 missiles at White Sands Proving Grounds.[43]

Thackwell and Porter proposed a solid-propellant version of the Hermes A-1, a smaller version of the V-2 then being considered for development as a weapon system. The new missile, Hermes A-2, would carry a 1,500-pound payload 75 nm using 3,000 pounds of polysulfide solid propellant burning for 40 seconds. R&D began in May 1950. Using

FIGURE 4.2 The Thunderbird test rocket at Inyokern, California, 25 November 1947. This was the first flight test of Charles Bartley's polysulfide propellant as well as the first flight test of John Schafer's 10-point-internal-star grain design. Al Richardson is on the left. *Photograph courtesy of Jet Propulsion Laboratory.*

Thackwell's knowledge of JPL's mistakes, Thiokol researchers decided on a modified star-grain design with more rounded points and a much thicker case, 0.200 instead of 0.065 inch. Phase I subscale tests with 6-inch-diameter, 1,000-pound motors were conducted, with nearly 300 successful tests using four different internal star grains (five, six, and seven points). These tests measured levels of unburned propellant (due to star-grain combustion geometry, slivers of unburned propellant remained at burnout, insulating the casing), the ratio of initial to final perimeter, and use of multiple propellant batches due to available mixers having a smaller capacity than the final motor volume.[44]

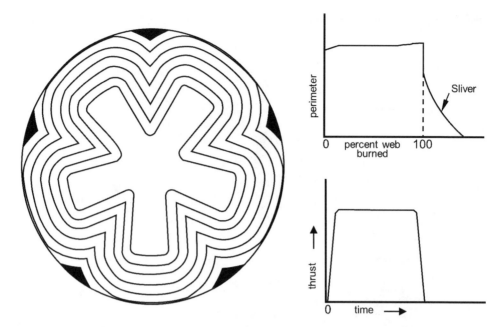

FIGURE 4.3 Illustration of the burn pattern of a five-point-star grain. The five dark areas are called slivers and represent unburned propellant at the end of thrust. The upper right figure illustrates the constant perimeter burning area with tailoff to the sliver. The lower right figure shows the neutral thrust curve versus time, the preferred profile for use with ballistic missiles. *Adapted from "Solid Propellant Grain Design and Internal Ballistics," NASA SP-8076.*

One unanswered question remained. Could a 31-inch-diameter propellant load support itself at elevated temperatures, or would it sag or slump? A 24-inch-diameter test motor, nicknamed "the lard tub," was poured, cured, and stored horizontally for two months while subjected to thermal cycling between 160 degrees F for six hours and room temperature for 18 hours. While slight shifting occurred, it appeared that propellant sagging would not be an issue.[45]

Phase II began with the first full-scale static firing with a five-point-star, 31-inch, 4,786-pound motor on 2 December 1951, only 20 months from the start of the program. A total firing time of 41.2 seconds generated a total impulse of 795,000 pound-seconds. This represented a milestone in solid propellant development, a powerful indicator of things to come.[46]

Laboratory research also had been focused on increasing the propellant's specific impulse. Consequently, for the second full-scale static test, oxidizer content was increased from 56 to 67 percent. Pressure fluctuations, called combustion instability, occurred, but reduction of the oxidizer content to 60 percent eliminated this problem.

In 1952 fabrication of 26 full-scale 31-inch motors under the Phase II program enabled successful static firing of 21 motors and successful flight testing (Hermes A-2 designation was changed to RV-A-10) of 4 in 1953, with 1 placed in storage until it was fired successfully seven years later. The RV-10-A program ended in September 1953 with a decision not to turn it into a weapon system.[47]

In early 1952 Redstone Arsenal contracted with JPL to review all Army missile programs. The initial study covered only liquid-propellant missiles but expanded in 1953 to include solid-propellant missiles. The conclusion was that solid-propellant missiles were equal or superior to their liquid counterparts with a 1,500-pound payload at ranges of 25 to 100 nms. JPL staff suggested further study and proposed designing and developing a Sergeant tactical ballistic missile (TBM) system. In 1954 both JPL and Thiokol received contracts to jointly develop the Sergeant missile as a direct outgrowth of the RV-A-10 program. The first test article in the Sergeant development program consisted of two RV-A-10 motor cases welded together and filled with the highly successful Thiokol TI7E1 propellant.[48]

On 16 June 1955, the secretary of the Army approved the Sergeant weapon system, assigning it the Army's highest priority. Sergeant would have a range of 25 to 75 nm, carry a 1,500-pound warhead, and measure 30 inches in diameter by 34 feet long.[49]

## Solid Propellant Breakthrough

While castable propellants facilitated the design of large-diameter, solid-rocket motors, the state of the art had reached maximum size with the Sergeant in 1955. In developing the final Sergeant motor design, Thiokol had worked through five propellant formulations and motor designs in its quest for higher performance with decreased missile size. The company encountered combustion instability that caused, at best, severe oscillations that the guidance system could not survive and, at worst, catastrophic motor failure. On the fifth try, Thiokol found a successful propellant formulation for the deployed system, but it proved insufficient for longer-range missiles.[50]

Atlantic Research Corporation (ARC) engineers made a breakthrough in 1950, when Les Weil adapted the plastisol process to formulation of a novel solid propellant. Plastisol is a mixture of a polymer such as polyvinyl chloride in a plasticizing oil to produce an emulsion polymerization process from which the product emerges as tiny spheres. When heated to the cure temperature of 350 degrees F, the polymer and plasticizer quickly fuse to form a tough, uniform plastic mass. The plastisol process was well known in industry, with a variety of uses, but it had not been applied to solid propellants. When mixed with a fine-grain oxidizer, this new type of propellant could be cured more easily than other composite propellants.[51]

The second part of the breakthrough involved addition of powdered metal to increase flame temperature and specific impulse. An article by Caltech's Hsue-Shen Tsien in 1945 briefly mentioned the potential use of powdered metal additives to increase combustion temperature and thus performance.[52] In 1947 a series of papers by JPL's Arthur S. Leonard had surveyed a large number of metal-powder, solid-propellant composition possibilities, including aluminum.[53] In late 1950 NOTS's Bill McEwan had investigated embedding thin aluminum wires in propellant to increase the burn rate and specific impulse. Earlier work at NOTS had developed a black powder–coated-magnesium mixture used in their subsequent rocket designs. This might have been the starting point for ARC's work.[54]

In 1952 ARC's Weil and Keith E. Rumbel demonstrated that adding powdered aluminum could increase the specific impulse of the company's Arcite polyvinyl chloride plastisol composite propellant. Rumbel presented their results at a classified meeting of the Joint Army–Navy–Air Force Solid-Propellant Group at Redstone Arsenal in 1952. At the time the plastisol process was not conducive to casting large-diameter motors, but the seed had been planted for using powdered metals to increase specific impulse. Replacing a portion of ammonium perchlorate (density 1.4 g/cm$^3$) with aluminum (density 2.7 g/cm$^3$) increased propellant density, thereby increasing the propellant mass loaded into the motor case. The resulting increase in total impulse, with no increase in the inert (hardware) mass, delivered a gain in performance.[55]

Apparently at the same time, or at least overlapping with their work, Aerojet-General's Schultz and Dekker, using mechanical calculators to solve simultaneously several equations governing combustion products, painstakingly calculated the theoretical effect of adding powdered aluminum to composite solid propellants their company manufactured. Schultz and Dekker found improvement in specific impulse would level off, then decline after addition of up to 7 percent powdered aluminum by weight. The theoretical results baffled the Aerojet engineers, but the company did not fire actual motors with aluminum levels above 7 percent, turning instead to other methods for improving specific impulse.[56]

In 1953 A. C. Scurlock, K. E. Rumbel, and M. L. Rice submitted a patent application for solid polyvinyl chloride propellants containing metal. They described adding up to 15 percent powdered aluminum or magnesium by weight as being almost equally effective for increasing the Arcite propellant's specific impulse from 226 to 243 seconds while decreasing organic fuel from 25 to 20 percent and oxidizer from 75 to 64.4 percent. Their application mentioned increases up to 25 percent aluminum powder but found 15 to 20 percent was optimal. If the data reflected that described in the meeting a year earlier, the 7 percent increase in specific impulse was interesting but apparently not worth pursuing at the time (Figure 4.4).[57]

In 1954 the Navy Bureau of Ordnance contracted with ARC for a method to increase the specific impulse of solid propellants to a level competitive with liquid propellants.[58] Armed with the earlier powdered aluminum–Arcite results, Rumbel and Charles B. Henderson, assisted by E. Lawrence and E. Rosholdt, reinvestigated the chemical equations. The heats of formation for lithium-to-zirconium metal oxides suggested lithium, beryllium, boron, magnesium, and aluminum oxides as the most likely candidates for further study. The researchers rejected lithium due to its reactivity and beryllium, initially, due to the toxicity of its dust.[59] Boron oxide proved to be an unstable combustion product, and magnesium powder was difficult to handle safely, which left aluminum as the best choice.

Aerojet engineers had assumed the combustion reaction products were nitrogen gas, carbon dioxide, carbon monoxide, and water, plus aluminum oxide. Lawrence and Henderson calculated that substitution of certain metals for a portion of the oxidizer would limit the oxidation of hydrogen to water and result in higher specific impulse (the metal reacted with water formed during combustion, resulting in the metal oxide and hydrogen gas).[60] They now had an explanation for the results Rumbel reported in 1952. Internal-burning grains—5 inches in diameter and 11 to 14 inches long—of a newly developed composite-modified, double-base propellant containing 24 to 27 percent aluminum powder by weight were prepared. Graphite became the nozzle material, based on the reducing conditions and higher temperatures of the combustion products. Efficient aluminum com-

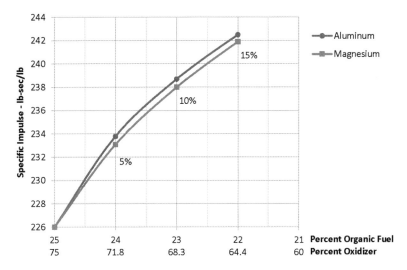

FIGURE 4.4 The effect on specific impulse of the addition of powdered aluminum or magnesium to a polyvinyl chloride propellant with ammonium perchlorate as the oxidizer. In 1952 this effect on specific impulse was insufficient to garner much attention. Six years later, the effect was the basis for the success of the Polaris and Minuteman missiles. *Image adapted by author from US Patent 3,107,186.*

bustion required several conditions, the most critical being a sufficiently energetic propellant mixture to ignite the aluminum particles. This resulted in a specific impulse of 260, which caught the Navy sponsors' attention. Aerojet engineers heard about it and quickly confirmed the results using 100-pound aluminized polyurethane-propellant motors.[61]

In many cases, but not all, adding a small percentage of aluminum helped eliminate some forms of combustion instability.[62] Not all current propellants, however, were compatible with higher percentages of aluminum. Thiokol engineers found, for example, that increasing the amount of ammonium perchlorate oxidizer in their polysulfide propellants had the desired effect of raising flame temperatures but also caused combustion instability. Addition of 2 percent aluminum eliminated this problem, and addition of up to 15 percent to the later polysulfide formulations increased specific impulse. Any further addition proved ineffective because the polysulfide binder did not have sufficient hydrogen content to make use of the higher aluminum levels. This consideration led Thiokol to search for a new class of binders that evolved into the polybutadiene propellants used for the Minuteman I Stage I motor (see Chapter 8).[63]

★ ★ ★

The breakthrough discovery of aluminized solid propellants was the key development that allowed the Navy to bypass a proposed liquid-propellant SLBM and replace it with a solid-propellant missile, Polaris. Development began well before Minuteman, much to the chagrin of the Air Force.

# 5

# FROM POLARIS CAME MINUTEMAN

A heretofore unexplored aspect of the Minuteman ICBM story is the role played by the Navy's development of the Polaris missile. Early on, the Navy's SLBM development was stymied by the need to work with the Army on the liquid-propellant Jupiter IRBM. Reluctant from the beginning to place a large, liquid-propellant missile on a submarine, the Navy sponsored research that led to a breakthrough in solid-propellant technology and development of the Polaris missile, which deployed before the first full-scale Minuteman test launch in 1961.

## Polaris

The Navy entered the strategic offensive missile picture in a somewhat roundabout way. On 8 November 1955, in response to an NSC directive, Secretary of Defense Charlie Wilson assigned specific responsibilities for ICBM and IRBM development to each of the services based on the Killian Report and NSC recommendations (see Chapter 1). Wilson assigned the Air Force sole responsibility for the ICBM and IRBM #1 (Thor). He gave the Army and Navy the "dual objective of achieving an early shipboard capability and of also providing a land-based alternative, IRBM #2 (Jupiter), to the Air Force program."[1] Two months earlier, on 8 September 1955, President Eisenhower had directed that the ICBM program be given the single highest priority among Defense Department programs, with the IRBM programs being second only to the ICBM.[2]

On 15 December 1955, the Army–Navy Joint Ballistic Missiles Committee submitted its initial development plan to Wilson. The plan included language about the desirability of developing a solid-propellant IRBM. Five days later, Wilson approved the proposal.[3]

The Navy already had experience with the problems of launching large, liquid-propellant missiles on land and at sea. In 1946 the Naval Research Laboratory had begun a high-altitude sounding rocket program, Project Viking, to investigate upper-atmosphere effects on high-performance aircraft. The Glenn L. Martin Company built Viking, about half the size and weight of the German V-2, using an aluminum airframe with monocoque tanks. Monocoque meant the tank skin formed the skin of the missile, unlike a V-2, where aluminum tanks were housed in a steel airframe. There were two Viking versions, both 45 feet long, but the first eight measured only 32 inches in diameter, and numbers 9 through 12 measured 45 inches in diameter. The XLR10-RM-2 rocket engine, built by Reaction Motors, was the first fully gimballed engine (the Atlas North American MX-774 engines were hinged, not fully gimballed) and used ethanol and liquid oxygen as the propellants.

The first launch, on 3 May 1949, at the Army's White Sands Proving Ground, was partially successful. Eleven more launches took place through 4 February 1955.[4]

Operation Sandy marked the Navy's first launch of a large, liquid-propellant missile from a warship at sea. The captured V-2 missile successfully launched from the USS *Midway* (CV-41) on 6 September 1947.[5] Operation Pushover, a follow-on test at White Sands Proving Ground, investigated the effect of a V-2, fully loaded with propellants, tipping onto a simulated aircraft carrier's armored deck. Buckling and cracking of the deck, while deemed repairable, signaled danger when using large, liquid-propellant missiles onboard ships.[6] Nonetheless, a successful launch of Viking 4 from the USS *Norton Sound* (AVM-1), near Christmas Island, on 11 May 1950—part of a research program on cosmic radiation and high-altitude atmospheric pressure at the equator—also benefited ongoing evaluation of the complexities of shipboard launches of large liquid-propellant missiles. The tankage requirement for storing large quantities of highly volatile propellants, particularly liquid oxygen, and the problem of a ship's motion (even in calm waters) during a liquid-propellant missile's slow acceleration at liftoff caused great concern. Still, a desire to remain competitive in ballistic missile system development caused the Navy to reluctantly partner with the Army.[7]

At the outset the Army's preference for a longer missile, originally designed to be over 90 feet long, was a problem. The Navy countered that maneuvering a missile that tall aboard a surface ship would be cumbersome, and, fully loaded on a submarine, it would seriously alter the vessel's metacentric height, thereby complicating its surfaced stability. The services compromised with a length of 50 feet by increasing the missile diameter to 105 inches.[8]

On 11 February 1956, a Secretary of Defense Ballistic Missile Scientific Advisory Committee report included both Navy and Air Force presentations on the advantages of solid-propellant missiles: ease of storage, quick ready time, and ease of transportation. Furthermore, "for Navy use, still other advantages appeared; short length of overall missile, elimination of the need for additional storage space on shipboard for fuel and the high initial acceleration so that the ship superstructure is not subject to blast for a long time." The committee recommended that the Navy solid-propellant IRBM program receive the same top priority as the other IRBM programs. At that time the Air Force had funded research on solid-propellant improvement but had not proposed such a weapon system.[9]

The committee observed that the Navy and Air Force had different approaches to using solid propellants for large rocket development:

> The Navy presentation envisaged an attempt to produce a very early IRBM by using modifications of existing components and postponing optimization until after a useable missile has been obtained. The Air Force presentation outlined an imaginative research program that would provide new basic information to be used for subsequent optimization of the Navy's missile, or possibly even for the design of a solid fuel ICBM.[10]

On 11 April 1956, the Navy Special Projects Office gave Lockheed Missiles and Space Company a letter of intent to investigate the use of solid-propellant motors for the fleet ballistic missile program. Captain Levering Smith from NOTS, China Lake, California, led the solid-propellant effort. Smith's experience with large, solid-propellant missiles included work on the short-lived Big Stoop program conducted five years earlier at NOTS. The Atomic Energy Commission (AEC) had requested evaluation of the ability of solid-propellant missiles to carry nuclear warheads. The result was Big Stoop, a two-stage, 51-foot-long, 1.9-foot-diameter missile crafted from two Bumblebee booster airframes manufactured by ABL for the Navy's Talos supersonic surface-to-air and surface-to-surface missile programs. The flight test program consisted of three successful flights with a range of 20 nm.[11] Smith recognized the feasibility of solid-propellant missiles with much greater range, but the Navy, in 1952, simply gave the Army the test results, which initiated development of the Honest John surface-to-surface missile (Figure 5.1).[12]

On 26 April 1956, the Defense Department approved the Navy solid-propellant motor proposal. Thereafter, Navy interest in adaptation of the proposed liquid-propellant Jupiter airframe for shipboard use diminished greatly.[13] One month later, in May 1956, the Navy reviewed a Lockheed concept for Jupiter-S, a solid-propellant missile that weighed nearly twice as much as the original Jupiter, primarily due to the lower performance of existing solid propellants and the need to loft a 3,000-pound reentry vehicle–warhead combination plus a heavy inertial guidance system. The proposed design included a cluster of six modified Sergeant airframes (enlarged to 40 inches in diameter) for the first stage and a single Sergeant airframe for the second stage. This resulted in the shorter missile the Navy desired for submarine deployment, 41.3 feet for Jupiter-S versus 50 feet for Jupiter, but with greater diameter, 10 feet compared to 8.75 feet for Jupiter. The estimated weight of the missiles was 160,000 pounds each, so a submarine displacing 8,500 tons could feasibly carry only four.[14]

Simultaneous with the Navy's study, Smith asked Dr. Frank E. Bothwell of the Weapons Planning Group at China Lake to initiate studies for improving every aspect of Jupiter-S. Bothwell and his team conducted Project Mercury from spring through September 1956. Areas of concern included solid-propellant performance, guidance-and-control system weight, reentry vehicle design, and thermonuclear warhead weight and yield. The team concluded it was foolish to build Jupiter-S with the current, nearly obsolete, components.[15]

In December 1955 the newly appointed chief of naval operations, Adm. Arleigh Burke, had asked the National Academy of Sciences Committee on Undersea Warfare to study the looming problem of protection against the growing capabilities of the Soviet submarine fleet. The result was Project Nobska, a summer-long conference in several parts, involving 60 Navy, government, and university laboratory participants at Woods Hole Oceanographic Institute, Woods Hole, Massachusetts. Bothwell's Project Mercury was well underway when the Project Nobska Conference began on 18 June 1956.[16] When the participants decided to include strategic aspects of undersea warfare to understand and exploit possible Soviet weaknesses, they invited nuclear weapon and missile experts to attend, Bothwell and Dr. Edward Teller from Livermore Radiation Laboratory among them. Bothwell persuaded the Project Nobska members to consider a 30,000-pound fleet ballistic missile based on recently demonstrated substantial improvement in solid-propellant performance coupled with a lighter guidance system.[17]

Driven by the earlier stipulation for Atlas accuracy of 5 nm or better at the proposed 5,500-nm range, military planners had settled on a one-megaton warhead weighing 1,600 pounds for ballistic missiles. The Army team had adopted this for both the Army and Navy versions of Jupiter.[18] When discussion turned to the feasibility of a nuclear warhead for

FIGURE 5.1
The only existing photograph of the Big Stoop flight test vehicle. Big Stoop was made up of two Talos boosters as part of an Atomic Energy Commission project. *Official US Navy photograph.*

FIGURE 5.2
The progression, left to right, from the liquid-propellant Jupiter to the solid-propellant Jupiter-S and finally to the operational solid-propellant Polaris A-1. The Minuteman IA is shown for comparison. The 1957-to-1958 breakthrough in solid–propellant performance made the Polaris program feasible. *Adapted from "A History of the FBM System."*

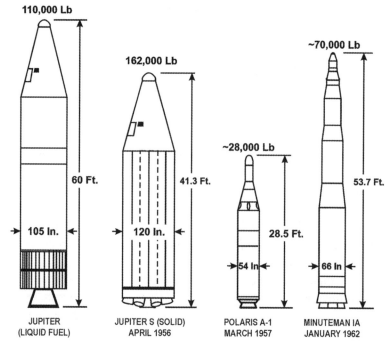

torpedoes, Teller asked why conference participants were discussing a 1965 weapon design using 1958 technology. He explained one could take the trend toward ever-lighter warheads and predict a warhead weight with one-megaton yield that would work on either a torpedo or the new missile. Although J. Carson Mark, the Los Alamos National Laboratory representative, disagreed with some of Teller's conclusions, he admitted the trend Teller mentioned was possible. Mark suggested that a lighter, half-megaton-yield warhead seemed more realistic.[19]

The Project Nobska final report impressed Secretary of Defense Wilson, who authorized the Navy, on 8 December 1956, to end all efforts for the navalized version of Jupiter or Jupiter-S and proceed with development of the new, lightweight Polaris missile, submarine application being its priority. Wilson abolished the Joint Army–Navy Committee at the same time (Figure 5.2).[20]

Thus, the Navy, well before any decision from the Air Force, opted to design and deploy a large-diameter, solid-propellant missile. It became one of the highest-priority programs in its budget. Would Minuteman have come about without the Polaris program? Surely. Did Polaris nudge the Air Force to reluctantly move forward with a solid-propellant ICBM? Very likely. Was there an Air Force preference for the cumbersome, liquid-propellant Atlas and Titan I missiles? They were the best fit at the time. In its rush to field ICBMs, the Air Force was learning painful lessons about guided missile development. Neither the Pentagon nor Congress wished to repeat those programmatic mistakes. Their reticence prolonged the Minuteman funding process.

## Minuteman

The convoluted path culminating in Col. Edward Hall's February 1958 briefing on System Q (the Minuteman program's first name) to the secretary of defense began three years earlier with an unsolicited proposal from a professor at Cornell University. Details about this early path are confusing and, in some cases, contradictory.

The first mention of large-diameter, solid-propellant rockets for an Air Force weapon system came on 15 June 1955 from Brig. Gen. Bernard A. Schriever, WDD commander, in response to a proposal for a solid-propellant ICBM from Dr. Dwight Gunder of Cornell University. While appreciating Gunder's proposal, Schriever indicated the Air Force had no requirement for a solid-fueled ICBM. Available nuclear weapon and guidance technology, plus propellant performance, meant the warhead would be too small and the dispersion at intercontinental ranges too great for such a weapon to be useful. Nonetheless. he invited Gunder to meet with Ramo-Wooldridge staff, systems engineering consultants to WDD, to discuss the possibility of a solid-fueled TBM.[21]

The next day, Hall, WDD's Propulsion Unit chief, sent Dr. Louis Dunn at Ramo-Wooldridge and Col. Charles Terhune at WDD a memo commenting on Schriever's reply to Gunder.[22] Hall's memo detailed the pros and cons of a solid-propellant missile and listed several remaining problem areas: ability to terminate thrust, structural integrity of grain bonding to motor case under acceleration, withstanding temperature gradient conditions that would occur with a TBM, and reproducibility of all major performance parameters from batch to batch of propellant.[23]

Hall suggested that WDD invite Gunder to visit for an update on Hall's current ideas. This resulted in a summer study proposal by Gunder and his associates at RamoWooldridge Corporation.[24] Eight days later, on 24 June 1955, Terhune admonished Hall to proceed no further pending comments from the RamoWooldridge staff. This began a contentious relationship between Hall and virtually everyone else at WDD and RamoWooldridge.[25]

Available records reveal that in July 1955, industry experts agreed with Hall on the virtues of solid propellants, but many drawbacks indicated the need for more work. The four major firms with significant solid-propellant expertise were Aerojet-General, Grand Central Rocket, Phillips Petroleum Company, and Thiokol. Representatives from each company briefed WDD on their respective assessments of the state of the art. As a result, in December 1955, Hall submitted a formal proposal to ARDC for an urgent study program.[26]

The next official mention of Air Force solid-propellant ballistic missile development came in the Secretary of Defense Ballistic Missiles Scientific Advisory Committee's first report on 11 February 1956. The committee recommended that solid-propellant IRBM development receive the same high priority as Thor and Jupiter. As previously described, the report noted differences in the Navy and Air Force approaches, with the latter feasibly leading to development of a solid-propellant ICBM.[27]

On 28 February 1956, at the request of Lt. Col. L. F. Ayres, WDD submitted a detailed Solid Rocket Development Plan authored by Hall to ARDC. It described the Air Force Large Solid Rocket Feasibility Program, which had the following objectives:

> To obtain the basic information necessary to determine the feasibility of developing a solid propellant rocket suitable for use as the primary power plants for missiles of the IRBM type and as boosters for nuclear powered rockets and ramjets and to develop design criteria for development of these rockets. Immediate objectives are to develop practical techniques for manufacturing high specific performance rockets of maximum reliability with overall impulses of five million to 20 million pound-seconds and burning times of between 50 and 150 seconds.[28]

The multiyear development plan listed additional problem areas for study over a two-year period: hardware, controls, propellants, interior ballistics, ignition, and support.[29] Contracts for the first year's study went to Aerojet-General, Grand Central Rocket Company, Phillips Petroleum Company, and Standard Oil Company of Indiana.[30]

The companies were assigned specific areas for nonproprietary investigation. Aerojet-General, for example, received the task of investigating optimal case materials but pursued work in all the problem categories. At the end of the first year, the Air Force dropped Grand Central Rocket Company and Phillips Petroleum from the program, replacing them with Hercules Powder Company and Thiokol.[31]

Early on Aerojet contributed a key element to the overall Minuteman design program: use of a depressed trajectory. Aerojet engineer John Fuller realized the greater robustness of a solid-propellant airframe compared to a liquid-propellant missile meant the former could pitch over toward the target much more quickly, thereby eliminating almost entirely the long, fuel-consuming vertical climb characteristic of liquid-propellant missiles. The key parameter is "body bending load," calculated by multiplying maximum dynamic pressure by angle of attack. A solid-propellant airframe could withstand significantly higher dynamic pressures and a much greater angle of attack because solid rocket motor casings (airframes) were built to withstand pressures of 600 to 800 psi and thus thicker than liquid-propellant airframes pressurized to 50 to 100 psi. Fuller recommended an immediate pitch over of 30 degrees from the vertical, giving a 20 percent increase in range compared to the standard Atlas or upcoming Titan I trajectory.[32]

At a meeting of the OSD-BMC on 13 March 1956, the Navy and Air Force presented their respective approaches to solid-propellant IRBM development. The Navy proposed using state-of-the-art technology in the Jupiter-S program.[33] Hall made a point-by-point counterpresentation arguing that state-of-the-art inadequacy for both propellant and case design rendered the Navy proposals premature. The Air Force, on the other hand, was pursuing several fundamental research lines in solid propellants and case design, plus studying cluster-rocket versus single-rocket launch techniques and solid-propellant thrust termination. He implied the Air Force had pulled ahead of the Navy in large solid-propellant missile R&D. Considering the clear differences between Navy and Air Force assessments of whether current solid-propellant technology could successfully support a full-scale IRBM program, the committee chose to postpone any decisions.[34]

At the next meeting of the OSD-BMC, on 20 March 1956, the committee approved the Navy Jupiter-S program as "limited to resulting system studies and program to be presented to the OSD-BMC for review before entering upon a full-scale missile development program." It "requested that the assistant secretary of defense for R&D review the Air Force solid propellant program to ensure that it is complementary to the Navy effort and that it continues at the exploratory research level."[35]

WDD's overriding mandate was to develop the Atlas and Titan liquid-propellant missiles as rapidly as possible. In June 1956, at Schriever's behest and to Hall's dismay, the Large Solid Rocket Feasibility Program moved from WDD to Wright Air Development Center in Ohio. Hall, still propulsion chief at WDD, continued to monitor developments and offer advice.[36]

The Air Force did not take the leap of faith Hall wanted when he had insisted that recently released information on higher-performance solid-propellant experimentation at ARC meant a solid-propellant ICBM was within reach (see Chapter 4). Given the complexity of managing Atlas and Titan ICBMs, plus the Thor IRBM, the major headaches associated with a solid-propellant missile—thrust termination and directional control—seemed best left for solution at a later time.

Hall saw these issues in a different light. Solid-propellant missiles had intrinsic benefits, such as extremely simple motors and easily stored propellants. Those factors enabled a swift countdown; loading propellants immediately before launch was not an issue. Although hypergolic propellants could be stored on a missile, the oxidizer—some form of nitric acid or nitrogen tetroxide—and the hydrazine-based fuel were highly toxic and the oxidizer highly corrosive. Recent advances in solid-propellant formulation, in addition to the use of aluminum powder, mitigated the explosive potential and significantly reduced the number of people needed for maintenance and launch tasks.[37]

Manufacturing issues, such as the ability to mix uniform batches of solid propellant in the volume necessary for the large rockets, caused concern. At the time, for example, the 31-inch-diameter Sergeant motor contained 6,000 pounds of propellant. Development of lightweight, high-strength motor casings also presented challenges. Hall and others, however, regarded these issues as straightforward manufacturing problems, relatively easy to resolve.[38]

Fortunately, Richard Young, a consultant for the Navy's ABL (operated by Hercules Powder Company since 1945) in Maryland, had potentially solved the problem of lightweight, high-strength casings years earlier. Young had developed a motor casing of wound fiberglass filament that flew successfully on a Nike sounding rocket test in the early 1950s, but the information may have been unavailable to Hall or the Air Force.[39]

The first official Air Force discussion of a large solid-propellant missile system occurred on 18 July 1956. Perhaps due to information coming from Project Nobska meetings that had begun a month earlier, Lt. Gen. William Tunner, commander-in-chief of United States Air Forces in Europe, released an operational requirement for a TBM to replace the subsonic TM-61 Matador cruise missile in the early 1960s.[40] Tunner wanted a mobile solid-propellant missile with a maximum range of 1,500 nm and a CEP of 1 nm or less that crews could move between surveyed launch positions. In addition, the requirements noted, "Component replacement type maintenance that would permit rapid exchange of defective electrical or mechanical assembly parts in shortest possible time before launch." This mirrored Hall's concept of basing for a simple solid-propellant ICBM.[41]

On 28 February 1957, the OSD-BMC's 15th Monthly Report on Progress of ICBM and IRBM programs sent to President Eisenhower included an update on progress by the Navy on solving the solid rocket motor grain-cracking problem with successful cycling of motors to minus 40 and minus 80 degrees F. Forty-inch-diameter motors 100 inches long and containing 7,500 pounds of propellant were being tested to evaluate whether the propellant held up under firing stresses.[42]

On 20 March 1957, Lt. Gen. Donald L. Putt, Air Force deputy chief of staff, development, in a memorandum to Lt. Gen. Thomas S. Power, commander, ARDC, drew attention to "the many significant advantages offered by the solid propellant rocket for IRBM propulsion and the early indications of success in the effort toward that end [which] make it timely to consider the establishment of a weapon system requirement." Putt reiterated the need to retain liquid-propellant IRBMs for use against targets requiring longer ranges and payloads. He emphasized difficulties in funding systems currently under development but suggested that while the more stringent size and weight requirements of the submarine would decrease its range or payload capabilities, Polaris might be compatible with Air Force needs. He asked Power to provide Headquarters USAF "with information necessary to determine the optimum role of the solid propellant IRBM and to aid in the proper planning for initiation of system development." He also requested an estimate for when development of the solid-propellant IRBM could begin without adversely affecting Atlas and Thor deployment. To avoid duplication of R&D efforts, Putt also sought arrangements for detailed examination of the Navy program by the Air Force.[43] On 1 April 1957, Power assigned WDD responsibility for solid-propellant IRBM management.[44]

One day earlier, on 31 March 1957, the 16th OSD-BMC monthly report to Eisenhower established Polaris parameters at 28.5 feet in length, 54 inches in diameter, and weighing approximately 29,000 pounds. Earlier that month, the first Aerojet 40 KS-100,000 motor fired successfully, generating 4 million pound-seconds total impulse, the highest yet achieved with a US solid-propellant motor.[45] Sixty-inch-diameter motors 80 inches long with 13,500 pounds of propellant had been tested at 70, 50, and 30 degrees F for five days without showing signs of grain cracking or liner separation. Navy progress drew Air Force attention.[46]

On 23 April 1957, Maj. Gen. James Ferguson, Air Force director of requirements, informed various Air Force commands that recent studies, presumably by the Navy, warranted his service's further consideration of a solid-propellant IRBM. Ferguson directed ARDC to undertake a comparative study of a solid-propellant missile compared to a supersonic cruise missile, with possible adaptation to the Thor program. Noting ARDC's ongoing solid-propellant work, Ferguson requested completion of the study as soon as possible.[47]

The Navy continued making significant progress on a solid-propellant formulation for Polaris. On 24 April 1957, Hercules Powder Company's ABL successfully demonstrated a new propellant formulation using a double-base propellant (nitrocellulose and nitroglycerine) with 21 percent ammonium perchlorate and 21 percent aluminum powder, giving a specific impulse of 255 seconds, significantly higher than required for the second stage. Although not used in A-1, the first Polaris variant, the A-2 used this propellant after Hercules Powder Company won the A-2 Stage II contract.[48]

On 27 June 1957, the Navy successfully demonstrated graphite-lined nozzles for the Stage I motor. Not a full-scale motor test, it used an Aerojet 30KS-50,000 motor, but the completely graphite-lined nozzle withstood a temperature of 5,150 degrees F for 30 seconds.[49]

Col. Benjamin Paul Blasingame, WDD's Titan I program director, in a 17 July 1957 memorandum to Col. Charles Terhune, deputy commander, weapon systems, WDD, cautioned against focusing solely on a solid-propellant IRBM. He perceived the IRBM program as a distraction from basing strategic missiles on US soil, where they were less vulnerable to attack than in Europe. While the cost of developing a first-stage solid propellant seemed prohibitive, Blasingame thought a second stage and possibly a third stage for Titan I seemed feasible. The Titan I second stage posed the greatest hazard to reliability because its engine required an at-altitude start. The hybrid vehicle might be a product improvement, with test facilities already in place.[50]

On 1 July 1957, Maj. Gen. Ferguson published General Operational Requirement Short Range Ballistic Missile Weapon System (GOR-161). He considered the previous qualitative operational requirements, such as simple replacement of critical components and fast reaction time—5 minutes indicated but 10 to 15 minutes initially acceptable—in the interest of early fielding of the weapon system. GOR-161 emphasized use of either solid propellants or storable liquid propellants, with the need for a self-contained (inertial) guidance system that could operate for extended periods being paramount. Also specified was a CEP of 1,500 feet with a three-megaton warhead and deployment to operational units by 1962. It made no mention of using solid propellants in an ICBM weapon system.[51]

In the 20th monthly OSD-BMC progress report, for July 1957, the Navy touted success in solving the problem of thrust termination, one of the most difficult technical challenges with solid-propellant motors. Thrust termination with solid rockets required either redirecting thrust or extinguishing the flame front. The Navy found that by opening ports in the forward section of the motor casing, it could redirect exhaust gases forward, thereby reversing thrust, and the sudden decrease in motor pressure quickly extinguished the flame front. It successfully demonstrated forward-end thrust reversal on 16 July 1957, with a change in directional force of 7.5 gs in 3.33 milliseconds after initiation.[52]

In July 1957 a panel of 18 scientists met, at Ramo-Wooldridge Corporation's request, to discuss future developments in ballistic missiles. Named after its chairman, Dr. Robert F. Bacher, a Caltech physicist, the Bacher Committee spent two weeks reviewing ICBM force survival, the reentry problem, propulsion systems, guidance systems, and glide rockets. The panel recommended serious consideration of a mobile ICBM to complement hard-base installations. Finding the Air Force solid-propellant program inadequate, the panel recommended the Air Force continue R&D on thrust termination and thrust control methods, along with propellants for larger advanced systems. Moreover, Bacher's committee encouraged Air Force pursuit of a solid-propellant ICBM based on the Navy Polaris fleet ballistic missile.[53]

Simultaneous with the Bacher Committee deliberations, on 1 July 1957, Ramo-Wooldridge's Guided Missile Research Division released a "Design Study and Systems Analysis of Solid Propellant Intermediate Range Ballistic Missiles." Essentially a point-by-point response to Putt's memo of March 1957, Ramo-Wooldridge's study confirmed the feasibility, in the 1960s, of employing solid-propellant IRBMs—with ranges of 250 to 2,000 nm, projected guidance system improvements, and existing state-of-the-art solid-propellant motors—against both hard and soft targets. Finally, the study estimated the cost of equipping and training operational squadrons with solid-propellant missiles as nearly 25 percent lower than with liquid-propellant missiles.[54]

The 21st OSD-BMC monthly progress report, for August 1957, showed Polaris continuing to demonstrate solutions to key technologies. Forward thrust termination had been demonstrated at an altitude of 92 nm. Thrust vector control using the jetevator concept, a movable ring placed at the end of each exhaust nozzle to deflect exhaust gases, was achieved; a rhodium-coated molybdenum ring showed the greatest resistance to erosion from the aluminized propellant. A six-point-star propellant grain configuration was demonstrated successfully with the first 54-inch-diameter Stage II motor. The polyurethane aluminized propellant yielded a specific impulse of 239 seconds, extrapolated to 1,000 psi, closing in on the required second-stage thrust. The Polaris program confirmed that large-diameter, solid-propellant missiles were feasible. At this point the main issue for the Air Force became extension to ICBM range.[55]

On 16 August 1957, Maj. Gen. John W. Sessums Jr., ARDC vice commander, wrote to Schriever expressing "considerable interest" in GOR-161. He advocated any "lash-up of existing components such as single or clustered solid units in one, two or three stages" from any of the services and requested that Schriever respond by the third week of September. Sessums's letter ended:

> The information is needed for a special purpose and unless a practical solution not requiring years of development time is readily available from existing components for guidance, control, logistics considerations, the exercise will be of no value.[56]

Clearly he did not envision an entirely new program but simply a "lash-up" solution. On 18 September 1957, Blasingame weighed in further on GOR-161. He cautioned against a development program modeled on Titan, emphasizing that it contained "not a single word in all the procurement data" describing the basing or ground support for the missile once deployed, the result being a considerable amount of expensive jury-rigging for the test program. A year and a half after contract award, he noted a severe lack of detail on what the ground support equipment would look like. Blasingame urged consideration of an eight-month to one-year system development program using the best talent from both the Air Force Ballistic Missile Division (WDD had become AFBMD) and Ramo-Wooldridge. In short, while not directly saying it, he was against Sessums's "lash-up" approach.[57]

Schriever's reply to Sessums, on 20 September 1957, contained the first mention of any kind of official Air Force proposal for a solid-propellant ICBM. While true that Ramo-Woolridge implied as much in a December 1956 study, and again in February and July 1957, those did not constitute Air Force policy. Schriever vigorously opposed a "lash-up" program, indicating Sessums would receive a comprehensive system development plan in November that outlined the use of solid propellants for a three-stage ICBM. This plan would include options for using the first and second stage, or second and third stage, or only the third stage as IRBMs. He mentioned ranges of 5,500 nm for the ICBM and 250 to 2,500 nm for the IRBM variants. Schriever also favored Blasingame's suggestion of a solid-propellant second stage for Titan I.[58]

On 4 October 1957, Lt. Gen. Samuel E. Anderson, the new ARDC commander, wrote to Gen. Thomas White, Air Force chief of staff, concerning progress on solid-propellant systems. Anderson detailed ARDC's significant state-of-the-art advances. Although current operational propellants had a specific impulse of 225 seconds, ARDC planned to fire, in December, experimental compositions with specific impulses of 235 to 240 seconds in large-diameter motors. Industry already had under development motors with specific impulses of 250 to 260 seconds. Further specific impulse increases would lead to smaller, more logistically flexible missiles. With current propellants, a two-stage ICBM weighing 87,000 pounds could be developed. By comparison, the liquid-propelled Thor weighed 110,000 pounds. A specific impulse increase to just 240 seconds meant the same two-stage missile would weigh only 50,000 pounds.[59]

On 7 November 1957, the Security Resources Panel of the Office of Defense Mobilization Science Advisory Committee released the Gaither Report (see Chapter 2). The report included a description of probable new Soviet and US weapon systems. It found that "second and later generation missiles, with solid propellants, CEPs measured in thousands of feet, instead of several miles, and with larger megaton warheads and quicker reaction time, will be put in production by 1960–1965."[60]

On 8 November 1957, in response to the Bacher Committee report and requests for investigation of Polaris missile utility as a land-based IRBM, Hall requested Lockheed's Missile Systems Division study how the Polaris program could meet GOR-161 objectives. By this time Hall had a clear vision of how to conduct the System Q development study.[61]

As Hall's group juggled the myriad aspects of System Q development, it investigated basing concepts with low-cost, simple ground support equipment and as few maintenance and support personnel as possible. Hall addressed silo basing in a 22 November 1957 memo to STL, formed from Ramo-Wooldridge's Guided Missile Research Division earlier in the month.[62] He emphasized minimum size and cost, suggesting elimination of a separate exhaust duct by designing the silo with an annulus, a ring-shaped region between the missile and the silo wall to serve as the exhaust path. At this time the missile parameters included three stages totaling 65 feet in length, with a maximum diameter of 6 feet and a gross weight of 65,000 pounds.[63]

Silo basing and in-silo launch eliminated the problematic missile elevators used with Atlas F and Titan I. Without the need for costly liquid-propellant storage and transfer facilities, silos could be simple, essentially stand-alone structures. Since solid-propellant missiles did not require constant checking or rapid loading of propellants before launch, Hall and his counterpart at STL, Barney Adelman, realized the silos could be unmanned and dispersed over remote areas. Monitoring from a distant control room would be similar to how oil refineries had transducers that reported up to 15 pieces of information from remote cracking towers to the refinery control room.[64]

Preliminary design of the missile airframe occurred through the Large Solid Rocket Feasibility Program at the same time as Polaris program successes continued to mount. The best Polaris propellant had a specific impulse of 239 seconds at 1,000 psi, more than sufficient. Early small-scale tests at ABL had demonstrated specific impulses as high as 255 seconds at 1,000 psi. Aerojet-General was working on three manufacturing processes for motor casings: rolled and welded, hydrospin, and spiral wound. Several successful tests with the spiral-wound process using 13-inch motors had occurred in the Polaris program, which sought to achieve the lowest possible casing-to-propellant ratio. Hercules Powder Company had a proprietary fiberglass filament-wound casing that weighed much less than metal casings and could withstand even higher pressures. In addition, the previously intractable reentry problem now had several solutions (see Chapter 9).[65]

Hall sent Terhune a memo on 26 December 1957 commenting on Ramo-Wooldridge's final report that proposed development of a highly accurate guidance system, one to permit assigning hard targets, such as missile sites, dams, and hardened command posts. Hall laid out the purpose of his System Q as a deterrent force aimed at enemy cities, which rendered the proposed guidance system unnecessary. The Ramo-Wooldridge report advocated a combined radio-inertial system employing Doppler. System Q, on the other hand, avoided the constraints imposed by ground-based guidance stations by employing all-inertial guidance. Ramo-Wooldridge also recommended a heavier warhead, which required a larger missile, thereby negating the small missile and lower operating costs of System Q. Perceiving that most Ramo-Wooldridge recommendations would further delay the solid-propellant ICBM program, Hall concluded:

> We can afford numerical superiority sooner by accepting an accuracy satisfactory for a city force deterrent mission by compressing the development cycle to the utmost, and by committing the Q Weapon System to production as soon as possible. As significant improvements in missile performance are achieved, the improved missiles can be phased into the force to the degree required.[66]

Hall's conservative approach on capabilities paralleled the Polaris program. Eventually both programs accepted lower performance for the first missile variants in return for earlier deployment. Carrying Hall's cost-conscious approach to an extreme, Lt. Col. F. K. Bagby, on 27 December 1957, priced a 70-foot-deep, closed-bottom "economy" model silo made of 12-foot-diameter corrugated metal culvert at $50,000 each, if produced in quantity.[67]

The OSD-BMC met for two days in December 1957 to discuss the direction of Air Force ballistic missile programs. The committee recommended that a second generation of long-range ballistic missiles proceed along two paths, the first directed toward "simplicity, small size, mobility, quick reaction and similar characteristics to further operational capability" and the second toward "larger size and higher performance, to be operated from relatively large fixed installations, which would lead to missiles of much larger payloads and to the first stages for advanced satellite and space vehicles."

Much to the dismay of the System Q development staff, the OSD-BMC report suggested "early attainment of an operationally much superior IRBM can best be achieved if the Air Force makes full use of Polaris, adapting early versions to land use. We recommend that this procedure be urgently explored and such a program be undertaken."[68]

The Sputnik launches on 4 October and 3 November 1957 caused the Navy to immediately borrow funds from other missile programs to advance the deployment of the first Polaris fleet ballistic missile submarines from 1963 to 1960. Progress in the Polaris program during December warranted this acceleration since the first- and second-stage motors had successfully test-fired with the candidate propellants and, on 30 December 1957, the motor for Polaris AX flight testing had passed its firing tests, with jetevators and nozzles operational throughout the test. The first static firing of a 40-inch-diameter, strip-wound motor case also succeeded, clearing the way for a full-scale, 54-inch-diameter motor test.[69]

A final discussion item in 1957 involved a production base for the various solid-propellant missile configurations—theater-based, short-range IRBMs and ICBMs. Hall envisioned multiple contractors for each variant, with an industrial base eventually capable of producing 2,000 missiles annually.[70]

On 7 January 1958, the Navy expressed its confidence in the Polaris developmental process to OSD-BMC:

> The missile which will be provided for the first submarine in October 1960 is not in any way a degraded missile. Its main characteristics will be as follows: range 300–1,200 nm; warhead yield - 3/10 to 6/10 of a megaton with a weight of 600 pounds; FBM CEP 3–4 nm. These characteristics are the same ones planned for the previous IOC date of 1 January 1963.[71]

Drs. Killian, York, and Kistiakowsky met with President Eisenhower and his staff secretary, Gen. Goodpaster, on 4 February 1958. They discussed the Air Force solid-propellant ICBM program. Kistiakowsky said a solid-propellant ICBM could not be expected until 1965 or 1966 and work on storable, self-igniting propellants should continue for what soon became Titan II.[72]

On 8 February 1958, Schriever and Hall briefed Secretary of Defense McElroy on the Minuteman program.[73] Envisioning a force of 1,616 missiles by the end of 1965, Hall described a three-stage, solid-propellant missile capable of delivering a one-megaton or half-megaton warhead 5,500 or 6,500 nm, respectively.[74] Each missile would be housed in a 200-psi hardened concrete silo and dispersed to prevent loss of more than one missile if targeted with a five-megaton weapon. Hall emphasized the need for minimal ground support equipment, no field maintenance, and one control facility per 50 missiles. All major missile components would be designed for extended operation, with a self-aligning guidance system to ensure each missile remained constantly available for launch. Attempting to circumvent the already circulating idea of adapting Polaris for land basing, Hall emphasized possibly using second- and third-stage combinations for low-cost tactical and intermediate-range missiles by 1961 (Tables 5.1 and 5.2).[75]

Hall based his estimate on an annual attrition rate of 50 percent through training launches, malfunctions, and aging. That enabled a rapid buildup to Hall's 1964 operational force projection of 4,000 missiles. He estimated the cost per missile at $350,000 after the first thousand. In addition, he estimated the cost of each LF at $175,000 and ground support equipment at $100 million. Renting commercial telephone lines from AT&T would provide redundant communications. Hall also estimated the R&D costs (Table 5.3).[76]

After Hall's briefing, McElroy tacitly agreed that Minuteman represented a second-generation system worth pursuing, OSD-BMC should hear Hall's Minuteman briefing, and overall, Air Force ICBM force objectives needed careful appraisal within the next 60 days.[77]

On 10 February 1958, Secretary of the Air Force James Douglas sent Secretary McElroy a memorandum stating, "now is the time to establish a firm program (Minuteman) which will provide a three-stage solid propellant ICBM." He expected "that following rapid buildup of the far less expensive and less demanding Minuteman force, it will overtake the requirement of additional liquid propellant missiles. It is requested that you approve immediate initiation of the Minuteman program."[78]

Apparently, the Air Force kept the Minuteman briefings closely held. At a meeting on 13 February 1958, the Ballistic Missile Panel subcommittee of the President's Science Advisory Committee summarized the "Technical Progress and Actions Required in the Long-Range Ballistic Missile Program." The discussion included continued Polaris program progress and a recommendation that the Air Force seriously consider Polaris as a land-based IRBM, provided demonstration of satisfactory performance [79]

TABLE 5.1   Status of Solid Propulsion Systems, February 1958[a]

|  | CURRENT | IN DEVELOPMENT | ICBM DESIGN CRITERIA |
|---|---|---|---|
| MOTOR GROSS WEIGHT (LBS.) | 9,000 | 22,000 | 9,000–48,000 |
| DELIVERED SPECIFIC IMPULSE (SEC.) | 230–270 | 260–276 | 275–280 |
| CASE MANUFACTURING TECHNIQUE | rolled, welded | strip winding | strip winding, Hydrospin |
| THRUST VECTOR CONTROL | jet vanes | jet vanes, jetevators swiveled nozzles | jet vanes, jetevators swiveled nozzles |
| THRUST TERMINATION SYSTEM | none | thrust reversal nozzle | thrust reversal |

a) "Minuteman Financial History, 1955–1963."

TABLE 5.2  Minuteman General Characteristics, February 1958[a]

| GENERAL CHARACTERISTICS | | | | |
|---|---|---|---|---|
| missile gross weight | 65,000 lbs. | | | |
| warhead weight | 330 lbs. | | | |
| warhead yield | 500 kiloton | | | |
| CEP 6,500 nm | 0.3 nm estimated | | | |
| 7,000 nm | 1.0 nm estimated | | | |
| AIRFRAME CHARACTERISTICS | | STAGE I | STAGE II | STAGE III |
| motor weight (lbs.) | | 46,200 | 13,500 | 4,300 |
| thrust (lbs.) | | 138,000 | 43,600 | 16,800 |
| duration (secs) | | 67 | 67 | 60 |
| propellant weight (lbs.) | | 41,500 | 11,800 | 3,660 |
| hardware weight (lbs.) | | 4,700 | 1,700 | 640 |
| specific impulse (secs) | | 260 | 276 | 280 |
| REENTRY VEHICLE CHARACTERISTICS | | | | |
| ballistic coefficient ($W/C_oA$, lb./ft$^2$) | 800 | | | |
| total weight (lbs.) | 550 | | | |

a) "Minuteman Financial History, 1955–1963."


TABLE 5.3  Minuteman R&D Cost Estimates (in Millions), February 1958[a]

| AREA | FY 58 | FY 59 | FY 60 | FY 61 | FY 62 | TOTAL |
|---|---|---|---|---|---|---|
| INSTRUMENTATION | 0 | 8.0 | 10.0 | 12.0 | 12.0 | 42.0 |
| SYSTEM INTEGRATION | 8.0 | 20.0 | 35.0 | 35.0 | 35.0 | 133.0 |
| GROUND SUPPORT EQUIPMENT | 3.0 | 10.0 | 10.0 | 8.0 | 8.0 | 39.0 |
| GUIDANCE | 5.0 | 20.0 | 45.0 | 70.0 | 83.0 | 223.0 |
| NOSECONE | 0 | 5.0 | 15.0 | 15.0 | 17.0 | 52.0 |
| PROPULSION | 14.0 | 77.0 | 125.0 | 140.0 | 140.0 | 496.0 |
| TOTAL | 30.0 | 140.0 | 240.0 | 280.0 | 295.0 | 985.0 |

a) "Minuteman Financial History, 1955–1963."

On 18 February 1958, Gen. White sent the JCS a memo concerning Evaluation of the Minuteman Ballistic Missile Program. He reiterated that in addition to the envisioned Minuteman ICBM capability, use of the second and third stages as an IRBM, or simply the third stage as a theater ballistic missile, made Minuteman all the more attractive. Due to the similarity of the Minuteman IRBM version to the proposed land-based Polaris, WSEG compared the two.[80] At the next OSD-BMC meeting, on 19 February 1959, Hall made another Minuteman "informational" presentation.[81]

A flurry of discussions and activity about budgeting and force levels occurred over the next several weeks. On 21 February 1958, Brig. Gen. Charles M. McCorkle, assistant chief of staff for guided missiles, responded to Director of Guided Missiles William Holaday's request to formalize how Minuteman might fit into overall Air Force ICBM planning objectives and program costs. Mirroring the Gaither Committee's recommendations, McCorkle projected a 600-missile force by 1964. The Air Force would need $26 million in augmentation funds in 1958 to initiate the final R&D plus manufacturing capacity, with an additional $223 million in 1959. Estimated Minuteman funding would become substantially lower per missile, compared to Atlas or Titan, over FY 1960 to 1963 (Table 5.4).[82]

On 24 February 1958, Hall countered McCorkle's memo with one of his own to Terhune. While admitting McCorkle accurately stated the general thrust of the early Minuteman program, Hall expressed concern about the implications of a 600-missile ICBM force by 1964, fearing that Minuteman opponents might forever limit the future number of Minuteman missiles deployed to the 400 scheduled for the 600 total ICBM force in 1964. Hall reminded Terhune that the enthusiasm LeMay and others exhibited during the initial Minuteman briefings stemmed from the proposed deployment of thousands, not hundreds, of Minuteman ICBMs.[83]

Also, on 24 February 1958, Schriever provided the new Air Force ballistic missile program recommendations to Gen. White. Referring to the earlier February briefings, Schriever now included the proposed Minuteman force in his Atlas and Titan calculations. He requested immediate release of $26 million to enable further Minuteman R&D in FY 1958, thereby ensuring Minuteman operational deployment by 1963 to 1965; for FY 1959, he requested $150 million.[84]

On 27 February 1958, Holaday approved funding for only Minuteman R&D, specifying "that the effort should be concentrated in those areas where critical technical problems might exist so that a firm development plan could be established at an early date." Furthermore, he stipulated that funding would remain at the R&D level until the Air Force was "prepared to present a complete evaluation of Minuteman versus Polaris as a follow-on land-based IRBM."[85]

TABLE 5.4  ICBM Force Objectives and Estimated Dollar Requirements[a]

| | ICBM FORCE OBJECTIVES | | | | | | |
|---|---|---|---|---|---|---|---|
| | FY 58 | FY 59 | FY 60 | FY 61 | FY 62 | FY 63 | FY 64 |
| ATLAS | -- | 10 | 30 | 70 | 90 | 90 | 90 |
| TITAN | -- | -- | -- | 10 | 50 | 110 | 110 |
| MINUTEMAN | -- | -- | -- | -- | -- | 100 | 400 |
| TOTALS | | 10 | 30 | 80 | 140 | 300 | 600 |
| | ESTIMATED DOLLAR REQUIREMENTS (MILLIONS) | | | | | | |
| | FY 58 | FY 59 | FY 60 | FY 61 | FY 62 | FY 63 | FY 64 |
| ATLAS | 653.7 | 665.4 | 588.8 | 443.1 | 142.0 | 107.0 | 72.0 |
| TITAN | 356.4 | 560.5 | 636.7 | 821.0 | 316.0 | 421.0 | 133.0 |
| MINUTEMAN | 26.0 | 233.0 | 276.0 | 308.0 | 290.0 | 240.0 | 185.0 |
| TOTALS | 1,036.1 | 1,458.9 | 1,501.5 | 1,572.1 | 748.0 | 768.0 | 390.0 |

a) Piper, *Development of the SM-80*, Document 53.

Finally, on 28 February 1958, Schriever received authorization, via telex, from Headquarters Air Force:

> You are authorized to proceed immediately with research and development of the Minuteman ICBM program as generally outlined in the 15 February 1958 AFBMD Ballistic Development Plan for Minuteman. Inclusion of the IRBM/TBM [tactical ballistic missile] requirements will be deferred pending further evaluation of Polaris potential. Request you submit a definitized development plan as soon as practicable. Program priority being determined. Action being initiated to provide necessary funds.[86]

On 4 March 1958, ARDC designated Minuteman as WS-133A (WS for weapon system) and advised Hall to select an SM (strategic missile) number. He chose SM-80. Although AFBMD had unofficially named Minuteman, the Defense Department did not make the popular name official until 18 June 1958.[87]

When the influential Ballistic Missile Panel of the President's Science Advisory Committee met on 4 March 1958, it concluded that Thor, Jupiter, Atlas, Titan, and Polaris would have limited retaliatory effectiveness. The liquid-propellant missiles all had long response times, and their complexity rendered them unreliable under prolonged operational use. Polaris was expensive because the submarine component required two hulls for every one on station, and limited communication with submarines on station risked slow response time. Compared to liquid-propellant systems, the panel found Polaris had "superior reliability and a high probability of surviving a surprise attack." Its members also noted that the solid propellant's specific impulse had reached 230 to 240 seconds, an improvement of nearly 40 percent since 1945. With focused effort, 260 to 290 seconds seemed feasible, possibly permitting longer ranges or heavier payloads.

Regarding the Minuteman program, the panel opened the door for what turned out to be several months of debate on details the Air Force thought were already settled. While agreeing in general with the Air Force proposal, panel members believed an initial operating capacity in 1963 would require another crash program, which the service should avoid. Furthermore, the available propellants and motor casing, plus prospects for a lightweight inertial guidance system, constituted a marginal design whose total weight nudged the limit of transportability on surface roads. If Minuteman deployed in 1963, however, further Atlas and Titan improvements might be avoided.

In a prescient suggestion, the panel offered:

> A second alternative plan would be initially to set a more modest range of about 4,000 nm for the Air Force's Minuteman. The bases for this Minuteman could be planned as far north on the North American continent as possible, which would permit coverage of many targets. Meanwhile a vigorous development of solid propellants should be undertaken, with the objective of changing propellants in the Minuteman motor by about 1965 and so extending the range to 5,500 nm in the subsequent missiles.

The panel's report ended with admonishment not to repeat the Atlas, Titan, Thor, and Jupiter mistakes of "plunging into rigidly fixed hardware development programs, without due allowance of time for advanced development."[88]

Finally, on 8 March 1958, Dr. James R. Killian, head of the President's Scientific Advisory Committee, gave President Eisenhower both an executive summary and a full copy of the committee report. Quite possibly this was Eisenhower's first real briefing on the Minuteman program. Even the executive summary contained trepidation about any kind of a crash program.[89]

Two days later, on 10 March 1958, Killian and Eisenhower spoke again about Minuteman. Killian summarized the conversation:

> With regard to the proposal for a well-conceived basic research effort on solid propellants, the President strongly stressed that an overall group, such as ARPA, should conduct this research. Otherwise, it would be done in bits and pieces. In fact, he thought that all research on fuels should be kept centralized, avoiding the wastes of duplication effort. Dr. Kistiakowsky reported that there has really been very little support for, or interest in, a solid propellant development program.[90]

Kistiakowsky either did not know about or was unimpressed by the Air Force and Navy R&D efforts. Eisenhower also opposed rushing into the proposed Minuteman program; he would not give approval until he received more carefully considered information.[91]

The next Air Force opportunity to present detailed Minuteman plans came on 14 March 1958, at a Secretary of Defense Scientific Advisory Committee meeting. The service emphasized "the lower operational complexity, higher reliability and lower cost of production" compared to current ICBM and IRBM programs. Discussion of operational use contrasted the slow response time of liquid-propellant systems to the nearly instantaneous solid-propellant response, which allowed "massive retaliation even under enemy attack," because the deployment plan dispersed the missiles at least three miles between silos hardened to 65 to 100 psi. A single LCC would monitor 50 or more missiles and could immediately launch all of them. With no on-site maintenance, crews would remove missiles needing repair, transport them to a repair base, and immediately install replacements.

In summary, the Air Force requested approval of:

> The basic concept of operation and initiation of the development phase by 1 April 1958, the development and test program can be completed and Minuteman ICBM missiles begin to flow to operational units by August 1962.[92]

On 18 March 1958, McCorkle reported to the JCS on funds needed for Minuteman in FY 1958 and 1959. The Air Force sought $11 million for R&D and $5 million for industrial facilities in FY 1958. The amounts increased to $113 million and $35 million, respectively, for FY 1959. None of the industrial facilities for producing the large amount of propellants and missiles currently existed, and a long lead time necessitated allocation of funding as soon as possible.[93]

Air Force higher headquarters remained opposed to a crash program. Ramo-Wooldridge agreed with the OSD-BMC that a later operational deployment date seemed more reasonable. Even the director of Defense Research and Engineering, Dr. Herbert York, expressed concern about the anticipated need for improved propulsion and missile weight.[94]

The debate about Minuteman funding and the missile's place in the Air Force ICBM program continued. On 31 March 1958, the service submitted a revised development plan that eliminated Minuteman "A" (the IRBM version) and limited FY 1958 and 1959 funding to $22 million and $236 million, respectively. The plan included 25 captive test missiles, 105 flight test vehicles, and 246 operational missiles by the end of FY 1964, with the first flight test occurring in December 1960.[95]

Possibly in response to the skepticism voiced by Minuteman's detractors, the Air Force released an updated preliminary Minuteman operational concept on 8 April 1958. The missile's basic design remained the same, but the possibility of fins on the first stage appeared for the first time. The LF would be reusable, supporting both operational and crew training launches. Any LCC could launch all of a squadron's missiles, either singly or in a salvo. Furthermore, alternative or higher-command headquarters could launch Minuteman if the local LCCs became disabled. Dispersal of LFs meant destruction of no more than one launcher by a 5-megaton weapon and no more than one LCC by a 50-megaton weapon.[96]

Beyond the confusion about funding and programmatic details, the issue of Minuteman program management remained unsettled. Would it be the same as Atlas, Titan, and Thor, where Ramo-Wooldridge or a similar company served as systems engineering and technical development (SETD) contractor, or would it be the earlier, more traditional "prime contractor" arrangement? Although not a new concept, having been used in industry for years, systems engineering in the Air Force began due to the complexity of diverse disciplines in the Atlas program. With SETD, a company like Ramo-Wooldridge served as program coordinator, overseeing and evaluating subcontractors' performance. In the case of Atlas, both industry and the Air Force had little experience with liquid-propulsion systems, missile guidance, or warhead development. By 1958, however, missile contractors and the Air Force had gained sufficient expertise to render a separate SETD firm unnecessary. Nonetheless, Terhune had indicated earlier in the year that Minuteman management would be similar to Atlas, Titan, and Thor, which meant reliance on an SETD contractor.

On 28 March 1958, in preparation for a 2 April meeting of the AFBMC to discuss the Ramo-Wooldridge/STL role for Minuteman, STL director Louis G. Dunn clarified STL's involvement. He noted that, in many ways, the relative simplicity of solid- versus liquid-propellant propulsion would make Minuteman SETD less complex. In Minuteman's case, however, weight margins were considerably narrower, making the SETD effort more critical. Much tighter control over the technical design of the subsystems and interactions between the missile and ground operations would be essential. Treating the completed stages as Class II explosives meant manufacturing and subsequent handling of the missiles would require new safety procedures for assembly and checkout. Finally, the SETD contractor would need to exert tighter control over captive and flight testing, plus experimentation with ground support and control.[97] On 9 April 1958, Schriever reemphasized that the STL division of Ramo-Wooldridge would handle Minuteman SETD. All levels of the Air Force, including the secretary, agreed.[98]

On 26 April 1958, the die was cast. Ramo-Wooldridge would divest itself of STL, as recommended by the AFBMC, and STL would continue as SETD contractor. This ended eight months of debate concerning Ramo-Wooldridge's inside track, through STL, for all missile development contract details. Henceforth, STL's connections would be scrutinized to ensure no unwarranted favoritism existed.[99]

On 28 May 1958, the Navy entered the Minuteman motor performance debate via a letter to Dr. Kistiakowsky from Dr. William S. McEwan, Chemistry Division chief at NOTS. McEwan had learned of Kistiakowsky's earlier solid-propellant research summary through a recent joint Army–Navy–Air Force meeting at Redstone Arsenal. Concerned that the significant breakthroughs at Navy-sponsored facilities—NOTS, ABL, Naval Powder Factory, and Solid Propellant Information Agency—had been overlooked or ignored, McEwan pointed out that NOTS's research department alone employed 40 chemists, 22 of whom had doctorates, and that NOTS laboratories were second to none. Research at NOTS had developed a castable solid propellant with a specific impulse of 255 seconds at 1,000 psi (not corrected for heat loss) and a new system with a calculated specific impulse of 275 seconds. McEwan ended his epistle by noting that better solid propulsion systems could come from a better understanding of combustion efficiency, since many of the test mixtures with promising higher specific impulse lacked the necessary mechanical properties for use in large propellant grains.[100]

On 17 June 1958, the president's Ballistic Missiles Panel met to assess overall progress on the eve of a new fiscal year. Considerable discussion centered on the Minuteman missile's total weight not exceeding 65,000 pounds to accommodate road transportation. The panel acknowledged the feasibility of a three-stage Minuteman missile, assuming slightly higher motor efficiencies than Polaris, development of a guidance system with half the weight of current systems and twice the accuracy of the all-inertial system planned for Atlas and Titan I for 1960 to 1961. Current reentry vehicle technology also required improvement to make the thermal shield's weight a much smaller fraction of the total reentry vehicle weight.[101]

The panel questioned whether the Air Force could resolve all the issues, including thrust termination, lightweight motor casings, and movable nozzles, in time to achieve the 1962-to-1963 operational goal, suggesting the 1964-to-1965 timeframe seemed more likely.[102] Elimination of the requirement for road transportation of the fully assembled missile seemed logical to the panel, whose members recognized that all too often, development of important weapon systems "suffered greatly from artificially imposed restrictions derived from operational considerations that subsequently proved not to be important." The panel recommended no authorization to proceed with the current Minuteman design.[103] Clearly, President Eisenhower's advisors still favored slower Minuteman development and a different operational concept.[104]

On 20 July 1958, award of prime R&D contracts went to Avco Manufacturing Corporation for the reentry vehicle and North American Aviation's Autonetics Division for the guidance system. Thiokol Chemical Corporation received the prime contract for Stage I and a competitive contract for Stages II and III; Aerojet-General Corporation was awarded the Stage II motor contract and competitive contracts for Stages I and III. Hercules Powder Company had the prime contract for Stage III (see Chapter 8).[105]

On 24 July 1958, Dr. Clark Millikan, OSD-BMC chairman, finally supported the Air Force plans for Minuteman. Responding to an earlier letter from Holaday, who sought a rationale for delaying Minuteman, Millikan explained that Schriever had requested formation of an ad hoc group of OSD-BMC to advise AFBMD's Minuteman team. That group met with STL and AFBMD staff members to review progress on solid propellants and various Minuteman subsystem components. According to Millikan:

> The Group is satisfied that feasibility of the concept has now been demonstrated as well as its great importance to the overall ICBM program of the country. These two factors lead the Group to the strongly held feeling that the "study" phase is now completed and that the program should now be established on a weapon system basis, retaining the flexibility mentioned above. This implies that decisions as to operational numbers of missiles required should not be made at present and that production tooling should not be considered until later freezing of major dimensions and characteristics. The potential importance of the program and of its early demonstration of an IOC are such that the level of financial support should now be materially increased over that appropriate of a study phase, although much less than that of a major production effort.[106]

Also, on 24 July 1958, a slightly revised budget submission included an estimated cost for deploying Minuteman on the basis of launch site, LCC, missile, ground support equipment, and communications for 100 R&D missiles ($1.1 million per missile) or extrapolated to $669,000 based on the cost of the 1,000th missile. Totals for FY 1958 and 1959 remained nearly constant compared to the June 1958 estimate—$20.6 million and $210 million, respectively.[107]

Not all group members shared Millikan's conclusions. In a letter to Kistiakowsky on 1 August 1958, Dr. H. W. Bode, director of research, Physical Sciences, Bell Laboratories, agreed that an 80,000-pound missile seemed more reasonable, considering propellants were already developed and allowed a wider margin of error for inevitable weight increases of subsystems. Bode, however, found the guidance system problem far from solved and, getting into details, wondered about public reaction to hundreds of fenced-in LFs and a high-yield thermonuclear weapon at each site. The cost of a constant armed patrol at each site also concerned him.[108]

Millikan's letter also failed to convince Holaday. On 7 August 1958, Holaday requested further information from the JCS. The number of ballistic missile systems then underway—Atlas, Titan, Thor, Jupiter, Polaris, and the new one, Minuteman—bothered him. He acknowledged Minuteman's relative simplicity, both the missile itself and its logistical support, as signaling the future direction for ICBM development. However, increased funding for Minuteman, given the Eisenhower administration's current austerity drive, meant taking money from the first-generation missile programs whose R&D were well underway and their deployment on the horizon. Holaday sought the JCS's reassurance that Minuteman would be more than a sufficient replacement for the first-generation missile systems. He sensed Minuteman's initial deployment would deliver reduced performance and, consequently, require a smaller warhead, meaning more missiles per target and greater costs. Holaday asked the JCS to thoroughly study the warhead criteria and advise him on the "necessity of achieving this mid-1962 IOC when considered in the light of our total strategic weapons system delivery posture at that time." Clearly, he wanted a reason to keep Minuteman in the development phase.[109]

On 4 August 1958, the second change to the June 1958 development plan appeared. It reduced missile production to 131 at the R&D production facility—84 for flight tests and 47 for captive firing—with no change in cost.[110]

On 6 August 1958, Headquarters Air Force issued GOR-171, the requirement for "a quick reaction intercontinental ballistic missile which uses solid or storable liquid propellants. Such a missile dispersed and hardened should greatly increase the national security by providing an improved defensive capability for the Strategic Air Command." The operational performance requirements were:

Readiness
The missile should be designed to permit it to be maintained in a constant state of readiness to provide an almost immediate launch capability from a hardened environment with a minimum exposure time.

Range
A range of 5,500 nautical miles is required. Longer range is desired to provide flexibility in site location and more complete target coverage

Guidance
The guidance system must be as reliable and accurate as is technically feasible in ballistic systems. It must be capable of remaining in a state of alert for extended periods without serious loss of accuracy or excessive replacement of components. The design of the guidance system should be directed towards the use of self-contained guidance which does not rely upon ground guidance stations after launch. The guidance system should be designed to permit target changes to be made in a minimum of time prior to launch.

Yield/CEP
The weapon system should be designed to provide the best initial yield and CEP capability achievable without delaying operational availability. The initial yield and accuracy will be improved at the state-of-the-art permits. The ultimate objective should be to provide the capability of destroying any target of known location. This includes counter-force targets as well as government control centers and other soft area targets. As a guide to the proper yield and CEP, the initial goal should be 1 megaton and 1 nautical mile.

GOR-171 went on to emphasize the need for utmost simplicity of all equipment. Operational performance could be varied to provide for earlier availability.

The overall size and weight of the assembled missile should be as small as possible in consonance with the requirements of transportability, simplicity and hardening. Technological advances which occur during the development of this weapon system will be used when practical to provide larger yields, reduce complexity, greater accuracy, increased range without increasing the missile size and weight.

The missile design would maximize storage life with minimum requirements for inspection and maintenance. Automatic checkout equipment should be capable of being operated from a control point remote from the actual launch site (Figure 5.3).[111]

To provide for changes made through R&D advances, on 29 August 1958, Schriever

~~SECRET~~

Weapon System 133A

# SM-80 MINUTEMAN

GOR 171, 6 August 1958

*Missile Characteristics*

(Wing II Configuration)

|  | Stage I | Stage II | Stage III |
|---|---|---|---|
| Length | 24 ft 10 in | 13 ft 1 in | 7 ft 5.5 in |
| Diameter | 5 ft 5 in | 3 ft 8 in | 3 ft 1.5 in |
| Weight | 50,100 lbs | 12,050 lbs | 5,800 lbs |
| Propellant Weight | 44,980 lbs | 10,400 lbs | 3,660 lbs |
| Total Weight at Liftoff | 69,980 lbs |  |  |
| Ratio of Propellant Weight to Total Weight | .876 | .913 | .907 |

*Nominal Trajectory*

(5,500 nautical miles)

| Item | Time Sec | Alt n.m. | Range n.m. | Velocity Ft/Sec |
|---|---|---|---|---|
| 1. Liftoff | 0 | 0 | 0 | 0 |
| 2. Begin Gravity Turn | 3 | 165 ft | 0 | 134 |
| 3. Stage I Burnout | 58.9 | 15.3 | 22.2 | 7045 |
| 4. Stage II Burnout | 116.7 | 52.3 | 108 | 14,300 |
| 5. Stage III Burnout | 180.6 | 117.8 | 286 | 23,000 |
| 6. Apogee | 862.4 | 710 | 1987 | 16,670 |
| 7. Re-Entry | 34.9 min. | 41.4 | 5384 | 23,400 |
| 8. Impact | 35.5 min. | 0 | 5500 | 800 |

**Flight Profile and Events**

~~SECRET~~

This document is classified; the extract used is declassified IAW EO13526

FIGURE 5.3 A reproduction of the original Minuteman General Operational Requirement 171 illustration describing the operating requirements for the Minuteman missile. *Adapted from "The Development of the SM-80 Minuteman."*

updated the entire Air Force ballistic missile program and included two Minuteman program plans. The first plan, for FY 1963 to 1964, had 100 R&D and 400 operational missiles, with a projected budget for FY 1959 of $194 million, growing to $265 million in FY 1960, $321 million in FY 1961, and down to $316 million in FY 1962. The second plan had 47 operational missiles from R&D excess by FY 1963 and 100 production operational missiles by FY 1964, with nearly the same expenditures for FY 1959 and 1960; for FY 1961 and 1962, the totals were considerably less, $242 million and $239 million, respectively. In neither plan did Schriever mention a higher-level request to plan for only $50 million in 1959, because he considered that funding level inadequate to satisfy the 100/400-missile objectives.[112]

On 4 September 1958, Headquarters Air Force, presumably responding to a JCS request, asked AFBMD to assess the impact of FY 1959 funding options—$50 million, $120 million, or $150 million—for a 100/400 Minuteman force becoming available in the FY 1963 to 1964 timeframe. Schriever responded that $50 million would cover just manpower and was therefore not a viable option. The $120-million option would cover R&D only, slipping an operational force one year or more without any guarantee of meeting the 100/400-missile force numbers; $150 million merely amounted to slightly better funding for the 100/400-force-level option.[113]

Another funding obstacle arose on 12 September 1958, when the Eisenhower administration announced a freeze on the $1.17-billion supplemental Defense Department funding voted by Congress. This meant deferring Minuteman procurement.[114] Consequently, on 17 September 1958, Holaday rejected a request, sent to him on 11 August 1958, for $210 million in FY 1959. He judged the newly submitted plan as resembling the crash programs for Atlas and Titan, which he did not want to repeat. In the interim he approved $100 million for FY 1959—$50 million in new funds and $50 million reprogrammed from current Air Force programs.[115]

On 9 October 1958, Boeing Aircraft Company won the contract for Minuteman final assembly and testing, but the battle for a viable program had only just begun. A month earlier, on 12 September 1958, Maj. Gen. Benjamin I. Funk, deputy director of Ballistic Missiles Directorate Procurement and Production, wrote to Schriever about delays in construction of Minuteman facilities. Funds allocated to Thiokol and Aerojet-General in June remained unreleased as of early September, which put funding for further propellant research efforts nearly on a day-to-day basis. Furthermore, Funk had learned that Minuteman was on the defense secretary's list of "deferred" items.[116]

Both the Navy and Army resisted supporting the Minuteman program. In late 1957 the Naval Warfare Analysis Group in the Office of the Chief of Naval Operations had released a study, "The Policy Implications of Atomic Parity," postulating "that all-out war is obsolete as an instrument for attainment of national objectives," and a "graduated deterrent" capability that relied on nuclear weapons—high-yield down to the lowest yields, complemented with conventional weapons—posed a critical new development. More pointedly, and seemingly indirectly addressing the proposed large-scale deployment of Minuteman, this concept of graduated deterrence "must not be hindered by the allocation of excessive resources to over-inflation of our strategic striking power based on a preventive-war or blunting capability."[117]

The Army argued against mobile Minuteman, asserting that a mobile system required mobile logistics support. Having been forced to relinquish its Jupiter IRBM system to the Air Force, the Army perhaps felt an Air Force mobile IRBM or TBM Minuteman variant would mean the latter service was overstepping its responsibilities and wasting money yet again, because the Army already had such a logistics system for supporting the soon-to-be-fielded Pershing solid-propellant IRBM.[118]

Inadvertent support for Minuteman came during a presentation by V. Adm. John H. Sides, director, WSEG, to the NSC on 13 October 1958. He reviewed US offensive and defensive weapon systems in detail and claimed even one Minuteman squadron of 100 missiles would "severely tax the estimated Soviet production capabilities if they were to attempt to produce enough ICBMs to permit neutralizing a Minuteman squadron." Some, however, might have interpreted his statement as a suggestion to limit Minuteman deployment to only 100 missiles.[119]

On 29 October 1958, Killian wrote to Gordon Gray, President Eisenhower's national security advisor. Commenting on Sides's presentation, Killian suggested a study to answer the question "How many nuclear weapons are needed to meet our strategic objectives?" It could evaluate the need for a deterrent force and a "counter-force" force. The study also might help determine the proper mix of bombers and missiles, plus "the desirability of having mixed forces of missiles, as for example, a combination in the future of Titan, Polaris and Minuteman."[120]

Delays in releasing already approved funding continued to plague the R&D program. Congress had added a rider to the FY 1959 appropriations bill requiring the Air Force to justify construction of new solid-propellant motor testing facilities at Edwards AFB, California, causing a further delay. Meanwhile, delayed release of funds threatened slippage in construction of the 0.30-scale Minuteman aboveground silo test facility at Edwards.[121]

On 25 November 1958, Schriever submitted a revised Minuteman Development Plan to Headquarters Air Force and the AFBMC. It had alternative deployment concepts in response to GOR-171. In addition to hardened, dispersed silos and the mobile concept, the plan described a "shell game" variant of the hardened, dispersed concept called "hardened mobility." Range and payload also changed to 5,500 nm carrying a 600-pound reentry vehicle/warhead or 6,500 nm with a 350-pound reentry vehicle/warhead. Guidance accuracy became one nm or less with a one-megaton warhead. Especially important to keeping the program on schedule, the Air Force needed $1.7 million to construct the full-scale silo testing facilities at Edwards. Schriever needed approval of 1959 funding by January to avoid schedule slippage.[122] By approving Schriever's plan, the AFBMC acknowledged "the program should proceed under a high Department of Defense priority in an orderly manner consistent with the state-of-the-art." The committee also agreed "continuation of the current program schedule and scope at the level of $184 million fiscal 1959 support was compatible with Air Force objectives and R&D requirements." It noted that current advances in the R&D effort justified FY 1960 support at the planned $260-million level.[123] The new plan went for OSD-BMC review on 11 December 1958.

To close out a seemingly productive, albeit frustrating, year for starting the Minuteman program, WS-133A received a 1A priority on 22 December 1958. Thus, 1958 ended with the

Minuteman still struggling to break out from purely R&D funding constraints but finally gaining full weapon system status and a higher priority rating.[124]

Calendar year 1959 began on an extremely positive note, with OSD-BMC approval of the latest Minuteman program plan on 8 January 1959 and a funding ceiling of $184 million. The approval included test facilities at Edwards AFB.[125] On 30 January 1959, Headquarters Air Force directed AFBMD to plan for Minuteman operational construction, training, production, and deployment consistent with the approved funding.[126]

On 25 February 1959, the AFBMC sent the secretary of defense a proposal to accelerate the Minuteman development program based on its review of the rapid development of associated technologies. Acceptance would enable limited Minuteman operational capability by July 1962, with a buildup to 150 operational missiles one year later. Air Force planners understood, however, that acceptance of this proposal would result in degraded range, warhead yield, and accuracy for the first 150 missiles as development continued toward achieving design objectives for the remaining missiles. The committee proposed increasing funds from $260 million to $347 million, but planners felt program acceleration would succeed only if Defense Department and Air Force elements gave their complete support. To ensure this level of cooperation, Minuteman needed a Defense Extraordinary Priority (DX) rating for unimpeded development of an urgently needed weapon system.[127]

On 13 May 1959, the OSD-BMC "approved in principle the plans for intensifying the development effort to ensure the technical progress necessary to provide an adequate basis for achieving the earliest practical production decision" but needed further justification before giving Minuteman a DX rating. It also recommended that JCS establish force levels and production rates to allow an acquisition decision between October 1959 and April 1960.[128] One week later, the secretary of defense approved accelerated Minuteman development, with $347 million reserved pending congressional approval.[129] On 30 June 1959, Minuteman finally received the $347 million allocation for FY 1960.[130]

On 21 August 1959, in response to the deputy secretary of defense's request concerning Minuteman force and production levels, the JCS issued a Directorate of Strategic Plans and Policy (J5) report that revealed disagreement among the Air Force, Army, and Navy about Minuteman plans. It said:

> production planning for Minuteman should be based on the force objective of 100 to 150 missiles for the first year following the delivery of the first operational missile to a unit deployed to an operational site and capable of launching the missile. Factors not yet available or assessed, bearing on the future size and composition of the U. S. Armed Forces, must be determined before Minuteman force levels can be established for the second and subsequent years.[131]

On 4 September 1959, Headquarters Air Force finally assigned Minuteman a DX priority rating. Amendments to all current Minuteman contracts gave the program's contractors the same high-priority access to materials enjoyed by contractors on other missile programs.[132]

★ ★ ★

The Navy's Polaris program played a pivotal role in the technological developments needed to make the Minuteman program feasible. Prodded by the success of the Polaris R&D efforts, the Minuteman program was born on 8 February 1958 with the briefing to the secretary of defense. By the end of 1959, the Minuteman program finally had solidified. Although many funding battles remained, along with unresolved details about deployment locations, number of missiles, and the rail-mobile concept, Minuteman's prospects soared. It no longer was a matter of whether and when but, going forward, a matter of where and how many at each location.

# 6

# SITING AND FACILITY DESIGN

After losing the WS-110A (XB-70) supersonic bomber contract to North American Aviation in December 1957, Boeing's Seattle division management were in a quandary. With the production of the B-52 Stratofortress being transferred to the Wichita, Kansas, facility, the Seattle division either had to lay off a large number of skilled personnel or quickly find another suitable program. Such a program appeared to be on the horizon, and this time Boeing would be fully prepared.[1]

In early 1958 the Air Force published *The United States Air Force Report on the Ballistic Missile: Its Technology, Logistics and the Strategy*, a review of ballistic missile programs, both intermediate- and intercontinental-range systems, which caught the eye of a Boeing systems analyst.[2] The report was a compilation of articles previously published in the summer 1957 issue of *Air Quarterly Review*, the professional journal of the United States Air Force. In the book Gen. Thomas D. White, Air Force chief of staff, briefly discussed the feasibility of conversion from liquid to solid fuels:

> Use of solid propellants would greatly facilitate maintenance and logistics problems, enhance movability, permit more extensive dispersal and hardening, reduce requirements for skilled technicians, and allow for greater automaticity. The relative simplicity of solid-fuel power plants would increase reliability and improve reaction capability. Moreover, it is anticipated that over-all cost of procuring and maintaining solid-fuel missiles will be considerably below that for the liquid-fuel type. For all these reasons solid-fuel power plants will undoubtedly find increasing use in future generations of ballistic missiles.[3]

Apparently the first national press mention of the Minuteman program was in *Time* on 10 March 1958. Calling it the "wildest blue-yonder project in Air Force history," the article went on to describe Minuteman in surprising detail, including length, weight, and probable deployment in the southwestern states, a 5,500-nm range, and an estimated 4,000 missiles to be produced for $3.5 billion. *Time* scooped the two major aviation journals of that time, *Missiles and Rockets* and *Aviation Week*.[4]

On 16 April 1958, Boeing senior management met with staff from the AFBMD to determine what, if any, role the company could play in the Minuteman program. As with other interested firms, the AFBMD staff gave Boeing a detailed briefing on the Minuteman concept. In a new approach to weapon system contracting, AFBMD would be the prime contractor, with system engineering and technical direction support from Ramo-Wooldridge via STL. Four major associate contracts would be offered by the Air Force for the missile component of Minuteman: motor, guidance and control, reentry vehicle, and assembly and test (A&T). A separate architect–engineer (A&E) contract would be given to a firm with major facility construction experience. The Air Force considered Boeing to be a strong competitor for the A&T component.

At this point the goals of the Minuteman program remained those originally proposed by Col. Hall two months earlier: (1) thousands of missiles, (2) inexpensive missiles and ground equipment easily mass produced, (3) three- to four-minute response time depending on economic considerations, (4) no on-site maintenance as entire faulty missiles would be removed and replaced, and (5) no highly trained personnel at the wing or squadron level. Further desired requirements were one LCC per 25 to 50 or perhaps 100 missiles duplexed to a second LCC for redundant launch capability, protective measures to prevent unauthorized launch to be present but not specified, approximately 15 variables to be remotely monitored using readily available commercial equipment, and operating ground equipment easily upgraded as new models of the missile airframe and components were deployed. A fence and electronic surveillance would be used instead of on-site security teams. Site locations would be dictated predominately by already available roads and communication lines.[5]

The A&T contractor would be tasked with design and fabrication of the missile interstage structures and A&T of the missile. The contractor would also function as the test conductor, including operation of test facilities at Patrick AFB, Florida, as well as operations at the assembly plants. Integration of the launcher, launch control systems, communication, transportation, and emplacement of the missile would also be part of the contract.

With the missile anticipated to be inexpensive compared to Atlas and Titan, manning Minuteman would be less complicated, with perhaps 5 men per missile rather than the 100 or more for the current systems. LFs were projected to cost between $50,000 and $100,000 each and be approximately 12 feet in diameter. LCC design was not mentioned.

On 13 May 1958, Boeing learned it was on the bidder's list for the Minuteman A&T contract. The request for proposals was issued on 24 July 1958. On 8 September 1958, Boeing submitted its Minuteman A&T proposal for $163.7 million, one of 17 proposals. On 10 October 1958, Boeing was awarded the Minuteman A&T contract, AF 04 (647)-289. As the A&T contractor, Boeing would work closely with the other associate contractors: Avco (reentry vehicle) Autonetics (guidance and control), and Thiokol, Aerojet-General, and Hercules (motor/airframe).[6]

## Siting

The Minuteman A&T contract included responsibility for system integration: developing the launcher, launch control, transportation of the missile, and installation and checkout of all facilities at the operational bases. A month after the award, Boeing began a study of the transportation system needed to support ground-and-air movement of the missiles to and from the operational launchers, the support base, and the missile assembly facility. Surveys were conducted in five states: Colorado, Iowa, Kansas, Nebraska, and Texas.[7]

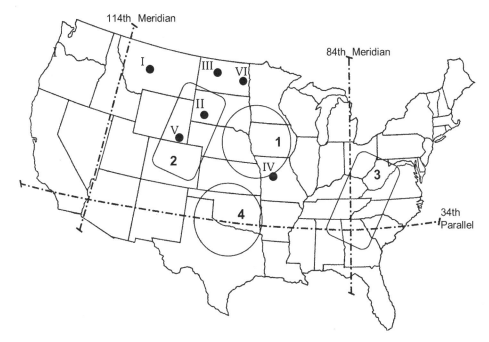

FIGURE 6.1 Original prospective deployment areas for Minuteman outlined as Areas 1 through 4. Subsequently Area 3 was eliminated due to geological limitations. Actual deployment locations indicated by Roman numerals. *Adapted from "Minuteman History Summary 1958–1959," courtesy of the Boeing Company.*

## Missile Assembly-and-Recycle Facility

In early 1959 the first major decision for the selection of Minuteman bases was the location of the missile assembly-and-recycle facility or facilities. Col. Hall's concept of limited maintenance of the missile at the LF required that the assembly-and-recycle facility be close to, or collocated with, the operational bases, unlike Atlas or Titan I or II.

On 5 May 1959, Boeing was directed to further study the economic, logistical, and technical issues dictating the number of assembly-and-recycle facilities required for either a 1,000-, 2,000-, or 4,000-missile force. The initial assumptions included missile production rates to support these three force sizes and combat training launch rates, which would consume 5 to 20 percent of annual missile production within a 36-month period projected to begin in July 1962. The possible locations for a single operational base were narrowed down to four multistate areas (Figure 6.1).[8]

Only two days later, on 7 May 1959, the assumptions had changed:

1. Force size of 500, 1,000, or 2,000 missiles with a buildup within a 36-month period.
2. Silo emplacement:
    a. Silo separation distances of 3, 5, 10, and 15 nm working outward from Hastings, Nebraska.

b. Areas to be avoided for silo siting included populations centers, areas subject to flooding, areas of poor geology, or buffer zones for Atlas and Titan sites.
c. Limits of emplacements were east of the 114th meridian and west of the 84th meridian, south of the Canadian border and north of the 34th parallel.
d. No additional roads.
e. Assumption of 10 silos per LCC.
f. Assumption that the weapon system would be operational from July 1962 to December 1972.
g. Ten percent of the operational force would be expended in verification firings with missiles chosen at random.
h. Assumption of the use of Hastings, Nebraska, as the missile assembly facility, with and without first-stage engine assembly capability.
i. Inclusion of all equipment costs in the assembly and recycle facility.
3. Inclusion in transportation costs of the costs of missiles and ground support equipment in operation during transportation. The cost of each missile with warhead would be $600,000 and the missile alone, $400,000.[9]

Boeing analyzed costs related to site separation distances, independent of weapons effects, and found a cost increase of $33,000 per missile facility for each additional mile of separation over three nm. The major factor was the buried cable system connecting the LFs and LCCs.

The result was another change in assumptions, released in June 1959:

1. Forces sizes of 1,000, 2,000, and 4,000 missiles with buildup over a 36-month period.
2. Assume rail-piggyback transportation mode and as an alternative, all-rail transportation.
3. Weapon system operation from July 1962 to December 1972.
4. Twelve percent of the operational missiles would be expended in verification firings, with missiles selected at random.
5. Silo emplacement:
    a. Three-nm spacing in 50 to 60 missile complexes.
    b. Avoidance of Atlas and Titan sites, population centers, and other areas based upon flooding, transportation, etc.[10]

By July 1959 the projected cost of construction of silos spaced 3 nm apart would be $1,786,040 per missile; at 15 nm apart, $2.5 million per missile.[11]

In early June 1959, SAC outlined its interest in the Omaha, Nebraska, area; the Appalachian part of Indiana; Wichita Falls, Texas; and Cheyenne, Wyoming. A decision on the location of the operational missile assembly-and-recycle facility had originally been scheduled for July 1959 but was postponed to September 1959, causing considerable concern since missile production to support 500 to 4,000 missiles at 30 missiles per month meant that delay in facility construction posed a serious limitation to Minuteman deployment.[12]

Locating the assembly-and-recycle facility was not as simple a task as it might appear. A committee composed of Aerojet, Thiokol, and Boeing evaluated possible locations. Sites evaluated included Hastings Naval Ammunition Depot, Hastings, Nebraska; Cornhusker Ordnance Plant, Grand Island, Nebraska; Nebraska Ordnance Plant, Wahoo, Nebraska; and Kansas Ordnance Plant, Parsons, Kansas. Selection criteria focused on economy—use or modification of existing government facilities, ready access to an airfield as well as highway and rail transportation, and location near the operational area, since planners still felt that more than one production facility would be needed. First-stage motor production would also take place there.

The Naval Ammunition Depot at Hastings was seen as an ideal choice, and survey work was conducted during the spring of 1959. In September 1959 the AMC vetoed the Hastings location due to a change in the support philosophy. Instead of leaving an inoperative missile in the silo until a scheduled recycle operation, now mobile maintenance teams would quickly repair or replace guidance and reentry vehicle subsystems. The original deployment plan called for up to 600 missiles in one deployment area; this was reduced to 150 to 200. Deployment optimizing use of existing facilities was now changed to deployment in sparsely populated areas. Lastly, while the original concept had been a contractor-operated plant, AMC now wanted to reserve the option of depot maintenance, which eliminated the Hastings site.[13]

Nonetheless, on 15 October 1959, the AFBMD and SAC recommended to the AFBMC that the assembly, recycle, and first-stage production plant be located in Hastings. Deployment of the first 150 missiles would be located north of Kansas City, with the option to expand to 600 missiles. Whiteman AFB would be the support base.[14] Instead, on 30 November 1959, the Air Force announced that Hill AFB in Ogden, Utah, would be the location of the first missile assembly-and-recycle facility, named Industrial Plant 77 and designed to produce 30 missiles per month. The close proximity of the Thiokol (Stage I) and Hercules Powder Company (Stage III) production facilities made Hill AFB a more logical choice. The Ogden Air Materiel Area Depot, located at Hill AFB, was selected by AMC as the manager for support of the Minuteman program on 6 January 1959.[15]

On 15 September 1960, construction began on Plant 77, which Boeing would operate. Two months later, Plant 78 construction began next to Thiokol's Utah Division facility at Brigham City for Stage I production. Construction of Plant 81 next to Hercules Powder Company production facilities at Magna, Utah, began in July 1961 for Stage III production. Aerojet expanded its production capabilities in Folsom, California, for Stage II at the same time. As construction of the new facilities neared completion on 29 June 1961, authorization for doubling production capacity to 60 missiles per month came from the Air Force due to a decision Secretary of Defense McNamara made in April 1961.[16] The Air Force accepted Plant 77 on 23 August 1961, and less than a month later, construction for the added capacity commenced. In the end Hill AFB was the only location for Minuteman missile assembly and recycle.[17]

## Base Location Consideration

Possible ICBM base locations were described in WSEG Report 26, Geographical Locations of Initial ICBM Units, on 17 October 1957. The report concluded:

a. In contrast to interior U. S. locations, all ICBM locations considered outside the Continental United States are excessively vulnerable to prelaunch attacks by enemy aircraft and sea-launched missiles. They greatly increase the cost of the system. Their greater vulnerability weakens the deterrent effect of the ICBM, and can increase the total hazard by making an attack more likely.
b. Within the United States, the north-central region, in and around North Dakota, is nearly optimal, and adjacent areas are not far from optimal.
c. The serious enemy ICBM threat can be greatly reduced by hardening our bases (missile facilities).[18]

This advice was ignored in the deployment of Atlas, somewhat followed with Titan I, ignored with Titan II, but followed closely with the Minuteman program.

In early 1959 SAC recommended the entire United States for ICBM sites since dispersal over the entire country complicated the targeting problem for the Soviet Union. SAC bases were already dispersed around the country, so no increase in danger to the population centers would be incurred.[19] On 9 July 1959, the Air Force established that operational sites would not be located within 15 nm of communities of 25,000 or more and that preferred locations would be west of 96 degrees longitude, approximately west of the Minnesota and Dakota boundaries down to the Gulf of Mexico.[20]

On 30 March 1960, Gen. LeMay testified before the Subcommittee on Military Operations of the Committee on Government Operations of the House of Representatives. Chaired by Representative Chet Holifield (D-CA), the subcommittee was investigating the country's civil defense program. LeMay described locating ICBM bases and launch sites so as to protect population centers from the direct effects of nuclear attack as much as possible while optimizing ability to cover enemy targets as well as utilizing existing facilities and geological formations amenable to hardening methods. Curiously, during his testimony, LeMay described how increasing the hardening of a launcher greatly increased the number of missiles needed to destroy it, from 2 for a soft launcher to as many as 40 for a higher level of hardening. When asked about the increase in fallout due to hardening, LeMay replied that it had been taken into account in the siting decisions.[21]

Clarification of the site selection policy was made by Secretary of the Air Force Eugene Zuckert a year later. A number of allegations had been made in public discussions about missile launch site locations. Most prominent was that the Department of Defense (DoD) appeared to be treating the result of an attack on launch sites as if conventional weapons would be used, ignoring the radioactive fallout component of a nuclear attack. Zuckert pointed out that in planning for the deployment with such widely dispersed missiles, it had to be assumed that an attack on the United States would be massive enough to ensure success—that is, destroying our ability to act as a nation. Such an attack would include not only military facilities but also industrial capability, transport and communication, the seats of local and national government, and educational institutions. Therefore, the fallout problem could not be treated simply on a local basis by choosing locations downwind of population centers. As a result, the Air Force had developed a comprehensive list of criteria for site locations: (a) locations from which targets can be reached, (b) avoidance of built-up areas of the eastern United States, mountainous locations, and coastal areas with shallow water tables, (c) favorable geological conditions, and (d) logistical support including roads, proximity of communities for housing, and established support bases to avoid expensive duplication.[22]

## Deployment Area Decisions

### WING I, MALMSTROM AFB, MONTANA

With selection of Hill AFB as the missile assembly and recycle plant, a new search was undertaken for the first operational wing. Malmstrom, Ellsworth, Hill, Larson, and Glasgow AFB were evaluated. The physical criteria for the Minuteman LFs were far less demanding than those for Atlas F and Titan I and II launch complexes. Minuteman launch tubes were only 12 feet in diameter and 83 feet deep, compared to the 80-foot-wide by 180-foot-deep silo for Atlas F and Titan I, or the 55-foot-wide and 145-foot-deep silo of Titan II.[23] The result was consideration of a greatly expanded geographical area. On 23 December 1959, the AFBMC and the Air Staff approved the selection of Malmstrom AFB, Montana, for the 150-missile Wing I (see Figure 6.2 for diagram showing wing locations and missile field layouts).[24]

The usual reason given by historians of ICBM development for the choice of Malmstrom AFB was the reduced range of what would be called the Minuteman IA, deployed only at Wing I: 4,800 nm versus the original 5,500 nm (later changed to 4,910 nm).[25] An additional

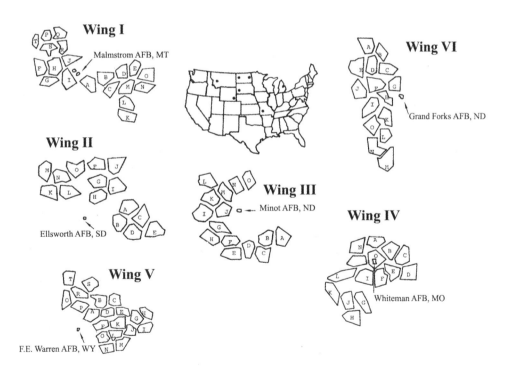

FIGURE 6.2 The location and distribution of the Minuteman missile flights at each wing. *Adapted from "Minuteman Familiarization."*

factor was its 3,972-foot elevation, which provided additional range due to the lower air density in the earliest stages of flight.[26]

The first discussion of deploying Minuteman IA was in the Lauritsen Committee report released on 31 May 1960. The committee members believed that the first consideration was to meet the accelerated deployment schedule, which could only be achieved by relaxing the operational range, reliability, and accuracy requirements. The report stated, "We feel that the urgency associated with the current schedule is great and must be maintained at the expense of performance."[27] On 2 August 1960, AFBMC accepted the recommendations, stating Minuteman IA would have a reduced range of 4,500 to 5,200 nm, with countdown and in-flight reliability at 50 to 70 percent and in-commission rate expected to be 75 to 85 percent.[28]

## WING II, ELLSWORTH AFB, SOUTH DAKOTA

The choices for the second wing were Ellsworth AFB, South Dakota, or Minot AFB or Grand Forks AFB, North Dakota. On 12 October 1960, SAC chose Ellsworth for Wing II, with Minot and Grand Forks tentatively selected for the third and fourth wings. On 19 December 1960, Ellsworth was approved as the second wing, and on 5 January 1961, Secretary Zuckert announced the decision.[29]

## WING III, MINOT AFB, NORTH DAKOTA

In November 1960 LeMay voiced concern over selection of any North Dakota base. He suggested Pease AFB in New Hampshire, which had better housing and community facilities. On 5 January 1961, Minot and Grand Forks AFB were removed from consideration for the third and fourth wings, and Pease tentatively was selected pending soil test results.[30] Initial results at Pease were not favorable, and Minot was selected for further investigation. Sufficient sites were located at Minot for a three-squadron wing, but the geological data indicated that the horizontal and vertical displacements from an attack would be larger than those predicted for Malmstrom and Ellsworth due to a high water table, which would enhance transmittal of blast effects. It would be necessary to design out changes in the LF and LCC so that the operating ground equipment would not be sensitive to the increased earth movement. On 9 June 1961, the Air Force announced the selection of Minot as the location of Wing III.[31]

## WING IV, WHITEMAN AFB, MISSOURI

On 9 June 1961, Whiteman AFB, which had been the original location for Wing I, was selected as the fourth Minuteman wing. In November 1961 concerns about the ability of the Whiteman area highway bridges, built in the 1930s, to support transportation of Minuteman missiles caused the Air Force to reconsider its decision. The cost of road improvements in the Malmstrom area had reached $18 million, and estimates for the Ellsworth and Minot areas were approximately $3.1 million and $2.5 million, respectively. The estimated cost to improve roads and bridges around Whiteman was $27 million. Eliminating Whiteman would delay the program six to eight months. On 12 January 1962, Secretary Zuckert reaffirmed Whiteman as the fourth wing but directed that 25 LFs be relocated, for a savings of approximately $3 million.[32]

## WING V, FRANCIS E. WARREN AFB, WYOMING

On 23 September 1961, Secretary of Defense McNamara issued a draft presidential memorandum that recommended increasing the Minuteman force to 800 missiles (see Chapter 2). In late 1961 the Air Force began looking for a location for a fifth wing of Minuteman missiles, which would have 20 LCCs and four squadrons, each with 50 LFs. Grand Forks AFB, North Dakota; Glasgow AFB, Montana; and Francis E. Warren AFB, Wyoming, were considered. On 27 February 1962, Secretary Zuckert approved F. E. Warren for the fifth wing, and on 10 December 1962, the public announcement was made.[33]

## WING VI, GRAND FORKS AIR FORCE BASE, NORTH DAKOTA

Wing VI base selection had begun in early April 1962. Anticipating deploying Minuteman II with a range of 5,500 to 7,500 nm, depending on reentry vehicle configuration, the Air Force expanded the search area. Prior to 15 April 1962, Grand Forks AFB, North Dakota; Tinker AFB, Oklahoma; and Reese AFB, Texas, had been surveyed, with 220, 165, and 165 sites selected, respectively. The Tinker and Reese AFB locations had problems similar to Whiteman. Added to the list in mid-April were Stead AFB, Nevada; Lowry AFB, Colorado; Schilling AFB, Kansas; and Sheppard and Carswell AFB, Texas. West Coast and additional East Coast locations were also investigated, without detailed surveys, including areas in Oregon and California, as well as Georgia, North Carolina, Virginia, New York, New Hampshire, and Maine. Shallow water tables, saturated sandy soil conditions, and projected higher construction costs, as well as high population densities, quickly eliminated the eastern states and Oregon. Three locations outside of the lower 48 states were considered and surveyed: Elmendorf AFB, Alaska; Anderson AFB, Guam; and the islands of Kauai, Maui, and Hawaii. High transportation and support costs eliminated them from further consideration. The first recommendation was Carswell AFB, with three squadrons, as it would cost $136 million for construction, while the second choice, Grand Forks AFB, would cost $160 million, with neither figure including equipment or missiles.[34]

Stressing strategic requirements as more important, the Air Force Ballistic Systems Division recommended Grand Forks AFB, as it was 900 nm closer to the Soviet Union.[35] On 8 February 1963, the Air Force officially announced Grand Forks as the sixth and final Minuteman wing, with 150 missiles.[36]

## 564TH STRATEGIC MISSILE SQUADRON, MALMSTROM AFB

On 24 July 1964, Headquarters United States Air Force announced that an additional squadron of 50 Minuteman II missiles would be added to the 341 SMW. Briefly designated as the 491st Strategic Missile Squadron (491 SMS), it was replaced by the 564 SMS, which was reactivated on 1 April 1966 (the 564 SMS had been an Atlas D squadron at F. E. Warren AFB and had been deactivated on 1 September 1964) and was the 20th and final Minuteman squadron.[37]

# Dispersal of Launch Facilities and Launch Control Centers

A hardening study in 1958 generated the Soviet threat estimates detailed in Table 6.1. These estimates indicated the need for over 10,000 Soviet missiles to destroy 1,359 aim points in the United States. The question now was how much hardening was necessary and would it be affordable (Table 6.1).

Super-hardening could make the missile complex nearly impervious to all but a direct hit. Deep underground facilities such as abandoned mines offered superior hardening possibilities but slow access to the surface. Natural caves or vaults were another alternative but, again, would result in slow surface access. Either option was counter to the basic concept of the Minuteman weapon system—rapid response. Wide dispersal would prevent multiple-site damage from a single missile hit but would be extremely costly due to logistical considerations.[38]

A study on missile site separation conducted for AFBMD, completed in October 1959, revisited the optimization of missile site separation and concluded that the best definition of "optimum separation distance" was one that considered both system effectiveness and cost.[39] Clearly the exposure time of the system was critical. Exposure time was defined as the time during the launch sequence and early trajectory that the missile would be vulnerable to the effects of two psi of overpressure, enough to cause severe damage to the missile skin. Another concern was the exposure to thermal energy, since 100 to 135 calories/$cm^2$ is sufficient to melt aluminum aircraft skin, or in the case of Minuteman, cause damage to the guidance section, since at the time, Minuteman I had steel Stage I and II motor casings, but the guidance section skin was aluminum. Exposure time for Minuteman was estimated to be 30 to 90 seconds. Radiation effects decrease at an inversely proportional rate to the square of the distance from the source, which meant a small increase in separation distance had a large effect in diminishing thermal radiation exposure.[40]

There were two attack scenarios used in the study. In the first scenario, the majority of enemy weapons would reach US forces prior to initiation of a retaliatory launch. In this case, each LF would still have its silo door closed and be protected to 100 psi overpressure, the initial design consideration for Minuteman given in the study. A minimum five-nm separation ensured that each site would have to be treated as a separate target. In the second scenario, a majority of enemy weapons would reach the US after initiation of a retaliatory launch. Silo doors would be open, with missiles vulnerable in their silos or in the early stages of flight. The optimum LF site separation for Minuteman at 100-psi overpressure in either scenario was determined to be five nm (Table 6.2).[41]

As constructed, the final LF separation distances were five to seven nm for Wing I, which changed to three to five nm for Wings II through VI for reasons of economy, as the increased hardening to 300 psi allowed for closer spacing. LCCs and adjoining LCFs were separated from the nearest LF by a distance varying between 5 and 25 nm. The downwind distance from population centers of 25,000 or more was increased to 18 nm.[42]

## Design Considerations

During the Cold War, volumes were written addressing the perceived capabilities and vulnerabilities of US and Soviet strategic forces. Reliability, probability of surviving to launch, and missile guidance system accuracies were all valid points to study. The designers of the hardened silos for Atlas, Titan, and Minuteman had to contend not with the probability of a Soviet nuclear weapon reaching its target but the effects from its detonation. In Table 6.3

TABLE 6.1  US Programmed Forces and Estimated Soviet Threat, 1961 and 1965[a]

|  | 1961<br>SOVIET FORCES 500 ICBM, 2 MT, 4 NM CEP; 170 SUBMARINE LAUNCHED MISSILES, 3 MT, 3-4 NM CEP | | | 1965<br>SOVIET FORCES, 750 ICBMS, 5 MT, 2 NM CEP; 250 ICBMS, 13 MT, 1.4 NM CEP; 280 SUBMARINE LAUNCHED MISSILES, 3 MT, 2 NM CEP | | |
|---|---|---|---|---|---|---|
| US FORCES | AIM POINTS | HARDENING (PSI) | SOVIET MISSILES REQUIRED | AIM POINTS | HARDENING (PSI) | SOVIET MISSILES REQUIRED |
| BOMBER BASES | 62 | 2.8 | 434 | 62 | 2.8 | 124 |
| ICBM SOFT SITES | 9 | 2 | 36 | 8 | 2 | 16 |
| ICBM HARD SITES | 3 | 25 | 72 | 1,289 | 25–100 | 10,386 |
|  |  | TOTAL | 542 |  | TOTAL | 10,526 |

a) "Ballistic Missile Hardening Study," 10 July 1958.

TABLE 6.2  Characteristics of Basing Systems with Recommended Separation Distances[a]

| SYSTEM | SITE HARDNESS (PSI) | MISSILES/ SITE | SITES/ SQUADRON | EXPOSURE TIME (MINUTES) | SEPARATION DISTANCE (NAUTICAL MILES) |
|---|---|---|---|---|---|
| ATLAS, TITAN I SILO LIFT | 100 | 1 | 12 Atlas F<br>9 Titan I | 5–7 | 14–18 |
| TITAN II, IN-SILO[b] | 100 | 1 | 9 | 2–3 | 7–10 |
| MINUTEMAN[b] | 100 | 1 | 50–100 | 0.5–1.5 | 5 |

a) Adapted from "Missile Site Separation," October 1959; b) original estimate was 100 psi; actual construction was to 300 psi.

selected calculated effects are listed from 2-, 5-, or 30-megaton weapons detonated as air or ground bursts, as those yields were used in many of the planning documents.[43] Many criteria can be used to evaluate how close a warhead would have to impact to prevent launch of the targeted missile. The two scenarios that follow illustrate how variable the answer is to the question of "How close is close enough?"

The first scenario is the effect of surface overpressure. The original Minuteman I launcher closure door was designed to withstand an overpressure of 100 psi. Table 6.3 lists the distance, as a result of an air or surface burst, at which the overpressure would be 100 psi or higher. Depending on warhead yield, the radial distance varies from 4,500 to 11,000 feet.

The second scenario is the physical disruption of the LF by placing it within a portion of the crater. Physical damage to the launcher would have occurred in the rupture zone, which would contain numerous soil cracks of various sizes, or the plastic zone, where soil is

TABLE 6.3  Weapon Effects for a 2-, 5-, and 30-Megaton Air or Surface Burst

| EFFECT | 2 MT | 5 MT | 30 MT |
|---|---|---|---|
| 100 psi radius overpressure contour[a] | 4,500[b] | 6,000 | 11,000 |
| Radius plastic zone, surface burst[c] | | | |
| dry soil, soft rock | 1,600 | 2,200 | 3,800 |
| wet soil, wet soft rock | 2,000 | 3,000 | 5,000 |
| wet hard rock | 2,000 | 2,800 | 5,100 |
| dry hard rock | 1,300 | 1,800 | 3,300 |

a) Adapted from "Air Force Manual for the Design of Hardened Structures"; at overpressures below 30 psi, an air burst can be significantly more effective than a ground burst in area coverage; above 30 psi, both air and ground bursts give nearly the same effect; b) dimensions are in feet; c) *Effects of Nuclear Weapons*.

compressed or deformed permanently.[44] This zone extends out to the edge of the continuous ejecta zone. From Table 6.3 it is readily apparent that the continuous ejecta zone would have a radius from 1,300 to 5,100 feet, depending on soil type and weapon yield. More than six inches of soil on top of the silo closure door for Titan II was sufficient to prevent door opening. The corresponding depth for Minuteman remains classified.[45]

In reality, the question of "How close is close enough?" becomes a difficult one to answer completely. Keeping in mind that these calculations are considered to be approximations and, in many cases, scaled from small-weapon-effect demonstrations, one can see that, at some point, the designers simply had to accept a cost-effective solution and proceed.

## Design of the Operational Facilities

In early 1959 design studies for technical aspects of LFs, LCCs, and support facilities were prepared by Ralph M. Parsons Company, the National Engineering Science Company, and Kaiser Corporation. The final design contract was awarded to Ralph M. Parsons Company, which had designed the Titan I and II facilities as well as the Minuteman launch tube evaluation facilities at Edwards AFB.

In the summer of 1960, the Air Force investigated the feasibility of significantly increasing the hardness of the system from an overpressure of 100 psi for the LF and 500 psi for the LCF to 300 and 1,000 psi, respectively.[46] These changes necessitated shock-mounting a portion of the upper level of the launcher equipment room (LER), increasing the thickness of the launcher closure door from 24 to 40 inches, redesigning the missile support system, and increasing concrete structural components of the LCC. In order not to delay construction of the Wing I facilities, an improved missile suspension system was delayed until Wing II. The geological conditions of the Malmstrom area permitted the use of the original design and still met the new hardening specifications.[47] The final plans were released by AMC to the Corps of Engineers Ballistic Missile Construction Office (CEBMCO) on 31 October 1960.[48]

Unlike the Atlas and Titan I and II launch complexes, where the LCC was adjacent to the missile silos, the Minuteman system facilities, as described earlier, were dispersed, with the LCC situated approximately nine nm from the nearest LF. The LCC was underground, adjacent to the aboveground launch control support building (LCSB).[49] While this relationship between the three main structures remained the same throughout construction of the Minuteman facilities, significant changes in design took place as the wings were built due to changes in strategy from one of massive retaliation during the first year of the Kennedy administration to controlled response (Table 6.4).

Controlled response meant not only launch selectivity and the addition of dual targeting but also increased backup power capability from six hours to nine weeks after an attack. Extended survival required an increase in the hardening of the missile suspension system, the launcher support building (LSB), the LCC, and the launch control equipment building (LCEB). This, in turn, had far-reaching effects in terms of equipment and facility redesign and concomitant cost increases with construction rescheduling. Redesign of the LCCs or LSB was not feasible for Wing I at Malmstrom, as construction was too far advanced.[50] When asked for a bid to change its construction contract to incorporate the new designs into Wing II, which was also already under construction, Peter Kiewit and Sons' Company responded with a price increase of $48.1 million for the entire wing or $31.9 million for the last two squadrons. The Air Force decided that extended survivability could wait until Wing III at Minot. For Wings I and II, the changes not easily accommodated as redesign would be added as modifications or "overlays" to the original design as was feasible, with launch selectivity being incorporated starting with Wing I and dual targeting beginning with Wing II (see Chapter 14).[51] Each of the major design changes will be described on a

TABLE 6.4  Minuteman Facility Hardness and Survivability Criteria, 1964[a]

|  | MINUTEMAN I | | | MINUTEMAN II |
| --- | --- | --- | --- | --- |
|  | WING I | WING II | WINGS III–V | WING VI AND 564TH |
| MISSILE | LGM-30A | LGM-30B | LGM-30B | LGM-30F |
| LAUNCH FACILITY | 300 psi | 300 psi | 300 psi | 300 psi |
| LAUNCH FACILITY EMERGENCY GENERATOR[b] | soft | soft | 25 psi | 300 psi |
| LAUNCH CONTROL CENTER | 1,000 psi | 1,000 psi | 1,000 psi | 1,000 psi |
| LAUNCH CONTROL CENTER EMERGENCY GENERATOR[c] | soft | soft | 300 psi | 1,000 psi |
| SURVIVABILITY | 6 hours | 6 hours | 9 weeks | 9 weeks |
| TARGETING | single | dual | dual | 8 |

a) "USAF Ballistic Missile Programs, 1962–1964"; b) the launcher support building for Wings I and II was considered soft; c) the emergency generator for Wings I and II was in the aboveground launch control support building.

wing-by-wing basis, but before discussing the construction of representative wing facilities, a brief description of each of the structures is necessary.[52]

## Launch Facility

A Minuteman LF consists of the launcher, LSB, security system, and service area situated on 1.8 to 2 acres of land surrounded by a seven-foot, six-inch security fence (Figure 6.3).

The design of the launcher is less variable than that of the LSB or LCEB. Major LF variations between the wings are primarily found in the missile suspension system. The LSB changed positions relative to the launcher and also evolved from a structure with its roof slightly above ground level in Wings I through V to that of a fully buried structure in Wing VI and 564 SMS at Malmstrom AFB (Figure 6.4).

As per the original concept by Col. Hall, the LF was to be unmanned and quite spartan in terms of support equipment since no on-site maintenance was to take place. Only the minimal equipment necessary to determine a system fault and its location would be at the LF. An entire malfunctioning missile would be removed and replaced by a functioning one, with the malfunctioning missile shipped to the support depot for repair.[53] This concept changed on 15 March 1959, when the Air Force announced that the missile maintenance concept was altered to allow for the removal and replacement of one or more sections of the missile at the LF for repair at the supply depot.[54]

### LAUNCHER

The Minuteman launcher structure is composed of the launch tube, enclosed at its upper level by a two-level LER and on the top by the launcher closure door. Compared to the massive silo for the Titan II, which was the first in-silo launch design, the Minuteman facilities are less complex since it is solid-fueled, thus not requiring a complicated propellant transfer system. Earlier work at Edwards AFB had shown there was no need for the separate exhaust ducts found in Titan II, as the launch tube served as an annular exhaust duct. These test results froze the basic design at a 12-foot-inside-diameter launch tube with a flat exhaust deflector (see Chapter 12). The operational LFs were aligned 8.5 degrees off true north—351.5 degrees—to accommodate collimator bench orientation limitations. The LFs at Vandenberg AFB were likewise oriented off true north approximately 48 degrees to accommodate the westerly and southerly azimuths for launching to Kwajalein, Eniwetok, Phoenix Islands, and Oeno Island (Figure 6.5; see Chapter 13).[55]

### LAUNCH TUBE

The launch tube has an inside diameter of 12 feet and is 80 feet, 5 ¾-inches deep measured from the top of the launcher closure to the top of the launcher foundation. Wings V and VI and the 564 SMS had a launch tube 90 feet in depth, possibly to accommodate a future missile after Minuteman III.[56]

The launch tube liner was fabricated from plate steel in two sections. The first section, the lower launch tube liner, is ¼-inch-thick plate, 12 feet in diameter, and extends 52 feet (62 feet at Wings V and VI and 564 SMS) from the top of the launch tube foundation, 3 feet,

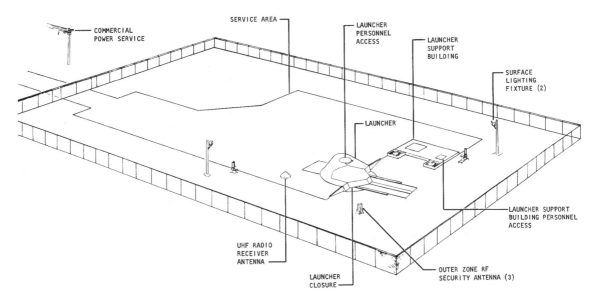

FIGURE 6.3 Typical layout of a Wing II through V LF showing the security fence, the early design intrusion detection system antennas, and the service area. The LSB position relative to the launcher was changed to the east side of the launcher after construction of Wing I. *Adapted from "Minuteman Familiarization."*

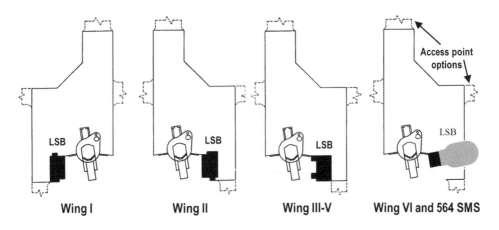

FIGURE 6.4 Plot plans of the LFs of each of the wings. The access point location varied depending on the topography of the site. The LSB roof was slightly above ground level for Wings I through V (black), while for Wing VI and the 564 SMS, it was completely buried (gray). The operational LFs were aligned 8.5 degrees off true north to accommodate collimator bench orientation limitations. *Adapted by author from "Minuteman Familiarization."*

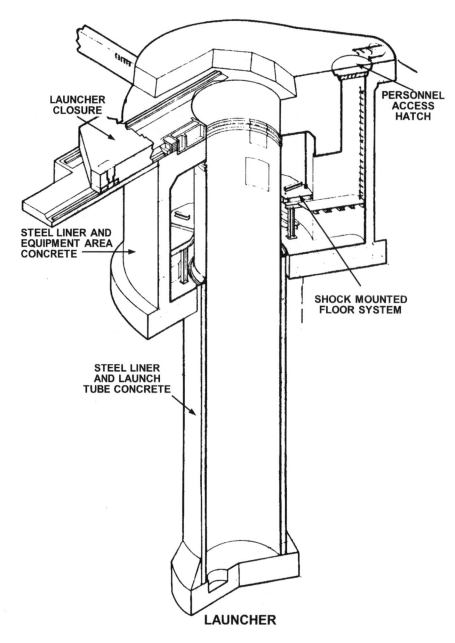

FIGURE 6.5 Typical LF underground layout showing the two-level LER and launch tube. Details of the Personnel Access System are not depicted. *Adapted from "Minuteman Familiarization."*

6 inches to 4 feet thick, depending on soil conditions, to the floor of the lower level of the LER in all wings. This portion of the launch tube liner has, depending on soil conditions, a 12- to 14-inch exterior reinforced concrete shell. At the bottom of the lower launch tube liner is the ½-inch-thick steel deflector plate at Wing I—2 inches thick at Wings II through VI and 564 SMS—with a depression for a sump pump.

The second section, the upper launch tube liner, is 12 feet in diameter and ½ (Wing 1) or 3/8 inch thick (Wings II through VI). The upper liner extends 17 feet, 10 inches to the upper rattle-space rubber boot fitting, which is 1 foot, 2 inches in height and ends at the ceiling of the upper LER for a combined length of 19 feet. The upper launch tube liner does not have a concrete shell, as it forms the interior wall of the LER on both levels. Normal access to the launch tube is through an opening in the upper launch tube liner. An additional 6 feet of ¼-inch-thick steel plate tube extending through the ceiling of the LER is not considered part of the launch tube, as it is separated from the upper launch tube liner by the rattle-space joint.[57]

There are two major equipment components within the launch tube: the missile suspension system and the guidance-and-control section umbilical retraction system (the lower umbilical releases due to missile motion as it leaves the base support ring at launch).

*Missile Suspension System*

The missile suspension system is designed to prevent damage to the missile due to vertical and horizontal ground motion from earthquakes or a nearby nuclear blast. An analogy is the function of shock absorbers in a motor vehicle. The shock absorber "absorbs" the up-and-down movement of the wheel, allowing the frame to remain relatively steady. In the case of a missile, the suspension system acts as the shock absorber and the missile acts as the vehicle frame; the result is that the missile stays in place as the launch tube moves around it. With Atlas F and Titan I and II, a hold-down system was necessary, locking the suspension system in place while the engines developed sufficient thrust for liftoff. The suspension system remained locked as the missile lifted off, preventing upward movement, known as lofting, of the thrust amount due to the absence of the weight of the missile. Since solid-propellant motors reach full thrust almost instantaneously, there was no hold-down system for Minuteman. The lofting issue was avoided by the suspension system spring return rates as well as a tether system.

There have been three variations in the missile suspension system. The Wing I system was deployed only at Malmstrom. As mentioned earlier, the geological conditions at Malmstrom were considered sufficient to permit use of the original design and not delay deployment. Wings II through VI and 564 SMS had the second system to provide additional protection against nuclear-blast ground shock. A third design with additional shock-absorbing capability was installed in the LFs for both Minuteman II and III in the 1970s as part of the Force Modernization/Command Data Buffer (CDB) modification. In each case the missile was located in the same position from the bottom of the launcher closure. Modifications were needed for Minuteman III to accommodate its greater length, meaning the guidance system alignment window was now lower (see Chapter 10).

*Missile Suspension System Wing I*

The design of the original Wing I system is shown in Figure 6.6.

The Stage I skirt rested in a receiver ring, which was supported by the base support ring attached to the missile suspension system. The receiver ring had four alignment ears to help guide the missile into the correct position on the receiver ring. The receiver ring could be rotated to a limited extent using the azimuth drive motor to rotate the missile so that the

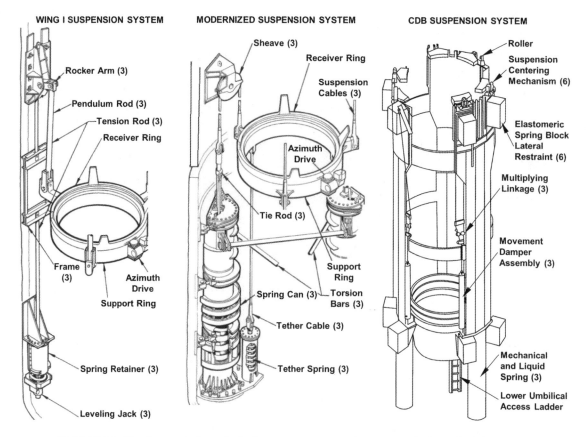

FIGURE 6.6 The three variations of missile support or suspension systems. Left to right: Wing I; Wings II–VI and 564 SMS; Wings I, III–VI, and 564 SMS after the CDB and ILCS modifications. *Adapted from "Minuteman Service News," courtesy of the Boeing Company.*

guidance system's optical alignment opening was positioned correctly in relation to the collimator on Level 1 of the LER. The original system is no longer installed at Malmstrom.

The original Wing I shock-isolation system consisted of three 100-inch-long pendulum rods connecting the base support ring to one end of the three rocker arms, with the other end connected via rods to vertical coil springs within cans anchored to the launch tube wall 300 inches below the rocker arms. The three pendulum rods had universal joints at each end and were spaced at 120-degree increments around the base ring. The brackets supporting the rocker arm fulcrums were attached to the launch tube wall. The vertical springs provided isolation against vertical movement, and the pendulum rods provided isolation against horizontal ground motion. The "windowpane" feature provided protection for the pendulum attachment point against impact with the launch tube wall. Vertical motion was limited to four inches and horizontal motion limited to six inches. Snubbers were added to the rocker arms to limit base support ring movement after the springs were unloaded during launch, preventing rebound damage to the Stage I nozzles.

Vertical oscillations were damped out in approximately six seconds, but lateral oscillations took considerably longer—approximately five minutes—limiting the ability to launch immediately after an attack. The design specifications required that after a shock, the suspension system would return the missile axis to within a 0.25-degree angle from the vertical, ½ inch from the launch tube centerline for lateral position, and the missile elevation to within ¼ inch.[58]

*Missile Suspension System Wings II through VI and 564 SMS*

The missile suspension system was redesigned for use in Wings II through VI and 564 SMS to allow for increased ground shock requirements as the Minuteman system was now to be used for controlled response, which required riding out an attack. The vertical spring stiffness was reduced to allow for higher shock levels, making it necessary to add torsional stiffener bars between the vertical spring cans to increase pitch movement resistance. The torsion bars were mounted on the vertical suspension spring cans and joined to tie rods by connecting rod assemblies so that the bars furnished no resistance to vertical motion as long as the three suspension springs moved in unison. Any force tilting the base support ring, and therefore the missile, exerted a twisting motion on the torsion bars that was resisted, thereby providing the pitch stiffness necessary to ensure stability. The rods and rocker arms of the Wing I system were replaced by a cable-and-pulley arrangement with the same vertical dimensions. The spring cans, attached to the deflector plate, were larger and more complicated, resulting in a substantial increase in cost. Vertical displacement was now limited to 12.5 inches and the horizontal displacement to 6 inches by snubber cables (Figure 6.6).[59]

*Changes with Deployment of Minuteman II*

The Force Modernization Program was established in 1964 to replace the Minuteman IA and IB missiles, at Wing I and Wings II through V, respectively, with Minuteman II. By the time Force Mod reached Wing III in 1970, the decision was made to replace the Minuteman IB missiles at Wings III and V with Minuteman III instead of Minuteman II. This required changes in the missile suspension system since Minuteman II, at 57.4 feet in length, was 3.9 feet longer than Minuteman IA and 1.5 feet longer than Minuteman IB.[60] For Wing I this required relocating the missile suspension system wall brackets and shortening the lower connecting rods; for the remaining wings, the change simply involved changing the cable lengths suspending the base support ring.[61]

*Changes with Deployment of Minuteman III*

With the decision to install the first flight of Minuteman III missiles at Wing III in June 1970, the missile suspension system again had to be altered so that either Minuteman II or III missiles could be installed. The modification was actually quite simple through the use of a 36-inch-long forged steel cable link. To place the base support ring in the lower position to accommodate Minuteman III, the link was placed between the end of the upper cable and the base support ring. An identical length of cable was removed from the lower tether cable. If the decision was made to emplace a Minuteman II in the LF after removal of a Minuteman III, the link would be removed from the upper cable and installed between the tether cable and bottom of the base support ring. This eliminated stocking two different sets of the upper and lower cables.[62]

*Missile Support System Upgrade Silo Program*

Beginning with Wing V Force Modernization/CDB deployment, an entirely new missile suspension system for use with Minuteman III replaced the earlier cable system found in Wings I, III through VI, and 564 SMS (Wing II was not slated for Minuteman III installation and was not modified) and now was called the missile support system. The new system provides much greater protection against ground motion. A large steel cage surrounds the missile as it sits on the base support ring within the cage. Instead of the spring assemblies being attached to the launch tube floor, they are now attached to the bottom of the cage structure at the level of the base support ring. Cables lead up from the spring cans through cable guides at the top of the cage and extend further up to attach to leveling jackscrews anchored to the launch tube wall at the level of the lower LER floor. The cable guides at the top of the cage afford horizontal motion stability and eliminate the need for torsion bars. Inside the cage the missile is protected from lateral movement by six rubber spring blocks that press against the upper surface of Stage I. Prior to launch these blocks are pulled away by articulating arms and do not interfere with launch.

The cage structure is prevented from contacting the launch tube wall by the pendulum suspension of the cage through the cable guides and by two sets of foam blocks located outside the upper and lower cage. During launch the cage's upward movement as it is relieved of the missile weight is arrested by a combination of the six articulating arms that open outward and press against the launch tube wall and a tether cable system that links the bottom of the cage structure to the floor of the launch tube (Figure 6.6).[63]

*Umbilical Retraction System*

The guidance-and-control section umbilical retraction system is unique to the Minuteman program. For Titan II all the umbilicals were flyaway, in that lanyards attached to flexible fittings on the launch duct wall served to disconnect the upper umbilical as the missile lifted off the thrust mount. The same system cannot be used in Minuteman since the guidance-and-control section umbilical carries not only guidance-and-control electrical connections but also two coolant hoses for the guidance system. The umbilical plug is secured to the guidance umbilical receptacle by a threaded shear stud screwed into the receptacle by rotating a handwheel on the umbilical head.

Retraction of the guidance-and-control section umbilical head immediately prior to launch involves an explosive charge that shears a retaining pin, disconnecting the umbilical from the guidance-and-control section of the missile. A simultaneous separation signal activates an explosive charge in the umbilical retraction mechanism, which reels in the umbilical head, rapidly retracting and locking the umbilical into a cavity in the launch tube wall in a position outside of the launch envelope.[64]

## Launcher Equipment Room

WINGS I THROUGH V

The LER is a two-story cylindrical structure completely surrounding the launch tube; it measures 29 feet, 5 ¾ inches in depth as measured from the top of the roof to the bottom of the foundation, with an outer diameter of 25 feet. The reinforced concrete walls and ceiling are two feet, three inches and six feet thick, respectively. The distance between the upper launch tube liner exterior and the inner surface of the equipment room outer wall is five feet, six inches. The upper-level rigid floor section is made of ¼-inch plate steel on 8-inch I-beams, while the lower-level floor is ¼-inch plate steel on top of a 4-foot-thick reinforced concrete foundation. The opening of the launch tube through the LER ceiling is covered by the launcher closure. The lower-level floor, walls, and ceiling of the room are lined with the ¼-inch plate steel to serve as an electromagnetic radiation shield as well as provide a surface for welding of equipment fixtures.

The LER's two levels are isolated from the launch tube liner by a 6-inch rattle-space gap between the two floors and the launch tube for lateral movement and a 14-inch rattle-space gap at the top of the launch tube, at the ceiling of the first level, for movement due to vertical ground motion (Figures 6.7 and 6.8).

Access to the LER is via the Personnel Access System (PAS) shaft, 3 feet, 6 inches in diameter and 21 feet, 3 ¾ inches deep, which protects against unauthorized entry with primary and secondary access doors. The original design of the PAS shaft had only a single door, called the A-plug, which weighed 10,000 pounds and was based on the same idea as a commercial safe security system. The report of the Fletcher Committee in September 1961, which had reviewed Minuteman flexibility and safety, recommended changes to the PAS. A review by the Nuclear Weapons Safety Group of the AEC revealed that the original design could be circumvented relatively easily by professional safecracking techniques, demonstrating that combination locks were insufficient security for an unmanned facility housing a nuclear weapon.

The solution for enhanced security was twofold. First was the addition of a secondary door, a 14,000-pound plug of concrete, steel, and copper 41 inches in diameter and 45 5/16 inches thick, called the B-plug. The B-plug served to block the PAS hatch shaft until lowered to the floor of the upper level of the LER by a jackscrew mechanism. The jackscrew mechanism lowering rate was timed such that security forces from the parent LCF could respond in time to prevent unauthorized entry.[65] The B-plug was designed with four layers: copper to absorb heat from a torch, hardened steel to resist drilling, Mylar composites to resist impact, and dust-creating devices to impair the vision of intruders.[66]

The second solution was a redesign of the collimator sight tube, which was originally 22 inches in diameter. Theoretically, a 200-pound individual could slide through the sight tube, bypassing the B-plug, and gain entrance to the LER. The sight tube diameter was reduced to nine inches by an insert welded into the original tube. The space between the original tube and insert was filled with grout.[67] Design changes were implemented beginning on 31 July 1962 and retrofitted into the Malmstrom and Ellsworth LFs that were already built (Figure 6.9).[68]

## WING VI AND 564 SMS

There were several changes to the design of the LER. The azimuth sight tube was relocated from its earlier true north orientation, accessible after opening the primary access door to the LER personnel access shaft, to a true south orientation, which allowed direct access through the LER wall to the collimator bench. The sight tube retained the narrow

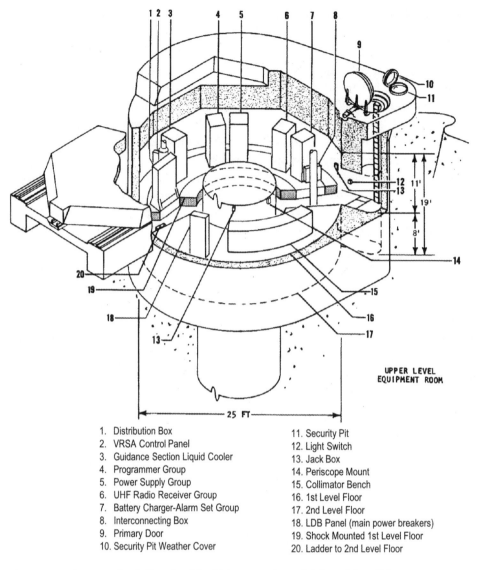

FIGURE 6.7 LER upper level. (Details of the PAS are not shown.) *Adapted from "Minuteman Familiarization."*

diameter to prevent unauthorized access. Only one alignment mirror was installed, and the collimator bench was much smaller due to the new guidance system for Minuteman II, first installed at Wing VI and 564 SMS. With the CDB modification, 2 of the emergency batteries were removed, and missile power for tests and launch came from the remaining 10 batteries. Later in the program, during the Minuteman Extended Survivable Power (MESP) modification, four lead–acid batteries were replaced by lithium-ion batteries at some sites. The shock-isolation platform, weighing 29,000 pounds, was suspended by eight springs and had a horizontal rattle space of 12 inches and vertical rattle space of 14 inches.[69]

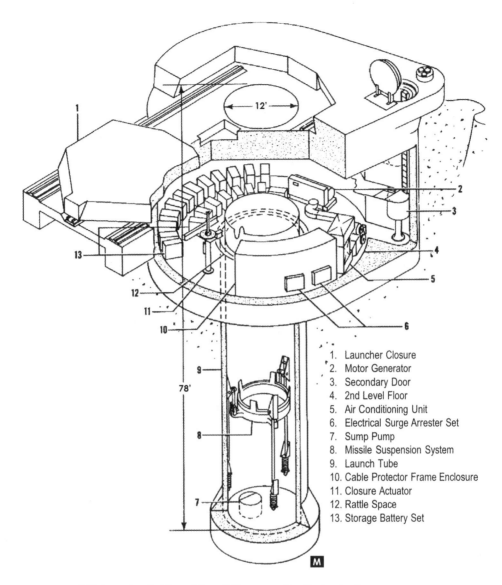

FIGURE 6.8  LER lower level. *Adapted from "Minuteman Familiarization."*

## Launcher Closure

The top of the LER roof is covered with a 3-foot, 5 ¾-inch reinforced concrete closure structure that surrounds the irregular-hexagonal-shaped launcher closure on three sides. The launcher closure is 20 feet, 2 ½ inches long, 18 feet wide at its widest point, 17 feet wide at the narrowest point, and weighs approximately 80 tons. The launcher closure sides are slanted so that in the closed position, the surround serves to limit upward movement. Horizontal movement is limited by the surround to the north, east, and west and an interior locking mechanism to the south. The launcher closure was originally 3 feet, 5 ¾ inches thick,

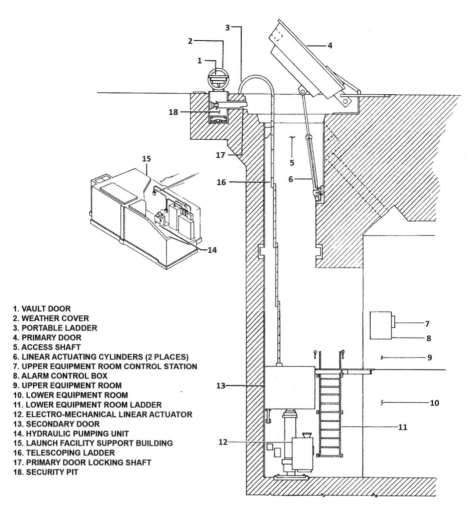

1. VAULT DOOR
2. WEATHER COVER
3. PORTABLE LADDER
4. PRIMARY DOOR
5. ACCESS SHAFT
6. LINEAR ACTUATING CYLINDERS (2 PLACES)
7. UPPER EQUIPMENT ROOM CONTROL STATION
8. ALARM CONTROL BOX
9. UPPER EQUIPMENT ROOM
10. LOWER EQUIPMENT ROOM
11. LOWER EQUIPMENT ROOM LADDER
12. ELECTRO-MECHANICAL LINEAR ACTUATOR
13. SECONDARY DOOR
14. HYDRAULIC PUMPING UNIT
15. LAUNCH FACILITY SUPPORT BUILDING
16. TELESCOPING LADDER
17. PRIMARY DOOR LOCKING SHAFT
18. SECURITY PIT

FIGURE 6.9 PAS cross-section. This is the original installation design, as the hydraulics for opening the door are inside the access tube. A ladder section was brought to connect to the retractable section attached to the wall and retracted or extended by motion of the B-plug. The unlocking mechanism for the hatch cover was secured with a combination lock in the weather-protected vault shown on the left. *Adapted from "Minuteman Familiarization."*

with sides and lower surface of ¼-inch plate steel. The interior is made up of reinforced concrete and housings for four 18-inch-diameter, 7-inch-wide steel wheels that move along a pair of 39-foot, 6-inch-long, 2-inch-wide, 1 ¼-inch-thick steel tracks embedded in concrete. In the closed position, the launcher closure rests on a 1-foot-wide, 14-foot-diameter bearing ring that contains an embedded 1-inch-wide electromagnetic shield gasket, which also extends to surround the door-actuating mechanism (Figure 6.10).[70]

The launcher closure has two methods for opening and one method for closing. The operational opening method is one-way. The ballistic actuator, a gas generator system driving a piston, is used to open the launcher closure lock and move the closure clear of the launch tube opening. The expanding gas from the four (originally two before the Upgrade Silo Program) ballistic gas generators enters the actuator piston, forcing it down. The down-

FIGURE 6.10  Launcher closure door tracks. The door opens and closes by rolling on the two outer tracks. The central track serves as a cogged rail for manually opening and closing the door for maintenance purposes. There is no provision for retaining the door on the tracks during operational opening, as each of the launchers at the operational bases were single-use. For the simulation tests, a barrier of up to 6,000 sandbags was created to prevent the door from rolling off the tracks. *Photograph courtesy of Chuck Penson.*

ward motion moves through a system of rods and pulleys to first open the locking mechanism and then, through a multiplying linkage, initiate launcher closure movement with a large piston, which causes the wheels to move out of a 5/8-inch detent in the track, lifting the door off the bearing ring. Further cable motion disengages the multiplying linkage in the door as it continues to move open and eventually roll off the rails during an operational launch. During tests of the door-opening mechanism, several thousand sandbags are used to prevent the launcher closure from leaving the rails. For maintenance purposes the launcher closure is opened and closed manually using a hydraulic pipe pusher and a cogged center track between the two closure rails. The original device was an electric mule, but early on it was unreliable, and so a hand-controlled hydraulic pipe pusher was used from Wing III on, and Wings I and II were retrofitted as the electric mules needed to be replaced.[71]

### Upgrade Silo Wings I, III through VI, and 564 SMS

The launcher closure and surround were modified by the addition of 10 inches of borated concrete to increase protection against neutrons as part of the CDB modification. The launcher closure now weighs approximately 107.5 tons.[72]

*Vandenberg AFB*

Repeated use of the Minuteman LFs at Vandenberg requires a launcher closure retaining system. This is accomplished by a revetment structure that incorporates spring-loaded buffers to restrict launcher closure movement once clear of the launch tube opening (Figure 6.11).

## *Launcher Support Building*

### WINGS I AND II

The LSB is located approximately 15 feet from the west side of the LER for Wing I and the east side for Wing II. The LSB is a single-story, nearly completely buried, underground structure with its roof exposed approximately four inches at grade. The building is 30 feet, 10 inches long, 15 ½ feet wide, and 11 feet deep, with a 5-inch-thick roof, 8-inch-thick walls, and floors made of reinforced concrete. Two removable steel covers in the LSB roof facilitate equipment emplacement and removal.

Personnel entry is through an access shaft covered by a metal grate at the south end of the building. The metal grate is padlocked, as is the entrance door, which uses the same key as the site gate. The Support Information Network (SIN) telephone is immediately to the right upon entering the LSB and used by maintenance teams to communicate with the flight's LCC as part of the LF entry procedure.

The LSBs at the Wings I and II LFs were not required to survive an attack and therefore not hardened. Support equipment within the building includes a diesel generator used to supply electrical power to the launcher in the event of commercial power failure; a brine chiller used to supply the air-conditioning unit in the LER; a dry-air compressor used to pressurize the underground hardened intersite cable system (HICS) and indicate cable damage; automatic switching unit for the diesel generator; instrument air compressor for the launcher environmental control system, which uses pneumatic controls; the hydraulic pump for LER A-plug opening (later removed when a manual opening system was installed); magnetic security switches; and the SIN telephone (Figure 6.12).[73]

### WINGS III THROUGH V

As discussed earlier, the decision was made partway through the construction of Wing II to increase the hardening of the LSB from soft to 25-psi overpressure. This was implemented for Wings III through V. The building is located 12 feet east of the LER and is still mostly underground, as in Wings I and II. The building is slightly larger, 24 feet, 9 inches wide by 28 feet, 4 inches long, with a depth-to-floor level of 15 feet, 8 inches. The hardening upgrade consisted of thicker walls, floor, and roof: 18 inches, 2 feet, and 5 ¾ inches, respectively. In addition, three-foot-diameter blast dampers were installed, and major equipment is mounted on a shock-isolation platform similar to the platform in the LER. The LSB platform is 5/16-inch plate steel suspended from 10 10-foot, 2-inch suspension springs. Horizontal movement is dampened with rubber bumpers.[74]

The lack of a significant cover to the LSB is due to the fact that the roof is horizontal

FIGURE 6.11 Launcher closure door revetment at LF-02, Vandenberg AFB. Since these sites were reused, the LF had a revetment with bumpers (left), which prevented the launcher closure door from leaving the tracks. *From Corps of Engineers, "Cold War Properties Evaluation."*

and nearly at ground level, so if there is an air blast, the shock wave will pass over the roof, and while it would deform some of the openings, they were reinforced to mitigate damage. The increased structural thickness and the shock-isolation platform provide increased protection against ground shock.[75]

An additional difference between Wings I and II and Wings III through V is the survivability issue. The earlier two wings had sufficient backup power for only six hours of operation for either the LF or the LCC. With the change in strategy for the use of Minuteman, the survivability was increased to nine weeks through improved hardening for Wings III through V and by additional fuel for the backup diesel generators.[76]

### WING VI AND 564 SMS

Hardening of the LSB to 300-psi overpressure was achieved in Wing VI and 564 SMS through a significant redesign of the structure. Instead of a box shape with its roof slightly above ground level, the new LSB was a completely buried, capsule-shaped structure with an access shaft and blast door.

The capsule interior was lined with ½-inch plate steel. The capsule was 22 feet in diameter by 32 feet, 6 inches long. The capsule was encased in a shell of reinforced concrete

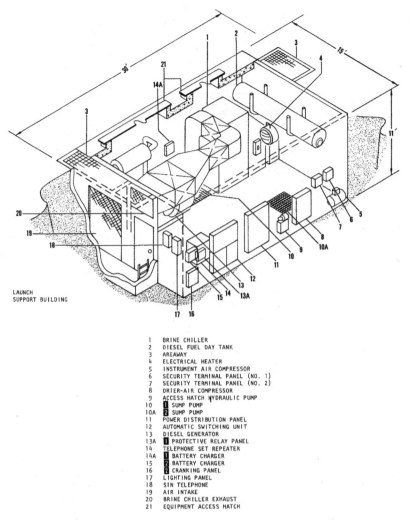

| | |
|---|---|
| 1 | BRINE CHILLER |
| 2 | DIESEL FUEL DAY TANK |
| 3 | AREAWAY |
| 4 | ELECTRICAL HEATER |
| 5 | INSTRUMENT AIR COMPRESSOR |
| 6 | SECURITY TERMINAL PANEL (NO. 1) |
| 7 | SECURITY TERMINAL PANEL (NO. 2) |
| 8 | DRIER-AIR COMPRESSOR |
| 9 | ACCESS HATCH HYDRAULIC PUMP |
| 10 | 1 SUMP PUMP |
| 10A | 2 SUMP PUMP |
| 11 | POWER DISTRIBUTION PANEL |
| 12 | AUTOMATIC SWITCHING UNIT |
| 13 | DIESEL GENERATOR |
| 13A | 1 PROTECTIVE RELAY PANEL |
| 14 | TELEPHONE SET REPEATER |
| 14A | 1 BATTERY CHARGER |
| 15 | 2 BATTERY CHARGER |
| 16 | 2 CRANKING PANEL |
| 17 | LIGHTING PANEL |
| 18 | SIN TELEPHONE |
| 19 | AIR INTAKE |
| 20 | BRINE CHILLER EXHAUST |
| 21 | EQUIPMENT ACCESS HATCH |

FIGURE 6.12   LSB, Wings I and II. The buildings were identical, but at Wing II, the building was rotated 180 degrees and placed on the east side of the launcher. The LSBs for Wings III through V were of similar design, with the roof still near ground level, but the equipment is on a shock-suspended platform. *Adapted from "Minuteman Familiarization."*

with a minimum thickness of one foot, six inches, which increased to a maximum of five feet at the blast lock entrance. The exterior dimensions were 37 feet, 6 inches long and 25 feet in diameter. The outside top surface of the capsule was 13 feet, 4 inches below grade.

Surface access to the LSB was via a 32-feet, 4-inch-deep access shaft with an opening 10 feet, 4 inches square. The access shaft had a two-foot foundation and walls varying in thickness from one to two feet. Access was by ladder after unlocking a padlocked weather cover. A ladder landing platform, 4 feet wide by 6 feet, 4 inches long, was located 16 feet, 10 inches below the access hatch for safety reasons, as there was no equipment on the platform. The blast door dimensions are not available. The blast door protected the eight-foot-high, five-foot-wide, three-foot-long entrance passage connected to the platform with a hinged plate.

Support equipment was mounted on a shock-isolated floor of ¼-inch plate steel on a metal frame 26 feet, 9 ¾-inches long and 16 feet, 11 ½ inches wide. There were four shock isolators and snubbers to dampen horizontal movement. Equipment on the platform included the service entrance for commercial power to the building, automatic transfer equipment to emergency standby, standby diesel engine generator, fluid-level indicator for the diesel storage tank next to the building, intake-and-exhaust blast-valve delay tube system, and environmental control system, which provided conditioned air in the building as well as chilled brine for the LER environmental control system (Figure 6.13).[77]

## Launch Facility Security System

The original LF security system was divided into outer and inner zones. All detected security violations are displayed in the LCC. The outer zone included a radio frequency system that detected surface activity within a specific area. The inner zone included two switches at the launcher closure, one switch on the primary door, one combination lock security switch in the A-plug door vault, security switches on the B-plug door, and penetration detection at the LSB vault door. In addition, vibration sensors were situated within the launch tube. There have been numerous upgrades to the system, which now include video surveillance. Additional details remain classified.[78]

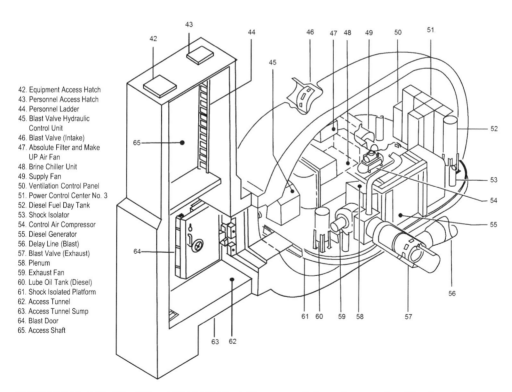

FIGURE 6.13 LSB, Wing VI and 564 SMS. The LSB was now fully underground and hardened with a blast door protecting it from the unprotected access shaft. *Adapted from "Minuteman Familiarization."*

## Launch Control Facility

The LCF consists of the LCSB, the underground LCC, and LCEB in Wings III through VI and 564 SMS, various above- and belowground communications antennas, and additional support structures on 4.75 acres surrounded by a 7.5-foot security fence with one entrance gate. Unlike the Atlas and Titan I and II launch complexes with adjacent LCCs, there is one LCC per flight of 10 missiles, usually located a minimum of nine nm from the nearest LF.

## Launch Control Support Building

### WINGS I THROUGH V

The LCSB is a 4,200-square-foot, single-story, standard frame structure that provides housing for the launch crew support personnel and the flight's security team and maintenance crews as needed. Seven bedrooms and a kitchen support up to 21 personnel as well as the missile combat crew. In the Wings I and II design, the emergency diesel generator and non-emergency air-conditioning for the LCC are located in the support building.

For Wings III through V, the emergency diesel generator is underground in the LCEB. The aboveground entrance to the LCC, and later the LCEB, is located in the Security Control Center portion of the building (Figures 6.14 and 6.15).[79]

The LCSB is larger, 6,800 square feet, and, depending on location, has seven to nine bedrooms and a larger garage (Figure 6.18).

## Launch Control Center

### WINGS I AND II

Access to the LCC is by an elevator located at the Security Control Center. The floor of the LCC normally varies from 29 to 31 feet below grade level. The LCC is a capsule-shaped structure with 10-foot, 3-inch-radius hemispherical ends and a cylindrical portion 21 feet in length. It is fabricated from ¼-inch plate steel. The steel structure is surrounded by reinforced concrete varying in thickness from 4 feet, 3 inches for most of the structure to 7 feet, 6 inches at the blast door entrance for a final exterior length of 53 feet, 9 inches and exterior diameter of 29 feet (Figures 6.16, 6.17, and 6.18).

The LCC is protected by an approximately eight-ton blast door 6 feet, 6 inches high on the outside face and 5 feet, 2 inches high on the inside face, 5 feet, 4 inches wide, and 2 feet, 9 inches thick, made of ¾-inch plate steel and filled with grout. The doorframe is set in reinforced concrete, which, at this location in the structure, is 7 feet, 6 inches thick. The door's hydraulic unlocking system can only be activated from inside the LCC. For additional security there is a manually operated door latch on the inside face of the door. Twelve four-inch-diameter locking pins lock the door to the frame and hold the door pressed against an environmental seal. An interior corridor five feet, two inches high, three feet wide, and five feet, five inches long leads to a hinged plate connecting the corridor to the shock-isolated platform.[80]

FIGURE 6.14  Aerial view of LCF Alpha-01, 10 SMW, 341 SMW Malmstrom AFB. *Official USAF photograph.*

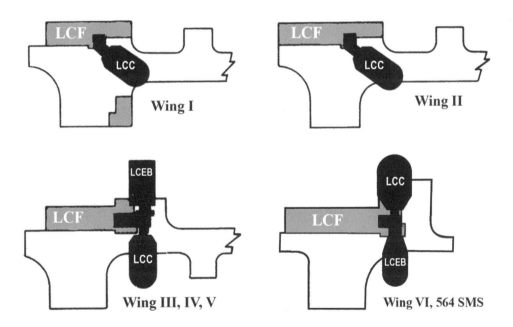

FIGURE 6.15  LCC and LCEB (black) positions relative to the LCF (gray) structure. Wings I and II did not have separate LCEB structures; the support equipment was housed in the LCF. Notice the much larger LCC at Wing VI and 564 SMS. *Adapted by author from "Minuteman Familiarization."*

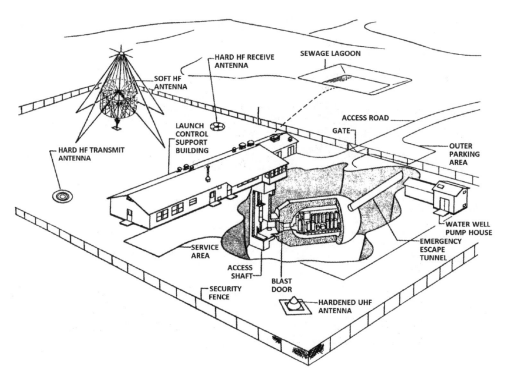

FIGURE 6.16 Typical site plan for Wings I and II LCF showing location of the LCC relative to the aboveground structures. Notably missing in this drawing are the intake and exhaust blast valves and delay lines to and from the LCC capsule. *Adapted from "Minuteman Familiarization."*

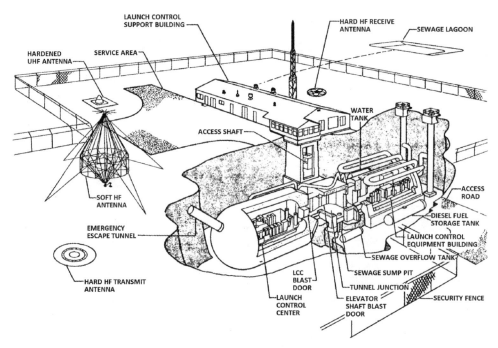

FIGURE 6.17 Typical site plan for Wings III through V LCF. The environmental control system and backup diesel generator for the LCC are now in the hardened LCEB structure. *Adapted from "Minuteman Familiarization."*

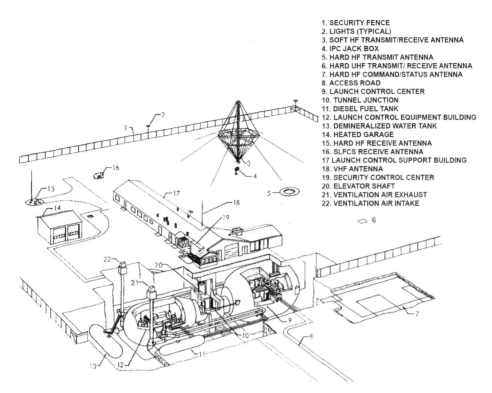

FIGURE 6.18 Typical site plan for Wing VI and 564 SMS LCF. The LCEB has changed position relative to the LCC. The WS-133B LCC was much larger than the WS-133A-M LCC due to increased survivability capacity. *Adapted from "Minuteman Familiarization."*

Access to the shock-suspended LCC floor is via a hinged section of 3/8-inch plate steel. The platform floor varies in width from 10 feet, 6 inches to 12 feet and is suspended from the capsule ceiling by four pneumatic shock-isolation springs. These serve to suspend the floor in a manner similar to a pendulum, and snubbers are located at each corner of the platform to dampen the platform motion after a ground shock.[81]

The original equipment contained on the LCC platform includes the launch control consoles, the communication control console, power supply group, data analysis central, primary alert system, radio set, telephone set repeater, and the SAC Automated Communications and Control System. A bed, oven, refrigerator, and toilet are also on the platform, as well as emergency oxygen-generation equipment and a drinking fountain. Batteries and the motor generator are located in compartments below the floor of the platform. Environmental control equipment is also located on the platform; chilled brine is circulated from the refrigeration unit in the equipment room of the LCSB for air-conditioning. The chilled brine is also used to chill water in an emergency backup chilled-water tank stored against the capsule wall. In the event of isolation from the LCSB, the chilled water tank provides cooling to the equipment in the LCC. The entire platform is enclosed in a sound-attenuating shell.

Air-conditioning for the LCC is provided from the LCSB via the two-foot-diameter blast-valve delay tube system installed in the seven-foot, six-inch walls at the entrance to the capsule. Air entering or leaving the capsule passes through a delay tube system that, when coupled with the blast valves, serves to shut off airflow from or to the outside as a result of overpressure from a nearby blast. A chemical, biological, and radiological filtration system, located in the environmental control equipment room in the LCSB, first filters makeup air from the outside. The cleansed air is chilled using a brine chiller and then recirculated to the LCC.[82]

The launch crew has an escape tube, 23 to 47 feet in length, depending on location, positioned at the far end of the capsule that leads up at a 30-degree angle to the exit hatch 5 feet below the surface. The escape tube is filled with sand to prevent collapse from ground shock. No shock mitigation equipment is involved because of the depth of the exit beneath the surface (Figures 6.19 and 6.20).[83]

### WINGS III THROUGH V

The major difference in the LCC configuration is that the source of conditioned air is no longer coming from the LCSB but instead from the added underground LCEB adjacent to the LCC (see below). A length of chain is inserted between the ceiling and the piston rod, replacing mechanical dampers. The rattle space is 34 inches up, 11 inches down, and 11 inches in any lateral direction.

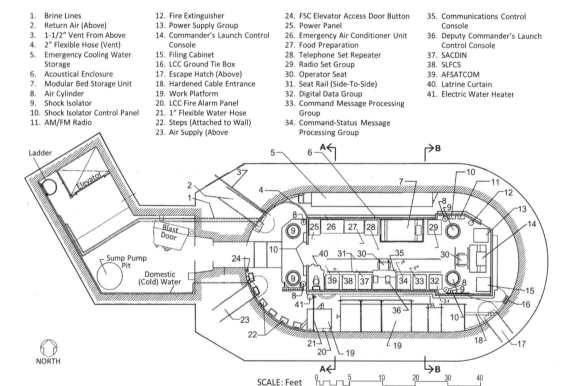

FIGURE 6.19  Floor plan of LCC Delta-01, Ellsworth AFB, circa 1991. *Adapted from Historic American Engineering Record.*

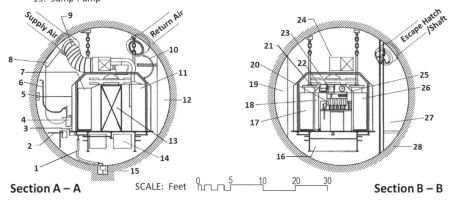

1. 3" Flexible Hose (Sewer)
2. Work Platform
3. Steps
4. Water Heater
5. Fire Alarm Bell
6. 1" Flex Hose (Domestic Water)
7. AFSATCOM
8. Shock Contactor Enclosure
9. Flexible Duct
10. 1-1/2" Vent Flexible Hose
11. Emergency Air Conditioning Unit EACU)
12. EACU Chilled Water Tank
13. To Blast Door/Tunnel Junction
14. Emergency Batteries
15. Sump Pump
16. KO2 Canister Storage
17. Radio Set Group
18. Power Supply Group Rack
19. Commander's Launch Control Console
20. Fire Alarm Bell
21. Acoustical Enclosure
22. Mirror
23. Television
24. Mechanical Equipment
25. Acoustical Headliner
26. Digital Data Group
27. Electrical Surge Arrestor (ESA) Cable Protection Frame Enclosure
28. Wall Thickness – 4'-3"

FIGURE 6.20 Sectional views of LCC Delta-01, Ellsworth AFB, circa 1991. *Adapted from Historic American Engineering Record.*

## WING VI AND 564 SMS

The Wing VI and 564 SMS LCCs were significantly larger than the Wings I through V design (Wing VI and 564 SMS facilities have been retired). The capsule shape remained the same—a cylinder with hemispherical ends. The cylinder was 24 feet in length, and the hemispheres had a 15-foot, 6-inch radius, which gave a total volume nearly three times that of the earlier LCCs.

The reinforced concrete shell varied in thickness from seven feet, six inches for most of the capsule to eight feet, six inches at the blast door entrance. For Wing VI the blast door was made of ¾-inch plate steel, 6 feet, 6 inches high, 5 feet, 4 inches wide, 2 feet thick, and filled with grout, for a total weight of nearly eight tons. For 564 SMS, the door was the same dimension but only partially filled with grout, resulting in a weight of approximately six tons. In either case the door was locked with 12 pins, 6 on each side of the door, operated hydraulically from within the LCC. Access to the shock-isolated equipment platform was via a corridor 5 feet, 4 inches tall, 3 feet, 6 inches wide, 6 feet, 9 inches long, connected to the equipment platform by a hinged panel of ¼-inch plate steel.

The shock-isolated platform was made of ¼-inch plate steel on a metal frame and was 42 feet, 9 inches by 20 feet, 3 inches, which gave a total area nearly 2.5 times that of the earlier system. The rattle space was the same as in the previous LCCs. The increase in volume mentioned above was a direct reflection of the increase in surface area of the shock-isolated platform to accommodate double the amount of emergency backup batteries—the Wing VI and 564 SMS configuration had 32. All LCCs in Wing I have 16 batteries stored underneath the platform, while Wings II through V had 10 or 12, depending on whether the LCC was also an alternate command post.

With the increase in size of the capsule interior came the ability to store 6,700 gallons of chilled brine for use in emergency cooling. Wings I through V can only store approximately 1,200 gallons of chilled water.[84]

The LCC was connected to the access shaft tunnel junction by a 13-foot-long, 18-foot-inside-diameter tunnel fabricated from 1-inch plate steel with a flexible joint on the LCC end. The floor is ¼-inch plate steel (Figure 6.21).

## Launch Control Equipment Building

### WINGS I AND II

There was no LCEB for Wings I and II; the LCSB housed the air-conditioning and backup diesel generator equipment.

### WINGS III THROUGH V

By the time construction began for Wings III through V, the backup diesel generator and associated equipment were no longer part of the LCSB. Instead, these were now located underground and hardened to withstand 300-psi overpressure. The air-conditioning system for the LCC was also relocated to the LCEB (Figures 6.22 and 6.23).

Access is through the Security Control Center in the LCSB, as the LCEB is located at the same level as the LCC. LCEB blast overpressure protection is provided with a blast door located between the bottom of the elevator shaft and the tunnel junction. The junction blast door is taller and wider, 8 feet, 10 inches tall, 7 feet, 5 inches wide, but not as thick, at 1 foot, 7 3/8 inches, as the LCC blast door. It is not filled with grout and is mechanically locked and unlocked via a handwheel on either side of the door, with locking pins located opposite the door hinge. There is also a latch mechanism located on the inside door surface that can be operated from either side. The junction blast door has to be much taller and wider because the diesel generator and brine chiller equipment must be able to pass through the opening during installation and repair or replacement.

Access from the tunnel junction to the LCEB on the left, or the LCC on the right, is by two tunnels, 10 feet and 8 feet long, respectively, and 12 feet wide, made of 3/4-inch steel, with a flexible connection to each building wall. The entrance to the LCEB is eight feet, six inches tall and five feet, three inches wide.

The LCEB is a modified arch made of reinforced concrete and lined with ¼-inch plate steel. It has a six-foot-thick foundation. The walls at the base of the arch are 4 feet thick and taper to a ceiling thickness of 1 foot, 6 inches at the apex of the arch, which is 15 feet, 6 inches above the top of the foundation. The end walls are five feet thick.

The emergency diesel generator, brine chiller, and air-filtration and air-conditioning

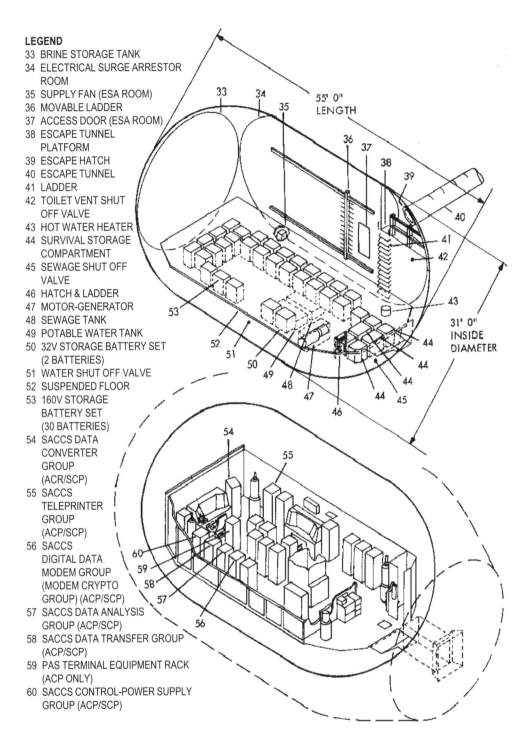

FIGURE 6.21 Details of a WS-133B LCC. *Adapted from "Minuteman Familiarization."*

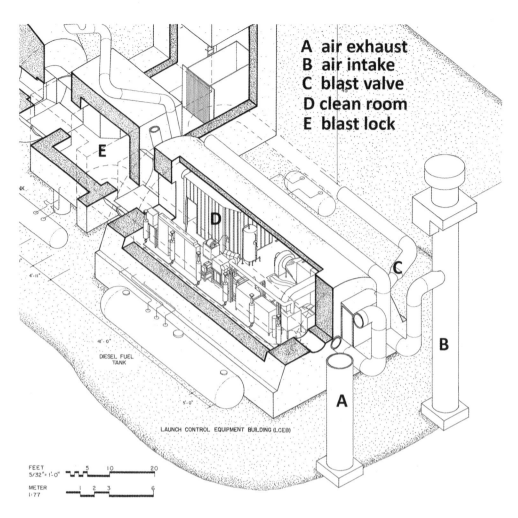

FIGURE 6.22 Cutaway drawing of LCEB Oscar-01, Whiteman AFB, circa 1991. The serpentine piping from the intake shaft to the blast valve was a delay tube used in conjunction with the blast valve to protect the LCEB from air shock. Blast valves were present on both intake and exhaust lines. Only a part of the delay tubing is shown in this illustration. *Adapted by the author from Historic American Engineering Record.*

equipment are all located on a 5/16-inch-thick steel platform suspended from the arch wall by 12 shock-isolation springs. Also located on the platform are a hydraulic pumping unit for the blast valves, the communications cable pressure unit, and telephone repeater equipment, as well as the diesel day tank and large ventilation fans. Horizontal movement is dampened by 16 leaf spring dampers located around the periphery of the platform. The air-conditioning equipment for the LCC is isolated from the rest of the room by a metal structure called the "clean room."

Airflow to the LCEB is via intake-and-exhaust shafts protected by a blast valve and delay tube system. Makeup air is drawn through a chemical, biological, and radiological filter system outside of the clean room and ducted inside, where it is mixed with the

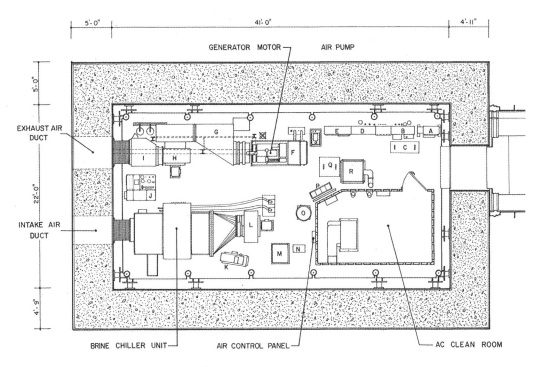

**LAUNCH CONTROL EQUIPMENT BUILDING KEY**

A. 440V JUNCTION BOX
B. TRANSFORMER
C. FLOOR PLATE
D. ELECTRICAL DISTRIBUTION
E. AUTOMATIC SWITCHING UNIT
F. GENERATOR PANEL
G. EXHAUST PLENUM
H. EXHAUST FAN
I. DAMPER CHAMBER
J. BLAST VALVE CONTROL UNIT
K. AIR COMPRESSOR
L. INTAKE FAN
M. TELEPHONE CONNECTING AND SWITCHING SET
N. AIR DRYER
O. DIESEL FUEL DAY TANK
P. BATTERIES
Q. FLOOR PLATE
R. CHEMICAL, BIOLOGICAL, AND RADIOLOGICAL (CBR) FILTER AND FAN

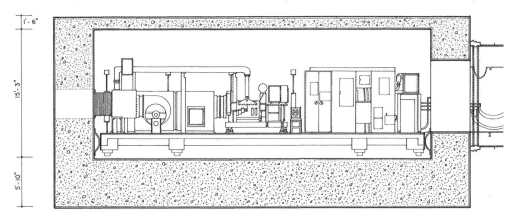

FIGURE 6.23  Top: Floor plan of LCEB Oscar-01, Whiteman AFB, circa 1991. Bottom: sectional view. *Adapted from Historic American Engineering Record.*

recirculating air as needed. Chilled brine is used with the heat exchanger in the clean room to provide conditioned air to the LCC. Chilled brine is also supplied to the LCC for chilling the cold-water emergency cooling backup tank.[85]

The LCC has additional blast protection for the air-circulation system. While seemingly redundant, blast protection for the LCEB and the LCC have different requirements due to what they protect. The LCEB blast valves are 48 inches in diameter and designed primarily to protect the environmental control system, ventilation ducts, and the emergency diesel and motor generator. The LCEB also acts as an expansion chamber for blasts that enter the building, thus lowering the initial force that might be transmitted through the ducting at the air-conditioning equipment and on through to the LCC blast valves. The LCC blast valves were smaller, 22 inches in diameter, and protected a harder structure. The combined LCEB blast valves, LCEB volume, and LCC blast valves allow the crewmembers, ducting, and survival systems to withstand attack and still function. It is a layered approach of blast containment.[86]

## WING VI AND 564 SMS

The Wing VI and 564 SMS LCEB was capsule shaped for increased overpressure protection. The capsule had a cylindrical portion measuring 25 feet, 10 inches and hemispherical ends with a radius of 11 feet, 6 inches. The liner was ¼-inch-thick steel plate surrounded by a reinforced concrete shell varying in thickness from seven feet, six inches for most of the capsule to seven feet, eight inches at the entrance to the blast door. Access to the LCEB was by the same access shaft used for the LCC (Figure 6.24).

There was no blast door at the tunnel junction. Instead, the door was now at the building itself. The door was 10 feet high, 8 feet, 10 inches wide, 2 feet, 3 inches thick, and partially filled with grout. The door weighed approximately 13.5 tons and was operated with a hand pump hydraulic system using 18 4-inch-diameter pins, 9 per side, to hold the door closed against the environmental seal. The door could be opened from either side.

The LCEB floor was suspended by four liquid shock isolators, one at each corner of the platform. These shock isolators were similar to the ones used in the LCC, but in the LCEB, they were jacketed with a heating system to maintain temperature at 125 degrees F. This was necessary because temperature changes in the LCEB due to equipment operation could appreciably change the volume of the isolator fluid, thus affecting the level of the platform.

The LCEB contained the LCEB environmental system, brine chiller unit, diesel enclosure, hardened cable air compressor–dryer unit, stored water tank, chemical, biological, and radiological filter, and pneumatic-control air compressor. The diesel generator and associated equipment were contained within a metal heat-shielded cabinet to isolate the equipment from the LCEB environment.

The air supply to the LCEB was via intake-and-exhaust supplies protected by a blast-valve delay system. The valves were programmed to reopen after 30 minutes. Air was supplied at approximately 6,300 cubic feet per minute and combined with recirculation air as needed to control the temperature above 40 degrees F.

Conditioned air for the LCC was not provided from the LCEB. Instead, chilled brine was circulated to a heat exchanger in the LCC. Makeup air was provided from the LCEB to the LCC after passage through the chemical, biological, and radiological filter system.

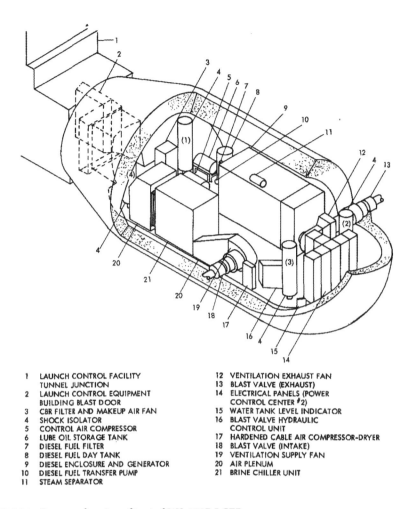

| | |
|---|---|
| 1 LAUNCH CONTROL FACILITY TUNNEL JUNCTION | 12 VENTILATION EXHAUST FAN |
| 2 LAUNCH CONTROL EQUIPMENT BUILDING BLAST DOOR | 13 BLAST VALVE (EXHAUST) |
| 3 CBR FILTER AND MAKEUP AIR FAN | 14 ELECTRICAL PANELS (POWER CONTROL CENTER #2) |
| 4 SHOCK ISOLATOR | 15 WATER TANK LEVEL INDICATOR |
| 5 CONTROL AIR COMPRESSOR | 16 BLAST VALVE HYDRAULIC CONTROL UNIT |
| 6 LUBE OIL STORAGE TANK | 17 HARDENED CABLE AIR COMPRESSOR-DRYER |
| 7 DIESEL FUEL FILTER | 18 BLAST VALVE (INTAKE) |
| 8 DIESEL FUEL DAY TANK | 19 VENTILATION SUPPLY FAN |
| 9 DIESEL ENCLOSURE AND GENERATOR | 20 AIR PLENUM |
| 10 DIESEL FUEL TRANSFER PUMP | 21 BRINE CHILLER UNIT |
| 11 STEAM SEPARATOR | |

FIGURE 6.24 Cutaway drawing of typical WS-133B LCEB.

Originally the LCSB supplied chilled brine to a refrigerant condenser, but during one of the force modifications, this equipment was removed, leaving only the brine chiller equipment in the LCEB.[87]

The LCEB was attached to the tunnel junction by an 8-foot-long tunnel, 13 feet, 9 inches in inside diameter, made of ¾-inch plate steel, with a flexible joint on the LCEB side.[88]

★ ★ ★

Col. Edward Hall's original concept of a simple, unmanned LF with minimal support equipment quickly evolved into a much more complex installation with the decision to service the missile as much as possible at the LF. It is not clear what Hall's vision of the LCC design was, but it also evolved considerably, as did its support facilities. Amazingly enough, a search of the most important construction projects in the United States results in coverage of civilian but not military projects. By the early 1960s, the American public had witnessed the construction of monumental facilities such as the Hoover and Grand Coulee Dams, the Empire State Building, and the Tennessee Valley Authority Project. It is doubtful that anyone could successfully argue that these were not amazing projects. As will be seen in the next chapter, the construction of the 1,100 Minuteman launch and launch control facilities over a six-year timeframe deserves a place in the pantheon of construction marvels in this country.

# 7

# CONSTRUCTION AND ACTIVATION

Fifty-eight years have passed since construction began on the 341st Strategic Missile Wing (Wing I), Malmstrom AFB, Great Falls, Montana. It is easy to forget just how monumental a task it was since 105 (24 Atlas F, 27 Titan I, and 54 Titan II) sites were still under construction. Construction was soon to start on the 44th Strategic Missile Wing (Wing II) at Ellsworth AFB, South Dakota, to be followed by the rest of the wings in rapid succession over the next five years (Table 7.1).

## Construction Issues

Construction began on the first operational Atlas D ICBM complex at F. E. Warren AFB, Cheyenne, Wyoming, in June 1958. While the Air Force had facility design responsibility, the Army Corps of Engineers was responsible for actual construction. The situation was a perfect storm for cost overruns and delays due to the frantic pace of construction. The conundrum was the concept of concurrency—R&D of the actual missile taking place simultaneously with the design and construction of the launch complexes. As new information became available on missile design and performance, facility design changes made change orders for the contractors the rule rather than the exception. One year later, the Atlas construction at F. E. Warren had slipped by six months. Contractors placed the blame for delays and cost overruns on the Air Force and the Corps of Engineers and made their grievances known to their representatives in Washington.[1]

In June 1960 Gen. LeMay, Air Force vice chief of staff, toured the missile sites under construction. LeMay immediately realized that the scope of construction and site activation had been badly underestimated by the Air Force and the Corps of Engineers. The missile manufacturers would give design specifications to an architectural engineering firm, which would draw up designs of the operational facilities. The design would be given to the Corps of Engineers, which then became responsible for the construction, drawing on the expertise of firms capable of large-scale construction projects. After the brick-and-mortar construction was complete, the Air Force stepped in with the installation contractors, who installed the equipment and the missile and brought the sites onto alert.

What LeMay discovered was that the missile manufacturers and the architectural engineering firms had stipulated specifications far more demanding than those previously seen for standard military construction. The result was that the Corps of Engineers was inundated by representatives of the Air Force and the contractors trying to ensure that their specifications were being met. The Corps of Engineers did not have a process at the construction site to deal with conflicting specifications. Further investigation by LeMay's staff

TABLE 7.1  Minuteman Wing Construction Timeline

| WING | 61 Q1 | Q2 | Q3 | Q4 | 62 Q1 | Q2 | Q3 | Q4 | 63 Q1 | Q2 | Q3 | Q4 | 64 Q1 | Q2 | Q3 | Q4 | 65 Q1 | Q2 | Q3 | Q4 | 66 Q1 | Q2 | Q3 | Q4 |
|---|---|---|---|---|---|---|---|---|---|---|---|---|---|---|---|---|---|---|---|---|---|---|---|---|
| 341 SMW MALMSTROM AFB | | ● | ━ | ━ | ━ | ━ | ━ | ● | | | | | | | | | | | | | | | | |
| 44 SMW ELLSWORTH AFB | | | | ● | ━ | ━ | ━ | ● | | | | | | | | | | | | | | | | |
| 455 SMW MINOT AFB[a] | | | | | | ● | ━ | ━ | ━ | ● | | | | | | | | | | | | | | |
| 351 SMW WHITEMAN AFB | | | | | | ● | ━ | ━ | ━ | ● | | | | | | | | | | | | | | |
| 91 SMW F. E. WARREN AFB | | | | | | | | | ● | ━ | ━ | ━ | ━ | ━ | ● | | | | | | | | | |
| 321 SMW GRAND FORKS AFB | | | | | | | | | | | | | | ● | ━ | ━ | ━ | ━ | ● | | | | | |
| 564 SMS MALMSTROM AFB | | | | | | | | | | | | | | | | | ● | ━ | ━ | ━ | ━ | ━ | ━ | ● |

a) The 455th Strategic Missile Wing was renamed the 91st Strategic Missile Wing on 25 June 1968, "Missileers' Heritage."

revealed an administrative nightmare with lines of communication needlessly jumbled. Contractors were receiving conflicting information from over a half dozen agencies, with decisions being delayed for weeks as the buck was passed back and forth.

On 9 July 1960, LeMay directed that each site would have a single manager responsible from site selection to turnover of the completed facility for operational use to SAC. This change did not affect the Atlas D and E facilities already under construction, as they were nearing completion. Site activation was assigned to the AMC. LeMay reassigned Maj. Gen. Thomas P. Gerrity from the Oklahoma City Air Materiel Area to the AMC Ballistic Missile Center and made it clear that he was the man in charge and would be the one LeMay held responsible for the ICBM program construction and activation. Next, LeMay had 18 colonels with reputations for administrative skills reassigned to specific missile construction sites. He let it be known that he was holding each of these officers directly responsible for site activation, construction, and turnover to SAC. The change in oversight for construction was just in time for the first Minuteman wing at Malmstrom AFB. This new concept became known as the Site Activation Task Force (SATAF).[2]

To meet the level of coordination required by LeMay, CEBMCO was established on 2 August 1960 as part of ARDC. Brig. Gen. Allan Welling was made the single point of authority over construction, and his office was located adjacent to that of Gerrity. The Corps of Engineers placed an officer at each site as the area engineer, serving as deputy for construction to the SATAF commander with complete on-site authority for all phases of construction. The AFBMD assigned an officer who was intimately familiar with the particular missile system as deputy for engineering. Thus, each site had on location the necessary command structure to make decisions quickly. Initially there was some conflict between the Air Force and Corps of Engineers officers, but the excellent working relationship between Gerrity and Welling soon permeated the two organizations.[3]

While the new organizations helped, problems continued to develop. Seven months later, on 13 February 1961, the House of Representatives Subcommittee on Military Construction, known as the Sheppard Committee, began hearings to "lay all the facts on the table and clean up this mess." There were 28 witnesses, including the major contractors and Air Force and Corps of Engineers officers. Joseph V. Charyk, under secretary of the Air Force, gave the committee a description of the concept of concurrency:

> We would like to discuss briefly with you the concept of concurrency in the ballistic missile program, which was established at the time this program was first initiated. By concurrency, we are producing missiles in quantity and constructing facilities before prototypes are carefully tested and proven. The Air Force made the decision on concurrency because of the rapid advancement of science in our modern-day weapons. The urgency of having an operational force in the inventory as soon as possible made World War II and Korea manufacturing methods and schedules obsolete. We entered upon this concept knowing that there would be an unusually large number of change orders as a result; however, we felt that the saving in time would be well worth the effort involved. Many of our problems in the ballistic missile program have been due to the extremely dynamic nature of all phases of the aerospace field. It has probably been highlighted more in the ballistic missile area because of the extreme urgency of this program and the fact that we were trying to develop a very large operational force in a minimum of time and in a scientific field where major breakthroughs are frequent. In the Atlas program these major evolutions came so rapidly that almost each missile site for the first seven or eight squadrons was substantially different in concept. The first squadron at Vandenberg has the missile in a vertical position at all times. The second squadron, which was placed at Warren AFB, is on the so-called 1x6 concept with six missiles in a horizontal position in one location. The third squadron at Warren and fourth at Offutt are on the so-called 3x3 soft configuration. The next three squadrons at Fairchild, Forbes, and again at Warren are in what we call the 1x9, semi-hard concept. The next six squadrons, which we consider the "heart" of our Atlas program, are in the 1x12, fully hardened configuration. It was with this rapid change in a site configuration and in the missile itself that it was extremely difficult to develop the usual learning curve that we have experienced in weapons systems in the past.[4]

On 3 March 1961, the committee's findings were published. The committee chairman, Harry R. Sheppard (D-CA), expressed concern with the upcoming Minuteman construction at Malmstrom AFB since the bids in December 1960 had been 55 percent in excess of the government estimate (see below). The report stated that:

Action must be taken by those concerned to bring order and proper direction to the program and provide this Nation with a ballistic missile construction program which will meet the operational needs of our defense forces in a realistic manner without squandering defense funds.[5]

The Air Force and the Army Corps of Engineers responded quickly to the committee's findings with a further reorganization of the construction program. The ARDC and AMC were replaced by the Air Force Systems Command (AFSC) and Air Force Logistics Command (AFLC), respectively. The SATAF was now part of the AFSC, delegated to the Ballistic Systems Division (BSD), which had replaced the AFBMD.

Construction specifications were much more stringent than many of the heavy construction contractors were used to, causing conflicts that had to be quickly resolved. Often equipment would arrive and not fit, leading to disputes as to who was at fault and whether the government or the contractors must bear the financial burden. The SATAF deputy for engineering and CEBMCO deputy for construction worked together to overcome these problems, only to face union strikes, weather delays, and some unusual construction challenges, but these problems had reasonably straightforward solutions.[6]

A standard Minuteman wing is composed of 150 LFs and 15 LCCs. The LFs are organized in groups of 10, known as flights, which are labeled alphabetically A through O, or A through T for F. E. Warren and P through T for 564 SMS. For Wings I through V within each flight, the LCF and LCC are designated, for example, as Alpha-01, and the LFs Alpha-02 through 11. The numbering sequence is clockwise, starting with the first site past 12 o'clock with the clock center on the flight's LCC. Five flights are organized into squadrons of 50 missiles. For Wing VI and the 564 SMS, the LCF, LCC, and LF nomenclature changed. LCF and LCC were now Alpha-0 to Oscar-0 for Wing VI and Papa-0 to Tango-0 for the 564 SMS. The LFs were now numbered by squadron: for Wing VI, LFs Alpha-01 to Echo-50, and so on. For the 564 SMS, the LFs were numbered Papa-01 to Tango-50 (see Table 7.2).

The Minuteman sites are widely dispersed, with the greatest dispersion occurring within Wing I, which encompasses an area of 18,000 square miles, with Flight Foxtrot 60 miles west of Great Falls and Flights Oscar and Kilo 100 miles east and southeast of Great Falls, respectively (this reflects current deployment). Approximately 3,000 miles of road connects the sites, with the longest travel time being to LCF Oscar-01 near Roy, Montana, which takes nearly four hours for a typical crew change, and over twice that for the transporter-erector used to install or remove a missile (Table 7.3).

## Construction Begins

On 15 September 1960, the final design of the Minuteman operational facilities for Malmstrom was given to the Minuteman Configuration Control Board created by the AFBMD. The board had one overriding responsibility: to ensure that any changes that were adopted would not alter the schedule. All major changes would have to be introduced into the system on a block basis, meaning in a unit of 150 missiles, or a wing.[7] One month later, on 26 October 1960, AFBMD released the final drawings to CEBMCO to be placed out for bid. On 15 December 1960, the bids were opened, and, much to the consternation

TABLE 7.2  Minuteman Launch Facility and Launch Control Facility Nomenclature

| SQN. | WING I | | WINGS II-III | | WING IV[a] | | WING V | | WING VI | |
|---|---|---|---|---|---|---|---|---|---|---|
| | LCF | LF | LCF | LF | LCF | LF | LCF | LF | LCF | LF |
| I | A-01 | A-02  A-11 | A-01 | A-02  A-11 | A-01 | A-02  A-11 | A-01 | A-02  A-11 | A-0 | A-1  A-10 |
| | B-01 | B-02  B-11 | B-01 | B-02  B-11 | B-01 | B-02  B-11 | B-01 | B-02  B-11 | B-0 | B-11  B-20 |
| | C-01 | C-02  C-11 | C-01 | C-02  C-11 | C-01 | C-02  C-11 | C-01 | C-02  C-11 | C-0 | C-21  C-30 |
| | D-01 | D-02  D-11 | D-01 | D-02  D-11 | D-01 | D-02  D-11 | D-01 | D-02  D-11 | D-0 | D-31  D-40 |
| | E-01 | E-02  E-11 | E-01 | E-02  E-11 | E-01 | E-02  E-11 | E-01 | E-02  E-11 | E-0 | E-41  E-50 |
| II | F-01 | F-02  F-11 | F-01 | F-02  F-11 | G-01 | G-02  G-11 | F-01 | F-02  F-11 | F-0 | F-1  F-10 |
| | G-01 | G-02  G-11 | G-01 | G-02  G-11 | H-01 | H-02  H-11 | G-01 | G-02  G-11 | G-0 | G-11  G-20 |
| | H-01 | H-02  H-1 | H-01 | H-02  H-11 | J-01 | J-02  J-11 | H-01 | H-02  H-11 | H-0 | H-21  H-30 |
| | I-01 | II-02  H-11 | I-01 | I-02  H-11 | K-01 | K-02  K-11 | I-01 | I-02  H-11 | I-0 | I-31  I-40 |
| | J-01 | J-02  J-11 | J-01 | J-02  J-11 | L-01 | L-02  L-11 | J-01 | J-02  J-11 | J-0 | J-41  J-50 |
| III | K-01 | K-02  K-11 | K-01 | K-02  K-11 | F-01 | F-02  F-11 | K-01 | K-02  K-11 | K-0 | K-1  K-10 |
| | L-01 | L-02  L-11 | L-01 | L-02  L-11 | I-01 | I-02  I-11 | L-01 | L-02  L-11 | L-0 | L-11  L-20 |
| | M-01 | M-02  M-11 | M-01 | M-02  M-11 | M-01 | M-02  M-11 | M-01 | M-02  M-11 | M-0 | M-21  M-30 |
| | N-01 | N-02  N-11 | N-01 | N-02  N-11 | N-01 | N-02  N-11 | N-01 | N-02  N-11 | N-0 | N-31  N-40 |
| | O-01 | O-02  O-11 | O-01 | O-02  O-11 | O-01 | O-02  O-11 | O-01 | O-02  O-11 | O-0 | O-41  O-50 |
| IV | P-0 | P-01  P-10 | | | | | P-01 | P-02  P-11 | | |
| | Q-0 | Q-11  Q-20 | | | | | Q-01 | Q-02  Q-11 | | |
| | R-0 | R-21  R-30 | | | | | R-01 | R-02  R-11 | | |
| | S-0 | S-31  S-40 | | | | | S-01 | S-02  S-11 | | |
| | T-0 | T-41  T-50 | | | | | T-01 | T-02  T-11 | | |

a) With the deployment of the Emergency Rocket Communication System (ERCS) at Wing IV, Squadron II flights were now G, H, J, K, and L, and Squadron III flights were now F, I, M, N, and O. The reallocation was due to proximity of the flights to LCC Oscar-01 to simplify wiring changes to accommodate the ERCS deployment.

of the Air Force, the lowest bid was $79 million, not the $58 million it had budgeted for the Malmstrom construction. The bids were formally rejected on 23 December 1960, and a review group formed to develop changes to the contract to bring it closer to the available funds. The group concluded that the high bids were caused by the tight schedule requirement due to the severe winter weather, high penalty clauses, strident security measures, and validation costs.[8]

Maj. Gen. A. M. Minton, chief of engineers, Department of the Army, recommended that instead of the lump sum contract used in the past, a fixed-price incentive-type contract be negotiated to provide contractors a bonus or penalty depending on their ability to reduce the cost of the work. He recommended expanding the number of prospective bidders and setting a target price of $52 million. He suggested that a conference be convened for prospective bidders where the type of contract would be explained, the relaxation of the time schedules and specifications discussed, and contractors' ideas for maximum cost reduction could be solicited.[9]

On 28 February 1961, CEBMCO awarded a fixed-price incentive contract for the construction of the Malmstrom operational facilities to George A. Fuller Company and Del E.

Webb Corporation, a joint venture from Los Angeles. Taking profits into consideration, the contract's target price was $61.8 million, with the government agreeing to pay all costs up to $58 million and share costs above $58 million to a maximum of $78 million. The contractor could increase his profits on a sliding scale based on how far below the $58-million target he could keep his costs (the rest of the Minuteman construction contracts were fixed price).[10]

There is a paucity of specific information for the construction of the sites at Wing I. Therefore, the general construction procedures for the 44th Strategic Missile Wing (Wing II), Ellsworth AFB, South Dakota, will be used as the overall example, with differences in construction at the various wings discussed. Following the general discussion, several examples of construction problems are described.[11]

## General Construction, 44th Strategic Missile Wing

On 1 August 1961, the operational site construction contract for Ellsworth was awarded to Peter Kiewit Sons' Company of Omaha, Nebraska, for the bid price of $56,220,274. The actual physical construction started on 21 August 1961, and on 14 September 1961, the initial excavation was begun at LF Bravo-05.[12] See Table 7.4 for a summary of construction start and completion dates.

### *Launch Facility*

LAUNCHER

The first excavation was an open cut to a depth of 12 feet, the base of the foundation for the LSB, resulting in a circular excavation approximately 70 feet in diameter. Two access ramps led to the floor of this cut. A second circular excavation for the launch tube shaft was then made to a depth of 32 feet, with a diameter of approximately 40 feet, the depth of the LER foundation, The center of the second excavation coincided with the center of the launch tube and was moved slightly to the west and north of the center of the first cut. This provided a working platform for cranes at the minus-12-foot level and allowed a platform to the southeast for the LSB foundation. With an experienced operator and no equipment failures or bad weather issues, this excavation could be completed in two 10-hour shifts, moving a total of 1,400 cubic yards of material.[13]

From the minus-32-foot level, a 52-foot, 5 ¾-inch-deep shaft was excavated, 14 feet, 4 inches in diameter for the wet sites and 14 feet for the dry sites. Fifteen feet from the bottom, the shaft belled out to sixteen feet in diameter. Although the contract allowed for blasting only in hard rock, greatly reduced charges were allowed for use in the shale and sandstone material. Fifty-eight holes were drilled in four concentric rings, with all but the outer ring slanted inward to varying degrees. The dynamite was detonated electrically with millisecond delays in an attempt to prevent overbreak at the edges of the shaft. The shattered material was removed and the process repeated until the correct depth was reached.

Specifications called for the exposed shale in the shaft to be protected from air within eight hours. After the walls were dressed with pneumatic spades, wire mesh was secured

TABLE 7.3  Minuteman Wing Geographical Statistics

| WING | # OF FLIGHTS | APPROXIMATE DEPLOYMENT AREA(SQ. MI.) |
|---|---|---|
| 341 SMW MALMSTROM AFB[a] | 15 | 18,000 |
| 44 SMW ELLSWORTH AFB[b] | 15 | 12,600 |
| 91 SMW MINOT AFB[c] | 15 | 9,300 |
| 351 SMW WHITEMAN AFB[d] | 15 | 7,500 |
| 90 SMW F. E. WARREN AFB[e] | 20 | 12,000 |
| 321 SMW GRAND FORKS AFB[f] | 15 | 10,000 |
| 564 SMS MALMSTROM AFB[g] | 5 | 2,500 |

a) Minuteman Technical Facilities, Malmstrom AFB, Great Falls, MT; b) "History of Minuteman Construction Wing II," Ellsworth Area Engineering Office; c) WS-133A Minuteman Missile Facilities, Vol. II, Minot AFB, Minot, ND; d) History of Construction Activities and Contract Administration Phases Encountered by Whiteman Area Office Corps of Engineers during Construction of Minuteman Strategic Missile Wing IV, Vol. I, Whiteman AFB, MO; e) "Warren Area Minuteman History"; f) History of Grand Forks Area during Construction of Wing VI, Minuteman II ICBM; g) History of Malmstrom Area during Construction of Collocated Squadron Number 20 Minuteman II ICBM Facilities.


TABLE 7.4  CEBMCO Start and Completion Dates

| WING | START | COMPLETION | DAYS |
|---|---|---|---|
| 341 SMW MALMSTROM AFB[a] | 16 Mar 61 | 21 Sep 62 | 585 |
| 44 SMW ELLSWORTH AFB[b] | 21 Aug 61 | 22 Apr 63 | 609 |
| 455 SMW MINOT AFB[c] | 18 Jan 62 | 11 Jul 63 | 539 |
| 351 SMW WHITEMAN AFB[d] | 2 Apr 62 | 7 Nov 63 | 584 |
| 90 SMW F.E. WARREN AFB[e] | 25 Oct 62 | 19 Jun 64 | 603 |
| 321 SMW GRAND FORKS AFB[f] | 12 Mar 64 | 10 Dec 65 | 638 |
| 564 SMS MALMSTROM AFB[g] | 8 Mar 65 | 26 Oct 66 | 597 |

a) Minuteman Technical Facilities, Malmstrom AFB; "Minuteman Historical Summary, 1961–1962"; b) "History of Minuteman Construction Wing II," Ellsworth Area Engineering Office; c) WS-133A Minuteman Missile Facilities II, Minot AFB; d) History of Construction Activities and Contract Administration Phases Encountered by Whiteman Area Office Corps of Engineers during Construction of Minuteman Strategic Missile Wing IV; e) "Warren Area Minuteman History"; f) History of Grand Forks Area during Construction of Wing VI Minuteman II ICBM; g) History of Malmstrom Area during Construction of Collocated Squadron Number 20 Minuteman II ICBM Facilities.

to the shaft walls. At intervals of approximately 7 feet, a ring of corrugated steel liner plate 16 inches wide was placed against the walls, suspended from the liner plate above with tie rods. Dry-pack grout was placed behind the liner plate, and gunite was sprayed against the walls, held in place by the wire mesh. Three eight-hour shifts per day were commonplace, with an average of seven full working days per launch shaft excavation.[14]

The foundation of the launch tube was four feet of reinforced concrete. Mats of reinforcing bars were placed near the top and bottom of the four-foot slab as well as around the sump pump opening. The deflector plate, 12 feet, 2 ½ inches in diameter, 2 inches thick, with the sump well attached and openings for concrete placement, was positioned using I beams spanning the top of the shaft and connected to the deflector plate via strongbacks. The plate was leveled using turnbuckles, and then braces extending to the shaft walls were welded to it. Approximately 36 cubic yards of concrete was poured through the holes, which were then welded shut.[15]

Concrete operations continued through the winter months of 1961 and 1962. To protect the concrete from freezing, water was heated to 90 degrees F, aggregate was heated and dried, and high early strength concrete was used. Completed concrete pours were covered with tarpaulins, and heat from forced-air butane-type heaters was blown between the concrete and tarps.

The lower launch tube liner, a cylinder 52 feet long and 12 feet in diameter, plus or minus ½ inch, was fabricated from ¼-inch steel plate in Gary, Indiana. When the liner tube reached the Rapid City fabrication yard, studs were welded to the exterior, and reinforcing rods tied in place around the outside. Five spider braces were spaced along the interior of the launcher to maintain its cylindrical shape during shipment from the factory (Figure 7.1).

The launch tube liner had embedded plates of various thickness already welded to the walls, with the most important being the missile suspension system mounting plates, which were 10 feet, 10 inches high by 2 feet wide and made of 2.5-inch-thick steel plate. The three missile suspension system attachment points were located at azimuths of 90, 210, and 330 degrees around the liner circumference, with the bottom of the plates 18 feet, 3 inches from the top of the deflector plate. An air-conditioning duct extended 38 feet, 6 inches down inside the lower launch tube liner.

Placement of the lower launch tube liner involved two cranes attached to an equalizer bar at the top of the liner and a third crane attached to the bottom. Once the lower launch tube liner was vertical, the third crane was detached, and iron workers inspected the exterior of the liner for detached reinforcing steel, which was then repaired as the liner was lowered into the shaft. As the liner was lowered into position, a crew rotated it into its final orientation and plumbed the liner to the tolerance of 1:500 before welding the liner to the deflector plate. Inspection ports were cut in the bottom of the liner and welded shut after inspection and cleaning of the construction joint from the previous pour for the foundation. Ring braces were installed at 3-foot centers for the lower 32 feet of the liner and at 9-foot centers for the upper 20 feet. The top of the liner extended four feet out of the shaft so that a concrete collar with a radius of seven feet from the launch tube liner center could be poured. This collar served to delineate the launch tube liner side of the LER rattle space at the lower level (Figure 7.2).

Concrete was poured through tremies, vertical pipes that extended to the bottom of

FIGURE 7.1 The lower launch tube was 52 feet long and 12 feet in diameter; the cylinder was fabricated from 0.25-inch plate steel. Here the lower launch tube has been delivered without the exterior reinforcing steel. Workers are welding studs and reinforcing tie rods to the exterior at the fabrication yard, Rapid City, South Dakota. *Library of Congress.*

FIGURE 7.2 The lower launch tube insertion is nearly complete at a Delta Flight LF. The lower spider brace can be seen on the right side of the tube. As the liner was lowered into the launch tube excavation, a crew rotated it into its final orientation and inspected it for detached reinforcing steel, which was repaired. *Library of Congress.*

the shaft between the shaft wall and the exterior of the launch tube liner. This method of placement floated any water coming into the shaft to the surface and did not allow it to be mixed with the concrete as it was being placed. Vibrators extended down the sides of the shaft and were raised with the tremies as the concrete level rose. Approximately 60 cubic yards of concrete was used for each shaft.

After the lower launch tube liner pour, an interface control team consisting of two surveyors from the Corps of Engineers, two surveyors from Boeing, and an Air Force officer conducted a post-pour check. The top of the lower launch tube liner had a one-foot-wide concrete collar with 40 one-inch-diameter bolts set on 12-inch centers for connection to the upper launch tube liner. As part of that check, the team established a certified benchmark on the bolt that was on the north–south line of the launch tube. This benchmark was used to set all the equipment bases on the lower floor of the LER.[16]

## LAUNCHER EQUIPMENT ROOM

Once the fine grading for the LER was completed, forms were set for the lower LER floor, providing a six-inch rattle space between the lower launch tube liner collar and the lower level floor. Two mats of reinforcing steel were set with radial and circumferential bars. The lower level of the LER had a floor of ¼-inch plate steel with holes provided for the concrete pour. Reinforcing bar protruded vertically at the edge of the foundation to link to the LER exterior walls at the second pour. The outside forms for the floor were placed in a 16-foot, 9-inch-diameter circle, which increased somewhat in the northeast section to provide for the personnel access shaft. Approximately 112 cubic yards of concrete was placed in the LER floor slab, which was four feet thick.[17]

Once the floor foundation cured, the ¼-inch plate steel liner for the inside walls was set with an interior radius of 12 feet, 6 inches from the launch tube center line. Next the upper nine feet of the upper launch tube liner was set in place on temporary stringers while the LER ceiling plate was welded in place. This was followed by the 12-foot-diameter, 6-foot-high launch tube throat through the LER ceiling. Survey work was then conducted to ensure accurate positioning of the launcher closure actuator box as well as the collimator alignment sight tube. Once completed, the rebar for the ceiling was emplaced, final forms were erected, and catwalks installed for concrete placement. The first four yards of concrete was a thin grout mixture to cover the previously sandblasted and cleaned construction joint connecting the LER foundation to the wall. The 6-foot-thick roof and 2-foot, ¼-inch-thick walls were placed in a monolithic pour (Figure 7.3 and Figure 7.4).[18]

## LAUNCHER SUPPORT BUILDING

The LSB foundation, a total of 20 cubic yards, was poured at the same time as the LER walls and ceiling. After two days of curing the wall and roof, forms were set and 41 cubic yards of concrete poured for the walls and roof. Approximately eight feet south of the LSB, a pad was placed for the diesel fuel tank to be installed later.

## BACKFILL OPERATIONS

The next operation was the first placement of backfill. The backfill process was done in four stages due to the need for placement of numerous electrical and environmental control

FIGURE 7.3  The LER foundation is complete, and the interior 0.25-inch steel plate liner sections for the LER are being placed at a Delta Flight LF. Above the ladder are sections of the LER roof. *Library of Congress.*

connections between the LER and the LSB, as well as additional conduits from the LER to the locations for the security system antennas and the HICS connecting the launcher to the other LFs and LCC. The first placement started at the minus-32-foot elevation and ended at minus 24 feet. A 2,500-gallon diesel fuel tank was set on six inches of compacted sand, which was placed on the previously poured concrete pad.

The second backfill operation began at the minus-24-foot elevation and stopped at minus 8 foot, 8 ¾ inches to permit installation of the launcher closure tracks and track beams, launcher closure wing walls, and transporter-erector jack pads. The track beams extended the closure tracks to the south to allow the launcher closure door to be rolled back from the launch tube, and they were built as cantilever beams with rebar embedded in the LER walls. The foundations for the retaining walls were set against the LER on the east and west sides to allow for the abrupt grade change of 3 feet, 5 ¾ inches between the top of the launcher closure surround and the top of the track beam. The retaining wall foundations were at two levels, eight feet, six inches and six feet, eight inches below the wall top, and poured monolithically with the wall itself.

The jack pads for the missile transporter-erector were designed to allow the hydraulic rams, which lifted the missile transporter-erector into the vertical position, to be safely

FIGURE 7.4 Rebar hoops are being installed on the exterior of the wall of the LER. The PAS shaft is seen in the foreground with the collimator sight tube projecting down to the left. *Library of Congress.*

anchored. The two pads were set on top of columns two feet by three feet, six inches in cross-sectional dimensions. The column lengths varied from site to site and were 6 to 12 feet deep and set on a common 3-foot-thick footing.

The third backfill brought the elevation to minus 3 feet, 5 ¾ inches to allow the launcher closure enclosure surround, maintenance track slab, and gear pad pylons to be built. Non-frost-susceptible material was placed under the launcher closure track beams and maintenance track slab for a depth of 18 inches.

The fourth backfill came up to final elevation; this allowed the launcher closer surround and door to be formed and poured. The launcher closure door frame was brought to the site complete with reinforcing bar. It was set on the tracks and rolled into place with the wheels sitting in their indents in the track rails while the door bearing ring rested on the lower bearing ring located on top of the LER roof. The door was pulled down tight to the bearing ring, checked for alignment, and then welded to the launch tube throat in the ceiling of the LER to prevent movement during the concrete pour. With the door in place, prefabricated steel shrouds were placed against the door on three sides, spaced the proper distance from the door with shims on one-foot centers. The shrouds served as the inside form of the closure surround (Figure 7.5).

FIGURE 7.5 Beginning of the concrete pour for the launcher closure door and surround. Just to the right of the concrete bucket is one of two transporter-erector pylons. The workers are standing on the door reinforcing steel, which was #8 rebar (one-inch diameter) on six-inch centers. The door was welded in the closed position to prevent movement during the concrete pour. *Library of Congress.*

The PAS hatch door, frame, and hinge box were preassembled off-site. The assembled unit was placed on top of the access shaft, and the hydraulic cylinder attachment lugs on the door were oriented precisely above the respective cylinder bases set in the access hatch. The hatch was enclosed, and the security pit was positioned relative to the lock recess and the access hatch, then welded into place. Remaining forms for the outside of the surround were set, and rebar put in place. The transporter-erector pylons were welded onto the rebar, jack pad steel inserts placed on top of the columns, and two gear pads—13-inch-square, ½-inch steel plate set on reinforced columns 4 feet, 6 inches deep—were set. The construction joints for the surround and the LER roof were sandblasted and cleaned, and then 91 cubic yards of concrete was poured for the launcher closure, surround, and gear pads (Figure 7.6).

FIGURE 7.6 Workers prepare the sealing surface for the Personnel Access System hatch. This is the original hatch design, as can be determined by the interior hydraulic mechanism for raising and lowering the hatch. The hatch had to be large enough for equipment to be transferred to and from the LER. *Library of Congress.*

After the fourth backfill operation, a nine-inch concrete slab extending north of the top of the LER to the jack pads was formed and poured. Similarly, the launcher closure maintenance track slab was poured.[19]

## Interior Structural and Mechanical Work

### LAUNCHER EQUIPMENT ROOM

The structural framing for the fixed and shock-mounted floors for the upper level of the LER was prefabricated, lowered into the LER through the launch tube throat in the ceiling, and rough-set into position. The upper half of the upper launch tube liner had been set on stringers spanning the lower launch tube liner prior to the installation of the equipment room roof liner plate. This upper portion was raised by cranes outside the equipment room, and the lower portion, in three pieces, was placed under it and welded to it. The seams between the lower portions were then welded together, and the entire upper liner was then oriented to true north and checked to see that the collimator slot was at the proper elevation. A mark on the ring at the lower end of the upper launch tube was aligned with the lower launch tube liner benchmark, the elevation set to align with the collimator bench on the upper level of the LER, and finally, the two sections were bolted together and then circumferentially welded on the interior face.

The collimator bench and theodolite stand rail were then set, surveyed for proper position, and then grouted into place. A water stop was placed in the rattle space between the launch tube liner and the LER floor. A wire mesh electromagnetic shield was placed over the water stop and rattle space and brazed to both sides of the rattle-space gap.[20]

## LAUNCHER SUPPORT BUILDING

For the LSB a backup diesel generator was lowered through the equipment hatch and bolted to the floor, as well as the diesel fuel day tank, a water chiller unit to provide chilled water for the LER environmental control system, and the hydraulic unit for operating the personnel access hatch. Hatch covers were installed, gratings for the air intake and exhaust were put in place, and the access door was hung.[21]

## *Launch Control Facility*

### LAUNCH CONTROL CENTER

The first phase of the LCC construction was excavation to the depth of 25 feet above the invert (bottom) elevation of the LCC capsule. The result was a rectangular excavation approximately 100 feet long by 50 feet wide with ramps at both ends. The second excavation made a 20-foot cut to an elevation 4 feet, 6 inches above the invert elevation of the capsule. The size of this cut was approximately 95 feet long by 40 feet wide, which left a working platform at the bottom of one ramp.[22]

The capsule wall thickness was four feet; another one foot of sand bedding and two inches of concrete and waterproofing made the invert excavation a total of an additional five feet, two inches in depth. The width of the excavation at the top was 29 feet, and the excavation radius was 15 feet, 9 inches. The length was 59 feet and rounded to conform to the final capsule shape.

Next, the sump pit for the LCC access shaft was prepared in the two-foot-thick foundation for the access tunnel. Inside and outside forms for the walls and ceiling of the access tunnel were then placed, and 51 cubic yards of concrete poured monolithically for the tunnel walls and roof. The access elevator shaft was poured in three or four lifts, depending on the depth of the LCC capsule, which varied somewhat with site conditions. All the forms for the tunnel walls and ceiling were left in place during the elevator shaft concrete placement.

A fairly difficult operation now took place in the capsule invert excavation. A minimum one-foot-thick sand cushion was placed in the capsule invert, lining the excavation in which the capsule would sit. The material was emplaced in four-inch layers, and each layer compacted. The unusual part was placement of the cushion material and compacting it. At the bottom of the excavation, this was relatively simple, but as the cushion material continued up the slope, it became impossible to hold it in place. The solution was to blow wet sand using a gunite machine, screed it to the proper shape, and then quickly cover it with a thin layer of gunite. On top of that was a two-inch layer of concrete and a five-ply waterproof membrane.

Next came the placement of capsule-wall-reinforcing steel, followed by the lower section of the LCC's ¼-inch plate steel liner, which was prefabricated with embeds welded to the skin. Strongbacks conforming to the interior shape of the lower liner plate were placed at six-foot intervals along the length of the capsule to prevent the lower capsule liner plate from floating when the concrete was poured. A total of 228 cubic yards of concrete was placed through pour holes in the liner plate and around the outside edge (Figure 7.7).

Following the curing of the lower capsule concrete placement, the strongbacks were removed and the structural steel floor for the shock-mounted LCC placed inside on top of temporary steel shoring to facilitate attachment of shock-isolation springs once the capsule was completely enclosed. Inside braces to support the ¼-inch plate steel capsule walls and ceiling were mounted on the floor. The sleeve for the escape tube and the conduits for the hardened communications were placed in the ceiling, as well as two 22-inch-diameter blast valve sleeve penetrations.[23] The two-foot-thick concrete-and-steel blast door was lowered down the elevator shaft and rolled through the completed access tunnel into position for attachment to the network of rebar at the capsule entrance. Door operation validation was performed; cold water, vent, and brine lines installed and tested; and the upper capsule wall

FIGURE 7.7 LCC capsule upper liner under construction at Grand Forks AFB LCC Oscar-0. The lower liner concrete has already been poured. *Library of Congress.*

forms placed. After a survey to ensure proper location of critical components, 505 cubic yards of concrete was placed for the upper capsule, the largest single pour on the job. After the concrete cured, waterproofing was applied in the same manner as the LER and the LSB (Figures 7.8–7.10).[24]

A considerable amount of work remained before backfilling the excavation; all of the electrical conduits, cable entrances and exits, hydraulic lines, brine lines, and air inlets and outlets had to be stubbed out in the proper locations so they could be connected to the surface-level LCSB. See Table 7.5 for a summary of the construction materials used at Wing VI.

## Interior Structural and Mechanical Work

### LAUNCH CONTROL CENTER

After installation of the blast door, structural steel work continued with the installation of the cable protector frame enclosure, which provided electromagnetic shielding of all oncoming cables from the rest of the capsule equipment. On the LCC platform, the walls and ceiling of the acoustical enclosure were positioned. Air lines and control panels for

FIGURE 7.8  Wing VI LCC Golf-0 blast door installation. The blast door and frame were assembled off-site. Unlike installation at Ellsworth, the door was positioned prior to completion of the access tunnel. The installation learning curve was steep, as tolerances proved difficult to meet. Notice the four tabs securing the door in the closed position. The door opening was six feet, six inches high and five feet, four inches wide. For scale notice the grease gun in the lower right-hand corner. *Library of Congress.*

FIGURE 7.9 LCC Delta-01 capsule concrete forms being placed, Ellsworth AFB. The forms were removed and reused. The blast door entrance is on the lower right. The large cylindrical pipes protruding from the form are part of the exhaust air system. *Library of Congress*.

the pneumatic shock-isolation spring system were installed, and the fine alignment adjustments for the blast door made. The brine piping for the emergency cooling water tank was installed, as well as the toilet and holding tank. The hydraulic system for the blast valves was also installed.[25]

LAUNCH CONTROL EQUIPMENT BUILDING WINGS III THROUGH V

There are no surviving CEBMCO documents describing the construction of the underground LCEB. There are, however, photographs (Figure 7.11).

FIGURE 7.10   LCC concrete pour is complete, showing the access shaft and intake (left) and exhaust air (right) delay lines. For scale, note the man on the ladder in the foreground. *Library of Congress.*

## *Launch Control Support Building*

### WINGS I THROUGH V

The LCSB overall length is 162 feet, 6 inches, with a width of 33 feet. Included in the structure are seven bedrooms, bathroom, kitchen, water treatment room, telephone equipment room, dining and recreation room, equipment room, security office, entrance to the elevator to the LCC, generator room, and garage. The perimeter footing is five feet deep and consists of 77 cubic yards of concrete. Inside the perimeter footing walls, pipes and conduits were installed for the chilled brine, drinking water, telephone lines, conditioned air, and all electrical circuits to the LCC. Then the reinforced concrete slab was poured. The frame was prefabricated and shipped to the site, as was the roof.

A 100-kilowatt emergency diesel generator was set in the generator room. The heating system was located in the equipment room; a chlorination system for drinking water was located in the water treatment room. A 2,000-pound-capacity service elevator was installed, which was controlled by passenger push button.[26]

TABLE 7.5   Summary of Construction Material Requirements for Operational Facilities for Wing VI, Grand Forks AFB[a]

| ITEM | STRUCTURAL STEEL (TONS) | REINFORCING STEEL (TONS) | CONCRETE (CUBIC YARDS) |
|---|---|---|---|
| LAUNCH CONTROL CENTER | 124 | 260 | 2,030 |
| LAUNCH CONTROL EQUIPMENT BUILDING | 65 | 155 | 1,106 |
| ACCESS SHAFT (INCLUDES TUNNEL) | 42 | 24 | 494 |
| TRANSMIT ANTENNA | 26 | 50 | 656 |
| RECEIVER ANTENNA | 19 | 28 | 240 |
| LAUNCH CONTROL SUPPORT BUILDING | -- | 3 | 236 |
| **TOTAL PER LCF** | 444 | 520 | 4,818 |
| **TOTAL FOR 15 LCFS** | 6,700 | 7,800 | 72,500 |
| LAUNCHER | 39 | 24 | 200 |
| LAUNCHER EQUIPMENT ROOM | 80 | 115 | 181 |
| LAUNCHER SUPPORT BUILDING | 48 | 38 | 625 |
| ACCESS SHAFT | 3 | 10 | 131 |
| CLOSURE DOOR AND TRACK | 25 | 20 | 154 |
| BLAST VALVE PIPING, TANKS, MISCELLANEOUS | 72 | -- | 19 |
| **TOTAL PER LF** | 267 | 207 | 1,310 |
| **TOTAL FOR 150 LFS** | 40,000 | 31,000 | 200,000 |

a) History of Grand Forks Area during Construction of Wing VI Minuteman II ICBM.

WING VI AND 564 SMS

The LCSB floor plan was essentially the same but included a larger garage and equipment storage area.

## Major Construction Issues

Despite the experience gained in building the Atlas and Titan facilities, significant problems due to weather, excavation, and structural and mechanical issues still occurred. The urgency of the deployment meant working through some of the worst winter weather in years, the widely dispersed facilities meant geological variations were a constant problem, and close tolerances were often at odds with what was achievable.

### *Weather*

Construction continued even in severe winter weather, stopping only when absolutely necessary. At Malmstrom, the severe winter between November 1961 and February 1962

FIGURE 7.11 Wings III through V LCEB rebar placement. The construction workers are assembling the rebar around the air intake and exhaust shafts. *Library of Congress*.

caused a 40-day delay due to the inability to backfill excavations successfully because of frozen soil. At Minot excavation started and progressed in temperatures that were lower than minus 30 degrees F with frost to 60 inches in depth. During the second winter, in order to keep up with the schedule, steam coils were buried in the sand and aggregate piles as well as in the water tanks. The result was a doubling in cost of concrete placement.

## *Excavation*

Most of the sites at Ellsworth were relatively dry, with water flow limited to one to two gallons per minute, if at all. At LF Bravo-10, work had to be delayed for 30 days due to groundwater intrusion at the rate of 20 to 40 gallons per minute. The standard dewatering system was unable to keep up with this flow rate. The solution was to install a well point system consisting of 44 wells approximately 15 feet deep in a circular pattern around the excavation. After pumping for 60 hours, excavation resumed and was completed two days later. The well point system continued operation through the end of construction to keep the water table below the bottom of the open cut, and waterproofing was completed without further problems.[27]

At Minot the soil conditions caused concern that settlement of the LER would occur due to the nature of the foundation soil. After numerous consultations within and outside the Corps of Engineers and the Air Force, the decision was reached to raise the affected site

LERs from 0.10 to 0.15 foot, and settlement rates were then monitored. A typical settlement rate was 0.054 foot in nine months, which then leveled off to 0.0033 foot per month. The assumption was made that the settlement rate would taper off, and design elevation would be achieved approximately three years later. This was indeed the case.[28]

Launch tube shaft excavation methods were considerably different at F. E. Warren AFB. The prime contractor, Morrison–Knudsen, developed a method using a large-diameter auger to excavate the shafts more quickly and precisely. Comprehensive study of the boring logs for each of the LF sites indicated that there was very little hard rock, and the water table was, for the most part, below 100 feet. Where rock was encountered, the layers were thin and could be broken through using jackhammers or minor drilling and blasting. In many cases the layer of rock was thin enough that a large auger could be simply dropped onto the layer several times, fracturing it, before continuing. Five truck-mounted cranes had their normal drilling attachments modified for use with a telescoping, high-strength Kelly bar, which was capable of operating while extended below the level of the crane to the depth of 70 feet or more. A Kelly bar uses a section of pipe with a polygonal or splined outer surface that passes through the matching polygonal or splined Kelly (mating) bushing and rotary table. The bushing is rotated via the rotary table, and thus, the pipe and the attached auger turn while the polygonal pipe is free to slide vertically in the bushing as the auger digs the shaft deeper. Each Kelly bar could mount one of three specially constructed augers: 6, 10, or 15 feet in diameter.

Each drilling rig had the three auger sizes on-site in anticipation of unexpected variations in soil type. If possible, the shaft was started and finished with a 15-foot auger. Routinely, however, a full-depth pilot hole was excavated using either the 6- or 10-foot auger, and then the shaft was reamed to final dimension using the 15-foot device, with the bell-shaped contour at the bottom of the shaft dug manually using pneumatic spades. When the 15-foot-diameter excavation reached approximately 6 feet, the operation stopped, and a corrugated collar was installed to prevent sloughing of the shaft. The operator would then continue digging until sufficient earth was on top of the auger, at which point he would pull the auger to the surface and, while still aligned with the excavation, spin the auger at high speed, casting the dirt away from the shaft, where it could then be loaded into dump trucks. While this was not particularly efficient in terms of casting the dirt, the method kept the auger aligned with the shaft, thus ensuring a clean, vertical shaft.

If unstable soil was encountered, steel liner plates were installed in the shaft as it deepened. Where the wall would stand on its own, no stabilization was necessary until the shaft was complete. At this point stabilization was accomplished via high-strength gunite, bituminous material, installation of steel wire mesh, or a combination of these.

The first shaft was started at F. E. Warren in the second week of November 1962 at LF Alpha-06, and the last shaft was completed on 4 June 1963 at LF Tango-11, for a total of 200 shafts completed in nearly 200 calendar days, with Sundays off. Round-the-clock operations with a six-day work week produced an average of better than one completed shaft every 24 hours. Under ideal conditions a crew could bottom out a shaft only 16 to 18 hours after arriving on-site. This technique also was used for many of the launcher shaft excavations at Grand Forks AFB and 564 SMS at Malmstrom AFB (see Figures 7.12 and 7.13).[29]

At Grand Forks a combination of the geology of the region, which was primarily glacial till in the upper layers, and the severity of the winter of 1964 to 1965 resulted in the significant disturbance of structure foundations due to frost heave. In the spring of 1965, four months after the start of subzero weather, the routine inspection of structure settlement revealed upward movement of several of the structures. These results prompted inspection of all the structures, and many of the LSBs were found to have uplifted as much as eight inches. The LERs had been less affected. The structures that had displaced only slightly were allowed to settle naturally. Where displacement was greatest, the decision was made to thaw the ground with steam jets, and most of the structures settled to the design level. When settlement after thaw was below design level, chemical grout was injected beneath the structures and the problem resolved.[30]

## *Structural and Mechanical*

Interference between reinforcing steel in the launcher roof and the closure track beams caused considerable delay in the construction of the first squadron at Ellsworth. Setting the steel liner on the lower level of the LER was also problematic. Technique changes in each instance made construction of the remaining squadrons much more straightforward.

The blast door design for the Minuteman LCC was significantly more complicated than that of the Titan II doors. The four blast doors for the Titan II blast lock structure, which isolated LCC from the silo in one direction and the access portal in another direction, each had 4 hydraulically operated pins; the Minuteman LCC blast doors have 12 pins, 6 on each side. At the beginning the linkage system was a major mechanical problem at LCC Alpha-01 due to binding of the pins. A redesign partially resolved the problem, and modifications made on a site-by-site basis completed the installation successfully.[31]

## *Inspection and Turnover of Completed Facilities*

In most cases each site received two formal inspections prior to turnover to the Air Force SATAF for equipment installation by Boeing and other contractors. The pre-final inspection gave the contractors an idea of how close they were to finishing the site. At the final inspection, the CEBMCO representative had the final word as to whether an item found on the inspection was a legitimate deficiency or not. If there was disagreement with the SATAF representative, the deficiency decision was referred to the SATAF wing office and perhaps to the CEBMCO area office. Construction costs for each wing's operational facilities are given in Table 7.6.

## Site Activation

Boeing had responsibility for developing the LF and LCC electronic equipment and the missile transporter-erector, as well as installation and checkout of all facilities at Wings I through V. Nine months after receiving the contract, Boeing began construction of the Weapon System Integration Laboratory in Seattle. Manufacturing activities began in support of the Edwards AFB Air Force Flight Test Center (AFFTC) activities such as test silos and "battleship" missiles (nonflight, thick-walled, steel motor case). Boeing also invested

FIGURE 7.12　Wing V geology permitted the use of large augers to excavate most of the LF shafts. The 6-foot auger was often used to dig a pilot hole for the 15-foot auger. *Photograph from Samuel Phillips Collection, National Archives and Research Administration.*

FIGURE 7.13 Upper photograph shows the 15-foot-diameter auger in operation. Lower photograph shows the uniformly smooth walls of the launch tube excavation, which greatly simplified construction at F. E. Warren AFB. *Photographs from Samuel Phillips Collection, National Archives and Research Administration.*

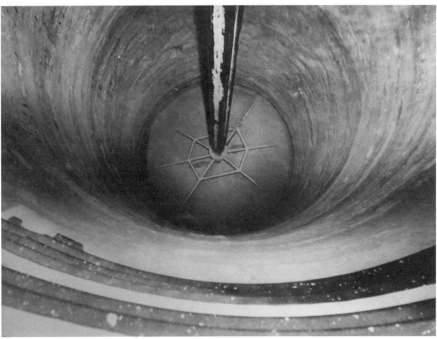

TABLE 7.6  Construction Contracts WS-133A and WS-133B Operational Facilities

| WING | CONTRACTOR | BID | FINAL | # OF MODIFI-CATIONS |
|---|---|---|---|---|
| I[a] | Fuller-Web, joint venture | $61,773,644 | $79,284,385 | 243 |
| II[b] | Peter Kiewit Sons' Company | $56,220,274 | $73,401,386 | 104 |
| III[c] | Peter Kiewit Sons' Company | $67,800,000 | $90,129,570 | 91 |
| IV[d] | Morrison-Hardeman-Perini-Leavell | $60,664,500 | $73,998,558 | 88 |
| V[e] | Morrison-Knudson Company Inc. | $83,950,000 | $90,082,024 | 204 |
| VI[f] | Morrison-Hardeman-Perini-Leavell | $121,290,000 | $160,961,018 | 204 |
| 564 SMS[g] | Morrison-Knudson Company Inc. | $46,585,000 | $53,995,154 | 148 |

a) Minuteman Technical Facilities, Malmstrom AFB, Great Falls, MT; b) "History of Minuteman Construction, Wing II," Ellsworth Area Engineering Office; c) WS-133A Minuteman Missile Facilities, Vol. II, Minot AFB; d) History of Construction Activities and Contract Administration Phases Encountered by Whiteman Area Office Corps of Engineers during Construction of Minuteman Strategic Missile Wing IV, Vol. I; e) Warren Area Minuteman History; f) History of Grand Forks Area during Construction of Wing VI Minuteman II ICBM; g) History of Malmstrom Area during Construction of Collocated Squadron Number 20 Minuteman II ICBM Facilities.

heavily in R&D of the necessary electronics for the operational bases, as well as certain components of the missile.[32]

The primary objective of the Weapon System Integration Laboratory was the development and testing of equipment and procedures to be used at the Atlantic Missile Range (AMR), Patrick AFB, Florida, and during operational development programs at the Pacific Missile Range (PMR), Vandenburg AFB, California. This required developing simulators and full-scale mockups of operational equipment. In January 1960 Boeing was notified that the launcher closure testing would be at Seattle instead of Edwards AFB. A site known as the Desimone property, which was adjacent to the Development Center building, was purchased as the location for a full-scale prototype launcher and one of two "soft" aboveground LCCs connected to the second mockup LCC within the Integration Laboratory. By 5 February 1960, planning for what was now to be called Seattle Test Program (STP) III was approved for the Desimone site. Boeing also financed an electronic manufacturing facility. While this was being built, an interim circuit board fabrication area was activated in an adjacent building for use in manufacturing electronic equipment for use at the AMR.

The launcher closure test facility construction was completed in April 1960. Testing began in May to determine the static and sliding friction values for the bearing rings and rolling friction values for the wheels and tracks, as well as debris resistance with different depths of debris. The tests quickly validated the Malmstrom design for the proposed actuation system using ballistic gas generators.[33]

Construction for the STP III facilities was well underway when the final drawings for Wing I at Malmstrom were released in October. Modifications included the size and configuration of the LER and the size of the LCC to accommodate a motor generator set and SAC communication equipment, which was as yet not defined. Between October and December 1960, STP III facility designs were revised to duplicate as closely as possible, within the confines of the Boeing facilities, the new Malmstrom configuration, which included changing the LER to a two-level structure to provide additional space for an electrical surge arrester (ESA) and a motor generator set.[34]

## Site Activation Task Force

The detailed SATAF records for each of the Minuteman wings have proven difficult to find. Fortunately, the 341 SMW SATAF history is nearly complete and serves as an example of the activation process. Tables in Appendix A summarize the available detailed activation information for each of the wings. Other important dates, such as first missile emplacement and achievement of alert status, are summarized in a short chronology for each of the wings, also in Appendix A. A comprehensive chronology for each wing can be found in the digital appendices.

### 341ST STRATEGIC MISSILE WING, MALMSTROM AFB, MONTANA

On 15 July 1961, 341 SMW was activated at Malmstrom AFB, replacing the 4061st Air Refueling Wing and supporting organizations. Construction of the missile facilities had commenced four months earlier on 16 March 1961 with the start of the excavation for the Alpha Flight LCC Alpha-01 located 3.4 miles northwest of Raynesford, as well as three nearby LFs.[35] Five months later, on 9 August 1961, backfill operations were begun on LF Alpha-02. Three months later, on 14 November 1961, LF Alpha-02 was turned over to the SATAF commander for Malmstrom, Col. Harry E. Goldsworthy.[36] By the end of November 1961, all of the LFs in Alpha Flight were complete and turned over to SATAF.[37] On 1 December 1961, the 10 SMS and the 341st Missile Maintenance Squadron (341 MIMS) were activated.[38]

The Alpha Flight LFs had been accepted in November by SATAF with several deficiencies. On 6 January 1962, a final inspection marked the completion of construction of all Alpha Flight facilities, a short 10 months since construction started. SATAF could now begin the installation of the air-conditioning equipment and checkout of the underground electrical cables connecting the LCC to the 10 LFs. Boeing began installation of the missile firing and guidance equipment. Missile installation, also part of Boeing's responsibility, was anticipated for midsummer.

Construction of the remaining 140 LFs and 14 LCFs was on schedule, with Bravo Flight more than halfway to completion. Concrete work was nearly finished at Charlie through Echo Flights. Foxtrot through Hotel Flights had the launcher and LER concrete work finished, and work was well underway on the LSBs. India through Juliet Flights were in the process of completion of LER reinforcing steel placement, while Kilo through Oscar Flights were completing the launcher lower annular ring and grout emplacement.

The incorporation of the Fletcher Committee recommendations concerning safety and operational flexibility were expected to cause a delay of 90 to 100 days for turnover of the first squadron to operational status (see Chapter 14). The recommendations included Selective Launch, Dual Targeting, and Launch Enable. Only Selective Launch and Launch

Enable would be added to Wing I, while the entire set of changes would be added to each of the other wings, starting with Ellsworth. A similar delay was expected for the second and third squadrons at Malmstrom and possibly the first squadron at Ellsworth, but as of 9 January 1962, the first operational flight, Alpha, was expected to be accepted on 10 October 1962.[39]

In mid-January 1962, the final report of Project High Climber was issued by Headquarters, 341 SMW. Malmstrom had been designated as one of four test sites to furnish information for determining the types and numbers of aircraft and ground support vehicles necessary for the widely dispersed operational area characteristic of the Minuteman program. Nearly 3,000 miles of roads connected the 165 dispersed Minuteman facilities with Malmstrom AFB.[40] The flight time ranged from 0.2 to 1.1 hours, while surface transportation ranged from 0.7 to 4 hours. Eight of fifteen LCFs required over two hours of surface travel time.[41] The magnitude of the problem was illustrated by nearly 33,000 miles of actual and simulated travel, including both air and surface transportation, during the first month of the program. The report concluded that short takeoff and landing aircraft (L-20 and L-28) operating from 1,200-foot austere landing strips located near the LCFs would be the most efficient method for launch crew changeover.[42] Surface transportation would work best for the maintenance crews, LCF support, and security staff. Use of nearby civilian airstrips would permit C-47 aircraft for rapid equipment movement as necessary. Helicopters (H-19) were also evaluated, with the recommendation that helipads be located at both LCFs and LFs. The report also recommended a wide variety of surface vehicles.[43]

On 1 March 1962, the 12th Strategic Missile Squadron (12 SMS) was activated at Malmstrom AFB. By the end of the month, Alpha Flight was in the assembly-and-checkout phase of equipment installation, while at Bravo through Echo Flights, all construction was complete and the facilities nearly ready for turnover to SATAF. Foxtrot through Juliet Flights had nearly complete LFs, with LCCs in the process of backfilling. Kilo through Oscar Flights were still in the process of pouring concrete for the LER, LSB, and LCC. November and Oscar Flights' lower launch tube annular ring concrete work was complete, and the reinforcing structural steel was in the process of being laid for the LER and LSB. LCC capsule and shaft concrete work was complete, and backfill was starting.[44]

As of the end of April 1962, Alpha Flight missile combat crews were scheduled to complete operational readiness training (ORT) in September 1962 and become combat ready by October 1962, with Bravo and Charlie Flights reaching the same status in November and December, respectively.[45]

Project Test Hop began 12 April and ended on 26 April 1962. This was a 14-day exercise to evaluate, as realistically as possible, the 24-hour missile combat crew duty shift schedule as proposed to Headquarters SAC by the Air Force Institute of Technology. Simultaneously, the project exercised the support functions of maintenance, supply, civil engineering, food service, security, central transportation control, communications safety, and the Wing Command Post to establish coordinated procedures and checklists designed to optimize the operational efficiency of the Minuteman weapon system. Since LCC Alpha-01 was still undergoing assembly and checkout, a bedroom in the LCSB was converted into a simulated LCC. The missile combat crews were composed of two officers working a 24-hour shift, with five participating crews. None were certified prior to the test but received training from a Standboard Crew during the test.[46]

By the end of April 1962, Alpha through Delta Flights were well into assembly and checkout status. Echo Flight had been turned over to SATAF. Foxtrot and Golf Flights' concrete work was finished, as well as backfill at all facilities. Hotel through November Flights' concrete work was finished, but backfill awaited conduit placement at the LCC capsules. Oscar Flight equipment room steel placement at all LFs was underway. LCC capsule reinforcing steel was in place, and the elevator shaft was complete.[47] On 1 May 1962, the 490th Strategic Missile Squadron (490 SMW) was activated. Alpha through Foxtrot Flights were all in the assembly-and-checkout phase, and Golf Flight was nearly ready for turnover to SATAF. At Hotel and India Flights, electrical and mechanical installation was proceeding, as well as the shock-mounted floor in the LCC capsule, with backfill nearly complete. For Juliet and Kilo Flights, backfill was complete at all LFs as well as LCCs, and backfill was to the top of the capsule, waiting for the final conduit work. For Lima Flight backfill was now to the level of the LSB floor, as all concrete work had been completed and installation of electrical mechanical equipment begun. Backfill was to the centerline of the LCC capsule. For Mike and November Flights, the LER concrete work was complete at all LFs as well as the LCC, and backfill was to the centerline of LCC Mike-01 capsule but not yet begun at the Mike Flight LFs. For Oscar Flight the LER steel liner was complete at two of the LFs, while, for the remainder, the reinforcing steel for the LER was complete. LCC capsule reinforced steel placement was complete, and the elevator shaft concrete had been finished.[48]

Project Test Hop was extended through August 1962 to further exercise and evaluate emergency war order (EWO) procedures as well as support requirements for the more distant and remote LCFs. The first phase of Project Test Hop had been highly successful in establishing operational procedures and support requirements.[49] Project Test Hop II was conducted from 10 to 19 June 1962, at LCC Echo-01 to evaluate a 48-hour shift alert, as well as support issues for the more distant site, 119 miles from Malmstrom AFB. Additional objectives were evaluation of programmed facilities and equipment to support the mission of the wing as well as exercise operational and maintenance functions, security forces, safety procedures, food service support, and other services. Four crews were utilized, with a fifth serving as emergency relief.[50]

By the end of July 1962, Golf through India Flights had been turned over to SATAF for assembly and checkout. For Juliet through November Flights, all major construction was finished, with electrical and mechanical installation in progress. For Oscar Flight LER concrete was complete, as was backfill at all LFs with launcher closure surround. Backfill of LCC Oscar-01 and elevator shaft concrete was complete. Foundation of the LCF support building was complete, and installation of underground piping was in progress.[51]

The first Minuteman IA missile, Serial Number 62-3601, was delivered by rail on 23 July 1962 at six o'clock from Hill AFB, Utah, and emplaced in Alpha Flight LF Alpha-09 on 27 July 1962.[52] By the end of August 1962, the 10 SMS and 12 SMS LFs and LCFs were all still undergoing assembly and checkout. Kilo Flight was accepted by SATAF, and six additional missiles had been installed in Alpha Flight LFs.[53]

In August 1962, as the operational date for the first flight of missiles at Malmstrom was drawing near, a new project was added for evaluation of procedures in the operation of the Minuteman force, Project Confirm. Its purpose was to determine the operating procedures for the maintenance control division of the 341st Missile Maintenance Squadron. Time schedules developed by Boeing were now evaluated by maintenance personnel. There were four types of maintenance teams: electromechanical, target and alignment, missile maintenance, and missile handling. At the end of the project, the Boeing estimates were confirmed as realistic.[54]

The results from the first two phases of Project Test Hop demonstrated that 24- and 48-hour alert shifts were feasible. The final phase, Project Test Hop III, was conducted from 10 to 22 September 1962. LCCs India-01 and Kilo-01 were used for both 48- and 24-hour alert tests for the first time, with two LCCs available, without interfering with the SATAF work.[55]

The HICS installation for Alpha Flight was completed in September. The system could not maintain air pressure within specifications due to leaks at faulty splice cases. Of the 237 splice cases in the system, 179 had been tested, with 71 still leaking; however, 61 had been repaired with potting material, which appeared to solve the problem.[56]

By the end of September, Juliet and Lima Flights were turned over to SATAF, and Mike through Oscar Flights were in the final stages of cleanup for inspection before being turned over to SATAF. Thirteen Minuteman IA missiles were emplaced in LFs (Table 7.6): ten in Alpha Flight and three in Bravo Flight.[57] Alpha Flight was ready for the demonstration and acceptance protocol where Boeing and SATAF would turn the 10 LFs and one LCF over to SAC after a thorough evaluation. There were three classes of demonstration and acceptance procedures:

> Class I demonstrations are defined as formal demonstrations of maintenance equipment utilizing validated and verified technical data and conducted once per each Wing prior to the delivery of the first flight.
>
> Class II demonstrations are defined as formal demonstrations conducted immediately prior to acceptance of each deliverable increment (flight or squadron) of the weapon system, using validated and verified technical data.
>
> Class III demonstrations are defined as semi-formal demonstrations and tests conducted by the contractor as part of the assembly and checkout of each LF and launch control facility equipment which is not feasible to duplicate after complete tie-in of the facility.[58]

The priorities of the Cuban Missile Crisis meant there was no time for a separate demonstration and acceptance program prior to turnover to SAC; instead, reconfiguration took place. Reconfiguration meant installation and targeting of guidance sets and installation of war reserve reentry vehicles. Four teams of personnel were required: missile handling and transportation, missile guidance and reentry vehicle, electromechanical, and target alignment. The 341 SMW did not have enough of these teams to bring 10 missiles to alert as quickly as the national emergency required, so Boeing and SATAF personnel were used as well (a detailed discussion of the Cuban Missile Crisis and the 341 SMW, as well as the Minuteman missiles of the 6595th Aerospace Test Wing, Vandenberg, AFB, can be found in Chapter 14).

## The Beginning of Minuteman Continuous Alert

On 11 December 1962, at a ceremony at Malmstrom AFB, SAC declared Alpha Flight of the 310 SMS combat ready and placed it in the nation's nuclear deterrent force inventory as part of the 341 SMW. Bravo Flight had been accepted by SAC from SATAF and Boeing on 30 November 1962 and was in the reconfiguration process, which would bring the flight to strategic alert status.[59] On 19 January 1963, all 20 Alpha and Bravo Flight missiles reached full strategic alert status, and the normal two-LCC launch control configuration was achieved.[60]

The remaining 13 flights were turned over to SAC in rapid succession, and on 13 June 1963, the last Minuteman IA, Serial Number 63-054, was delivered to Malmstrom. The 341 SMW reached operational status on 3 July 1963, nine days ahead of schedule. The Air Force determined 15 July 1963 as the operational date for Wing I.[61] In October 1963 the 490 SMS reached full alert status, completing the first Minuteman wing. During its first month of full alert status, the 341 SMW averaged 127 missiles on alert, which was lower than the required 148 due to programmed maintenance and required engineering change proposal modifications to the system.[62]

One of the most vexing problems facing the SATAF effort during Wing I activation had been finding an economical solution to the HICS air-pressure leaks. With 1,754 miles of cable connecting LFs and LCCs, 127 stream and 170 major highway crossings, 34 river and 74 railroad crossings, let alone the countless farm fields subject to plowing and the 3,415 splice casings, the problem had become apparent in March 1962. The solution involved digging up the splice cases and filling the void between the inner and outer casings with potting compound, at a cost of $250 per splice.[63]

★ ★ ★

A little more than six years from the start of construction, all 1,000 LFs and 100 LCFs had been built and brought to strategic alert (see Appendix A, Tables A.1 through A.7). While the level of Minuteman missiles would fluctuate slightly between 1963 and 1991 due to upgrades and missile maintenance, combined with the 54 Titan II missiles, the 1,054-ICBM component of the strategic triad was now in place. This undertaking was truly remarkable and easily forgotten in the haze of history.

Even more remarkable is the longevity of the program. The original projected service life of Minuteman I was a mere 10 years. Each variant in succession had a similar approximate projected service life. Minuteman IA was deployed from October 1963 to February 1969, with replacement by Minuteman IB anticipated as soon as possible. Minuteman IB was deployed from June 1963 to January 1975. Minuteman II was deployed from October 1965 to October 1991, nearly tripling its expected lifetime.[64] Minuteman III was first deployed in June 1970, and as of November 2019 was still deployed. Obviously, much has had to be done to extend the original longevity from 10 to 49 years and counting (see Chapter 16).

The development of the missile itself is as impressive as the construction of the facilities that housed it. From the first briefing by Col. Hall in February 1958 to the first flight test of the Stage I motor at Edwards AFB on 15 September 1959 was only a matter of 19 months.[65]

# 8

# MOTORS AND AIRFRAME

With approval of the Minuteman development plan in March 1958, the Air Force acted quickly to move forward with R&D of missile hardware. In July 1958 propulsion contractors were selected. The task now was to overcome the technological hurdles that the recently completed Large Solid Rocket Feasibility Program had identified.

## Technological Hurdles to Minuteman I

### *Motor Cases*

Facilities for producing the large diameter (66 inches, originally 74 inches), thin-walled metal motor case proposed by Thiokol for Stage I and Stage II (58 inches) did not yet exist. The largest metal cases flown to date were the 31-inch-diameter, 0.103-inch-thick, 4130 steel Sergeant and X-17 motors manufactured by Thiokol. Work at Aerojet-General on the Polaris 54-inch-diameter rolled and welded AMS (Aerospace Material Specification) 6434 steel case began in 1957 and culminated with the first AX full-scale Polaris launch on 24 September 1958, which was a "partial success." The flight test program ended on 2 October 1959 after 17 flights, of which 5 were successes, 11 partial successes, and 1 was a failure.[1] The outlook was slightly better for the Stage III glass filament–wound case proposed by Hercules, since a 54-inch glass filament case for the Polaris A-2 missiles had been successfully hydrotested. While success was not guaranteed, at least feasibility of the glass filament–wound case had been demonstrated.[2]

### *Nozzles*

Lightweight movable nozzles capable of withstanding the higher temperatures, pressures, and erosive nature of aluminized propellant had not yet been demonstrated for the expected 60-second powered flight regimen for each stage. The Sergeant missile had the longest thrust duration at the time, 25 to 30 seconds, with a fixed 1020 alloy steel nozzle with a graphite throat insert and a nonerosive propellant. Thiokol's proposal called for all three stages to have the same nozzle materials: a fiberglass shell lined with phenolic resin, impregnated with quartz fiber and zirconium oxide, and molded to the required shape. Graphite would serve as the nozzle throat material to resist erosion.[3] Nozzle throat insert erosion was due to the presence of reactive gaseous compounds in the exhaust stream. In the case of graphite inserts, the most damaging gases were those that reacted with carbon to form carbon monoxide.[4] Aluminum particles were erosive as well.

## *Thrust Vector Control*

Three thrust vector control options were available in 1958: jet vanes, jetevators, and gimballed nozzles. The Navy chose the jetevator concept, where a ring with a spherical inside surface was rotated into the exhaust plume for thrust vector control for both stages of Polaris A-1. The aluminum oxide particles caused considerable developmental problems for the Navy, and Polaris A-2 used movable nozzles for second-stage thrust vector control (Figure 8.1).

The Air Force eliminated the jetevator system because its impingement into the exhaust caused a drop in specific impulse by as much as 9 percent at full deflection.[5] Designing nozzle-actuating equipment able to withstand the radiant heat caused by the much higher aluminum content of the Minuteman propellant proved to be a challenge that took until 1960 to solve.[6]

## *Thrust Termination*

Thrust termination for solid-propellant motors at first appeared to be a difficult problem since, unlike a liquid rocket engine, there were no valves to instantly shut off propellant flow. By July 1957 the Polaris program demonstrated successful thrust reversal within milliseconds of actuation using blowout ports in the forward dome of the motor.[7] Combustion was terminated by the sudden decrease of chamber pressure. In addition, forward-venting pressure was sufficient to create a net reverse thrust, resulting in separation from the reentry vehicle. Flight testing of the Polaris forward-facing system achieved full-scale success by mid-1959.[8] The Minuteman Stage III design modified this concept to allow for less-than-full-range targets (see below).

## *Propellant Processing*

By 1958 the largest solid-propellant motor in production was the Sergeant missile (and the X-17 Stage I derived from it), which contained approximately 6,000 pounds of propellant.[9] The early full-scale Polaris AX Stage I, at 54 inches in diameter and a propellant weight of 15,200 pounds, indicated large-scale production of the 45,000-pound Minuteman I Stage I was feasible.[10]

## *High-Altitude Testing*

Simulation of flight test conditions was not possible at the Minuteman motor contractors' facilities. All tests were done at the high-altitude test facilities at Air Force Arnold Engineering Development Center (AEDC), Arnold AFB, Tennessee. Testing at AEDC covered virtually all aspects of motor development, from sub- to full-scale experiments of

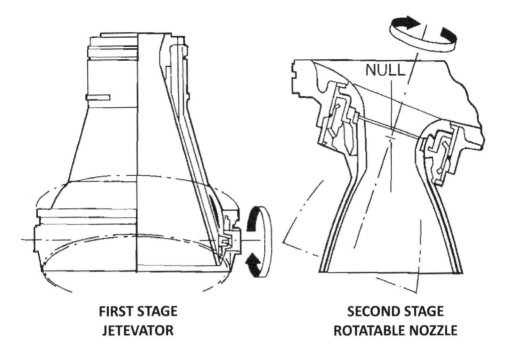

**FIRST STAGE JETEVATOR**

**SECOND STAGE ROTATABLE NOZZLE**

FIGURE 8.1 Examples of the jetevator and axial nozzle design used in the Navy's Polaris A-1 (both stages) and A-2 (second stage) fleet ballistic missiles, respectively. The jetevator was an inefficient thrust vector design and, with the development of the A-3 airframe, was replaced by the axial nozzle design for the first stage and liquid-injection thrust-vector control for the second stage. *Adapted from "A History of the FBM System."*

nozzle and casing design as well as propellant performance and proposed thrust termination concepts. Nozzle design was empirical in nature. Tests used subscale models ranging from 2,500 to 6,000 pounds of thrust at simulated altitudes of 80,000 to over 120,000 feet. Total and specific impulse, nozzle temperature, and pressure evaluation began in March 1959 and continued through 1960.[11]

## Stage I

### *Thiokol Stage I M55, M55A1*

DEVELOPMENT

Thiokol was selected as prime contractor for Minuteman Stage I development and a competitive contractor for Stage II and III motors in July 1958.[12] It had participated in the second year of the Air Force's Large Solid Rocket Feasibility Program. Thiokol's experience in solid-propellant motors began in 1947 when the Army Air Forces requested the company begin the design and manufacture of solid-propellant rockets. Thiokol agreed and, in late 1947, constructed its first facility at Elkton, Maryland. The first Thiokol rocket was fired in

1948. Thiokol moved to the Army's Redstone Arsenal in 1949 and began development of a variety of polysulfide composite–propellant rockets, culminating in the 31-inch-diameter Sergeant XM-100 booster using the T17E1 (TRX110A) polysulfide rubber–ammonium perchlorate propellant with a star-shaped internal grain (see Chapter 4).[13]

Thiokol's researchers followed the other solid-propellant manufacturers' investigations into aluminized propellants but found that the hydrogen content of the polysulfide was insufficient to utilize the higher amounts of aluminum. A switch to polybutadiene polymers with higher hydrogen content allowed for incorporation of increased levels of aluminum with a concomitant lowering of the molecular weight of the exhaust, thus significantly increasing the specific impulse.[14] The use of polybutadiene acrylic acid (PBAA) polymer propellant was a large step forward, as PBAA polymer had a theoretical specific impulse of 267 seconds, compared to the best polysulfide composition at 237 seconds.

Early in the PBAA development process, Thiokol found that PBAA-based propellants had poor tear strength compared to polysulfide. Thiokol engineers solved the problem in 1957 by adding 10 percent acrylonitrile as a third monomer to the PBAA copolymer formula to form the polybutadiene acrylic acid–acrylonitrile polymer (PBAN).[15]

The Stage I motors for Minuteman I through III used a six-pointed-star grain configuration. Final development and the operational missiles used PBAN 1.3 TP-H1011 for the main motor and PBAN TP-H1043 (slower burning) for the aft closure load. The aft closure load was cast as a 15-inch molded surface to direct exhaust flow to the four nozzles and provide insulation for the aft closure.[16]

CASE

The 18.6-foot-long, 5.5-foot-diameter Stage I motor case application was unexplored territory for Thiokol. A thin-walled structure of this size, designed to a minimum yield strength of 220,000 psi at temperatures ranging from 200 to 400 degrees F, was a significant design challenge.[17] Many candidate metals were tested before the D6AC steel alloy was selected (also known as Ladish D6AC after the Ladish Forging Company that developed it), a low-alloy steel that had been in use for high-strength applications in the aircraft industry.[18]

Fabrication experiments with small case specimens revealed that girth welds (welds on the circumference of the cylindrical case) were twice as strong as longitudinal welds (welds the length of the cylindrical case). The use of seamless rolled rings, girth-welded to form the case body, was the solution. Further testing showed that a welded nozzle outlet was the next point of failure. Cross-forging of the closure dome, followed by counterblow hammer forging, was the solution. The enclosures were contour-milled to final dimensions, with the forward closure girth welded and the aft closure attached with a threaded breech lock system. The case thickness in the cylindrical section was between 0.144 and 0.149 inch (approximately 1/8 inch, twice the thickness of a United States quarter).[19]

Experimentation also revealed that tempering the case to maximum hardness led to a significant increase in notch sensitivity, a term used to describe the sensitivity of a metal to small notch flaws. Known for many years, it was not thought to be a problem for solid-propellant motor cases, but early failures pointed to its importance. The solution was counterintuitive: tempering the metal to a strength lower than the calculated maximum minimized notch sensitivity.[20]

## ATTITUDE CONTROL

The greatest hurdle was that the Minuteman flight profile required a 58- to 64-second burn time for each stage; thus, new nozzle actuator designs were necessary. Full-thrust vector control had been decided early in the program, with four gimballed nozzles of convergent–divergent de Laval design on each stage, enabling pitch, yaw, and roll control.[21]

Each nozzle has a blast tube, which serves to straighten out the exhaust gas flow and is connected to the exit nozzle with a flexible bellows. The nozzles pivot in four mutually perpendicular planes parallel to the centerline of the motor and are capable of moving through an angle of plus or minus eight degrees. Yaw control is achieved by moving Nozzles 1 and 3 in the same direction. Roll control is achieved by moving Nozzles 1 and 3 in opposite directions. Pitch attitude is maintained by moving Nozzles 2 and 4 in the same direction. The nozzle control unit (NCU) hydraulics are insulated from base heating by Buna-N rubber, mastic insulation, and high-density graphite parts as well as a phenolic–Refrasil base plate shield. Base heating is a phenomenon where, at supersonic speeds, there is a backflow of the exhaust gases into the engine compartment.[22] Each nozzle has an exterior insulating boot.[23]

## MOTOR TESTING

Motor testing at the AEDC began in May 1959 with 16 Thiokol TU121F firings to measure specific impulse and thrust performance at an altitude of 90,000 feet, ending on 26 August 1960. Nozzle geometry was also studied extensively, as well as the effect of aluminum particle size. A unique test motor, called the nozzle entrance simulator, utilized a single nozzle offset from the motor case to approximate the exhaust flow typical in one quadrant of the full-scale four-nozzle configuration (Figure 8.2).[24]

Thiokol's first static firing of a large-scale Stage I motor containing 44,000 pounds of propellant, constituting the largest solid-propellant motor fired to date in the free world, took place on 13 April 1959, lasting 42 seconds before a nozzle attachment joint burned through. Test data showed the specific impulse requirement was met. On 26 May 1959, Thiokol successfully hydrotested a flight-weight Stage I motor case to pressures in excess of flight requirements. One month later, two subscale Stage I motors were successfully fired, further proving Stage I design elements and specific impulse. On 1 September 1959, Thiokol fired the first full-scale Stage I motor. The aft closure separated from the motor after two seconds of operation due to aft dome insulation failure. On 26 October 1959, the first full-duration Stage I motor test was successful, with nozzles and aft closure remaining intact. Continued nozzle and aft dome insulation problems caused the BMD to halt the Stage I testing while these issues were reviewed. The solution was to proceed with two simultaneous test programs. A battleship (thick-walled) steel motor case was used to evaluate movable nozzle design, while flight-weight cases with fixed nozzles were used to evaluate insulation solutions.

On March 1960 a Stage I flight-weight motor case with four heavy fixed nozzles was successfully tested, though the nozzles showed excessive throat erosion. On 19 March 1960, a full-scale Stage I movable nozzle withstood a full-duration firing without excessive nozzle throat erosion. The nozzle used a tungsten–molybdenum alloy throat insert to retain the graphite throat material. In April 1960 three of four full-scale Stage I motors with movable nozzles were successfully fired. In May 1960 five Stage I motors with fixed nozzles were

FIGURE 8.2 The offset nozzle simulator used an Aerojet 5KS4500 designed to approximate the exhaust flow typical in one quadrant of the four-nozzle Minuteman Stage I motor. The offset nozzle resulted in a 5 percent loss of specific impulse but has withstood the test of time as it continues to be used with Minuteman III Stage I. *Official USAF photograph.*

successfully fired for full duration. On 14 July 1960, the first flight-weight Stage I motor with movable nozzles was successfully fired for the full 62-second test. The specific impulse was measured at 250 seconds, 2 percent above the contracted requirement. On 31 October 1960, Thiokol completed its first preliminary flight rate test for the Stage I motor. This was the fourth successful full-duration test of the month, with all systems functioning properly. In November 1960 3 of 4 firings were successful, bringing the Stage I motor development to 11 consecutive firings without structural failure. On 8 November 1960, the first Stage I flight test motor was delivered to the AMR for assembly into the first flight test missile.[25]

Improved aft closure insulation methods and lightweight nozzle designs from Arde Inc. and the Allison Division of General Motors, with assistance from STL scientists, proved to be the solution to the seemingly intractable failures. Final nozzle design includes a steel shell up to the nozzle actuator linkage point. Beyond this is a phenolic resin–impregnated Refrasil liner backed by plastic. The nozzle throat inserts are tungsten with graphite retaining rings (Figures 8.3, 8.4, and 8.5).[26]

Minuteman IB through III Stage I motors differed from Minuteman IA motors primarily through improved aft closure insulation and the modification of the nozzle geometry to Prandtl design parameters to increase efficiency (Table 8.1 summarizes motor nomenclature changes; Table 8.2 summarizes Stage I particulars).[27]

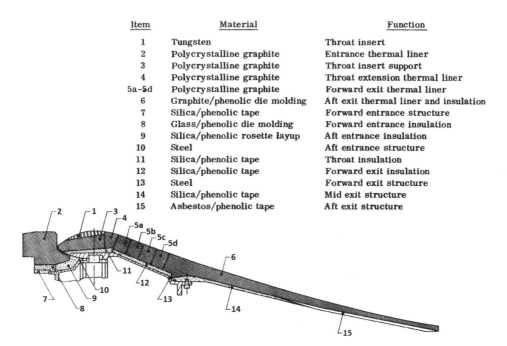

FIGURE 8.3 Design detail of Minuteman Stage I nozzle. *Adapted from "Solid Rocket Nozzles, NASA SP-8115."*

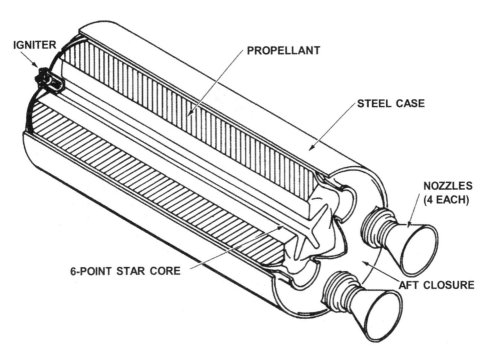

FIGURE 8.4 Cutaway illustration of Minuteman I Stage I motor. The igniter is in the center of the forward dome and ignites the full length of the six-pointed-star grain. *Adapted from "Minuteman Familiarization."*

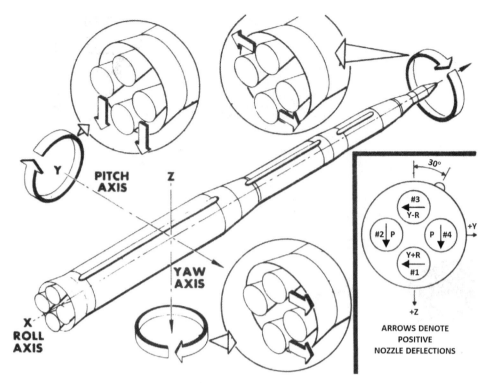

FIGURE 8.5 All three stages of the Minuteman IA and IB used four steerable nozzles for thrust vector control through differential gimballing of pairs of nozzles. *Adapted from "Minuteman Familiarization."*

TABLE 8.1 Summary of Minuteman Motor Designations, 1968[a]

| STAGE | MINUTEMAN I LGM-30A | MINUTEMAN I LGM-30B | MINUTEMAN II LGM-30F | MINUTEMAN III LGM-30G |
|---|---|---|---|---|
| I | M55 Thiokol 2K-SRM55-3 | M55A1 Thiokol 2K-SRM55-3 | M55A1 Thiokol 2K-SRM55-3 | M55A1 Thiokol 2K-SRM55-3 |
| II | M56 Aerojet 2K-SRM56-3 | M56A1 Aerojet 2K-SRM56-3 | SR19-AJ-1[b] Aerojet 2K-SR19-3 | SR19-AJ-1 Aerojet 2K-SR19-3 |
| III | M57 Hercules 2K-SRM57-3 | M57A1 Hercules 2K-SRM57-3 | M57A1 Hercules 2K-SRM57-3 | SR73-AJ-1[c] Aerojet 2K-SR73-3 |

a) Adapted from "New Minuteman Motor Designations," *Minuteman Service News*, no. 38; b) Rocket Motor SR19-AJ-1: Overhaul Instructions and Illustrated Parts Breakdown, T. O. 2K-SR19-3;
c) Minuteman III Stage III was also manufactured by Thiokol and Chemical Systems Division of United Technologies Corporation.

TABLE 8.2  Minuteman I through III Stage I M55A1 Motor Specifications[a]

| | |
|---|---|
| WEIGHT, TOTAL | 51,251 pounds[b] |
| WEIGHT, PROPELLANT | 45,670 pounds[b] |
| WEIGHT, NOZZLES | 1,056 pounds[c] |
| WEIGHT, CASE | 4,525 pounds |
| LENGTH (INCLUDING NOZZLES) | 24.6 feet[c] |
| DIAMETER (EXCLUDING EXTERNAL INSULATION) | 5.5 feet[b] |
| THRUST | 200,400 pounds[b] |
| PROPELLANT | TP-H1011(case), TP-H1043(aft closure)[b] |
| MOTOR CASE | D6AC steel[b] |
| NUMBER OF NOZZLES | 4 |
| MANUFACTURER | Thiokol |

a) M55A1 refers to the Minuteman IB, II, and III Stage I motor; b) "Minuteman Weapon System History and Description," 2001; c) T. O. 2K-SRM55-3, 1 May 1967.

## Stage II

### Aerojet Stage II M56, M56A1, SR 19-AJ-1

DEVELOPMENT

Aerojet was selected as the prime contractor for Stage II and parallel contractor for Stages I and III in July 1958.[28] Aerojet had been a key part of the Air Force's Large Solid Rocket Feasibility Program from 1956 to 1958 and in 1956 had become the lead motor contractor for the Navy's Polaris fleet ballistic missile program.[29]

In 1960 Robert L. Duerksen and Joseph Cohen, Aerojet engineers, filed a patent describing the use of cross-linked polyurethanes that, combined with powdered aluminum, produced propellants with excellent physical properties and performance characteristics. The aluminum content could be as high as 40 percent. The polyurethane formulation could be cured at low temperature with no measurable heat of reaction while curing. Additionally, this new propellant could bond directly to the case lining.[30]

A number of different polyurethane–aluminum formulations were tested during Minuteman motor development, and ANP-2862 and ANP-2864 were used in the operational Stage II for Minuteman I (M56A, M56A1). A bipropellant system was chosen to get a sliverless burnout. The bipropellant system was cast with a four-point-star grain of the fast-burning propellant. The initial thrust was provided by the fast-burning grain and then transitioned into a slower-burning grain, providing a neutral thrust profile. The transition to the slower-burning propellant ensured the simultaneous burnout of the flame path from the initial fast-burning grain surface, which resulted in complete consumption of propellant. The grain shape changed at the forward dome into a six-radial-fin structure to provide a more gradual acceleration tailoff.[31] Minuteman II and III Stage II (SR-19-AJ-1) used a cylindrical grain of the higher-energy, carboxyl-terminated polybutadiene propellant (Figure 8.6).[32]

CASE

The original Minuteman IA Stage II (M56) motor case design used a unique fabrication technique called Aerowrap. Aerojet engineers were looking for a way to make aluminum motor casings. They adapted a British process using thin bands of high-strength aluminum strip wrapped around a mandrel and bonded together using adhesive. Aerojet decided to use high-strength Sheffield strip steel instead, in a process marketed as Aerowrap. Both Polaris and Minuteman Stage I experimental casings were made, but development problems ended this approach. The M56 case was made of D6AC steel using methods similar to those at Thiokol.

Aerojet switched to Ti-6A1-4V titanium alloy for Minuteman IB (M56A1) and Minuteman II and III (SR-19-AJ-1). The titanium case was 55 percent lighter (353 versus 639 pounds), while the titanium casing thickness was actually greater, 0.096 to 0.101 inch versus 0.070 to 0.075 for the Ladish D6AC cases. Since the Minuteman Stage II motors had to withstand the greatest buckling forces during flight (the highest dynamic pressure, also called max q), the increase in thickness with titanium alloy, while still yielding a large overall decrease in weight, was a design bonus. The titanium case allowed for a large part of the increase in range found with Minuteman IB.[33]

ATTITUDE CONTROL

Stage II attitude control for Minuteman IA and IB used four movable nozzles in the same pattern as Stage I, but the deflection was only plus or minus six degrees.[34] Stage II blast tube and nozzle designs were similar to those of Stage I and were provided by Bendix, Straza Industries, or Aerojet's Downey Division. The nozzles had tungsten throat inserts with graphite backing. The blast tubes were graphite-backed, with resin-impregnated glass fiber.[35]

There were three major modifications to Stage II attitude control for Minuteman II and III: the use of a single submerged nozzle, liquid-injection thrust vector control (LITVC), and the new roll control system necessitated by the single nozzle design. A submerged nozzle increased the amount of propellant in the motor without increasing the length or diameter. Submerged nozzles are de Laval nozzles with a short converging section protruding into the rocket motor chamber.[36]

LITVC was first described in a patent by A. E. Wetherbee Jr., who was working at United Aircraft Corporation, filed on 30 December 1950 and issued on 5 July 1960. By injecting fluids, a series of shock waves are selectively produced on the inner surface of the exhaust nozzle, which redirects the exhaust stream.[37] Thrust vector control by fluid injection, in this case gas, was first successfully demonstrated in 1952 by George F Hausman at United Aircraft for use with jet engines, a concept he patented on 23 July 1957.[38] LITVC was first successfully used in large solid rocket motors with the 120-inch-diameter Titan IIIC.[39]

Searching for a lighter-weight and simplified thrust vector control as opposed to movable nozzles, L. T. Bankston began work in 1958 at NOTS. Bankston concluded that

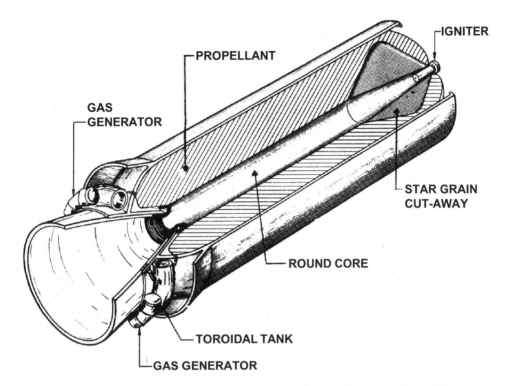

FIGURE 8.6 Cutaway illustration of Minuteman II Stage II. The grain design was changed from a bipropellant star grain to a circular grain in the main propellant, which then changed to a star grain toward the end of the burn. *Adapted from "Minuteman Familiarization."*

injecting a fluid into the expansion cone of a supersonic nozzle appeared to be the most promising technique.

Minuteman II and III Stage II LITVC initially used Freon 114B2, but due to environmental concerns, Freon was replaced with perfluorohexane. The liquid is injected at four locations 90 degrees apart, downstream of the titanium exhaust cone nozzle throat but before the exit cone extension attachment ring. Injection of the fluid through one of these valves into the exhaust stream causes a shock wave that alters the main thrust vector. Deflection of the main thrust vector changes missile pitch or yaw attitude. The Freon or perfluorohexane tank is pressurized by a gas generator. Early in the flight test program, it became apparent that an excess amount of Stage II liquid injectant had been designed into the LITVC system to cover control contingencies during the flight test program, which proved to be unnecessary. Rather than expend money to redesign the system, it was decided to use a software change to dump the excess early in the Stage II burn. An equal flow of injectant is expended from Pintle Injectors 2 and 4 so that there is no net control force. The removal of the excess injectant enabled the missile range to be extended, allowing more targeting flexibility.[40]

During Stage II flight, the roll control system uses two valves located diametrically opposite each other on the skirt of the motor case. Each valve has two opposing exhaust ports. The roll generator provides pressurized gas that is exhausted equally to all four ports during normal flight, resulting in zero roll. If roll is required, the valves unbalance, exhausting gas to produce roll thrust in the desired direction (Figure 8.7).[41]

MOTOR TESTING

On 19 April 1959, Aerojet fired its first full-scale interim-weight Stage II motor for 56 seconds. On 25 May 1959, the second full-scale Stage II engine test took place using a bipropellant mixture. The motor failed after 25 seconds due to a crack in the propellant grain. In June another Stage II test was only partially successful due to failure of a nozzle throat section after 34 seconds. On 22 July 1959, in another full-duration firing, nozzle failures occurred at 46 and 48 seconds. On 23 December 1959, Aerojet fired its first flight-weight Stage II engine with movable nozzles. Burn through of the nozzle bellows joint occurred just three seconds into the burn.

On 14 September 1960, the 10th consecutive full-duration firing of a Stage II full-scale flight-weight motor took place. The string of successes indicated that nozzle and insulation problems had been overcome. In October 1960 four Stage II full-duration firings were successful, with two being preliminary flight rating tests and two testing the prototype motors with the proposed lightweight nozzles, which would add 200 nm of range. On 3 October 1960, Aerojet delivered the first flight test Minuteman IA Stage II motor to the AMR to be assembled on the first flight test missile (Table 8.3).[42]

# Stage III

## Hercules Powder Company Stage III M57, M57A1

DEVELOPMENT

Hercules Powder Company had participated in the second year of Air Force's Large Solid Rocket Feasibility Program and was already working on the Polaris program. Three contractors bid on the Stage III motor: Thiokol, Aerojet, and Hercules. Unlike the conservative approaches of Thiokol and Aerojet, which proposed the use of composite propellants and steel motor cases used in Stages I and II, Hercules proposed a high-energy, double-base propellant not previously used in a motor this large, as well as a glass filament–wound, epoxy resin–impregnated, lightweight casing, which held great promise for increased range. In June 1958 Hercules's ABL had been selected by the Navy to develop a glass filament–wound, double-base-propellant, 54-inch-diameter second stage for the Polaris A-2 missile.[43]

Maj. Ralph Harned, chief for Minuteman Propulsion, AFBMD, found the Hercules approach promising, noting:

> Because the Hercules approach was more advanced, there was even less known about many of its features. The other contractor approaches were more conservative. We had less assurance that Hercules could deliver a high-quality reliable engine in time to satisfy Minuteman's needs. The additional performance promised by the Hercules approach was too attractive to pass up, but the risk was too great to have the complete weapon system dependent on that approach.[44]

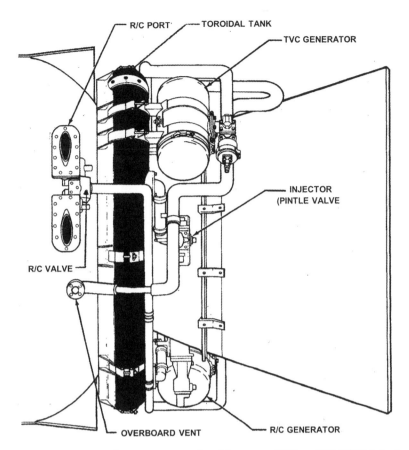

FIGURE 8.7 Minuteman II and III Stage II as well as Minuteman III Stage III LITVC. Injection of fluid, initially Freon, into the exhaust stream, allows pitch and yaw control. A separate system, R/C, uses a gas generator and four nozzles for roll control. *Adapted from "Minuteman Familiarization."*

TABLE 8.3  Minuteman Stage II Motor Specifications

|  | M56 | M56A1 | SR19-AJ-1 |
|---|---|---|---|
| WEIGHT, TOTAL | 12,072 pounds[a] | 11,530 pounds[d] | 16,057 pounds[a] |
| WEIGHT, PROPELLANT | 10,380 pounds[a] | 10,343 pounds[d] | 13,680 pounds[a] |
| WEIGHT, CASE | 639 pounds[b] | 353 pounds[b] | 353 pounds[d] |
| LENGTH (WITH NOZZLES) | 13.2 feet[c] | 13.3 feet[c] | 13.5 feet[d] |
| DIAMETER | 3.7 feet[c] | 3.7 feet[c] | 4.3 feet[d] |
| THRUST | 45,600 pounds[a] | 45,600 pounds | 60,700 pounds[a] |
| PROPELLANT | ANP-2862, ANP-2864[a] | ANP-2862, ANP-2864[a] | ANB-3066[a] |
| MOTOR CASE | D6AC steel[b] | Ti-6Al-4V[b] | Ti-6Al-4V[d] |
| NUMBER OF NOZZLES | 4 | 4 | 1 |
| MANUFACTURER | Aerojet | Aerojet | Aerojet |

a) "Minuteman Weapon System History and Description," 1990; b) "Design and Fabrication of Titanium Rocket Chambers"; c) T. O. 2K-SRM56-3; d) T. O. 21M-LGM30B-1-3; T. O. 2K-SR19-3.

As insurance against insurmountable problems with the Hercules approach, Thiokol and Aerojet were also given development contracts for Stage III.[45]

Hercules had decades of experience with double-base propellants reaching back to World War I. In late 1945 the Navy engaged Hercules to operate ABL. Work began immediately on the Bumblebee anti-aircraft missile booster, 17 inches in diameter and 120 inches in length. Extrusion manufacturing techniques had worked for smaller double-base propellant grains, such as the bazooka, but were not suitable for larger grains. Clusters of small-diameter rockets were an obvious solution, but Hercules scientists Dr. John F. Kincaid and Dr. Henry M. Shuey perfected a method that permitted production of double-base propellant grains of nearly unlimited diameter. Extruded grains were plasticized using a mixture of nitroglycerin with stabilizers to form a uniform propellant that could be cast into large-diameter grains. The first successful test of a booster using the double-base casting process occurred in August 1957. This development eventually led to ABL being part of the development of the Honest John, Nike, Snark, Talos, and Terrier booster rockets with diameters up to 30 inches.[46]

In the late 1940s, Richard Young at Young Development Laboratories began research into lightweight materials and fabrication techniques for pressure vessels such as solid rocket motor cases. In theory glass fiber–wound cases could have a girth strength of 100,000 psi and a strength equivalent to mild heat-treated steel but with a density nearly equal to magnesium. Young patented a winding technique that enabled dome end closures to be fabricated monolithically with the cylindrical portion of the vessel. The process was given the trade name Spiralloy (Figure 8.8).[47]

Unlike Stages I and II, which used an internal burning star grain, Minuteman Stage III was known as a core-and-slotted-tube modified end burner. The design was elegant and unique, maximizing mass loading of propellant while affording potential thrust termination for short-range targets by exposing the thrust termination ports early in flight. The slots, widening from ¾ to 2 inches, opened up fully as the propellant burned, reaching the thrust termination ports at approximately 30 percent burn. The cone shapes were to allow initial burning in the slots to flow without restriction to the nozzles.[48] Stage III also utilized a bipropellant system; both propellants were composite, modified double-base formulations that afforded a highly efficient, high-energy system, thus improving range, payload capability, or both (Figure 8.9).

The propellant load was in three layers. The CYH 77 burned to the DDP 80 layer, which had a higher burn rate. The burn front sped up momentarily so that it reached the second layer of CYH 77 uniformly, with the result being a relatively sharp thrust cutoff. The thrust–time curve for main CYH 77 load was slightly regressive, with the diminishing thrust allowing for uniform acceleration as the weight of propellant decreased due to consumption.

TESTING

On 5 December 1959, the first full-scale flight Stage III motor test failed after six seconds due to bond separation at the thrust termination ports. On 23 December 1959, the second of two successful full-scale Stage III motor tests took place for the full flight duration of

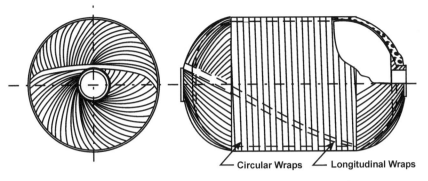

FIGURE 8.8 Filament winding pattern similar to that used for the fabrication of Minuteman Stage III motor casings. *Adapted from "Hercules Chemist."*

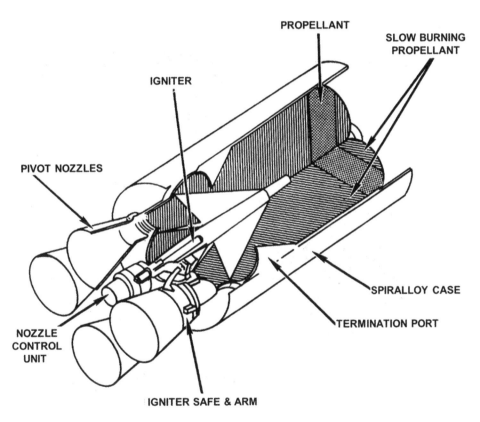

FIGURE 8.9 Minuteman I and II Stage III. The Stage III propellant charge was called a slotted core. Four slots connected the center core with conical ports extending to each nozzle. The propellant configuration was arranged so that a slow-burning propellant was separated by a thin layer of fast-burning propellant at the forward end of the motor case. This configuration provided a relatively constant thrust and prevented a long tailoff period. *Adapted from "Minuteman Familiarization."*

60 seconds. Both tests were with casings that did not have the thrust termination ports in place. On 17 February 1960, a full-scale Stage III motor with actuator-controlled nozzles was successful, the first time a Minuteman motor with four movable nozzles functioned for the full 60-second duration. Five days later, two full-scale motor tests were successful, one with nozzles in a fixed position and one with the movable nozzles not actuated. A month later, five flight-weight Stage III motors were successfully tested with nozzle actuation and with the first thrust termination system successfully operated. On 12 August 1960, parallel Stage III development by Thiokol and Aerojet was terminated, and Hercules was selected for further R&D as well as production of Stage III. Between October and December 1960, four preliminary flight rating tests on full-scale Stage III motors were successful and included testing of the thrust termination ports. On 10 October 1960, Hercules delivered the first Stage III flight test motor to the AMR.[49]

CASING

Development of the large Spiralloy glass filament–wound case for Minuteman IA, IB, and II Stage III was based on earlier work on the Polaris A-2 second stage. The Stage III cases were supplied by Black, Sivalls, and Bryson, Ardmore, Oklahoma, and the Hercules plant at Rocky Hill, New Jersey.[50]

ATTITUDE CONTROL

Nozzle design for Stage III was similar to the Stage I and II nozzles. As with all aluminized propellants, the benefit of increased specific impulse and reduction of combustion acoustic instability was somewhat offset by the erosive nature of the aluminum oxide particles in the areas of highest gas flow, the nozzle throat and the aft case closure insulation. The nozzle throat erosion was finally overcome for Stage III by using tungsten inserts. Another cause for nozzle failure was ejection of the phenolic graphite nozzle liner. Hercules engineers discovered that nozzle bonding material was outgassing, causing delamination and ejection. The solution was to drill vent holes in the liner to permit release of the gas. The holes were through the phenolic liner to the inside of the structural aluminum cone. As with Stages I and II, the four nozzles pivoted and provided pitch, yaw, and roll control.[51]

## *Aerojet SR73-AJ-1*

On 25 July 1966, Aerojet-General received the R&D contract for the improved Stage III motor (SR73-AJ-1) for Minuteman III. The diameter was increased to 52 inches, but the casing remained a glass filament material. The propellant composition was carboxyl-terminated polybutadiene with a cylindrical grain and progressive thrust curve. The thrust termination system ports were increased to six and relocated to the Stage III forward dome.[52]

The Aerojet Minuteman III Stage III uses a LITVC mechanism similar to that of Minuteman II and III Stage II. The injectant was changed to a solution of strontium perchlorate, which has higher performance than Freon (Table 8.4).[53]

TABLE 8.4  Minuteman Stage III Motor Specifications[a]

| | M57A-1 | SR73-AJ-1 |
|---|---|---|
| WEIGHT, TOTAL | 4,443 pounds[a] | 8,197 pounds[a] |
| WEIGHT, PROPELLANT | 3,668 pounds[a] | 7,292 pounds[a] |
| WEIGHT, CASE | -------- | --------- |
| LENGTH (INCLUDING NOZZLE) | 7. 1 feet[b] | 7.6 feet[c] |
| DIAMETER (EXCLUDING EXTERNAL INSULATION) | 3.2 feet[b] | 4.3 feet[c] |
| THRUST | 17,100 pounds[a] | 34,000 pounds[a] |
| PROPELLANT | ANB-3066[a] | ANB-3066 type I[a] |
| MOTOR CASE | S-901 filament-wound glass fiber, –epoxy resin–impregnated[a] | |
| NUMBER OF NOZZLES | 4 | 1 |
| MANUFACTURERS | Hercules | Aerojet; Thiokol; Chemical Systems Division, United Technologies Corporation[a] |

a) "Minuteman Weapon System History and Description," 2001; b) T. O. 2K-SRM57-3; c) ICBM Handbook, 2004.

## *Thiokol and Chemical Systems Division SR73-AJ-1*

The Aerojet Minuteman III Stage III contract was terminated after 343 were purchased due to cost overruns early in the Minuteman III program. Thiokol took over production of the remaining 392 motors in October 1968, with the designation remaining the same. The Chemical Systems Division of United Technologies Corporation also produced SR 73-AJ-1 motors.[54]

## *Stage III Flight Reliability Problems*

MINUTEMAN II

The M57 motor was known to have combustion instability problems. The oscillations induced by the instability were low, and the NCU and guidance system operation were not affected, so the instability was deemed acceptable. In November 1968 the oscillation phenomena increased in magnitude and caused Stage III NCU failures in two consecutive Minuteman II flights and a third in April 1969 (see Chapter 13, Minuteman II Flight Programs, Special Test Missiles). Investigation by Hercules engineers found that the aluminum powder used in the propellant formulation was from a new plant, though from the same company and sold under the same designation.

Two research efforts were undertaken by the Space and Missile Systems Organization (SAMSO) in December 1968. The first was to find out how the instability oscillations affected the NCU and the second to determine how the new batch of aluminum powder, Lot 1-11, changed the magnitude of the oscillation. Bench testing indicated that the new oscillation frequency was resonant with one of four hydraulic lines in the NCU. When the oscillation became too large, a piston in the NCU became dislodged and stuck, preventing return to the neutral position. The solution was to increase the length of the piston, negating its response to the oscillation. The change in the characteristics of the aluminum powder was determined by high-speed photography, which revealed significant differences in combustion between the new and old aluminum powder, most likely caused by a difference in the oxide coating of the particles. The problematic powder had been quenched at 400 degrees F instead of the original 900 degrees F. The lower quench temperature had left the oxide coating thinner, allowing the aluminum particles to ignite sooner and liberate more energy, which led to oscillations of greater magnitude.[55]

Six years later, a new problem developed with Stage III during the operational test launch of Minuteman II GIANT MOON 7 on 5 September 1975. The Stage II chamber saddle pressure, the pressure present in the motor when the main motor grain burning begins, fell from a normal pressure of 200 to 230 to 108 psi. Although the flight was successful, Ogden Air Logistics Center engineers were concerned that motor flameout was a possibility. The problem was traced to the failure of the aluminum grains in the Stage III propellant to ignite. The issue was restricted to those Minuteman II Stage III motors containing propellant from Lot 16 produced by Hercules. This finding was reported to the SAC ICBM Systems Panel meeting on 3 December 1976, and the decision was made to immediately replace the 32 defective motors.[56]

MINUTEMAN III

In April 1975 the Minuteman Propulsion Long-Range Service Life Analysis Program conducted by Ogden Air Logistics Center revealed that the liner bonding the propellant to the motor case in both the Minuteman II Stage II and III Stage III had changed from its normally firm, pliable state to a viscous, sticky fluid that was designated "sticky liner." By January 1976 Ogden had determined that only Minuteman III Stage III capability was affected.

While the Aerojet contract for production of Minuteman III Stage III motors had been terminated early in the production history of Minuteman III, 172 of the defective motors were produced, and 29 remained in stored inventory as of January 1976. Ninety-nine of these motors were on deployed missiles. In March 1976 Headquarters Air Force directed that the Stage III motors be replaced with motors from Thiokol, which had taken over the contract; this was accomplished during the Upgrade Silo modification program.[57]

## Minuteman III Post-Boost Control System

*Autonetics*

DEVELOPMENT

In the fall of 1957, partially as a response to the launch of Sputnik, the DoD set up the Reentry Body Identification Group to study the feasibility and practicality of developing an

ABM defense system. The committee issued its report on 2 April 1958, concluding that an ABM defense system was feasible, but it could be easily overcome by decoys, chaff, reduced radar cross-sections for the reentry vehicle, tank fragments, or multiple warheads.[58]

In September 1959 the Navy's Fleet Ballistic Missile Program adopted the idea of multiple warheads, deciding to arm the Polaris A-3 with three reentry bodies (the naval terminology for reentry vehicle) in a concept the Air Force called multiple reentry vehicles (MRV), which would be released simultaneously and impact as a cluster. In 1961 Air Force Ballistic Systems Division began studying the idea of MRVs for Minuteman based on the new Mark 12 reentry vehicle design. The Air Force was reluctant to fractionate the payload into smaller-yield weapons, so at this time, the MRV concept was not adopted.[59]

The original use and validation of simple post-boost control system (PBCS) vehicles was the first multiple satellite deployment, which took place on 22 June 1960, with the launch of a Thor Able-Star booster carrying two satellites, Transit IIA and Solar Radiation/Galactic Radiation and Background (GRAB).[60] The Able-Star second stage successfully started and stopped twice to place the satellites in their proper orbits. On 16 October 1963, an Atlas-Agena B combination was used to launch a pair of Vela satellites into two different orbits.

The first true PBCS, Transtage, was the upper stage of the Titan III space launch vehicle. The concept used a restartable stage to precisely place multiple satellites in varying orbits using one launch vehicle, saving the expense of multiple boosters. While development at Martin Marietta began in August 1962, there was no immediate need for the capability. After a delay of nearly four years, on 16 June 1966, Transtage successfully delivered eight Defense Department Initial Defense Communications Satellite Program satellites into eight dispersed orbits.[61]

In early 1963 engineers at North American, working to improve Minuteman accuracy in response to increased hardening of Soviet targets, independently conceived the Vernier Velocity Unit, a combination of the Minuteman guidance system and a simple pressurized gas propulsion unit that could accelerate or decelerate the reentry vehicle a fraction of a foot per second before release.[62] In mid-1963 North American made an unsolicited proposal to the Air Force for the Precision Deployment System (PDS).

In early June 1964, the Air Force called a meeting to discuss problems with the Mark 12 reentry system (Mark 12 was the name of the reentry vehicle system as well as the reentry vehicle itself). There were design problems concerning deployment of decoys and other penetration aids more effectively. The original concept was to emulate the MRV concept of the Navy, but that meant the first reentry vehicle to detonate on the target would likely destroy any other reentry vehicles in the area unless the targets were further apart than desired. The PDS represented a solution to the fratricide problem as it had the capability to loft or depress the trajectories of each of the reentry vehicles enough to spread out their time of arrival over target. Simultaneous with these developments was the discovery that the guidance system of the Mark 12 reentry system was susceptible to x-rays generated from ABM warhead detonation and required shielding, which meant additional weight.[63]

On 29 December 1964, Headquarters Air Force approved the development of the Mark 12 multiple independently targetable reentry concept (MIRV) using a PBCS (the new name for PDS) for Minuteman III. Originally meant to be deployed on Minuteman II, the increased weight of the guidance system shielding meant that the Stage III performance would have to be improved. When Secretary of Defense McNamara approved development of Minuteman III on 8 December 1965, the new missile had a larger-diameter Stage III, matching that of Stage II, to accommodate the Mark 12 MIRV reentry system.[64]

## Minuteman III Propulsion System Rocket Engine

The Propulsion System Rocket Engine (PSRE) provides propulsion as well as attitude control to the Minuteman III post-boost vehicle (PBV), carrying up to three reentry vehicles after Stage III thrust termination. The PSRE is located directly behind the Missile Guidance Set (MGS). When mated, the PSRE and the MGS are referred to as the PBCS, which is manufactured by the Autonetics Division of North American Rockwell Corporation, with Bell Aerospace as a subcontractor. When combined with the reentry vehicle system, they comprise the PBV.

Eleven small liquid rocket engines are used in the PSRE—ten attitude-control engines and one axial thrust engine. There are four pitch engines and two yaw engines mounted rigidly on the skin of the PSRE aluminum wafer section, which is 52 inches in diameter and 18 inches in height. The six engines are interchangeable and deliver 24 pounds of thrust each. During storage and launch, the throats of engine nozzles are closed by covers that are flush with the missile skin and blown off when PSRE begins operation. The four roll control engines are similar in design but deliver only 18 pounds of thrust. The axial engine is much larger and delivers 315 pounds of thrust. Unlike the fixed attitude control engines, the axial engine is gimballed. The purpose of the axial engine is to compensate for the shifting center of gravity as propellants are consumed and reentry vehicles are released.

The use of liquid propellants in the PSRE was a contentious subject. The hypergolic propellants, nitrogen tetroxide and monomethylhydrazine, are both highly toxic; nitrogen tetroxide is also highly corrosive. At the time of the development of the PSRE, restartable solid-propellant motors were not available, and the weight penalty of a simple pressurized gas system was too great. Approximately 160 pounds of liquid nitrogen tetroxide is stored inside a stainless steel bellows tank housed in a cylindrical stainless steel tank 13.8 inches in diameter and 30.25 inches long. Approximately 100 pounds of liquid monomethylhydrazine is stored in an identical configuration (Figure 8.10).

Prior to PSRE ignition, there are two preparatory actions that take place within the PBCS. First, the propellant lines are purged of air during the Stage III burn by cycling the bipropellant valves of the attitude control engines at the command of the MGS computer. This action is followed by the opening of all the isolation valves, allowing helium pressure to act on the stored cylinder bellows to force the oxidizer and fuel from the storage tanks and fill the distribution lines up to the closed bipropellant valves.

Separation of the reentry vehicles from the PSRE is accomplished through the use of the two forward-firing pitch engines and two similar yaw engines also functioning as retrorockets to slow the PSRE and pull it away from the reentry vehicles. The reentry vehicles are released from the PSRE bulkhead by a spring system.[65]

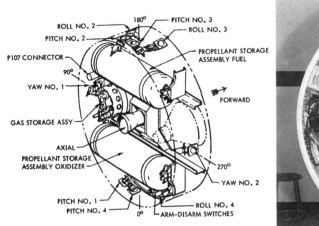

FIGURE 8.10  Right: Aft view of post-boost system rocket engine. Left: Identification of components. *Adapted from "Minuteman Service News," courtesy of Boeing Company.*

## Missile External Insulation

Unlike the Titan II silo, which had dual exhaust ducts to divert engine exhaust from the launch duct and protect the missile during emergence, the Minuteman missile is briefly surrounded by 3,000-degree F exhaust.[66] Acceleration is rapid when compared to Titan II, but measures had to be taken to protect the missile prior to emergence as well as during atmospheric flight, as aerodynamic heating in the lower atmosphere is also a problem due to Minuteman's rapid acceleration. Three approaches were taken during the early design stages of the program: casing material that could tolerate the expected heat load, use of insulation to protect the skin, or use of an insulative and ablative material that would burn away during the ascent phase of flight, both protecting the skin and, as the material ablated, decreasing the weight of the missile for extended range. The insulative and ablative approach was chosen. Boeing chose Avco Corporation's Avcoat I as the exterior insulation for Minuteman IA. From Minuteman IB on, cork was also used.[67] Avcoat I (white) and II (light green) are both epoxy-polyamide resins, with Avcoat II being more flexible. The Avcoat material is sprayed in a thin layer on the exterior of the skirt, stages, and interstage areas. The cork used to coat Minuteman is ground prime cork with a thermosetting phenolic resin, which is formed into sheets that are then bonded to the missile surface with adhesive. A second cork material that can be troweled into small imperfections is also used. The distinctive green color is due to the paranitrophenol fungicide used to prevent fungus growth during prolonged exposure in the launcher. A second fungicide, Hypalon, is opaque (Table 8.5).

Unlike the early developmental Minuteman I pad launches at Cape Canaveral, where the missile was a pristine white during flight, missiles emerging from LFs are a uniform charcoal black. On launch films the exterior coating is actually on fire, which extinguishes

TABLE 8.5  Minuteman Missile Ablative Coating Configurations[a]

| AREA | LGM-30A[b] | LGM-30B | LGM-30F | LGM-30G |
|---|---|---|---|---|
| SKIRT | Avcoat I | cork[c] | cork | cork |
| STAGE I | Avcoat I | Avcoat II | Avcoat II | Avcoat II |
| 1–2 INTERSTAGE | Avcoat I | cork | cork | cork |
| STAGE II | Avcoat I | Avcoat II | cork (white) | cork (white) |
| 2–3 INTERSTAGE | Avcoat I | cork | cork | cork |
| STAGE III | Avcoat I | cork (white) | cork (white) | cork |
| GUIDANCE AND CONTROL | fiberglass | cork[d] | cork (white) | cork (white)[e] |
| REENTRY VEHICLE | Avcoat I | Avcoat II | Avcoat II | cork[f] |
| RACEWAY | Avcoat I | cork | cork | cork |

a) Adapted from *Minuteman Service News*, August 1963; b) LGM-30A missiles removed from deployment February 1969; c) cork areas are green except where noted; d) fiberglass on Wing II installations only; e) includes PSRE; f) shroud before dust hardening replaced the aluminum cork-covered shroud with one fabricated from titanium, requiring no cork covering.

as the missile accelerates. Beginning with Minuteman IB, cork material replaced some areas of the missile previously protected by Avcoat I, and the remaining areas are protected by Avcoat II. A red sealant material is also used at several locations on the missile exterior.[68]

## Additional Airframe Components

### *Interstages*

The missile stages are connected by aluminum alloy structures called interstages, manufactured by Boeing. The interstages are frustrum-shaped, except for the interstage on Minuteman III Stage II and III, which is a cylinder; all are of semi-monocoque (ring-stiffened skin) construction. The interstages join the three motors aerodynamically and structurally, providing protection to Stage II and III nozzles during the severe launch tube egress environment as well as providing separation and flight stability for each stage during powered flight. The interstage skin thickness varies from 0.22 to 0.23 inch on Minuteman III.[69]

Stage separation and interstage removal in flight are accomplished by linear-shaped charges. Each interstage is equipped with an electronic arm–disarm mechanism for the stage separation system and a mechanical lanyard to arm the fracture charges. The stage separation charge, approximately 0.20 inch in width, is contained in a circumferential aluminum enclosure that is not a primary load-carrying member. The enclosure is covered with a rubber-like material, PR-1910, which serves to contain any particle fragments from the charge as well as helping to absorb the shock caused by the separation charge. For increased reliability, detonators are attached to each end of the stage separation charge and electrically connected to the arm–disarm mechanism.

For Minuteman I, Stage I and II staging took place at approximately 60 seconds into flight, with a signal from the MGS cutting the interstage off at the Stage I separation plane (Stage II nozzle exit plane) with the simultaneous ignition of Stage II. A lanyard attached to Stage I tripped a 20-second delay timer on Stage II upon stage separation, followed shortly thereafter by the detonation of four longitudinal, linear-shaped charges to cut the interstage into four sections, which fall away from Stage II.[70] When Minuteman I was designed, solid-propellant ignition at altitude was not well understood. Maximum dynamic pressure in the flight profile occurred at Stage I and II separation and Stage II ignition, so interstage panel fragmentation and separation were delayed until maximum dynamic pressure had passed. A similar set of events happened at the ignition of Stage III approximately 117 seconds into flight, except this time, the interstage detachment from the separation plane of Stage II and fragmentation into four panels was delayed only 2 seconds.[71]

### *Stage I Skirt*

While the terms interstage and skirt are often used interchangeably, skirt refers to the support structure at the aft end of Stage I that supports the entire missile in the missile support ring component of the missile suspension system as well as providing support during transportation and emplacement. The skirt also provides aerodynamic shielding of the Stage I motor nozzles, thermal shielding of the aft Stage I motor Y joint, and support for the lower umbilical connector. With Minuteman IA the lower portion of the 36.24-inch-tall skirt was a frustrum. Flight testing with Minuteman IB showed that the conical shape was not necessary, and modification to a cylindrical shape gave increased range as aerodynamic drag was reduced.[72] The skirt is semi-monocoque, and skin thickness is a straight taper from 0.133 inch at the forward end to 0.177 inch at the aft end (Figure 8.11 compares all four missile variant dimensions; Table 8.6 gives total production numbers).[73]

TABLE 8.6  Minuteman Airframe Production[a]

|  | # PRODUCED | YEAR(S) |
|---|---|---|
| MINUTEMAN I | | |
| R&D | 54 | 1961–1964 |
| OPERATIONAL IA | 185 | 1963 |
| OPERATIONAL IB | 744 | 1963–1965 |
| MINUTEMAN II | | |
| R&D | 20 | 1965–1967 |
| OPERATIONAL | 648 | 1965–1970 |
| MINUTEMAN III | | |
| R&D | 31 | 1969–1970 |
| OPERATIONAL | 739 | 1971–1978 |

a) "Air Force Statistical Digest, Fiscal Years 1962–1978."

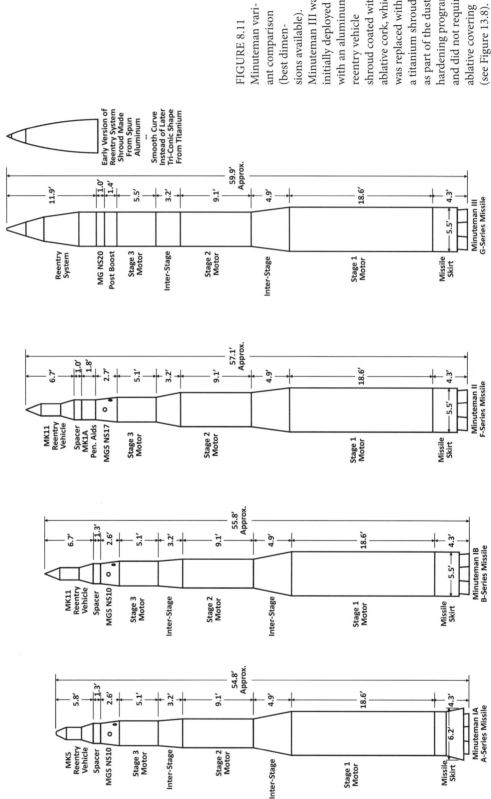

FIGURE 8.11 Minuteman variant comparison (best dimensions available). Minuteman III was initially deployed with an aluminum reentry vehicle shroud coated with ablative cork, which was replaced with a titanium shroud as part of the dust-hardening program and did not require ablative covering (see Figure 13.8).

★ ★ ★

A myriad of technological advances coalesced in a three-year period culminating in the Minuteman IA airframe and propulsion system. This was but one of three challenges that needed to be overcome to bring the Minuteman program to fruition. Just as the Minuteman program required advances in solid-propellant technology, the higher speed of the ICBM reentry vehicle, when compared to IRBMs, required a new approach to thermal protection of the warhead against the aerodynamic heating from the atmosphere as the vehicle emerged from the vacuum of space. The second key to the success of the Minuteman program was the timely development of the second-generation reentry vehicles that used ablative technology rather than the heavier heatsink system.

# 9

# MARK 5 AND 11 SERIES REENTRY VEHICLES

The need to quickly design and field the reentry vehicle system for a relatively large warhead using readily available materials led to the first-generation heatsink concept used with the Air Force Thor and Atlas D (Mark 2) and Navy Polaris A-1 and A-2 (Mark 1).[1] The second-generation reentry vehicle system, ablation, was first demonstrated by the Army with the Jupiter reentry vehicle. The Minuteman Mark 5 and Mark 11 series reentry vehicles represented the culmination of the pyrolytic, or charring, method of ablation with their smaller size and greater accuracy compared to heatsink reentry vehicles. Due to classification issues caused by current world events, the third generation of Air Force reentry vehicles (Mark 12, 12A, and 21) is not discussed, though it has been described in some detail by Lin (see Table 9.1 for a summary of Air Force reentry vehicle designations).[2]

## Early Research

Arming a guided missile with an atomic warhead was an obvious next step in strategic warfare. Concerned with the vulnerability of the eastern United States to long-range missiles from the Soviet Union, in 1945 NACA saw an urgent need to begin studying the problems of hypersonic flight (defined as greater than five times the speed of sound, which is the speed at which aerodynamic heating begins to be significant). By the late 1940s, two major NACA facilities, Ames Aeronautical Laboratory, Moffett Field, California, and Langley Aeronautical Laboratory, Hampton, Virginia, responded by expanding their aeronautical work to study aerodynamic issues involved in ballistic missile flight.[3]

Theoretical research into the problem of aerodynamic heating of ballistic missiles during reentry was first published in 1949 by Carl Wagner.[4] The first comprehensive theoretical work began in 1951 with the work of H. Julian Allen and A. J. Eggers Jr., engineers at Ames. They studied the problem of reentry heating for ballistic, glide, and skip-entry trajectories. Their investigation of these three types of trajectories was driven by the need to find a flight path that could best utilize the thermal protection materials then available. At the start Allen and Eggers dismissed the pointed nose shape, a carryover from rifle bullet design, instead focusing their calculations on a blunt, hemispherical shape, recommending that "not only should pointed bodies be avoided, but that the rounded nose should have the largest radius possible." (Figure 9.1)

It is important to note that these calculations were made with "light" and "heavy" missile options, and no mention was made of a reentry vehicle as such. The light missile's optimal nose shape from a heat-transfer standpoint was a blunt shape; for the heavy missile, a slender shape was best. Their calculations showed that high drag caused a detached shock

TABLE 9.1 Air Force Reentry Vehicle Designators through Minuteman III[a]

| DESIGNATOR | WEAPON SYSTEM | CONTRACTOR |
|---|---|---|
| Mark 1 | Atlas D, Thor | General Electric (development, not flown) |
| Mark 2 | Atlas D, Thor | General Electric, Avco (not flown) |
| Mark 3 | Atlas D | General Electric |
| Mark 4 | Atlas E, F, Titan I | Avco |
| Mark 5 | Minuteman IA | Avco |
| Mark 6 | Titan II | General Electric |
| Mark 7 | Skybolt | General Electric (canceled) |
| Mark 8–10 | not assigned | — |
| Mark 11, 11A, 11B, 11C | Minuteman IB, Minuteman II | Avco |
| Mark 12 | Minuteman III | General Electric |
| Mark 12A | Minuteman III | General Electric |
| Mark 13, 14 | Titan II, Atlas | never authorized |
| Mark 15, 16 | unknown | — |
| Mark 17, 18 | Titan II, Minuteman III | never authorized |
| Mark 19, 20 | unknown | — |
| Mark 21 | Peacekeeper, Minuteman III | Avco |

a) Barry Miller, "Studies of Penetration Aids Broadening," *AWST*, 20 January 1964; D. Ruchonnet, *A Brief History of Minuteman and Multiple Reentry Vehicles*, 1976.

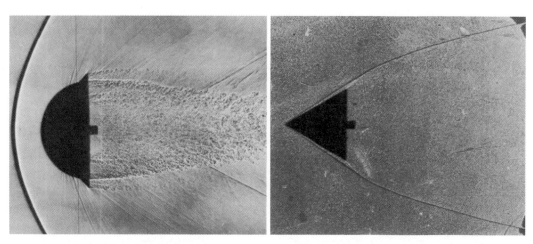

FIGURE 9.1 Wind tunnel Schlieren photographs illustrating the detached bow shock wave generated by a blunt reentry vehicle body (left), compared to the attached shock wave with a pointed reentry body. The detached bow shock wave dissipates heat well away from the reentry body surface. *Adapted from "Wind Tunnels of NASA."*

wave; thus, the majority of the heat generated dissipated into the atmosphere, leaving only radiated heat to contend with, unlike a sharply pointed body, where the shock wave was attached to the tip, causing heat transfer and probable destruction of the body. Additionally, the heat reaching the blunt body would be more evenly distributed.[5]

In order to reach targets at a range of 4,000 to 6,000 nm, ballistic missiles need to accelerate to speeds of up to approximately Mach 20 (15,223 miles per hour, just short of orbital velocity), 10 times the speed of a high-powered rifle bullet.[6] Reentry into the atmosphere at these speeds generates a shock wave, heating the atmosphere to temperatures approaching 12,000 degrees F, which exceeds the melting point of tungsten, the metallic element with the highest known melting point: 6,116 degrees F.[7] At this temperature the air plasma is also highly chemically reactive. There is a transport of heat by mass conduction from the air plasma to the vehicle surface, which is dependent on both the temperature and density of the air in the plasma. At high altitudes, where the air density is low, the mass transport of heat is low, in spite of the very high shock wave temperature. Conversely, at lower altitudes, the higher density plasma results in a higher heat flux for equal reentry vehicle velocities (Figure 9.2).[8]

## *Solutions to the Reentry Problem*

Theodore von Kármán, perhaps the leading aerodynamics expert of his time, described what he called "the reentry problem" at a symposium in Berkeley, California, in June 1956. Reentering the atmosphere at speeds of Mach 12 to 20 was "perhaps one of the most difficult problems one can imagine . . . a challenge to the best brains working in these domains of modern astrophysics."[9]

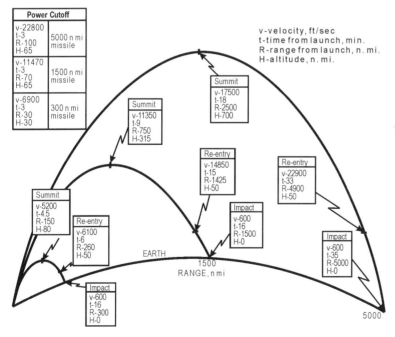

FIGURE 9.2 Short-, medium-, and long-range ballistic missile trajectory characteristics. *Adapted from Ordnance Engineering Design Handbook, Ballistic Missile Series: Trajectories.*

While the workers at Ames, Langley, and other facilities had partially met the challenge via theoretical calculations about vehicle shape, which led to the design of testing facilities, what was the solution to the remaining aspect—taming the thermal load encountered at these high speeds?

Four categories of cooling were considered: (a) radiant cooling via emittance from the vehicle surface, (b) solid heatsinks having sufficient mass to absorb the heat, (c) transpiration and film cooling for heat removal by material phase change, and (d) ablative heat dissipation via the many protective processes associated with surface removal. Radiant cooling was best for long-duration reentry environments where heat load was relatively low and constant and, in practice, worked best at temperatures below 2,000 degrees F. Solid heatsinks could accommodate higher temperatures as long as the heating rate was not so rapid as to melt the material. Additional large structural mass was necessary to store the heat and protect the payload. Transpiration and film cooling would work over a wide thermal environment but were mechanically complicated, which might reveal reliability issues. Ablation worked well for short-duration, high-temperature environments. The question was which materials to select and how to test them.[10]

A key description of a reentry vehicle is its ballistic coefficient, beta ($\beta$). $\beta$ is defined as $W/(C_d A)$, where W is the weight of the reentry vehicle, $C_d$ is the coefficient of drag, and A is the cross-sectional area. With reentry vehicle weight being held constant, reentry vehicles with a low $\beta$ (high coefficient of drag and cross-sectional area and thus high air resistance) decelerate at a relatively high altitude, where the density of the atmosphere is low and heat fluxes are lower but reentry times are longer, facilitating radar detection and resulting in decreased accuracy. Medium-$\beta$ vehicles decelerate at a medium altitude with higher heat fluxes but shorter detection times and increased accuracy. High-$\beta$ vehicles decelerate at much lower altitudes, encountering much denser air and, hence, higher heat fluxes, but for a shorter time, thereby allowing less time for radar detection and also greatest accuracy. Obviously, these considerations were critical to mission requirements but were constrained by both the materials and testing facilities available at the time.

## *The First Generation—Heatsink*

The work of Allen and Eggers had clearly shown the importance of selecting a relatively blunt nose shape to minimize aerodynamic heating. Their research showed that most of the aerodynamic heating would be outside the boundary layer and not in direct contact with the reentry vehicle, provided that the boundary layer remained laminar. A considerable amount of radiative heat still had to be dissipated. Since radiation varies as the fourth power of the temperature, it was likely that the reentry vehicle would not be an efficient radiator, with the result that surface temperature would rise beyond either the structural stability of available materials or the tolerance level of the enclosed equipment, the fusing and actual warhead. Both the Navy (Mark 1) and Air Force (Mark 2) elected to use the heatsink concept for their first-generation reentry vehicles but quickly moved on to the second-generation ablation technology, which had become available during the heatsink reentry vehicle design and production (Figure 9.3).[11]

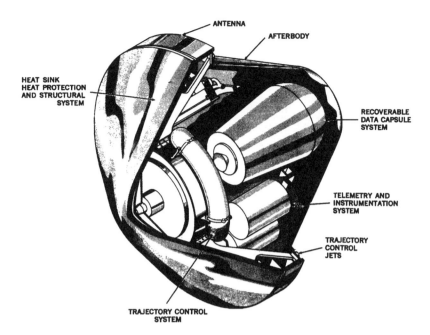

FIGURE 9.3 The General Electric Mark 2 reentry body components (top) and the Avco Mark 2 version (bottom), which never flew but was a competing contract to General Electric. The Avco Mark 2 heatsink material was beryllium with a highly polished face to prevent uneven heating during reentry. *Top courtesy of Donald Schmidt; bottom courtesy of Phil Fote.*

## THE SECOND GENERATION—ABLATIVE

The process of ablation during reentry is described as follows:

> As heating progresses, the outer layer of polymer may become viscous and then begin to degrade, producing a foaming carbonaceous mass and ultimately a porous carbon char. The char is a thermal insulation; the interior is cooled by volatile material percolating through it from the decomposing polymer. During the percolation process, the volatile materials are heated to very high temperatures with decomposition to low molecular weight species, which are injected into the boundary layer of air. This mass injection creates a blocking action, which reduces the heat transfer in the material. Thus, a char-forming resin acts as a self-regulating ablation radiator, providing thermal protection through transpirational cooling and insulation. The efficiency, in terms of heat absorbed per weight of material lost, is about 40 times that of the earlier copper heatsink design.[12]

### *Army*

Ablation provided thermal protection for the Jupiter reentry vehicle. The Army's ablative approach came from the fortuitous result of research begun in 1953 investigating materials to replace graphite for jet vane application during the development of the Redstone missile (jet vanes were used for directional control instead of gimballing the engine). In the search for a replacement, researchers tested a jet vane made of commercial-grade, fiberglass-reinforced melamine. Exposure to the Redstone rocket motor exhaust eroded the vane as expected, but ¼ inch beneath the surface, the material was not only undisturbed, but the embedded thermocouples revealed no heating had taken place. While the tested material was not used as a jet vane, the results led the Army Ballistic Missile Agency (ABMA) researchers to bypass the heatsink concept and go straight to ablative reentry vehicle materials.[13]

Scientists at ABMA evaluated five candidate materials: Refrasil–phenolic, fiberglass–melamine, unfired ceramic, and beryllium and copper to provide thermal protection for a proposed heat shield design. Refrasil, fiberglass–melamine, and ceramic were found to be the materials of choice. An expedient method for evaluating candidate materials was to expose flat plates of the material to rocket exhaust at a heat flux of 100 BTU per square foot per second and a velocity of 6,700 feet per second. The plates were four inches square and tilted at a 45-degree angle in the exhaust stream. Further research in resin-based ablative materials revealed that asbestos-reinforced phenolic resin would be the best overall material for the Jupiter reentry vehicle environment. After initial evaluation of the plate material, reentry vehicle shapes were tested both with the rocket exhaust technique and via shock-tube studies by Arthur Kantrowitz at Cornell University.[14] Using a variety of rocket motors, researchers were able to simulate heating rates up to 2,500 BTU per square foot per second. Transonic wind tunnel tests of a half-scale model Jupiter reentry vehicle were conducted at the Air Force's AEDC, Arnold AFB, in June 1957 and at the hypersonic test facilities of the Naval Ordnance Laboratory, White Oak, Maryland, in September 1957, confirming the full-scale design.[15]

For flight testing of the one-third-scale Jupiter reentry vehicle designs, the Army's Redstone TBM was modified into a three-stage booster. The first stage had an elongated fuel tank and used a more powerful fuel called Hnyde (unsymmetrical dimethylhydrazine). The forward section of the first stage was strengthened to support the new upper stages. The second stage was made up of a cluster of 11 scaled-down Sergeant solid-propellant missiles, six inches in diameter, housed in a cylindrical fairing called the "tub." The third stage was located in the center of the second stage and made up of three additional scaled-down Sergeant missiles. Atop the third stage was a 300-pound, one-third-scale ablative (one-tenth surface area) reentry vehicle composed of a welded steel shell supporting the heat shield. While fabrication techniques were being perfected for the resin–asbestos material, a heat shield made of layered disks of melamine was flown first. The tub was spun up by electric motors at launch to provide ballistic stability. The resulting vehicle was called Jupiter-C (Jupiter Composite) and had a range of over 1,500 nm with an apogee of over 175 nm.[16]

Only 3 of 13 scheduled flights were necessary for the Jupiter-C program. The first launch was on 20 September 1956, Jupiter-C Missile RS-27, with the missile reaching an altitude of 600 nm and a speed of Mach 18. This was a test of the modified propulsion and staging system and was successful. The second flight, Jupiter-C Missile RS-34, was launched on 15 May 1957. The missile pitched up at 134 seconds into flight. While the planned range was not reached and the reentry vehicle was not recovered, telemetry indicated that the fiberglass–melamine ablative material functioned as expected. The first subscale operational Jupiter reentry vehicle design using a phenolic resin–asbestos ablative material was flown on Jupiter-C Missile RS-40 on 8 August 1957. The booster and high-speed upper stages worked well. Failure of the reentry vehicle to separate from the third stage changed the reentry trajectory, reducing the angle of attack at the point of maximum heating. Nonetheless, the reentry vehicle traveled 1,168 nm, achieving a velocity of 13,000 feet per second and withstanding a temperature of over 2,000 degrees F, conditions similar to those expected for an IRBM reentry vehicle. While the reentry vehicle did not separate as planned, the heat of reentry melted the magnesium ring of the separation system, and the recovery system deployed successfully. Analysis of the recovered reentry vehicle ablative surface showed only a 1.5 percent loss. The reentry vehicle was displayed in President Eisenhower's office and is in storage at the National Air and Space Museum in Washington, DC. Ablation technology was proven with the ultimate test—full IRBM range and velocity.[17]

Full-scale Jupiter reentry vehicles were successfully recovered on three flights: Jupiter AM-5, launched on 18 May 1958, the first recovery of an IRBM reentry vehicle; Jupiter AM-6B, 17 July 1958, which also carried a lightweight high-explosive warhead; and Jupiter AM-18, 28 May 1959, which carried two monkeys, Able and Baker, which survived unharmed. While the reentry vehicle flown on AM-5 showed an ablation depth of 3/8 inch at the greatest point of loss, the remaining flights showed considerably less, validating the ablative results of the subscale model flown and recovered earlier.[18]

The deployed reentry vehicle, built by Goodyear Aircraft Corporation, was a hermetically sealed conical aluminum shell with a 12.5-inch radius spherical tip attached to a cone frustrum with a base 65 inches in diameter and an overall length of 9 feet. The molded nose cap was composed of 30 percent, by weight, phenolic resin, with 70 percent Type E glass;

the frustrum material was a layer of a mixture of 45 percent phenolic resin and 55 percent chrysotile asbestos.[19] A key design feature, also found in other reentry vehicle designs, was a convex aft cover that conferred the ability to recover from any attitude to the correct reentry alignment. The ablative material was much thinner than the subscale fiberglass–melamine heat shield. The complete reentry vehicle with warhead weighed 2,617 pounds; the W49 nuclear weapon weighed 1,600 pounds.[20]

## Air Force

The concept of ablation was not new to the Air Force. The two contractors selected to develop the Atlas Mark 2 heatsink reentry vehicle, General Electric and Avco, were directed to look at all methods for solving the reentry problem. Wright-Patterson ARDC was also evaluating ablation materials, as were Langley and Ames. The decision to include work with the ablation concept stemmed from recommendations by a number of scientific advisory committees and panels.[21]

On 31 August 1957, in the OSD-BMC "21st Monthly Report on Progress of ICBM and IRBM Programs," a shift in reentry vehicle design was noted. While the heatsink design for Atlas and Thor was sufficient, developments in materials and testing capability indicated that ablation reentry vehicles could have ballistic coefficients five to eight times greater than the Mark 2 heatsink, which would lead to greater reentry speed with increased accuracy and decreased vulnerability.[22]

On 28 August 1958, after only two Atlas B flights with the Mark 2 and just before the start of the Thor Mark 2 flight testing, almost exactly one year after the highly successful conclusion of the Army's Jupiter-C reentry test vehicle program, Brig. Gen. O. J. Ritland, vice commander of BMD, notified ARDC of the decision to reorient the ICBM reentry vehicle program from heatsink to ablative technology. The decision was based on "recent developments aimed toward improving the solution to the ICBM reentry problem." The Mark 2 heatsink reentry vehicles would be supplied for all WS-315A (Thor) and early operational WS-107A-1 (Atlas D) missiles at the two operational sites at Cooke AFB (Cooke had not been renamed Vandenberg yet). All Avco work on heatsink development was to be discontinued. General Electric was now assigned development responsibility for a lightweight, second-generation reentry vehicle capable of carrying a 1,600-pound warhead, to be flight tested on the Series D Atlas missiles with deployment starting at F. E. Warren AFB. This was the Mark 3. Avco was assigned responsibility for a heavyweight, second-generation reentry vehicle capable of carrying a 3,000-pound warhead, to be flight tested on Lot J Titan I missiles. This was the Mark 4.[23]

### ABLATIVE MATERIALS

As early as 1956, plastics had been examined for use in the high-temperature environment of ramjet engines. Researchers at the Marquardt Aircraft Company exposed model ramjet inlet cones made from three fiberglass-reinforced plastics, Conolon 505 (phenolic), DC 2106 (silicone), and Vibrin 135 (polyester) for 20 minutes at temperatures up to 600 degrees F at a speed of Mach 2. They found that all three materials showed little or no detrimental effects, concluding that reinforced plastics might have a role in missile reentry vehicle development.[24]

Researchers at General Electric's Missile and Ordnance Systems Division in Philadelphia expanded on the Marquardt work by estimating a candidate ablative material's ability to absorb heat up to 8,000 degrees F under equilibrium conditions. The results showed that plastic materials had the highest theoretical heat-absorbing capacities, more than twice that of beryllium. The more gas a material generated upon heating, again under equilibrium conditions, the better the material. The higher the melting point and the more viscous the resulting liquid, the more optimal the thermal effects. Phenolic resin plastics were found to decompose slowly at high temperature and did not liquefy, instead forming gaseous by-products and a char layer that protected the base material. Exposure of phenolic–glass cloth with 65 percent resin to 12,000 degrees F in a high-temperature arc showed only 1.4 percent erosion; phenolic–Refrasil with 41 percent resin only 2.1 percent erosion, and phenolic–nylon cloth with 57 percent resin only 1.0 percent erosion. Key variables were type of resin, orientation of the fibers, type of fiber, and ratio of resin to fiber. Phenolic resins gave a higher yield of carbon char. Orientation of the fibers had a significant effect on performance, with random orientation giving the best results. At temperatures above 5,000 degrees F, amorphous silica fibers were superior to ordinary glass, and organic fibers were superior to glass fibers. Clearly, ablative materials had come of age for use in ICBM reentry vehicle heat shields.[25] The result was General Electric's Mark 3 reentry vehicle, deployed only on Atlas D.

Avco was also studying and developing ablative and heatsink materials. Unlike the engineers at General Electric, who had studied ceramics and dismissed them as too difficult to work with compared to reinforced plastic resins, Avco engineers decided to pursue the use of ceramics for the nose section of the reentry vehicle, where the heating was the most severe.

Expanding on ceramics research conducted at the Georgia Institute of Technology and Battelle Memorial Institute for the Jupiter program, Avco researchers focused on solving the brittle fracture problem that was preventing the fabrication of large and complicated ceramic reentry vehicle shapes light enough to be practical. The decision was made not to search for new materials but rather to focus on new fabrication techniques. The eventual solution was to use a metal honeycomb structure to hold small "pencils" of ceramic that did not easily fracture. By orienting the pieces in honeycomb cells at 90 degrees to the surface, optimum thermal protection and structural strength was obtained. In 1959 Avco's Research and Advanced Development Division announced the development of Avcoite, a magnesium honeycomb reinforced ceramic for use on the nose of the Mark 4 reentry vehicle originally destined for Titan I but also deployed on Atlas E and F.[26]

## FLIGHT TESTING

Once the feasibility of ablative material had been experimentally determined, flight testing of subscale reentry vehicles began. The primary R&D flight testing operations for evaluating the early Air Force sub- and full-scale ablative ICBM reentry vehicles were the Thor-Able 0 and II, Atlas D, and Titan I Lot J programs.

*Thor-Able*

The first in a series of boosters used for Air Force reentry vehicle development was the Thor-Able launch vehicle. Use of the Atlas ICBM was considered but rejected at this point, as integration of reentry vehicle testing would interfere with the early weapon system development objectives. In October 1957 BMD and STL began the design of an advanced reentry test vehicle (ARTV) that could be ready for use within six to eight months using existing hardware. The critical capability of the ARTV would be to reach ICBM reentry speeds of approximately 24,000 feet per second carrying a one-half-scale reentry vehicle. The result was the Thor-Able booster, a Thor first stage and Vanguard second stage modified with eight spin rockets.[27]

*Able RTV*

The Thor-Able 0 program flight tested three General Electric reentry vehicle development models, designated Able RTVs. The RTVs were biconic spheres, 34 inches long, with a base 38 inches in diameter, weighing 620 pounds, fabricated with ablative material, and flown from Cape Canaveral from 23 April to 23 July 1958. There was one failure due to booster malfunction and two partial successes. All three flights carried biomedical experiments with mice, and while the two successes clearly demonstrated the efficacy of ablation at ICBM ranges and speeds, the reentry vehicles were not recovered as planned. The data provided by these tests helped determine how much the heat shield weight could be decreased and still be effective, as well as verifying the superior performance of ablative materials compared to the heatsink materials. A description of the RTV series vehicle's ablative materials has proven elusive.[28]

*Able RVX-1*

For the Thor-Able II program (the Thor-Able I program was used for Operation Mona, three unsuccessful lunar probe launches), a modified Thor booster was used with its inertial guidance package removed and replaced with the radio-inertial guidance system for the Titan I installed in the RVX-1 reentry vehicle. These six flights were designated as Precisely Guided RTV launches with two goals: evaluating the new guidance system, which would also indicate the exact point of impact, and continuing to evaluate new ablative materials.[29]

Instead of using the recovery concept that had been used unsuccessfully with the Thor-Able 0 flights, General Electric developed a more robust system to handle the much heavier RVX-1 vehicles. The data capsule system used in the Mark 2 heatsink program was used to record the telemetry during the flight and reentry phase when ionization phenomena prevented telemetry transmission. Earlier experiments with the X-17 sounding rocket research program had demonstrated that an ionized air layer surrounding the vehicle during the highest-temperature period of reentry caused a telemetry blackout. For full-scale flight testing of the Mark 2, General Electric engineers had developed a buoyant data capsule. The capsules were 18-inch spheres made from two hollow hemispheres of polyurethane foam that housed a tape recorder, radio beacon, battery pack, dye pack, and sound fixing and ranging (SOFAR) bomb device for locating the capsule. The bottom half of the capsule was coated with shark repellent after a test capsule was recovered with a shark bite mark.

The capsule was attached to a small rocket to boost it free of the reentry vehicle. The urethane sphere was encapsulated in an ablative shell, which shattered on impact, releasing the buoyant capsule. Contact with saltwater triggered the release of dye, the SOFAR device, and activation of the radio beacon.[30]

General Electric provided the RVX-1 internal frame used to test both the General Electric and Avco ablative materials. The RVX-1 was a conic sphere–flared cylinder configuration 67 inches long, with a cylinder diameter of 15 inches and a flare diameter of 20 inches, and weighed 645 pounds (Figure 9.4).

The Avco RVX-1 vehicles, 68 inches in length, with a nose cap of 11 inches, a cylinder diameter of 17 inches, cylinder length of 39 inches, a flare length of 18 inches, and a flare base diameter of 28 inches, had Avcoite on the nose and phenolic–Refrasil tape covering the midsection and flare.[31] On 8 April 1959, the Avco RVX-1-5 successfully flew 5,000 nm downrange with a maximum altitude of 764 nm and a reentry speed of 15,000 miles per hour. The nose cap easily withstood the heat of reentry, as did the Refrasil material coating on the cylinder and flare sections. The Avcoite ceramic melted and flowed back asymmetrically a short distance down the cylindrical body, as expected. Telemetry results indicated no effect on aerodynamic stability (Figure 9.5).

The RVX-1 flight program, even with the failures due to not recovering all the vehicles (complete telemetry was obtained via the data capsules), further confirmed the maturity of ablative materials for use in high-speed reentry where the reentry vehicles were exposed to temperatures exceeding 12,000 degrees F. The RVX-1 test vehicles were the

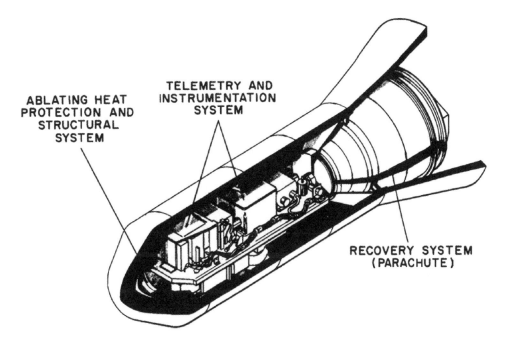

FIGURE 9.4 The RVX-1 components. Avco tested a variety of heat shield materials on the RVX test vehicle. *Courtesy of Donald Schmidt.*

FIGURE 9.5  The recovered Avco RVX-1–5 on display at Avco facilities. The nose cap was removed for further inspection and replaced with a mockup prior to display to the press due to security concerns. The RVX-1-5 is now in storage, along with the removed nose cap, at the National Air and Space Museum's Garber facility. *Courtesy of Phil Fote.*

direct progenitors of the General Electric Mark 3 (Atlas D) and Avco Mark 4 (Atlas E and F and Titan I) reentry vehicles.[32]

*RVX-2 Series*

By mid-1960 Atlas D missiles were available for use in the final phase of ablative material testing: flights of full-scale reentry vehicles at operational ranges and reentry speeds. The RVX-2 series involved tests of newer plastic ablative materials. Ranges flown varied from 4,000 nm to the Ascension Island impact area to 6,400 nm off the coast of Cape Town, South Africa, and further yet, 7,900 nm to the South Atlantic off the Prince Edward Islands. The reentry evaluation portion of the program commenced on 8 March 1960 and ended on 23 January 1961.[33]

Three General Electric RVX-2 reentry vehicles were flown to test a new type of ablative material, unreinforced phenolic resin, General Electric Series 100, for the proposed Titan II Mark 6 reentry vehicle.[34] The RVX-2 was a conic sphere configuration, 12 feet tall and 5 feet in diameter, weighing over 2,000 pounds, the largest reentry vehicle yet flown. The RVX-2 had a phenolic resin–chopped nylon nose cap with unreinforced phenolic resin frustrum side panels. The first two flights suffered guidance and booster failures on 17 and 18 March 1959, respectively, but the last flight, on 21 July 1959, was successful, and the reentry vehicle was recovered intact after a flight of 5,000 nm. Photographs of the recovered vehicle show a close resemblance to the General Electric Titan II Mark 6 reentry vehicle, which also incorporated these materials.[35]

The RVX-2A program had three flights during the Atlas D test flight program, 12 August, 16 September, and 13 October 1960. The RVX-2A vehicle had the same dimensions as the RVX-2 and weighed slightly more than 2,700 pounds. The main difference between the two was the instrumentation. The RVX-2A was used for extensive scientific experiments as well as reentry. A recovery system similar to that of the RVX-1 program was used on all the flights, with successful recovery on only the final flight.[36]

Avco flew one RVX-2A flight on 16 September 1960. The nose cap was RaD 58D, followed by a 26-inch frustrum section of RaD 58B and 100 inches of tape-wound Refrasil. Test plugs of Avcoat X3007 and RaD 58E were inserted at alternating 90-degree intervals in the forward portion of the tape-wound Refrasil section. RaD 58E was a candidate material for the Minuteman missile reentry vehicle, and Avcoat was a proposed low-temperature ablation coating for the boost phase of the Minuteman trajectory. Telemetry problems prevented transmission of thermal and ablation data.[37]

## MARK 3

The Mark 3 reentry vehicle was designed for the Atlas F missile, as mentioned earlier, but deployed only on Atlas D. The Mark 3 was a direct descendant of the General Electric RVX-1 program. Measuring 114.8 inches in overall length, there were two Mark 3 shapes. Both had the conical sphere nose shape, 29.22 inches in length, and a cylindrical midsection 20.7 inches in diameter and 40.6 inches in length. The Mark 3 (Mods I, IX, and IA) had a single biconic frustrum flare, 35.9 inches in diameter, that blended smoothly with the reentry vehicle adapter spacer atop the missile. The Mark 3 (Mods IB and IIB) had a biconic-2 shape with a second, wider flare at the base, 42.8 inches in diameter, resulting in a characteristic conical ring "skirt" slightly outward above the spacer, which was not modified to a more streamlined appearance. The second flare aided in reentry stability by moving the aerodynamic center of pressure toward the rear of the reentry vehicle. Available photographic evidence indicates the biconic-2 modification was the deployed version. The nose section was thermally protected by molded phenolic nylon and the midsection and flare by tape-wrapped phenolic nylon (Figure 9.6).[38]

Eleven full-scale Mark 3 reentry vehicles were flight tested as part of the Atlas D R&D program from 8 March 1960 to 23 January 1961, with ten successful and one failed due to booster failure prior to launch.[39] The Mark 3 was deployed on Atlas D missiles from 1960 to 1965.[40] The Mark 3 (Mod 3) operational reentry vehicle weighed 2,200 pounds, of which 1,600 pounds was the warhead.[41]

## MARK 4

Unlike the General Electric Mark 3, the Avco Mark 4 design required additional experimental flights, designated RVX-3, a 0.72-scale model, and the RVX-4, 0.94-scale model, due to modified Air Force requirements. The RVX-3 was flight tested on five Titan I Lot C missile flights from 12 December 1959 to 28 April 1960. The RVX-4 was to have been the full-scale model, but the diameter of the warhead was changed slightly, leading to the actual full-scale Mark 4. The RVX-4 was flight tested on one Atlas D, 27 January 1960, and seven Titan I Lot G flights, 24 February to 29 September 1960.[42]

The Mark 4 was a sphere-cone-cylinder-biconic flare shape, 126.7 inches long, 33 inches in diameter at the cylindrical midsection, and 48 inches in diameter at the base. The Mark 4 flare varied from 7 to 22 degrees, with two very small spin fins at the base of the flare. The nose cap was made of Avcoite varying from 1.32 to 0.82 inches thick, the cylindrical body and flare protected by oblique tape-wound Refrasil at a thickness of 0.61 to 0.32 and 0.44 to 0.66 inch, respectively, and the afterbody was protected with fiberglass. The Mark 4 with warhead weighed 3,800 pounds.[43] A second reference cites the operational Mark 4 as weighing 4,100 pounds, of which 3,100 pounds was the warhead (Figure 9.6).[44]

The Mark 4 was flight tested on one Atlas D, seven Atlas Es, seven Atlas Fs, and 28 Titan I Lot J and M missiles from 11 October 1960 to 1 May 1963. The Mark 4 was deployed on Atlas E and F and Titan I from 1962 to 1966.[45] One Mark 4 was flown on a Titan II during the R&D program, Titan II N-11, 6 December 1962.[46]

## *Minuteman Reentry Vehicles*

### MARK 5

On 13 January 1958, in discussions within the Nose Cone Division of Space Systems, BMD, a decision was made that Avco would have initial design responsibility for the advanced reentry vehicle for Minuteman due to the heavy technical load already assigned to General Electric. On 5 February 1958, a letter was issued to Avco confirming the request for an advanced reentry vehicle design study, which included design specifications. This was not a sole-source contract for the reentry vehicle production; as with other reentry vehicles, the development contract would be a competitive one.[47]

With contractor bid proposals to be evaluated in late June, on 28 May 1958, the Nose Cone Division clarified the desire, previously discussed in the proposed preliminary operational concept of Minuteman dated 8 April 1958, for two reentry vehicle designs and two warheads.[48] One vehicle would have a weight of 790 pounds, including a 600-pound warhead, for a range of 5,500 nm; the second would have a weight of 550 pounds, including a 350-pound warhead, for a range of 6,500 nm. The two designs would permit the preliminary operational plan target coverage from bases located in the southwest portion of the United States. The designs would be optimized for maximum-range target coverage, permitting each missile to fly to the maximum range estimated for the payload, negating the need for lesser-range targeting for a given missile. Phase I and II warhead feasibility studies had already been completed, permitting reentry vehicle dimensions to be established. Requests for proposals were issued to 10 contractors for the reentry vehicle studies covering either or both of the reentry vehicles.

The Nose Cone Division explained the need for two reentry vehicles:

> There are several reasons which in our opinion make it imperative that we continue with development of both vehicles. These relate primarily to the warhead development itself. At the present time this country is considering a moratorium on testing,

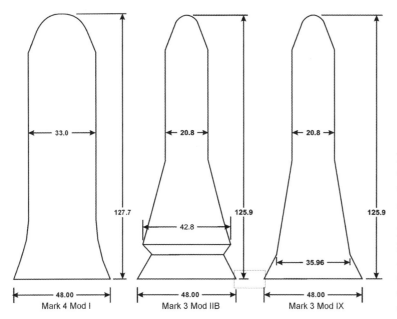

FIGURE 9.6 Comparison, left to right, of the Avco Mark 4 Mod 1, General Electric Mark 3 Mod IX, and Mark 3 Mod IIB. *Adapted from Flight Summary Report, Series D Atlas Missiles.*

the duration of this being somewhat indeterminate. For this reason, AEC laboratories are endeavoring to carry out during Hardtack all tests which appear to them of importance in development of weapons for which requirements have been stated. At the same time the AEC is attempting to get acceptance by DOD of the concept of multi-use warheads. Under this concept, which has been favorably received, a weapon system requiring a warhead of a particular weight will be forced to use an already available or planned weapon, which in some instances will have been developed for quite different requirements. In our case, for example, the smaller warhead will be that now under development for Nike-Zeus. Provided that the Minuteman requirements are incorporated into the weapon design initially, which can be done if we establish our need for this weapon, there will be no difficulty in obtaining maximum performance of the system (the same is not true for the 600 pound Polaris warhead which must be modified to a considerable degree to meet Minuteman requirements). If on the other hand we do not today establish a firm requirement for a second, lighter warhead, it will be designed on the basis of Nike-Zeus requirements and will be completely incompatible with the Minuteman system in the event that we choose to use the second reentry vehicle at some later date.

We feel emphatically that development must continue on both warheads, and hence, both reentry vehicles since a requirement for one cannot be established without the other.[49]

On 20 July 1958, AFBMD announced that Avco had been selected from a group of seven proposals (Aerophysics Allison, Avco Corporation, Ford Aeronutronic, General Electric, McDonnell, Republic Aviation, and Douglas/Goodyear) to develop the two Minuteman reentry vehicles. The contract required development of a light and heavy reentry vehicle to accommodate two possible warhead designs weighing 350 and 600 pounds, respectively, with warhead dimensions to be forthcoming.[50] The contract was formally awarded to Avco on 19 September 1958.[51]

The light version was canceled on 4 December 1958 to reduce costs (Avco was directed to continue studying the light version on a lower priority basis). The decision was based on the lower yield available for the light vehicle warhead as well as complications introduced into the missile flight test program by multiple combinations of reentry vehicles and the missile airframe. The result was a 790-pound reentry vehicle, of which 600 pounds was due to the warhead. The larger reentry vehicle could also more easily accommodate changes in warhead dimensions.[52]

After nearly a year of indecision on design, on 1 September 1959, the Minuteman warhead was finally authorized. Avco's sphere-cone-cylinder-flare design was based on the Mark 4 shape but was considerably smaller due to weight constraints (Figure 9.7).

Extensive wind tunnel and light-gas gun evaluations of ablative material composition and thickness, as well as studies of attitude control and structural design to withstand deceleration forces of 20 to 50 g, were undertaken. Mark 5 (Mod I) flight test vehicles were in production by the end of 1960. Like the Mark 4, the Mark 5 had a nose cap of Avcoite, in this case Avcoite-1, bonded to the top of the cylindrical and flare sections, which were machined out of a block of RaD-58B phenolic resin–Refrasil material. Reformulation of the ceramic material reduced the melting and flowing that occurred with the Mark 4. The aft closure was configured to stabilize the reentry vehicle during early reentry and coated with Avcoat. The Mark 5 did not have an active attitude control system. It tumbled upon entering the atmosphere until two small fins induced a stabilizing spin before the fins ablated early in reentry (Figure 9.8).[53]

The full-scale R&D flight test program began on 1 February 1961, with the successful launch and flight of FTM-401, a fully configured Minuteman IA, from the Launch Complex 31A pad, Cape Canaveral Air Force Station, Florida. Two more pad launches took place, 19 March (failed) and 27 July 1961 (successful). Silo R&D launches at Cape Canaveral began on 30 August 1960 with a spectacular failure and ended on 20 February 1963 with 6 failures out of 21 launches. Mark 5 flight tests also utilized Atlas D (one), E (four), and F (three) missiles, with one failure. The Atlas flight tests commenced on 13 May 1961 with a Mark 5 Mod I flown on an Atlas E and ended on 31 July 1963 with a successful Atlas D flight.[54]

The deployed Mark 5 Mod 5B weighed approximately 200 pounds, not including the warhead. The Mark 5 was deployed on only 150 Minuteman IA missiles, beginning in 1962 and ending in 1969 (Figures 9.9 and 9.10).[55]

## Mark 11 Series

In October 1960 the DoD and the AEC authorized development of an advanced version of the XW-56X1 warhead. In December 1960 the Air Force requested development of a lighter and higher-yield warhead, designated the XW-59. One month later, it was decided to have Avco develop a new reentry vehicle, the Mark 11, able to carry either of the new warhead designs and to be deployed starting with the second Minuteman wing, equipped

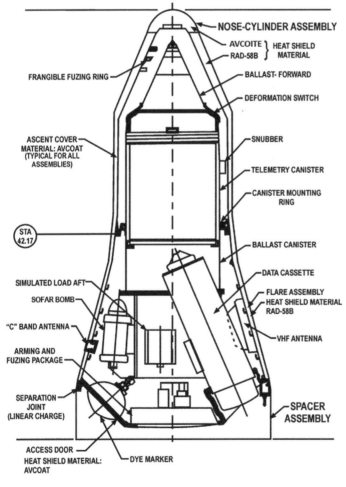

FIGURE 9.7 (above) Comparison of the size and shape of the Avco Mark 5 on the left and the Mark 4 on the right. The technician standing behind the Mark 5 gives a sense of scale. *Courtesy of Phil Fote.*

FIGURE 9.8 (left) Major components of the Mark 5 flight test vehicle. It is not clear if the spin fins were on the flight test vehicles. *Adapted from "The Development of the SM-80 Minuteman."*

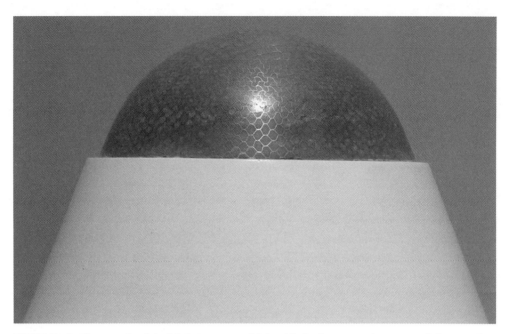

FIGURE 9.9 The nose tip of the Mark 5 reentry vehicle with the Avcoat coating removed showing the cells of Avcoite ceramic material that took the brunt of the reentry heating. *Photograph by the author, courtesy of the National Museum of Nuclear Science and History.*

FIGURE 9.10 Effect of reentry on heat shield material of a recovered Mark 5 reentry vehicle. Approximately 11/16 inch thick at this point on the reentry vehicle, reentry caused ablation of less than 1/8 inch. The fracture damage was caused by water impact and retrieval of the reentry vehicle. *Photograph by the author, courtesy of the National Museum of Nuclear Science and History.*

with Minuteman IB, at Ellsworth AFB. The Mark 11 series reentry vehicle had an operational requirement for a reduced radar cross-section during the exoatmospheric portion of its trajectory.[56]

The Mark 11 series, 11, 11A, B, and C, had a somewhat similar size and shape to the Mark 5 but was slightly longer and had a distinctively sharper nose tip. Avcoite-1 was again used in the nose forecone. For the Mark 11, the body of the vehicle was made of RaD-60, a molded silica phenolic using chopped silica fibers, which was machined to fit over an airframe made of 50 magnesium ribs that were covered with a spin-formed magnesium skin for both the cylindrical and flare sections (formed separately). The two assemblies were bonded with epoxy, the final machining completed, a radar-cross-section–reducing mesh applied, and a final layer of Avcoat 2 applied. The pointed tip was made of glass fiber resin–impregnated cloth formed on a mandrel and epoxied to the nose. It was used to provide protection to the nose radar cross-section–reducing material from boost-phase heating. Once the Mark 11 entered the atmosphere, the nose fairing, as well as the radar cross-section–reducing mesh, were removed by ablation. At this point in reentry, the vehicle was producing a highly ionized and readily detectable wake, which was unavoidable. While the Mark 5 tumbled at the start of reentry and thus provided a large radar return, the Mark 11 series was spin-stabilized. Two pitch-and-spin rockets were fired, at a time established by the arming and fusing–attitude control assembly, to properly orient the vehicle for reentry.

The Mark 11A, B, and C had a different fabrication process from the Mark 11. The new aluminum frame was heavier than the Mark 11 magnesium frame but stronger, a feature required for the nuclear hardening of the vehicle—a new operational requirement due to anticipated advances in the Soviet ABM system. The flare, cylindrical body, and nose cap frames were bolted together, and then the heat shield applied using oblique tape-wound Refrasil by a unique process developed by Avco. After curing the heat shield was machined to specification, and the radar cross-section–reducing material was applied and covered with a final layer of Avcoat 2. The aft fairing was specifically designed to reduce the radar cross-section.[57]

The Mark 11 deployed from the third stage with only a slight increase in velocity, so the third stage served almost like a radar beacon for Soviet ABMs. For the Mark 11, 11A, and 11B, Avco developed a retrorocket spacer that had 10 small thrusters, which fired in pairs to provide a random velocity to the third stage after release of the reentry vehicle. Before firing the retrorocket thrusters, a tumbler motor fired perpendicular to the centerline of the third stage to impart a rotation rate. This combination randomized the third-stage position relative to the reentry vehicle and reduced the problem of the third stage serving as a radar beacon.[58]

For the Mark 11C, the retrorocket spacer was replaced with a penetration aid spacer, which carried a number of Mark 1A chaff dispensers, each equipped with different-level impulse thrusters. This was in response to the low-frequency Soviet ABM radars. The chaff dispenser was connected to the Mark 11C via a lightweight spacer made of beryllium rather than aluminum, as this configuration was up against a weight limit due to the chaff system and beryllium was 30 percent lighter than aluminum. After Mark 11C release, the chaff dispensers were fired up and down the range-insensitive axis to generate a train of chaff clouds spaced far enough apart that the defensive systems would have to target each cloud.[59]

Virtually indistinguishable in outer appearance, the Mark 11 series was approximately 100 inches in height, with a cylindrical section 19 inches in diameter and a base diameter of 32 inches. All used the same ablative materials (Figure 9.11).

The Mark 11 R&D program included six flights on Atlas D missiles beginning on 28 August 1963 and ending 12 February 1964, with one successful flight; the failures were due to booster malfunctions. Minuteman IB flight tests began on 7 December 1962 and ended on 8 December 1967 after 41 flights with 6 failures. The Mark 11C penetration aid's capability was tested on the final six flights, which began on 28 April 1967.[60]

The Mark 11 was deployed on Minuteman IB. All four variants were deployed on Minuteman II. The weight of the Mark 11 was 200 to 250 pounds. The Mark 11A, B, and C were approximately 25 percent heavier than the Mark 11.[61] The deployment history from 1962 to 1975 is shown in Figure 9.12.[62]

★ ★ ★

The development of the second-generation ablative material reentry vehicle was the second key technological advance driving the Minuteman program. The reduction in payload capacity, compared to Titan II, was offset by the large number of Minuteman missiles deployed but made even more important the need for a successful solution to the last technological challenge, a lightweight and accurate inertial guidance system.

FIGURE 9.11 Comparison of Minuteman IA and IB reentry vehicles. While similar in size, the Mark 5 (left), FTM 423A, 7 January 1963, had a blunt nose compared to the Mark 11 (right), FTM 424, 14 March 1963. The small spin fins on the Mark 5 are indicated by the two white arrows. The single white arrow points out the telemetry antenna fairings located on the instrumentation wafer found on the flight development missiles at the Cape and on the combat training launch instrumentation wafer on the flight test missiles at Vandenberg AFB. (Arrows inserted by the author.) *Official USAF photograph, courtesy of the Air Force Space and Missile Museum, CCAFS.*

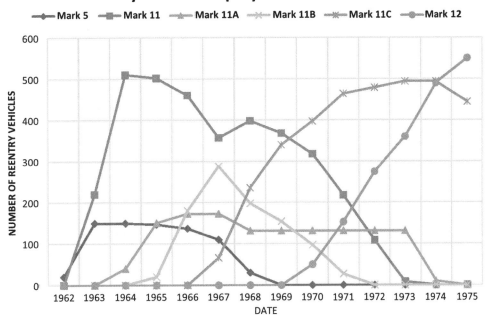

FIGURE 9.12 The deployment history of Minuteman Mark 5, Mark 11 series, and Mark 12 reentry vehicles, 1962 to 1975.

# 10

# GUIDANCE AND CONTROL

The first demonstration of an all-inertial guidance system in an IRBM took place on 19 December 1957 at the AMR, Florida. The successful flight of Thor 113 concluded with reentry vehicle impact nine nm from the target. Available records do not list the range for this flight; however, earlier successful flights had been over 1,000 nm. The flight was considered a success, despite the large miss distance.

The flight of Jupiter AM-6B on 17 July 1958, for 1,241 nm, with reentry vehicle impact 1 nm short and 1.5 nm cross-range, further confirmed the feasibility of an all-inertial guidance system.[1]

The Jupiter ST-90 guidance system was based on the Redstone ST-80 all-inertial system, which had first flown at a much shorter range on 12 September 1955.[2] While the range of the Redstone, Jupiter, and Thor missiles was considerably shorter than that proposed for Minuteman, in addition to flying a significantly different trajectory, the length of the powered flight portion was very close to that of Minuteman, indicating pure inertial guidance was feasible under ICBM flight conditions. Atlas all-inertial guidance tests did not begin until 11 June 1960.[3]

Now concern shifted to three major questions. Would the launch environment of Minuteman result in excessive flexure of the stable platform support structure? The relatively slow acceleration of liquid-propellant missiles had not proved a problem, but Minuteman accelerated much more rapidly. Could Minuteman's guidance system and computer be designed to run constantly while the missile was on strategic alert instead of being brought into operation just prior to launch as with Jupiter, Thor, Atlas, and Titan I? Titan II faced this same hurdle. Could this all be done with a weight of no more than 470 pounds?[4]

## Minuteman I NS10Q Series Missile Guidance System

The 8 April 1958 Proposed Preliminary Operational Concept for Minuteman listed as a requirement a CEP of 1 to 3 nm at a range of 5,500 to 6,500 nm. Missiles would be in hardened launchers, unattended, for a minimum of six months except for replacement of missile-borne target information. The guidance system would have single-target capability, with retargeting to be accomplished in two to three hours.[5]

STL staff had played a key role in writing the Minuteman specifications. Dr. J. Robert Burnett, manager of Minuteman Guidance and Control for STL, capitalized on lessons learned with the earlier ICBM programs' guidance-and-control system design and construction. He determined that with Minuteman, the guidance-and-control contractor would be responsible for not only the stable platform, computer, and the guidance-and-control section of the missile, but also all of the downstage (below the guidance system) components.[6]

The first the staff of the Autonetics engineering department heard of Minuteman was on 10 March 1958, when Joseph Boltinghouse, an engineer instrumental in the development of the G6 free-rotor, gas-bearing gyroscope, brought in that week's issue of *Time*, which contained an article describing the program. What caught the attention of the engineers was the statement in the article, "development at MIT of a new type guidance gyroscope that can be kept running continuously inside the underground silo for as long as two to three years."[7] Autonetics engineering staff were confident that their G6 series free-rotor, gas-bearing gyroscope would be the determining factor in favor of Autonetics.[8]

The Air Force request for Minuteman guidance-and-control proposals was issued in May 1958, with a due date of 12 June 1958. Preproposal work at Autonetics incorporated the G6 free-rotor, gas-bearing technology, VM1 induction velocity meters, a three-axis platform derived from the inner three axes of the N5B platform, a magnetic logic digital differential analyzer (DDA) computer, a digital flight control system, and a set of digital angular accelerometers for airframe angular acceleration measurement. The engineers proposed continuous operation as the only mode that would satisfy the instantaneous-response requirement. Experience gained from the Nike air defense missile system revealed a higher failure rate from repeated starting and stopping of the guidance system compared to continuous operation.[9] The final system was called the NS10: a three-axis platform, two free-rotor, gas-bearing G6 gyroscopes for platform stabilization, a single three-axis VM4 induction velocity meter, and an alignment block carrying mirrors for optical azimuth alignment, as well as two bubble levels for local vertical and the Autonetics magnetic logic computer, called QUAntized INtegrating Torquer (QUAINT).[10]

On 18 July 1958, Autonetics was awarded an R&D contract, AF04(647)-256, for development of the inertial guidance-and-control equipment for Minuteman.[11] The main reason Autonetics received the contract was its proven track record with the free-rotor, gas-bearing type gyroscope, an early model of which had been running continuously since May 1952. The fact that Autonetics had a large, broad-based engineering department with expertise in nearly all of the required disciplines for this new, expanded approach to the guidance-and-control system was probably also a factor.[12]

Development of the free-rotor, gas-bearing concept into an operational system had been funded by the Navy starting in 1954 for use in the Ships Inertial Navigation System for the fleet ballistic missile submarine program. Further development continued at Wright-Patterson AFB, as the Air Force was interested in application of the concept for aircraft, ballistic missiles, and cruise missiles. The research culminated in 1957 with the design and manufacture of the first production free-rotor, gas-bearing gyroscope, as described in US Patent 3,251,233, filed on 21 February 1957 and issued to Donal Duncan and Joseph Boltinghouse on 17 May 1966:

> This invention relates to gyroscopes, and more particularly to a gyroscope with a free rotor supported by a spherical gas bearing. A free rotor gyroscope is one which spins on a bearing which permits three degrees of angular freedom of the rotor about its support. A gas bearing is most readily suited for the support of such a free

rotor. This gas bearing is, in some instances, pressure fed from an external compressor, and, in others, it is self-lubricated by hydrodynamic action. The free rotor gyro has several advantages over conventional gyroscopes. Vibration and bearing noise can be made practically nonexistent with a gas spin bearing, and since little or no wear occurs, high reliability and long life are possible. In addition, a free rotor gyro permits outputs on two axes, namely those normal to the spin axis and normal to each other. Therefore, to obtain outputs for all three axes in space, only two such gyroscopes with their rotor spin axes normal to each other are required.[13]

Development elsewhere had been dropped due to manufacturing difficulties. The Autonetics innovation was to make the bearing small, the size of a large marble, with a gap between the rotor and the bearing on the order of a few ten-thousandths of an inch.[14] The result was the G6B4 two-axis, free-rotor, gas-bearing gyroscope, which was utilized in the Minuteman I, II, and III guidance systems.[15]

Autonetics's estimated cost for the system was $37 million, while the BSD estimate was $123 million. After one year of work, Autonetics estimated it would require $280 million if the mobile Minuteman system was to be incorporated.[16]

## Reliability

Unlike Thor, Jupiter, Atlas, or Titan I, whose guidance systems were brought into operation as part of the countdown, Minuteman's guidance-and-control equipment would ideally operate unattended for at least months, and hopefully years, at a time. In 1958 available missile guidance-and-control equipment routinely needed adjustment after a matter of days or weeks.

### *Electronic Components*

The high-reliability electronics design concept proposed by Autonetics was revolutionary, as vacuum tubes were replaced by the relatively new solid-state components and adjustable components such as potentiometers were eliminated. Extensive circuit analyses were conducted in the design phase, with every circuit required to pass a worst-case analysis of all its parameters. In the development program, part derating was used (derating refers to operation of the device at less than its rated maximum capability), and a thermal management mechanism for the guidance system was designed to reduce thermal stress during operation. Semiconductors underwent accelerated life testing combined with physics-of-failure analyses, manufacturing process development, and yield improvement. Burn-in and extensive environmental testing were also used to force improvement of reliability. It was anticipated that this intense oversight, previously unheard of in the industry, would achieve a 100-fold improvement in failure rates. By the late 1950s, the mean time between failure (MTBF) for individual components had already reached more than a year. The problem was the relationship between component and system reliability. A system containing 3,000 components, each with an MTBF of 100 years, would theoretically have a system MTBF of only two weeks.[17] As a consequence, Autonetics began an intensive program for reliability improvement, not only in its own development and manufacturing process, but those of its suppliers. These steps included batch control of material used, detailed operator instructions for the assembly line workers, and statistical sampling under continuous operational conditions as well as marginal operating conditions.

Suppliers had to submit details on how they were going to conduct their reliability improvement program. The companies that appeared most promising had their facilities inspected by Autonetics reliability engineers, and if they passed inspection, actual components were tested to evaluate their product's current reliability. When a supplier was finally selected, it had to agree to a set of 15 tasks, which comprised the framework for an unheard-of increase in reliability. Perhaps most onerous was Task 13: Serialization, "Each device manufactured for the test requirements of Tasks 4, 7, and 8 shall be serialized and the data will be suitably recorded for machine processing." Task 13 facilitated the tracking of components and batches of components for failure analysis. The practice of discarding faulty units was eliminated, and instead, components failing tests were analyzed for the mode of failure so that the manufacturer could adjust its processes.[18]

While companies balked at the initial increase in expense due to the strident manufacturing conditions, as the programs took hold, significant reduction in rejected parts mitigated the cost increases, with prices actually lowered for high-reliability parts. General Electric was able to reduce prices on its High Reliability foil tantalum capacitors an average of 30 percent after Autonetics notified it that the capacitors attained the ultimate reliability goal needed in the Minuteman program.[19]

For the computer component of the Minuteman guidance system, the results of these procedures were readily apparent. Prototype models of airborne computers usually operated for 10 hours before failure. The prototype Minuteman computer operated for 585 hours. The failure was not in the electronic components but a mechanical component of the computer.

## *Electromechanical Equipment*

### GYROSCOPES

Application of the high-reliability program to the inertial guidance system involved the selection of an improved model of the G6 series free-rotor, gas-bearing gyroscope, the G6B4, which had demonstrated the potential for extremely high reliability as well as accuracy. Since the G6B4 was a two-axis-of-freedom gyroscope, only two were necessary, simplifying the stable platform design, therefore increasing reliability.

### VELOCITY METERS

The three-axis VM4 integrating accelerometer (velocity meter) was expected to provide high reliability. The VM series integrating accelerometers were an Autonetics invention and much less complex than alternative designs. However, problems developed with the VM4A instruments in the final design when the magnet component supplier could not provide consistent materials.[20]

STABLE PLATFORM

The use of an internal gimbal system on a central core knuckle joint meant that slip ring assemblies, with their inherent lack of reliability, were eliminated. This gimbal system had been successfully used in the Jupiter guidance system. While considerable weight was saved, the drawback was limited rotational freedom in the yaw-axis, only plus or minus 20 degrees, with 70 degrees in roll and 90 degrees in pitch.[21] With the original NS10Q1 single-target capability of Minuteman IA or the restricted dual-target capability of the NS10Q2 of Minuteman IB, this was not a limiting factor (Figures 10.1 and 10.2).[22]

In order to simplify the guidance-and-control in-flight computations, and therefore reliability, the inertial platform was aligned on the launch azimuth in the launcher instead of acquiring the launch azimuth after liftoff. With the NS10Q2, the NS10Q1 modification used for Minuteman IB, the ability to select a second target was added, provided that the second azimuth was no more than 10 degrees off the primary target heading. Though some accuracy was lost with this method, dual targeting was more simply achieved via additional computer programming rather than a major modification of the stable platform.[23]

## *NS10Q Major Component Description*

The NS10 missile guidance system was housed in missile body Section 42, a frustrum with a base diameter of 37.5 inches, a top diameter of 32.5 inches, and a height of 31.5 inches. The housing provided the support structure between the top of the Stage III motor and the spacer supporting the reentry vehicle. Two half-toroid sections surrounded the stable platform. One toroid section comprised the D17B computer. The second housed the electronics for the stabilized platform, Stage III flight control, thrust termination, and stage separation ordnance control, as well as penetration aid dispensing (Minuteman IB) and reentry vehicle prearm and release (Figure 10.3).[24]

G6B4 FREE-ROTOR GAS-BEARING GYROSCOPE

The G6B4 was a two-axis displacement-type gyroscope fabricated entirely from beryllium except for the electrical components and rotor sleeves. The rotor, which was driven by means of an induction drive motor, was supported by a spherical, self-lubricating gas bearing. The gas bearing gave the rotor three degrees of angular freedom, permitting the definition of a rotor spin axis and two displacement axes. The free-rotor gyro design eliminated the problems of buoyancy balance and convection torques associated with liquid-floated gyroscopes. When the gyroscope was powered down, there was contact of the rotor with the bearing, so continuous operation was preferred.

The first Minuteman G6B4 free-rotor, gas-bearing gyroscopes were fabricated from 2024 aluminum alloy. Concern arose about certain performance characteristic results from high-speed sled tests that revealed excessive rotational distortion of the rotor. Partial correction of this distortion was achieved by elongating the cavities along the spin direction of the rotor. Additionally, higher-viscosity gas and higher fill pressure were implemented. While this minimized the effect of the rotor distortion, the higher-viscosity gas density

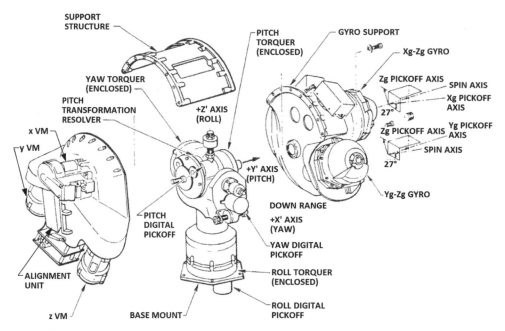

FIGURE 10.1 The exploded-view diagram of the NS10 stable platform identifying system components. The knuckle joint is in the center of the illustration. *Adapted from "Function of Guidance and Control System, Student Study Guide."*

FIGURE 10.2 Minuteman I NS10Q series stable platform. The two G6B4 gyroscopes are on the right, and two of the three velocity meters are visible on the left. The design of the velocity meters was such that they did not have to be orthogonal to each other. This is one of two known remaining examples of the NS10Q1series stable platform. *Photograph courtesy of Monte Watts.*

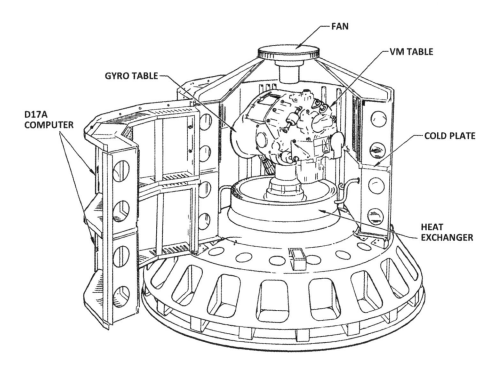

FIGURE 10.3 The major components of the Minuteman I NS10 guidance system. Half of the torus surrounding the stable platform was composed of the D17B computer. The other half held the flight control electronics. Enclosed in a hermetically sealed housing, the guidance system was kept between 66.5 and 68.5 degrees F using coolant circulated from the LER via the upper umbilical. *Adapted from "Function of Guidance and Control System, Student Study Guide."*

resulted in an increase in random drift, resulting in marginal performance (Figures 10.4 and 10.5).

In 1958 Brush Beryllium Company contacted Autonetics about the availability of hot-pressed beryllium blocks. Two engineers at Autonetics, John Slater and Joseph Boltinghouse, realized that the rotor distortion issue could be corrected by using a beryllium rotor. Converting the rotor to beryllium required that the support ball and shaft would also have to be fabricated from beryllium or another metal with a similar thermal expansion coefficient. The decision was to again use beryllium for both the ball and the support shaft. Finally, the gyroscope housing was converted to beryllium, which eliminated the need, when aluminum was used, for complex flexure assemblies between the motor and torquer assemblies, as well as enabling the motor and torquer to be cemented directly to the gyroscope housings.

The decision to convert to beryllium was made in early 1959, and basic design work was started. Two critical problems remained: how to plate the beryllium in the rotor cavity with hard nickel and how to chrome-plate the support ball. Two separate laboratories within North American Aviation were given different approaches to this problem. One of the approaches proved feasible just as prototype hardware had reached the point where plating would be necessary.[25]

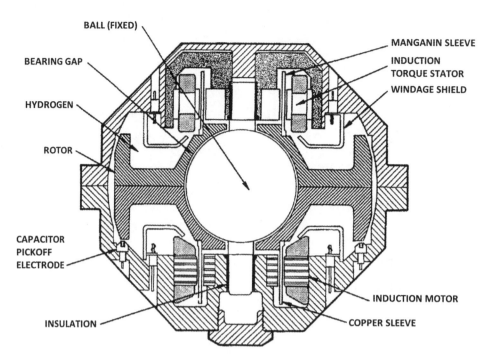

FIGURE 10.4 Diagram of an early model of the Autonetics G6B4 two-axis, free-rotor, gas-bearing gyroscope showing major components. *Adapted from "20 Years of Inertial Navigation."*

FIGURE 10.5 Cross-section of the operational G6B4 gyroscope. *Adapted from "A Brief History of Minuteman Guidance."*

With self-generated gas-bearing gyroscopes, the greatest point of wear and tear was at startup or shutdown. Autonetics's solution to minimizing this problem was to use a much higher than normal voltage to literally jolt and accelerate the rotor quickly to generate the required gas cushion. The G6B4 gyroscope was used in all three of the Minuteman variants' inertial guidance systems, with over 4,500 manufactured as of 1970. A large share of the success is due to the choice of beryllium as the major fabrication material.[26]

## VM4A VELOCITY METER

The production models of the NS10Q1 and NS10Q2 used three VM4A singly integrating velocity meters "scrunched" around an approximate boosted trajectory line.[27] They were an advanced version of the original VM1 design. While the design was simple enough, the difficulty with the VM4A was twofold. The first was that a delicate heat-treatment process was necessary to stabilize the magnetic component. The second was that in early 1962, Autonetics unknowingly received a shipment of commercial-grade manganin, a component in the velocity meter, instead of military grade. The higher iron content in the commercial product was not discovered until post-fabrication testing. The material was reordered, but the delay seriously affected production.

As the difficulties with the VM4A continued, the Air Force asked Autonetics to evaluate the American Bosch ARMA Corporation vibrating string accelerometer (VSA), which was being used with success on the Atlas E and F inertial platforms. Extensive testing revealed that the VSA was unable to withstand the Minuteman vibration environment. Another alternative was the Draper Instrument Laboratory's pendulous integrating gyro-accelerometer (PIGA). It was much more expensive to build, as well as not being available in time to keep the Minuteman I deployment schedule. The final proof was in-flight test results, and the VM4A proved sufficiently accurate to keep Wing I deployment on track. The production problems were ironed out by the end of February 1963.[28]

## D17B COMPUTER

The final Autonetics proposal offered a modified version of its VERsatile Digital ANalyzer (VERDAN) computer. The VERDAN was a complex computer that contained approximately 10,000 diodes and 1,500 transistors yet only needed approximately 350 watts of clean, three-phase 400-Hz power. It was also the first production computer with electronics mounted on both sides of the circuit boards. It weighed 82 pounds and occupied 1.4 cubic feet. It consisted of a 128 integrator DDA section supervised by a serial general-purpose computer and was used successfully in the Air Force Hound Dog cruise missile. After the award of the Minuteman contract, it became clear that the VERDAN system had significant rate and memory limitations in performing calculations with the dynamic ranges and complexities of the Minuteman I guidance task. Major modifications would be necessary, as the decision had been made, after the contract award, to have the guidance computer directly in the control loop rather than having the flight control hardware itself respond to the raw steering commands. This added requirement was beyond the capability of the VERDAN system.

STL was adamant that the VERDAN system could not be modified properly and offered its own design, called Blue Book. The Autonetics computer team evaluated the Blue

Book design and pointed out that any change in trajectory constants would force the need for major reprogramming. Claude King, the STL Minuteman guidance manager, gave the Autonetics team 24 hours to come up with an alternative. Led by Robert Knox, the chief scientist for Minuteman guidance at Autonetics, a team of computer designers including Marshall McMurran (computer tasks and programming) and Al Jennings (computer logic design) worked through the night to design a new general-purpose computer, the D17. The gas-bearing magnetic disk memory was a given in the design because of its many proven qualities. The major objective was to attain the required speed without incurring an unacceptable complexity of electronics. That hurdle was finally cleared by adopting two key design elements: (1) agreement to employ fully the "look ahead" logic (pipelining) that an engineer, Bob Booher, had employed in the FADAC, a field artillery digital computer designed by Autonetics for the Army; and (2) the incorporation of several short-loop registers on the magnetic disk to ease the programming difficulties. With these measures the D17 computer was born. While some complexity was added to the electronics and magnetic disk memory by these measures, the penalties were deemed acceptable in order to attain the required speed. With a serial computer and the non-random-access memory, a nightmarish burden was placed on the software programmers to attain the required speed of computation. But it was clear that the architecture of the D17 had growth potential, which would surely be needed downstream in the program. Later that morning, briefings were given to STL and the Air Force about the new D17 design, which was then accepted for incorporation into the NS10 guidance system.[29]

The D17B, the operational version of the computer, was a general-purpose airborne digital computer operating in a serial mode—the first ballistic missile computer to extensively use semiconductor components. The computer functions included solving guidance equations and generating missile steering commands, solving flight control equations and controlling staging and thrust termination, supplying prearm signals to the reentry vehicle, controlling ground alignment calibration, and periodic testing of the missile systems. Diode-resistor logic was used extensively, while the less reliable diode transistor logic was used only when necessary due to the lack of reliable transistors at that time.[30]

The D17B was 20 inches in height by 29 inches in diameter, occupying one half of the 12-sided right polygonal shell (toroid) 5 inches in depth and surrounding the stable platform in the NS10Q guidance compartment. The computer weighed approximately 62 pounds and had a magnetic disk memory rotating at 6,000 rpm with a capacity of 2,562 (24-bit) words. The computer contained approximately 14,711 parts, including 1,521 transistors, 6,282 diodes, 1,116 capacitors, and 5,094 resistors (at one point the Minuteman I missile guidance system production absorbed over 90 percent of all transistor production in the United States).[31] These components were mounted on 74 double-sided, copper-clad, engraved, gold-plated, glass fiber–printed circuit boards. All were coated with polyurethane for environmental protection. Early MTBF was approximately 1,200 hours and, by the end of 1961, reached 3,000 hours. The D17B eventually achieved an MTBF of 5.5 years. The record for continuous operation of an NS10 series guidance set was 2 minutes short of 63,990 hours (7.3 years) after having been in operation from 2 June 1965 to 20 September 1972 at Wing V.[32]

NS10Q ALIGNMENT

The NS10 series guidance system had neither the instrument precision nor the computer capacity to perform self-contained gyrocompassing (north reference) azimuth alignment without external reference. Coarse alignment was done at emplacement by rotating the missile on the missile suspension system receiver ring so that the guidance section optical alignment window was facing the autocollimator bench. Prior to this the autocollimator had been aligned to the launch azimuth using external references. Fine alignment was achieved through interaction between the guidance system optical alignment mechanism and the autocollimator (see Chapter 11).

ENVIRONMENTAL CONTROL

The MGS was enclosed in a welded housing that served to provide a foolproof seal from the LF environment and ensured that the equipment could only be accessed at the factory or a properly authorized depot. The nominal NS10 temperature was between 66.5 and 68.5 degrees F. Temperature control was via an umbilical from the guidance-and-control cooler equipment located in the LER. The coolant was circulated through cold plates located in the computer structure while the missile was on strategic alert and, once in flight, via heat exchangers and helium gas that was circulated by an axial fan located in the top of the hermetically sealed cover.[33]

FLIGHT CONTROL GROUP

The flight control group instruments included an angular accelerometer unit, three NCUs (one for each stage), an ordnance switching unit for stage separation, and missile cabling. The flight control group's function was to stabilize missile flight in order to follow the computer-commanded trajectory.

*Angular Accelerometer Unit*

The P68 angular accelerometer, located on the forward dome of Stage II, contained two inertial accelerometers for the pitch and yaw axes. The accelerometers provided bending rate information to the guidance computer to ensure that the missile airframe could withstand the bending modes generated from the trajectory at maximum dynamic pressure. At the end of Stage II flight, the angular accelerometers were no longer necessary, as there was no further comparable bending rate limitation in Stage III flight.[34]

*Nozzle Control Units*

Minuteman IA and IB had four exhaust nozzles on each stage. The NCUs were of the same design, differing primarily in size and weight. Each NCU had four hydraulically operated arms with hydraulic power provided by an auxiliary power unit. The Stage I nozzles could travel up to eight degrees on either side of neutral, while Stage II and III could swivel six and four degrees, respectively. Each NCU was thermally protected by a phenolic–Refrasil baseplate cover.[35]

The downstage (below the guidance system) flight control hardware was not normally energized. While the failure of the MGS during routine testing did not require removal of the entire missile, a failure of an NCU required the removal of the reentry vehicle and guidance section, followed by removal of the downstage for shipment to Hill AFB Depot for repair and replacement. Monitoring the NCUs was a routine function of the missile guidance computer.

*Airborne Power*

Airborne power was supplied by three 28-volt DC batteries: two SE-12, one each for Stage I and II, and one SE-13, which supplied power to the MGS as well as the Stage III NCU. The battery electrolyte was stored out of contact with the battery cell plates. Five seconds into the terminal countdown, electrolyte was released into the battery by computer command. After full internal power had come up (a few seconds), the upper umbilical was released and retracted, and the computer proceeded to go through the self-test, the test of all three NCUs and various guidance systems, before placing the system in the in-flight control mode just prior to Stage I ignition. Battery activation was an irrevocable action, which necessitated replacement of the batteries should the launch be aborted. Ground power through the upper and lower umbilicals enabled MGS monitoring and downstage flight control group operation during strategic alert (Table 10.1).[36]

In order to achieve strategic alert or return to it after maintenance, the MGS had to undergo an approximately 2.5-hour test of the velocity meter and missile. The missile would routinely remain on strategic alert for approximately 30 days until the next velocity meter calibration, at which point it would again be off alert for 2.5 hours. This shortcoming was addressed in the upgrade to the Autonetics Minuteman II NS17 Missile Guidance System.

## *Minuteman II NS17 Missile Guidance System*

In July 1962 Autonetics received a contract for R&D of the Minuteman II NS17 guidance system. Four months earlier, Boeing had received a letter contract to proceed with conceptual development of Improved Minuteman, WS-133B, which would have a 7,500-nm range, passive and semiactive penetration aids, and, most important for Autonetics, eight target options for each missile up to 20 degrees off the primary launch azimuth. The new guidance system would have improved response following a severe seismic event that could degrade the optical azimuth reference. Improved Minuteman was also to have greater accuracy for use against hardened targets, a necessity with the change from the massive retaliation strategy to that of counterforce. In order to accommodate these changes, a new guidance computer would have to be developed as well.[37]

The Minuteman I guidance-and-control concepts and mechanizations that were retained were:

1. The magnetic disk memory for the computer, providing an efficient source of alterable/programmable nonvolatile memory. The downside remained the serial access and the special skills and time required to "optimize" the software.
2. Preferred orientations for the stable platform and inertial instruments to optimize accuracy for a standard boost trajectory/orientation.

TABLE 10.1  NS10Q2 Operation during Flight, Minuteman IB Trajectory[a]

With the receipt of a valid launch command, the missile guidance computer sequenced through a guidance-and-control test, then, after successful completion:

1. After launch, a pitchover angle to 20 degrees was commanded after vertical velocity reached the programmed value. If the launch was to Target 2, a yaw steering command was issued at pitchover to the new target azimuth.
2. When the acceleration rate decreased from peak acceleration, Stage II ignition and Stage I separation occurred.
3. When Stage II approached burnout and a decrease from peak acceleration was sensed, the computer issued Stage III ignition and Stage II separation.
4. Reentry vehicle mechanical disconnect was issued when the computer determined that cutoff velocity would be reached in three to four major cycles (a major cycle was 45 revolutions of the disk memory, 0.45 second).
5. The computer entered fine countdown when cutoff velocity was one to two major cycles away and changed flight program solution rate to a shorter time.
6. Thrust termination was issued when velocity to be gained equaled zero. Thrust termination was achieved by explosively opening four vent ports on the side of Stage III, which extinguished the propellant.
7. Ten to twenty milliseconds after thrust termination, Prearm 1 was issued if a flight safety check was successfully passed. If the flight safety check was not passed, the computer issued electrical disconnect and terminated the flight program.
8. If the Prearm 1 signal was issued, 89 milliseconds later, Prearm 2 was issued.
9. Twenty milliseconds after Prearm 2, the computer issued electrical disconnect. Separation springs in the reentry vehicle mechanical disconnect system pushed the reentry vehicle away from the Stage III–guidance package.
10. Three to five seconds after electrical disconnect, Retrorocket 1 fired, giving the guidance-and-control section–Stage III motor an alternative ballistic path.
11. Within a programmed time interval, Retrorocket 2 fired, causing a programmed maneuver of the Stage III motor–guidance-and-control section. This provided greater separation between the reentry vehicle and the Stage III–guidance section trajectories.
12. Final arming of the reentry vehicle occurred upon reentry into the atmosphere.

a) Minuteman Weapon System Familiarization: Flight Control Group, Student Study Guide.

3. Primary platform alignment references were the two-axis level detector and an autocollimator optical transfer to a mirror on the alignment assembly.
4. Limited rotational freedom gimbal system allowed interconnection to the stable member by means of the gyro stabilized platform wire harness (no slip rings).
5. Two G6B4 gyroscopes for platform stabilization.
6. Actuators and downstage electronics for Stages I and III were essentially the same as Minuteman I except for electronics update.
7. The liquid injection system for Stage II used hydraulic power supply for pintle actuators.
8. Airborne electrical power supply by a battery mounted on Stage I (SE13G) and a battery mounted in the missile guidance system upstage body section (SE12G).
9. The missile ordnance system utilized dual-redundant hot bridge wire-initiated ordnance trains.
10. The downstage cabling system utilized ordnance initiated staging connectors and circular connectors for interconnect of boxes/components.
11. System designed with the goal of detecting (by pre-launch operations/testing) launch critical failures but with minimal increase in complexity/launch facility failure modes to accomplish this.
12. The launch/countdown sequence for MM II was similar to that for MM I; the flight profile was the same length as MM I.

The guidance-and-control operations changed from Minuteman I to Minuteman II were:

1. Rather than using the targeting van for guidance and control tape loading, punched mylar tapes were loaded using a portable tape reader, the C-164 Tape Transport Unit.
2. The guidance and control tape load included the operational ground/flight program and depot derived inertial measurement unit parameters unique to the particular inertial instrument and targeting data and codes for secure launch command processing. The Code Inserter Verify Facility was used to prepare Wing-developed tapes and to derive a computer memory security check number unique to the particular missile guidance system/launch facility involved. The same 24-bit number had to be derived by the missile guidance system at the launch facility and manually verified/matched by the code team prior to completion of the guidance and control startup process. Computer memory security check generation ensured data loaded into the guidance computer was complete and unaltered.
3. Each missile/missile guidance set could accommodate eight different targets with any particular target tape load. These targets could be changed to a new set by a targeting tape load at the launch facility; no reorientation of the missile or autocollimator was required to change to a new target.
4. Targeting flexibility was achieved by means of a platform-mounted GI-T1-B azimuth gyroscope which provided angular freedom (in azimuth) between the platform and inner body of the gyroscope. The inner body of the gyroscope

contained two level detectors and two opposed mirrors which provided the means for precision alignment and pendulous integrating gyro accelerometer (PIGA) calibration. The precision azimuth alignment was provided by the autocollimator which was always west-pointed, independent of the orientation of the plane of the trajectory (azimuth) to the target. In essence the mirror normals were always pointed East-West while the platform was moved under it to whatever orientation was appropriate for its assigned target. Accurate indexing of the platform with respect to the GI-T1-B body was provided by an optical electrical resolver which had holding positions every 316 seconds of arc throughout the 360 degrees of relative motion. A computer-controlled servo was used to change the optical electrical resolver nulls and then to hold onto the desired null.

5. All missiles were aligned in roll so that zero-degree missile roll was pointed North, independent of the direction to their assigned targets. This allowed optical access (through the window in the missile guidance body section) between the azimuth gyroscope and the west-pointed autocollimator. Since the stable platform was in a sealed box, it also had a window to accommodate this optical access.

6. Two types of target change methods could be used when commanding a particular missile/missile guidance set to change to a different target:

    a. Minimum Reaction Time (MRT) target change did not reorient the platform to the preferred orientation for the new target. Such a target change was accomplished without any delay. In this case the expected accuracy was degraded, more so as the "mis-orientation" increased. In order to limit the possible accuracy degradation due to MRT and to avoid gimbal angle freedom problems, the maximum allowable "mis-orientation" was approximately plus or minus 10 degrees. Degradation in accuracy estimates were maintained for MRT type launches as a function of the "mis-orientation angle."

    b. Optimum Accuracy Circular Error Probable (CEP) target change resulted in the platform being rotated to the preferred orientation of the new target. If MRT was commanded and the "reorientation" exceeded 10 degrees, the missile guidance system responded with a CEP target change. The missile off-alert time due to reorientation was somewhat variable with the size of the "reorientation" angle with typical times being on the order of 12 to 15 minutes.

7. The NS17 included a GI-T1-B azimuth gyroscope housed in the inner (rotatable) gimbal element. The azimuth gyroscope utilized a two-position mechanization to achieve a high accuracy secondary azimuth alignment mode for conditions where the autocollimator might have failed or might have been moved as a result of a seismic disturbance. This two-position mode utilized periodic 180-degree rotations of the GI-T1-B, alternately pointing the input axis nominally east and then west. This required two opposed mirror surfaces to allow two-position mechanization while referencing to the autocollimator light beam.[38]

8. In order to improve accuracy, compensation schemes were incorporated into

the operational flight program or targeting program to compensate for the predictable effects produced by the more significant acceleration dependent error terms, both for the gyroscopes and accelerometers. This was made possible by the increases in throughput speed and memory capacity for the D37B and C computers. These error terms included two acceleration and two acceleration-squared dependent drift rate coefficients for the G6B4 gyroscopes, and an acceleration dependent scale factor coefficient for the three 16 PIGA-G accelerometers.

9. The early Stage I flight sequence included "a roll to the target plane" subsequent to pitch over to the desired pitch angle. This accommodated limited gimbal freedom, the fixed orientation of the missile and all-azimuth target coverage.

10. The missile guidance set electronics included provisions to recover from in-flight hostile environment effects. A PIGA pulse "double counting" scheme was used to (approximately) compensate for the velocity increments which were lost when the computer was turned off in the "semi-Somnus" circumvention/recovery mode due to the detection of increased radiation levels. Due to inaccuracies in the "double count" scheme and other uncompensated effects of the hostile environments, some degradation in guidance and control accuracy was expected if the flight was accomplished in a hostile environment. The expected accuracy was a function of the particular hostile environment scenario being considered.

## *NS17A1 Equipment Changes*

The packaging of the NS17 MGS changed from that of the NS10Q sealed container to a set of four major "black boxes" consisting of the D37C computer, P92 amplifier for downstage flight control and ordnance electronics with a code device called the P-plug (used for launch command processing, with each MGS having its uniquely coded P-plug), an MGS control box—which was the inertial guidance electronics—and the gyro-stabilized platform. This change was intended to add flexibility and efficiency to the depot repair/retest process as well as improved efficiency in manufacturing.

STABLE PLATFORM

The major design modification for the NS17 stable platform was the change from the inside-out knuckle joint design to the more traditional outside-in gimbal ring design, but slip rings were again avoided. The need to accommodate eight possible launch azimuths required greater gimbal-axis angular freedom, and this was accomplished most easily with the change in gimbal design. Electrical connections were achieved through flexible wiring harnesses with gimbal stops restricting movement to prevent damage to the wiring. The gimbal stops also defined the limit for the MRT.

The change in platform design necessitated a move from aluminum to beryllium as the platform structural material to provide greater rigidity for increased accuracy while simultaneously mitigating weight increases due to changes in inertial instrumentation located on the platform. The gimbal rings had a hollow box cross-section machined from blocks of beryllium. The structure weight was approximately 6 pounds per gimbal ring and approximately 8.5 pounds for the completed instrument mount, a total of approximately 20.5 pounds versus approximately 29 pounds if made of aluminum.

The new stable platform was now a three-axis, two-gimbal platform with a gimbal axis order of roll, yaw, and pitch from outermost to innermost gimbal. The stable platform supported two G6B4 gyroscopes, three 16 PIGA-Gs, the GI-T1-B azimuth gyroscope assembly, platform electronics, and a vane-axial cooling fan. The total weight of the new platform with inertial instruments was approximately 70 pounds (Figure 10.6).[39]

## GI-T1-B GYROSCOPE

The addition of GI-T1-B, a high-precision, single-axis gyroscope, provided the ability to remotely command and change platform azimuth orientations while maintaining accurate referencing to the autocollimator. It also provided a method to maintain azimuth alignment if the autocollimator reference was disturbed as a result of a seismic event. This issue became apparent as a result of the effect of the earthquake in Alaska on 27 March 1964, which caused the NS10Q guidance systems to lose acquisition of the autocollimator. Now all missiles were installed with the same roll orientation, independent of the orientation of the target plane. This simplified targeting because now the autocollimator simply had to be placed accurately at the east–west position; missile rotation in the launcher for specific targets was no longer necessary.

## 16 PIGA-G ACCELEROMETER

Autonetics continued to work on the velocity meter design with modifications that evolved into the VM4B, but improved accuracy and reliability were elusive. Robert Knox, the chief scientist for Minuteman II guidance development, finally called a "come to God" meeting with Al Grant, the deputy to the head of the Navigation Systems Division at Autonetics and the inertial instrument engineers. The staff did not want to give up on the VM4B but could not commit to the increased performance required by Minuteman II specifications. The problem remained stability of the permanent magnet. Finally, Knox announced that they would have to seek Air Force approval to switch to the Draper Instrument Laboratory PIGA instrument. This decision was crucial to the program. It offered the potential of meeting, even exceeding, the accuracy requirements, but it did incur serious penalties of reliability, weight, cost, and time.[40]

The PIGA concept had originated with the German V-2 guidance system. Draper Instrument Laboratory had been working on PIGA designs for many years prior to this decision. The first major refinement by Draper was to "float" the pendulous gyroscope to virtually eliminate gimbal bearing friction and its effect on accuracy. The result was an accelerometer that was difficult and expensive to build but accurate enough for use in the form of the 16 PIGA-F (16 refers to the 1.6-inch diameter of the end housing of the device) in the Polaris A-3 missile.[41]

The first version of the 16 PIGA destined for Minuteman utilized ball bearings in the gyroscope. A continuing problem that compromised accuracy at the level of instrument sensitivity required for Minuteman II was known as the jog anomaly. In the laboratory this phenomenon repeated with regularity over many hours, which was clearly not acceptable for

FIGURE 10.6 Stable platform for the Minuteman II NS17 and Minuteman III NS20 missile guidance systems. The design now changed to an external gimbal system with the velocity meters replaced by PIGAs. The dome-shaped object in the center of the picture is one of two G6B4 gyroscope enclosures. Above it and to the right is the top of one of three PIGAs. Directly above the G6B4 is one of many ceramic-printed circuit boards. *Adapted from Air Force Association of Missileers Collection.*

missile guidance. Detection was reasonably simple. The instrument trace made by a recording milliwatt meter in the wheel motor circuit showed an abrupt jog, which was immediately followed by a gradual return to the normal trace. The source of the jog was traced to the asymmetrical buildup of lubricant in the ball bearing retaining ring. Workers at AC Spark Plug were able to measure the lubricant thickness down to a level of 10 micro-inches and confirm this was the case. After over 70 attempts to design a retaining ring that would prevent this phenomenon, Michele Sapuppo, the program manager, had to admit defeat.

Coincident with this was the desire of the Air Force to "simply" replace the ball bearing support with a gas bearing. Likely based on the success of the G6B4 gyroscope, the Air Force finally won over Sapuppo, and the conversion was begun in 1962, resulting in the 16 PIGA-G. Sapuppo recalls that what appeared to be straightforward was not simply a matter of replacing the ball bearing's support with a gas bearing. This single change in design involved a multitude of other changes, one of which was a change to a different flotation fluid, which involved considerable effort. A second modification was the use of electromagnetic suspension to support the residual unfloated mass of the float assembly.[42]

Once the Draper design was successful, the matter of production of the device became apparent. Draper was known for exacting design specifications that proved difficult to transform into production drawings and successful fabrication. Bendix, AC Spark Plug, and Honeywell won contracts to build the 16 PIGA-G. While difficult and expensive to build, it eventually reached MTBFs of 10,000 to 15,000 hours.[43]

## D37 SERIES COMPUTERS

The design of the D37 series computer incorporated the first extensive use of integrated circuitry by the Air Force. This decision to use a nascent technology was not without its risks in terms of reliability and ease of manufacture. On the plus side, it allowed for a substantial increase in computer capability with a significant decrease in weight. It was now feasible to store eight target sets in the computer and have the missile retargeted from the LCC. Additionally, the new design enabled programming the computer to assume a greater role in the day-to-day monitoring and functioning of the missile and ground facility systems. The equivalent of 16 drawers of ground equipment containing over 10,000 parts were eliminated from the LER as the D37 assumed their function. The application of microelectronics reduced the missile computer electronic parts count from 14,672 to 4,507, and in the process, the volume of the computer decreased from 1.6 to 0.43 cubic feet, while the weight decreased from 61.5 to 38.3 pounds.[44] The rotating magnetic disk memory from the NS10 was retained, but capacity was significantly increased with denser magnetic media. Through the use of both sides of the disk, the read-only memory was increased from 2,562 to 7,222 24-bit words. As in the D17, a 24-bit word could be divided into two 12-bit words and processed simultaneously to treat flight control and other nonglobal parameters (Figure 10.7).

During D37 development the Air Force became concerned with the possibility of in-flight radiation events and directed Autonetics to develop radiation protection. The D37B computer was the soft version, and the D37C was the hardened version. The principal differences were that the D37C incorporated an ionizing radiation detector system that temporarily disabled the write amplifiers so no changes could be made to the memory. The

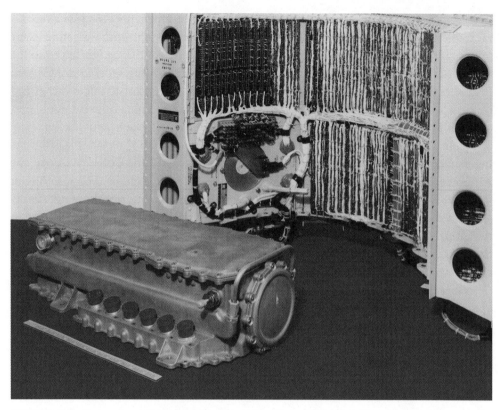

FIGURE 10.7 Comparison of the size of the Minuteman II and III D37 (left) and Minuteman I D17B (right) computers. A 12-inch ruler is in front of the D37. The disk drive is beneath the circular cover at the end of the D37 computer. *Adapted from Association of Air Force Missileers Collection.*

computer turned off the secondary power supplies for one or more integral revolutions of the disk. When ionizing radiation was no longer present above the threshold, the computer turned back on and picked up where it left off. Extrapolation of variables was then made for the time the computer had been asleep. Some degradation of accuracy occurred because of the small errors and extrapolation, but this was not considered a serious problem.[45]

Parts and materials that gave unacceptable performance due to radiation events were replaced with less susceptible material. All semiconductor devices were screened to verify acceptable performance in such an environment. Shielding made of high Z (originally zircalloy and then changed to 0.025-inch tantalum sheeting) material was added to protect electronics from damage, and missile cabling was encased in a convoluted metal shield to minimize transient current on the conductors due to electromagnetic pulse (EMP).[46] The D37B was initially deployed on the first 60 Minuteman II missiles; the D37C was deployed on the remaining 490 missiles and eventually replaced the remaining D37Bs as well.[47]

NOZZLE CONTROL UNITS

The reduction in weight of the NS17A1 MGS, as well as a redesign of the downstage flight control circuitry, meant that the NCU electronics could be moved to the MGS compartment (P92 amplifier). This change meant that maintenance on the NCU electronics, which previously had required removal of the missile, could be done by removing the reentry vehicle and then the guidance section. The Stage I and III NCU hydraulics and actuators remained the same as Minuteman I, while the Stage II NCU was smaller due to the new thrust vectoring control system (see Chapter 8).[48]

## *Elusive Reliability*

There had been relatively little time to establish integrated circuit reliability. The consensus within the industry was that since integrated circuits were fabricated in a manner similar to the silicon transistors already in use, the testing protocols from the Minuteman I high-reliability component program would be applicable to the new devices. For Minuteman II the high reliability goals had to be achieved in a much shorter time. Additionally, most of the components for the NS10 guidance system had been single-sourced, while for the NS17, there would be multiple contractors for many of the components. Autonetics developed the Component Quality Assurance Program (CQAP) with each distributor. There were two major areas in the CQAP effort. The supplier conducted a high stress test program for detection and isolation of failure modes, while the physical analyses on the subcomponent level would be done by Autonetics to provide identification of the mechanisms of failures.[49]

Long-term reliability tests conducted by Westinghouse and Texas Instruments at elevated temperatures had accumulated over 10 million device-hours with no new modes of failure identified. These encouraging results indicated that failure rates of 0.0002 to 0.001 percent per thousand hours could be achieved under the Minuteman operating conditions. Data published in 1963 indicated that integrated circuits would have failure rates closely approximating those of individual silicon transistors.[50]

The final report on the NS17 guidance system integrated circuit failures, issued in May 1970, summarized the major failure modes as being excessive surface leakage of current, cracked die, oxide dielectric defects, foreign particles, and die-to-header bond failures.[51] The "purple plague" problem, the development of an $AuAl_2$ compound where gold wires are bonded to aluminum metallization regions in silicon transistors and integrated circuits, was also identified, and solutions were developed by the manufacturers.[52]

The LF environment reliability goal for the NS17 MGS was an MTBF of 7,722 hours. There were 11 R&D flights, all successful, but they were not indicative of the extended operation times of the operational environment. With installation of the first Minuteman II at Grand Forks on 31 October 1965, reliability problems began to appear. The plug rate was 4.5 to 1 (bringing a flight of 10 missiles to strategic alert took installation of a total of 45 guidance sets instead of the assumed 10). By the time the last flight at Grand Forks was brought on alert on 22 November 1966, the plug rate was down to 1.1 to 1, but the MTBF was approximately 1,200 hours, or 50 days.

In-flight effectiveness was high; the R&D program had 15 successful launches and flights to impact from the Eastern Test Range (ETR) and 10 from the WTR (see Chapters 12 and 13). Thus, the weapon system effectiveness was not a matter of accuracy and in-flight performance but rather an inability to provide an adequate supply of long-term reliable, operational NS17 MGSs. In addition to the deployment of 150 Minuteman II missiles to

Wing VI at Grand Forks and the 50 Minuteman II missiles at 564 SMS at Wing I, the Force Modernization Program, replacing Minuteman IA and IB with Minuteman II, was to begin in mid-1966, so the need for serviceable guidance-and-control sets was going to increase markedly.

By February 1966 178 NS17 MGSs had been received at Wing VI, and of these, 127 had been rejected and returned for repair or replacement. Even the assembly and checkout phase of missile preparation at Plant 77 at Hill AFB was experiencing a high rejection rate, so it was not just the LF environment that was causing the problem.[53]

The limiting factor to fielding Minuteman II was the ability to rapidly repair and replace the failed guidance systems. The cause was the lack of test equipment at the Aerospace Guidance and Metrology Center Depot, Newark Air Force Station, Ohio. Additionally, Autonetics test stations became overtaxed by repair activities, which impacted production deliveries. The pressure to provide rapid turnaround of field returns delayed work on failure analysis and corrective actions, further compounding the problem. Finally, the lack of serviceable NS17 MGSs forced SAC to remove LFs from strategic alert status as Wing IV Force Modernization guidance sections were diverted to Wing VI in order to bring it to full strategic alert on schedule. In addition, SAC removed some LFs from operational status to allow Autonetics and the Newark Air Force Station Depot to get the recovery proceeding well enough to allow the production of guidance systems on a planned, rather than reaction, basis.[54] The number rose from 40 missiles off alert on 31 December 1966 to 72 on 30 April 1967 and then dropped to 61 on 30 June 1967.[55]

By the end of 1966, the NS17 MGS was experiencing an MTBF of approximately 2,000 hours, and the preoperational plug rate was 1.5 to 1. It became clear that the remaining problems had to be identified and solved quickly. There were three approaches to finding a solution. One was to get the New Jersey Air Force Station Depot output accelerated. The second was the start, in January 1967, of the Minuteman II Guidance and Control Recovery Program at Autonetics. The third, which also began in January 1967, was the authorization by the Air Force for Boeing to conduct an investigation of the LF environment that might be contributing to the guidance system failures. This program was known as Determination of Impact of Autonetics Boundary on Logistics and Operation (DIABLO). Four sites were selected, two of which experienced high failure rates—two had low failure rates. Completed in September 1967, the final report found no correlation between the electronic environment in the selected LFs and the failure rate of the guidance systems.[56]

Many reliability improvements had been developed, and some had been incorporated into the guidance system design prior to the start of the recovery program. A block change point was also established and served as a culmination point for the many studies and resulting design changes that could not be readily introduced piecemeal in the assembly lines. Approximately 200 of these NS17A1s were produced.

By January 1968 guidance section availability had improved significantly from that of a year before, with spare systems readily available, and MTBFs were 1.5 times as long as they had been a year earlier. Available data for guidance section failures through November 1967 showed 778 failures in LFs were attributed to electromechanical parts and 376 to electronic parts. The majority of the electromechanical failures were due to the new 16 PIGA-G and

GI-T1-B instruments that fell short of their expected reliability. The electronic failures were viewed as an example of failure to fully appreciate the difficulties with integrated-circuit technologies (Figure 10.8).

## *Minuteman II Guidance and Control Improvements after Initial Deployment*

The WS-133B Operational Ground/Flight Program underwent 13 changes from 1965 to 1971, with the final version incorporating the cancel launch in progress mode—the ability to cancel or delay a launch—new biasing and alignment routines for the inertial components, and, for the first time, provision of inertial instrument data to the LCC.

The WS-133A-M Force Modernization program, the upgrade of Wing I through V LCCs and LFs to accommodate Minuteman II missiles, had eight Operational Ground and Flight Program updates, which incorporated similar changes to those in the WS-133B system. Unlike the WS-133B Minuteman II system, which was replaced by the Minuteman III, the Force Modernization missiles remained in their older facilities until Minuteman II deactivation in 1991. The more significant changes included the following:

### SOFTWARE STATUS AUTHENTICATION SYSTEM

To provide greater security against unauthorized activities, this modification to the LCC hardware and software and the operational ground program incorporated the ability to authenticate critical status messages between the LFs and LCCs. This modification was incorporated at Wings I, III, and IV in the 1974-to-1975 timeframe via the Software Status Authentication System (SSAS) 1 Operational Ground Program, with the SSAS-1 upgraded to the SSAS-11 in 1975 to preclude a software glitch whereby the azimuth gyroscope alignment reference could be lost upon exit from inertial measurement unit calibration.

### IMPROVED LAUNCH CONTROL SYSTEM

The Improved Launch Control System (ILCS) modification upgraded all LF electronics equipment to be identical to the new Minuteman III CDB (see below) configuration. The principal design change was deletion of the LF autocollimator, which was replaced by the perturbation self-alignment technique (PSAT) calibration. The PSAT utilized two-position gyrocompassing for azimuth alignment of the platform. This modification was incorporated in Wings I and IV starting in 1977 and concluding in the early 1980s. Two complete electronic racks in the LF equipment room were eliminated. Sensitive command message processing, programmer group, and coupler racks were combined into one through the use of integrated circuits. The result was a tremendous increase in reliability and internal fault detection, which eliminated manual tests with their checkout equipment. Wing II stayed in the SSAS configuration.

### ACCURACY, RELIABILITY, SUPPORTABILITY IMPROVEMENT PROGRAM

As the NS17 MGSs reached 100,000 hours total operating time in the late 1970s, the electronics hardware developed signs of aging, which raised concerns regarding flight reliability. This problem also began to affect Minuteman III NS20 spares, since many of the parts

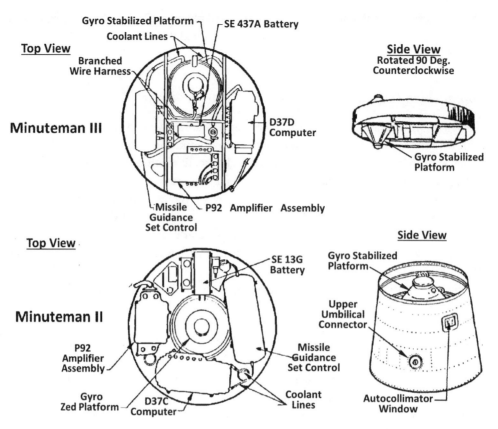

FIGURE 10.8 The Minuteman II missile guidance system was configured to be housed in the same missile section as that of Minuteman I, which had a height of 31.5 inches. Changes in design and advances in technology permitted the Minuteman III missile guidance system to be housed in a section 13.25 inches high. *Adapted from "Minuteman Familiarization."*

were common to both systems and some were no longer easily obtained. The LF reliability of the Accuracy, Reliability, Supportability Improvement Program (ARSIP) electronics eventually exceeded goals, with the MTBF of the new electronics over twice the MTBF with the pre-ARSIP equipment. The seven flight tests of the ARSIP upgrade were highly successful.

### Minuteman III NS20 Missile Guidance Set

The idea of multiple warheads for use on Minuteman had been first discussed in March 1964 by Dr. Alexander Flax, the assistant secretary of the Air Force for R&D. The Mark 12 reentry vehicle then under development would be used in a cluster mode as a penetration aid, similar to the Navy's Poseidon system. In July 1964 the decision was made to sacrifice the yield of the warhead in the Mark 12 in order to meet weight specifications for the cluster idea.[57]

On 15 July 1965, the Air Force issued a contract, AF04(694)-791, to Boeing for the research, development, testing, and evaluation of improvements to Minuteman II, as the need of an entirely new missile had not yet been agreed upon within the DoD. The major improvements proposed for the Minuteman II at this point were a PBCS for a MIRV concept and a trajectory accuracy prediction system.[58] The MIRV concept had been considered for use with an improved Minuteman II, but a delay in the development of the proposed reentry vehicle, the Mark 12, resulted in further refinement of the Mark 11 series reentry vehicles for Minuteman II and the postponement of the MIRV concept. By January 1965 the R&D effort had produced a satisfactory PBCS. Now what was needed was the functional Mark 12 reentry vehicle, and this proved to be the limiting factor, as early flight testing was disappointing (see Chapter 13).[59]

On 8 November 1965, Secretary of Defense McNamara, in a memo to Secretary of the Air Force Harold Brown, recommended adoption of the Mark 12 reentry system with its Mark 12 reentry vehicle, which logically meant a new name for the improved missile:

> The Minuteman missile equipped with the multiple reentry vehicle and other changes associated with that warhead is so much more effective and carries much more destructive power that, I believe, we should give serious consideration to giving it a new name.[60]

On 8 December 1965, Secretary McNamara approved development of Minuteman III, the decision having been made to further advance the basic Minuteman concept as a new airframe rather than make modifications to Minuteman II.[61]

On 28 March 1966, the Air Force issued System Management Directive 6-61-133B (10) defining the Minuteman III weapon system. It designated the system the LGM-30G.[62] The Stage III diameter was to be increased from 37 to 52 inches to accommodate the new PBCS payload—the Mark 12 reentry system composed of three Mark 12 reentry vehicles—as well as increase overall range with the larger Stage III motor. Ground equipment changes were to be made as well. A month later, Autonetics received the contract for the R&D of the Minuteman III guidance system.[63]

The PBCS would provide the capability to individually target the three higher-$\beta$ Mark 12 reentry vehicles (350 pounds each). The higher-$\beta$ reentry vehicles could reenter at a steeper angle, thereby spending less time in the atmosphere, resulting in less exposure to high-altitude winds and other atmospheric dispersive forces. Using the PBCS, the guidance system could disperse the three Mark 12 reentry vehicles in an elliptical area 50 nm across and extending 200 nm downrange.

The CEP estimates approved by the JCS for use in SIOP Revision G, 1 January through 30 June 1970, for Minuteman IB with the Mark 11 and Minuteman II with the Mark 11C were 1.6 and 1 nm, respectively. The CEP estimates for Minuteman III's first, second, and third Mark 12 reentry vehicles were 0.2, 0.3, and 0.4 nm, respectively, within that ellipse. The increased accuracy due to the PBCS, combined with the lighter warheads compared to the Mark 11C on Minuteman II, made the three-reentry-vehicle MIRV concept viable. The MIRV concept permitted an increase in target coverage without the cost of an increase in the number of Minuteman LFs, as the existing LFs would be retrofitted to accommodate the longer Minuteman III airframe.[64]

## NS20 MISSILE GUIDANCE SET

The NS20 was a modification of the Minuteman II MGS and associated black boxes. The major modifications were:

1. Expansion of the P92 flight control functions to accommodate the post-boost mission. The post boost propulsion system (PBPS), also referred to as the propulsion system rocket engine (PSRE), was designed and manufactured by Bell Aerospace.[65]
2. Increase in computer memory and throughput to handle the additional calculations of the MIRV trajectories; the computer designation was changed to D37D.
3. Further refinement of the nuclear hardness and survivability features.
4. Addition of precision azimuth alignment by the gyrocompass to potentially replace the need for the autocollimator.

Autonetics was also responsible for the Stage III P116 LITVC and hot gas roll system, which would be similar to those used with the Minuteman II Stage II motors. The significant differences were the use of strontium perchlorate instead of Freon, and the valve action was electromechanical rather than hydraulic. Through the use of new software, the angular accelerometer assembly that had been part of Stage II for Minuteman I and II was removed, which further simplified maintenance as well as increasing range.

The Minuteman III MGS and flight control requirements that changed from Minuteman II included:

1. An increase in accuracy.
2. Additional nuclear hardness and survivability design changes.
3. Flight reliability requirements to include the post boost mission.
4. Guidance reliability requirement was initially reduced to 4,000 hours MTBF, which applied only to the early deployed NS20 units.

An increase in accuracy was considered to be fairly straightforward as additional refinement of modeling compensation for inertial instrument errors was incorporated into the software. The nuclear hardness and survivability design changes involved changing from junction-isolated to dielectrically isolated integrated circuits. The major in-flight reliability concern was that of possible shock and vibration effects on the NS20 from operation of the PBPS/PSRE axial and attitude control systems. Part of the program for increased reliability was modification to the PIGA wheel bearing assembly. LF reliability was somewhat optimistically thought to have been solved with the solutions to the NS17 reliability issues, but this problem was to also haunt the initial NS20 deployment. As a result of the NS20 problems, the Air Force established a second source program, which was won by Honeywell but later phased out.

## NS20 DEVELOPMENT PROGRAM

The NS20 MGS development program was conducted in four stages, Blocks I through IV. Block I involved minor changes from the NS17 design and nonflight hardware. Block II systems were flightworthy or new build hardware and software but lacked many of the reliability improvements being developed for the Minuteman II NS17 production instruments. Block III hardware and software included the current Minuteman II reliability improvements as well as those for Minuteman III. Block IV was essentially production systems and included the transition to the newer integrated circuits. The transition from Block III to IV turned out to be the most troublesome.

### Stable Platform

The changes to the stable platform were relatively minor and involved increasing angular freedom by removing gimbal stops and other small mechanical modifications to accommodate the changes in targeting. The gimbal access–no slip ring configuration of the NS17 was retained. Provisions were also made for the new gyrocompass azimuth self-alignment protocol. This was a four-position calibration scheme called self-alignment technique calibration (SATCAL). This calibration mode was performed during the initial startup sequence and subsequently approximately every 90 days.[66]

### Azimuth Alignment

The Minuteman III autocollimator viewing window vertical position was offset from that of Minuteman I and II, so a periscope device was used to transfer the light beam from the autocollimator to the sighting window on the guidance section. The autocollimator system was retained until the SATCAL system had been demonstrated at actual LFs and was then removed.

### Nozzle Control Unit

The Aerojet Stage III design replaced the four nozzles of Stage III with a single nozzle and liquid-injection thrust vectoring. Unlike the downstage NCUs, Stage III uses electrical rather than hydraulic control of the valves, resulting in a lighter-weight system. Roll control is via a hot gas generator similar to that of Stage II (see Chapter 8).

### 16 PIGA-91

Extensive work was done to improve the MTBF of the 16 PIGA-G accelerometers, with modification to the 16 PIGA-91 the final result. Additional work was done to improve wheel bearing reliability in a design that was called the 16 PIGA-111, but it was eventually abandoned as unsuccessful.

### D37D Computer

There is little information on the D37D computer other than software changes to accommodate the increased calculation load from the PBPS/PSRE system. The computer was an upgrade of the D37C. A major change for the D37D was the use of double density for the disk memory, which doubled the memory capacity in light of the added complexity of the ground and flight programs as well as deployment versus simple release of the next reentry vehicle (see Hybrid Explicit program discussion later in this chapter).

The Air Force's decision to evaluate a second source for the "build to print" of the NS20 included a new computer. Honeywell won the competition with its proposal to design and develop the Improved Digital Computer Unit to replace the D37D. The new computer would incorporate random access memory to speed throughput and improve the efficiency of the software development cycle. The Honeywell system was deployed in limited numbers at Wing III and later replaced with the Autonetics NS20A after the development of the Hybrid Explicit software (Figure 10.9).[67]

POST-BOOST PROPULSION SYSTEM

The system releases each reentry vehicle with no residual thrust and then backs away and makes the necessary maneuvers for the release of the next reentry vehicle, as directed by the targeting program (see Chapter 8).

RELIABILITY

Although the Minuteman II flight mission was completed at thrust termination, telemetry data accumulated from over 50 flights indicated that the MGS was able to withstand the shock of Stage III separation as well as release of the reentry vehicle. It was, therefore, surprising to find in-flight reliability problems in the post-boost phase of the Minuteman III guidance-and-control system. The shock resulting from the pyrotechnic separation of the PBPS from Stage III was responsible for the failures. Concurrent studies by Raytheon for the Apollo Program showed that tiny particles within integrated circuits, small signal devices, and electronic assemblies were typically bound by electrostatic forces that could only be overcome by static discharge (unlikely in assembled hardware) or very sharp accelerations (shock). The fixes to achieve acceptable in-flight reliability were a combination of device-and-assembly design changes as well as process-and-screen testing. The NS20 MGS branched wiring harnesses had 28-gauge (0.0126-inch) wires. Although the wire harnesses were covered with braided shields, the termination areas were found to be susceptible to vibration, shock, and particle contamination. Replacing the 28-gauge with 22-gauge (0.0254-inch) wire helped improve reliability. A large part of the problem was that as the devices were assembled, screening tests became less and less effective. Extensive, expensive, and time-consuming efforts finally resolved many of the problems.

When installation of the Minuteman III missiles began at Minot AFB on 19 June 1970, the plug rate per flight was 1.5 to 1 and varied between 1 and 1.8 to 1 over the 18-month installation period. The predominant failure modes were with the PIGA and GI-T1-B instruments. The missile guidance system MTBF went from approximately 7,000 hours in 1970 to a high of 17,000 hours in 1976 and then averaged approximately 12,000 hours through 1994, the end of the available information.[68]

## Improvements after Initial Deployment

COMMAND DATA BUFFER PROGRAM

The CDB program was implemented in 1973 at Wing V as Minuteman III missiles were being installed during the Force Modernization Program. This guidance-and-control upgrade included a new post-attack operating mode called PIGA leveling, which removed

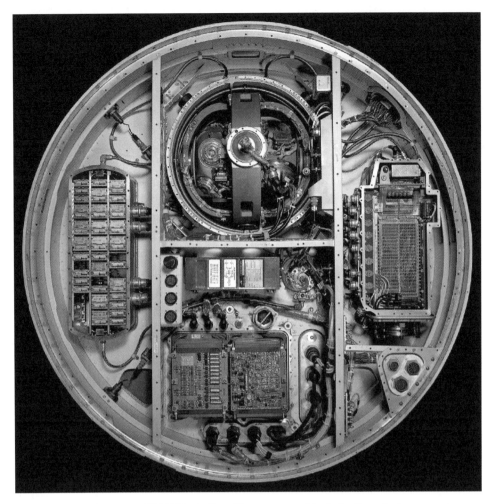

FIGURE 10.9 Minuteman III NS20 missile guidance system. Clockwise from the top, gyro stabilized platform, D37D computer, P92 amplifier assembly, MGS control. *Photograph courtesy of National Air and Space Museum.*

level reference and leveling loop system errors resulting from seismic disturbances. The short-term stability of the G6B4 drift rates was also improved by the addition of turbulence-suppression baffle hardware within the gyroscopes. These changes were retrofitted into the entire Minuteman III force.

The CDB program also included a significant change in the self-alignment procedure used for the azimuth orientation of the stable platform. In the initial deployment of the NS20, the autocollimator was used as a short-term holding reference for periodic calibration of the GI-T1-B gyrocompass by means of a four-position gyrocompassing technique that measured the misalignment and azimuth between the input axis of the gyrocompass gyroscope and the axis defined by the two gyrocompass mirrors. Post-deployment research later indicated that the autocollimator could be removed completely, which eventually took place.

Changes in the command-and-control system between the LCC and the LF enabled remote changes of target data as well as changes in program routines and the computer disk memory.

## HYBRID EXPLICIT GUIDANCE MECHANIZATION/HYBRID EXPLICIT FLIGHT PROGRAM

On 30 November 1972, Headquarters SAC issued ROC 19-72, which established the need for improved Minuteman III guidance performance—either the replacement of the computer or development of a new flight programming tape. This tape was designated the Hybrid Explicit Guidance software.[69] This change involved simplifying and reducing the number of targeting parameters for a particular launch site and target set. The new program reduced the computer memory words for each target set from 760 to 70, enabling the option of the retention of the D37D computer. Prior to this change, the Minuteman system used the Delta Guidance sequence, with control of the missile to a predetermined trajectory and then correcting between the actual trajectory and the programmed trajectory during flight. The Delta Guidance scheme involved lengthy and complex computations in the targeting program. The Explicit, or Q, Guidance scheme used by the Thor, Polaris, and Poseidon systems relied on the concept of "velocity to be gained." The inertial guidance system sensed the velocity history of the missile in flight so that the boost phase ended when the difference between missile velocity in the X, Y, Z coordinates and that needed to reach the target at the calculated time reached zero.[70] The Hybrid Explicit Flight Program (HEFP) modification permitted rapid targeting computations for new targets at the LCC, with subsequent reduction in time to transfer from the LCC to the LF and the missile's computer. The retargeting time was reduced from 30 minutes per target to 7 minutes. The preloaded target set was increased from three to four per reentry vehicle, and the steering mechanization was modified to allow a much larger off-azimuth launch angle, increasing from 14 degrees for the MRT launch to 45 degrees.[71]

## GUIDANCE IMPROVEMENT PROGRAM

The Guidance Improvement Program (GIP) was implemented after completion of the initial R&D flight testing, which revealed errors that were included in the guidance error budget at that time. The expectation was that the CEP would improve by approximately 15 percent. For a variety of reasons, this decrease in CEP was not realized until the guidance upgrade program (GUP).[72]

## GUIDANCE UPGRADE PROGRAM

The GUP was a result of work done to analyze the primary causes of the downrange miss and cross-range dispersion errors observed in the Minuteman III R&D and operational flight tests starting in 1978. Errors such as gravity model discrepancies were considered the prime reason for the dispersions, but further studies concluded that the errors were caused by the guidance-and-control instrumentation. The downrange bias was found to be largely the result of problems with the PIGA, a result of errors in the accelerometer calibration. The effect was predictable and consistent between guidance sets and, once understood, compensated for through targeting computations. The program culminated with the development of improved test equipment to accommodate the measurement of gyroscope coefficients, allowing change to the targeting program for each guidance set.[73]

## MINUTEMAN III EXPANDED EXECUTION PLANS PROGRAM

The Minuteman III operational ground programs were modified in 1987 (WS-133B) and 1988 (WS-133 A-M) with the Expanded Execution Plans (EEP) program to enable increased flexibility in execution plans. The execution plan message provided each LF with the launch delay time, the target set number, and whether the target set would be targeted as a CEP or MRT launch. Previous to this upgrade, each LF had stored in the D37D memory 100 active and 100 inactive execution plans, which were selectable by the launch crew via the Preparatory Launch Command-A. The active/inactive partitioning allowed a new set of execution plans to be remotely loaded from the LCC and then made instantly active as needed. The EEP program eliminated partitioning and coding, now allowing a total of 2,400 different options. This was done to accommodate changes in war plan scenarios.

Another part of the EEP dealt with the sensitivity to earth motion from even distant earthquakes, which had been observed for many years in both the Titan II and Minuteman guidance systems. This was a major reason for the azimuth alignment gyroscope assembly in Minuteman II and the self-alignment method for Minuteman III. The solution was to make use of the most sensitive motion detector in the guidance set, the azimuth alignment gyroscope assembly. If sufficient noise of the right signature and duration was detected in the azimuth alignment gyrocompass determination of azimuth, entry into the PIGA leveling protocol would occur. If an EMP event was detected by the operational ground equipment, the crew in the LCC could also initiate the PIGA leveling protocol.[74]

## GUIDANCE REPLACEMENT PROGRAM NS50A MISSILE GUIDANCE SET

By the early 1990s, the Air Force decided that the Minuteman III system deployment would need to be extended to the 2020 timeframe. Simultaneous with this decision, concerns regarding the extended support of the NS20 MGS led the Air Force to initiate the Guidance Replacement Program (GRP). The GRP modernized guidance system electronics; improved radiation and EMP protection; adopted a standard-language, solid-state, random access computer; and increased memory storage. The modified MGS was designated as the NS50A and retained the existing NS20 interfaces (see Chapter 16).[75]

★ ★ ★

The development of a lightweight inertial guidance system for the Minuteman program, like the other technological challenges of solid-propellant development and the second-generation ablative reentry vehicle, took place in a remarkably short period of time. The key innovation of the free-rotor, gas-bearing gyroscope and advances in semiconductor technology were critical factors.

One aspect of ballistic missile inertial guidance that has been lost in all the technology is that of the guidance system knowing where it is and where it is going. Targeting Minuteman from the early 1960s to the present day has also evolved.

# 11

# TARGETING MINUTEMAN

A conundrum facing advocates of the ICBM was the question of accuracy. The first ballistic missile used as a weapon, the German V-2, had a reputation for inaccuracy that was driven in part by guidance system and structural issues. At least as important were launch site and target coordinate errors. The comprehensive geodetic surveys (a system of coordinates or datum) of Britain and Europe were ineffectively linked at the start of World War II. Most industrial nations had generated their own geodetic systems to the detail and accuracy needed for commercial and government applications by combining local government surveys. At this point there were well over 100 different geodetic datums, each with slight differences in their parameters sufficient to make hitting targets at intercontinental distances a difficult task.

Until the age of the ICBM, there was no apparent need for consistency in the geodetic surveys between neighboring countries. For instance, there were nine geodetic datums for Southeast Asia, and they were not tied together. The only international or continent-spanning surveys were those done by parties interested in determining the shape of the Earth. A partial solution to the multiplicity of systems was to select several major systems as the preferred geodetic systems for a particular area and make adjustments to the subsumed systems. This approach resulted in selection of the North American 1927, European (apparently not including England), Tokyo, and Indian datums, but they were not tied together.[1] By the mid-1950s, this was still not sufficiently accurate for use with intercontinental weapon systems such as air-breathing guided missiles or ICBMs using inertial guidance systems. In April 1960 the World Geodetic System 1960 (WGS 60) was introduced. Crafted from separately developed US Army Map Service and US Air Force Aeronautical Chart and Information Center (ACIC) geodetic systems, WGS 60 enabled the first reasonably accurate targeting of ICBMs.[2]

## Locating the Launch Facilities

Aiming an ICBM accurately to hit a target at intercontinental distances required minimizing errors at both ends of the trajectory as well as the variable errors encountered in flight due to differences in gravity along the flight path. At the beginning of the ICBM program, little was known of the geodetic data at the target, so the focus was to make the geodetic data at each launch site as accurate as possible. To this end, in 1959 the Air Force formed an entirely new organization, the 1381st Geodetic Survey Squadron (Missile; 1381 GSS), tasked with the highly important mission of performing astronomic, geodetic, and gravimetric surveys in support of accurately positioning each LF. The 1381 GSS was initially located at Orlando AFB, Florida, as part of the Air Force's 1370th Photomapping Group but was relocated to F. E. Warren AFB, Cheyenne, Wyoming, in 1965. The unit was a mix of military and civilian technical staff providing accurate geospatial and geophysical information for the ICBM programs, but later the responsibilities were expanded to provide for targeting needs of the US military on a worldwide basis.

After the Air Force and the Army Corps of Engineers agreed on Minuteman wing locations, the Corps of Engineers would locate the individual LCCs and LFs, marking property boundaries, access points, and easements. This included stipulating where the security fence, survey pedestal monuments, and temporary perimeter fence would be located, as well as the intercomplex communication cabling. Once the facility locations were known in general, a survey team from the United States Coast and Geodetic Survey would tie the locations to the nearest geodetic control station in North American Datum 1927 (NAD 27). Corrections were made, shifting the NAD 27 geodetic positions to WGS 60, since NAD 27 was a nongeocentric national geodetic system and thus unsuitable for targeting ICBMs.[3]

Next, either Boeing contractor personnel or a team from the 1381 GSS would install the guidance alignment survey monuments. These were concrete pedestals 2.5 feet in diameter, 5 to 7 feet above ground, and positioned 10 feet deep, with a fitting on the top for a surveyor's target. The monuments normally were 1,000 to 1,500 feet from the LF personnel access hatch and covered an arc of approximately 90 degrees from the leftmost to the rightmost targets.[4]

The sequence for establishing the geodetic coordinates of each monument varied somewhat, but in general, an astronomic survey team would use astrogeodetic methods for precise location of one monument, which would then be used as a reference for the remaining monuments, whose number varied. A gravity survey team would take gravimetric readings both at the monuments and in the surrounding area. After the LF was built, the Station A point marker was placed near the LF personnel access hatch. Its precise geographic location was determined from sighting on the reference monuments.[5]

Another component needed for accurate targeting of ICBMs is the determination of gravity values in the region surrounding the LF. There are two components to the need for gravity data. The first is to determine the gravity data directly behind the launcher's target azimuth since initial flight is relatively slow. The effect of gravity decreases as a function of the inverse square of the altitude. This means that errors in the measurement of gravity have the greatest effect at launch and lessen quickly as the missile gains altitude. Gravity issues at the target are not nearly as important for the target portion of the trajectory due to the altitude and speed of the reentry vehicle near the target and its brief time to impact.[6] At first glance gravity surveying is not obvious. How can one detect variations in the Earth's gravity that would mean anything? Current gravimeters can detect changes in gravity as small as one hundred millionth of the total gravity force.[7]

Another reason for measuring local gravity variations is to determine what is called the deflection of the vertical, which has a latitudinal and a longitudinal component. Two methods are available for the determination of deflection of the vertical at a site: the astrogeodetic and gravimetric methods. The astrogeodetic method utilizes astronomic latitude and longitude coordinates obtained from star observations made at the site where geodetic latitude and longitude are also known. The latter must be geocentric geodetic latitudes and longitudes to be useful for ICBM guidance support. These two sets of coordinates (astronomic and geodetic) are used in simple formulas to create the north–south and east–west components of the deflection of the vertical. With the gravimetric method, the site deflection of the vertical components is computed using the Vening-Meinesz formula with a regional set of gravity values as input data. Both the geodetic coordinates (on the ellipsoid) and the astronomic coordinates (on the geoid) were determined at each LF to generate the deflection components so that the proper corrections could be applied. The geoid is defined as that particular equipotential surface of the Earth that coincides with the mean sea level over the oceans and extends hypothetically beneath all land surfaces.[8]

### *Locating an Individual Launch Facility*

Once the construction of the LF was complete, an initial astronomic survey took place to precisely locate it. Normally at night, a field survey team would set up a T3 theodolite over the Station A point. Illuminated targets were placed in holders on the external monuments (the actual number varied from two to three depending on the wing) and the astronomic azimuths for each monument relative to the Station A point recorded. This was done 16 times with the T3 oriented in the direct mode and 16 times with it in the reverse mode. This was done to control systematic errors due to instrument misadjustment. In addition to the observations on the targets mounted onto external monuments, the observer also included direct and reverse sightings on the North Star (Polaris) and the exact time of the observation. The recorded times were checked and corrected via the time signals broadcast via shortwave radio on National Bureau of Standards Station WWV at Fort Collins, Colorado. The computational reduction of the measured directions to the targets and the Polaris observations yielded the true azimuths from north for each line between the Station A point and the external monuments, precisely locating the Station A point.[9]

## Minuteman I Guidance Alignment

The missile guidance system computer of the Autonetics NS10Q series did not hold the target astronomic azimuth. The missile was actually launched on an azimuth that would culminate in the target location after the programmed flight time. Thus, the launch astronomic azimuth was not a direct heading to the target at time of launch. Instead, the computer held the flight equations necessary to reach a particular target on a particular astronomic azimuth, having taken into account rotation of the Earth during the flight.

The final step in targeting a Minuteman missile was the process of generating a reference point for the launch astronomic azimuth. This required optically transferring the LF astronomic azimuth position from Station A to the autocollimator bench on Level 1 of the LER. This enabled determination of the precise location of the autocollimator on the autocollimator bench, which was the guidance system alignment reference point.

Carrying the launch azimuth and alignment to the missile guidance system was the job of the Target and Alignment Team (later the name was changed to Combat Targeting Team, or CTT). Prior to initial missile installation, the three-man team, one officer and two enlisted men, would conduct a reference mirror alignment verification (RMAV). First, they would check the angles of the external monuments with the Station A point near the launch tube using a theodolite (Figure 11.1).

If they were in tolerance, meaning a farmer had not hit one with a tractor or harvester, then the reference angle from either one was sent down the autocollimator sight tube using the theodolite at the Station A point and a second theodolite positioned on the autocollimator bench on Level 1 of the LER. The second theodolite was then used to measure the angle to the nearest of two reference mirrors located on the LF wall, one at either end of the autocollimator bench. Eight measurements, or sets, were done, each one consisting of a direct and reverse reading from each theodolite as well as using different parts of the theodolite angle scale to eliminate systematic errors. The operators also took a theodolite level reading, which was used to compensate for leveling errors. The measurements were then averaged, and each of the eight had to have an error less than plus or minus five arc seconds. If the error was larger, the team would have to reshoot the sets. Once one of the reference mirrors was verified, the other was verified via the downstairs theodolite (Figure 11.2).

These measurements were preferably done at night so that sun heating was not an issue for the Station A theodolite and operator, as distortion of the monument targets due to

FIGURE 11.1 Members of a CTT measuring the angle (Angle A) between two external alignment monuments as part of the RMAV process. If the measured Angle A is within three arc seconds of the geodetic survey angle, they could continue and establish the azimuth of the two reference mirrors in the LER and then accurately align the autocollimator to the missile's launch azimuth. *Official USAF photograph.*

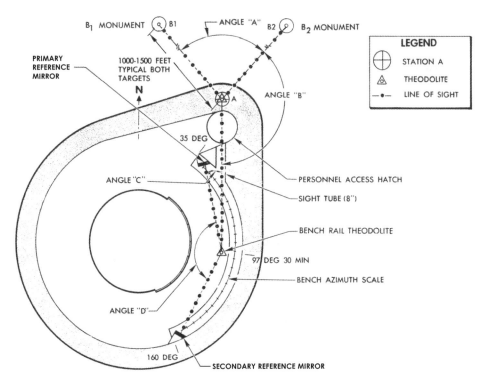

FIGURE 11.2 Targeting and alignment procedures diagram for establishing reference mirror azimuths using external reference monuments. *Adapted from Student Handouts, Missile Alignment Diagrams, courtesy of Jim Barnard.*

ground heat waves was a problem. If it had to be done during the day, the Station A theodolite was sheltered from the sun and the weather. In the winter the personnel access hatch was covered by a metal cap with an optical glass insert so the light beam between the Station A and LER theodolites was not distorted due to the warmer LER temperature air currents rising up the hatch.[10]

## Autocollimator Alignment Procedures

Once the reference mirrors had been checked and their azimuth confirmed or reestablished, the autocollimator positioning could take place (Figures 11.3 and 11.4).

The Minuteman I flight launch azimuth could be anywhere in a 307- to 67-degree arc. Due to guidance stable platform gimbal hardware limitations, the LF internal reference point for the guidance system, the autocollimator, was positioned on the autocollimator bench (which had a protractor strip arc of 35 to 160 degrees, with 90 degrees being due east) at the launch azimuth plus 90 degrees (launch azimuth of 310 degrees plus 90 degrees gave 400 degrees; 400 degrees minus 360 gave 40 degrees, the location on the autocollimator bench). The bench was divided into sectors that defined which reference mirror would

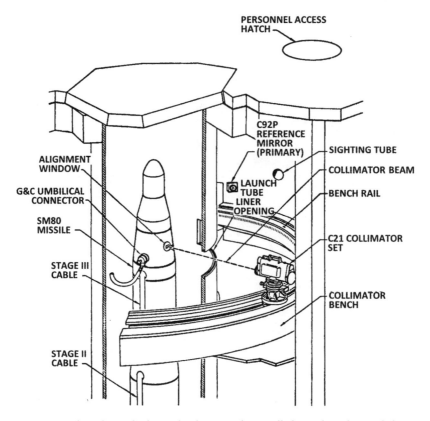

FIGURE 11.3 Typical mechanical relationship between the installed missile and optical alignment equipment on LER Level 1 for aligning Minuteman I and II guidance sets. The Minuteman II bench was shortened at some wings since the autocollimator was now always located at 90 degrees. *Courtesy of Jim Barnard.*

be used and how many theodolites were needed to accomplish the alignment tasks. For Minuteman II and III, the autocollimator was in the same position on all sites, known as Angle X (90 degrees on the bench). Minuteman III used autocollimators for a time until the computer self-aligning technique process (gyrocompass; see Chapter 10) was validated.

The launch astronomic azimuth determined from the reference mirror azimuth plus 90 degrees yielded the position for the autocollimator on the autocollimator bench. The theodolite was then moved to that position. The autocollimator was then coarsely positioned directly behind the theodolite. Using a gunsight on the theodolite, the autocollimator was moved so its centerline marker was lined up with the gunsight. Then the letter on the autocollimator cover was noted for determining the vertical position.

## Missile Indexing

Next, the centerline of the missile had to be determined. The missile might have a slight bend in the airframe or slight lean from the vertical due to the suspension system. This meant the centerline of the guidance-and-control alignment window might not be the true

FIGURE 11.4 Two members of the Targeting and Alignment Team, A1C Duane Bowser (left) and A1C Ronald Sammons (right), verify the secondary reference mirror azimuth alignment prior to positioning the autocollimator. The reference mirror can be seen on the wall at the upper right. The autocollimator bench is the arc stretching from the lower left to the upper right. *Official USAF photograph, courtesy Andy Doll.*

horizontal centerline of the missile in the launch tube. The theodolite, already at the launch azimuth plus 90-degree position on the bench, was rotated to face the missile and pointed to the left edge of the guidance-and-control section. The operator would take a reading, point to the right edge, take a reading, raise the theodolite telescope a small amount, align to the right edge, take a reading, point to the left edge, and take a reading. The guidance section was white, and to facilitate the centering task, the launch tube was painted black directly behind the guidance system section. A worksheet computation would determine if the rough theodolite position was within tolerance. If not, the theodolite was moved slightly to the new worksheet value, which was the true horizontal centerline of the missile.

## Missile Rotation

The missile had been previously rotated by the Missile Maintenance Team (MMT) so that the guidance system alignment window was already coarsely opposite the launch azimuth plus 90 degrees on the autocollimator bench. The targeting team now precisely rotated the missile to launch azimuth so the guidance system optical alignment window was now directly across from the theodolite and autocollimator. Next, the bench theodolite was turned to the launch azimuth plus 270 heading (turned 180 degrees away from the autocollimator). Above the outer guidance system alignment window on Minuteman I was a decal with three vertical lines (on Minuteman II and III, an inverted T decal was used). Using a power controller box that interfaced with a gear drive at the base of the missile, the operators could rotate the missile left or right to position the middle line for Minuteman I, and for Minuteman II and III, to align the vertical of the inverted T with the theodolite reticle.

At the same time, a vertical measurement was taken to determine missile vertical centerline in relationship to the autocollimator. The technical order had a chart that, with this measurement, in conjunction with the autocollimator cover letter from above, indicated if autocollimator spacers needed to be added or removed to adjust the autocollimator height.

### *Autocollimator Coarse Positioning*

Next, while looking through the bench theodolite telescope, the autocollimator mount was moved until the autocollimator light beam was visible in the telescope. The autocollimator was then tightened down to the bench. The autocollimator mount fine adjustment knobs were used to move the autocollimator left or right to align the light beam to the theodolite reticle.

### *Autocollimator Fine Alignment*

The final and most critical procedure was to accurately adjust the autocollimator light beam. This was accomplished using an optically flat, double-sided mirror called a transfer mirror installed on the theodolite bench mount between the autocollimator and the theodolite. The transfer mirror was aligned so the theodolite gauss was centered on the theodolite reticle.[11] Then, using the autocollimator mount fine adjustment knobs, the autocollimator was aligned to the transfer mirror so the Autocollimator Test Set indicators AC1 and AC2 indicated the beam was reflected straight back into the autocollimator. This was exactly what the guidance system stable platform mirror reflection would look like to the autocollimator.

Once verified, the transfer mirror and theodolite were removed, and the inertial guidance system went into alignment-and-adjustment mode when commanded from the targeting console (see below). The oscillating light beam from the autocollimator was reflected from the inertial guidance system azimuth alignment mirror back to the autocollimator, which then automatically calculated one half of the angle error from direct reflection and sent a signal to the inertial guidance system, which adjusted the guidance platform to the new setting, repeating this until the reflected beam was coincident with the outgoing beam. This took approximately 30 minutes. A pair of metal barriers were clamped to the bench to prevent anyone from inadvertently walking through the light beam.

### *Targeting*

Targeting Minuteman IA or IB was done by means of a targeting console (C-24) located in a targeting van on the surface of the LF. The console and a portable battery power supply

(C-95) were connected to the Programmer Group in the upper equipment room by a large cable. The targeting console operator then started up the missile guidance system, which required the supplemental power from the battery power supply to boost the power available in the guidance-and-control coupler. The extra power was necessary to quickly start the gyroscopes (see Chapter 10). Once the guidance system was powered up, the operator ran three tapes. The first was the maintenance tape, which checked out the missile and associated aerospace ground equipment that ran various tasks prior to launch. The second tape was the alignment tape, which aligned the guidance platform in the missile to acquire the light beam on the autocollimator. This was followed by the 60-second test, at the end of which the Stage I nozzles moved through their full range of motion, making a distinctive whine that varied in pitch as they moved. If everything checked out, a "calibrate" light came on to tell the operator to start the approximately 190-minute calibration sequence, during which the guidance platform slewed away from the autocollimator window to begin calibration.

While this was taking place, the targeting team was packing up and backing out of the inner and outer security zones. Once all the way outside the LF fence, the team called the missile combat crew in the LCC to reset the inner and outer security zone detectors. If the alignment was successful, the LF was back on strategic alert. As long as the oscillating return beam was centered and detected by the autocollimator, the missile remained on strategic alert.[12]

## Minuteman II Guidance System Alignment

The 321 SMW at Grand Forks AFB, North Dakota, was the first to receive the Minuteman II missile. At the beginning of deployment of Minuteman II, the new inertial guidance system, the Autonetics NS17, was aligned using the Minuteman I RMAV process. However, after approximately six months, a new alignment system was fielded, the azimuth laying set (ALS), also known as the portable gyrocompass assembly (PGA).[13] The ALS required only a two-man team to operate, eliminating one man that previously worked the aboveground optical alignment from the Station A position. This process was called the reference mirror azimuth determination (RMAD).

The ALS used a geosensor apparatus, which was the gyrocompass being used in the more advanced, soon-to-be-deployed Minuteman III NS20 missile guidance system. The ALS could sense the Earth's rotation and autonomously determine which way was true north to an extremely accurate value. By utilizing mirrors attached to the gyrocompass, a theodolite, and the reference mirror, the angle between the ALS mirror and the reference mirror was calculated. This gave the horizontal astronomic azimuth of the reference mirror in relation to true north. This is what the RMAV had done, but now it was all below ground, and the old system was no longer necessary. The purpose of the RMAV and RMAD was exactly the same: calculation of the reference mirror astronomic azimuth with respect to true north. The name change just distinguished how the measurements and calculations were made. Once theodolite measurements were complete, and the worksheet calculations proved everything was within tolerance, the reference mirror astronomic azimuth could be used to accomplish any other optical alignment procedures necessary in the LF unless the launcher closure was opened.

In 1982 problems with the heading sensitivity testing (HST) program at the 91 SMW's Minuteman LF-08 at Vandenberg AFB, California, brought to light issues with the accuracy of the ALS system. The HST program involved putting Minuteman II and III missile guidance systems through simulated flight testing to ascertain the position of the impact of the warheads. At LF-08 the tests revealed a 12–arc second gap between what the missile guidance system and the ALS were reporting.

SSgt. John Mills and his team were sent to the Autonetics facilities in Anaheim Hills, California, where the LER environment was recreated for extensive testing of the ALS over a three-week period using the original ALS Technical Order procedures and then the current procedures. The differences between the old and new procedures were evaluated, and the engineers found an inherent bias that had been created by continued changes to the procedures over time by the Air Force. This internal bias partially contributed to the 12–arc second deviation discovered at LF-08. Another factor involved the hollow autocollimator bench at LF-08, where the temperature was not constant. The ALS was very temperature-sensitive.

The SAC was not impressed with the results of these tests and insisted on one more validation test: reopening the autocollimator sighting tube and reshooting the old monuments from the Station A marker. The 1381 GSS came out and resurveyed the sites, and the CTT shot the RMAVs using the original technical data provided by Autonetics. The results of the RMAVs were less than stellar due to the horrendous weather conditions at Vandenberg. However, several good optical sequences were accomplished, which validated the Anaheim Hills tests and the need to return to the original ALS procedures.

Instead of following the recommendations of Autonetics, SAC disallowed all RMADs at the operational Minuteman wings effective by the end of 1983. Following the elimination of RMADs at Ellsworth, the target-and-alignment teams went out and did target tape loads only but still utilized the autocollimator to provide the east–west astronomic azimuth heading.

The NS17 computer calculated the launch astronomic azimuth for the selected target coordinates. In its memory were the geodetic latitude and longitude coordinates for the LF and the eight targets. The autocollimator position indicated where east was, and thus the computer could calculate north. All three pieces of information enabled the computer to calculate the target flight geodetic azimuth, and it rolled the missile to that azimuth shortly after launch. Now essentially all the missiles were aligned the same and launched on a due north heading before rolling to acquire the launch geodetic azimuth shortly after clearing the launch tube.[14]

The NS17 used the CDB and the ILCS upgrade that was used exclusively at Whiteman and Malmstrom AFBs. Later on, Ellsworth AFB adopted the same targeting techniques of the ILCS alignment but still kept the autocollimator, with minimal alignment being accomplished by the CTT. The ALS was shelved (except at Vandenberg AFB) and the autocollimator left at the last verified position. From that point on, the few optical alignments that were completed were the autocollimator azimuth check, alignment, and checkouts, along with missile indexing following an emplacement.

In an eight-month period, the Air Force went from shooting RMADs at Ellsworth AFB to totally eliminating the task and sending the ALS to depot for destruction. The M1 reference mirror was still utilized for theodolite setup and calibration, as the last known position was accepted and assumed it never moved from then on. While Vandenberg AFB continued to use the ALS for the HST program, it was now becoming difficult to find technicians qualified in optical alignment. Eventually Autonetics took over the procedures in 1985 and 1986.[15]

The Targeting and Alignment Team name was changed to CTT in the late 1960s, while remaining a three-person team, one officer and two enlisted men. In 1973 SAC asked the 44 SMW to develop procedures that would reduce the team to one officer and one enlisted man. The issue was how to maintain positive control on the coded target tape materials that the team used to load the target sets into the missile inertial guidance computer. Since one man had to be topside to lower equipment to the other team member in the personnel access hatch area, two-man control would not easily be accomplished.

The methods were worked out and ready to be confirmed via the semiannual visit of the 3901st Strategic Missile Evaluation Squadron (3901 SMES) from Vandenberg AFB. Rik Gervais recalls all too well the day the evaluation took place as his two-man team carried out the new procedure under the inspector's watchful eyes. The procedure was completed without any errors, and Gervais was pleased as they waited with the evaluator outside the LF security fence for the sortie to come back on strategic alert. His mood quickly changed when the evaluator told him that their performance was excellent save for one important point: They had failed to maintain proper two-man control, a violation of the SAC Two-Man Policy. Gervais recalls being devastated, as they had shown it was possible to carry out the alignment and targeting tasks efficiently with a two-person team. Apparently the issue was debated between the 3901 SMES and higher headquarters for several days, and Gervais's team's rating was changed from "unqualified" to "highly qualified."[16]

## Minuteman III Guidance System Alignment

Precision alignment by the guidance system gyrocompass eventually eliminated the need for the autocollimator and the support system for it, resulting in substantial program savings. Retargeting now takes far less time, as input and verification can all be done at the LCC (see Chapter 10).

## Minuteman Target Locations

Prior to 1970, the estimated accuracy of some target locations in the Soviet Union was at best 2 to 3 miles and at worst as much as 30 miles. At the onset of Project 117L, also known as Corona, in 1955, the Air Force ACIC was the center for reduction of geodetic and cartographic information related to photographic imagery. The interpretation of the Corona imagery for ICBM targeting purposes was called Project Shoe Lace, which ran from 1958 to 1967 and under the name Sentinel Lace through 1972. The goal of Project Shoe Lace was to reduce targeting error due to location error to a circular error (CE) of 1,000 feet with 90 percent probability by 1970. According to William C. Mahoney, Project Shoe Lace's leader, by 1965 the team was certain that a CE of 450 feet in position, with 90 percent probability and a linear error of 300 feet in elevation with 90 percent probability, had been obtained. To be clear, this was a mapping accuracy and not a reflection of the inertial guidance system accuracy at the time.[17]

SAMSO and ACIC maintained a Minuteman Geodetic and Geophysical (G&G) Error Budget. The error budget listed in three major headings (Launch Site, Inflight, and Target Position) the G&G data required by Minuteman for testing and deployment. Accuracy values were also listed for this data. The G&G Error Budget was used by the ACIC to identify the data instrumentation, data acquisition programs, and resources required to achieve the listed data accuracy values. The target position category, determining accurate WGS horizontal and vertical positions for potential targets in politically inaccessible areas, was the most challenging of the G&G tasks, requiring extensive efforts by ACIC technical personnel.

As the accuracy of Minuteman inertial guidance systems and other missile-related factors improved, to fully exploit this improvement, the G&G-related activities of ACIC and the GSS also increased to support achievement of the improved impact accuracy for Minuteman. Similarly, such G&G refinements have continued through the years, as evident from the development and application of WGS 84 and its revision in 2004.[18]

★ ★ ★

In this age of global positioning satellites and the ability to track a cell phone's location to a physical address, it seems archaic that optical surveying techniques were used to locate the Atlas, Titan, and Minuteman LFs. Even more astounding was the extension of those techniques across continents for the initial target locations in the Soviet Union and other countries.

The concept of concurrency meant that the flight test program and operational facility development were taking place simultaneously. The first flight test of the Minuteman Stage I took place in September 1959, just 19 months after Col. Hall's initial briefing to the secretary of defense. This was the beginning of the integration of all the parts of the Minuteman program into an operational system, starting with the determination of full-scale launcher design at Edwards AFB.

# 12

# RESEARCH AND DEVELOPMENT FLIGHT PROGRAMS

The Minuteman hardware R&D programs began in early 1959. A new facility, the Weapons System Integration and Vibration Laboratory, was designed, and modifications were made to existing structures at the Seattle Boeing Development Center, with completion planned by early 1960. These facilities housed test equipment and full-scale replicas of equipment and structures for use in design, modifications, and integration as the program progressed.[1]

## Silo Development

In-silo launch experiments had been underway in Britain at the Rocket Propulsion Establishment, Westcott, Buckinghamshire, since 1956 as part of its liquid-propellant Blue Streak IRBM program (theoretical studies were also carried out at Aerojet-General in 1956). The feasibility of in-silo launch was first demonstrated by British scientists through extensive subscale research, which culminated in a horizontal one-sixth-scale aboveground test structure. Tests were conducted to evaluate sound energy suppression utilizing sound-absorbing panels and water deluge with a J-shaped exhaust duct (the launcher was described as U-shaped, a combination of the launch duct and the exhaust duct). British scientists visited the United States to share their results with the Air Force in 1958. The Blue Streak program was canceled before an actual underground LF was built.[2]

A month after Boeing was awarded the Minuteman contract, the in-silo launch option was discussed at BMD for Titan II. The launch environment for Titan II was much more complicated than that anticipated for Minuteman. The Titan II launch sequence involved engine ignition, followed by approximately 1.1 seconds to reach 77 percent of full thrust, at which point three 1.8-second timers started, allowing the engine to come to full thrust (experience at Aerojet, the engine manufacturer, showed that if the engine reached 77 percent, it would continue to full thrust). When any one of the three timers timed out, explosive nuts fired, releasing hold-down bolts, and the missile left the thrust mount, taking another four seconds to clear the silo. Thus, the missile was exposed to enormous acoustic and thermal energy during the approximately six to seven seconds from ignition to clearing the silo. Water deluge and sound attenuation panels lessened the energy to acceptable levels. Upon engine ignition, entrained outside air forced the exhaust fumes, heat, and the steam generated by the deluge water down and out the two exhaust ducts located on either side of the launch duct. Angled cascade vanes directed the exhaust fumes away from the silo opening, preventing recirculation.[3]

One of the many advantages of solid-propellant missiles was the nearly instantaneous development of full thrust, which meant a much shorter exposure to the in-silo launch environment. With this advantage in hand, Boeing engineers, building on the British experience, conducted a series of 1/20-scale tests in Seattle of the in-silo launch environment in December 1958 and, one month later, were conducting cold flow tests to provide supplemental data on flame deflector shapes prior to the start of one-third- and full-scale tests. The results indicated that because of the rapidity of missile liftoff (approximately 1.1 seconds after Stage I ignition) and the stronger airframe due to the solid propellant, the silo could serve as the annular exhaust duct, considerably simplifying construction time and expense, just as suggested by Col. Hall at the beginning of the program.

## Edwards Air Force Base

The one-third- and full-scale test programs using Stage I, with several seconds of propellant, were conducted at the AFFTC, Edwards AFB, California, with the first four one-third-scale tests conducted by the Air Force and the remainder conducted by Boeing engineers (Figure 12.1).

The Boeing one-third-scale test series began at Edwards AFB in California on 15 May 1959 using a scaled 12-foot-diameter test silo (4.2 feet actual diameter). Similar to what had

FIGURE 12.1 Aerial view of the Edwards AFB Minuteman silo launch test facilities. Two silos were built due to the urgency of design decisions. Silo 1 is on the right. *Official USAF photograph.*

been done in Britain, this test silo was horizontal and above ground, facilitating modification and instrumentation placement. A cluster of four ARC rocket motors were manifolded together and inserted in the first-stage section of the one-third-scale missile, firing for 3.25 seconds (Figure 12.2).

The facility design was such that various lateral and emerged missile positions could be evaluated, as well as heat transfer to the nozzle compartment, heating of the missile skin, and the in-silo acoustic environment. Early results permitted concurrent construction of the full-scale silos at AFFTC to allow for testing of three diameter configurations (Figure 12.3).[4]

Two full-scale silos were constructed at the AFFTC to determine the optimum silo configuration. Each silo was 26 feet in diameter, approximately 76 feet deep, with provision to insert steel liners of 12-, 14-, and 16-foot diameter for use to verify the optimum silo design results from the subscale tests (Figure 12.4).

Eleven ground test missiles (GTM) were built in Seattle. The so-called battleship missile motor casings were thicker than the flight-weight structures. The proposed operational thrust-to-weight ratio was achieved using dummy ballasted Stage IIs and IIIs. Thiokol

FIGURE 12.2 Horizontal in-silo launch test structure at Edwards AFB, September 1959. The deflector at the left end of the structure could be exchanged for different designs. This test appears to be set up for the water deluge deflector, as evidenced by the water line connected to the deflector. The one-third-scale model test missile could be moved horizontally to various positions within the test structure during the test. Deflector design and the silo environment during launch were investigated using this structure for the design of the LFs at Cape Canaveral. *Adapted from "Air Force Flight Test Center History, July–December 1959."*

FIGURE 12.3 Left: One-third-scale model used in the horizontal test structure. This test article is frequently misidentified as a wind tunnel model. Right: STM 101 being lowered into the silo launch test facility at Edwards AFB. Note the two heavy nylon tethers attached to the Stage II to III interstage. The reentry vehicle is a mockup for weight and balance. The white triangular structures are telemetry antennas. *Official USAF photographs, courtesy of John Hilliard.*

R&D Stage I motors were used with sufficient propellant to boost the missile clear of the silo in approximately 1.5 seconds and propel it several hundred feet into the air. A nylon tether served to ensure that the missile landed nose first to preserve the Stage I nozzles and actuating system for post-flight inspection.[5]

Silo Test Missile (GTM was renamed STM) 101 was successfully launched on 15 September 1959, with Silo 1 configured at a 16-foot diameter with a sombrero-type flame deflector. The test objectives were focused on validating the silo design, the STM, and the instrumentation. In the original design, the missile was to be guided out of the silo with a mechanical guidance system in contact with the silo wall, but by mid-1959, the design had been changed to free flight out of the silo with missile position determined by onboard accelerometers. The remaining STMs had an Autonetics gyroscope installed. The first flight lasted 2.6 seconds, with excellent data recorded. It was clear that the test silo's design would function as planned without any major modifications. Additionally, the question of whether flight control should commence at ignition or after the missile cleared the silo was clearly answered: Flight control could commence just prior to Stage I ignition. Damage to the silo was minimal (Figure 12.5).[6]

STM 102, launched successfully on 2 October 1959 from Silo 2, again configured with a 16-foot diameter but this time with a flat-plate flame deflector. Attitude sensors showed that STM 102 had no appreciable yaw, but there was some pitch movement. The tether system functioned as planned, and the missile was recovered 160 feet from the silo.

STM 103 was fired on 22 October 1959 from Silo 2, with the same configuration as STM 102. The flight objectives for STM 103 included collecting acceleration, acoustic, pressure, and temperature data. STM 103 had a fixed nozzle configuration, and this flight was used to determine the free-flight characteristics of a missile with fixed nozzles.

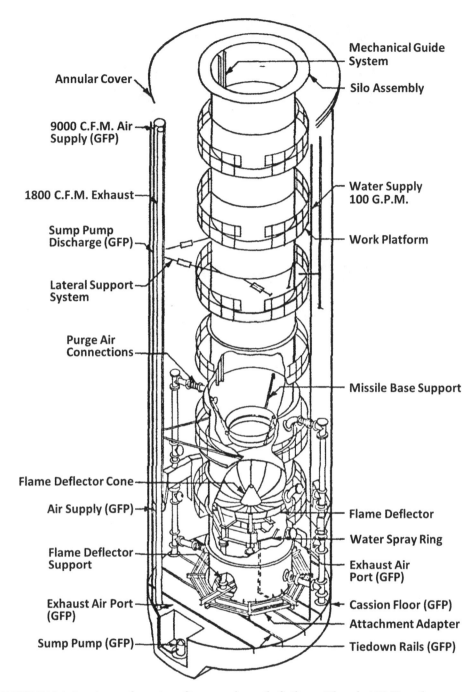

FIGURE 12.4 Interior configuration of in-ground test silo facility at Edwards AFB. Two silos were constructed in case of damage to one during testing. Both silos could have different diameter liners installed. The liners were installed in segments to facilitate working on the missile prior to launch. *Adapted from "Minuteman Historical Summary, 1958–1959," courtesy of Boeing Company.*

FIGURE 12.5
The first tethered launch on 15 September 1959 was highly successful. STM 101 has reached the apex of the flight profile, restrained by the tether system. *Official USAF photograph.*

The first three flights demonstrated that the subscale predictions were conservative estimates. The temperature and acoustic data were within 10 percent of the subscale values, confirming the validity of using the subscale models for further testing as the program progressed. Because of the success of the first three launches, the remaining tests used a 12-foot-diameter silo insert. The launch of STM 104 was postponed for approximately a month and the test objectives modified to include water spray into the exhaust to reduce the temperature and the acoustic environment in the silo, as had been tested in Britain.

STM 104 was launched successfully on 22 December 1959. The water spray resulted in an unexpectedly stronger acoustic environment, and no further tests with it were conducted. To evaluate materials for external insulation of the missile airframe, six white acrylic panels were installed on the missile, 10 inches square and 0.04 to 0.06 inch thick.

STM 105, the first missile with an autopilot to control the newly installed movable nozzles, was successfully launched on 27 January 1960 from Silo 2, with a flat-plate flame deflector. Twenty-four insulation panels were placed on the airframe: six white acrylic, six black acrylic, six cork, and six polypropylene. The missile rotated 270 degrees as planned during the short flight, and a nearly flat landing preserved the motor nozzles for evaluation.

STM 107 was the first missile that utilized flight-weight hardware. It successfully launched on 3 March 1960 and utilized a slightly slower-burning propellant, which caused it to emerge more slowly from the silo. STM 107 also had an autopilot and movable nozzles, and the missile rotated 310 degrees, as expected. Unfortunately, STM 107 landed on the Stage I skirt, and the nozzles were not available for inspection. Still, the flight was considered highly successful.

STM 109 successfully launched from Silo 1 on 5 April 1960. The silo had been converted to the 12-foot-diameter configuration with a concrete liner, which appeared to suffer little damage, although flaking did occur and would have to be dealt with since the flakes could damage the missile before it fully emerged from the silo.[7] This time the missile landed nose first, with the nose, the instrumentation section, the guidance-and-control section, and half the Stage III motor buried. The Stage II ballasted motor had split open, but the Stage I motor remained intact. The nozzles and hydraulic actuators were shipped to Thiokol for evaluation.

The last silo test launch was that of STM 111, launched on 6 May 1960 with essentially the same configuration as STM 109, except that the concrete interior of the silo was now lined with steel to prevent flaking. For the first time in the test program, the Stage I motor failed approximately 10 feet above the silo lip. This did not interfere with in-silo launch evaluation, which was the primary goal of the test series. The AFFTC in-silo launch test program was so successful that it ended earlier than planned, with the return of facilities to Edwards AFB control on 16 May 1960.[8]

## *Air Force Missile Test Center, Atlantic Missile Range*

Concurrent with the silo design program at Edwards AFB was construction of Minuteman assembly, storage, and launch facilities at the Air Force Missile Test Center (AFMTC), AMR, located at Patrick AFB, Florida. The objective of Minuteman testing at AFMTC was to determine the practical aspects of the weapon system program as well as to evaluate operational ground equipment design and procedures. This would involve a detailed R&D program using both pad and silo launches. Missile storage and assembly facilities, a launch control blockhouse, pad, and silo LF construction were necessary, as well as development of missile handling equipment for the operational bases (Figure 12.6).

The flight test program was designed to be conducted between December 1960 and December 1962, utilizing approximately 50 fully operational configuration XSM-80 missiles launched using two aboveground pads and two silo launchers. Flight objectives were (a) evaluation of functional operation of the missile in flight, (b) demonstration of in-silo launch of a fully configured missile, and (c) demonstration of guidance and control of the missile as well as the performance of the reentry vehicle.

The flight test program was divided into three blocks. Block I would be eight missiles fired between December 1960 and June 1961, with emphasis on evaluation of the propulsion system. Block II would be 22 missiles fired between June 1961 and April 1962, with emphasis on evaluating guidance-and-control performance as well as that of the reentry vehicle. The Block II launches would be primarily from the silos, with most of the launches to a range of 4,400 nm, but with single firings from 2,000- to 6,500-nm ranges. Block III would be 20 missiles fired between April and December 1962 from the silos. These would be used to evaluate system compatibility and would include 10 launches approximating operational procedures within the R&D data-gathering requirements. A fourth block of 10 to 15 missiles could be added to achieve more statistically relevant accuracy and reliability information.[9]

Facility construction began in July 1959, making use of several structures remaining from the Launch Complex 9/10 used with the Navaho cruise missile and Jason and Alpha Draco sounding rocket programs. The launch pads and facilities other than the silos were completed by November 1960. The dual pad, silo, and blockhouse facilities had

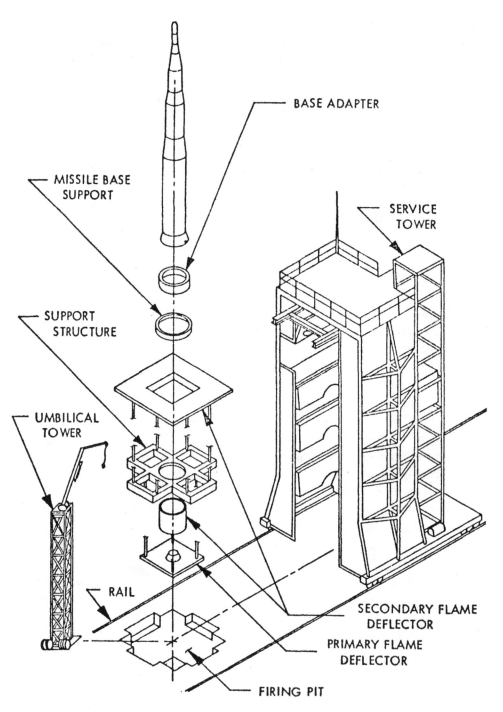

FIGURE 12.6 Configuration of pad LFs at Patrick AFB, Florida, circa 1961. There were two pad LFs, Launch Complex 31A and 32A. *"Minuteman Historical Summary, 1960–1961," courtesy of Boeing Company.*

been approved to permit launch operations on the high-priority program to continue if a test failure caused significant damage to either one. Construction on the two silos had awaited the results of silo tests conducted at Edwards AFB, which had ended in May 1960. All Launch Complex 31/32 facilities were officially turned over to the Air Force on 18 April 1961, though the first launch had taken place on 1 February 1961.[10]

*Minuteman I Research and Development Flight Test Program*

MINUTEMAN IA

*1961*

The Category I flight test program was at first conducted by Boeing personnel and then by Air Force personnel from the 6555th Guided Missile Group (6555 GMG) at the AMR. After two months of delays due to a heavy test schedule and unresolved radio interference from some of the ground equipment, the successful launch of Flight Test Missile (FTM) 401 took place on 1 February 1961 from the pad at Launch Complex (LC) 31A. The launch and flight were a resounding success, the first time a multistage missile had been launched with all stages operational on its first flight. In context, it had been only three years since the initial briefing on Minuteman to the secretary of defense. The missile flew 3,999 nm, impacting 2.3 nm long and 0.4 nm to the left of the target in a broad ocean area near Ascension Island, as planned. This launch was referred to by Col. Samuel C. Phillips, Minuteman program manager, as "December 63rd, 1960," in reference to the originally scheduled launch date of 1 December 1960 (Figure 12.7).[11]

Unfortunately, the next launch, FTM 402, LC-31A, was not as successful. Launched on 19 May 1961, first-stage operation was normal, and the second stage appeared to ignite properly, but shortly thereafter, the missile began to pitch and roll, and at 98.7 seconds into flight, the missile broke up. Subsequent investigation showed that the NCU for Stage II had malfunctioned.

The flight of FTM 403, launched on 27 July 1961, LC-31A, was nearly as successful as FTM 401, except that the thrust termination system activated prematurely, and the reentry vehicle impacted 425 nm short of the intended target. This launch signaled the end of the pad launches, which were generally felt to have been an excellent beginning to the flight test program.[12]

The next flight was that of FTM 404, the first full-flight test in-silo launch at LC-32B on 30 August 1961. All appeared normal with Stage I ignition, but less than one second later, with the missile emerging from the silo, the Stage I and II interstage separation device ignited simultaneously with that of the Stage II motor. The missile impacted six seconds later, spreading burning propellant across the launch complex. Amazingly, there was little damage to the silo and launch equipment. The guidance computer initiated all staging commands, so it was clear the problem must be a guidance-system malfunction. Since the tests at Edwards AFB had shown the guidance system could withstand the in-silo launch vibration environment, what could have gone wrong with the guidance system? Fortunately, a significant amount of the guidance system was recovered, and the problem

FIGURE 12.7   Minuteman IA FTM 401 at the instant of Stage I ignition on 1 February 1961, LC-31A. The two umbilicals, one for the guidance system and one for the instrumentation wafer, have just released. The bright orange color at the base of the missile is a reflection of the Stage I igniter flame. The white triangular structures just below the top black section are telemetry antennas located on the instrumentation wafer below the guidance section. *Official USAF photograph, courtesy of Air Force Space and Missile Museum, CCAFS.*

was discovered to be one of quality control. The soldering tabs in the torus from which signals were distributed had vibrated loose, shorting the connections, causing a loss of power to the guidance-and-control system, which caused the premature ignition (Figure 12.8).[13]

There was a delay of nearly three months while Autonetics worked to improve quality control in the fabrication of the guidance-and-control section. A decision needed to be reached on whether to launch from the pad or the silo, as the launch schedule was getting tighter and tighter in order to meet the proposed operational date of December 1962. As an added precaution, for the next launch, LC-32B was lined with fiberglass material to serve as sound attenuation.

FTM 405 launched successfully from LC-32B on 17 November 1961. The launch and flight were successful, with all three stages of powered flight and thrust termination performing as planned. The reentry vehicle impacted 2,993 nm downrange. Data indicated the vibration environment was lower than for FTM 404. A nozzle control perturbation occurred on Stage II, but the autopilot successfully corrected the problem.[14]

With the successful silo launch of FTM 406 on 18 December 1961 from LC-31B, the in-silo launch concept was confirmed. Further research on launcher hardware and

FIGURE 12.8 Unsuccessful launch of FTM 404, the first full-flight test of in-silo launches at LC-32B on 30 August 1961. All appeared normal with Stage I ignition, but as the missile emerged, excessive vibration caused the missile guidance system to fail, which led to premature Stage I and II separation and Stage II ignition. The Stage I forward dome and Stage II engine nozzles were destroyed, leading to a spectacular explosion. *Adapted from official USAF photograph, courtesy of John Hilliard.*

positioning would be one of the goals for 1962. Six launches had been attempted with four successes, two each for the pad and silo launchers. When compared to the first six launches of the Thor (one success) and Atlas programs (three successes), the Minuteman program showed essentially the same growing pains. While it might be argued that the liquid-propellant missiles were more complex and not a reasonable comparison, neither the Thor nor Atlas missiles had to survive the in-silo launch environment.

*1962*

Boeing and the Air Force had an ambitious 30-launch schedule for 1962. The program included high-trajectory minimum and midrange launches, offsetting the position of the launch mount, removal of the Stage I skirt, and evaluating the proposed Malmstrom missile suspension system equipment.[15]

Flight testing resumed on 5 January 1962, with the successful launch and flight of FTM 407 from LC-31B to a midrange target. This was the most successful flight to date, with a complete set of telemetry. On 25 January 1962, FTM 408 was successfully launched and completed a midrange, high-trajectory flight profile.

Missile processing delays caused by a shortage of flight-ready guidance systems, repair to silo launch damage, and incorporation of operational equipment to be used in Wing I threatened the ambitious schedule. The late delivery of missiles and launcher hardware from Boeing, as well as reliability issues with telemetry equipment, also jeopardized the schedule. To further complicate matters, a lightning strike near LC-32B resulted in extensive damage shortly after removal of FTM 410 for inspection due to moisture accumulation.

The first use of the emergency power cutoff procedure took place on the aborted launch attempt of FTM 410 on 9 February 1962. After umbilical disconnect, the digital data programmer lost lock-on with the computer, and the launch was aborted. On 15 February 1962, FTM 410, the initial flight of the Skirt Removal Test Program, launched successfully from LC-31B. While all flight objectives were met, the reentry vehicle impacted 12 nm long. There are no further details concerning the results of the skirt removal flight tests.

On 24 April 1962, FTM 412 was successfully launched, but only seconds into flight, the missile strayed off course, and the range safety officer had to activate the destruct mechanism. The quick look data analysis showed a reduction in Stage I chamber pressure, which most likely had been caused by a burn through in the aft closure dome. This was hard to reconcile with the ground test results, which had not demonstrated that mode of failure during the entire static test program. Salvage ships were sent out to recover as much of the missile as possible, which was then pieced together in a vacant hangar. A sufficient number of fragments were recovered to show that the drop in pressure had indeed been caused by a burn through located in the center of the aft closure dome.[16]

By April 1962 ground preparation time per missile had been reduced from 60 days to an average of 27. The complexity of missile assembly was not as much of an issue as the delays in delivery of components from manufacturers. By July 1962 missile launches were taking place on an average of every 19 days (Figure 12.9).

FTM 417 was launched on 12 July 1962, with normal flight until approximately 45 seconds, when the missile experienced abnormal pitch, roll, and yaw maneuvers. The flight

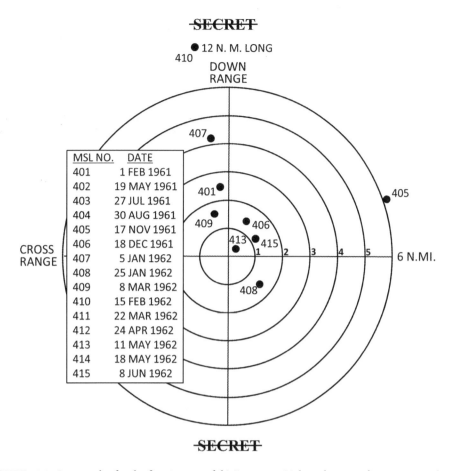

FIGURE 12.9 Impact plot for the first 9 successful Minuteman IA launches out of 15 attempts. Though an insufficient number of flights for the determination of a CEP, the results showed that the Minuteman IA accuracy requirement could be met.

terminated at 52 seconds. The failure was attributed to the Stage I No.1 nozzle. FTM 417 was the first flight of interim Wing II motors prior to their qualification tests.

On 17 October 1962, FTM 422, a Wing I–configured Minuteman IA, was successfully launched from LC-31B on a minimum-range high–reentry vehicle heating trajectory. Just seconds into flight, the missile veered off course and had to be destroyed. This was a night launch, and the destruction of the missile created a spectacular Roman candle–type fireworks display. Thiokol had modified the Stage I motor nozzles in both thickness and length to increase thrust and thus range. If the problem was in the nozzle design, this had to be quickly addressed and the nozzles replaced on subsequent test vehicles. There was no indication of a drop in chamber pressure, indicating a burn through of the aft dome, as had been seen on FTM 412, was not likely the cause.

Thiokol engineers requested permission to enter the silo and inspect the launch ring to see if possibly the longer nozzle design had resulted in contact with the missile support ring and been damaged at liftoff. The ring showed evidence of contact, but it was not possible to conclude if it was a guidance system problem or a result of the guidance responding to a damaged nozzle. The launch ring was scalloped out in the area of the four nozzles to provide greater clearance, and the problem was not experienced again (the modified launch ring was installed in the operational launchers).[17]

FTM 426, launched on 20 December 1962, was the first of a series of experiments to evaluate the effect of both missile offset and tilt that might result from launcher movement during an attack. While the missile suspension system was designed to be self-centering, this test would evaluate the worst-case scenario. The missile centerline was offset 1.25 inches from the silo centerline and did not prevent a successful launch. Flight was normal through most of Stage II operation, but a malfunction occurred just before Stage II to III staging that caused the guidance stable platform to tumble, and the missile impacted short of the target.[18] The Minutemen IA flight test program did not achieve the 30-flight schedule for 1962. There were 17 launches with 4 failures, for a 76 percent success rate.

## 1963

FTM 423A was launched from LC-31B on 7 January 1963. The flight was normal during Stage I and II operation, but midway through the Stage III operation, a burn through of the aft closure at the ignition port occurred, and impact was 1,990 nm short and 40 nm miles left of the target (unlike Stages I and II with the igniter at the top of motor, the Stage III igniter port was in between the four nozzles).

FTM 419 was launched on 23 January 1963 from LC-32B. The flight was successful, but the reentry vehicle did not separate from the Stage III motor. However, 90 percent of the test objectives were met.

FTM 421B was launched successfully from LC-32B on 20 February 1963. Powered flight and staging events were normal, with the reentry vehicle following the short-range profile of 3,100 nm, with impact 0.32 nm short and 0.41 mile right of the target. FTM 421B was the first missile launched with a deliberate tilt, 0.35 degree from the silo centerline, and no detrimental effect was seen (later experiments with tilt up to 0.7 degree also were successful). These tests were important because they validated not only the ability to launch after attack with a displaced missile suspension system but also the missile-bending moment tolerance calculations by Boeing engineers.

The Minuteman IA flight test program concluded at AMR on a positive note. All the program's flight test objectives had been met. Of the 24 flight tests, 16 were considered successful, 4 were partially successful, and 4 were failures (Minuteman IB FTM 423 with three Wing II motors is not counted as a Minuteman IA failure). See Appendix B, Table B.1, for a complete list of Minuteman IA R&D launches.

## MINUTEMAN IB

### 1962

FTM 424 was successfully launched on 7 December 1962 from LC-32B; all aspects of the flight appeared normal, including the first flight of the Mark 11 reentry vehicle.

FTM 423 was successfully launched from LC-31B on 14 December 1962. The missile was the first programmed for impact in the Ascension Island Station 2 Missile Impact Locator System (MILS) Net. The flight was a successful test of the interim Wing II configuration, except for the range instrumentation and Wing I reentry vehicle. FTM 423 was the first missile that had extensive use of cork insulation in place of Avcoat. The base areas of the Stage I and I to II interstage, as well as the exterior of the Stage II to III interstage and all of the raceway covers, were insulated with AC 2755 cork.[19] Temperatures recorded at launch and in flight were significantly lower than predicted. Samples placed on the silo wall and exposed to missile launch conditions showed ablation of 0.04 to 0.06 inch, indicating a substantially thinner layer could be used in the future, with a significant savings in weight.

*1963*

The fifth Minuteman IB missile, FTM 425, launched from LC-31B on 18 March 1963. FTM 425 was the first test of a cylindrical Stage I skirt. Flight was normal until a burn through of the Stage III motor. The use of a cylindrical aft skirt appeared to be successful, as Stage I performance was normal in all aspects.

FTM 419A was launched from LC-32B on 27 March 1963. This was the last missile scheduled to use the Mark 5 reentry vehicle. Flight to the target area was successful.

On 19 April 1963, the 6555 GMG, Minuteman Operations Branch, was assigned to operate the Minuteman launch complexes at AMR with Boeing personnel serving in a support capacity, supplying the missiles, and supporting flight operations. Four prior launches, FTM 416, 421, 424, and 417, from LC-32B, had been all-military operations.[20]

FTM 432, launched 24 July 1963, continued the deliberate offset tests with the centerline of the missile mount deliberately offset three inches from silo centerline. The launch was successful, and the missile followed the planned short-range trajectory, impacting in the target area, Broad Ocean Area Zone 2, approximately 1,860 nm downrange.

The Minuteman IB guidance system had dual target capability. The first test of this capability was on 16 July 1963, but due to an electrical malfunction in the NCU of Stage I, the launch was unsuccessful. The next attempt was on 5 August 1963; the launch of FTM 433 was successful, and the yaw maneuver to align to the second target trajectory was completed as programmed before a Stage III malfunction.[21]

In January 1963 the Air Force authorized, through STL, the modification of three Wing II–configuration flight test missiles, FTM 446, 447, and 448, with WS-133B (Minuteman II) interim Stage II motors, which had a single nozzle and a LITVC system (see Chapter 8). The flight of FTM 438 was also used to continue evaluation of a cylindrical rather than flared Stage I skirt. This modification provided a substantial gain in range but made the missile more aerodynamically unstable; however, the guidance system was able to successfully compensate. Furthermore, the results showed that missiles with the cylindrical skirt would be able to fly a hot, or high–dynamic pressure, trajectory, so the skirt design change was incorporated into the Minuteman IB and II operational configurations as well as Minuteman III.[22]

Between 1 July 1963 and 30 September 1964, which was the last Minuteman IB launch at AMR, the 6555 GMG, Minuteman Operations Branch, launched 19 Minuteman IB missiles, 9 from LC-31B and 10 from LC-32B. Fourteen of the nineteen flights were highly successful, with a string of twelve consecutive successes.[23] See Appendix B, Table B.2, for a complete list of Minuteman IB R&D launches.

## *Minuteman II Research and Development Program*

On 28 March 1962, Boeing received advanced notice from the Air Force that Contract Change Number 388 to the original Minuteman contract, AF04 (647)-289, would be authorized for a limited study effort on Improved Minuteman (Minuteman II). The resulting study was favorably received by the Air Force in May 1962.

Minuteman II overall capabilities were to include an increase in range to 7,500 nm; two warhead options, 0.5 MT for long-range targets or 5 MT for shorter-range targets; eight target options in each missile, with up to a 20-degree off-target azimuth; 63-day survivability at the present hardness; 1,024 target programs, with war plan options (salvo or ripple fire) included; infinite connectivity radio (low-frequency); and secure status information with an increase in status information to 50 parameters for drawer-level malfunctions. Additionally, Minuteman II could be launched by the Airborne Launch Control System (ALCS: see Chapter 15). The new missile configuration would involve a new guidance system, new reentry vehicle, a new Stage II motor with an increase in diameter to 52 inches and a single fixed nozzle with LITVC, Stage I and III unchanged, missile weight at 73,500 pounds, and an optional penetration aids wafer, 13 inches long, would be inserted between the reentry vehicle and the guidance-and-control section.[24]

The first flight test was tentatively scheduled for January 1964 at AMR, with the only significant modification to those facilities being the upgrade of the existing silos at LC-31B and LC-32B to accommodate Minuteman II. This meant repositioning the umbilical retraction mechanism and increasing the silo depth to accommodate the longer missile.[25]

### *1964*

Modifications to LC-32B were completed on 14 May 1964. FTM 449, the first Wing VI configuration missile, was emplaced in LC-32B on 14 September 1964. On 24 September 1964, the Minuteman II R&D launch program began in earnest with the successful launch and flight of FTM 449. All the Minuteman II launches were conducted by the 6555 GMG with Boeing contractor support for missile assembly and checkout (Figure 12.10).[26]

FTM 450 was emplaced in LC-32B on 14 October 1964. After being delayed due to guidance system problems, FTM 450 was successfully launched on 29 October 1964. The flight plan involved a programmed turn followed by the rollout onto the target azimuth, which took place as planned. At Stage II ignition, nozzle flow separated asymmetrically during the ignition transient, which caused bending moments in excess of design loads; however, the missile did not break up, as the guidance control corrected the problem. The flow separation had been triggered by Freon injection from the LITVC system. Early Stage II programmed roll perturbations were accomplished successfully, and the reentry vehicle impacted on target. Part of the test program was the guidance system–directed roll-and-yaw perturbations, which were successfully overcome.[27]

FIGURE 12.10 The first Minuteman II flight test missile, FTM 449, undergoing final preparation. The cork layers covering the interstages are the natural color. The guidance section to the left of the technician has startup and shutdown instructions taped to the missile airframe. *Official USAF photograph.*

On 17 November 1964, FTM 451 was emplaced in LC-32B and scheduled for launch one week later. The launch was rescheduled four times due to equipment malfunctions, but on 15 December 1964, 11 days behind schedule, FTM 451 was successfully launched and guided by the GI-T1-B internal missile alignment reference for the first time. All systems operated satisfactorily through powered flight, and the reentry vehicle impacted as planned in the Ascension Islands MILS Net.

FTM 452 successfully launched on schedule on 18 December 1964. This fourth launch was the first missile programmed to demonstrate the Wing VI high roll capability, with roll commanded through a 156-degree azimuth change. The flight was completely successful, with reentry vehicle impact as programmed.[28]

*1965*

FTM 453 was launched on 28 January 1965 and successfully carried out four identical roll perturbations programmed during Stage II flight to evaluate the torque capability of the roll control system.

FTM 455 was successfully launched from LC-32B on 7 May 1965. Reentry vehicle impact was as planned in the Ascension Broad Ocean Area target. Introduction of the pitch perturbations during Stage I flight to examine missile loads approaching wind shear design values was successful. This was the second successful flight demonstrating the launch capability from the internal azimuth reference mode.

FTM 456 launched from LC-32B on 25 May 1964 with a successful flight to target. FTM 456 implemented several firsts in the Minuteman II program. It was the first to carry radiation-protected, hardened subsystems, including the guidance-and-control system; to have the LITVC and roll control gas generators ignited by separate guidance-and-control discrete commands; and to fly with the Autonetics D37C computer using the fine countdown computation method.

FTM 457 had multiple launch delays due to difficulties with the guidance and control section but was successfully launched on 3 August 1965 from LC-32B. This was the second launch of a complete Wing VI configuration that included hardness provisions for all components except the reentry vehicle. Pitch, roll, and yaw perturbations to examine missile responses and loads near design values were successful, and the reentry vehicle impacted as planned in the target area.

FTM 454, which was somewhat of a hangar queen, was finally successfully launched from LC-31B on 23 August 1965 with a successful flight to the target area. The missile successfully demonstrated the first high counterclockwise roll maneuver during its Stage I flight and responded successfully to four roll perturbations programmed in Stage II flight. Instrumentation to determine the temperature at launch of missile suspension system spring cans revealed a high of 900 degrees F.

FTM 458 was successfully launched on 23 September 1965, with successful flight to impact. This was the first flight to test the semi-Somnus radiation protection system, where the guidance systems went to sleep for a short period as protection from radiation from a nearby nuclear detonation and then resumed operation (see Chapter 10).[29]

The Minuteman II flight test program for 1965 concluded with the successful launch and flight of FTM 459 on 1 October 1965, with impact on target.

## 1966

FTM 461 successfully launched on 5 January 1966, after nearly a month of delays due to guidance system problems.

The remaining six Minuteman II R&D flights, beginning with FTM 463, carried single Mark 12 reentry vehicles as part of the developmental program for the Minuteman III Mark 12 reentry vehicle system. The Mark 12 reentry vehicle heat shield, built by General Electric, was an entirely new technology based on a carbon filament design.

Launched on 8 July 1966, FTM 463 had a successful boosted flight, but the Mark 12 reentry vehicle disintegrated upon reentry due to spin reversal (see Chapter 13). The last launch, on 6 February 1968, FTM 468 from LC-32B was a successful boosted flight, but no further information is available concerning the Mark 12. Five of the six launches in the program successfully demonstrated the MRT targeting program. It is not clear if the last flight in the program used the MRT capability. The program stretched from July 1966 to February 1968, due mainly to difficulties with Mark 12 reentry vehicles.[30]

The last launch of a Minuteman II at Cape Kennedy was a demonstration of the Shelter-Based Minuteman concept where Minuteman III missiles would be on mobile transporters/launchers supported by a large number of hard shelters sited on a military reservation. After strategic warning the transporter/launcher would drive to a selected shelter and wait out the attack, emerging upon a signal from the Airborne Launch Control Center (ALCC), erect the launcher canister, and launch the missile. LC-31A primary and secondary flame deflectors were removed, and the ground instrumentation equipment was relocated to the Pad 31 Equipment Room. Minuteman 64-18024 was reconfigured with seven seconds of Stage I fuel and inert Stage II and III, along with a dummy reentry vehicle. This unique launch took place on 14 March 1970 and successfully demonstrated the concept (it was a test of the cold launch technique eventually used in the Peacekeeper Program).[31] See Appendix B, Table B.3, for the full Minuteman II R&D flight record.

## Minuteman III Research and Development Program

The earliest description of what was to become Minuteman III was a June 1961 inhouse study designated by Boeing as Growth Minuteman (Minuteman IIB). Growth Minuteman was a Category II(B) business plan. At Boeing business plans in this category were efforts desired by both the company and the customer, in this case the Air Force, but were funded solely by the company. The Growth Minuteman configuration included an increase in missile length of approximately 10 feet, enlarging the Stage I diameter by approximately 6 inches, the use of Fiberglass motor cases for Stages I and II, cork thermal insulation, the use of magnesium for the Stage I to II interstage, and the use of beryllium for the Stage II to III interstage. Total weight of the new missile would be approximately 110,000 pounds, which would necessitate significant changes in assembly, transportation, and handling. Due to the increase in size, the missile would be handled in two major sections, Stage I and Stage II and III, already joined with the guidance system, requiring final assembly to take place vertically within the LF. The study was begun on 5 June 1961, six months before the November announcement of the Minuteman II program by Secretary of the Air Force Zuckert. Based on the proposed weight of 110,000 pounds, Growth Minuteman was clearly not a description of Minuteman II but rather Minuteman III.[32]

Secretary of Defense McNamara's draft presidential memorandum of 3 December 1964, "Recommended Fiscal Year 1966-1967 Programs for Strategic Offensive Forces," described the planned improvements to Minuteman II. These included a new guidance system and the ability to deliver three reentry vehicles to geographically separated targets through the use of a PBCS (see Chapter 8).[33] Further analysis showed that in order to accommodate the increased weight of the PBCS equipment, the need to provide shielding against the effects of nearby nuclear detonations, and additional penetration aids, an improved Stage III for Minuteman II would be necessary. AFSC had made a similar recommendation in June 1965. The Air Staff rejected the request out of fear it would jeopardize the funding for the Advanced ICBM being planned to replace the Minuteman system.[34]

On 15 July 1965, Headquarters USAF issued the first Minuteman III R&D contract to Boeing. Minuteman III Stage I and II would not change from Minuteman II, but Minuteman III would employ an improved Stage III booster, carry more penetration aids to counter anticipated deployment of a Soviet ABM defense system, and be equipped to carry the Mark 12 MIRV system with three Mark 12 reentry vehicles.[35]

On 8 December 1965, Secretary of Defense McNamara approved development of Minuteman III. McNamara described the Minuteman III as being based on the Minuteman II and growing out of the requirement to improve the Minuteman II Stage III motor to carry the Mark 12 MIRV system. The term Minuteman III would apply to those missiles equipped with the Mark 12 reentry vehicles, while those carrying the Mark 11 or the proposed Mark 17 reentry vehicle would continue to be called Minuteman II.

On 28 March 1966, Headquarters USAF issued System Management Directive 6-61-133B (10), formally defining the Minuteman III weapon system, which would have the improved Stage III and be capable of carrying the Mark 12 MIRV system over a 5,500-nm range. The Air Force designated the new missile LGM-30G.[36]

*1968*

On 14 June 1968, modifications to LC-32B to accommodate Minuteman III launches were completed, and ground test missile GTM 071 was ready to be emplaced for compatibility testing. Verification and integration tests were completed on 23 September 1968 using GTM 071.[37]

The Minuteman III flight test program involved implementing improvements in the missile guidance system in a series of four steps, Blocks I through IV (see Chapter 10). The transition from Block III to IV, which were essentially the production systems, turned out to be very difficult and caused repeated delays in the flight test program at the ETR.[38]

The first launch in the Minuteman III program, FTM 201, took place on 16 August 1968 with a successful launch and flight to target. A modified Minuteman II guidance system was used with the new Stage III and Mark 12 MIRV system, which carried three Mark 12 reentry vehicles. Two of them were instrumented and, as on earlier Mark 12 developmental flights on Minuteman II, the reentry vehicles had spin rates that were unpredictable. Designed to spin at 60 revolutions per minute for greater accuracy, Mark 12 reentry vehicles would slow to zero spin rate and then begin spinning again in the same or opposite direction. In this case the two instrumented vehicles spun down to zero and then began spinning in the opposite direction. If at the zero point, the reentry vehicle had any trim angle, it would behave like a lifting body and potentially be 3,000 feet off the programmed trajectory. On this flight two different scoring methods were used to determine the CEP, so it was not possible to make a comparison between the instrumented and uninstrumented reentry vehicles.[39]

On 24 October 1968, the second launch was successful, but at Stage III thrust termination seal failure on one of the thrust termination ports caused damage to the PBV, resulting in failure to impact in the target area.

On 27 May 1970, a three-flight test series of Minuteman III special test missiles (STM) took place. As a result of the difficulties with the new NS20 guidance system, the Air Force had established a second source for the NS20 that also included an improved digital computer unit. Honeywell won the competition, and a limited number of the NS20 MGSs with the improved computer were built and flown at the ETR. The final flight of the Special Test

Missile Program conducted at Patrick AFB took place on 14 December 1970 and marked the 91st and final Minuteman missile full-range flight test to be launched from the ETR. The Minuteman program had spanned nearly nine years and had an 82 percent successful launch rate.[40] The available details for most of the ETR Minuteman III flight test program are sparse and summarized in Appendix B, Table B.4.

★ ★ ★

The ETR is more highly instrumented for the boost phase of ballistic missile flight when compared to the WTR, affording more complete coverage of missile performance during the development flight program. Once satisfied that the missiles were performing as expected, all launch operations moved to Vandenberg AFB, California. Unlike the facilities at the ETR, the Vandenberg LFs were in a near-operational configuration for each of the Minuteman variants. The only difference was the safety and telemetry equipment necessary for the collection of missile performance data. After the R&D flights were completed, operational modifications to the launcher equipment or missiles were evaluated at Vandenberg.

# 13

# OPERATIONAL FLIGHT AND EVALUATION PROGRAMS

## Launch and Launch Control Facilities

Boeing's Minuteman-related activities at Vandenberg AFB began in May 1960. The contract stipulated it would include an Engineering Test Program (Category I) that would continue evaluation of the weapon system, but now in the nearly operational configuration; Category II testing using operational configured facilities operated by Air Force personnel in as realistic and complete an environment as possible; and an Operational Readiness Training and Combat Training Launch (ORT/CTL) program conducted by SAC to train Minuteman crews. The ORT was the training program, and the CTL component would conduct missile launches by SAC crews under simulated operational conditions so that reliability and crew experience could be evaluated. The final program, which would be ongoing, was Category III, which would be operational system test and evaluation.[1]

There was considerable debate over the number of LFs and LCCs, both the hardened and soft aboveground test configurations needed to support these programs. Since Minuteman IA would be deployed only at Wing I, Minuteman IB at Wings II through V, and the already planned Minuteman II at Wing VI and 564 SMS, LFs and LCCs specific to the missile and wing configuration would be necessary. Construction began on 15 January 1961 for LF-01 through -05 for Minuteman I, with turnover to Boeing for equipment installation in January 1962.[2]

Category I flight test facilities would consist of one hardened LCC and LF-03 and LF-04, which would be ready by August 1962. The Category II flight program would use LF-01 and LF-02 along with one hardened LCC, and these would be ready by September 1962. Later a second soft LCC was added to the original plan, along with LF-06. LF-01 through -04 would be electrically connected in an operational configuration with the hardened LCC, while LF-05 and LF-06 were connected to the soft LCC in a test configuration (before the launch program began, all the LFs and LCCs were tied together in the operational configuration).

In February 1962 the Air Force made a proposal for additional facilities in the Wing III configuration with options for two additional hardened LCCs and 10 LFs, two additional hardened LCCs and 20 LFs, or five additional hardened LCCs and 50 LFs. In March 1962 approval was given for one additional hardened LCC in the Wing III configuration. A total of 15 Minuteman LFs were constructed between 1961 and 1965 for R&D and training purposes. The orientation of the LFs was changed from that of the operational launchers to permit launches toward the Pacific target areas (see Chapter 7). Unlike the operational

LFs, the topside design was modified to prevent the launcher closure from moving past the end of the launcher closure rails (see Chapter 6, Figure 6.11, Table 13.1, and Figure 13.1).[3]

Months before the first launch at Vandenberg, Boeing was tasked with determining the most desirable target areas in the Pacific Missile Range for Minuteman IA. Factors considered were target areas where instrumentation was already in place for accuracy evaluation and determination of CEP, maximum range, and any test plan restrictions due to the range environment. Boeing recommended that the firings be made to the Midway and Eniwetok Island splash nets for both the full-weight and deballasted reentry vehicles (Table 13.2).[4] The majority of the operational flight test targets for Minuteman I, II, and III were Eniwetok and Kwajalein Atolls, located in the Marshall Islands (Figures 13.2 and 13.3).

## Minuteman IA Flight Programs

### *Category I and II*

The Minuteman IA Category I and II launch program at the Pacific Missile Range, Vandenberg AFB (the name changed to Air Force Western Test Range in 1964, shortened to WTR in 1979, and finally to Western Range in 1991), began on 28 September 1962 with the successful

TABLE 13.1. Minuteman Launch Facilities at Vandenberg AFB: Total Launches[a]

| DESIGNATIONS | MM IA | MM IB | MM II | MM III |
|---|---|---|---|---|
| 394 A01, LF-02, LF-2 | 3 | 26 | 5 | 24 |
| 394 A02, LF-03, LF-3 | 3 | 85 | 8 | -- |
| 394 A03, LF-04, LF-4 | 16 | -- | 34 | 47 |
| 394 A04, LF-05, LF-5 | 10 | -- | 33 | 2 |
| 394 A05, LF-06, LF-6 | 18 | 57 | -- | 2 |
| 394 A06, LF-07, LF-7 | -- | 48 | 16 | -- |
| 394 A07, LF-08, LF-8 | -- | 10 | 12 | 34 |
| 394 A08, LF-09, LF-9 | -- | 41 | 1 | 61 |
| LF-10[b] | -- | -- | -- | 32 |
| LF-21 | -- | -- | 10 | 21 |
| LF-22 | -- | -- | 13 | 5 |
| LF-23 | -- | -- | 1 | -- |
| LF-24 | -- | -- | 13 | -- |
| LF-25 | -- | -- | 14 | 8 |
| LF-26 | -- | -- | 9 | 57 |
| TOTAL | 50 | 267 | 169 | 293[c] |

a) McDowell, JSR Launch Vehicle Database; Air Force Space and Missile Museum;
b) Designation changed from 394 AXX to LF-XX in January 1964; LF-22 became LF-10 in 1986; c) as of November 2019.

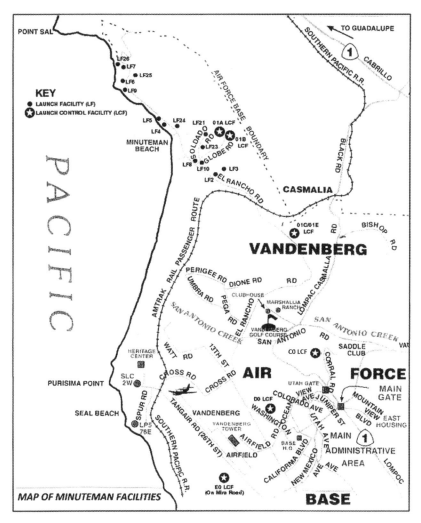

FIGURE 13.1  Map of the locations of the Minuteman LFs and LCFs (later the name was changed to missile alert facilities) at Vandenberg AFB, California. LF-10 was LF-22 until 1986, when it was redesignated LF-10 after its conversion from a WS-133B LF to one that could support missiles from WS-133A-M, Wings I through V. *Courtesy of Greg Ogletree.*

launch of FTM 503, nicknamed AIR CRUSADE, from 394 A01 by two crews from the 394th Missile Training Squadron (394 MTS).[5] This was the first Category I launch and had as primary objectives the demonstration of an operationally configured Wing I Minuteman IA missile and LF. This was the first use of the operational two-launch-vote protocol, which required agreement between two LCCs (one was the hardened LCC, with the first vote, and the other the soft LCC, with the second vote). Additionally, evaluation of the Pacific Missile Range tracking and telemetry equipment would be conducted. The programmed flight time was 1,527 seconds, with a range of 3,904 nm. Launch azimuth was 284.2445 degrees (true), and the impact point for the Mark 5 Mod 5B reentry vehicle was a broad

TABLE 13.2  Major Air Force Western Test Range Target Areas, 1980[a]

| TARGET AREA | LATITUDE | LONGITUDE | RANGE (NM) | LAUNCH AZIMUTH (DEGREES) |
|---|---|---|---|---|
| ENIWETOK ATOLL | 11°26'53.52"N | 162°13'37.20"E | 4,400 | 272 |
| GUAM ISLAND | 13°17'27.59"N | 144°42'13.59"E | 5,100 | 283 |
| KWAJALEIN ATOLL | 9°4'55.39"N | 167°33'16.22"E | 4,200 | 267 |
| MIDWAY ISLAND | 28°13'36.21"N | 177°2121'35.23"W | 2,800 | 278 |
| OENO ISLAND | 23°55'46.07"S | 130°44'32.96"W | 3,500 | 191 |
| PHOENIX ISLANDS | 3°31'41.05"S | 171°49'53.21"W | 3,600 | 242 |
| WAKE ISLAND | 19°11'37.53"N | 166°32'29.29"E | 3,900 | 276 |

a) Originally the Navy Pacific Missile Range, the designation was changed on 1 July 1964 and then shortened to Western Test Range in 1979; approximate locations determined from Google Earth.

ocean area target at 27.699 degrees north and 162.001 degrees east, approximately 580 nm north-northwest of Wake Island).[6]

The missile easily cleared the LF as expected, but 45 seconds into flight, the missile broke up due to excessive aerodynamic loads caused by a guidance system malfunction. Brig. Gen. Samuel C. Phillips, Minuteman Program director, summed up the launch as a good news–bad news event:

> The first Minuteman launch at Vandenberg AFB on 28 September 1962 was a major achievement when considered in the perspective of the overall Minuteman Program. The principal objective, to demonstrate that the ground system would check out and launch a missile and that all the systems of the total weapon system were compatible, was achieved. The airborne system malfunction which resulted in a missile destruct was an in-flight reliability-type failure, and though untimely, is secondary in importance to the significant success of the sequence of operations to that time. The test team in the field and all supporting associate contractor echelons have my sincere thanks and congratulations for the completion of this important milestone successfully and on schedule. As you are aware, we have many significant events yet to be successfully achieved in this program and I am sure you and your people are encouraged by this significant step forward in the integration and demonstration of the Minuteman weapon system as a major element of the United States defensive capability.[7]

The partial success in place, the next Category I launch, AMERICAN BEAUTY, FTM 501, was scheduled for launch on 29 October 1962. AMERICAN BEAUTY had test objectives that included those for AIR CRUSADE as well as demonstrating that the second Launch vote correctly determined the launch mode of either ripple or salvo and to test the Inhibit Launch command. The ripple fire option meant that two or more missiles would be fired sequentially with varying time delays after the second successful launch vote. The

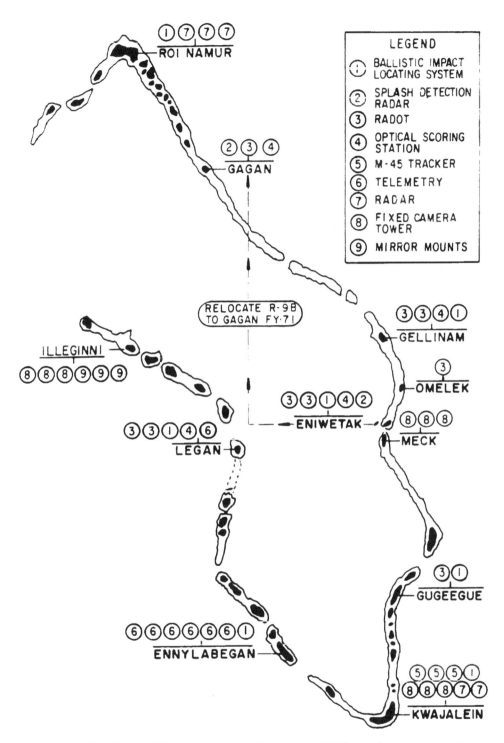

FIGURE 13.2 Kwajalein Atoll. The average depth of the lagoon is 200 feet, and recovery operations involve a two-man submarine for search operations and a SCUBA team and barge for actual recovery. *Adapted from "History of the 1st Strategic Aerospace Division, July–December 1965."*

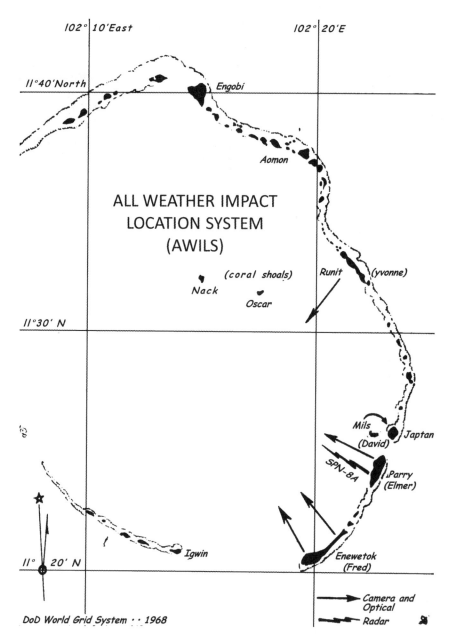

FIGURE 13.3 Eniwetok Atoll. The average depth of the lagoon is 120 feet, and recovery operations involve SCUBA and hardhat diving operations accompanied by a barge with recovery and decompression equipment. *Adapted from "History of the 1st Strategic Aerospace Division, July–December 1965."*

salvo fire option meant that two or more missiles would be launched nearly simultaneously after the second launch vote.

The Inhibit Launch command was not an original option in the launch sequence. This function had been installed as one of the recommended changes due to the findings of the Fletcher Committee Report submitted to the Pentagon in September 1961 (see Chapter 15). Due to concern of a rogue crew action or inadvertent launch even with the two-vote launch command system, the Inhibit Launch command, issued by either LCC, started a timer that would cancel the first launch vote if the second launch vote was not received within a preset delay period. If it was not received, the missile would exit the launch-commanded condition and return to strategic alert status as if nothing had happened. Thus, a rogue or inadvertent launch command could be easily canceled and would have been a nearly instantaneous response from other launch crews in the squadron.

FTM 501 would be launched from 394 A04 with the launcher closure door open. A ground test missile was in 394 A02 with the launcher closure door in the closed position to verify the modified launcher closure opening mechanism worked under these conditions. The launch mode would be ripple fire, with both LFs performing their own countdown. Flight time, target area, and payload for AMERICAN BEAUTY would be the same as for AIR CRUSADE. The launch was delayed due to the Cuban Missile Crisis, as all available LFs had fully operational missiles installed and were placed on strategic alert (see Chapter 15).

The launch was rescheduled for 21 November 1962 but was delayed several times due to malfunctions in the guidance system and unfavorable weather conditions. On 3 December 1962, the countdown for FTM 501 proceeded smoothly to T minus zero, but the missile did not respond to the launch command because the upper umbilical failed to retract due to a mechanism squib failure. The interrupted Stage I ignition was due to a flight test interlock circuit unique to Vandenberg. The result was a launch safety interrupt that prevented launch. If the umbilical did not release, it would act as a tie-down that would tug on that side of the missile during exit from the launch tube. This would pull the missile sideways, resulting in the upper stages contacting the launch tube wall. Additionally, the guidance computer would signal the Stage I to swivel as it tried to keep the missile vertical, which would also cause Stage I to impact the launch tube wall, the missile suspension system support, or both, in any case resulting in catastrophic damage to Stage I and the launch tube.[8]

One week later, on 10 December 1962, FTM 502, which had been substituted for FTM 501 because of guidance system issues, was successfully launched but had an even shorter flight time than FTM 503, again due to a guidance system failure. At 35 seconds into flight, the missile pitched up abruptly, and the range safety officer destroyed the missile. While the flight objectives were not met, the test of the Inhibit Launch command function was successful. The two crews waited the delay period caused by the Inhibit Launch command, and the missile subsequently returned to strategic alert status. The first crew restarted the launch procedure, continuing the countdown as normal. The second crew initiated their launch vote with ripple fire selected, which initiated a delayed simulated launch for the ground test missile in 394 A02 according to preset instructions in the missile's guidance computer.[9]

It soon became apparent that the Minuteman program would suffer the same problems as the other missile programs at Vandenberg in that delays in the Category I test program would cause delays in the Category II programs. The first Category II launch, VELVET TOUCH, FTM 546, had been scheduled for mid-December and was now postponed indefinitely. The change in schedule now permitted the missile combat crews from Wing I, Malmstrom AFB, to participate in the test program.[10]

After several months of investigation by Autonetics, the probable cause of the in-flight problems of FTM 503 and 502 were isolated to the Vandenberg-specific Combat Test Launch Instrumentation wafer, the telemetry equipment used for transmission of in-flight data. Radio frequency interference generated by corona discharge (static electricity) caused guidance-and-control system failures in exhaustive ground testing. Comparison of flight test failures at the AMR and Vandenberg also indicated that unanticipated vibrations caused by external missile configuration changes due to added range safety and telemetry equipment were also a contributing factor. Modifications were made to the next scheduled missile, FTM 546, and the launch scheduled for mid-March 1963. On 15 March 1963, FTM 546 failed to fire at T minus 0. Subsequent investigation revealed that the reason for the abort was a faulty thrust termination switch. FTM 546 was replaced with FTM 534 and launch rescheduled for April. Two more launch attempts occurred on 20 March and 4 April but failed due to LF and telemetry issues.[11]

The first successful launch and flight of a Minuteman IA from Vandenberg AFB was AFGHAN RUG, FTM 534, from 394 A05 on 11 April 1963 as part of the Category I program. AFGHAN RUG was the third launch in the Category I test program, which was completed on 5 July 1963 with the successful launch and flight of GRAND TOUR, FTM 518, 394 A04.[12]

The Minuteman IA Category II flight program began on 12 April 1963 with the successful launch of FTM 565, VELVET TOUCH, from 394 A01. This was the first launch where the primary LCC was manned 24 hours a day on strategic alert until launch. Missile combat crews in the primary LCC were from the 3901 SMES, a Headquarters SAC unit based at Vandenberg AFB, while the second LCC was manned by the crews from the 394 MTS. Launch procedures were as close to the SAC command-and-control procedures as possible. The Mark 5 Mod 5B reentry vehicle impacted in the Eniwetok Lagoon as planned, at a range of 4,385 nm.[13]

Many of the early Vandenberg flight tests included running various scenarios in the command-and-control procedures for Minuteman IA. On 8 May 1963, FINE SHOW, FTM 604, the fifth Category II test, was launched from 394 A05. Emergency conditions were simulated, including loss of all LFs except 394 A05 and all LCCs except hardened LCC-01, loss of the autocollimator in 394 A05 (this test evaluated guidance system ability to launch with the target azimuth stored in memory), and loss of commercial power. Emergency power was started, a single launch command was given, and after the War Plan Timer ran out, FTM 604 had a successful flight and impacted in the Eniwetok Lagoon as planned.[14]

## Demonstration and Shakedown Operation

The Demonstration and Shakedown Operation (DASO; Category III was the earlier program term) consisted of launch operations conducted by the operational commands in an operational environment using operational procedures. The Minuteman IA DASO program utilized missile combat crews from Malmstrom AFB. During this program procedures were to be refined, system capabilities and limitations demonstrated, and the determination made that the system was ready to perform its mission on a continuing basis.[15]

The program consisted of five launches from June to August 1963. The first launch, WAR AXE, FTM 521, 394 A03, took place on 18 June 1963 and was successful, with reentry vehicle impact in the target area (the target area was defined as a circle with a radius of 3.5 times the official CEP for the weapon system).[16] Ten days later, TRIM CHIEF, FTM 546, 394 A01, was successfully launched but was considered only a partial success in that Stage III failed to ignite. On 11 July 1963, TRIPLE PLAY, FTM 650, 394 A02, had a successful launch and flight to Eniwetok Lagoon, as did DIAL RIGHT, FTM 514, 394 A05, on 27 July 1963, and finally WELL DONE, FTM 520, 394 A01, on 8 August 1963. With an 80 percent successful launch-and-flight record, the system was ready for operational testing (see typical Minuteman IA flight events, Table 13.3).[17]

## Operational Testing 1963 through 1964

The next test program was called Operational Testing (OT). This phase was used to determine weapon system reliability and accuracy under representative operational conditions. This was in contrast to the objectives of the R&D (Category I and II) or DASO programs, which were conducted with the purpose of improving these factors prior to the operational tests.[18]

The OT program was comprised of 26 launches and began on 17 October 1963 with the successful launch of CEDAR LAKE, FTM 672, from 394 A03 by two missile combat crews from Wing I. This was the first flight operation where the missile was removed from a Wing I LF and transported to Vandenberg for the flight test program. Late operation of the thrust termination system caused the reentry vehicle to land 781 nm past the target area and 24.3 nm to the right.[19]

The second OT launch was on 31 October 1963. DRAG CHUTE, FTM 664, was selected on 3 October 1963 and emplaced in 394 A05 on 15 October 1963. One week later, the missile was placed on strategic alert. Nine days later, FTM 664 was launched, with impact as planned in Target Option 19, 20 nm northeast of the Eniwetok Lagoon, at a range of 4,370 nm after a flight of 27.5 minutes. Approximately eight minutes elapsed between the predicted impact time and the actual SOFAR bomb detonation that indicated the Mark 5 Mod 5B reentry vehicle point of impact. The reentry vehicle had apparently floated on the surface for eight minutes before sinking.[20]

The first ripple fire launch of two missiles took place on 29 February 1964 with the successful launch and flight to target of BRASS RING, FTM 581, from LF-04, and BOX SEAT, FTM 686, from LF-05. While the primary objective for the OT program was to determine system reliability and accuracy, a key secondary objective was to exercise the War Plan BRAVO (ripple fire) option. BRASS RING was launched first, and after the programmed delay period of 19 minutes and 58 seconds, BOX SEAT lifted off as planned. The BRASS RING Mark 5 Mod 5B reentry vehicle impacted in the Option 19 target area of the Eniwetok Lagoon, but due to the failure of the SOFAR bomb to detonate, the point of impact was calculated by computer as 0.51 nm left and 1.45 nm long. BOX SEAT was

TABLE 13.3  Typical Minuteman IA Flight Sequence[a]

| | MM IA | | NAUTICAL MILES | |
|---|---|---|---|---|
| EVENT | PLANNED | TIME (SEC) ACTUAL | DOWN RANGE[b] | ALTITUDE[b] |
| IGNITION | -0.20 | -0.27 | | |
| LIFTOFF | 0.00 | 0.00 | 0.00 | 0.00 |
| PITCHOVER | 3.00 | 3.13 | 0.17 | 0.17 |
| I/II STAGING DISCRETE | 60.75 | 58.96 | 24 | 17.3 |
| STAGE II SKIRT JETTISON | 80.75 | 77.52 | 39 | 22 |
| II/III STAGING DISCRETE | 121.95 | 120.51 | 131 | 59 |
| STAGE III SKIRT JETTISON | 121.95 | 121.07 | 131 | 59 |
| R/V MECHANICAL DISCONNECT | 170.00 | 169.38 | 292 | 114 |
| THRUST TERMINATION | 171.33 | 170.93 | 298 | 116 |
| R/V PREARM #1 | 171.68 | 170.96 | 298 | 116 |
| R/V PREARM #2 | 171.76 | 171.05 | 298 | 116 |
| ELECTRICAL DISCONNECT | 171.87 | 171.07 | 298 | 116 |
| APOGEE[b] | 808 | ---- | 2206 | 750 |
| IMPACT | 1648.46 | 1648.21 | 4367 | 0 |

a) Operational Test Exercise Report: Drag Chute, Sortie Number 99-64-M-3, Minuteman A, LGM-30A AFSN 63-124/FTM 664, 31 October 1963;
b) Downrange, altitude, and apogee values estimated from powered flight data by Jonathan McDowell.

targeted for Option 18, which was also within the lagoon, and the Mark 5 Mod 5B impacted 0.06 nm left and 0.01 nm long.[21]

The OT program ended with the successful launch and flight of QUICK JUMP, FTM 661, on 9 November 1964 from LF-05. The Mark 5 Mod 5B reentry vehicle impacted in the Option 18 target area, 0.325 nm left and 0.036 nm long. The OT program had 26 launches with 5 failures, 2 partial successes (launch and Stage I flight successful), and 19 fully successful flights for a flight-to-target success rate of 73 percent.

### Follow-on Operational Test Program 1965 to 1966

The Follow-on Operational Test Program (FOT; the flight test reports give the program name as Follow-On Operational Test Program, FOOT; other Air Force documents use

FOT) conducted tests on a continuing basis to ensure that the established reliability and accuracy factors were preserved during the lifespan of the weapon system.[22] FOT consisted of 13 launches and commenced on 24 August 1965 with the successful launch and flight of SHUTTLE TRAIN, FTM 677, from LF-06, with the Mark 5 Mod 5B reentry vehicle impacting in the target area. A unique feature of this launch and the launch of PILOT ROCK, FTM 509, LF-04, one day later was that both were timed to coincide with the overhead passage of the Gemini V spacecraft for observation by astronauts Pete Conrad and Gordon Cooper. One of their mission tasks was to determine if they could photograph launches from Vandenberg or Cape Kennedy as part of the D-7 terrestrial photography observation program. With an orbit varying in altitude from 86 to 188 nm, the astronauts briefly watched the missiles climb toward the spacecraft since a typical Minuteman IA flight profile had a maximum altitude of approximately 850 nm for a full-range flight.[23]

With the successful flight of SHUTTLE TRAIN, the Minuteman IA FOT program appeared to pick up where DASO had left off, but such was not the case. Five launches were attempted, with only three successful: SHUTTLE TRAIN, PILOT ROCK, FTM 509, LF-04, 25 August 1965, and GRAND RIVER, FTM 516, LF-04, 14 December 1965. PILOT ROCK had an in-flight failure when Stage III failed to ignite. The two failed launch attempts were ARCTIC HOLIDAY and ROCKY POINT on 16 November 1965 and 7 December 1965, respectively. The ARCTIC HOLIDAY abort was due to the failure of the downstage checkout of the Stage II NCU during countdown due to incorrect wiring. The ROCKY POINT abort was caused by the failure of the D17B guidance computer to receive the launch command. Two successful flights out of five launch attempts caused considerable concern within SAC.[24]

Fortunately, the program quickly recovered with the first salvo launches of Minuteman IA missiles on 24 February 1966, SEA DEVIL, FTM 529, LF-04, and BROAD ARROW, FTM 629, LF-06, with impact at Target Option 18 in the Eniwetok Lagoon. The operation plans for both missiles included mandatory recovery of the reentry vehicle/warhead combination (see the digital appendices for reentry vehicle recovery operation details). The salvo launch option had been an objective since the beginning of the Minuteman program at Vandenberg. Only recently had the ability to monitor the telemetry with two launches at the same time become available. The earlier ripple launches during the OT program had been spaced sufficiently apart so that the available single-channel instrumentation could be used. The limitation had not only been at Vandenberg, the downrange facilities also had to be able to handle two in-flight missiles with the need for separate destruct signals as well as tracking the reentry vehicles to impact. Both reentry vehicles successfully impacted in the Eniwetok Lagoon within 0.8 second of each other, 3.3 nm apart; SEA DEVIL was 1.51 nm short, and BROAD ARROW was 1.82 nm long, with a calculated CE of 1.82 nm for BROAD ARROW and 2.1 nm for SEA DEVIL (the only unclassified data for accuracy is given as circular error, or CE, the radial distance from the point of impact to the center of the target, and is not to be confused with the circular error probable, or CEP).[25] Target Option 18 is only 4 nm from the nearest land.[26]

The Minuteman IA FOT program was completed on 3 October 1966, with a total of 13 launches in 15 attempts, 11 successful launches and flights to target, 2 partial successes, and 2 aborts. The 77 percent flight-to-target success rate was acceptable, but of interest was the decrease in accuracy over the period of the test through the first half of 1966. The center of impact changed from 0.241 nm long and to the left of the target with the 1965 flights to 2.015 nm long and to the left of the target from January to June 1966 (see Appendix C, Table C.1).[27]

## Minuteman IB Flight Programs

### Research and Development Flight Program

The Minuteman IB was a substantial improvement over Minuteman IA, in booster performance with Stage I, a new, lighter Stage II titanium casing, the improved guidance system (NS10Q2) with dual target capability, and the new Mark 11 Mod 5B reentry vehicle.

The four-flight Minuteman IB R&D flights (Categories I and II combined) began on 24 May 1963, with the successful launch and flight of HEY DAY, FTM 658, launched from LF-07. This was the first flight from Vandenberg using an operational Stage III retrorocket system. The retrorocket operated normally, and the reentry vehicle impacted in the Eniwetok Lagoon as planned (see Chapter 9). Of the remaining three flights, GLASS WAND, FTM 625, LF-07, 30 August 1963, had an in-flight failure at T plus 117 seconds; STATE PARK, FTM 770, LF-08, 27 September 1963, was successful; but the fourth, GOLD DUKE, FTM 695, LF-07, 4 October 1963, suffered a Stage II failure.[28]

### Demonstration and Shakedown Operation 1963 to 1964

The Minuteman IB DASO program consisted of nine launch attempts with five successful launches and flights, which commenced with the flight of BIG CIRCLE, FTM 842, on 27 November 1963 from LF-07, and ended with the flight of ECHO HILL, FTM 808, LF-07, on 30 January 1964. Three of the four aborted launch attempts were due to malfunction of Vandenberg-specific equipment. The remaining launch abort was due to a procedural error; the second launch console key would not turn, and while the launch would have occurred with expiration of the War Plan Timer, the launch was aborted because the test program specified two launch commands. The Air Force scored the DASO program as a five-for-five success.[29]

SAC specified 13 test objectives to be achieved during the Minuteman IB DASO Test Program:

1. Verify the capability of the missile combat crew to receive and execute a launch message. This objective was attained.
2. Verify the capability of the missile to fire to the primary target. This objective was attained.
3. Verify that the second launch command selects the war plan and initiates the automatic launch sequence. This objective was attained.
4. Determine the spatial position of the third stage relative to the reentry vehicle at reentry. This objective was not obtained as the Pacific Missile Range did not

have the necessary midcourse tracking capability at the time. This objective was recommended to be transferred to the Operational Test program.
5. Determine the actual circular error. This objective was attained.
6. Recover the reentry vehicle. This objective was attained where specified.
7. Verify the capability of the missile to fire to the secondary target after performing a maximum yaw left maneuver. This objective was attained.
8. Verify the capability of the missile to fire to the secondary target after performing a maximum yaw right maneuver. This objective was attained to the extent of the existing platform azimuth alignment capability for launches into the Eniwetok Lagoon, as target location in the Eniwetok Lagoon was required to enable recovery of the reentry vehicle. A yaw right maneuver of 2.1 degrees was performed compared with maximum design yaw capability of 10 degrees.
9. Identify deficient operational procedures and recommend improvements in technical data, personnel subsystems, and equipment. This objective was attained.
10. Verify the capability of the weapon system to launch after being in emergency alert mode for a minimum of two hours. This objective was attained.
11. Verify the function of the "Inhibit Launch" command to neutralize a single launch command. This objective was attained.
12. Verify that "Sensitive Command Network test" and "test" commands do not delay a launch reaction when introduced after the first launch command. This objective was attained.
13. Verify the weapon system capability to launch on a single launch command. This objective was attained.

The dual target capability for Minuteman IB meant that the missile guidance system had to be able to yaw the missile a maximum of plus or minus 10 degrees if the secondary target was selected. While this capability had been demonstrated in the Category I and II flights at both Vandenberg and Cape Kennedy, one of the objectives of the Vandenberg program was to verify the ability of the missile to fire to the secondary target after performing a maximum right yaw maneuver. Due to the launch azimuth limitations from Vandenberg to the Eniwetok Lagoon target area for recovery of the reentry vehicle, which was also a program objective for several of the flights, the maximum yaw right to Target Option 18 in the Eniwetok Lagoon was only 2.1 degrees. In order to achieve this flight azimuth, the Minuteman LFs at Vandenberg were built rotated approximately 40 degrees counterclockwise from the usual 8.5 degrees counterclockwise from true north found in the operational facilities.[30]

The first launch in the program, BIG CIRCLE, impacted as planned at Target Option 19, 20 nm northeast of the Eniwetok Lagoon. The test requirement for this flight was for a nighttime reentry, which precluded the use of the cinetheodolites used for targets in the lagoon. Impact was determined by detection of a SOFAR bomb detonation picked up by missile impact sound fixing and ranging system (MISS) hydrophones placed at 3,000 feet deep. Normally a Mark 11 Mod 5B reentry vehicle would sink to that depth in five minutes, but due to the reentry vehicle configuration for the test with a weight of 450 pounds versus the normal 750 pounds, it was likely that the reentry vehicle floated before sinking and drifted an undetermined amount, thus negatively affecting the point-of-impact measurement.[31]

Perhaps the most important finding in the DASO program was what happened on the flight of ECHO HILL, FTM 808—the last flight, on 30 January 1964, launched from LF-07. While the reentry vehicle impacted within the target area, it failed to prearm. A preliminary investigation indicated that a flight safety check of velocity versus time (one of several prearm checks) by the airborne guidance system was not satisfied, and thus prearming did not occur. With obvious implications for the operational use of the reentry vehicle and warhead, further investigation was needed. The first three test exercises had used the Mark 11 Mod 5B reentry vehicle configuration, which weighed approximately 55 percent (450 pounds) of the operational Mark 11 Mod 3 configuration weight (830 to 845 pounds). These reentry vehicles had not been ballasted to make up the difference because of the added weight of the Combat Training Launch Instrumentation wafer and airborne destruct system. The Mark 11 Mod 5B reentry vehicle used with ECHO HILL was to have weighed 707 pounds due to special test needs but was later found to have weighed 745 pounds. Since the target tape had the trajectory factors for a 707-pound reentry vehicle, the safety check of velocity versus time was not met, and the warhead prearm signal was not sent by the missile's guidance system. Normally the guidance computer would have corrected for this problem, but an additional factor played a large part, important for future launches from Vandenberg—the role of propellant temperature.

Solid-propellant performance is closely tied to propellant temperature, with loss of performance as temperature decreases. The target tape program value was for 80 degrees F. Due to Vandenberg-specific LF configurations, specifically the closure of the opening for the autocollimator alignment door, which normally allowed for air circulation and temperature control, launch tube temperatures were much lower. To aid in the rapid refurbishment of the LFs between launches, the launcher environmental protection system (LEPS) closed the autocollimator opening at the beginning of the Peacetime Requirement and Preparations countdown, which typically lasted more than an hour and was not something that would have taken place with an operational launch. Additionally, the launch duct heater was routinely turned off prior to attaining missile ready status. As a result, the launch tube temperature around the missile was near the ambient temperature of 60 degrees F. Thus, the missile sat for a considerably longer time at a lower temperature than would have occurred at an operational LF. Booster performance was therefore affected, with the result being a longer-than-programmed flight time, which contributed to the flight safety velocity issue. The conclusion was that there were sufficient arming safety checks for the warhead; thus, the acceleration performance check via the flight safety velocity computation might result "in an unnecessarily narrowed weapon system capability." Warhead performance was not evaluated.[32]

The report emphasized that the small sample size of five flights presented limitations to interpretation of the test results, especially in terms of accuracy. All five of the reentry vehicles impacted within the target area. The DASO CEP was calculated to be 0.66 nm (see Appendix C, Table C.2).[33]

## Operational Test Program 1964 to 1965

The Minuteman IB OT program started with the successful launch and flight of SNAP ROLL, FTM 813, LF-02, on 25 February 1964, and was completed on 6 July 1965 with the launch and successful flight of STAR DUST, FTM 1069, LF-07. Fifty-three launch attempts were made, with fifty successful; the three aborts were weapon system failures. Thirty-nine missiles impacted in their target areas, and eleven had in-flight failures. All launches were accomplished on the first attempt except PURPLE LIGHT, which was successful on the second attempt.

Wing II, Ellsworth AFB, had 13 successes and 5 failures; Wing III, Minot AFB, had 10 successes and 5 failures; Wing IV, Whiteman AFB, had 16 successes and 1 failure. With nearly twice the number of launches as in the Minuteman IA OT program, the Minuteman IB OT program had a comprehensive list of test objectives from both the Air Force and the Army.[34]

TEST OBJECTIVES

*Air Force*

SAC objectives were focused on reliability and accuracy. Three specific test exercises were planned:

1. To boost an EWO-weight Mark 5 reentry vehicle with a denuclearized Mark 59 warhead to target Option 18 in the Eniwetok Lagoon for verification of total weapon system reliability (Mixed Marble II SAC OPS PLAN 36–6I4). This objective was assigned to Exercises GEORGIA BOY, FTM 973 and ORANGE CHUTE, FTM 848, with successful flights for both, 9 July 1964 and 6 November 1964, respectively.
2. To boost an EWO-weight Mark 11 reentry vehicle with a denuclearized Mark 56, Model 1, warhead to Target Option 18 in the Eniwetok Lagoon for verification of total weapon system reliability (Mixed Marble II SAC OPS PLAN 36–6I4). This objective was assigned to Exercise PRONTO ROSE, FTM 1142, 8 February 1965, which was a successful flight.
3. To boost an EWO-weight Mark 11 reentry vehicle with a denuclearized Mark 56, Model 1, warhead to Target Option 18 in the Eniwetok Lagoon for verification of total weapon system reliability. This was not a Mixed Marble objective as defined in confidential message from SAC (DPLR/DOOTM 03166) April 1965. This objective was assigned to Exercise VIOLET RAY, FTM 1003, but it was not achieved due to an in-flight failure on 10 May 1965. It was later assigned to Exercises SILVER CLOUD, FTM 1115, and STAR DUST, FTM 1069; both launches were successful, 18 May 1965 and 6 July 1965 respectively.[35]

The Mixed Marble II test program was a response to the Comprehensive Test Ban Treaty that had gone into effect on 10 October 1963. Mixed Marble II was designed "to confirm in practice the nuclear weapon reliability data which results from laboratory quality assurance tests." The program fulfilled a SAC requirement for evaluating all components of an operational missile. Mixed Marble II went beyond the testing of arming and fusing to actual detonation of high explosives, which replaced the nuclear materials, to verify functioning of the firing mechanism. Originally only the Titan II Mark 6 W53 warhead would have had the explosives, while Minuteman was to use an instrument package to signal successful detonation. In June 1966 the decision was made to use the explosive option for both systems. Height-of-burst data was determined by the All-Weather Impact Location System cameras on Runit, Eniwetok, and Parry Islands. There were eight launch attempts in the Minuteman IB OT Mixed Marble program: two aborts, one in-flight failure, and five successful flights. The Mixed Marble II exercise was considered highly successful, with the GEORGIA BOY and ORANGE CHUTE Mark 5 detonations taking place at 5,561 and 5,440 feet, respectively—an average of 180 feet above the planned altitude. The PRONTO ROSE, SILVER CLOUD, and STAR DUST Mark 11 detonations took place at 6,362, 6,759, and 6,261 feet, respectively, an average of 400 feet above planned altitude (see Appendix C, Table C.3).[36]

The BSD's objectives were directed primarily toward evaluation of the Mark 5 and Mark 11 reentry vehicle design and Mark 11 decoy deployment:

1. To verify reentry vehicle drag characteristics. This objective was assigned to Exercises NICKED BLADE, FTM 1163, 9 December 1964, which was an in-flight failure, and SWEET TALK, FTM 799, which was a successful flight on 2 July 1965.
2. To determine by optical means the visible and cinespectographic characteristics of the reentry vehicle plasma and wake. This objective was assigned to Exercises SPEED KING, FTM 669, 10 June 1965, and NICKED BLADE. Both exercises were failures. The objective was later assigned to Exercise SWEET TALK which was a successful flight.
3. To obtain reentry vehicle signature data. This objective was assigned to Exercises PAINTED WARRIOR, FTM 994, 29 September 1964; FIVE POINTS, FTM 1007, 7 July 1964; NICKED BLADE and SPEED KING. These four exercises were failures. Later, Exercise SWEET TALK was assigned this objective and the flight was successful.
4. To determine the effects of reentry on the Mark 11A heat shield. This objective was assigned to Exercises DOCK BELL, FTM 1132, 8 March 1965; QUICK NOTE, FTM 756, 25 March 1965; and WINTER BREW, FTM 730, 30 April 1965, all of which were successful flights.
5. To boost an EWO-weight Mark 5 reentry system to Target Option 18 in the Eniwetok Lagoon for verification of the reentry vehicle heat shield to include ECP-111 (spin fin modification) and recovery after impact in the lagoon. This objective was assigned to Exercise FIVE POINTS which incurred an in-flight failure. The objective was later assigned to Exercise OLD FOX, FTM 861, 13 July 1964, which was a successful flight with reentry vehicle recovery.
6. To boost an EWO-weight Mark 11 Mod 5B reentry vehicle to Eniwetok Lagoon, Target Option 18, to determine effects of reentry on the RV heat shield. This objective was assigned to Exercise ROSY FUTURE, FTM 1057, 18 December 1964, which was a successful flight.

7. To demonstrate proper operation of the Mark 11A arming and fuzing mechanism. This objective was assigned to the following Exercises, all of which were successful: DOCK BELL, QUICK NOTE, and WINTER BREW.
8. To demonstrate proper operation of the Mark 11 arming and fuzing mechanism. This objective was assigned to Exercise ROSY FUTURE which was a successful flight.
9. To determine spatial position of the third stage relative to the reentry vehicle at reentry. This objective was assigned to Exercises PAINTED WARRIOR, FIVE POINTS and NICKED BLADE. These exercises incurred in-flight failures. The objective was later assigned to Exercise SWEET TALK which was successful.
10. To verify the predicted decoy pattern. This objective was achieved by Exercises SMOKY RIVER, FTM 1150, 10 April 1965; SURF SPRAY, FTM 1054, 2 June 1965; WHITE GLOVE, FTM 1033, 23 June 1965; and MAPLE GROVE, FTM 1148, 29 June 1965.
11. To verify the survival altitude of the decoys. The objective was achieved by Exercises SMOKY RIVER, SURF SPRAY, WHITE GLOVE and MAPLE GROVE.
12. To determine, by optical means, the visible and spectrographic characteristics of the decoys during each flight. This objective was achieved by Exercises SMOKY RIVER, SURF SPRAY, WHITE GLOVE and MAPLE GROVE.
13. To determine the radar cross-section of the decoys during each flight. This was achieved by Exercises SMOKY RIVER, SURF SPRAY, WHITE GLOVE and MAPLE GROVE.
14. To verify the ballistic characteristics of the decoys. This objective was achieved by Exercises SMOKY RIVER, SURF SPRAY, WHITE GLOVE and MAPLE GROVE.[37]

Tests of the Mark 5 (3), Mark 11 (43), and Mark 11A (4) reentry vehicles were conducted during this program, though only the Mark 11 and 11A were deployed on Minuteman IB. The combination of an EWO Mark 5 reentry vehicle containing a denuclearized W59 warhead and the Combat Training Launch Instrumentation wafer was too heavy for a Minuteman IA to boost to the Option 18 target; therefore, Minuteman IB was utilized for two launches, with one successful—OLD FOX. Recovery of each of the reentry vehicles was an objective in two of these tests, which required Target Option 18 in the Eniwetok Lagoon with a depth of 150 to 200 feet to facilitate recovery (see Appendix C, Table C.4).[38]

One test objective left over from the DASO program was the effort to determine the spatial position of Stage III relative to the reentry vehicle at the time of reentry. Unlike the Mark 5, which tumbled at first during reentry and thus provided a large radar return before small fins imparted a stabilizing spin, the Mark 11 was spin-stabilized exoatmospherically in order to present a reduced radar return for as long as possible. The Mark 5 deployed from Stage III with only a slight increase in velocity, so the third stage would serve as a radar beacon for Soviet ABM systems. For the Mark 11, 11A, and 11B, Avco developed a retrorocket spacer that had 10 small solid-fuel thrusters that fired in pairs to impart a random velocity and position to Stage III. A lack of midcourse tracking capability had prevented evaluation of the spatial separation issue during the DASO program.[39] This objective was finally reached with Exercise SWEET TALK, FTM 799, launched on 2 July 1965. The test was conducted at night. Observation of the Stage III and reentry vehicle separation was done by support aircraft capable of continuous observation of the vehicle as it descended from 175,000 to 75,000 feet.[40]

*Army*

The Army Nike-X program objectives were:

1. To obtain reentry data on the reentry vehicle with Discrimination Radar. This objective was assigned to Exercise NICKED BLADE, which resulted in a failure.
2. To obtain radar data on the reentry vehicle plasma and wake during reentry. This objective was assigned to Exercise SPEED KING, which resulted in a failure. It was later re-assigned to Exercise SWEET TALK, which was successful.
3. To obtain infrared/optical data on the reentry vehicle during reentry. This objective was assigned to Exercise SPEED KING which resulted in a failure. It was later assigned to Exercise SWEET TALK which was successful.
4. To obtain reentry data on the reentry vehicle with both the TTR-L and TTP-5 radar. This objective was assigned to Exercise NICKED BLADE which resulted in a failure.
5. To obtain radar data on the reentry vehicle and decoys during reentry. This objective was achieved by Exercises: SMOKY RIVER, SURF SPRAY, WHITE GLOVE, and MAPLE GROVE.
6. To obtain infrared/optical data on the reentry vehicle and decoys during reentry. This was achieved by Exercises: SMOKY RIVER, SURF SPRAY, WHITE GLOVE and MAPLE GROVE.[41]

IN-FLIGHT PROBLEM AREAS

As might be expected in a 50-launch program, several in-flight problems occurred. Three flights, PAINTED WARRIOR, LONG SHOT, and VIOLET RAY, had in-flight failures at T plus 117 seconds. Looking back at the Vandenberg R&D program flights, GLASS WAND had suffered a nearly identical flight-time failure. The Air Force, Thompson-Ramo-Wooldridge (TRW), and Boeing conducted a joint investigation and concluded a Stage II forward dome rupture due to inadvertent destruct package firing was the likely cause. Since operational missiles did not carry the range destruct package, no operational changes were necessary.

Two other problem areas involved the Stage I nozzles, with three failures, and one failure involving NCU hydraulics. The Stage I Nozzle Reliability Improvement program resulted in redesigned Stage I nozzles for Minuteman II, but no changes were made for Minuteman IB. The hydraulic failure was resolved by modifying a shear pin in the hydraulic unit.[42]

IMPACT ANALYSIS

While all 39 successful flights impacted within the target area, which was a circle 6.30 nm in diameter (3.5 x CEP), only 19 landed within the area defined by the system CEP, a circle with a radius of 0.90 nm.[43] Most of the remainder landed to the left (maximum 2.75 nm) and long (maximum 2.45 nm) of the target. Starting with the flight of TOP RAIL, FTM

1101, 9 December 1964, LF-08, a change was made in the Mark 11 drag model. Prior to the change, 40 percent of the Mark 11 impacts were within the CEP, while after the change, the percentage increased to 73 (this does not include the Mark 11A flights). Two of the three Mark 5 flights to target were within the Mark 5 system CEP.[44]

Detailed analysis of reentry vehicle impact points is somewhat of an art, as distance and grouping errors imply different problems. If more of the impacts were short or long, then this suggested a problem with the reentry vehicle drag model. Outliers from the majority of the pattern could be from higher-than-normal winds at high altitude, reentry vehicle separation problems, or booster irregularities, among others. Right or left groupings implied a guidance bias. When multiple groupings of any of these occurred, then it was a problem that needed attention.[45]

## Follow-on Operational Test Program

The FOT program began on 27 January 1966 with the successful launch and flight of ANCHOR POLE, FTM 1243, LF-09, and ended on 10 June 1966 with the successful launch and flight of EBONY ANGEL, FTM 1261, LF-09, for a total of 14 successful launches and flights to target. Seventeen launches had been attempted, giving the highest launch success ratio—82 percent—and a 100 percent flight-to-target rate, the best so far in the program. The program was carried out by task forces from the 90 SMW, F. E. Warren AFB.

There were three aborts. The first was on 6 March 1966, Exercise BIG BARK, FTM 1287. The countdown proceeded as planned until reaching Stage I ignition. The LEPS inhibited the firing signal test because the Stage I to II interstage ordnance arm–disarm switch had not been rotated to the armed position during preparation for the launch. In the unarmed condition, the closing of a relay in the LEPS system was inhibited, and the firing signal was not sent to the missile. Arm–disarm switches were not the same as the safe–arm switches in the missile. The arm–disarm switches were used in the thrust termination and interstage separation circuits. The safe–arm switches were used in the Stage I, II, and III ignition systems.

In retrospect, a similar problem had occurred in two other launch attempts, one in the Minuteman IA OT program, Exercise BULL MARKET, and another in Minuteman IB OT, Exercise THUNDER VALLEY. An arm–disarm malfunction had prevented ignition of Stage III during the Minuteman IA FOT Exercise PILOT ROCK. Earlier, eight other missiles had to be recycled due to the safing shaft assembly not functioning properly. Further investigation discovered that the safety mechanism could be damaged during the installation and removal of the safing pin. Consequently, technical orders were revised and associated equipment modified to prevent recurrence of this problem.[46]

The second abort, NIGHT TICKET, FTM 1263, on 28 June 1966, was not due to a weapon system failure but rather an LF malfunction peculiar to Vandenberg. This had happened before due to the failure of the LEPS guillotine door to descend completely.[47]

The third abort, Exercise CLEAN SLATE, FTM 1244, was due to a downstage failure, date not given, and the missile was replaced with FTM 1211, with successful launch and flight to target as Exercise CLEAN SLATE, LF-03, on 11 March 1966.[48]

*Impact Analysis*

Eight of the fourteen impacts fell within the system CEP of 0.9 nm. The first and thirteenth flights in the FOT program, Exercises ANCHOR POLE and FOUR ACES, impacted within the target area but had excessive downrange errors of 2.383 nm long and 2.269 nm short, respectively. The cross-range errors were 0.080 and 0.526 nm left, respectively. With no clear answer for these large downrange errors, the 1st Strategic Aerospace Division (STRAD) Directorate of Materiel Launch Analysis Group recommended instrumenting the reentry vehicles to determine the cause for errors in future testing.

Eight of the impacts had a CE within 0.559 nm from the target, with the remaining four having CEs from 0.990 to 1.321 nm. Two different scoring systems were used, Metric Optic scoring and the All-Weather Impact Location System (AWILS), but all 14 flights were to Target Option 4 in Eniwetok Lagoon, so a target change source of error was not the cause for the impact dispersion. Of the impacts, 71 percent were to the right of the target, but the range errors were nearly equally long (57 percent) and short (43 percent; see Figure 13.4).

Five of the eight Mark 11A impacts and three of the four Mark 11 impacts were within the system CEP, excluding two flights whose reentry vehicles were not identified (see Appendix C, Table C.5).

## Major Test Programs

The formal Minuteman IB FOT program ended on 10 June 1966. Minuteman IB flight testing continued with an additional 192 operational and test launches from 11 July 1966 to 15 June 1993. Fourteen flights had in-flight failures, for an impressive 92 percent success record.

### OPERATIONAL TESTING

There was a total of 92 purely operational test flights, with 84 successful and eight in-flight failures, from 11 July 1966 to 10 December 1971. These launches were not part of a test program but were an ongoing sampling of the operational fleet to maintain confidence. The program achieved a 91 percent success rate. Through 30 June 1970, the test range CEP for Minuteman IB was 0.49 nm with the Mark 11 and 0.93 nm with the Mark 11A.[49]

The boosted launch portion of the program was highly successful, but impact accuracy with the Mark 11A reentry vehicle was troubling. Analysis showed that the Mark 11A accuracy deteriorated as the test reentry vehicle approached the operational weight. The program was suspended in January 1968, and three instrumented Pacer Kite program (Exercise OLYMPIC TRIALS) missiles were flown during the first half of 1968. OLYMPIC TRIALS B-1, B-2, and B-3 were flown on 2 February (Mark 11B), 10 April (Mark 11A), and 22 May 1968 (Mark 11A), respectively. On 12 July 1968, OLYMPIC TRIALS B-4 (Mark 11A) was launched, with a successful flight to impact, concluding the first phase of the Pacer Kite program. Unfortunately, no significant results were obtained from these flights. The next launch took place on 30 October 1968, but OLYMPIC TRIALS B-5 (Mark 11A) failed in flight due to burn through of the aft dome of the Stage III motor. The next three flights were successful: OLYMPIC TRIALS B-6 (Mark 11A) on 7 December 1968, B-7 (Mark 11A) on 21 January 1969, and B-8 (Mark 11A) on 21 February 1969. These flights

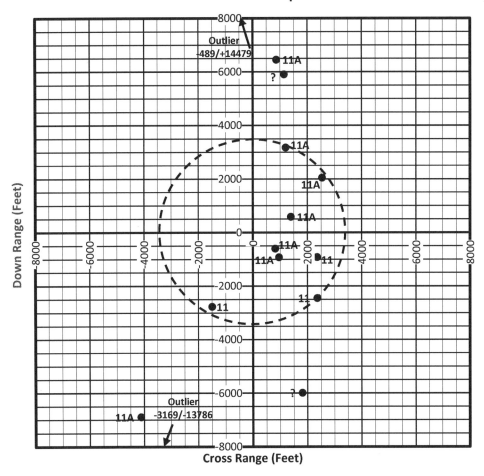

FIGURE 13.4  Impact plot for the Minuteman IB Follow-on Operational Test Program. The CEP for the 14 successful launches was 0.50 nm, the dotted circle, well within the system CEP of 0.9 nm.

produced significant results. The data did not reveal a single large problem; rather, small unrelated errors were found in the guidance system programming, reentry vehicle separation, and flight performance. The first group of program corrections was expected to improve the Mark 11 and 11A CEP by as much as 0.34 and 0.58 nm, respectively, and a new target tape reflecting the results was expected to be in place by April 1969. Additional corrections to be added to the target tape at a later date were expected to improve the CEP for both reentry vehicles by another 0.3 nm.

With three flights left in the program, the decision was made to slow the launch rate and authorize the use of Minuteman IB FOT flights with Mark 11A reentry vehicles and full retrorocket loads to be flown to the Kwajalein targets. The goal was not just to get the Mark 11A CEP under one nautical mile but also to continue to determine sources of error in accuracy.

One possible explanation of the failures was the placement of the raceway that carried cabling between the stages on the exterior of the missile. Due to range safety equipment, the raceway sizes and positions were unique to Vandenberg. The proximity of the raceway to one of the Stage III thrust termination ports was causing impingement of the exhaust and imparting a spin. This was thought to result in the inaccurate deployment of the reentry vehicle since there was no way to counteract the spin at this point in the trajectory. Modifications were made with apparently little or no effect.

Partway through the OLYMPIC TRIALS series, it is not clear after which flight, engineers at TRW—systems engineering technical advisors for Minuteman—were reviewing the trajectory tape launch site gravity values and found a clerical error had been made. Review of the discovery by the Air Force Air Staff Board concluded that the major source of impact error had been found. The final two flights in the OLYMPIC TRIALS program prior to the resumption of the DASO program confirmed the findings, as the point of impact was now within the CEP.[50]

With the successful flight to impact of OLYMPIC TRIALS B-9, the flights of B-10 and B-11 carrying Mark 11A reentry vehicles were retargeted to either Midway Island or Johnston Island in order to use a 24-degree reentry angle. This angle was used in the ETR flights, as opposed to the 19-degree angle used on the WTR and in the operational fleet. Minuteman flight program computer instructions on reentry vehicle ablation and shape changes had been developed from the early R&D flights at the ETR. Reentry vehicles using the 19-degree reentry angle experienced up to 30 percent higher aerodynamic heat loads and possibly a significantly different airflow pattern.[51]

Detailed analysis at the end of the Pacer Kite program demonstrated that approximately 0.6 nm of total miss was caused by reentry modeling errors due to nonrepresentative data from the ETR program. These modeling errors resulted in the reentry vehicle landing short of the target. Additionally, differences between the predicted density of the atmosphere at the target and the actual density had caused further reentry error. As a result, density calculations were revised, and new guidance information was added to the targeting tapes, with the result that the Mark 11 targeting CEP decreased to 0.41 nm and that of the Mark 11A to 0.65 nm.[52]

COLD/HEAT SOAK SPECIAL TEST PROGRAM

Only one description of the Cold/Heat Soak program conducted by SAC and the AFLC has been declassified. Exercise GLOWING SAND, FTM 728, LF-06, was launched on 16 January 1968, the third in a six-launch program evaluating the effect of prolonged cold temperature on Minuteman propellant performance. The primary objective of these tests was to determine the effects on propellant structural stability of temperature conditioning of the Minuteman motors below the 80 degrees (plus or minus 20 degrees) F stipulated in the weapon system's specified temperature limits. The Stage I selected for this particular test had cracks on the core surface of the propellant, but it had not been subjected to the temperature conditioning. Stages II and III had been conditioned separately for 120 hours. The performance of all three stages was within specification limits, and the ballistic parameters were similar to earlier Minuteman IB launches. The data indicated that Stage I motors with small cracks could be successfully launched, as all six launches in this program were successful.[53]

## MINUTEMAN REENTRY SYSTEM LAUNCH PROGRAM

At the end of June 1969, the AFLC announced that the last Minuteman IA LF at Vandenberg, LF-06, would be converted to a Minuteman IB configuration for use in support of the Army's Safeguard System Target Test Program. The modification program was called Minuteman Reentry System Launch Program (RSLP). Modifications were also made to LF-03 at the end of the operational testing of Minuteman IB in 1973. Both LFs were controlled from the Launch Support Center, the former soft LCC used with LF-03 and LF-06. The major modifications included raising the missile support ring 16 feet to elevate the payload above the top of the launch to remove it from the launch tube and launch environment effects (Figure 13.5).[54]

RSLP flights, in addition to the Safeguard System Target Test Program flights, took place from 8 October 1982 to 15 June 1993. There were 15 launches with 2 in-flight failures. The last launch, on 15 June 1993, was known as the Casmalia Express due to the fact that immediately after launch, the missile veered in an easterly trajectory toward the town of Casmalia. The range safety officer destroyed the missile, which resulted in a spectacular fireworks display akin to a Roman candle. Several fires were started and fanned by the Santa Ana winds. The accident investigation revealed that the probable cause for the errant flight was a simple miswiring of the guidance system (see Appendix C, Table C.6, and Figure 13.6).[55]

## ARMY SAFEGUARD SYSTEM TARGET TEST PROGRAM

Twenty-eight launches providing targets to the Kwajalein Safeguard test site were conducted from 23 September 1969 to 1 August 1974, with two in-flight failures.[56]

## ADVANCED BALLISTIC REENTRY SYSTEM

On 14 May 1963, the Advanced Ballistic Reentry System (ABRES) Program, 627A, was established by the DoD as a joint-service program covering research in reentry vehicle development and testing. Areas of research included investigation of optical and radar observable wakes during reentry, deployment of penetration aids such as decoys and chaff, and evaluation of maneuverable reentry vehicle designs. By the end of 1964, launch programs were being conducted at two locations, Vandenberg AFB and the Utah Launch Complex, Green River, Utah. At Vandenberg the program utilized surplus ICBM boosters for flights of full-scale reentry vehicles to Kwajalein Atoll. Minuteman IB missiles were used as boosters for the program from 9 March 1973 to 4 October 1981, for a total of 27 flights with 1 failure.[57]

## REENTRY BODY SUPPLEMENTAL FLIGHT TEST PROGRAM

Minuteman IB boosters were used for eight flight tests of the Navy's Trident C-4 SLBM Mark 4 reentry body (three) and Mark 500 Evader (five) from 6 March 1974 to 1 November 1975, while the C-4 missile was being developed. The Mark 4 was similar to the Air Force

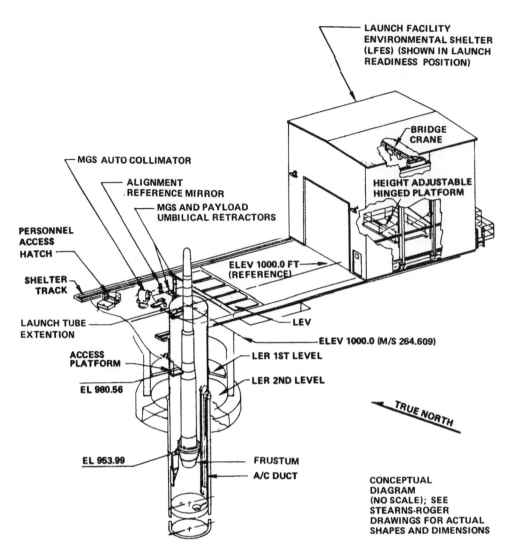

FIGURE 13.5 Modifications made to Vandenberg LF-06 and LF-03 in support of the Reentry System Launch Program. Further modifications were made in support of the Minuteman II Multi-Service Launch System program with flights from January 1987 to July 2001. *Adapted from "Minuteman I Reentry System Launch Program."*

Mark 12 reentry vehicle, while the Mark 500 Evader was a fully maneuverable reentry body that was not deployed. All were successful flights.

ARMY SPECIAL TEST PROGRAM HK

The Army Special Test Program HK (homing kill) was part of the Army Ballistic Missile Defense Program, a series of four flights from 28 October 1972 to 11 December 1973 that provided four Minuteman IB targets for the program. All four flights were successful.[58]

FIGURE 13.6 The Casmalia Express, Minuteman IB TDT-1, shortly after destruction by the range safety officer on 15 June 1993. The payload was a Target Development Test vehicle as part of the Ground-Based Interceptor Program. *Official USAF photograph.*

## *Minuteman II Flight Programs*

On 2 April 1962, the Air Force authorized Boeing to define the WS-133B Weapon System, Minuteman II, technical and scheduling requirements. Specifically, Boeing was to assist BSD and STL in preparing requirements for the weapon system functional analysis and facility concepts, design criteria for test sites and the Minuteman II operational sites, and test program requirements for the test sites.

The $7,486,000 contract for the construction of six Wing VI LFs and three LCFs at Vandenberg was awarded to Allied-Schaffer Construction Company on 11 December 1963, with the completion date scheduled for 16 November 1964. Due to program delays, Boeing did not take possession until 3 February 1965. Final acceptance by the Air Force took place for LF-021 and LCF-C0 on 28 July 1965, LF-022, LF-023, and LCF Delta-0 on 27 August 1965, and LF-021, LF-024, and LCF Echo-0 on 29 September 1965. Boeing began Force Modernization upgrades to LF-05 (Wing I) and LF-08 (Wing V) as well as LCF-01A and LCF-01C on 19 July 1965. Work was completed in January 1966. Eventually, 13 LFs were built or modified for use in the Minuteman II flight program (Table 13.1)[59]

FLIGHT TEST PROGRAMS

A number of factors resulted in a somewhat convoluted Minuteman II flight test program at Vandenberg. Minuteman II was being deployed at LFs and LCCs designed specifically for it at Wing VI, Grand Forks AFB, and the 564 SMS at Wing I, Malmstrom AFB (WS-133B). Simultaneously, the Force Modernization Program had begun (WS-133A-M), replacing the 150 Wing I Minuteman IA and the 300 Minuteman IB missiles at Wing IV, Whiteman AFB (150), and Wing II, Ellsworth AFB (150), with Minuteman II while modifying the LFs and LCCs to accommodate the new missile. Therefore, two test programs had to be run simultaneously to evaluate equipment and procedures for the two substantially different operational facilities. On top of this effort was the development and flight testing of the Mark 11A, 11B, 11C, and Mark 12 reentry vehicles.

Unlike Minuteman IA and IB flight programs, which were consecutive, the Category I and II flight programs (now called R&D flight tests) that tested the Mark 11A through C reentry vehicles overlapped with the Minuteman III Mark 12 development flights. Additionally, there were the two DASO programs that evaluated equipment and procedures for both the Wing VI and Wings I through V Force Modernization facilities present at Vandenberg. Problems with the Minuteman II NS17 guidance system and accuracy issues with the new reentry vehicles prolonged the test programs significantly. Unfortunately, the detailed flight test reports remain classified for Minuteman II. Available details of specific flight tests are discussed below.

WS-133B

*Wing VI*

The launch program at the ETR had evaluated the functioning of the components of the system. With the program transfer to Vandenberg, complete weapon system testing began for Wing VI with a successful series of two Category I and four Category II R&D flights starting on 18 August 1965 and ending on 16 February 1966. Two additional Category I flights took place in late 1967 as part of the Mark 1 penetration aid development flight program (see Appendix C, Table C.7).[60]

*Mark 12 Reentry Vehicle Development Flight Tests*

While the Air Force remains sensitive to a detailed discussion of the development of the Mark 12 reentry vehicle for use in the Minuteman III program, there is some unclassified information pertinent to the flight testing at Vandenberg. The first flight was FAINT CLICK, FTM 2049, launched on 17 March 1966 from LF-08. The boosted portion of the flight was successful, but the reentry vehicle nose tip fractured during flight, and the reentry vehicle disintegrated before impact. Data collected from the flight revealed that the fabrication of the heat shield was the problem. The carbon filament material was wrapped at an angle, and when the resin that held the material in place melted during exposure to the high temperature of reentry, ridges were exposed, which acted as turbine blades and caused the reentry vehicle to spin up to 2,000 revolutions per minute, nearly 33 times faster than designed. The result was a centrifugal force of 1,000 gs, which literally tore the reentry vehicle apart. To help dissipate the heat, a Teflon cap was installed and the internal structure strengthened. The first successful flight of a Mark 12 reentry vehicle took place on 11 May 1967: BUSY FELLOW, FTM 2095, LF-21. Unfortunately, the problems were not solved completely, as spin reversal continued, and lofting became an issue (see Appendix C, Table C.7, also Minuteman III, below).[61]

*Demonstration and Shakedown Operation*

The reentry vehicle configuration for the Wing VI DASO flight program was the Mark 11A Mod 5 BH ballasted to EWO weight. The target was Eniwetok Target Option 18, with the possibility of flights to Kwajalein as well. The retrorocket spacer would be stripped of retrorockets. The trajectory would be Shape One, with the option to use Shapes Two and Three.

The Wing VI DASO flight-test program had an auspicious start on 1 August 1966 with the successful flight to target of STAR BRIGHT, FTM 2061, LF-24. Remote target change was achieved from the LCC, and the missile rejected a Launch command deliberately sent before the Launch Enable command (see Chapter 15). Followed by a second successful flight to target with CAREER GIRL, FTM 2026, on 26 August 1966, LF-23, the Minuteman II program seemed to be on solid ground. Such was not the case, as on 2 November 1966, GOLDEN AGE, FTM 2068, LF-22, was successfully launched, but the range safety officer had to destroy the missile at T plus nine seconds into flight. This in-flight failure, combined with the fact that the Mark 11A reentry vehicles were consistently falling short of the target, led to suspension of the DASO program and the start of a nine-flight special test series called OLYMPIC TRIALS (see Appendix C, Table C.8, and previous discussion of Minuteman IB).[62]

The Wing VI DASO resumed with the flight of GIANT BLADE 1, FTM 2503, on 30 April 1968, with a successful launch, only to have the Stage II safe–arm circuit fail, resulting in the second suspension of the program.[63] Investigation determined that surges of electrical power coming from the operational ground equipment in the interval between first-stage ignition and umbilical disconnect were causing the Stage II safe–arm system to return to safe. Instrumentation was added to the next flight that confirmed the problem. The program resumed on 21 November 1968 with the successful launch of GIANT BLADE 2, LF-25, only to have a Stage III flight control failure due to excessive vibration. The Wing VI DASO program ended with this flight.

*WS-133A-M*

The Force Modernization Category I flights began on 22 January 1966 with a successful launch, but FTM 2022 had a Stage II control anomaly and did not impact on target. This was followed by four successful flights to target. While the boosters performed properly, the Mark 12 reentry vehicles broke up on reentry (also see Minuteman III below). The program ended on 11 January 1967.[64]

Because the GIANT BLADE 1 failure had been in operational ground equipment, meaning an LF equipment problem, SAC opted to begin the Force Modernized DASO program. In this case it was essentially a continuation of the Wing VI program in that the same missiles were being flown, so the demonstration was really that of the LFs. The first flight, GIANT FIST 1, FTM 2589, LF-04, had a successful launch and flight to target on 8 July 1968. Three days later, the first attempt to launch GIANT FIST 2 resulted in an abort, and investigation of the problem caused a delay in the program for three months. The successful launch and flight to target of GIANT FIST 2A on 24 October 1968, LF-04, seemed to put the program back on track, but the Stage III failure of GIANT FIST 4, FTM 2537, LF-05, on 13 November 1968 caused the third suspension in the program (see Special Test Missiles below and Minuteman II Stage III problems, Chapter 8).

The Force Modernization DASO program resumed on 2 February 1969 with the successful launch and flight of GIANT FIST 5, FTM 2208, LF-04, and concluded with the successful launch and flight of GIANT FIST 3, FTM 2642, LF-04, on 12 March 1969 (see Appendix C, Table C.9).[65]

EMERGENCY ROCKET COMMUNICATIONS SYSTEM

The Emergency Rocket Communications System (ERCS) Category I flight tests took place concurrently with the Force Modernization Category I flight program, with three successful launches beginning on 13 December 1966 and ending on 17 April 1967 (see Appendix C, Table C.9, and Chapter 14).

SPECIAL TEST MISSILES

Investigation by SAMSO of the GIANT FIST 4 Stage III failure revealed that recently fielded Minuteman II Stage III motors had irregular lots of propellant, which caused intense acoustic vibrations and loss of steering control as the NCU was unable to counter the vibrations. Modifications were made to the NCU to reduce sensitivity to vibration. Three special test flights confirmed the changes were successful: SAMSO special test missile (SSTM) 050, FTM 2504, 29 January 1969, LF-02; STM 051, FTM 2252, 7 March 1969, LF-05; and STM 052, FTM 2287, 18 April 1969, LF-25. While SSTM 052 had a problem with Stage III operation, it did not involve the NCU fix (see Chapter 8, Minuteman II Stage III problems).[66]

OPERATIONAL TEST AND EVALUATION PROGRAM

The Operational Test and Evaluation (OTE) flight tests consisted of two phases totaling 99 launches from April 1969 to November 1987, with only 3 in-flight failures and 2 aborts. Since the radiation-hardened guidance control sets and reentry vehicles were in limited supply, those missiles were excluded from the operational base selection process. The first OT launch, GLORY TRIP 19M, FTM 2339, LF-05 (GLORY TRIP XXM signified Force Modernized Minuteman, Wings II through V, while GLORY TRIP XXF signified Wing VI and 364 SMS, Improved Minuteman), took place on 16 April 1969. The launch was successful, but the missile had an unmodified Stage III that was still carrying the mixed-propellant motor, and the severe acoustic vibrations caused Stage III failure. Testing continued on 25 April 1969, with successful launch and flight to target of GLORY TRIP 05F, FTM 2096, LF-22. The program concluded on 10 November 1987 with the successful launch but mission failure of GLORY TRIP 149M, LF-07. The reentry vehicle flight problem had been resolved, as the CEP was now better than 0.5 nm.[67]

For variable range and azimuth testing to simulate the operational launch direction, a set of three targets were selected, one on Oeno Island and two located south of the island, designated Broad Ocean Areas B1 and B2. Flights to Oeno Island would be on a 191-degree azimuth from Vandenberg and therefore represented the ability to launch in an operational direction, that is, instead of nearly directly west, on a more polar trajectory to verify guidance capability on a significantly different azimuth. While not specifically stated in the report, this may have been to mollify those critics of the MX program who were arguing that US ICBMs had never flown anything other than a westerly trajectory. Obviously, a northerly direction was out of the question. Owing to the remoteness of the island, a Broad Ocean Scoring Systems ship with Splash Detection Radar and radar reflectors on the island would be the scoring method. SAC planned to launch four Minuteman II OT flights to Oeno in September and October 1969.[68]

One flight to the Oeno Island target has been partially declassified. GLORY TRIP 55F, FTM 2486, was successfully launched from LF-25 by ALCS on 21 May 1970. The reentry vehicle successfully impacted at Broad Ocean Area Target 33, Oeno Island, with a CE of 0.48 nm. As far as SAC was concerned, the success of this flight put to rest the issue of the ability of Minuteman guidance systems to accurately fly a route other than the nearly westerly flight path normally used to the Kwajalein and Eniwetok target areas.[69]

On 3 April 1974, GLORY TRIP 119M was successfully launched from LF-02 with a successful flight to impact with a CE of 0.085 nm (516 feet). All of the five Minuteman II OT flights in FY 1974 were targeted to Kwajalein and were successful (see Appendix C, Table C.10).[70]

SPECIAL FLIGHT PROGRAMS

There were a number of test programs conducted during the Minuteman II flight test program in addition to the flights specifically for Minuteman II operational force booster and reentry vehicle evaluation.

*Airborne Launch Control System*

Eleven of the OT launches were commanded from ALCS aircraft between 18 April 1969 and 18 December 1970 (see Chapter 15 for a discussion of ALCS and Minuteman II).[71]

*Emergency Rocket Communications System 494L Operational Tests*

The nine-flight ERCS program began on 4 August 1970 and ended on 16 July 1979, with all launches and flights successful. There was one operational test flight (see Appendix C, Table C.10, and Chapter 15 for discussion of ERCS).[72]

*GIANT PATRIOT Operational Base Launch Safety System*

GIANT PATRIOT required an in-flight safety system in case of malfunction during boosted flight between Malmstrom AFB and the Pacific Coast. Two flights from Vandenberg AFB successfully tested the system: GIANT PATRIOT 1, LF-07, on 13 June 1972, and GIANT PATRIOT 2, 26 July 1972, LF-07 (see Appendix C, Table C.10, and Chapter 14).[73]

*Mark IA Penetration Aids*

Concerned with the developing Soviet ABM system capability and its increased coverage between 1965 and 1968, SAC recommended development of penetration aids for the Minuteman and Titan II programs in the form of chaff or decoy dispensers, respectively. Two systems were tested, the Mark I and IA, the major difference being the type of chaff dispensed—bags of chaff or foil dispensers, respectively. The Mark I version had been flight tested with Minuteman IB, but only the Mark IA chaff dispensers were deployed with the

Minuteman II Mark 11C reentry vehicles in a wafer between the guidance section and the reentry vehicle. There were nine chaff clouds created, one of which could mask the reentry vehicle when released into it. The Mark IA penetration aid flight development program consisted of six flights, five successful, three concurrent with the DASO program, and three during the OT program. The flight testing began on 20 December 1968, with the flight of FTM 2418, LF-08, and ended on 25 July 1969, with the flight of FTM 2357, LF-08 (see Appendix C, Table C.10, and Chapter 9).[74]

*Space and Missile Systems Organization*

The final SAMSO Stage III modification test, SSTM 053, was flown during the OT program on 18 April 1969, LF-25. SSTM 060, FTM 2510, LF-21, launched on 1 October 1969, and SSTM 061, FTM 2508, launched on 19 November 1969, LF-21, were for evaluation of the Minuteman II all-up retrofit configuration. Both were successful flights.[75]

*Ogden Special Launch*

In April 1975 engineers at the Ogden Air Logistics Center discovered, as part of the Minuteman Propulsion Long-Range Service Life Analysis Program, that the liner that bonded the solid propellant to the Aerojet Minuteman II Stage II and Minuteman III Stage III motor cases had transformed from a normally firm and pliable state to a viscous, sticky fluid designated "sticky liner." Further investigation determined that the problem was probably not a flight reliability issue for Minuteman II Stage II motors. After extensive ground testing, the Ogden Special Launch (OSL in the flight lists) missile was successfully launched on 23 September 1975, LF-04, with flight to impact. Ogden concluded that the degraded liner had no significant effect on Stage II performance.[76]

MULTI-SERVICE LAUNCH SYSTEM

Surplus Minuteman II boosters remaining from the deactivation of the Minuteman II in 1991 were modified to serve as launch vehicles in the Multi-Service Launch System program (MSLS). These launches took place from the Vandenberg RSLP facilities previously used with surplus Minuteman IB boosters. Missile modifications included a new inertial guidance system, the Litton Systems LN-100 ring laser gyroscope, GPS, and replacement of the reentry vehicle with a 615- to 771-pound target vehicle. All eight flights, from 17 January 1997 to 15 July 2001, were successful in boosting the target vehicle to the required 800-nm suborbital altitude.[77] The MSLS program provided targets for the development of the Ground-Based Interceptor Program after the Homing Overlay Experiment Program successfully demonstrated the concept.

## Minuteman III Flight Programs

Launch complex integration and simulated flight tests with Minuteman III ground test missile GTM 073 at LF-02 were completed on 27 March 1969. FTM 301 (5001; two numbers are given in several reports, so the second is included for historical accuracy) was launched with successful flight to target on 11 April 1969, LF-02.[78]

Hope for a flawless flight test program proved short-lived, for while the launch of the next test, FTM 302 (5002), LF-02, on 29 May 1969, was successful, an in-flight staging anomaly occurred. Problems with the missile guidance system checkout and the derailment of the train car carrying FTM 303 (5003) to Vandenberg AFB caused a several-month delay in the flight test program. FTM 304 (5006) was successfully launched from LF-02 on 13 September 1969, with successful flight to target. FTM 305 (5004) was returned to Plant 77, Hill AFB, due to a suspect component of the Stage III LITVC system and was replaced in the launch schedule by FTM 305 (5007), which was successfully launched on 15 October 1969 from LF-02. An integrated circuit failure in the PBPS prevented successful flight to impact. FTM 304 (5003) was successfully launched on 31 October 1969 from LF-02; further information about impact is not available.[79]

The first series of flights at the WTR had been with the Block III NS20 guidance system (see Chapter 10). The first flight with the Block IV production guidance system was FTM 309 (5009) on 2 April 1970, from LF-08. This was the first flight using the periscope alignment system with the autocollimator in the LF (see Chapter 10). Launch was successful; further details are not available. FTM 308 (5010) was successfully launched on 22 April 1970 from LF-02, with highly successful flight to impact. FTM 310 (5011) was successfully launched on 8 May 1970, but further flight information is not available.[80]

The first Category II flight test was the successful launch of FTM 312 (5012) on 17 June 1970 from LF-02. Further details are not available. Six days later, FTM 311 (5016) was successfully launched from LF-02 on 28 July 1970, the last launch of the R&D Program. No further details on either flight are available.[81]

## DEMONSTRATION AND SHAKEDOWN OPERATIONS

The DASO program began on 27 August 1970 with the successful launch from LF-08 of OLD FOX 01M, FTM 5034; further flight details are not available. On 25 September 1970, OLD FOX 02M, FTM 5036, was successfully launched from LF-02. This flight had been planned for launch by the ALCS, but when LF-02 failed to respond to the launch message, the launch was commanded by LCC missile combat crew. Investigation revealed that there were no equipment failures at the LF. Therefore, the problem originated from the aircraft and was found to be a frequency deviation problem that had been properly rejected by the LF equipment. No further data on the flight is available. On 27 January 1971, OLD FOX 04M, FTM 5042, was successfully launched from LF-08. The flight was successful through thrust termination, when all telemetry was lost. The fifth flight in the DASO program took place on 16 February 1970 with the successful launch from LF-02 of OLD FOX 05M, FTM 5069, with successful flight to impact. The final flight in the program was OLD FOX 06M, FTM 5075, successfully launched from LF-26 on 23 April 1971, with flight through thrust termination successful; impact data is not available.[82]

## OPERATIONAL TEST

*Phase I*

The OT program was conducted in two concurrent phases. Phase I began on 23 March

1971 with the successful launch and flight to impact of GLORY TRIP 01GM from LF-08 (M signifies Force Modernization and B signifies Wing VI or 564 SMS). Phase I consisted of 39 successful launches, 1 partial success, and 2 in-flight failures.

During GLORY TRIP 17GB, launched on 16 October 1972, LF-25, a malfunction in the guidance-and-control unit resulted in failure to deploy the third reentry vehicle and chaff. During the flight of GLORY TRIP 22GB, launched on 3 May 1973, LF-22, the failure of the chaff dispenser subsystem prevented chaff deployment. SAC considered both failures to be random malfunctions, and GLORY TRIP 17GB was rated a partial success, while GLORY TRIP 22GB was rated a success.[83]

Phase I ended on 10 July 1979 with the successful launch and flight to target of GLORY TRIP 40GM from LF-09 and GLORY TRIP 68GM, LF-08, 12 seconds later as part of a ripple launch by crews from the 319 SMS, 90 SMW.[84]

*Phase II*

Phase II began on 5 December 1972 with the successful launch of GLORY TRIP 41GM from LF-02. However, three minutes into flight, Stage III underwent an uncommanded attitude change, and the range safety officer issued the destruct command. Investigation revealed the failure of the mechanical bond between the solid propellant and the motor case insulation of Stage III. The result was an uneven burning of the propellant, which caused a burn through of the aft dome and the subsequent attitude aberration. Of the 148 Aerojet Stage IIIs in the operational fleet, 6 were determined to have voids in the propellant sufficiently large enough to cause the burn-through problem, and these were replaced.[85]

Beginning in 1967 SAC had been helping AFSC develop new range evaluation and scoring systems for Minuteman III, as well as variable range and azimuth targets for both Minuteman II and III. Minuteman III missiles configured with three reentry vehicles and chaff, as well as the range safety equipment, were unable to reach the WTR target areas at Eniwetok and Kwajalein Atolls.

The Phoenix Islands Group was selected as the new target area. Located 3,600 nm southwest from Vandenberg AFB (227-degree azimuth), the group consists of six islands: Phoenix, Canton, Sydney, Enderbury, Birnie, and Hull, separated from each other by between 38 to 104 nm. Programmed targets were near Canton Island (three), Enderbury (two), and Hull (one), representing a triangular target area approximately 110 nm long with a base of 42 nm.[86] The Phoenix Islands were selected for a variety of reasons, perhaps most important of which was that support facilities were already present on Canton Island due to its use as a tracking station for the Gemini Program. Additionally, the British had used the airfield on Canton Island to operate Shackleton heavy bombers in support of the Christmas Island hydrogen bomb tests in 1957 and 1958.[87]

There was, however, one large problem. The Phoenix Islands Group was under joint US-British administration, and there would need to be negotiations for its use as a Minuteman III target. Negotiations became troublesome in July 1969 when the British Foreign Office refused to consider the American proposal due primarily to the possibility that MIRVs would be banned as part of the ongoing Strategic Arms Limitation Talks between the Soviet Union and the United States. An additional concern was for the welfare of the inhabitants of the islands and the bird sanctuaries found on Birnie Island. SAC developed a backup plan that would use Howland and Baker Islands as well as a broad ocean area near Kwajalein. In June 1970 a Conservative British government was elected, and soon thereafter, an agreement was reached.[88]

A Reentry Management System V phased-array radar and a TPQ-18 tracking radar were installed on Canton Island, as well as splash detection radars on Canton, Hull, and Enderbury Islands (installation on Birnie did not take place because of the bird sanctuaries). Additional support was provided by an Airborne Astro Graphic Camera System aircraft from Hawaii and a tracking ship. Radars in California, Hawaii, and on tracking ships between Vandenberg and the Phoenix Islands would be able to provide midcourse tracking information. The Midcourse Optical Station on Maui, Hawaii, would be used for midcourse analysis using long-wave infrared telescopes and television trackers.[89]

The splash detection radars were used to triangulate the water column resulting from reentry vehicle water impact. The AN/TPQ-18 measured target position in terms of slant-range azimuth and elevation, while the deployment position radar provided data for evaluation of deployment reliability of the reentry system by detecting the number of objects deployed, size and centerline of each chaff cloud, number of hard objects dislocated from chaff cloud by a specific distance, and velocity history of the reentry vehicles. The deployment position radar could detect and acquire the chaff cloud at a slant range of more than 800 nm, approximately four minutes prior to impact of the reentry vehicle. At a slant range of 250 nm, one minute prior to impact, target detection and tracking would take place on all targets. Then, approximately 45 seconds prior to impact, the lead object would penetrate the 400,000-foot altitude zone at a slant range of approximately 155 nm. At this point the radar would start to track the hard objects as they emerged from their respective pancaking chaff clouds (Figure 13.7).[90]

Most of the Minuteman III flights that carried three Mark 12 reentry vehicles and the Mark IA chaff dispenser would have been flown to the Phoenix Islands as described earlier. The available Minuteman III launch lists for the WTR do not include a description of the number of reentry vehicles or penetration aids flown and so do not permit determination of the number of flights to the Phoenix Islands. WTR histories for the dates of interest remain classified.

For the most part, details of the Minuteman III flight test reports remain classified; however, the first land impact mission has been partially declassified. GLORY TRIP 69GB, part of OT Phase II, was launched on 26 July 1979 from LF-26. The purpose of the mission was to demonstrate the Mark 12 reentry vehicle arming and fusing performance over land versus water. The first mission was targeted for Illeginni Island in the Kwajalein Atoll. Illeginni Island had been the site of a remote Spartan and Sprint launch complex, with the first Sprint launch occurring in March 1972. The facility reverted to inactive status at the end of 1973, although an unmanned communications relay station remained. Illeginni Island is approximately 33 acres in size, 2,800 feet long, and 1,200 feet wide at the widest point, though most of the island is 400 feet in width. The island is located within the Mid-Atoll Corridor (40 miles wide) of the Kwajalein Missile Range. Islands within the corridor are maintained without inhabitants to provide maximum range safety.

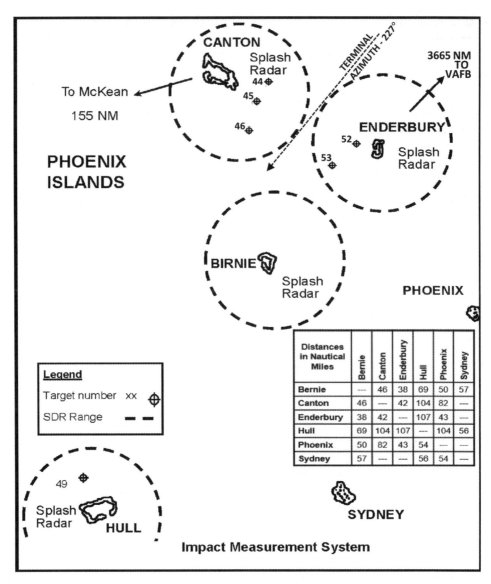

FIGURE 13.7 The configuration of the Phoenix Islands target area. Fully instrumented Minuteman III with range safety equipment could easily reach this target area, which was approximately 500 nm closer than Kwajalein. The islands were on a 227-degree azimuth from Vandenberg AFB. *Adapted from "History of Strategic Air Command 1974."*

The reentry system was composed of three Mark 12 reentry vehicles and no chaff dispenser. All three reentry vehicles were deployed successfully. Only Reentry Vehicle 2 was instrumented, and all arming and fusing requirements were confirmed for this vehicle. Reentry Vehicle 2 successfully impacted on the island. No CE data is available.[91]

The Phase II program ended on 18 June 1997 with the successful launch and flight to impact of GLORY TRIP 165GM from LF-10. The OTE program consisted of 165 successful launches and 7 in-flight failures.[92]

FLIGHT DEVELOPMENT EVALUATION PROGRAM

Flight Development Evaluation (FDE) was essentially a continuation of OTE Phase II but is listed separately to be consistent with the flight test records. FDE began on 20 February 1998 with the successful launch of GLORY TRIP 166GM from LF-04 but suffered an in-flight failure. The flight test record through October 2019 shows 67 successful launches and flights to target and 3 in-flight failures.[93]

TEST FLIGHT PROGRAMS

Through November 2019 there have been 44 test flights conducted for 10 development programs: GRP, Advanced Inertial Reference Sphere (AIRS), Propellant Replacement Program (PRP), Product Verification Missile/Special Test Missile, Dust Modification Performance, Honeywell NS20A Missile Guidance System Evaluation, Hybrid Explicit, PAVE PEPPER, Reentry Vehicle Performance/Nose Tip Redesign, and Safety-Enhanced Reentry Vehicle (SERV; see the digital appendices).[94]

*Guidance Replacement Program*

In August 1993 Boeing's Autonetics and Missile Systems Division received a contract to provide a modified MGS, the NS50A, which would have the same functionality as the NS20A, the original Minuteman III MGS, but with improved reliability and sustainability as more modern technology replaced the aging NS20A components. There were seven flights in the program. Two pertained only to the GRP: Integrated Demonstration Flight (IDF)-1, launched on 24 June 1998, LF-09, and IDF-2 on 18 September 1998, LF-26, both successful flights to target. Two more tests were required and were combined with the PRP. The reentry vehicle miss distances were considerably larger than expected. The decision was made to go into limited production in December 1999 and continue testing. Testing resumed during the FDE program with two flights on 28 September 2000: GLORY TRIP 173GM, LF-09, and 174GM, LF-04, were successfully launched with flight to impact. Combined with the results from the successful final two flights on 7 February 2001, GLORY TRIP 175GM, LF-10, and GLORY TRIP 176GM, LF-04, on 7 November 2001, the Air Force found that the NS50A accuracy did not meet requirements.[95]

The Air Force conducted an accuracy investigation that identified two primary sources of bias error in the guidance system software. One was the erroneous implementation of computational precision; that is, for several of the navigational computations, truncation was implemented where roundoff was intended. Likewise, for some of the guidance calculations, better approximations were needed to attain adequate precision. The other primary error source was a small, undesired residual velocity introduced into the calculations that governed the attitude of the reentry vehicles at deployment from the PBCS. The problem was complex, and as it happened, the bias errors reinforced one another in the westerly flights to Kwajalein. In other trajectories the errors would have increased dispersion but not contributed significantly to the weapon delivery error. Nonetheless, operational trajectories would still have been less than optimal if the situation had not been uncovered in the westerly flights. The Air Force took corrective action through an Accuracy Upgrade Program, and the first flight of the Minuteman III with the software corrections occurred on 7 June 2002 with the successful launch and highly accurate flight to impact of GLORY TRIP 179GM, LF-26. Now the guidance system accuracy met force requirements. Full production commenced in September 2001.[96]

## Advanced Inertial Reference Sphere

The first flight of the AIRS took place with the successful launch and flight of STM-11W on 15 July 1976, LF-21. This flight was the first test of the Missile Performance Measurement System, which was composed of the Honeywell Improved Digital Computer Unit and the AIRS. The AIRS was not part of the active guidance system on this flight. The flight demonstrated the technical feasibility of the AIRS system planned for the MX (Peacekeeper) advanced ICBM.[97]

## Propellant Replacement Program

With the goal of extending the life of Minuteman III boosters to the 2020 timeframe, the PRP was initiated in 1992 due to the prediction that the propellant would age out by 2002. Earlier programs had replaced the Stage II and III motors, but this was the first time all three stages were replaced at once, as well as the interstage structures. There were two flights in this program. The first took place on 13 November 1999 with the successful launch of FTM 01, LF-26. While the flight to target was successful, there was a higher-than-predicted use of injectant in the Stage III thrust vector control system, which was determined to be due to the misalignment of the Stage III nozzle. The second flight, FTM 02, was successfully launched on 24 May 2000, LF-09, but failed to reach the target due to the failure of the PSRE to separate from Stage III. The cause was a failure in the arm–disarm switch, and the Air Force determined that the anomaly was an isolated problem unrelated to the PRP, GRP, or PSRE programs.[98]

## Product Verification Missile/Special Test Missile Programs

Product Verification Missiles (PVM) were originally allocated to SAC for operational testing. The PVMs were used by SAMSO because they, like an STM, could be equipped with instrumentation not permitted on the OT missiles. The PVM program was designed to determine the ability of Minuteman III to launch from upgraded or modified ground facilities of the operational Minuteman force. The STM program's primary objective was to evaluate the performance of either new components of the weapon system, such as the new missile suspension equipment, or modifications incorporated to correct past missile malfunctions. Both the PVM and STM programs provided Headquarters SAC with valuable test data that was unattainable from the OT program.[99]

The PVM program began on 31 May 1972 with the successful launch and flight to target of PVM-1, LF-21, and ended after 19 successful flights on 27 March 1980 with PVM-19, LF-21. The STM program continued from launches at the ETR and began on 4 November 1970 with the successful launch and flight to impact of STM 1W, LF-21. The program ended after 17 flights with the successful launch and flight to impact of STM 18W on 30 August 1979, LF-21.

## Dust Modification Performance

Minuteman III was first deployed with a Mark 12 Reentry System shroud fabricated from an aluminum alloy that was easily formed into an ogive shape (a segment of a circle with the missile body tangent to it). The shroud was coated with a layer of fungicide-impregnated cork, which served as an ablative surface for protection against heating during egress from the launch tube and aerodynamic heating during the ascent stage of flight. The fungicide was a 15 percent solution of paranitrophenol dissolved in secondary butyl alcohol and was a clear liquid. Malachite-green oxalate was added to the solution to assist in application and gave the characteristic green color to the shroud skin (Figure 13.8).

On 18 December 1969, Headquarters Air Force authorized dust-hardening of the Minuteman III missiles starting with the 250th production model. Retrofit of LFs containing the unmodified missiles would begin in the first quarter of 1973 and be completed by early FY 1977.[100] The dust-hardening modification included replacing the ogive-shaped aluminum shroud with a biconic titanium shroud (see above). Dust-hardening was also considered as a retrofit for Minuteman II, including a shroud for the reentry vehicle, but was not implemented due to excessive costs.[101] The dust-hardened Minuteman III was 64 pounds heavier than the earlier production models, resulting in a range decrease of 40 to 90 nm (Figure 13.9).[102]

Information on the Dust Modification Performance flights is sparse. The first flight was on 2 August 1972, with the successful launch and flight to target of STM 7W, LF-21. Fifteen

FIGURE 13.8 Left: FTM 202 receives its reentry vehicle shroud on 16 October 1968 at the Cape. This is the original aluminum Minuteman III reentry vehicle shroud coated with ablative cork. Right: SSgt. Stephen Kravitsky inspects a Minuteman III at the 321 SMW in 1989. This is the dust-hardened shroud made of titanium, which did not require the ablative cork coating. *Photograph by SSgt. Alan Wyche.*

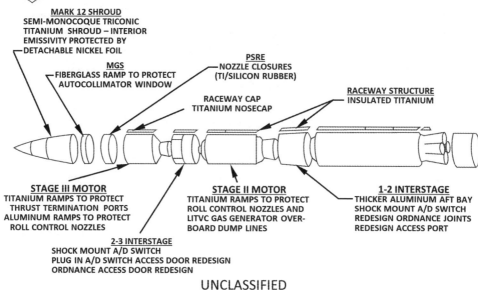

FIGURE 13.9 Minuteman III Dust-Hardening Program design change detail. Titanium shields replaced the aluminum counterparts as part of the nuclear-hardening program. Abbreviations: Arm–Disarm (A/D), Liquid-Injection Thrust Vector Control (LITVC), Missile Guidance System (MGS), Propulsion System Rocket Engine (PSRE), and titanium (TI). *Adapted from "History of Strategic Air Command, FY 1973."*

flights took place over the next four years; twelve were part of the ongoing PVM program (PVM-2 through PVM-13), and three were in the STM program (STM-8W through STM-10W). Detailed records indicating participation stop with the successful launch and flight of PVM 13 on 15 March 1976, LF-26.[103]

Deployment of dust-hardened Minuteman III missiles began in December 1972 at Wing V, F. E. Warren AFB, during the combined Force Modernization and Integrated Modification Programs. Dust-hardened missiles were then retrofitted into the force beginning on 17 September 1973 at Minot, 3 February 1975 at Grand Forks, and 20 January 1975 at Malmstrom. The program was completed on 11 July 1975 with the turnover of the 564 SMS at Malmstrom AFB (see Chapter 16).[104]

### Honeywell NS20A Missile Guidance System Evaluation

The STM flight program was used to evaluate the Improved Digital Computer Unit (IDCU), a component of the alternative source program for the NS20 guidance system, as part of the R&D program, which also took place at the WTR. The first flight, STM-1W, FTM 5040, was successfully launched from LF-21 on 4 November 1970, with successful flight to target.

STM-2W, FTM 5071, was successfully launched from LF-26 on 23 April 1971. Further flight information is not available. STM-6W, FTM 5004, was successfully launched from LF-21 on 11 June 1971; further flight details are not available. STM-3W, FTM 5143, was successfully launched from LF-08 on 20 October 1971, with successful flight to impact. The last flight in the program was STM-11W, 15 July 1976, LF-21, with successful flight to impact. The Honeywell system was deployed in small numbers at Wing III but eventually was replaced with the Autonetics system when the Air Force canceled production due to cost.[105]

### Hybrid Explicit

The successful launch and flight to target of PVM-10 on 6 May 1975, LF-26, was the first test of the CDB and development version of the Hybrid Explicit software. The second flight, PVM-11, was successfully launched on 2 July 1975, LF-26, to evaluate Stage III thrust termination and the HEFP. Flight to impact was successful. On 9 January 1976, the first flight test of the operational HEFP, PVM-12, LF-26, was successfully launched and completed flight to target. PVM-13 was successfully launched on 15 March 1976, LF-26, and completed flight to target. This was the final launch in the program prior to deployment (see Chapter 10).[106]

### PAVE PEPPER

The PAVE PEPPER program consisted of two flights, STM-9W and STM-10W (also part of the Dust Modification Performance program) on 16 May and 26 July 1975, respectively, both launched from LF-02. The primary objective of the PAVE PEPPER program was demonstration of the feasibility of the Mark 12 reentry system to deploy up to seven small Advanced Ballistic Concept reentry vehicles. The first flight was successful, but on the second flight, shortly after deployment of the fourth reentry vehicle, the MGS failed to process velocity values for the remaining reentry vehicles. This resulted in a continuous burn of the PSRE until fuel depletion, and reentry vehicles five through seven were not deployed. The Air Force assessed this as a random problem that would not affect the operational Minuteman III force. The PAVE PEPPER concept was not deployed with Minuteman III.[107]

### Reentry Vehicle Performance/Nose Tip Design

Both PVM and STM assets were involved in the Mark 12 reentry vehicle performance evaluation flight test effort. Eight PVM and three STM flights began on 31 May 1972, LF-21, with the flight of PVM-1, FTM 5255, and ended on 26 November 1974 with the flight of PVM-9, FTM 5552, LF-08. The Mark 12 reentry vehicle performance had been problematic from the beginning, when the first three test vehicles, flown singly on Minuteman II, broke up during reentry (see Minuteman II discussion above).[108]

By 1972, during the flight testing of the Mark 12 reentry system, attention was focused on the manufacturing technique for the three-section (forward, mid- and aft) heat shield, assembled using a breech lock mechanism. The sections were wrapped with two-inch-wide strips of rayon tape. Investigation showed that due to problems with General Electric's quality control procedures, the various sections of the heat shield were produced from rayon tape sewn together at both a positive and negative angle depending on the section being manufactured. It turned out that the manner in which the sections were wrapped had a significant effect on flight performance and accuracy. Upon separation from the PBV, the reentry vehicle was spun in a clockwise direction at 120 revolutions per minute using two small spin rockets. The spin was designed to stabilize the reentry vehicle before reentry. A positively wrapped heat shield section tended to increase the clockwise roll rate, while a negatively wrapped heat shield section tended to decrease the roll rate through zero; that is, slow down and change the direction of the spin. This resulted in the reentry vehicle impacting further downrange than intended. A year earlier, General Electric, the Mark 12 contractor, had changed its manufacturing procedures so that new Mark 12 reentry vehicle heat shield sections were produced with positively wrapped tape.

On 21 June 1973, SAMSO concluded that "it appears impossible to draw a positive conclusion about the vehicle roll rate characteristics based on the knowledge of heat shield wrap alone." There had been two PVM flights and one STM flight investigating reentry vehicle performance without conclusive results. Unlike the OT flights, these flights had carried instrumented reentry vehicles. Eight months later, in February 1974, after two more PVM flights and one STM flight evaluating reentry vehicle performance and one each PVM and STM flights evaluating nose tip redesign, SAMSO revised its earlier estimate and indicated negatively wrapped reentry vehicles had a 59 to 100 percent probability of rolling through zero. The range in probability depended on the location of the negatively wrapped section of rolling through zero, with resulting inaccuracy. Positively wrapped reentry vehicles had shown no inaccuracy problems. By 20 May 1974, a schedule for removal of the negatively wrapped reentry vehicles from the operational force was implemented.[109]

The nose tip redesign problem continued to be investigated with the flight of PVM-2 on 30 January 1973, LF-08. The nose tip of one of the two test reentry vehicles fractured, and the reentry vehicle was destroyed during reentry. Two earlier flights, GLORY TRIP 06GM on 15 October 1971, LF-25, and PVM-1, launched on 31 May 1972, LF-21, had suffered similar fates. Not only was the nose tip structurally weaker than expected, weather conditions at the target were believed to play a role. On 13 August 1973, AFSC was directed to proceed with the design and development of a new nose tip. During FY 1974 a new tungsten nose plug was developed to replace the aluminum insert that was currently deployed.[110]

Comprehensive weather data acquisition as part of the Minuteman Natural Hazards Program Mark 12 reentry vehicle tests began with the flights of PVM-4, FTM 5376, and PVM-3, FTM 5357, which were launched on 31 May 1973, LF-09, and 24 August 1973, LF-08, respectively. The test program was extensive, involving weather radar facilities for the near-surface weather conditions as well as a Lincoln Laboratory radar specifically monitoring the reentry corridor, lidar facilities, two WB-57F aircraft, a WC-135B aircraft, and Defense Meteorological Satellite Program real-time imagery. The PVM-4 reentry vehicles flew through a low-density cirrus layer but not the lower convective clouds that were in the area. By the time the PVM-3 reentry vehicles reached Kwajalein, the skies were clear. Coordinating weather events and missile flight times from Vandenberg turned out to be quite difficult (Figures 13.10 and 13.11).[111]

The next flight in the program, STM-8W, FTM 5309, launched on 22 December 1973, LF-08, carried the tungsten insert. This mission required minimum weather, which was defined as optically translucent clouds no greater than 2,000 feet thick. The surface weather

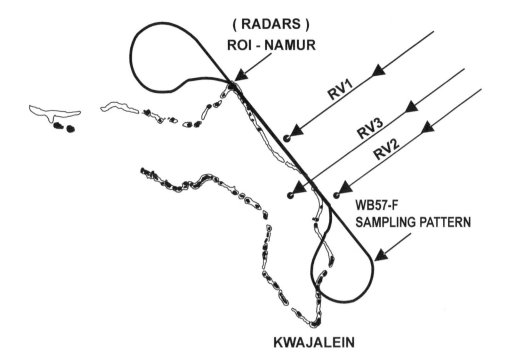

FIGURE 13.10 Cloud sampling pattern by WB-57-F aircraft minutes prior to reentry vehicle impact for flights PVM-3 and PVM-4. A repeat sampling run was made after impact. Even with the aid of real-time imagery from the Defense Meteorology Satellite Program, synchronizing reentry vehicle passage through cloud cover proved difficult. *Adapted from "Final Report of PVM-4 and PVM-3 Weather Documentation."*

at the time of impact was two-tenths coverage of scattered cumulus clouds at 1,400 feet and less than one-tenth coverage of stratocumulus clouds at 5,000 feet. Two of the reentry vehicles flew through low-density clouds of ice crystals 6 miles and 3.5 miles in altitude, followed by flight through low-level cumulus clouds near the surface. No further details on the integrity of the reentry vehicles have been declassified.[112]

PVM-5, FTM 5452, launched on 4 April 1974, LF-21, also carried the tungsten insert and encountered the first significant weather at Kwajalein, as planned.[113] The next two missions included Minuteman III GLORY TRIP 45GM on 17 August 1974, LF-02, and PVM-8, FTM 5507, on 5 October 1974, LF-04. Both were intentionally flown into minimum weather.[114] PVM-6, FTM 5485, and PVM-7, FTM 5454, launched on 11 October 1974, LF-21, and 12 October 1974, LF-25, respectively, with the objective to pass through heavy weather at Kwajalein. The weather objective was met as a major system moved through the Kwajalein target area during the impacts of PVM-6 reentry vehicles and had only partially cleared out of the area when the reentry vehicles from PVM-7 impacted three hours later. Both sets of reentry vehicles passed through clouds containing snow and ice crystals.[115]

One additional flight, PVM-9, FTM 5552, 26 November 1974, LF-08, took place, after which SAMSO concluded that weather was not a significant issue for Mark 12 reentry vehicle accuracy. Detailed information for the PVM and STM reentry vehicle performance program ends at this point.[116]

*Safety-Enhanced Reentry Vehicle*

The SERV program involved Minuteman III launches carrying a single Mark 21 reentry vehicle from the retired Peacekeeper missile force or the Mark 12A. The reduction to a single warhead was a result of the START II Treaty signed by President George H. Bush and Russian President Boris Yeltsin on 3 January 1993.[117]

Three flights were flown in the program: SERV-1 on 21 July 2005, LF-10, carrying a single Mark 21 reentry vehicle, which impacted east of Gagan Island, Kwajalein; SERV-2 on 25 August 2005, LF-26, again carrying a single Mark 21, which impacted at Kwajalein; and SERV-3 on 16 February 2006, LF-10, carrying two Mark 12A Mod 5F reentry vehicles to impact at Kwajalein. The fourth flight in the series was canceled due to the highly successful results of the first three flights. The third flight had demonstrated the successful integration of each of the major Minuteman III modification programs underway at the time. Five hundred Mark 21 reentry vehicles had been retained from the deactivated Peacekeeper program, which carried the modern W87 warhead compared to the much older W-62 (Mark 12) or W-78 (Mark 12A) warheads. In 2006 the Air Force purchased 140 modification kits to install single Mark 21 reentry vehicles on a portion of the Minuteman III force. The first modified missile was deployed at F. E. Warren AFB in December 2006.[118]

★ ★ ★

The flight programs at Vandenberg AFB evaluated not only reliability of missiles that had been deployed for years under operational conditions but also served as test beds for modification, such as the dust-hardening program and a variety of reentry vehicle evaluation programs.

There was a constant need for evaluating the reliability of the entire weapon system. In the early 1960s, the concept of operational base launch (OBL) programs began to be discussed. Within SAC there was an urgency to launch missiles from operational LFs at the 341 SMW to evaluate the reliability of the deployed missiles. Additionally, plans were developed to launch a Minuteman armed with a nuclear warhead from Vandenberg AFB. Realizing that such testing was problematic at best, OBL programs that simulated launch were developed in parallel.

FIGURE 13.11   Salvo launch of Minuteman III Glory Trip 40GM and 68GM from LF-09 and LF-08, respectively, on 10 July 1979. The launches were in conjunction with the annual Strategic Air Command Global Shield exercises. There was a 12-second delay in the salvo launch. These were the 118th and 119th Minuteman III launches, of which 117 had been successful, for a launch record of 98 percent. As of 7 November 2019, there have been 293 Minuteman III launches from Vandenberg AFB, with 12 failures, for a record of 95 percent successful launches. *Official USAF photograph.*

# 14

# OPERATIONAL BASE MISSILE TEST PROGRAMS

Flight testing at Vandenberg AFB was just one of several methods used to assess Minuteman weapon system reliability. Short- and full-range OBL programs were developed, and the former were actually conducted. In addition, two operational base ground test programs were developed and implemented: the modified operational missiles (MOMS) and simulated electronic launch Minuteman (SELM).

## Operational Base Launch

The OBL concept had an off-again, on-again existence from 1963 to 1974. Planning for a full-range launch from the 341 SMW at Malmstrom took place in 1963 but was canceled by Secretary of Defense Robert McNamara due to political concerns (Exercise ON TARGET; see below). While the full-range OBL program planning continued in the background, a short-range program was approved by McNamara in November 1964. Project Long Life entailed short-range launches from Ellsworth AFB, South Dakota, using modified Minuteman IB missiles. The objective was to validate the operational base configurations, which were sufficiently different, in the minds of the Air Force, from the test equipment–laden LFs at Vandenberg to warrant the tests. During its 11-year life, the OBL program was expanded to include planning for full-range launches with inert warheads from operational bases and a launch from Vandenberg with a nuclear warhead if the Nuclear Test Ban Treaty was lifted.[1]

### Short-Range Operational Base Launches

#### PROJECT LONG LIFE

Project Long Life had its origins in SAC OP PLAN 81-64, published in March 1964. Headquarters Air Force requested funds soon after, and the program was approved by McNamara on 27 November 1964. Development work on the technical data, instrumentation, and flight hardware had progressed well enough to support a launch date in the spring of 1965. A Stage I motor modified for a reduced burn time was successfully static-fired on 27 October 1964, with two more successful firings following in December.

Although a nonoperational Minuteman IB was to be launched, it was felt the configuration realistically represented the Minuteman IB fleet. Major objectives were to (1) demonstrate the effects of launch on an operational launcher, (2) document the ignition-through-egress environment in an operationally configured launcher, and (3) demonstrate the capability to launch with the operational launch control system.[2]

LONG LIFE I

The 44 SMW, Ellsworth AFB, South Dakota, was selected as the launch site in February 1965. Modifications to the Minuteman IB missile included reducing the Stage I propellant load to a seven-second burn time while maintaining the Stage I overall weight and balance, as well as weight and balance for the inert Stage II, Stage III, and reentry vehicle (the Stage I propellant load consisted of an inner core of active propellant surrounded by an outer layer of inert propellant). Modifications were completed on 20 January 1965 at Air Force Plant 77, Hill AFB, Utah. The missile was delivered to Ellsworth on 9 February 1965 and emplaced in the 68th Strategic Missile Squadron (68 SMS) LF November-02 on 12 February 1965. Installation of the operational guidance-and-control section and a Mark 11 Mod 5A test reentry vehicle, ballasted to operational weight and center of gravity, was accomplished on the same day. Subsequently, LF November-02 was electrically isolated from all but one of the squadron's LFs to prevent inadvertent launch of the fully operational missiles that remained on strategic alert. The missile was placed on modified strategic alert on 16 February 1965. A minimum of telemetry test equipment was installed and became part of a preliminary countdown, but the actual launch would be conducted by two combat crews in two LCCs as per normal operating procedures.[3]

LF November-02, ten miles north of Newell, South Dakota, was selected due to its relatively remote location, as the planned trajectory would have the missile rising approximately one mile and traveling no more than two miles downrange (Figure 14.1).

It was adjacent to Highway 79, which would be closed during the 30-minute launch window. A control-and-observation complex, the Long Life Command Post, was constructed approximately 8,000 feet from the LF, with a press and observation area for the public nearby. Two LCCs, LCC November-01 and LCC Oscar-01, were disconnected from the Sensitive Command Network, and test-unique launch codes were installed in both LCCs to ensure against inadvertent launch of a nuclear-armed missile.

On Monday, 1 March 1965, the weather was cold, with a high of 20 degrees F and a slight wind. At T minus 35 minutes, Col. Henry R. Cushman, the SAC/1 STRAD Task Force leader for Project Long Life, notified SAC Headquarters that all was ready, and the preliminary countdown was begun. With its successful completion, coded launch orders were sent from SAC Headquarters to LCC November-01, the primary LCC, located approximately eight miles from the launch site, manned by Missile Combat Crew Commander (MCCC) Capt. Dale B. Moneyhon and Deputy Missile Combat Crew Commander (DMCCC) Lt. Allan R. Martens. LCC Oscar-01, located 12 miles southeast of Castle Rock, South Dakota, manned by Capt. Herbert S. Schaefer, MCCC, and Lt. Gerald E. Hampton, DMCCC, also received the message in the secondary LCC (Figure 14.2).[4]

Martens clearly recalls the launch event 53 years later. After receiving and authenticating the message from SAC Headquarters, then participating in the conference call to the rest of the LCCs in the squadron, Moneyhon confirmed with the crew in LCC Oscar-01

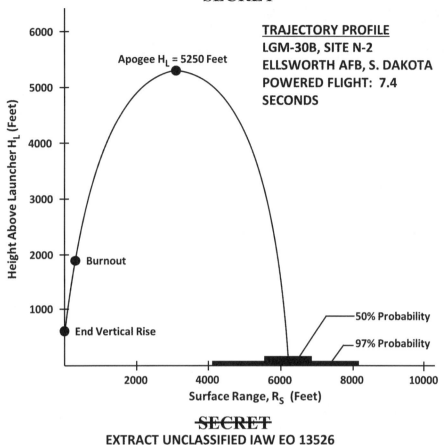

FIGURE 14.1 Programmed trajectory profile for Long Life I. The Stage I propellant was removed and replaced with an outer layer of inert propellant and an inner core of active propellant with a star grain designed to burn 7.3 to 7.4 seconds. *Adapted from "History of 44th Strategic Missile Wing, January–March 1965."*

that a valid message had been sent. Martens then selected and enabled LF November-02, readying the missile for the start of the terminal countdown and launch. At the time specified in the message, Moneyhon and Martens simultaneously turned their launch keys to initiate launch. At the same time, Schaefer and Hampton turned their keys for the second vote, and the countdown began. Six seconds before launch, the launcher closure door opened, the upper umbilical retracted, and then Stage I ignition took place; the missile was launched at 11:00 a.m., precisely on schedule.[5]

There was a slight breeze, which cleared the smoke from the LF, enabling photographs of the missile leaving the launch duct unobscured. The missile rose to 5,300 feet, pitching over slightly as per guidance system commands, and with the propellant exhausted, fell back to Earth tail first, impacting 6,100 feet downrange. Pieces of the interstages were

FIGURE 14.2　Primary launch crew for Long Life I. In the background is MCCC Capt. Dale B. Moneyhon, and in the foreground is DMCCC Lt. Allan R. Martens. *Courtesy of Alan R. Martens.*

scattered widely, some to a distance of 190 yards from the impact point. The Stage I motor split open longitudinally, with propellant burning.[6]

Cost for refurbishment of the launcher was estimated at $575,000. Originally Project Long Life at Ellsworth was to include three launches, at a cost of approximately $7 million, but the complete success of this first test resulted in cancellation of the remaining launches.[7]

## LONG LIFE II

Since the Long Life I test with Minuteman IB had been a complete success, the decision was made in November 1965 to extend the Long Life program to include the short-range OBL of a Minuteman II missile from the 321 SMW, Grand Forks AFB, North Dakota, in 1966 and another Minuteman II in 1967 from the Force Modernized facilities at the 351 SMW, Whiteman AFB, Missouri. At first the test missile was a Minuteman IB converted to the Minuteman II configuration with a Mark 11A Mod 5BH test reentry vehicle, but on 9 March 1966, SAC and BSD decided to use a Minuteman II instead.

The seven-second propellant load was again used, with the modified airframe and reentry vehicle ballasted as with Long Life I. To prevent inadvertent launch of the remaining missiles in the squadron, the test LF and two LCCs were again electronically and electrically isolated from the remaining squadron facilities. In addition, the nontest facilities would be placed on an Infinite Time Delay (see Chapter 15). The test launch codes were selected to be completely different from operational codes, and all nontest facilities were to be monitored to ensure no inadvertent message was accepted by the operational system.

The task force commander for the exercise was Col. Joe M. Schonka, 1 STRAD director of operations, with overall command control assigned to Maj. Gen. Harold E. Humfeld, commander, 1 STRAD. Launch director was Lt. Col. James A. Simons, 394 SMS, Vandenberg AFB, California. Preparations began on 31 May 1966, with an anticipated launch date of 2 June 1966, which was postponed due to the added support burden to the 321 SMW, which was in the process of full activation. Resumption of the program took place on 31 August 1966 with the activation of the Long Life II Task Force.[8]

The test missile arrived at Grand Forks AFB on 15 September 1966 and went on alert on 2 October 1966 at LF Hotel-24, 447th Strategic Missile Squadron (447 SMS). The first launch attempt was scheduled for 12 October, but on 5 October, a Stage II power supply failure occurred. The missile was removed and sent to Plant 77, where technicians replaced the faulty components. The missile arrived back at Grand Forks on 14 October and was on modified alert three days later in preparation for the test launch on 19 October.

Forty minutes before launch, faulty diodes in a power supply drawer in the LER halted the countdown. A third attempt was made on 28 October, but seven minutes prior to launch, the guidance-and-control system failed, as did the NCU in the inert Stage III. The missile was removed and the cause of the failure determined to be leaking wet tantalum capacitors. This had been a problem during the Minuteman II R&D at the Cape and the DASO program at Vandenberg. Further test launches were postponed until 1967.[9]

The third phase of Project Long Life was to be conducted at the 351 SMW, Whiteman AFB, Missouri. Whiteman had been the first base to undergo the Force Modernization LF and LCF update to reconfigure the previous Minuteman IB facilities to operate with

Minuteman II. A survey of the launch sites at Whiteman AFB revealed that there were no sites where a launch could be conducted safely, so the decision was made to return to Grand Forks and try again, this time with the program renamed Project Giant Boost.

On 14 August 1968, less than seven seconds before launch, the missile guidance computer shut down for the third consecutive failure in the Minuteman II limited-range launch program. The missile was immediately shipped to Hill AFB, and a ground test missile was installed in LF H-24.[10] Five consecutive successful countdowns were run on this missile, the flight test missile was reinstalled, and all connectors were x-rayed. The LER electronic equipment worked properly, as did the missile guidance section and the ordnance devices. Further investigation by SAMSO determined that a connector pin in the downstage umbilical cable between the missile and the LF power distribution box was found to be depressed 0.085 inch. This particular pin was the connection for the ordnance arming circuits. The length of the pin itself was within specifications, but if the umbilical connecting ring was loosened two turns and the cable flexed, as might be done during the installation, the result was an open circuit.[11]

No further attempts for a limited-range Minuteman II launch were made at Grand Forks. Instead, a Giant Boost modified missile, nicknamed SHORT ROUND, was successfully launched on 1 September 1968 from Vandenberg AFB. This successful test restored confidence in the Minuteman II among members of Congress who had been concerned with the failures at Grand Forks. On 6 September 1968, Secretary of Defense Clark Clifford and Secretary of the Air Force Harold Brown agreed to cancel the Giant Boost program at Grand Forks, and on 19 October 1968, the Giant Boost Task Force was inactivated.[12]

Concerned that other downstage umbilicals within the Minuteman II fleet had the same defect, SAC conducted an investigation and found 3 of 90 umbilicals showed the potential for a similar problem but, as installed, would not have prevented launch. Modifications were made to provide a routine continuity check capability for the ordnance circuits.[13]

While the Giant Boost program had been canceled, SAMSO, SAC, and the Air Staff agreed that the seven-second limited-range launches should still be made at each base by mid-March 1969, if funding was approved in time. On 19 October 1968, the Air Staff tasked SAC to develop specific plans for a series of six simultaneous limited-range launches or a series spaced over a 30-day period. On 1 December 1968, SAC issued a new plan for limited-range launches, and SAMSO proposed the use of an entire flight of a representative wing for each of the three Minuteman configurations: WS-133A, B, and A-M. These launches would be in conjunction with MOMS ground test missile operations (see below). This program was named Giant Roar.[14]

While a variety of different plans for seven-second limited-range launch programs were proposed in 1969, no further limited-range launch tests were conducted. The Air Force decided that the $10-million to $15-million cost for the program would yield little additional reliability data, and from a public relations standpoint, another seven-second missile failure, no matter what the malfunction, would be a disaster. The seven-second launches were replaced with ground test missile programs such as MOMS and SELM, discussed below.[15]

## *Full-Range Operational Base Launches*

EXERCISE ON TARGET

In February 1963 SAC, following instructions from the JCS, began developing a full-range OBL program. A preliminary study was developed and presented to Secretary of Defense McNamara on 11 March 1963, concluding that peacetime launches from operational sites, if properly selected, were indeed feasible. By 15 April 1963, the plan was complete for the launch of a Titan II from one of the westernmost launch complexes of the 390th Strategic Missile Wing, Davis–Monthan AFB, Arizona; the exact launch complex remains classified. The Titan II launch was to be the third in a series of four, the other three being a Titan I launch from 851st Strategic Missile Squadron (851 SMS), Beale AFB, California; an Atlas F launch from the 556th Strategic Missile Squadron (556 SMW, Plattsburgh AFB, New York; and a Minuteman IA from 341 SMW, Malmstrom AFB, Montana. There was considerable concern about the possibility of demonstrators protesting the Titan II launch since the flight path would extend over land for approximately 285 nm in both the United States and Mexico before arching out over the Pacific Ocean to a broad ocean area target. The Minuteman launch involved overflight of Canada to an impact area off the coast of Greenland, and it also ran into political opposition. The program was canceled by McNamara in January 1964.[16]

EXERCISE OCEAN VIEW II

The next discussion of full-range OBLs took place in 1966, with the plans for Exercise OCEAN VIEW II, SAC OP PLAN 3263.[17] Anticipating the possibility of the removal of the Nuclear Test Ban Treaty restraints preventing detonation of a nuclear weapon in the air or in the water, SAC developed a plan to launch an operational Minuteman IB with a Mark 5 reentry vehicle armed with a nuclear weapon to a broad ocean area target in the North Pacific. Due to the passage of time, the plan changed to use a Minuteman II armed with a Mark 11A, or if the ban was lifted sufficiently far in the future, a Minuteman III. Hazard analysis, risk evaluation, and local government approval were already being investigated for Exercise WIRE NET (see below), a full-range OBL with an unarmed reentry vehicle. Analysis of the possible impact locations for the first two stages would apply as well to OCEAN VIEW II and could be predicted from previous launches at both the Atlantic Test Range and Vandenberg. However, OCEAN VIEW II had the unique considerations of a nuclear warhead component. Not only were range safety issues, such as self-destruct without causing a nuclear explosion or the dispersion of nuclear materials, taken into account, but the probability of a tidal wave from the surface or air burst needed to be considered. These concerns were valid, not only for a full-range OBL but also if the OCEAN VIEW II test was conducted from Vandenberg. Exercise OCEAN VIEW II was not conducted.[18]

EXERCISE GIANT FOOT

Exercise GIANT FOOT, originally called BARE FOOT, was an operation to determine the accuracy and reliability of an unmodified Minuteman launched from Vandenberg in place of an operational base.

## EXERCISE WIRE NET

Exercise WIRE NET, an extension of Long Life with the plan to launch a Minuteman II from F. E. Warren AFB to a target in the Pacific, also encountered difficulties and after two and a half years was canceled in February 1968.[19]

## GIANT PATRIOT PROGRAM

Operational realism was always the goal for the Minuteman operational test launch programs at Vandenberg. For the most part, missiles were removed at random from one of the six operational bases and brought to Vandenberg, where the Combat Training Launch Instrumentation section (wafer) was added, which contained the command-and-destruct equipment. Additional telemetry instrumentation wafers might be added depending upon the test being conducted. The guidance-and-control raceways leading downstage from the guidance section were removed and destruct packages inserted for each stage. Raceway covers, now larger and heavier due to the range safety explosive charges, were then reinstalled. All of this was necessary for gathering in-flight information and for range safety, but the trade-off was that the missiles as launched were not true representations of the operational airframes (by 1973, this additional equipment increased operational test airframe weight nearly 300 pounds).[20]

On 13 September 1967, Headquarters SAC issued Required Operational Capability (ROC) 13-67 for a proposed system to replace the nuclear components in the reentry vehicles of Minuteman II or III operational test missiles with a new range safety device. The new Self-Contained Range Safety Abort System (SCRSAS) would restore the weight and balance of the missiles to the original operational configuration by moving the range safety equipment into the reentry vehicle, as well as relocating the destruct charges. The usual Vandenberg Minuteman missile destruct package consisted of Primacord placed in the raceway, which, when detonated, opened the booster stages like a tin can.

The concept was debated between SAC and Headquarters USAF over the next several months and finally validated by the Air Force with the issue of a Requirements Action Directive to AFSC to develop the SCRSAS. Since AFSC was already working on a similar system, there was optimism that the system would be ready for use within nine months.

Funding for the system was tied to the full-range OBL program in late 1968, with a design change from a relatively short-range system to that of a full-range OBL safety system, which would take an estimated 12 to 24 months to develop. As the need for the full-range OBL was debated, the SCRSAS program was modified and renamed the Operational Base Launch Safety System (OBLSS), which would consist of two shaped charges, located in the reentry vehicle, designed to propel a destructive force through the guidance system and into the third stage, thereby destroying the missile (Figure 14.3).[21]

The Minuteman full-range OBL concept culminated in the Giant Patriot program. The new SAC plan called for an extensive launch program from Vandenberg as well as operational bases. The program was expanded to include variable-range and azimuth launches from Vandenberg, including southerly launches, as well as uninstrumented launches from Vandenberg as a continuation of the Giant Foot program. The Vandenberg program would commence with one R&D launch for Minuteman II by September 1969 from a Vandenberg

## OBL SAFETY SYSTEM OPERATION

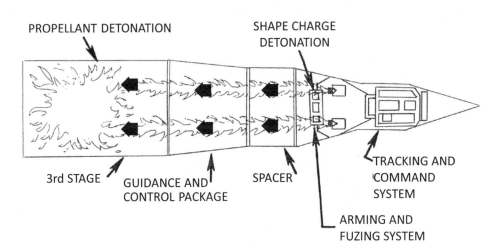

FIGURE 14.3 Profile of the OBLSS. The detonation of the third-stage propellant by the shaped charge in the reentry vehicle aeroshell would terminate missile flight. Placing the destruct charge in the aeroshell meant the missile was now as near operational configuration as possible. The OBLSS was available only for Minuteman II, and no more than 10 were built. *Adapted from "GIANT PATRIOT Minuteman Operational Base Launch Program."*

LF that was in an OBL configuration. This would be followed by a series of operational test launches from Malmstrom beginning with one launch in November 1969 and progressing to launches from two 564 SMS launch facilities (WS-133B) and two Force Modernization launch facilities (WS-133A-M) between January and June 1970. Subsequent launches would continue with one OBL from each of the other bases each year. It was hoped a similar program would begin for Minuteman III in 1971. The Air Force Scientific Advisory Board fully supported the OBL program and strongly recommended that only minimal changes to the weapon system configuration be permitted throughout the program. If the DoD required a range safety system for these launches, the Scientific Advisory Board recommended one that could be contained within the reentry vehicle as it was being developed with the SCRSAS/OBLSS. A major objective of the OBL program was the ability to fully test the national command-and-control system through launches at each of the bases.

On 9 November 1968, Deputy Secretary of Defense Paul H. Nitze approved the September 1968 proposals for both limited-range launches and the OBL program. The OBL would be conducted with a minimum of changes but would be preceded with a series of launches testing the range safety system. The Air Staff directed SAC to conduct further studies of possible destruct systems for the OBL program, including one proposed by the AEC, the C-band Airborne Tracking Safety System.

By the end of March 1969, the OBL program was expanded to include eight full-range flights from Malmstrom during the winter of 1969 and 1970, using available range safety systems or none at all. A month later, Secretary of the Air Force Dr. Robert C. Seamans

Jr. proposed a new OBL program. He pointed out that while the least expensive and most operationally realistic configuration was the uninstrumented launch program, there was too much risk and probable political opposition. Instead, he proposed a program consisting of six to eight full-range flights from Malmstrom during the winter of 1969 and 1970 using the current Vandenberg range safety system. In the interim SCRSAS would proceed to development for use in a full-range program in 1970 to 1971.

Three months later, on 6 June 1969, Secretary of Defense Melvin Laird requested further details on the full-range OBL program. Several weeks later, SAC was made lead command for the Malmstrom launch program, now named Business Leader. The new plan called for three WS-133B launches from 564 SMS LF Sierra-32 in January, February, and March 1970. Three WS-133A-M launches would take place in February, March, and April 1970 from Malmstrom LF Hotel-11, with all flights using the current Vandenberg range safety system. Simultaneous with the full-range launches would be four MOMS tests.[22]

On 26 January 1970, Secretary Seamans added to his earlier recommendations with a strong endorsement of the reentry vehicle self-destruct capability, which would now require approximately 19 months for development. Now the test missile aerodynamic characteristics would be identical to a fully armed operational missile. The OBL program would be conducted during the winter months or during the rainy season to keep fire danger to a minimum under the planned trajectory.

The OBL program plan, SAC OP 82-70, called for the launch of two missiles, with one serving as a backup. Tau Island, near American Samoa, 4,800 nm distant, was the impact area. Launched from LF Sierra-32, 564 SMS, Malmstrom AFB, the missile would exit the continental United States 92 miles north of San Francisco (42 miles south-southwest of Eureka). The reentry vehicle would be a Mark 11C Mod 6 containing an OBLSS in place of the nuclear warhead. The earlier hazards study for Exercise GIANT FOOT had concluded that a nighttime launch in the winter had the highest probability of success. By the end of FY 1970, the OBL program's remaining obstacles were approval by President Richard M. Nixon of the flight plan over Montana, Idaho, Oregon, and California and subsequent approval by Congress.[23]

A revised OBL schedule called for five launches from Malmstrom in FY 1972, with two more from another base in the winter of 1973 and 1974. All of these launches would be preceded by two OBLSS test launches from Vandenberg in the summer of 1971, continuing with eight additional launches scheduled through 1976 for comparison with the launches from operational bases. On 19 November 1971, the OBLSS design was successfully tested at White Sands Missile Range. In December 1971 a Senate–House conference committee deleted funding for the OBL program but did allow development of the safety system for the program to continue.[24]

As a critical part of the Giant Patriot program, flight testing of the OBLSS commenced at Vandenberg AFB on 13 June 1972, with the successful launch and then subsequent tracking of the Minuteman II by the mobile range safety system (MRSS). The MRSS would be deployed to the operational base as needed for temporary range safety tracking capability. The self-destruct charge was purposefully not detonated during this flight. On 25 July 1972, the second successful test launch took place. At 51 seconds into the Stage I burn, the range safety officer triggered the OBLSS, and the missile disappeared into a bright orange fireball over the Pacific. Funding was soon released by AFSC for the purchase of six additional OBLSS sets.[25]

The successful testing was timely in that the Soviet Union's OBL program, which also dated back to the mid-60s, had recently increased its launch rate, which caught the attention of Dr. Henry A. Kissinger, assistant to the president for national security affairs. Kissinger's inquiry into the state of a similar program within the Air Force spurred Headquarters SAC to respond on 23 February 1973 with justifications for the OBL program in the hope of encouraging Congress to approve funding for the program, as an additional $3.7 million was necessary. The OBL justifications included (1) provides the most realistic demonstration of Minuteman as a credible deterrent, (2) demonstrates complete end-to-end test of the weapon system, (3) permits evaluation of operational launch region gravity and geodetic modeling with respect to accuracy, and (4) provides the opportunity to evaluate the results of more than 14 years of ICBM operational testing.

A significant advantage of the OBLSS equipment was that the missile did not have to be removed from the LF in preparation for the test, as the only work necessary was that of removing the operational reentry vehicle and replacing it with the test reentry vehicle. Some modifications would be made to the launch duct to facilitate repair after launch. By September 1973 it appeared that it would be possible to begin the program in the upcoming winter months.[26]

On 28 December 1973, Secretary of Defense James R. Schlesinger announced the intent of SAC to conduct an eight-missile Minuteman II OBL program, named Giant Patriot, subject to congressional approval. There would be four launches in the winter of 1974 and 1975 and four launches in 1975 and 1976.[27]

A detailed description of the proposed program, which had been updated from the earlier programs on 15 November 1973 to become a two-year, eight-launch program, was released publicly through the 31 January 1974 issue of *Commanders Digest*. The greatest concern was the danger posed to the public by the spent 28-foot-long Stage I motor casing and the interstage panels that would be released upon ignition of Stage II. Experience from 123 prior Minuteman II launches from Vandenberg (122 successful) allowed the Air Force to estimate that the empty 4,800-pound booster would impact on uninhabited land near the Montana–Idaho border, 130 nm (150 statute miles) downrange, and the four interstage panels, each weighing 68 pounds, would land an additional 44 nm (50 statute miles) west near the intersection of the Idaho, Washington, and Oregon borders. The missile would cross the coast approximately 90 nm (103 statute miles) northwest of Medford, Oregon. The empty Stage II would impact in the Pacific Ocean approximately 530 nm (600 statute miles) off the coast.

One of several squadron configurations involved three missiles for the full-range launch, one serving as backup, and the remaining seven missiles in the selected flight configured for a SELM test (see below). Upon receipt of the test execution order from Headquarters SAC, one missile would be launched, and the seven SELM-configured missiles would complete a simulated countdown. Once the SELM results had been collected and the test range cleared, the second full-range missile would be launched several days later. Then a second squadron would repeat this same process. To verify the missile was performing properly at launch, the MRSS tracking system would be augmented by two sky screen devices, which were optical tracking systems for observing the earliest stages of flight prior to radar acquisition. Upon radar acquisition flight progress would be monitored by the instantaneous impact protection system.[28]

After several years of study and consideration of operational and safety criteria, 564 SMS and the 341 SMW, Malmstrom AFB, were selected as the first OBL test site. LFs Sierra-37 and Sierra-40, 564 SMS, were selected for the WS-133B launches, and 12 SMS LFs Golf-03, Golf-04, and Golf-05 were selected for the WS-133A-M flights. Problems existed with each of these launch sites. The missile launched from LF Sierra-37 had a Stage I probable impact area that included a large lumber camp with a population of 300 people. Missiles launched from either LF Sierra-37 or LF Sierra-40 would have the Stage I and II interstage skirt panels impact areas that encompassed sections of the Nez Perce Indian Reservation and the towns of Winchester and Rubens, Idaho. To circumvent this problem, Headquarters SAC proposed that the OBL missiles be inhibited from releasing the panels during flight and that LF Sierra-37 be excluded as a launch site. Headquarters USAF approved both proposals and instructed SAC to conduct two test launches from Vandenberg as quickly as possible to evaluate such an adjustment on missile reliability and accuracy.[29] The planned impact area was now 4,300 nm distant, near Canton Island in the Phoenix Islands target area. Canton had been selected because the Phoenix Islands target complex was already instrumented for use with the Minuteman III R&D and OTE launches (Figure 14.4).[30]

Immediately upon release of the public description of the program, Governors Cecil Andrus of Idaho and Tom McCall of Oregon were both highly critical of the idea. Senators Mike Mansfield (D-MT), Lee Metcalf (D-MT), Frank Church (D-ID), and Mark Hatfield (R-OR) signed a joint letter of opposition addressed to Secretary of Defense Schlesinger. Their opposition did not bode well for congressional approval of the necessary funding for land acquisition, hardware procurement, and test site preparation. With the pending conversion of the 564 SMS at Malmstrom AFB from Minuteman II to III, SAC set a deadline for the funding decision of 25 March 1974 for the four-launch program for the coming winter. On 29 March 1974, the $6.3 million was approved, and SAC decided to proceed with preparations for the full four-launch program even though the deadline had been exceeded by four days. Final approval for the actual launches required additional congressional action on the FY 1975 Defense Appropriations Bill, which was currently under consideration.

On 6 June 1974, the worst fears of the Air Force were realized when Senator Mansfield introduced an amendment to the FY 1975 Defense Appropriation Bill, S. 3000, deleting funds for OBL. Not only did the amendment delete funds for the operations at Malmstrom AFB, it went further and made clear:

> None of the funds authorized by this or any other Act may be used for the purpose of carrying out any proposed flight test (including operational base launch) of the Minuteman missile from any place within the United States other than Vandenberg AFB, Lompoc, California.[31]

In the subsequent debate on the amendment, Mansfield made the point that while there would be a welcome influx of funding to Montana's economy, the potential dangers

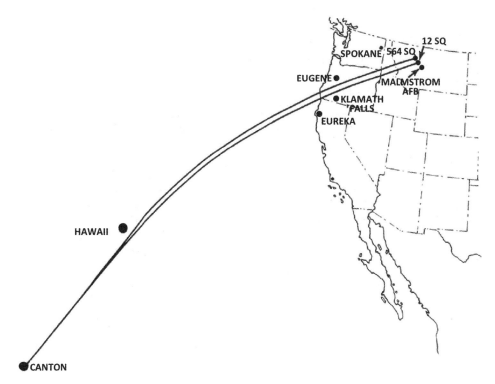

FIGURE 14.4 *Upper:* Proposed trajectory and target area for the first launch of GIANT PATRIOT Operational Base Launch Program. *Lower:* Flight profile showing probable impact distance from the LF for the Stage I motor casing and interstage panels. By 102 seconds of flight, the trajectory was such that the remaining stages and interstage panels would clear the continental United States prior to impact. Adapted from "GIANT PATRIOT Minuteman Operational Base Launch Program."

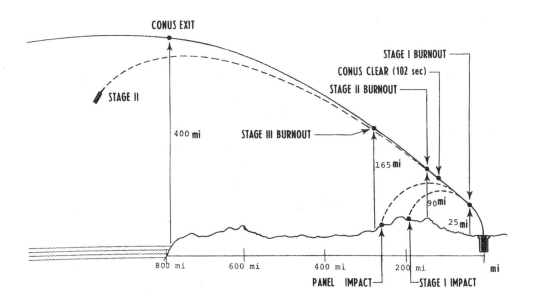

and international implications far outweighed any benefit of the tests for national security. Furthermore, it seemed to Mansfield that the logic for the test was flawed. Would a successful test program at Malmstrom then be enlarged to cover all the other Minuteman wings? In addition, Mansfield pointed out that the DoD had not described the possible consequences of such a test on the Strategic Arms Limitation Talks that were currently in progress.

Six months earlier, Mansfield had sent Secretary of Defense Schlesinger a letter requesting further details on the need for the OBL program. The response, from Deputy Secretary of Defense William P. Clements Jr., pointed out that two of the Malmstrom launches would be from LFs that did not have the protective shielding in the launch duct that was used at Vandenberg to facilitate repair after launch. Launches from these unprotected facilities would verify that the absence of the protective equipment did not adversely impact an operational launch. Landmass gravitational anomalies were presently calculated as an extrapolation for overwater flights. The land overflight involved with OBL would provide further verification of the present landmass gravitational trajectory models.[32]

Unfortunately for the Air Force, a recent book, *The High Priests of Waste,* by A. Ernest Fitzgerald, which was an exposé of the Lockheed C-5 Galaxy cost overruns, also included a scathing description of "inherent reliability problems in the Minuteman II advanced guided system which had an exceedingly high failure rate." What Fitzgerald did not point out was that the in-flight performance of the NS17 MGS was quite high, while the accuracy issues involved precision versus accuracy, not gross trajectory problems.[33]

The Mansfield amendment was approved, and the bill, as amended, was passed by the Senate on 11 June 1974. On 23 July 1974, a joint conference committee of the Senate and House Armed Services Committees approved the Senate amendment deleting the funding, and the OBL program ended.[34]

## Simulated Missile Launch Programs

### *Modified Operational Missiles Program*

Planning for the first simulated launch program, MOMS, began in November 1969, with SAC OP ORD 115-70. Two months later, the interim technical orders, ITO 21M-LGM30A-103-44, for replacing operational Minuteman II missiles on strategic alert with MOMs that would be able to respond to launch orders but be prevented from launch, had been written and reviewed by the Nuclear Weapons Systems Safety Group. The primary objectives of the MOMS program were twofold: (1) to increase the confidence in launch reliability of the Minuteman force and (2) to identify and correct possible weapon system deficiencies. A secondary objective was to evaluate, for the first time, the capability of single- and multiple-launch options of the Minuteman ground electronic systems using a complete flight of missiles at an operational base. The MOMS modification was successfully evaluated at Vandenberg in November 1969, where it had involved only one Minuteman II.[35]

Modifications to the missiles were kept to the minimum required for safety so the test would reflect missiles in as near an operational configuration as possible. The MOMS airframes were modified by Boeing at Plant 77, Hill AFB, Utah. A scoring device was attached to the Stage I igniter to indicate activation at the time of simulated launch. Each stage's motor igniters were replaced with ignition prevention devices, and all missile ordnance was disconnected or removed. Each missile had a large "MOMS" label stenciled on each stage to ensure proper identification.

EXERCISE GIANT PROFIT

Golf Flight, 447 SMS, 321 SMW, Grand Forks AFB, North Dakota, was chosen for the first operational base MOMS program test, Exercise GIANT PROFIT. Modification of the first missile began at Hill AFB on 19 May 1970, and the first MOMS missile arrived at Grand Forks on 28 May 1970.

LF modifications were also kept to a minimum. Depending on the program plan for a specific LF, the launcher closure door would be opened as per a normal launch or, as an alternative, an indicating squib was installed to indicate receipt of the launcher closure door opening signal at the ballistic gas generator but isolating it from activation and subsequent door opening. For the test where the launcher closure door would be opened, a buffer of sandbags was used to prevent the door from traveling off the guide rails. The only modification in the launch duct was the addition of padding at the upper umbilical launch duct receptacle to prevent damage upon umbilical retraction. The lower umbilical was pulled away as the missile lifted off and did not need protection as it would not be released during the test.

Of critical importance was the complete isolation of the test LCC and 10 LFs from the other 40 missiles in the squadron. This isolation would be achieved by physically disconnecting the command-and-status line hardened intersite communication cable, thus isolating Golf Flight from any nontest elements. An additional safety consideration was changing the radio frequency of the medium-frequency radio. These precautions allowed the remaining 40 missiles of the squadron to continue in their strategic alert posture. Only one LCC would be used, which meant two launch panels would be necessary as part of the ECC procedures (see Chapter 15).[36]

The GIANT PROFIT preparations began 15 May 1970. The first MOMS airframe was emplaced in LF Golf-14 on 29 May, and missile installations were completed on 31 July with the emplacement of the last missile in LF Golf-18. A dry run was made on 13 August 1970, with all aspects of the test sequence performed. Only minor problems occurred.[37]

Exercise GIANT PROFIT was conducted over 17 and 18 August 1970. Verification of flight isolation was completed at 1440Z without difficulty. The execution message was transmitted by the SAC Command Post controller with an execution time of 1745Z; it was received, verified, and acknowledged by the missile combat crew in LCC Golf-0 and another crew in the squadron at 1747Z. The simulated launches were executed in four increments in order to test several different launch scenarios.[38]

> Increment One
> After receipt of the execute message, the launch crew issued Preparatory Launch Command A (PLC-A). This sent a message to all the missiles in the isolated flight (further explanation of the launch commands is given in Chapter 115). The PLC-A

message contained a specific execution option which the missile guidance computer interpreted, based on a preloaded missile guidance set database, to select the target and delay time specific to the particular missile. Sorties 17 and 19 had their Infinite Hold delay, the default, change to a "hot" time (non-infinite) as a result of the PLC-A with delay times of three seconds and one second respectively. A second PLC-A message was sent to change the delay time for Sortie 16 from infinite to zero seconds resulting in Sorties 16, 17 and 19 having launch delay times of zero, three seconds and one second respectively. The Enable command was then issued, resulting in all test sorties (11–20) entering Enabled mode. These sorties remained Enabled until they successfully received an Inhibit command. Next, using the Emergency Combat Capability launch process, Execute Launch was commanded using Launch Control Panel A, followed by Launch Control Panel A replaced by Launch Control Panel B and the Second Execute Launch command sent to all test sorties. Only the sorties with "hot" delay times, Sorties 16, 17 and 19, accepted the two votes and processed the Execute Launch command. The three sorties indicated successful "launch" in the order of their delay time.

Increment Two
PLC-A was commanded and Sorties 11, 13 and 20 were given "hot" times of one, two and three seconds respectively while Sortie 15 had a delay time of zero seconds. The Execute Launch commands was given as before with the two panels, this time Panel B was first, and the sorties successfully "launched" in the order of their delay times.

Increment Three
Similar to Increment Two but this time Sorties 12 and 14 were selected. After the Execute Launch command was issued with Launch Control Panel A, a seven minute hold was called until a farmer in a field adjacent to Sortie 12 was verified to be outside the designated caution area. The second Execute Launch command was issued with Launch Control Panel B and the two sorties successfully "launched."

Increment Four
Due to a planned radio pre-emption, Sortie 18 did not enable during the all call Enable command issued during Increment One. Enable was now commanded for Sortie 18. A PLC-A command for a "hot" time of zero was generated for Sortie 18 followed by an Execute Launch commanded using Launch Control Panel B. Inhibit was then commanded and Sortie 18 entered Dis-enabled. If the second vote was not received within five minutes, the first vote would be deleted and the Enabled command needed to be resent. Nine minutes after the Inhibit command, Enable was commanded for Sortie 18, the two Execute Launch commands sent as before, and Sortie 18 indicated successful "launch."

The exercise LF Support Team inspected each LF for verification of successful simulation after each increment. Verification of the launcher closure opening was made on the surface. Verification of the simulated launcher closure opening was made upon penetration into each LER and visual inspection that the ballistic gas generator squib-simulator had fired. The launch tube inspection involved verifying the retraction of the upper umbilical and inspecting the Stage I ignition firing system to see that the indicating piston had extended.

Three anomalies were reported:

1. The upper umbilical retraction mechanism at LF Golf-15 did not hold the umbilical in the retracted position. Experience with a similar problem during operational test launches at Vandenberg indicated that this condition would not have prevented a successful launch but the problem needed to be addressed. Sortie 15 was scored a success.
2. The LF Support Team entering LF Golf-11 reported that the Stage I ignition indicator was not protruding and thus the missile launch was thought to be a failure. Further investigation showed that the indicator pin had activated but had been propelled out of the missile housing and was found on the floor of the launch duct. Therefore, the launch at LF Golf-11 was reported as a success.
3. LF Golf-17 failed terminal countdown. The LF support team reported that the Stage I ignition indicator was not protruding, the upper umbilical had not separated from the missile and the launcher closure squib simulator had not fired. The cause of the failure was that the Safety Control Switch (SCS) was not armed, and therefore the failure of the other indicators was explained. The Failure Analysis Team determined that the malfunction was due to a faulty SCS which could not have been detected except by commanding Launch and attempting to complete the terminal countdown. Subsequent testing revealed that the switch failed to remain in the armed position, which resulted in an open circuit and prevented the missile battery from activating, terminating the countdown. The switch was replaced and a second launch countdown initiated on 20 August 1970, which was a complete success.

Analysis of the SCS, which was removed in the failed condition, revealed that an improper ratchet wheel assembly had been installed, which did not engage a pawl, preventing the switch from staying closed. The switch had closed, sending the appropriate signal to the LCC, and then reopened before the launch signal had been sent, thus causing the No-Go condition. Replacement with the correct ratchet wheel restored proper functioning of the repaired unit. All operational SCS units were inspected to determine the installation of similar faulty units, 2 out of 200 (all the Minuteman II LFs at the time), which were then replaced. The first MOMS test with Golf Flight simulated Stage I ignition of 9 of the 10 missiles.[39]

Launcher closure doors were successfully opened at LF Golf-18 and Golf-20. The 804th Combat Support Group Civil Engineering Squadron constructed launcher closure door buffers at each site using approximately 4,500 sandbags, leaving an access hole large enough for the pipe pusher along the center rail until three days prior to the test day. The A alignment monument was dismantled to accommodate the sandbag buffer. The launcher closure door face penetrated only the first two rows of sandbags, but it was concluded that moist sand would have provided a more effective barrier. The A monument was rebuilt and resurveyed by Detachment 3 of the 1st Geodetic Survey Squadron prior to return to strategic alert on 25 October 1970, five months after the start of the test program.[40]

Recommendations for future GIANT PROFIT tests included (1) Develop a method to test the SCS at operational launchers without entering terminal countdown so as to establish confidence in the proper operation of the switch; (2) Redesign the upper umbilical head protection devices specific for the MOMS tests, as three umbilical heads were damaged, requiring repair or replacement; (3) Provide in the Test Operational Plans authority to reenter test configuration after test day for the purpose of failure analysis. Furthermore, failure analysis should continue until all failed missiles have a successful MOMS launch.[41]

During FY 1971 and 1972, four MOMS tests were completed, two per year. A fifth and final test was conducted in May 1973 at Minot AFB. The results of these tests were encouraging; out of the 50 simulated launches commanded by various LCCs and the ALCC (see Chapter 15), 48 had been successful. This 96 percent reliability demonstrated the effectiveness of the Minuteman system. Additionally, as had been the goal all along, the program identified two potentially serious problems affecting the system. The first had been the improperly built SCS. While the vast majority of the fielded switches were properly constructed, the test had served its purpose. The second problem involved the ALCC and a software S tape that had failed to transmit the necessary launch commands. Investigation revealed that the tape had been double-encrypted. Since the backup aircraft would have been able to launch the flight, the test was scored a success, but the discovery of the double-encrypted tape led to a $2.5-million software modification to prevent recurrence.[42]

At the end of the program, the cost of a 10-missile MOMS test was approximately $500,000 to $800,000, and the test could keep a flight of missiles off alert from 2.5 to 5 months. In an effort to develop a more cost-effective simulation program and prevent such long-term alert degradation, SAC sought to develop a simulated launch capability that did not involve removing the missile from the LF for conversion as an alternative to the MOMS program.[43]

## SIMULATED ELECTRONIC LAUNCH MINUTEMAN

Even as the MOMS program was being finalized, Headquarters SAC issued ROC 12-69 on 22 April 1969 for a Minuteman simulated combat launch capability (SCLC). This new requirement reflected the concern that attempted limited-range launch from operational bases had revealed that launch reliability was a significant area of weapon system uncertainty, as Long Life II and Giant Boost had unfortunately demonstrated. While it could be argued that the missiles in these two programs were highly modified and thus not representative of operational missiles, SAC needed a more economical, but still thorough, ground testing program. Such a program would conduct terminal countdown checks of critical equipment on a full-flight basis and not require missile removal for modification, as with the MOMS program.

On 29 July 1969, Headquarters Air Force validated ROC 12-69 and instructed the AFSC to study the feasibility of the SCLC concept. SCLC would permit testing of missiles in a nearly operational configuration while keeping safety considerations in mind and minimizing disconnection of LF equipment, all of which would be directed toward not requiring removal of the missile from the LF. On 17 August 1970, as the first MOMS test began at Grand Forks, SAMSO recommended canceling the SCLC program due to the expected $10-million to $15-million development costs and nuclear safety considerations. Both the Boeing and Sylvania SCLC proposals were promising, but both involved activation of the safe–arm devices, bringing the missile only one step from possible accidental Stage I ignition.

On 18 March 1971, Headquarters SAC Test Support Division released a cost comparison of SCLC versus MOMs: SCLC, $100,000 per test, three tests per year, time off alert 15 days; MOMS, $819,000, two per year, 77 days off alert. Clearly a simulation-directed test would be more cost-effective. A week later, SAMSO and the Ogden Air Materiel Area decided that the SCLC design's inherent safety issues prevented it from filling the needs of ROC 12-69, and on 30 April 1971, SAC formally canceled the SCLC requirement, deciding instead to develop a simulation program based on the ongoing Minuteman Bench Test Program.[44]

Utilizing reusable portable test sets, the Minuteman Bench Test Program evolved into SELM, which required minimal LF modifications and would allow the missile combat crews to turn their keys and go through the combat launch sequence with the test sets recording successful simulation of terminal countdown events. All that was needed for statistical evaluation of Minuteman reliability was the data from 40 missile tests a year. Since SAC ballistic missile evaluators needed to use an entire flight of missiles to test LCCs in conjunction with LFs, SELM looked like the ideal test program to accomplish both goals. The economics of the plan would hopefully permit annual evaluations of one flight at each wing.[45]

The SELM test set was designed to intercept certain launch signals issued by the test LCCs or ALCC while letting others pass to the operational ground equipment and then record whether the signals had been successfully transmitted. The equipment was to be installed in the LERs of the LFs to be used. There were three test boxes: the power control and load simulator serving as a circuit tester and signal router, the missile guidance power supply replacing the operational power source for the guidance-and-control section, and an ordnance load simulator. The operational reentry vehicles would be removed from the missile and replaced with reentry vehicle simulators, as had been done with the MOMS tests.[46] This was the only major modification to the missile and did not require its removal from the LF. Maintenance teams would disconnect the communication lines that linked the test flight to the rest of the squadron, as with the MOMS program. A test at Grand Forks or 564 SMS at Malmstrom would require turning off the test LCCs' medium-frequency radios or switching them to a test frequency to preclude signals from going to operational launchers on strategic alert. Another precaution was the use of special excluded test codes for the test sites. Each test missile would be equipped with the normal maintenance safing pins for each stage's igniters, and the lower Stage I umbilical would be disconnected to ensure prevention of Stage I ignition (Figure 14.5).

After review of the SELM test protocol by the Nuclear Weapons System Safety Group, approval was given to utilize, in real time, complete EWO command-and-control procedures, from message transmission from Headquarters SAC to missile combat crew commit with documents and key-turn procedures. Included in this approval was the first utilization of operational ultra-high frequencies (UHF) and squadron address tones by the ALCC to commit test missiles. While the MOMS test program had used ECC procedures since only one LCC sent the launch commands, in the SELM tests, the normal two-vote system was used.[47]

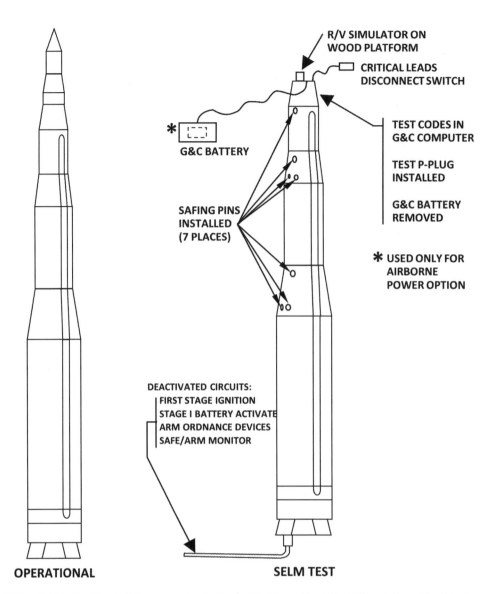

FIGURE 14.5  Profile of a Minuteman II missile modified for a GIANT PACE SELM test. The SELM test program continues to serve a vital role in maintaining operational capability with the Minuteman III force. *Adapted from "Strategic Air Command History, FY 1974."*

### Exercise GIANT PACE

GIANT PACE was the name given to the SELM tests. The first three GIANT PACE exercises were held in the last half of FY 1974, all involving Minuteman II missiles: the 44 SMW, Ellsworth AFB (11 missiles), 341 SMW, Malmstrom AFB (10 missiles), and the 351 SMW, Whiteman AFB (11 missiles). Launch commands were issued from both LCCs (17) and the ALCC (15). The test results showed 32 of 32 successful simulated launches.[48]

The average cost to SAC per SELM test was $40,000 to $50,000, which was considerably less than the $500,000 to $819,000 spent for each MOMS test. More importantly, the time involved in a GIANT PACE exercise averaged 6 weeks versus the 22 weeks associated with the MOMS program.[49]

The five GIANT PACE exercises in FY 1975 were nearly as impressive, with 54 of 55 successful simulated launches. By the end of 1976, 168 of 170 SELM tests had been successful, for a 98.8 percent success rate, convincing evidence of the reliability of the Minuteman force.[50] SELM testing continues to be performed as of November 2019 (Figure 14.6).[51]

★ ★ ★

The OBL concept evolved from planned full-range launches from Malmstrom to the SELM program still in use in 2019. The result was a much safer and less expensive evaluation of the reliability of major components of the launch operation. Additionally, missiles are not expended, a critical point with production having ended in 1977.

Another important area that needed to evolve with technological advances was that of the command and control of a 1,000-missile force. The move from a strategy of massive retaliation at the inception of the Minuteman program to control response required modifications to the system while at the same time maintaining the instant response that was the trademark of the Minuteman weapon system.

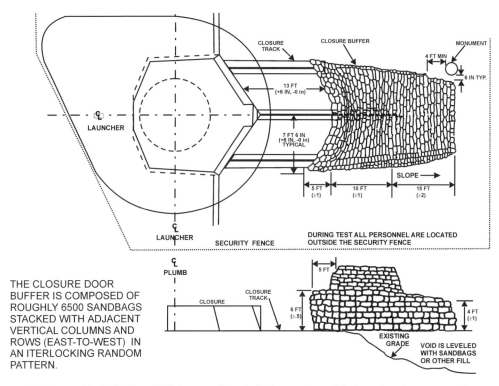

FIGURE 14.6 The MOMS or SELM test could include the opening of the launcher closure using the ballistic gas generators as in an operational launch. To prevent the launcher closure from leaving the rails, approximately 6,500 sandbags were positioned as a buffer. They did not always stop the door. *Adapted from "Minuteman Service News," courtesy of the Boeing Company.*

# 15

# ASPECTS OF COMMAND AND CONTROL

## Early Concerns

Col. Edward Hall's original concept for Minuteman was to launch a squadron of 50 missiles at a time. This massive strike was born from the prevalent massive retaliation strategy of the Eisenhower administration. An inadvertent or rogue launch of 50 Minuteman, when compared to potential single Atlas or Titan I or II launches, worried many inside and outside the Pentagon.[1]

On 12 February 1960, Eisenhower's science advisor, Dr. George B. Kistiakowsky, summarized his concerns about the Minuteman program. These included first-stage engine problems with insulation and movable nozzles as well as problems with development of the inertial guidance system required to run continuously for years with the missiles on alert. Kistiakowsky noted, "It is my considered opinion that all these problems will be eventually solved." His major concern, however, was the correct design of the overall system, especially the command-and-control aspects.[2]

Coinciding with Kistiakowsky's concerns was the work of WSEG, which provided analytical support to the OSD. Formed in 1949, WSEG produced reports that were seen as independent, objective, and authoritative. At the request of the JCS, WSEG began work on Report 50: Evaluation of Strategic Offensive Weapons Systems in September 1959. The focus of the report was "to evaluate the weapon systems and directly related functions which make up the strategic offensive posture of the US in the 1964–1967 period." The JCS and other agencies had input into the report.

By 15 September 1960, a preliminary copy was being reviewed by the Office of the JCS due to concern over the report's finding of severe deficiencies in the national command-and-control structure. On 26 September 1959, the concern was sufficient enough to brief the JCS themselves, as well as Secretary of Defense Robert Gates.[3] The report pointed out that the simplest way to eliminate the Minuteman missile force was to target LCCs, completely overcoming command and control of the force before its use.[4]

Also brought up in the report was the issue of inadvertent or rogue launch. Up to this point, the original 50-missile launch capability had been retained. Three months earlier, on 3 June 1960, the AFBMD expressed confidence that the design of the existing launch control system for Minuteman had sufficient safeguards to prevent inadvertent or rogue launches, and no mention was made concerning the minimum 50-missile launch response.[5]

On 15 December 1960, Brig. Gen. Bernard A. Schriever briefed Air Force and Defense Department officials on possible modifications to Minuteman in response to the questions raised by the WSEG 50 report. The recommended modifications—selective targeting, multiple-target capability, and stop launch—would be expensive and time-consuming, all of which would certainly delay deployment. The cost was estimated to be $40 million to $60 million, and it would take three years.[6]

The WSEG 50 report was released on 27 December 1960. For Minuteman the report recommended a standard wing size of 150 missiles composed of three squadrons of 50 missiles each with one LCC per 10 missiles. If needed, additional squadrons would be added. While the original configuration of three LCCs with 50 missiles each was no longer part of the operational plan, no mention was made of the preferred option to launch a full squadron of missiles at once. The response to the concerns expressed during its preparation and review, specifically those addressed by Schriever, were left to the new Kennedy administration to address.[7]

## Kennedy Administration

The WSEG 50 report proved a useful transition document for the Kennedy administration. The report covered the major strategic weapon systems in place or soon to be deployed and incorporated strategic input from the JCS. Secretary of Defense Robert S. McNamara and Deputy Secretary of Defense Roswell L. Gilpatric were briefed on WSEG 50 within days of taking office. Of particular interest was Enclosure C, Command and Control of Strategic Offensive Weapons in the Period 1964–1967. A second portion of the report also attracted McNamara's attention, Enclosure G, Weapons Systems Characteristics, which included a critical examination of the estimates provided for each strategic missile system's reliability, essential factors for formulating force structure, and operational planning.[8]

In April 1961 Schriever asked the Lauritsen Committee, formed a year earlier to independently review the progress of the entire ICBM program, to reconvene and conduct a follow-on review of the Minuteman program in light of the WSEG 50 report. Schriever specifically requested that they address the program's continuing technical issues and the operational concerns being voiced in terms of command and control. The committee's report was given to Schriever on 15 June 1961. The committee found that significant progress had been made on the propulsion and guidance issues, validating the decision a year earlier to proceed with a 4,700-nm-range missile, designated Minuteman IA, with less accuracy and reliability for the first, and possibly second, wing, in order to keep the operational deployment schedule. Selection of Malmstrom, AFB, Montana, near the Canadian border, ensured the ability to cover targets in the Soviet Union near the latitude of Moscow. The missile fields further south than 52 degrees north latitude in the Soviet Union or China could not be reached by Minuteman IA (see Figure 3.2, Chapter 3).[9]

As for accidental launch, the committee members expressed concern that the "highly automated and unattended" Minuteman launch sequence was controlled by a "sophisticated digital data processing system," which contrasted sharply with Atlas and Titan I and II in terms of human control. Furthermore, the committee members doubted that "the presently constituted reviewing agencies are technically qualified to perform the detailed analysis and evaluation which the system requires." Furthermore:

The committee is informed that 50 Minuteman missiles will be connected so that they can be salvo or ripple fired if desired. The committee feels most strongly that regardless of the precaution taken against inadvertent launch, the threat to world security implicit in such a system requires application of the highest level of approval and control.[10]

The lack of a stop-launch capability had been raised in 1960 by R. H. Rubel, assistant director, Strategic Weapons, Defense Research, and Engineering.[11] Rubel was aghast at the apparent ease of an accidental launch as well as the only options being a ripple or salvo launch of no less than an entire squadron of 50 missiles. Any two LCCs could launch all 50 missiles of a squadron, and if under attack with only one LCC left, all 50 could be launched after a time delay.[12]

The Air Force responded quickly to the report. The range issue was in the process of being solved, and missiles of the second wing at Ellsworth AFB, South Dakota, would have an estimated 5,700- to 6,200-nm range, which would cover all of the Soviet Union targets. The Air Force already had begun studies pertaining to unauthorized or inadvertent launch and found that the system was completely safe as designed. No mention was made of the 50-missile launch option, only that work in applying "flexibility features" was in progress. The Air Force felt that any additions or changes should be adopted only as they could be incorporated without delaying the operational program.[13]

On 6 July 1961, the Air Force forwarded a report written by Brig. Gen. Phillips, director of the Minuteman Program, and Col. R. T. Hemsley, chief, Minuteman Development Branch, to Dr. Herbert York, director of the Directorate of Research and Engineering in the Defense Department. The report addressed the Lauritsen Committee findings in greater detail, including budget estimates for their implementation, as well as the desire for more flexible targeting ability and selective launch. The earliest feasible implementation would be for the fourth wing, but from the view of cost and associated technical issues, delay to the fifth wing was suggested.[14]

The report's recommendations triggered another round of review by outside experts, as the DoD was in complete disagreement with the Air Force findings. On 27 July 1961, McNamara asked Secretary of the Air Force Eugene Zuckert to organize a committee to examine more closely the findings of the Lauritsen Committee report, the concerns mentioned in WSEG 50, and Rubel's specific concerns, with a goal of increased flexibility and safety for the Minuteman program. Formed on 30 July 1961, the group was formally known as the Minuteman Flexibility and Safety Group but informally as the Fletcher Committee after its chairman, Dr. James C. Fletcher. The main committee had nine members representing industry, academia, Pentagon staff, the RAND Corporation, and the JCS. The Engineering Subcommittee had five members from relevant aerospace companies and the RAND Corporation. The Fletcher Committee submitted its report on 26 September 1961.[15]

The response to the recommendations was swift. On 15 October 1961, Boeing received a 30-day contract for many of the committee's recommended design implementation changes and, just two weeks later, Contract Change Notice 299 was released, which required major changes in virtually all the launch control equipment. Boeing grouped the changes in two categories:

1. Enable/Retargeting overlay which would allow launch enabling by individual launch control facilities by means of remote control.
2. Unauthorized Launch Safety to be improved by seven separate changes:
   a. Sensitive Command Network (SCN) test enabling only the test message addressee to receive security status information.
   b. Volatile LF codes which included modification of the SCN decoder at the LFs so that code information would be destroyed upon detection of tampering.
   c. Extended No-Go Shutdown loops (LF) consisted of extending the No-Go continuity loops in the Sequence and Monitor racks so that removal of certain elements would cause a shutdown condition.
   d. Launch Enable Logic Unit (LELU) would provide for a reset of the LELU to a no-signal condition five minutes after receipt of an Inhibit Launch command.
   e. Monitoring conditions of the LELU Reset consisted of providing an LCC with a device to indicate that the launch signal had been received and registered.
   f. Monitor mechanical decoder shaft for rotation and report back to the LCC as a launch acceptance alarm.
   g. Secure the J-Box cover so that it would require an unauthorized person at least 30 minutes to remove.

Boeing was required to incorporate the changes without delay for the delivery of the first flight of missiles to Malmstrom. The Air Force was still reluctant to implement all the changes in Wing I, preferring to delay until Wing IV or later. While details of several of the Fletcher Committee recommendations remain classified even after 58 years, the Air Force responded to several key items, one of which was the ability to select eight targets remotely with the push to develop what the Fletcher Committee called "Improved Minuteman."[16]

In December 1961 McNamara made it clear that all wings would have the improved launch safety and selective launch features, while remote dual targeting would be added for Wing II and all subsequent wings, thus removing some of the complications in modifying the already built facilities. At the same time, excessive costs and deployment delay would be prevented.[17]

## The Evolution of Minuteman Command and Control

There were many concerns about the early design of the Minuteman command-and-control system. These concerns were addressed prior to deployment and put to the test in October 1962 during the Cuban Missile Crisis (see below). Over the nearly 60-year time span of Minuteman, two major command-and-control system additions were made: the ERCS and the ALCS. ERCS was a backup to broadcast the Emergency Action Message (EAM) to all SAC forces, while the ALCS was a backup to the ability to launch Minuteman if a Soviet first strike had rendered LCCs inoperative.

Minuteman command-and-control is the basic launch process; that is, receive valid Launch message, select target and launch time, Enable Launch, and Execute Launch. The launch process has evolved considerably since the initial deployment in 1962. Modifications included more target options, the ability to retarget missiles more quickly, and increased ability to monitor missile and LF functions, all the while maintaining launch promptness and, most importantly, improving weapon system safety.[18]

## *Readiness Verification*

To ensure the optimum launch readiness and performance for the current Minuteman system, a mixture of continuous status monitoring and periodic testing is employed. Launchers are continuously monitored for readiness by use of automatic fault detection, automatic status interrogations, and reports. Changes in system status are alerted to launch crews by audible alarms, status light changes, and printouts in the LCC. Details related to a general fault notification are obtained from the launcher using additional inquiries and responses: in some configurations, a voice reporting signal assembly (VRSA), and in later configurations, a dedicated, expanded fault report response—the ground maintenance response (GMR). Tests are run to exercise areas of the weapon system that are normally dormant until activated, as commanded by the launch crew. These areas include security monitors, internal missile flight controls, UHF communications, ordnance firing circuits, and the SCS, also known as the launch circuit safety device.

Launch readiness also includes maintenance response times. All types of maintenance teams are on alert to quickly troubleshoot and repair system malfunctions. Depending on the team and equipment involved, this could mean a team is on the road within an hour, though travel time to the more distant sites is considerable.

## *Target Selection and Launch Timing*

As Minuteman grew from having a single stored target on Minuteman IA to a maximum of eight for each reentry vehicle on Minuteman III, procedures to quickly select the desired target for 1 to 50 missiles in a squadron became essential. Each target has an associated launch time (actually a delay from initial launch execution), which is essential to coordinate the arrival of weapons at the target to avoid destruction by airborne debris created by previous explosions. Even small pebbles impacting the reentry vehicle exterior at Mach 10 can change the terminal flight path or destroy the weapon. In some cases weapon arrival time has to also consider other delivery systems such as aircraft, cruise missiles, and even submarine- and surface ship–launched missiles. Quickly defining the launch time, or delay, and target from the preloaded target set for each missile is, therefore, essential. As the world threat situation changed, the number of launch plans and subsequent launch timing complexity grew, resulting in weapon system modifications to maintain Minuteman's fast response time (see Chapter 16).

LAUNCH COMMANDS

*Enable*

The weapon system is normally in a fully safed configuration to ensure that even multiple potential equipment failures will not result in launch. The missile exits this positive safe condition when the Enabled configuration is selected: Electrical paths to ordnance devices are completed to prepare the launcher and the missile for all launch and flight actions. The ordnance devices provide the actions to open the launcher closure door to allow missile flyout, activate missile and guidance internal battery power systems, disconnect the umbilical cables from the missile, and ignite Stage I. The Enable command sets the stage for launching missiles within 32 seconds after key turn by the missile combat crews. The missiles can maintain the Enabled configuration for hours, or even days, safely, until it is reversed by an Inhibit command. Procedures only allow commanding Enable when launch is imminent.

*Execute Launch*

Execute Launch is the final step required to launch one or more missiles. This command sends a message, unique to each LCC, to all missiles in a squadron. Initiating this action requires precisely coordinated actions by both missile combat crew members using launch keys, which are secured in a double-locked container with access by each crew member opening their separate combination lock. When all the pieces of the process are correctly performed, a Launch command for that LCC is transmitted and is expected to be received by all missiles in the squadron. This single action can launch the missiles after a delay unless countered by an Inhibit command. If a second Execute Launch command from a different LCC is received, all selected missiles are committed to launch, and launch cannot be terminated by any of the LCCs in the squadron. The final steps for launch, the terminal countdown, will begin when the missile's specified launch delay time, if any, has expired.

*Inhibit*

The Inhibit command is a major safety feature that allows prelaunch actions to be canceled. Each squadron's LCCs has this capability. If, for any reason, the Inhibit capability is not immediately available from at least one LCC in the squadron, all squadron launch codes must be dissipated as a safety standard. Missiles will also be manually safed by visiting each launcher and installing a physical safing tool.

The Inhibit command requires the action of only one launch crew member and is immediately performed for a number of specific missile status conditions. Once initiated, the command is automatically repeated 30 times and floods the missile communication network to ensure each launcher receives it. The command contains a squadron-unique code value that must match a securely stored value at each missile launcher to be deemed authentic and acted upon. This is a primary nuclear surety requirement.

Since Inhibit will reverse some launch actions, additional safeguards prevent a lone crewmember from stopping launch by at least two launch crews. After Force Modernization conversion, a received Inhibit cycles an Enabled missile to Dis-Enable Commanded for five minutes. During this five minutes, the missile would accept valid Launch commands from at least two LCCs and respond as if Enabled. After five minutes, if no LCCs respond with Execute Launch commands, the missile cycles to the nonlaunch safed configuration.

If a missile has received only one valid Execute Launch command, the Inhibit command cycles the missile to the Inhibit Launch Commanded mode for five minutes and then reverts to the nonlaunch safed configuration if another valid Execute Launch command is not received.

The above processes ensure that two independent launch crews, upon receipt of a valid Launch message, can successfully launch missiles even if there are dissenting crew members in other LCCs in the squadron. Inhibit is tested periodically to ensure it is operating properly. In a nonlaunch condition, the code is checked for authenticity, and a status reply is generated to confirm that every missile can and will respond to the Inhibit command.

## Command-and-Control Changes by System Configuration

### Minuteman IA 1962 to 1969

Minuteman IA was the original deployment configuration for the weapon system and, as such, defined the concept for central LCCs controlling remote, unmanned missile LFs. The unmanned LFs were unique to Minuteman at the time.

TARGET AND MISSILE LAUNCH TIME SELECTION

Minuteman IA target and launch time selections were extremely simple. First, with only one target, the launch crew had no input for target selection. Launch timing was either salvo (immediate launch of all missiles) or ripple (sequenced launch of missiles based on a time value loaded during the targeting process at the LF). Salvo or ripple was selected on the Commander's Console Launch Control Panel using the War Plan switch. Position A was salvo, and Position B was ripple. This was selected before issuance of the Execute Launch command, and the setting was part of that command data. The start of both the salvo and ripple processes depended upon when Execute Launch was complete. This time was synchronized by crew actions at specific coordinated timing directed in the EAM from higher headquarters. This message included everything needed to authorize launch, plus which missiles to launch and when (Figures 15.1 and 15.2).

ENABLE

The Enable command for Minuteman IA was achieved by removing a safing tone from the missile. The safing tone was provided from a panel on the DMCCC's console and sent to the LF, where it was monitored, over a voice circuit in the hardened, buried cable network. When loss of tone was detected at the LF, circuitry was energized to complete initial arming of the missile ordnance and other launch-essential equipment. The deputy's console had a toggle switch, protected by a cover, for each LF in his flight (10 missiles) and allowed for individual missiles, or the entire flight, to be Enabled within a few seconds.

The electrical interconnect of LCCs to LFs restricted Enable control to only the parent flight; one capsule could not enable another LCC's missiles. This restriction resulted in an Enable failed armed design. If an LCC was destroyed during attack, its missiles would be Enabled by the default loss of tone generation from the destroyed parent LCC. The downside to this design was that certain equipment failures inside the LCC, or even damage to

FIGURE 15.1   341 SMW LCC India-01, May 1963, WS-133A. MCCC's Control Console. This console had three major functions: continuous surveillance of the 10 LFs under its control, generating commands to the LFs, and serving as a communications terminal for all operations. *Official USAF photograph.*

the buried communications cable, could result in a missile becoming Enabled. The response to this situation was to have a Minuteman maintenance team enter the Enabled LFs and manually insert an SCS lock pin assembly into the launcher distribution box to return the system to a safe condition and lock it in that state until the safing device pin was removed. When the safing tone was restored, maintenance reentered the launcher and removed the lock pin assembly. While the missile remained on strategic alert throughout this process, the safing process by the Minuteman maintenance team was a high-priority dispatch.

EXECUTE LAUNCH

Execute Launch sent the Launch command to the entire squadron. The command was sent to the parent launchers, where it was detected and relayed to other missile launchers until it fully permeated the entire squadron. The receiving LFs checked the command for validity (code matching) and stored the results in the guidance computer if the missile was in Enabled mode. When two unique Launch commands, normally sent from different LCCs, were received, the guidance computer authorized launch. This mode could not be

FIGURE 15.2  341 SMW LCC India-01, May 1963, WS-133A. DMCCC's missile combat crew Communications Console. This console had four major functions: enabling the 10 missiles (terminating the safe tone), voice reporting signal assembly used to detect faults and target status of each LF, monitoring the safe tone signal (which, when present, prevented arming and hence launch at each LF), and serving as a communications terminal. The DMCCC's Cooperative Launch switch was in the cabinet panel padlocked by the white bar to the left. The bar did not need to be removed to activate the switch. The tape recorders on the shelf above were used to record the Primary Alert System messages; the tapes had to be changed every 24 hours. *Official USAF photograph.*

remotely canceled. The only exit was entry into terminal countdown culminating in Stage I ignition and flight or upon guidance power removal. Critical fault detection could also cause guidance power removal and exit of terminal countdown.

The Execute Launch command was initiated from the MCCC's console using the Launch Control Panel Launch switch and coordinated with switch activation by the DMCCC. Each switch activation required a unique key. These keys were kept with the authentication documents in the double-locked safe. They were inserted into the keyed switches just prior to launch. The keyed switches were required to be simultaneously activated (clockwise rotation) to successfully initiate the command. Simultaneous activation and 12-foot physical separation were safety features to keep a single person from performing both key turns.

INHIBIT

The Inhibit command was issued from the MCCC's console on the Launch Control Panel from a rotary switch with a uniquely shaped handle. The Inhibit command, like the Execute Launch command, was detected at connected facilities and retransmitted to their interconnects until the Inhibit command reached every launch and LCF. Inhibit only affected the Execute Launch command. If received at a launcher with only one valid Launch vote received, the launcher cycled into a 205-second period waiting for a second valid Launch vote. If a second vote was not received within 205 seconds, the first Launch vote was canceled, and the launcher reverted to a no-launch condition. However, Inhibit had no effect on Enable arming of the missile.

EMERGENCY COMBAT CAPABILITY

It must be remembered that during the initial years of deployment, as the first flight in each squadron was completed, it was placed on strategic alert despite there being temporarily only one operational LCC. An Encoder Cavity Plug took the place of the operational Launch Control Panel in the MCCC's Launch Control Console to complete the command circuits.[19] When SAC control was established over the first flight in each squadron, that LCC was to be continuously manned. Guidance sections and reentry vehicles with War Reserve warheads would be installed on all 10 missiles of the first flight. Since the Launch Control Panel held both Execute Launch and Inhibit codes, which used the same connector, the LCC did not have Inhibit capability when the Encoder Cavity Plug was installed.

Launch would be initiated by removing the Encoder Cavity Plug, inserting the first operational Launch Control Panel in the Launch Control Console, generating the first vote, or Launch command, by key turn, then removing that panel, inserting the second Launch Control Panel, and generating, by a second key turn, a second Launch command or vote.[20] While it could be argued that this was not the instantaneous response of normal conditions, nuclear surety was the more important concern. Under peacetime conditions, the two panels were not located in the LCC but rather in a secure vault on the base. The ECC condition was only in place until a second LCC was activated (for further explanation, see the Cuban Missile Crisis discussion below).

OTHER PERTINENT INFORMATION

The communication structure in Minuteman IA did not allow LCCs in the squadron to detect the status of most of the rest of the squadron. While this did not impact launch, it did prevent another LCC from acting as a second set of eyes to observe status that required a crew response.

*Minuteman IB 1963 to 1974*

TARGET AND MISSILE LAUNCH TIME SELECTION

Launch timing selection remained either salvo or ripple. Minuteman IB was fielded with a slightly modified guidance set that permitted storage of a second target without reorienting the missile, as was the case with Minuteman IA (see Chapter 10). The commander's console received a modification to the Program Control Panel to permit selection of Target 1 or 2 for any missile in the flight. The target change was confirmed with a status change from the guidance set at the launcher.

LAUNCH COMMANDS

There were no changes to the individual launch commands. The ECC protocols remained the same as well.

## *Minuteman II 1967 to 1991*

Minuteman II introduced a new missile as part of the Force Modernization program (see Chapter 16). Force Modernization was a revolutionary change to both the LCCs and LFs, upgrading the Minuteman I facilities to those of the more modern Minuteman II found at Wing VI, 321 SMW, Grand Forks AFB, and the 564 SMS, 341 SMW, Malmstrom AFB. It introduced new electronics with greatly expanded capabilities, allowing new commands, target and launch timing control, and a vastly improved Enable system. A major communications architecture change allowed lost LCC capabilities to be assumed by another LCC, providing uninterrupted monitoring, testing, control, enabling, and executing launch (Figures 15.3 and 15.4).

TARGET AND MISSILE LAUNCH TIME SELECTION

Along with eight possible targets, launch time selection was vastly improved and included a set of 100 preplanned AOs for each missile, as well as the capability to set launch delay for an individual missile. The preplanned attack option was acted on by every missile in the squadron. Launch delay could range from zero to 9 hours, 59 minutes, 59 seconds. Additionally, launch delay could be set to infinite. An infinite setting would prevent missile launch even if valid Enable and Execute Launch commands were received.

Target and launch delay settings were initiated from the DMCCC's console. Receipt and action by the selected missile was confirmed by a message from each affected missile noting target selected and launch delay time (or an attack option number).

PREPARATORY LAUNCH COMMAND

The Preparatory Launch Command (PLC) capability was introduced with Force Modernization beginning in 1967. There are two types of PLCs: PLC-A, an all sortie–addressed command, and PLC-B, which is addressed to a single sortie in the squadron.

Normally the guidance set is aligned to its primary target, and the launch delay time is set to infinite (nonlaunch). The PLC-A command includes an all-call address code, which means the message will be received by all the missiles in the squadron. The other variable code within the command defines an attack option value between 00 and 99. The guidance computer uses the attack option and a preloaded database to set a target and delay time. The attack option database is unique to the target set for each missile and can include some null sets, meaning the attack option value may be associated with no target and an infinite

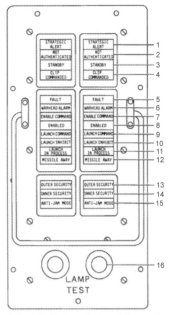
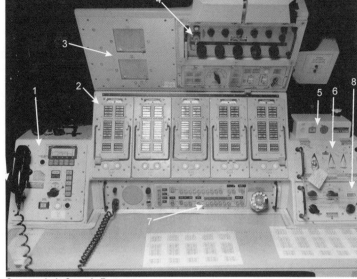

**Commander's Console Features**
1. Alarm Control Panel: Time Slot control and alarm notification/reset
2. Launch Status Missile Indicator Panel (4)
3. Primary Alerting System: Speakers and Acknowledge button
4. HF Transmitter/Receiver
5. Electro-Magnetic Pulse notification panel
6. Launch Control Panel: Launch and Inhibit Commands; War Plan(MM1 only)
7. Communication Panel: See DMCCC console \
8. Program Control Panel: Upper dials selects commands to missile and initiate. Lower dials select missile(s) to receive command. Commands: Missile Test, Sensitive Control Network Test, Calibrate, Sat Calibrate, Test (console), Tgt 1 (MM1B only), Tgt 2 (MM1B only). Modifications added: RWC - remote weather command, LECG - enable test, Enable, CLIP - cancel launch in progress, PIGA-LVL - PIGA leveling

1. Strategic Alert
2. Not Authenticated*
3. Standby
4. CLIP Commanded*
5. Fault
6. Warhead Alarm
7. Enable Command*
8. Enabled*
9. Launch Command
10. Launch Inhibit
11. Launch In Progress
12. Missile Away
13. Outer Security
14. Inner Security
15. Anti-Jam Mode*

FIGURE 15.3 Minuteman II MCCC's Launch Control Console, circa 1991, WS-133A-M, Whiteman AFB. *Adapted from Historic American Engineering Record.*

launch delay time. With the PLC-A, an entire squadron can be set to desired target and delay times with one command. A single PLC-A can also put all missiles into an infinite launch delay situation (the normal day-to-day configuration).

The PLC-B command results in the same launch delay options as the PLC-A but for an individual sortie. The DMCCC commands a PLC-B by selecting a flight and LF address, a target, and a delay time. When the command is sent, only the specifically addressed LF will take the actions defined by the deputy. A PLC-B can be used to set a finite launch delay sortie commanded by an earlier PLC-A to infinite delay. The maximum finite launch delay time value permitted is 9 hours, 59 minutes, 59 seconds (Figures 15.5 and 15.6).

Both the PLC-A and PLC-B have an option to select either CEP or MRT launch via a message bit. CEP set (1) causes the guidance set to realign to the new target azimuth before launch, which requires as much as a 15-minute delay before launch but is more accurate. MRT set (0) causes the guidance set to keep the current platform azimuth heading and fly to the new target using compensating calculations, resulting in some accuracy degradation (see Chapter 10).[21]

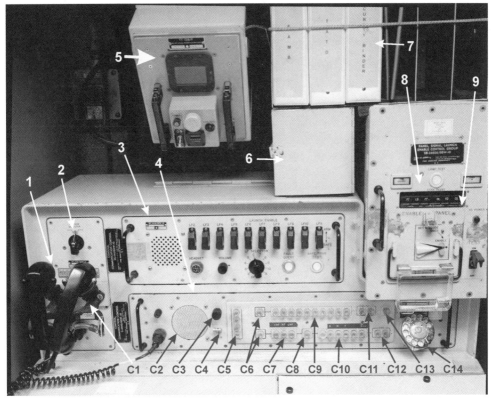

**DMCCC CONSOLE ITEMS**
1. ERCS Message Voice Playback Handset (510 SMS only)
2. ERCS Sortie Select Message Switch
3. MM1 Enable, converted VRSA during Force Modification (70s)
4. Communications Control Panel (see detail list)
5. SLFCS Receive Terminal/Printer
6. Storage
7. Documents Storage: quick access materials
8. Message Translate Thumbwheel Switches: Remote Weather (MM2), Enable Unlock and Selective Enable values insertion
9. Enable Panel: code storage and activation

**COMMUNICATION PANEL ITEMS**
C1 Communication Panel Handset
C2 Communication Panel Speaker
C3 Volume Control
C4 Microphone Switch: radio (up) or phone (down) handset connection
C5 Speaker Source Select: phone or radio (VHF, HF, UHF)
C6 Operation Switch: selected row of switches that were active-normally only one active
C7 Radio Set Select: VHF, HF, UHF
C8 Squadron Command Center select
C9 LF Connecting Buttons: fully depress for connect and ring, second depress for off
C10 HVC and LCC connect select
C11 Dial Line select: 1 or 2
C12 EWO 2 select/ring (ACP and SCP)
C13 EWO 1 select ring (ACP and SCP)
C14 Rotary Number Dialer

FIGURE 15.4 Minuteman II DMCCC's Communication Console, circa 1991, WS-133A-M, Whiteman AFB. *Adapted from Historic American Engineering Record.*

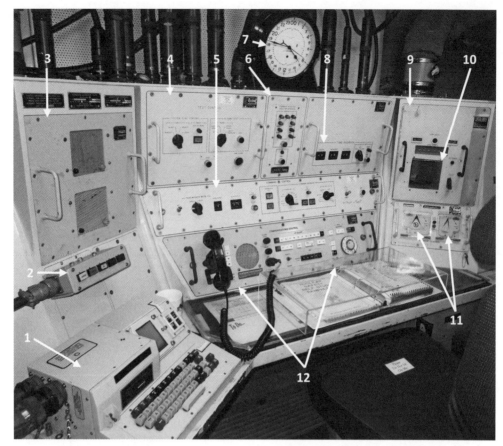

**MCCC CONSOLE FUNCTIONS**
1. **Keyboard Printer**: send/receipt of alphanumeric data to/from the LCC computer
2. **AMIIDS Control Monitor Panel**: entry detection, alarm and reset
3. **Primary Alert System (PAS) Monitor Panel**: voice alerts from higher authority source
4. **Test Control Panel**: Synchronizes message timing, tests timing monitor circuits and monitors Radio timing and sync; EMP is inoperative.
5. **Command Control Panel**: command type, variables, address, encryption and send actions; some message monitor functions
6. **Command Message and EMP Monitor Panel**: displays command type and source when non-status commands are detected
7. **24-Hour Clock**: mechanical, Zulu time
8. **Time Insertion Panel**: time setting and initiate to synchronized computer clock with coordinated universal time (WWV)
9. **Launch Enable Control Panel**: code storage, thumbwheels to alter code and clear values, Enable initiate switch
10. **Enable Selector**: initiates Enable command
11. **Launch Control Panel**: store codes and initiate Launch and Inhibit commands
12. **Communications Control Panel**: see DMCCC console

FIGURE 15.5   564 SMS MCCC's Command Console, WS-133B. The Command Console had four major functions: command generation for the LFs under its control, alerting the missile combat crew when commands were initiated by other LCCs, control of automatic interrogations of the LFs, and serving as a communications terminal. *Adapted from Historic American Engineering Record.*

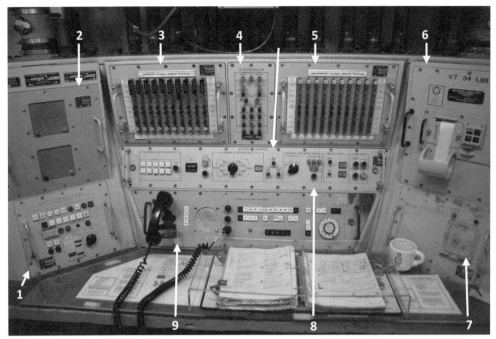

**DMCCC CONSOLE FUNCTIONS**
1. **External Communication Panel**: HF Radio transmit, receive and antenna controls
2. **PAS Monitor Panel**: (see MCCC figure)
3. **Primary Flight Group Status Panel**: Status for the LCC's primary 10 LF responsibility; see detail to right
4. **LF Status Call-Up Panel**: displays critical command status and detailed LF status display selected by Status Control Panel
5. **Secondary Flight Group Status Panel**: Status for the LCC's secondary 10 LF responsibility
6. **Signal Data Recorder**: system printer for LF status and some LCC monitors
7. **Launch Control Panel**: DMCCC launch initiation switch; operated in coordination with MCCC launch switch initiation
8. **Status Control Panel: Select primary/secondary LF status interrogation/monitoring** and display/print; include manual LF query select and LCC monitors
9. **Communications Control Panel**: voice (other than PAS) send and receive over landlines, radio, LCC and LF voice circuits. Indicates circuit in use, includes ring and number dial (mechanical) functions

**STATUS PANEL INDICATORS**
**Strategic Alert** (green) sortie is ready to launch
**Hold** (amber) infinite launch delay (typical)
**Standby** (white) not launch ready (typical for retargeting)
**Enabled** (amber) Enable command accepted
**LCH CMD** (amber) Launch Commanded, first launch vote accepted
**LCH IN PROC** (green) Launch in Progress, second launch vote accepted
**Missile Away** (green) missile has departed silo
**Inhibit** (amber) Inhibit accepted
**Fault** (red) A malfunction detected in facility
**Security** (red) security event active
**Status Change** (white) change in status has occurred

FIGURE 15.6  564 SMS DMCCC's Status Console, WS-133B. The Status Console had four major functions: receiving status from the 50 LFs in the squadron, control of cable and radio automatic flight interrogation by the LCC, permitting cooperative missile launch, and serving as a communications terminal. *Adapted from Historic American Engineering Record.*

ENABLE

The Enable function was completely redesigned. Enable became a new command similar to the Execute Launch command. The parent-only limitation disappeared with the command being relayed to all missiles in the squadron. At the missile the coded portion of the message was used as input to a code device (Command Signal Decoder-Missile, CSD-M) located in the interstage area between Stages I and II. If the code matched the stored CSD-M code, the CSD-M armed, completing the firing signal path to the Stage I motor and the safe–arm devices on each ordnance device.

The Enable command code, received as part of the EAM, was initiated from the DMCCC's console on a new Launch Enable Control Group (LECG) panel. The code was stored in the panel in a volatile mechanical device unit (MCU) identical to the ones used in the Launch Control Panel for Execute Launch and Inhibit codes. There were two forms of the Enable command. When the actuating knob was turned (not key secured), the code was presented on the detection lines, read, and formulated into either an individual LF-addressed message or addressed to the entire squadron as selected by the DMCCC following command direction. This allowed one LCC to Enable all missiles in the squadron with a single command or one or a few missiles for a limited launch.

The new Enable capability was resident in every LCC. The code was squadron-unique and handled only by maintenance code handlers. During semiannual code changes, maintenance installed the newly coded panels in the LCCs and updated the CSD-M to the new squadron code during simultaneously occurring LF recode operations.

EXECUTE LAUNCH

The Execute Launch command function did not change during the Force Modernization Program.

INHIBIT

A function was added to allow the Inhibit function to be tested. The squadron-specific code for the Inhibit command could be sent to all squadron missiles. If missiles were in a no-launch mode (not Enabled, not Launch commanded), the coded portion of the Inhibit command was compared to the authentication value stored in the guidance computer. If the values matched, the launcher sent this status to the LCC, reflecting a successful Inhibit test.

## *Command Data Buffer/Improved Launch Control System*

The CDB and ILCS modifications were deployed from 1973 to 1980 and had impacts similar to those achieved by the Force Modernization Program. CDB was the Minuteman III version, while ILCS was unique to Minuteman II. Both provided very similar improvements in hardness, electronics upgrades, and reliability. Only the Minuteman II command-and-control impacts will be discussed.

REMOTE TARGETING

CDB/ILCS provided increased computing capability within the LCC that supported a new data transfer protocol. Now, new target data was sent to the missile guidance computer directly from the LCC and avoided a maintenance/security forces dispatch to open the launcher and install new data at the guidance interface. New targets were received over existing communication systems as a compact target case (target location information). The target case was entered into the local weapon system computer, and a targeting program used the case to generate a target set, which was transmitted to the selected guidance set in incremental segments. When complete, the target was authorized for use in that guidance set by a separate monitoring LCC upon verification of the target set contents.

SECURE ENABLE

Secure Enable introduced a separate Enable unlock value to permit successful execution. The unlock value was entered into the thumbwheel switches on the LECG. When Enable was initiated by the DMCCC, the thumbwheel values and the value stored in the LECG MCU were combined into a new code that was transmitted to the squadron. The all-call or single-LF addressing features were retained. Single-LF Enables would require a different unlock value than a squadron-wide Enable. Unlock values were received in the launch, initiating EAM. Secure Enable was deployed separately or as part of the CDB upgrade if schedules overlapped.

RAPID EXECUTION AND COMBAT TARGETING

The Rapid Execution and Combat Targeting (REACT) Program was the response to the need to update the aging CDB command-and-control system and incorporate new features that would reduce crew launch reaction time, automate many command and monitoring tasks, and increase hardness and survivability. The CDB system had been designed in the early 1970s. The CDB LCC technology had become dated, with many part technologies becoming obsolete and unprocurable, along with sources going out of business or significantly changing their design and manufacturing focus. REACT was developed to upgrade both the Minuteman and Peacekeeper forces. Peacekeeper, deployed in Minuteman III LFs, used the same equipment with some modifications. REACT-A was installed in the WS-133A-M system, and REACT-B was installed in the WS-133B system.[22]

On 9 September 1988, Headquarters USAF issued a program management directive for the REACT system and the corresponding Missile Procedures Trainer (MPT). Two development contracts for REACT were awarded on 10 April 1989. The first was to Ford Aerospace, which merged with Loral Command and Control Systems and is currently under Lockheed Martin management. Ford Aerospace/Loral was to develop the crew console, rapid retargeting capability, and the new computer system to replace the weapon system controller equipment (basically a computer) commonly referred to as WSCE (pronounced "whiskey"). The second was to General Telephone Electronics (later General Dynamics) for the rapid message-processing equipment and communication systems, fielded as the Higher Authority Communication/Rapid Message Processing Element (HAC/RMPE) (pronounced "hack rempy").[23]

A significant change in LCC configuration was replacing the two separate launch consoles (commander and deputy positions) with one dual position console, providing improved crew communication, coordination, and resource management (less duplication). The new REACT system was also designed to automate communication systems, message reception, and the retargeting process by extracting and passing data directly from the communication system integrator suite across a one-way HAC/RMPE interface (one-way for security). Because targeting data was directly available to the crew, they could direct new missile targeting data to an LF via a CDB Remote Data Change Target (RDCT) function. The new system reduced by 50 percent the time needed to retarget the missile by manually keying in new target information from target data received via paper from the communication systems. This function potentially allowed retargeting of the entire Minuteman fleet in a number of hours. The automated RMPE process also reduced the time to decode and process encrypted EAMs, again shortening the launch process (Figure 15.7).

Automated data consolidation, sorting, automated logging, categorical alarming, and integration of facility equipment indications at the REACT console also reduced the physical and mental burdens placed on operations crews when compared to the indicator lamps, thermal paper record output, and analog alarm systems of CDB.

Automation replaced many of the manual CDB functions, including database-driven abilities to rapidly select missile target, timing, and alignment modes via preplanned PLC-A and manual PLC-B. LF response to the potential array of 2,400 different PLC-As was automatically monitored and checked by REACT for proper indications, with only errors being passed to the crew. Eventually the capability for crews to build PLC-B libraries, capable of supporting an additional 1,200 PLC-B execution options of up to 50 sorties each, with similar response monitoring and error notification, greatly increased Minuteman launch response capabilities while significantly reducing crew processing time.

The automation and increased capabilities required considerably more data processing capabilities than the original CDB LCC computer. Besides being underpowered, the LCC computer used obsolete technology, creating a looming spares shortage. That and an inability to operate through a nuclear event cemented the need for a totally new computer approach. Replacing this unit with a state-of-the-art computer with much larger memory and storage capacities was the solution.

The old plated-wire memory storage and large-volume drum memory (basically a crude hard drive and very slow) were replaced with RAM chips and an up-to-date hard drive. These two hardware changes (and associated circuitry) gave Minuteman twentieth-century data processing capability while adding nuclear event hardness.

Automation was carefully implemented to ensure the missile combat crews still fully controlled essential launch actions, including Preparatory Launch, Enable, and Launch commands sent to the squadron. The process changed from the assembling of commands and messages from the LCC to the LF via a series of switches and thumbwheels in the CDB system to a series of menu-driven selections and keyboard entries from the co-located operator workstations supported by a direct data feed from HAC/RMPE.

HAC/RMPE's new functionality was permitted by writing the WSCE software in approved ADA and FORTRAN programming languages, nuclear-certifying WSCE software product through extensive testing and an isolated, one-way data exchange of HAC/RMPE process data, also software certified with an irreversible data path. The product ensured a secure command-and-control processing capability that is resistant to modern cyber vulnerabilities through its antiquity when compared to IP-based, modern computing systems. Sometimes older is better.

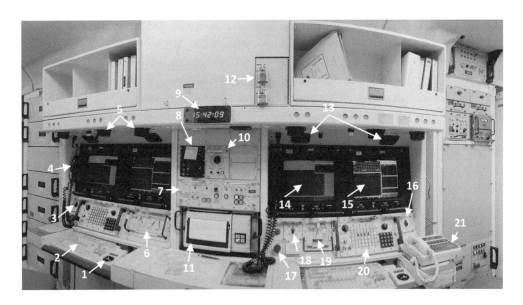

## OPERATOR POSITIONS
### MCCC Left                          DMCCC Right

**CONSOLE FEATURES**

1. Trackball (each station)
2. Keyboard (each station)
3. Cooperative Launch Switch (MCCC)
4. Telephone Handset (each station)
5. Console Lights (2 per station)
6. Launch Control Panel (Launch/Inhibit, MCCC only) operated in conjunction with #3
7. Auxiliary Control Panel
8. AFSATCOM Control Panel
9. Zulu Clock
10. UHF Control Panel
11. Printer
12. Secure Storage – 2 Person Access
14. Console Speakers (2 per station)
15. Higher Authority Comm. Display Panel (each station)
16. Weapon System Display Panel (each station)
17. Cooperative Launch Switch – Left (DMCCC)
18. Master Alarm Reset Button (each station)
19. Cooperative Launch Switch – Right (DMCCC)
20. Enable Panel (DMCCC only)
21. Voice Control panel (each station)
22. Secure Voice Telephone

FIGURES 15.7 REACT-A Console at LCC November-01, F. E. Warren AFB. To initiate a single launch vote, the two officers have to coordinate the turning of the Launch Control Panel Launch switch and the Cooperative Launch switch within two seconds of each other. The proximity of the two consoles, compared to the earlier configurations, still does not allow one crew member to execute the Launch command. *Adapted from Historic American Engineering Record.*

The new hardware and software architecture also facilitated periodic updating. As the longest-deployed command-and-control system for Minuteman, REACT continues to function as of this publication and is routinely updated to adapt HAC/RMPE to USSTRATCOM war plan changes (typically every six months) under contracts with Northrop Grumman, General Dynamics, and the US government HAC/RMPE Software Support Facility (HSSF). Nuclear-certified WSCE, Operations Ground Program (OGP), and planning and targeting support programs are also continually updated (typically every 18 months) for evolving operations procedures under the ICBM Operations Software Sustainment Program (IOSSP) under contracts with Northrop Grumman, Lockheed Martin, and Boeing.

## Emergency Rocket Communications System

In the late 1950s, SAC examined the vulnerability of communications with the strategic nuclear forces during or after a nuclear conflict. The ability to send the EAM to strategic nuclear forces was problematic since the current landline system invariably passed through primary target areas, and both very high-frequency (VHF) and high-frequency (HF) radio communication were vulnerable to the effects of nuclear detonations. On 19 December 1958, Headquarters SAC sent a Qualitative Operational Requirement to Headquarters Air Force for a UHF radio system that would utilize a rocket to launch a UHF transmitter in a suborbital trajectory, thus enabling communication with both the missile and bomber forces. This was the beginning of ERCS.[24]

### *Blue Scout Junior Program 279L (MER-6A)*

Three years later, on 29 September 1961, Headquarters Air Force issued Specific Operational Requirement 192 for ERCS, designated Program 279L. On 27 December 1961, Headquarters SAC clarified the capabilities and mission of Program 279L. Initial deployment would be an interim configuration of three modified Blue Scout Junior boosters, designated as Mobile Electronic Rocket (MER) 6A, in three locations near Offutt AFB, Nebraska. The system was to be operational by 1 January 1963. The planned final configuration would consist of four complexes of three rockets each, all located in the Midwest.[25]

The Blue Scout Junior rocket used in the early ERCS was based on the Ford Aeronutronic XRM-91 Blue Scout Junior solid-propellant, four-stage, small satellite booster; however, the ERCS booster was only three stages.[26] The payload housed a high-powered UHF transmitter and a storage/playback system for a voice message preloaded before launch. After the payload was deployed, it flew an arching trajectory with an apogee of 540 nm, broadcasting the voice message to all units within line of sight of the transmitter. Also known as BEAN STALK, the first launch on 31 May 1962 from Point Arguello, California, was successful and followed by six more successful launches through 17 December 1963 (Figure 15.8).[27]

Program 279L attained IOC on 11 July 1963 and was deployed at three launch points in Nebraska 60 to 75 miles northwest of Offutt. These were Site A-1, Wisner, Site A-2, West Point, and Site A-3, Tekamah. Each site consisted of a missile launch trailer housing an erectable missile, a launch control trailer, and various security and personnel support

FIGURE 15.8 The 13 March 1963 launch of MER 203 from Naval Missile Facility Point Arguello, Launch Complex A. This was the fifth launch in Program 279L, the progenitor to ERCS. The payload transmitter activated as programmed and transmitted a clear signal for 17 minutes before reentry. Adapted by the author. *Official USAF photograph, courtesy John Hilliard.*

vehicles and structures. The program was scheduled to be replaced by Program 494L, using Minuteman II boosters, during FY 1967, but delays in the 494L program required extending the 279L capability into FY 1968. The system became obsolete with the activation of the Minuteman II system at Whiteman AFB and was deactivated on 1 December 1967.[28]

## *Minuteman Program 494L*

On 16 November 1963, Deputy Secretary of Defense Roswell Gilpatric provided DoD approval for the beginning of a formal Minuteman ERCS. On 1 June 1964, the DoD directed the Air Force to simultaneously develop the 494L and Minuteman ALCS programs.[29]

In a move to increase availability, survivability, and connectivity, ERCS was redeployed as the 494L program, with a new communications package of two 1,000-watt transmitters atop six Minuteman II missiles in the 510th Strategic Missile Squadron (510 SMS) at Whiteman AFB, Missouri. The Minuteman variant of ERCS used the Minuteman II ICBM as the boost platform and the Minuteman II LF as a launch site, with launch control handled from the LCC. The ERCS transmitter payload replaced the Mark 11 reentry vehicle atop the missile with an aeroshell closely resembling it that contained the ERCS transmitter equipment. If necessary, SAC wanted to be able to easily replace ERCS equipment with the Mark 11 reentry vehicle containing the nuclear warhead. Flight parameters were adjusted for the ERCS payload weight and trajectory direction and profile to achieve a 701-nm apogee. ERCS utilized existing Minuteman missile guidance control signals to perform needed actions before and during flight. Little was required to accommodate the ERCS equipment and mission in Minuteman II, although some interface equipment was added to both the LF and LCC for ERCS-specific functions. Except for wiring changes, ERCS-specific equipment was only installed in LFs when the missile was configured with an ERCS payload.[30]

ERCS EQUIPMENT ADDITIONS TO THE LAUNCH FACILITY

To support the ERCS payload, the LF received an ERCS Data Transfer Unit (DTU), UHF ground transmitter drawer, and minor wiring changes to support ERCS voice, status reports, and transmitter setting functions. The DTU was mounted to the upper level suspended floor immediately to the right of the distribution box. Output of the existing UHF receiver (ALCC data and voice) was also routed to the DTU. The UHF receiver frequency and address tones were changed to unique settings to facilitate ERCS-only ALCC actions for the ERCS equipment and Minuteman launch. The UHF ground transmitter drawer was added to the equipment rack just below the existing UHF receiver. Minor wiring was incorporated in the existing interconnect box, ESA vault, and interface cabling (Figure 15.9).

ERCS EQUIPMENT ADDITIONS TO THE LAUNCH CONTROL CENTER

All ERCS-capable LCCs received a new ERCS Control-Monitor console adjacent to the crew commander's console on the left. Crews immediately dubbed it the "knee knocker," as it was easy to catch one's knee on it as the operator chair was swiveled. Circuits were added to cabling and the ESA vault to support the ERCS console interfaces with ERCS LFs, other ERCS-equipped LCCs, and command posts. All of these changes were minor (Figure 15.9).

The SAC airborne command post ALCC suite was modified to include controls similar to those in the ERCS LCCs to interface with the ERCS ground equipment and payload to load voice messages, receive voice playback via the new UHF ground transmitter drawer, and change or monitor ERCS payload transmitter channels.

ERCS GROUND OPERATIONS

The only Minuteman missile squadron with ERCS capability was the 510 SMS. All five LCCs and 50 LFs in this squadron could support ERCS-configured Minuteman II missiles. Each LCC could control and monitor two ERCS sorties (later increased to four during ILCS deployment at Whiteman. ERCS-configured LF locations were classified SECRET for most of its operating history. This was to prevent them from being preferentially targeted by the Soviets or sabotaged by ground attacks. Special procedures were employed for all missile payload operations (ERCS or Mark 11 reentry vehicle) in the 510 SMS to mask ERCS locations. Similar arrangements were made for the installation of ERCS sup-

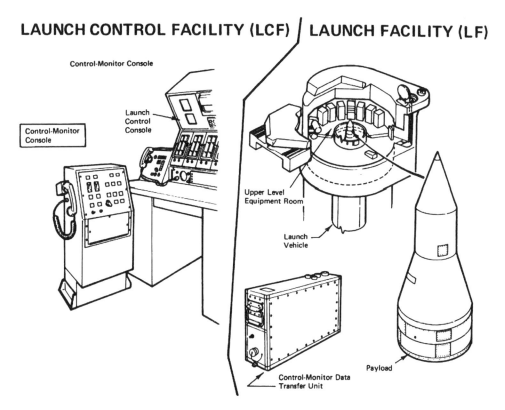

FIGURE 15.9 ERCS equipment additions to Minuteman facilities. ERCS-specific equipment was only installed in the LFs when the missile was carrying an ERCS payload. *Adapted from ERCS Orientation Handbook.*

port equipment. Communication protocols barred the use of any ERCS-distinguishing language over unsecured communication circuits. SIN, a voice line between the LFs and parent LCCs, and the standard telephone lines between the LCC and support base agencies (primarily maintenance job control and the local command post) were the primary focus of these restrictions.

Revealing ERCS locations resulted in a shift of the ERCS location using a shell game process. A spare ERCS payload was installed at a different LF during planned or unplanned (equipment failure) guidance or reentry vehicle maintenance. Early in the Minuteman II deployment, missile guidance system failures were almost daily (or one every three days in the 510 SMS), so this was easy to accommodate without showing unusual activity. The revealed ERCS payload was removed and a Mark 11 reentry vehicle installed. Maintenance actions also removed or reconfigured the LF to a non-ERCS condition. Inadvertent ERCS exposures were infrequent, but in the late 1970s, secure voice equipment was added to 510 SMS maintenance dispatches to avoid disclosure.

The newly configured ERCS sortie was initially checked out by the LCC crew using daily performed ERCS status checks via the ERCS Control-Monitor console. A self-test of the DTU was commanded and missile status monitored. Go/No-Go transmitter channels (A and B) were checked and, if necessary, stepped to the required settings. Finally, a voice message was read into the ERCS payload recorder, then monitored for quality and completeness during a separate playback action.

During daily ERCS function checks, the voice to the payload recorder was normally provided by the crew commander via the ERCS console telephone-type handset using the local message insertion push-button. However, the capability to load a message via the command post circuit was regularly exercised to confirm proper operation of that circuit. Also, all other 510 SMS LCCs could monitor ERCS playback from their consoles if the call-up push-button was activated. This function would allow all 510 SMS LCCs to confirm proper message insertion prior to an actual ERCS payload launch.

ERCS was considered a high-value asset and consequently had the highest priority for maintenance repair should it fail to respond to ERCS-unique checks or any other Minuteman system fault that would prevent launch. The most frequent of these were missile guidance system failure, but they also included communication and ground power problems. To mask the high-priority repair status for these actions, all similar tasks in the 510 SMS received similar responses, creating a unique maintenance team configuration for Whiteman. Extra missile maintenance and electromechanical teams were kept on call for fast-response maintenance should priority repair actions be required.

Periodically the ALCC ERCS operation was exercised to check out the UHF radio link between the airborne command post and the ERCS payload. To accomplish this, the ALCC aircraft would fly in the Whiteman area. With radio voice communications between the aircraft and the test LCC, the LCC commander would allow ALCC radio access to the LF UHF receiver by pressing the ALCC Exercise button on the ERCS Control-Monitor console. The DTU at the ERCS-connected LF would send a signal to the UHF receiver to allow ALCC access. This access was normally blocked unless the LF lost communication with all squadron LCCs. At the same time, it prevented any transmissions from the UHF transmitter, as a radio signal would compromise the ERCS sortie location. With access allowed the ALCC crew would insert a voice message into the ERCS payload recorder, then command playback, which would be monitored by the LCC crew for quality. After the recorder test, the ALCC would command payload channel changes, and the LCC would confirm proper response from the payload by monitoring ERCS Control-Monitor panel indicators. At the end of the test, the LCC would cancel the ALCC exercise function on its console and restore transmitter channels to operational settings. Any faults detected in the test would be evaluated and repaired by maintenance.

COMMUNICATIONS PAYLOAD

ERCS flight equipment consisted of the communications payload atop the Minuteman II booster. ERCS payloads were stored and maintained at the local base weapon storage area (WSA) to support the appearance of identical actions with the Mark 11 reentry vehicle, which was also stored and maintained at the WSA. The payload was the same diameter as the Mark 11 to allow connection to the Minuteman booster. The overall shape of the payload was very similar to the Mark 11 but could be readily distinguished by a silver-colored metal point at the top of the aeroshell. The payload was approximately 8 feet high, 32.6 inches in diameter, and weighed 875 pounds.

The payload was comprised of the following major assemblies: forward, aft, and spacer.

It separated from the Minuteman booster at the end of powered flight (approximately three minutes). The spacer section remained with the booster, and the forward section was also quickly jettisoned, leaving only the aft section to complete the mission.

## PAYLOAD OPERATIONS

While almost all payload operations occurred during flight, one preflight event was essential for operation. Battery activation began at 45 seconds before launch. The Minuteman guidance system launch terminal countdown sequence initiated a signal to the payload to pressurize a battery electrolyte tank in the forward section. This forced potassium hydroxide electrolyte into the large batteries located in the bottom of the aft section. Early payload battery activation ensured full power availability at the end of powered flight and allowed use of existing guidance system software and discrete signals instead of requiring a duplicate function.

During boosted flight the payload did nothing, much like the Mark 11 it replaced. Payload events began with Stage III thrust termination. At thrust termination the missile guidance system, using the Mark 11 discretes, fired gas generators in the spacer section to release ball locks located at the spacer–aft separation plane to mechanically disconnect the aft and forward sections. The spacer section stayed with the missile. Upon release of the ball locks, four compressed separation springs pushed the aft–forward combination away from the spacer section.

Shortly after physical separation, the missile guidance system fired tumble rockets mounted in the spacer section to cause Stage III to move out of the payload trajectory path. This caused a growing separation between the payload and Stage III to prevent a radio signal reflection and signal reception shadow when the payload transmitters began broadcasting. This was considered to be a Minuteman function rather than an ERCS function since the event was controlled by missile guidance system software and signals.

At 238 milliseconds after payload separation, spin rockets mounted on the aft section fired, inducing a one-revolution-per-minute rotation of the aft and forward sections. This rotation provided attitude stabilization during the remainder of the flight, keeping the transmitted signal pointed at the intended reception area.

At 4.52 seconds after payload separation, forward section separation occurred, with disconnection of tubing and signal cables and ball lock activation. At ball lock release compressed separation springs pushed the forward section away from the aft section (aft had considerably more mass). This separation achieved the final operation configuration for the aft section.

At 31 seconds after payload separation, the transmitter cooling system was activated to force water through cooling circuits in both A and B transmitters. This was followed at 31.02 seconds by transmitter high-voltage application to start signal transmissions from the payload via the antenna system. Operational ERCS trajectory flight ended with reentry and impact in broad ocean areas (Figure 15.10).

## OPERATIONAL TEST LAUNCH

Operational test launches were performed to verify the performance characteristics of the ERCS under actual launch and flight conditions using the same equipment configurations

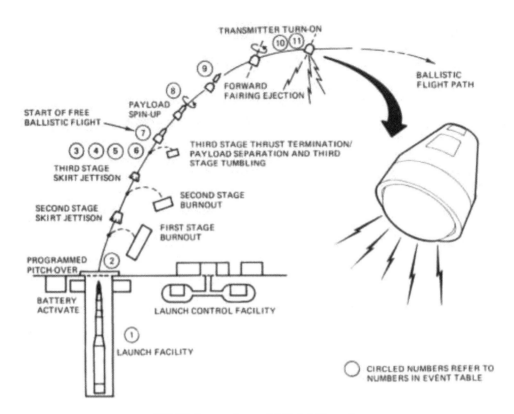

## ERCS Launch and Flight Events

| Key | TIME (Sec.) | EVENT | ACTION |
|---|---|---|---|
| 1 | T-45 | Battery Activate | ERCS batteries activated. |
| 2 | T-0 | Lift-Off | Minuteman first stage ignition. |
| 3 | TT-1.8 to -1.35 | Programmer Arm/Aft Mechanical Disconnect | Arm Programmer. Fire ball locks to release ERCS payload from Minuteman 3rd stage. |
| 4 | TT-1.0 to -0.5 | Aft Electrical Disconnect | Cut cable harness between 3rd stage and payload. |
| 5 | TT = 0 | Thrust Termination/ Payload Separation | 3rd stage thrust termination and payload separation. |
| 6 | TT | Programmer Start | ERCS Programmer timing starts. |
| 7 | TT+0.12 | Battery Activator Tube Cutting | Cut battery activate tube between forward and aft sections of payload. |
| 8 | TT +0.238 | Spin Motor Ignition | Aft section spin motors are fired. |
| 9 | TT + 4.52 | Forward Section Separation | Cut cable harness and release ball locks securing forward section to aft action. Forward section separates. |
| 10 | TT + 31 | Coolant Section Activation | Fire coolant system explosive valves to initiate coolant flow. |
| 11 | TT + 31 +0.02 | Transmitter Turn-On | Initiate high voltage to turn high- power transmitters on. |

FIGURE 15.10 ERCS flight profile. ERCS was one component of the Department of Defense Minimum Essential Emergency Communications Network. If ERCS failed to transmit within six minutes after launch, additional ERCS sorties would be launched as necessary to ensure coverage of required trajectories. *Adapted from ERCS Orientation Handbook.*

found at the operational locations. The only small difference was the addition of a telemetry unit to the payload. The telemetry unit replaced a weighted connector cap and dummy telemetry unit found on operational ERCS payloads to precisely match weight and balance for flight realism. The telemetry capability was implemented without altering the operational characteristics of the payload.

Flight testing began on 13 December 1966 with the successful launch of WATER TEST, FTM 2036, from LF-05 at Vandenberg AFB. After three additional flights, the last one involving message insertion via an ALCS aircraft, the test phase was completed, and on 10 October 1967, the first two sorties were placed on alert at Whiteman AFB. By the end of November, an additional four ERCS sorties were on alert, and on 1 December 1967, full operational capability was attained.[31] Nine additional successful operational flight tests were conducted, ending on 16 July 1979 with the launch of GIANT MOON 9, FTM 2225, LF-04.[32]

On 6 September 1975, signals from GIANT MOON 7 were picked up at Malmstrom AFB and Yokota Air Base, Japan. On the last operational flight test, signals from GIANT MOON 9 were received at Hawaii, Kwajalein, and Guam (Figure 15.11).[33]

DEACTIVATION

The Minuteman II ERCS configuration was active from 1967 until the deactivation of the Minuteman II system in 1991. Continuation of the ERCS mission by refitting it atop the Minuteman III booster was seriously considered but ultimately dropped, based on pending availability of the Air Force MILSTAR communication satellite to fulfill the same mission.[34]

## Airborne Launch Control System

The first public mention of the concept of the Minuteman ALCS took place during the testimony of Secretary of Defense Robert McNamara before the Senate Committee on Foreign Relations on 13 August 1963. McNamara was testifying in support of the Comprehensive Test Ban Treaty. His support focused on the military benefits and risks as a result of signing the treaty.

In reviewing a comparison of the United States and Soviet Union nuclear technology capabilities, McNamara reassured the committee of the continued capability of our strategic retaliatory forces to destroy Soviet targets. The recent tests by the Soviet Union of a 50-megaton nuclear weapon clearly demonstrated its superiority in the high yield-to-weight range of nuclear weapons. McNamara emphasized that below several megatons, the United States was clearly superior. Furthermore, McNamara stressed that the United States had purposely focused attention on the lower yield-to-weight weapons to facilitate deployment in large numbers of missiles such as Minuteman in order to saturate Soviet ABM defenses.

McNamara reiterated that the Minuteman facilities had been designed to withstand thermal, pressure, and ground motion effects typical of Soviet weapons detonated at relatively close quarters. He reassured the senators that the Soviets did not have anything like the number of missiles necessary to knock out the Minuteman force. McNamara went on

FIGURE 15.11 ERCS signal coverage. ERCS provided the ability to disseminate the Emergency Action Message to all SAC-controlled forces and other SIOP command elements in the European, Pacific, and North American areas. *Adapted from "History of Strategic Air Command, July 1975–December 1976."*

to place the 50- to 100-megaton weapon in perspective, noting that the Soviets did not currently have a missile capable of carrying such a warhead and it would be hard to hide the development of such a missile. McNamara noted that "we have duplicate facilities which will in the future include the ability of launching each individual Minuteman by a signal from an airborne control post."[35]

On 28 October 1963, Headquarters Air Force requested that SAC begin an operational study to evaluate the utility of using an ALCS for the Minuteman force as part of the Post-Attack Command-and-Control System (PACCS) that had been operating since 3 February 1961.[36]

This new study included the number of ALCS-equipped PACCS aircraft, flight patterns, altitudes, and the ALCS deployment concept that would extend the already existing EAM capabilities of the Looking Glass EC-135 aircraft operating out of Offutt AFB. The ALCS would be the post-attack replacement for destroyed Minuteman LCCs.[37]

On 1 December 1964, Headquarters SAC established a revised operational concept for the Minuteman ALCS that called for the PACCS aircraft, including the relay aircraft, to be ALCS capable (Figure 15.12).[38]

FIGURE 15.12  ALCS, EC-135 aircraft. EC-13A, C, G, and L models operated as PACCS/ALCS aircraft as well as serving as tankers. *Adapted from official USAF photograph, courtesy of Greg Ogletree.*

On 29 September 1965, Headquarters SAC completed a reevaluation of its command-and-control communications systems, which indicated that the Minuteman ALCS within the PACCS system was a matter of the highest urgency due to the pending deployment of a new Soviet ICBM. The Soviet R-36 (8K67, NATO designation SS-9) Mod 1 had begun flight testing on 3 December 1963. The SS-9 was a two-stage storable liquid–propellant missile that could carry a single 13,500-pound reentry vehicle 5,300 nm. With an estimated accuracy of 0.4 to 0.5 nm and carrying a 12- to 15-megaton warhead, the SS-9 Mod 1 was seen as placing the Minuteman LCCs at an unacceptable risk.[39]

On 26 October 1965, Headquarters USAF issued a system management directive to expand the ALCS program to include both Minuteman I and II.[40] In November 1965 McNamara directed an acceleration of the ALCS program and agreed to expand it to include the capability for the entire Minuteman fleet. The result was an increase in the number of PACCS aircraft from 27 to 32. The deployment of the ALCS would now force the Soviet Union to once again target the individual Minuteman LFs.[41]

On 7 July 1966, Boeing Company received an Airborne Command Post (ABNCP)-configured EC-135 aircraft for conversion to an ALCS-capable aircraft. The first test flight with the equipment installed took place on 22 September 1966. The first launch using the ALCS took place on 17 April 1967 with the successful launch and flight of Minuteman II

BUSY MISSILE, FTM 2143, LF-03, targeted for a broad ocean area. This flight was also the third and final flight test of the Minuteman ERCS combination, as the remaining scheduled flight was canceled due to the success of the first three.[42] There were two more successful ALCS launches and flights to target: Minuteman IB BUSY MUMMY, FTM 1019, LF-02, on 28 April 1967, and Minuteman II BUSY FELLOW, FTM 2095, LF-21, on 11 May 1967. The ALCS reached IOC on 31 May 1967.[43]

After four consecutive successful launches with ALCS, the first failure took place on 29 March 1968 with the attempted launch of OLYMPIC TRIALS 9, FTM 2522. Two launch attempts were made, with 17 Execute Launch commands transmitted during the first test and 21 during the second. From analysis of the telemetry, it was found that the problem was likely due to an intermittent air-to-ground data link and not in the code structure or computer operation. The test report included a recommendation that in future launches the ALCC command transmission be recorded, along with the signal strength. Equipment was installed for this purpose at Vandenberg AFB.[44]

By 20 June 1968, 600 of the 1,000 Minuteman launchers had the UHF installations completed. Aircraft-to-LF transmission tests were conducted to assess the reliability of the UHF systems. From January to June 1968, three tests of the Whiteman AFB installations were accomplished daily, and two weekly tests were flown for the Ellsworth, Malmstrom, and Warren LFs. Signal reception at the LFs was indicated at the respective LCCs, which then reported the success or failure to the ALCC aircraft. These tests were 98.7 percent effective. At the end of June 1969, 850 installations were completed, with the remainder at Grand Forks to be modified by mid-1970.[45]

## *Deployment*

From 1967 to mid-1970, the six Minuteman wings were covered by ALCS-equipped EC-135 aircraft operating from three locations: Looking Glass, flying out of Offutt AFB, covered Wing IV at Whiteman; one aircraft flying from Ellsworth AFB covered Wing II at Ellsworth and Wing V at F. E. Warren AFB, with a second aircraft covering Wing I at Malmstrom AFB; and one aircraft from Minot AFB covered Wing III at Minot and Wing VI at Grand Forks AFB. The ALCC missile combat crews for the Ellsworth AFB aircraft were assigned from the 67 SMS, and those for the Minot AFB aircraft were assigned from the 741st Strategic Missile Squadron (741 SMS). The ALCC missile combat crew positions were MCCC-A and DMCCC-A (for "Airborne"; Figures 15.13 and 15.14).

Early in 1970, the consolidation of PACCS aircraft and personnel from March AFB, California, as well as ALCS aircraft and personnel from Minot AFB, formed the 4th Airborne Command Control Squadron (4 ACCS) at Ellsworth. ALCC-1 covered Wings II and V as well as the duties of the Western Auxiliary Command Post functions of the PACCS; two aircraft were forward-deployed to Minot AFB, and their crews performed alert duty alongside the B-52 crews. ALCC-2 covered Wing I, while ALCC-3 covered Wings III and VI. The two ACCSs operating out of Offutt AFB covered Wing IV as part of the LOOKING GLASS mission.[46]

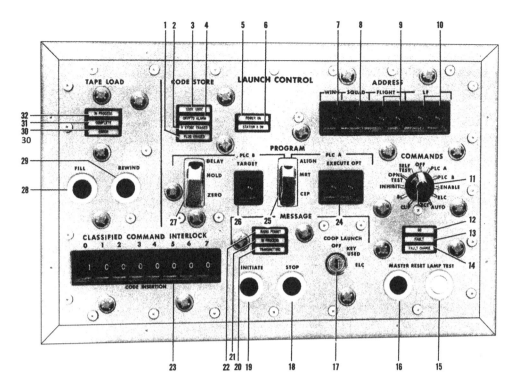

**ALCC LAUNCH CONTROL PANEL SWITCHES AND INDICATORS**

1. PLUG ERASED warning light (red)
2. V STORE ERASED warning light (red)
3. CODE USED warning light (red)
4. CRYPTO ALARM warning light (red)
5. POWER ON indicator (green)
6. STATION 3 ON caution light (amber)
7. WING selector
8. SQUAD selector
9. FLIGHT selectors (two)
10. LF address selectors
11. COMMANDS select switch
12. GO indicator (green)
13. FAULT warning light (red)
14. FAULT CHANGE caution light (amber)
15. LAMP TEST pushbutton
16. MASTER RESET pushbutton
17. COOP LAUNCH key switch
18. STOP pushbutton
19. INITIATE pushbutton
20. TRANSMITTING indicator (white)
21. IN PROCESS indicator (white)
22. RADIO PERMIT indicator (green)
23. CLASSIFIED COMMAND INTERLOCK/ CODE INSERTION selectors
24. PLC-A EXECUTE OPT selector
25. PLC A/B ALIGN switch
26. PLC-B TARGET selector
27. PLC-B DELAY switch
28. FILL pushbutton
29. REWIND pushbutton
30. ERROR warning light (red)
31. COMPLETE indicator (green)
32. IN PROCESS indicator (white)

FIGURE 15.13 ALCC MCCC-A Launch Control Panel illustration. See Appendix D for detailed explanation of panel controls. Missing from the illustration is the ERCS panel, located above the Launch Control Panel. *Adapted from Airborne Launch Control System Operation Instructions, courtesy of Greg Ogletree.*

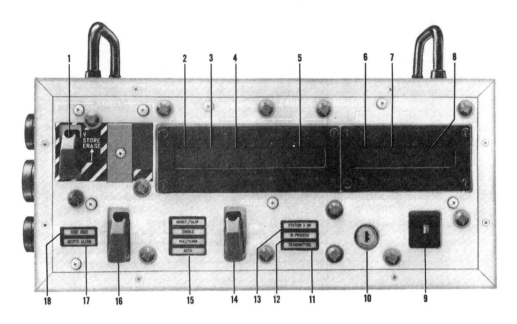

**ALCC CONTROL MONITOR PANEL SWITCHES AND INDICATORS**

1. V STORE ERASE switch
2. WING display
3. SQUAD display
4. FLIGHT displays
5. LF displays
6. DELAY display
7. TARGET display
8. PLC-A EXECUTE OPT display
9. LAMP TEST switch
10. COOP LAUNCH key switch
11. TRANSMITTING indicator (white)
12. IN PROCESS indicator (white)
13. STATION 3 ON caution light (amber)
14. RADIO switch
15. INHIBIT/CLIP, ENABLE, ELC/
16. CRYPTO ALARM OVERRIDE switch
17. CRYPTO ALARM warning light (red)
18. CODE USED warning light (red)

FIGURE 15.14 ALCC DMCCC-A Launch Monitor Panel illustration. The Launch Monitor Panel displayed output from the Launch Control Panel. The COOP Launch key switch had to be turned within 2.5 seconds of the key on the Launch Control Panel to initiate terminal countdown for the selected missile or missiles. See Appendix D for detailed explanation of panel controls. *Adapted from Airborne Launch Control System Operation Instructions, courtesy of Greg Ogletree.*

## Operations

The ALCC crew alert duty at Minot in the 1980s was a weeklong ground alert 24/7 and then off for a week, and the cycle repeated. Due to the need for rapid response, parts of the base were off limits to crew members, who had to reach the aircraft within five minutes after receiving the alert. When a klaxon sounded, all traffic on the base stopped except for the alert crews going to their aircraft.

Just as with the ground system procedures concerning codes, which required the codes to be carried separately in two parts, two officers would pick up the first half of the operational or test codes at the Codes Vault and then proceed to the aircraft. Once on board, they radioed to the Codes Vault, and the second team would bring the second half of the codes to the aircraft.

When the ALCS crew heard the alert alarm, a klaxon that could be heard across the base, they would immediately go out to the aircraft and board. While the engines were started and the pilot prepared to take off, the crew would decode the EAM. The preamble to the message indicated whether it was an exercise or actual alert message and whether they were to shut down the engines and return to quarters, taxi to the hold line, or launch.[47]

The ALCC command capability consisted of eight basic commands: (1) the launch facility radio test (LFRT) would simply transmit a UHF signal to all LFs at a wing for test purposes to ensure a link could be established; (2) PLC-A identified the launch execution option to an entire squadron; (3) PLC-B identified target, delay time, and retargeting mode to an individual missile, modifying the PLC-A; (4) Enable; 5) Execute Launch command (ELC)—one ELC was equal to one launch vote, placing the missile in the Launch Commanded Mode; (6) the second vote entered the missile into the Launch in Progress Mode, and the terminal countdown began if there was no hold stipulated by the PLC-A/B; (7) automatic (Auto) combined the Enable command and the two ELCs; (8) Inhibit nullified a single ground ELC and returned an Enabled missile to the Dis-Enabled state, or normal strategic alert status, after five minutes.

An additional safety measure was the ALCC switch, operated by the pilot, which controlled the transmission of the ALCC-classified commands. This switch prevented classified commands such as Auto, Enable, Inhibit, or Execute Launch to be transmitted to the LF until it was placed in the on position. The LFRT, PLC-A, and PLC-B were the only commands that did not require the ALCC switch to be in the on position.

The ALCS aircraft had to be at or above 33,000 feet and within plus or minus five nm of the designated track in order to remain in communication with the LFs. When turning was necessary, aircraft banks were limited to 15 degrees or less.[48]

There were four types of test exercises: a GIANT BALL mission, which was an LFRT; a GIANT PACE mission, which meant a SELM; an HETF I or II mission, which involved flying to Hill AFB, which had two training LFs with equipment that would record signal strength and other parameters—something that could not be done at the operational sites; and GLORY TRIP missions, an actual missile launch at Vandenberg. In each case test codes were used as a safety measure. The frequency of exercises varied from monthly to quarterly to annually for each wing, depending on the type of test.[49]

## GIANT BALL

The GIANT BALL LFRT missions involve LF UHF radio receiver tests. Each LCC and LF has a hold-off timer that prevents the LF UHF receiver from accepting a command message from the ALCS aircraft. This was a safety feature to prevent outside access to LFs via a UHF and conflict between the LCC and the ALCC. The ground missile combat crew receives a 10-minute warning prior to timer expiration. If the timer is not reset, a second advisory is generated at the two-minute point. Failure to respond before the two-minute advisory could be "career inhibiting."

During a GIANT BALL LFRT, the ALCC sends a UHF signal, not a command, to the LFs. Unlike the Peacekeeper system, where the LF replies to the aircraft, in the Minuteman system, each LCC reports the results of their flight of missiles to their Squadron Command Post, and the Squadron Command Post reports squadron results to the Alternate Command Post, which is usually designated as the exercise monitor and reports the results to the ALCC.[50]

## GIANT PACE

ALCC crews also participate in GIANT PACE SELM exercises using commands with test launch codes (see Chapter 14). Using test codes and radio frequencies, the ALCC missile combat crews go through the preparatory launch procedures and terminal countdown. The PLC-A command is sent to LFs with missiles in the SELM configuration. PLC-B often follows the PLC-A command and is used to select individual missiles in the selective launch procedure and modify the delay time. The ALCC's primary use of the Inhibit command is to nullify a single ground ELC in a degraded squadron so that the LFs in that squadron can be retargeted with airborne PLC commands. The Enable command arms the missile ordnance. The ELC command is transmitted in two parts, vote one and vote two, after key turn by the MCCC-A and DMCCC-A. After the transmission of the first ELC is finished, the second ELC is sent. The Auto command is unique to the ALCS and combines the Enable and ELC commands. It also requires simultaneous key turns by the ALCC missile combat crew.[51]

### *Hill Engineering Test Facilities, Strategic Missile Integration Complex*

After 20 January 1971, training flights issuing launch commands no longer had to be flown only at Vandenberg AFB, or up at Seattle at the Boeing STP III facility, or during SELM tests. The Hill Engineering Test Facilities (HETF) I, II, and III, also referred to as the Strategic Missile Integration Complex, were equipped with simulators on the ground to receive and process the airborne commands. Initially HETF I was configured as a Wing V unmodernized facility using the Minuteman IB missile and guidance set but was reconfigured to the Force Modernization upgrade in mid-1971. HETF II was configured as a 564 SMS facility with a Minuteman II missile. HETF I was reconfigured for Minuteman III and declared operational on 15 November 1971. HETF III was built for the Peacekeeper program.[52]

The advantage of the HETF facilities is that they are instrumented such that multiple simulated launches can be run in succession without having to set up again after each aircraft pass. The Minuteman facilities operate on different frequencies, which are also different from the frequencies used at the operational bases or at Vandenberg. The testing is conducted at altitudes above 30,000 feet and at a range of over 150 nm. On each flyover the launch commands are broadcast separately through five UHF transmitters carried in the aircraft and recorded both in the aircraft and on the ground to analyze for signal-to-noise ratio, field strength, and frequency deviation, among other parameters.[53]

## GLORY TRIP

Approximately twice a year, an ALCS aircraft participates in Follow-on Operational Test and Evaluation launches, nicknamed GLORY TRIP, of Minuteman missiles from test launch facilities at Vandenberg. These verified the reliability and capability of both the missile and the ALCC system.[54] As of May 2019, there have been 109 ALCS-commanded launches, with one failure.[55]

### Deactivation

On 27 September 1991, ALCC aircraft 24-hour alert operations were discontinued with the announcement by President George H. Bush that the Minuteman II force was being deactivated immediately. The squadrons continued flying training missions and retained the capability for alert deployments for another year.[56]

On 1 June 1992, SAC was inactivated, and the multiservice USSTRATCOM was activated. By 31 December 1992, all the ALCS aircraft had been upgraded to the Common/PACER LINK satellite communication system. On 1 October 1998, the Air Force retired the EC-135 aircraft, and the Navy's E-6B Mercury aircraft took over the USSTRATCOM LOOKING GLASS and ALCS missions, now called the USSTRATCOM ABNCP, which is on alert around the clock. It retains the ALCS capability and randomly flies airborne alert, periodically testing the ALCC.[57]

## Minuteman and the Cuban Missile Crisis

The most detailed analysis to date of the Minuteman program and the Cuban Missile Crisis is the work of Scott Sagan in his book *The Limits of Safety*. Sagan emphasized what he considered to be dangerous procedures used to bring to strategic alert the Minuteman missiles of Alpha Flight of the 310 SMS, 341 SMW, Malmstrom AFB, Montana. A review of the documents he cites, as well as additional official histories, reveals a significantly different version of the events at Malmstrom at that critical time in our country's, and indeed the world's, history. Since Minuteman missiles were placed on alert at both Malmstrom AFB and Vandenberg AFB, the discussion that follows will detail actions at both locations separately.[58]

### Prelude

On 14 October 1962, a SAC U-2 reconnaissance aircraft flying over western Cuba obtained photographs that provided conclusive evidence that 24 launchers for the Soviet SS-4 medium-range ballistic missiles, with a range of 1,129 nm, were in the last stages of construction and missile assembly was underway. These missiles placed Washington, DC, New Orleans, Houston, Dallas, and Cincinnati, as well as the Panama Canal and Mexico City, within range of an attack with a one-megaton nuclear weapon. Additionally, 38 percent of the B-52 bases and 8 percent of the Atlas ICBM bases were in jeopardy. Further intensifying the crisis was the fact that flight time was only 13 minutes to reach Washington, DC.[59]

On 22 October 1962, at 2300Z, President Kennedy, in a nationwide address lasting 18 minutes, informed the nation of the deployment of Soviet missiles with nuclear warheads in Cuba. After explaining the details of the Soviet actions, at 9 minutes, 10 seconds into the speech, President Kennedy announced seven initial steps to be taken immediately. The first step was that an immediate "quarantine on all offensive military equipment under shipment to Cuba is being initiated." Enforcement of the quarantine began at 1400Z, 24 October 1962.[60]

### 341st Strategic Missile Wing, Malmstrom AFB, Montana

On 22 October 1962, the battle staff and key contractor personnel assembled in the 341 SMW Command Post to listen to President Kennedy's speech. Simultaneously, the JCS declared defense readiness condition (DEFCON) 3 for United States military forces worldwide (at the time most American forces were kept at DEFCON 5 in normal peacetime with one exception, SAC, which was kept at DEFCON 4).[61] For the 341 SMW, this meant invoking OP 402–63, which had been updated on 10 October 1962 in support of the SAC/Air Force Systems Command Agreement for Emergency Combat Capability of Ballistic Missile Launch Complexes, written 4 May 1961, and Proposed Safety Rules for the HSM-80A (Minuteman)/MK 5 RV Weapon System, 2 October 1962.[62] OP 402–63 stipulated that upon declaration of DEFCON 3, the Malmstrom LFs would be assigned ELC status by the SATAF commander based on technical considerations. ELC status was defined as the "condition of the missile and the supporting ground environment wherein a launch is possible using available on-site personnel and equipment." SATAF activity was to continue in a timely manner in order to bring as many launchers as possible to ELC status. The operations plan also stipulated that at DEFCON 3, for all SAC-owned ECC flights, a condition under which, with strategic warning, a missile could be launched on an EWO mission, a standby missile combat crew for each flight would be stationed in the Standby Crew Room in the LCF. Specifically, for Minuteman, the ECC status was to exist after completion of one flight of missiles and an associated LCC. This designation only existed if the facility had not been designated as combat ready; that is, in emergency situations.[63]

On 23 October 1962, SAC placed the 341 SMW at DEFCON 2. An Alternate Command Post was established at Alpha Flight LCC Alpha-01. Thirteen missiles were already in LFs and being brought to demonstration-and-acceptance status by SATAF: 10 missiles in Alpha Flight and 3 in Bravo Flight. Turnover maintenance activities were focused on the Alpha Flight missiles. Alpha Flight was turned over to SAC on 24 October 1962 before completion of all required contractor system demonstrations. The demonstrations were completed by SAC and SATAF personnel in the process of bringing the Alpha Flight missiles to alert status.[64]

When the quarantine went into effect, the JCS ordered the generation of the remaining SAC forces, thus establishing DEFCON 2. One hour later, Gen. Thomas S. Power, commander-in-chief, SAC (CINCSAC), addressed all SAC forces via the Primary Alerting System:

> This is General Power speaking. I am addressing you for the purpose of reemphasizing the seriousness of the situation this nation faces. We are in an advanced state of readiness to meet any emergencies and I feel that we are well prepared. I expect each of you to maintain strict security and use calm judgment during this tense period. Our plans are well prepared and are being executed smoothly. If there are any questions concerning instructions which by nature of the situation deviate from normal, use the telephone for clarification. Review your plans for further action to ensure that there will be no mistakes or confusion. I expect you to cut out all non-essentials and put yourself in a maximum readiness condition. If you're not sure what you should do in any situation, and if time permits, get in touch with us here.[65]

At this point the 341 SMW had 10 ECC missile combat crews that had completed their training but had not yet been evaluated by the 3901 SMES for combat-ready status. All had been through a month of ORT at Vandenberg AFB, where the two-man crews were formed and trained together. There were six sections to the ORT evaluation at the end of training: maintaining alert, simulated EWO performance, readiness checkout, safety/emergency exams, proficiency exercise, and final rating. The rating key ranged from highly qualified, qualified, conditionally qualified, and unqualified. During a one-day visit to Malmstrom on 24 October, members of the 3901 SMES administered the safety/emergency examination to two crews as an abbreviated part of their combat readiness evaluation. All four crew members passed with highly qualified scores. The 341 SMW was found to have sufficient qualified crews for support of launch capability for Alpha Flight under ECC status. Furthermore, between 26 and 29 October, two Senior Standboard Crews were certified, and an additional 15 missile combat crews were ECC certified.[66]

With the declaration of DEFCON 2, all ELC flights that could be assigned ECC status were turned over to SAC for operational decisions and control. "The SAC squadron commander will provide a certified capable SAC launch control team which can receive, decode and authenticate all execution messages and conduct the launch countdown with assistance as necessary."[67] Furthermore, when SAC control was established over the first flight in each squadron, the LCC was continuously manned. Guidance sections and reentry vehicles with War Reserve warheads were installed in all 10 missiles of Alpha Flight. During this time frame, an Encoder Cavity Plug was installed in the Launch Control Console to complete the command circuits but without launch codes. Since the Launch Control Panel held both Launch and Inhibit codes, which used the same connector, LCC Alpha-01 did not have Inhibit capability. This is almost certainly what drove keeping the LF SCS safed and the launcher closure disconnected from its opening mechanism.

The operations plan also stipulated that at DEFCON 3 for all SAC-owned ECC flights, two operationally encoded Launch Control Panels were to be issued to the Standby Missile Combat Crew upon declaration of DEFCON 2. The Launch Control Panels were not to be taken to the LCC until both missile combat crews had authenticated and validated a launch execution message. Additionally, whenever an operational launch panel was present in an LCC, two missile combat crews had to be on duty within that LCC. The launcher closure ordnance was disconnected and the SCS used to safe the launcher during maintenance placed in the maintenance position, thus preventing a firing signal from reaching the missile guidance system. The launcher would remain in this condition until a launch execution order was received, validated, and authenticated by both missile combat crews.[68]

The procedures stipulated that upon receipt of a valid launch execution message, the Standby Missile Combat Crew was to take both Launch Control Panels to the LCC. Simultaneously, a Missile Maintenance Team would penetrate the selected LFs, reconnect the launcher closure ordnance in accordance with standard checklists, and arm the SCS. Execution of the launch would take place using the ECC checklists (see previous discussion on ECC).[69]

There were two launch modes for Minuteman IA. The normal mode would result from the generation of two separate Launch commands, normally one from each of two LCCs. The delay mode would produce a delayed launch and could be accomplished by a single Launch command generated by a missile combat crew in one LCC. Two war plans were available for execution. War Plan A was salvo fire; War Plan B was ripple fire.

With the normal launch mode, different war plan codes were stored in MCUs at each of the LCCs. The encoder and decoder designs were such that any attempt to disassemble the mechanisms would dissipate the code and thus prevent launch. If War Plan A was selected, all enabled missiles would be launched within 32 seconds, the length of the terminal countdown, upon receipt of the second command. If War Plan B was selected, all enabled missiles would be launched, with each missile being fired after a preselected time interval following receipt of the second command.

With the delayed mode, the Launch command started a delay timer in each of the launchers. When the delay timer ran out, the selected war plan would be executed in the same manner as if two Launch commands had been received. For Minuteman IA the delay timer could be set for a time delay ranging from 54.8 minutes to 5.69 hours.[70]

In response to the findings of the Fletcher Committee, an Inhibit Launch switch had been added to the Launch Control Panel that could be used, in conjunction with the Inhibit Launch code contained in the Launch Control Panel, to prevent launch in both the normal and delayed launch modes. If War Plan B has been selected, an Inhibit Launch command would be effective at any launcher where the 32-second countdown had not yet started, even if the delay timer had run down. When a single Launch command was initiated, followed by an Inhibit Launch command, initiation of a second Launch command within 3.4 minutes from a different LCC would launch the enabled squadron's missiles. If a second Launch command was not initiated within this time, the system returned to strategic alert status. Thus, a single crew and LCC could not produce an unauthorized launch, and no crew could destroy the operational capability of the squadron by using the Inhibit command.

Missile selection for launch was accomplished through the Launch Enable System (LES). Ten LES switches were located on the DMCCC's Communications Control Panel in the LCC. Each switch activated circuits that would arm or safe the SCS in the corresponding LF. When in the safe position, the SCS isolates the Stage I ignition and launcher closure ordnance circuits, thus preventing launch. Through this switch launch could be prevented even if the countdown was started, up to the time of Stage I ignition. The SCS could be manually locked in the safe position with a key, as was done during maintenance or targeting operations, making it impossible to arm the switch by means of the LES switch on the LCC console or accidentally during maintenance in the launcher. The LES was designed to fail in the enabled condition, thus permitting launch of the missiles on command from another LCC in the event that the parent LCC was destroyed.[71]

With the large number of contractor personnel working in and around the LCF support building and LCC, SAC made the decision to keep the coded launch panels and standby missile combat crews off-site at the Wing Command Post. This meant that once the launch execution order was given, it would take approximately 90 minutes to launch missiles.[72]

At 2100Z hours on 24 October, work began to bring LF Alpha-10 to alert status, followed by LF Alpha-06 at 2200Z hours and LF Alpha-07 at 2330Z hours. Work started on LF Alpha-09 at 0330Z on 25 October. While the maintenance crews were well trained, a lack of operational equipment and experience, as well as the technical orders still being refined, meant that Boeing personnel and their equipment became involved in bringing two LFs to alert and helping with the other sites. Boeing personnel, under the supervision of SATAF personnel, started work at LFs Alpha-05 and 08 on 26 October at 1900Z.

On 26 October SAC requested that BSD, through SATAF, conduct an immediate safety evaluation of the Alpha Flight in order to evaluate the potential of accidental launch since the normal demonstration-and-acceptance procedures had not been carried out. The next day, the BSD Minuteman System Program Office replied that:

> Technical evaluation of the turnover configuration of Alpha Flight Malmstrom did not reveal any hazardous conditions relative to achieving strategic alert. Those Alpha Flight Class 3 demonstrations unaccomplished by integrating contractor were considered to be adequately covered by checkout test performed prior to turnover. Launch message propagation test is covered by checkout procedures which assure the system is complete and capable of accomplishing the formal test of propagating test launch message from LF to LF when initiated at the LCC. Launch net verification test is functionally identical to operational routine sensitive command network test prescribed by the technical order procedures. These T. O. procedures are executed during the initial phase of post-delivery operational readiness preparation and will achieve the same confidence level as would have been accomplished by the checkout tests. Class 2 demonstrations unaccomplished prior to Alpha Flight turnover were either functionally covered by the contractor checkout tests or will be performed by SAC in the process of achieving strategic alert. By following prescribed T. O. procedures SAC can achieve operational strategic alert with no degradation of weapon system launch capability or safety. To ensure achieving this posture, the best qualified site activation and contractor personnel (Boeing Company and STL) are working with the 341 SMW personnel in an integrated team effort. Part Two. SAC requirements for an additional launch net verification test and command test did not have any adverse effects on the

technical performance of the weapon system. Part Three. The additional safety precautions prescribed by SAC do in fact provide additional safety at DEFCON Three, Two and One. The safety measures may be applied with equal effectiveness to both Vandenberg and Malmstrom.[73]

On 30 October 1962, BSD reassured SAC that its extraordinary safety precautions, which included disconnecting the launcher closure ordnance and safing the SCS, were not necessary. The precautions were indeed extraordinary since placing the LES switch in the safe position achieved the same purpose as manually safing the SCS (see Chapter 14).[74]

At 1816Z on 26 October, LF Alpha-06 achieved strategic alert status. Twelve hours later, the Sensitive Command Network monitoring system detected a failure in launch enable unit (LEU) No. 2, which resulted in a shutdown. The drawer was replaced, and LF Alpha-06 was back on alert on 28 October at 2330Z.[75] By 30 October LFs Alpha-05, 06, 07, 09, and 10 were all on alert at Malmstrom. The last missile to reach alert was at LF Alpha-02 at 1811Z on 10 November. At no time after 10 November were more than three missiles off alert at once. There was only one point when all 10 missiles were briefly on alert on 22 November 1962.

The 341 SMW remained at DEFCON 2 until 21 November, when SAC lowered the level to DEFCON 3. Six days later, SAC returned to its normal peacetime DEFCON 4 status.[76] The 10 launch facilities and missiles of Bravo Flight were accepted by SAC on 30 November, and reconfiguration of the missiles to strategic alert status was begun immediately.[77]

The bottom line is that LCC Alpha-01 had no launch capability until a valid launch execution message was received and validated in that LCC. Only then were launch codes permitted in the LCC. Until DEFCON 1 launch codes in the LCC would have been useless because the LFs were configured to deny all ordnance actions supporting launch. This included launcher closure opening, upper umbilical release, upper umbilical retract, battery activation in the guidance set and Stage I, and Stage I ignition.

## 6595th Aerospace Test Wing, Vandenberg AFB, California

At the time of the Cuban Missile Crisis, the Minuteman facilities at Vandenberg AFB consisted of six LFs with one hardened (belowground) and one soft (aboveground) LCC. Training of SAC instructor crews had begun in March 1962. The crew Missile Procedure Trainer had been delivered in July, and training was initiated by SAC. The various tests to tie together the LFs and LCCs had begun in August. Category II testing, to "determine if theories, techniques, personnel skills and material are practicable, or if equipment and component items are technically sound, reliable, safe, and meet established specifications or requirements," had begun in August. By this time five of the six LFs had demonstrated ability to achieve strategic alert using ground test missiles. Three of the LFs were in the Wing I configuration, and three were in the Wing II configuration, which differed mainly in the missile suspension system equipment.[78]

On 28 September 1962, at 16:59 PDT, the first Minuteman launch, AIR CRUSADE, FTM 503, took place from LF-04 (394 A03). The launch was a partial success in that the missile cleared the launcher, but 45 seconds into flight, the missile broke up due to excessive aerodynamic loads caused by a guidance system malfunction. On 12 October 1962, operations in support of the second launch, AMERICAN BEAUTY, this time from LF-05 (394 A04), were in progress with FTM 501 installed. SAC missile combat crews were training in three of the LFs at this time.[79]

At 2300Z on 22 October 1962, the 6595th Aerospace Test Wing (6595 ATW) Command Post was notified by AFSC that the Air Force had gone to DEFCON 3. Aware that DEFCON 2 would likely soon be declared, the various ICBM prime and associate contractors were alerted and tasked with evaluating the various LFs at Vandenberg to ascertain the work needed to bring them to ECC status. Under the supervision of the 6595 ATW, personnel from the 1 STRAD and the contractors began a focused effort to bring as many LFs as possible into ECC status as quickly as possible.

For Boeing Company this meant working on six LFs near Point Sal. Three of the LFs, LF-02 (394 A01), LF-03 (394 A02), and LF-07 (394 A06), required removal of ground training missiles followed by installation of operational missiles, guidance-and-control sections, and Mark 5 reentry vehicles. At LF-05 (394 A04), FTM 501 had to be removed and taken to the Destruct Ordnance Installation Building for removal of the destruct ordnance package, raceways, and the crew training launch instrumentation wafer before converting it to an operational missile, reinstalling it in LF-05, and installing the guidance-and-control section and Mark 5 reentry vehicle.

Each launcher had its autocollimator aperture in the launch tube wall welded shut and new ones opened to facilitate target alignment for the more northern launch azimuth toward the Soviet Union instead of the southwesterly azimuth normally flown on test or training flights. The two LCCs at Vandenberg were interconnected, so launch control was under normal two-vote control.[80]

At 1635Z on 27 October, LF-04 (394 A03) was the first Minuteman LF at Vandenberg to achieve strategic alert. What normally took five days had taken 51 hours using Air Force and contractor personnel. LF-05 (394 A04) reached strategic alert at 0700Z on 28 October, followed on 29 October by LF-06 (394 A05). On the following day, 30 October, two more LFs reached strategic alert, LF-02 (394 A01) and LF-03 (394 A02), bringing the total to five Minuteman missiles by the end of October.

LF-02 (394 A01) was taken off alert on 1 November for maintenance on a faulty guidance-and-control section, and the next day LF-04 (394 A03) was taken off alert for the same reason. On 4 November all Vandenberg LFs were returned to testing posture.[81]

★ ★ ★

Just as the command-and-control operation evolved with modernization efforts, so did the physical facilities at both the LCCs and the LFs themselves. This has been an ongoing process that started with the Force Modernization Program beginning in the mid-1960s and continuing in a variety of forms into 2019. The upgrade process also involved actual exposure of the LFs and LCFs to ground motion and overpressure at the levels expected in an enemy attack.

# 16

# KEEPING PACE: MODERNIZATION AND UPGRADES

The original design service life for each variant of Minuteman was a goal of 10 years. Minuteman IA was deployed from October 1962 to March 1969, Minuteman IB from June 1963 to January 1974, Minuteman II from October 1969 to October 1991, and Minuteman III has been deployed since June 1970. Excluding Minuteman IA, which the Air Force had intended to replace as soon as possible with Minuteman IB, each of the variants exceeded the 10-year goal.[1]

There have been a myriad of upgrade and modification programs to both the missile and the facilities since deployment began in 1962. The major programs are presented in chronological order where possible, although in some cases programs overlapped.

## High-Explosive Simulation Technique

Hardness design parameters for the Minuteman LF and LCC originated from the extensive data collected on the response of above- and belowground structures to overpressures above 50 psi obtained during the Operation Plumbbob series of nuclear weapon tests in 1957 at the Nevada Test Site (now named Nevada National Security Site).[2]

With the 1963 signing of the Limited Treaty Banning Nuclear Weapon Tests in the Atmosphere, Outer Space, and Under Water by the United States, Britain, Soviet Union, and 125 additional countries, the use of air- or ground-burst nuclear weapon detonations to test the hardness of components of the missile LFs and LCCs was no longer possible. In February 1964 the Air Force Special Weapons Laboratory began a nuclear blast effects simulation program. Two simulation techniques were studied: Project Gas Bag, the continuation of the gaseous mixture technique of several years earlier, and a second technique using Primacord (detonating cord).

The first experiments repeated the use of the gaseous mixture with the explosion confined by a 14,300-gallon water overburden. This was the first high-explosive simulation technique test (HEST). Difficulties were encountered in meeting the desired simulation properties: detonation rate, peak pressure, and the detonation waveform. The use of the gaseous mixture of hydrogen and oxygen had serious operational and safety problems, as well as requiring a large quantity of water, so a search was started for a more practical, solid explosive source and less troublesome overburden material.

A contract was awarded to General American Research Division of the General American Transportation Corporation based on its proposal to use a weave of Primacord in an air cavity over the structure to be tested. Covered by an earth overburden to contain the explosion and prolong its duration, the result would be a more accurate simulation of nuclear blast overpressures. Two tests were conducted on an area of 15 by 30 feet, but these efforts were not successful in producing the desired environment of a 300-psi overpressure wave. Seven tests to refine the technique were run over the summer of 1964 at Kirtland AFB, New Mexico. In the fall of 1964, the new system was successfully used to evaluate the hardness of the Minuteman LF using a small structural model (Figure 16.1).

The final configuration for the first large-scale test of the HEST system at Kirtland was a grid of assemblies of Primacord attached to five-by-seven-foot wooden frames of two-by-four-inch lumber. A continuous strand of Primacord was laced on each frame, thereby approximating the uniform properties of a solid sheet explosive. The wrap angle of the Primacord determined the rate at which the combustion products were formed along the length of the cavity. This was necessary because the detonation velocity for Primacord was higher than needed for the desired shock front simulation. The combustion products from the explosion acted like a piston by loading the cylinder of air in front of the detonation, which then formed a shock wave closely simulating the passage of the shock wave from a nuclear detonation. As the overburden moved upward as a result of the detonation, the cavity volume was increased and caused a corresponding decrease in pressure, as would be seen with a nuclear detonation blast wave passing over an LF.

At the end of the HEST development program, the system was able to simulate overpressures up to 3,000 psi for approximately the first 200 milliseconds of air blast for simulation of yields up to 10 megatons. HEST was recommended for testing shallow-buried and surface-flush structures since their principal failure mode was directly related to overpressure loading. Because the peak overpressure was uniform over the entire test area, structures with large surface areas could be more realistically tested (Figure 16.2).[3]

In 1965 a Hardness Verification Review Panel, first formed in 1963, reported to the BSD that 27 items did not meet the original design specifications in Wings I through V. The basic LF and LCC structures were not the problem; it was support equipment such as blast valves, pipes, electrical conduits, and similar items that studies had shown to be vulnerable to motion generated by ground shock. As a result, Minuteman LCC and LF hardness was estimated to be only about 10 percent and 25 percent, respectively, of the design specification. The proposed solutions would cost approximately $30 million and first be concentrated on improving the LCCs by 1972. On 5 November 1965, Secretary of Defense McNamara approved a plan to accelerate the corrective actions so that all the work would be completed by October 1968. With the development of the HEST capability, full-scale testing to evaluate proposed solutions to the problem could now be conducted.[4]

The first use of the HEST system on a full-scale military structure took place in December 1965 at the 90 SMW, F. E. Warren AFB, Wyoming. HEST I required the isolation of LF Quebec-04 from the rest of the squadron and the replacement of the operational missile by a ground test missile specially instrumented for the test. The LF was received by Boeing from SAC in August 1965, and the modified facility was returned to the Air Force Weapons Laboratory on 22 November 1965. The test took place on 1 December 1965 and was a complete success, with minimum damage sustained. The estimated overpressure was 300 psi. The refurbished site was returned to the Air Force on 10 November 1966 (Figure 16.3).[5]

HEST II, to test the hardness of the LCC, was also conducted at F. E. Warren AFB. LCC Delta-01 had an optical survey conducted to determine orientation and position prior to

FIGURE 16.1 Minuteman LF scale model used in HEST program at Kirtland AFB, New Mexico. *Adapted from "Proceedings of the Nuclear Blast and Shock Simulation Symposium."*

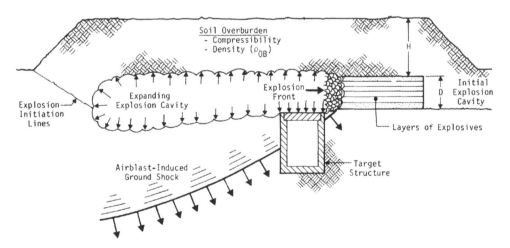

FIGURE 16.2 Illustration of the sequential detonation of explosives to create the moving shock wave. *Adapted from "Proceedings of the Nuclear Blast and Shock Simulation Symposium."*

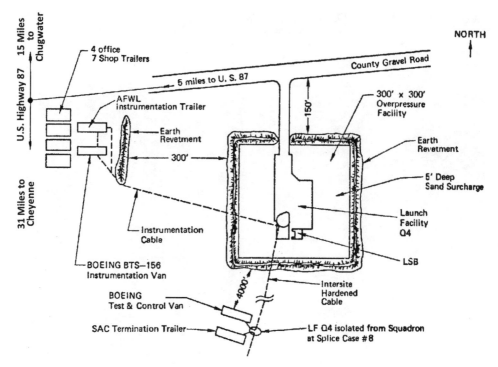

FIGURE 16.3 Layout of the first HEST at F. E. Warren AFB. *Adapted from "Minuteman Historical Summary, 1966–1967," courtesy of the Boeing Company.*

the test. The site was isolated from the rest of the squadron, and modifications began on 16 February 1966. Removal of all aboveground structures was completed on 27 February 1966, and instrumentation of the LCC began on 1 March 1966. The test structure was slightly modified to accommodate lessons learned from HEST I. HEST II used 80,000 pounds of Primacord. The test took place on 22 July 1966 and was again successful, as the LCC and LCEB continued to function despite damage to the LCEB from the blast that had generated an overpressure of 1,000 psi. Reconstruction of the LCEB began on 15 November 1966 and was completed on 11 August 1967 (Figure 16.4).[6]

The HEST program moved to Grand Forks AFB in September 1966, to investigate the hardness of the newly completed Minuteman II LFs as well as to evaluate possible degradation of the guidance-and-control system on a ground test missile due to blast effects. HEST III took place on 22 September 1966, at LF Mike-28, with a blast that exerted a force of 1,000 psi. While the LF remained operational for 72 minutes following the blast, it suffered significant damage. The loose soil and high water table at the site contributed to the displacement and flooding of the lower level of the LER as well as flooding in the launch tube. The launch tube flooding would normally have been taken care of by a sump pump, but the movement of the lower level of the LER had been sufficiently violent to break the emergency power line, preventing the pumps from operating. The blast also forced mud into the air-conditioning system, and mud covered the emergency power batteries located on the lower equipment room level. That the facility remained operational for over an hour

FIGURE 16.4 Preparation of layers of Primacord explosives for HEST II showing the specific angle on wooden frames. Note the size of the workers on the left side of the photograph and the overburden on top of the frame. *Adapted from "Minuteman Historical Summary, 1966–1967," courtesy of the Boeing Company.*

after the blast was encouraging, while the amount of damage validated the value of the test in revealing problems in the hardness of the system (Figure 16.5).[7]

During construction of the facilities at Grand Forks, two LERs settled beyond acceptable limits and had to be repositioned. Boeing and the H. C. Smith Construction Company, therefore, developed a technique for repositioning the structure. In the case of LF Mike-28, the 3-million-pound LER had to be releveled, laterally translated, and raised approximately 12 inches using 396 cubic yards of concrete under the foundation. Twenty-five 100-ton hydraulic jacks were used to raise the LER to the required elevation for placement of the lateral movement system. The next step was to place 12 lateral movement assemblies under the LER footing and wedge them firmly in place. The LER was then lowered onto the lateral movement assemblies and four sets of horizontal jacks used to move the LER into position. After an optical survey to ensure the building was in the proper location, steel wedges were positioned between the bearing surfaces, locking further movement. The wedges were welded into position and the bearing assemblies left permanently in place. All the hydraulic jacks were then removed and the space filled with concrete to within four inches of the foundation. The remaining space was filled with a nonshrinking pressure grout mixture. LSB repairs included slightly rotating and lowering the 1-million-pound building into a level position in a similar manner. Repairs to LF Mike-28 were completed in September 1967.[8]

FIGURE 16.5   HEST III took place on 22 September 1966 at Grand Forks AFB. Milliseconds after the detonation, note the uplift of the overburden at the left of the photograph when compared to the right side, where the explosives have just been detonated. *Adapted from "History of the 321st Strategic Missile Wing, July–September 1966."*

While the third test had dramatically revealed several major weaknesses in the Wing VI and 564 SMS Minuteman II LFs, the combined results from the three tests pointed to the need for improvements for all six wings. Areas for hardening improvement included air ductwork and improved design of the brine chiller to prevent dirt being forced into the strainers, which compromised the ability to cool the electronic equipment in the LER. Flooding was clearly a threat, as demonstrated in the last test, since other wings had similar soil and high–water table conditions. Flooding and subsequent generation of toxic gases from the backup batteries under the LCC floor, in addition to actual loss of power, were major concerns. The Air Force began a shock improvement effort that consisted of design changes of pipes, blast valves, and conduits but, for the most part, was directed toward the elimination of faulty welds, ensuring sufficient slack in cable runs, and adjusting the shock-isolation equipment on the various platforms.[9]

The AFSC recommended abandoning the program after the third test, but Air Force Chief of Staff Gen. John P. McConnell directed that it should continue. HEST IV was deferred, with the next test, HEST V, scheduled again for Grand Forks AFB. In October 1967 the Air Force conducted a scale-model test to correct a flaw in the simulation technique. The problem was a secondary shock wave caused by the collapse of the earth overburden onto the test site once the explosive gases, which lifted it up into the air, had escaped. The revised design caused the overburden to scatter, reducing the secondary jolt without interfering with the desired rolling shock wave.

HEST V took place on 5 September 1968 at LF Lima-16, which withstood the shock far better than LF Mike-28 two years earlier—the air-conditioning system continued to function after the explosion. The test launch crew shut down the ground test missile for inspection, returned it to alert for 11 hours, and then conducted a successful simulated launch exercise. The LF was then put on diesel power and returned to alert on 7 September 1968 for seven days, after which a second electronic launch was attempted. While the launcher closure opened as commanded and the upper umbilical retracted, there was no ignition signal. The results of the HEST program led to hardness modifications carried out during the Force Modernization Program.[10]

## Force Modernization (WS-133A-M)

On 27 January 1964, during DoD budget hearings before Congress, Secretary of Defense McNamara announced the Minuteman Force Modernization Program (Force Mod), which would replace all Minuteman IA and IB missiles with Minuteman II.[11] On 13 April 1965, Headquarters Air Force issued a Minuteman System Program Directive for implementation of Force Mod to reconfigure the entire Minuteman fleet with Minuteman II missiles by FY 1972.

As originally conceived, Force Mod upgraded Wings I through V LFs and LCCs to function and operate much like Wing VI and the 564 SMS, which already housed Minuteman II. After the program Wings I through V had a new weapon system designation, WS-133A-M (M for "modernization"), to accommodate what turned out to be either Minuteman II or III. Wings III and V went directly from Minuteman IB to Minuteman III. Wing VI and 564 SMS retained their WS-133B designation, even though the program resulted in their Minuteman II missiles being replaced with Minuteman IIIs.

Existing system equipment was used or adapted whenever practical. At 57.6 feet in length, the Minuteman II missile was 3.9 and 1.7 feet longer than Minuteman IA and IB, respectively.[12] This meant lowering the missile support ring in the launch tube to provide clearance for the reentry vehicle. For Wing I this required lowering the entire missile support system, while for Wings II and IV, the change was much simpler and involved using adapter links to increase cable lengths in the missile support system (see Chapter 6).

Since Minuteman II was aligned to true north rather than to a specific target azimuth, the autocollimator bench was shortened now that the autocollimator would always be at 90 degrees east. While the alignment window for Minuteman IB and II was at the same height, a new access opening in the launch duct liner had to be made at Wing I for the relocated autocollimator window, as well as relocating the autocollimator bench. An additional change in the launch tube liner was the repositioning of the new upper umbilical retraction system. The Minuteman II guidance-and-control section had thermal characteristics that required additional coolant capability, requiring modifications to the brine chiller equipment pumping system. Modification to the launcher closure opening cables had to be made to prevent chafing on the reentry vehicle, in the case of Minuteman II, or the reentry vehicle shroud of Minuteman III. To bring the survivability period for Wings I through V to the standard of Wing VI and 564 SMS, the LF emergency battery backup capacity was increased. Hardening improvements were also made to a variety of equipment installations.

LCC upgrades included increased emergency battery backup capacity, modifications to the LCC equipment to accommodate the increased target capacity for Minuteman II, a larger water tank to serve as a heatsink for Wings I and II, and standby power and controls moved into the LCEB from the LCSB on the surface for the wings with LCEBs. Command-and-control changes included the capability of a single LCC to interrogate not only its own flight but any or all flights in the squadron as well as set targets and missile launch delay times and provision for radio launch capability from the ALCS aircraft. Maintenance ground equipment changes totaled 27 new, 25 modified, and 19 deleted equipment items.[13]

Originally designed to take place on a squadron-by-squadron basis, the program was changed to a wing-by-wing basis beginning on 7 May 1966 with Wing IV, Whiteman AFB (see Table 16.1 for a summary of the Force Modernization dates for each wing). For Wing III, Minot AFB, and Wing V, F. E. Warren, the decision was made to upgrade directly to Minuteman III from IB.[14]

In July 1968 SAMSO proposed to SAC a change in the Force Modernization Program for the remaining three wings. Rather than relocate the autocollimator bench to accommodate the Minuteman III with a lower optical alignment window for the autocollimator, a periscope device could be used to displace the collimator beam downward 28 inches, saving several million dollars.[15] The feasibility of this was demonstrated in November 1968, and the periscope modification was adopted. In March 1969 SAMSO recommended that since Ellsworth was not at the time scheduled to receive both Minuteman II and III, the periscope would not be installed at Ellsworth, and so only some of the compatibility changes would be made. The decision was then made that Ellsworth and the first squadron at F. E. Warren would not get the periscope, missile suspension system, or umbilical retraction mechanism changes. Originally, three squadrons at F. E. Warren were scheduled for Minuteman III, and the decision for compatibility with Minuteman II was deferred (all four squadrons at F. E. Warren were converted to Minuteman III).[16]

The Force Modernization Program evolved into a nine-year effort that ended on 11 July 1975 with the completion of the conversion of 564 SMS at Malmstrom to Minuteman III. By the end of the program, Wing VI and 564 SMS were upgraded from Minuteman II to III, with Ellsworth receiving Grand Forks' Minuteman IIs. The result was a Minuteman force consisting of 450 Minuteman IIs and 550 Minuteman IIIs. SAC had planned to replace the entire aging Minuteman II force with Minuteman III, but with the January 1975 decision of Secretary of Defense James R. Schlesinger to limit deployment of Minuteman III to a total of 550 and cease Minuteman III production past FY 1976, the force mix was now set.[17]

## Upgrade Silo

The Upgrade Silo program was a result of the cancellation of the Hard Rock Silo Program (HRS), one of several programs the Air Force had looked at to protect the Minuteman force from improvements in the Soviet offensive strategic missile force that were taking place in the mid-1960s. Thirty concepts were evaluated, with HRS selection approved by Secretary McNamara in November 1967. HRS would develop new LFs, drilled out of solid rock, that would be able to withstand 3,000-psi overpressure, while the LCC would be able to withstand 6,000-psi overpressure. Initially they would house Minuteman III, but the LF design would be large enough to incorporate future ICBM systems.

A nationwide search located several feasible areas, ending up with the selection of

TABLE 16.1 Summary of Minuteman Force Composition by Wing Resulting from the Force Modernization Program

| WING | MISSILE VARIANT | START DATE | TURNOVER |
|---|---|---|---|
| WING IV, 351ST STRATEGIC MISSILE WING WHITEMAN AFB, MISSOURI | 150 Minuteman II | 7 May 1966[a] | 19 Oct 1967 |
| WING I, 341ST STRATEGIC MISSILE WING MALMSTROM AFB, MONTANA | 150 Minuteman II | 11 Aug 1967[b] | 27 May 1969 |
| WING III, 91ST STRATEGIC MISSILE WING MINOT AFB, NORTH DAKOTA | 150 Minuteman III | 12 Jan 1970[c] | 13 Dec 1971 |
| WING II, 44TH STRATEGIC MISSILE WING ELLSWORTH AFB, SOUTH DAKOTA | 150 Minuteman II | 4 Oct 1971[d] | 13 Mar 1973 |
| WING VI, 321ST STRATEGIC MISSILE WING GRAND FORKS AFB, NORTH DAKOTA | 150 Minuteman III | 6 Dec 1971[e] | 8 Mar 1973 |
| WING V, 90TH STRATEGIC MISSILE WING F. E. WARREN AFB, WYOMING | 200 Minuteman III | 4 Dec 1972[f] | 21 Jan 1975[g] |
| 564TH STRATEGIC MISSILE SQUADRONMALMSTROM AFB, MONTANA | 50 Minuteman III | 20 Jan 1975[g] | 8 Jul 1975 |

a) Ballistic Missile Organization Chronology, 1945–1990; b) "25 Years of Minuteman Deterrence: Malmstrom's Missiles of October Celebration"; c) "Ballistic Missile Organization Chronology, 1945–1990; d) *Aggressor Beware: A Brief History of the 44th Strategic Missile Wing, 1962–1994;* e) Strategic Air Command Weapon Systems Acquisition: 1964–1979; f) History of the 90th Strategic Missile Wing, October–December 1972; g) "The Development of the Strategic Air Command 1946–1981, A Chronological History."

Laramie Range near F. E. Warren AFB. There was enough suitable land to allow 1,000 LFs to be built, though initially only 150 to 450 would be constructed. On 21 November 1968 and 26 March 1969, Rocktest I and II, respectively, were conducted, simulating the air blast and direct induced ground motion characteristics of nuclear explosions using HEST on both quarter-scale model and full-size LF structures. On 30 June 1970, the Air Force Council recommended that the hard-point defense approach be changed to a relatively low-cost program, with the attributes of Minuteman III being the deciding design factor. Elements of the LF upgrades could then be approved in increments. HRS was canceled, and instead Upgrade Silo was implemented.[18]

The Upgrade Silo program was directed toward LF improvements to provide increased radiation protection and resistance to ground shock and air blast, thereby increasing post-attack survivability. The major components were the addition of launcher closure debris bins, a new missile suspension system, upgrade of the LER flooring shock-isolation system, strengthening the launcher closure-actuating and locking mechanism, increased capacity for the missile guidance-and-control cooling system, LF environmental control system improvements—including modifying the launch tube heater to accommodate the new missile suspension system, improvements to the UHF antenna, incorporation of the new hardened EMP antenna, and addition of a 10-inch borated concrete overlay to the launcher closure and surrounding structure to improve radiation protection.[19]

### Launcher Closure Debris System

The addition of the launcher closure debris system bins limited the amount of accumulated blast debris on the launcher closure door that would fall into the launch tube upon opening after an attack. The bins, located on the northern side of the launcher closure door, opened as the launcher closure door began movement, collecting debris that would fall as a result of the angle of repose of the soil deposited on the surface of the LF by a nearby blast. The corresponding headworks positions also had the bins installed at Wing II—at Wing V the bins were located only on the launcher closure. The capacity of the bins was based on the maximum depth of deposited soil with which the launcher closure could still open. The major components of the debris system were the collector assemblies, thermal shield, debris shields, modifications to the pylon spacer assembly, and an improved water seal (Figure 16.6).[20]

### Hardness and Overpressure

Hardness and overpressure improvements included lengthening the launcher closure track wheel detents, providing seals to reduce pressurization under the launcher closure structure, and providing a closure door track beam–controlled failure point with a cut in the rails. The launcher closure door wheels sat in approximately one-inch deep detents in the track so that the door-shielding ring had full contact with the radio frequency–EMP shield and overpressure seals when in the closed position. To prevent damage to the seals when the door was opened for maintenance, each track had a ramp from the detent to the surface of the rail that caused the door to lift above the seals upon opening. An additional 10 inches of neutron-absorbing concrete was added to the launcher closure and surrounding apron. The weight of the concrete added to the top of the launcher closure door and apron required a more gradual slope from the detent up to the surface of the rail. Additionally, the detent was lengthened slightly to allow the door to gather momentum before ascending the ramp. The closure-actuating and locking mechanisms were rebuilt to handle the heavier launcher closure. The ballistic power of the actuating mechanism was doubled with the addition of two gas generator cartridges, for a total of four. The seal on the leading edge of the launcher closure was also improved.[21]

### Missile Suspension System

Five wings, Grand Forks, Malmstrom (including 564 SMS), Minot, F. E. Warren, and Whiteman, had a new missile suspension system installed that enclosed the missile in an

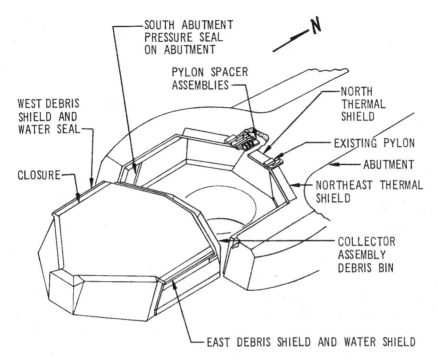

FIGURE 16.6 Launcher closure upgrade installation details. The debris bins open with launcher closure movement and are designed to capture soil and debris at the first moment of launcher closure opening. The thermal shields cover the closed debris bins for protection against thermal effects from a nearby blast and are made of ablative fiberglass material. *Adapted from "Launch Facility Upgrades Wing VI and Squad XX."*

elaborate shock-isolation cage. The major assemblies for the new system were the cage structure, articulating arms with an elastomeric spring and lateral restraint assembly, suspension leveling jacks, a tether assembly, and the NCU umbilical retractor (see Chapter 12). The new missile suspension system was not part of the Force Mod–Upgrade Silo effort at Ellsworth since there were no plans to place Minuteman III missiles there (Figure 16.7).[22]

## *Launcher Equipment Room*

The LER modifications included extending the shock-isolated floor on Level 1 to a full 180 degrees, upgrading the suspension system from mechanical to liquid springs, improving the environmental seal at the launch tube rattle space, miscellaneous modifications to the inner LER wall and ceiling, fixed-floor modifications, and changes to the ESA room.

The shock-isolated floor on the western side of the LER was essentially rebuilt using a welded box-beam structure. New liquid springs provided the shock isolation, and foam blocks provided lateral restraint between the outer wall, the platform, and the launch tube. The emergency power batteries and motor generator were removed from the Level 2 floor and suspended from the new shock-isolated floor (Figure 16.8).[23]

FIGURE 16.7  The current missile suspension system being lowered into an LF during the Rivet MILE program. *Adapted from Association of Air Force Missileers Collection.*

## Command Data Buffer

On 24 July 1968, Headquarters Air Force approved development of an LF processor and status authentication system that would upgrade the link between the LCCs and the missile guidance system. On 30 December 1968, this idea was incorporated into the Minuteman Command Control System (MICCS), a program to coordinate Minuteman II and III LFs and LCCs with the proposed ABM system so that offensive and defensive launches did not conflict. First deployment was set for March 1972 at Ellsworth and May 1972 at F. E. Warren. On 11 April 1969, budget cuts resulted in a reduction of funding and delay of IOC until March 1973. A year later, on 1 March 1970, Headquarters Air Force reduced the scope of the MICCS to remote targeting only and redesignated it as the Remote Retargeting/CDB system.

Further research at SAMSO indicated that the goal of remote targeting could be accomplished via a weapon system computer and data storage memory located solely at the LCCs, eliminating the need for new equipment at the LF at a significant reduction in program costs. At this point the program was limited to the Minuteman III facilities. Installation would take place concurrently with the Integrated Improvement Program, part of Force Mod, beginning at F. E. Warren.

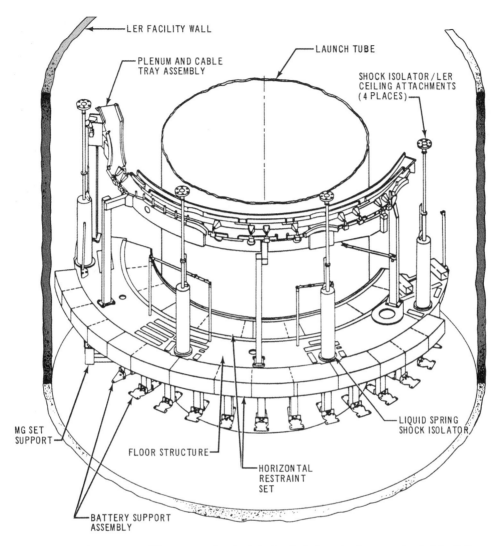

FIGURE 16.8 The major modifications to the shock-isolated platform in the upper level of the LER. The major protection was now for vertical shock, as the foam inserts on the interior and exterior of the platform prevented lateral movement. *Adapted from "Minuteman Service News," courtesy of the Boeing Company.*

The CDB upgrade meant that a target database would now be stored in the LCC computers instead of at the LF. This gave missile combat crews the ability to retarget the missiles in their squadron from the LCC instead of having to send a targeting team out to each LF. Now a missile no longer had to be taken off of strategic alert to accomplish retargeting. The new equipment at the LCC included a memory controller group to hold target information and a weapon systems controller that would prepare the new targeting data for transmission to the missile. In each LF and LCC, a secure data unit would ensure the transmission of the data in encrypted form. Retargeting was now simply a matter of data entry, taking approximately 15 hours for the entire 550 Minuteman III force.[24]

On 4 December 1972, CDB installation began at F. E. Warren, and the program was completed on 15 August 1977 at the 564 SMS. The equipment was also installed at Minuteman II LCCs as the ILCS.[25]

## Hybrid Explicit Flight Program

The Minuteman III HEFP decreased the number of memory words per target from 760 to 70 and decreased the time to retarget an individual sortie to seven minutes. HEFP simplified the development of, and significantly reduced the number of, targeting parameters for a specific launch site and target set. HEFP also increased the number of target sets and, by the modification of the missile steering mechanism, allowed a much larger off-azimuth launch angle, from 14 degrees to 45 degrees for an MRT launch (see Chapter 10).[26]

Initial deployment of HEFP took place at F. E. Warren and Minot between 19 July and 26 September 1976. The first installation was at F. E. Warren, beginning on 19 July 1976, but two days later, 12 missiles assigned to the 319th Strategic Missile Squadron (319 SMS) signaled a power shortage in their guidance system inertial measurement units and were taken off alert. By 24 July investigators had traced the malfunction to an incorrect sampling instruction in the operational ground software program, which falsely indicated a power outage. Corrected sampling instruction tapes were sent to F. E. Warren, and deployment was completed by 26 September 1976. Installation at all Minuteman III units was completed on 15 August 1977 at 564 SMS.[27]

## Integrated Improvement Program

The Integrated Improvement Program was the name given to the combination of Force Modernization, Upgrade Silo, EMP radiation hardening, HEFP modifications, and the CDB upgrade during the later stages of Force Modernization and encompassed Wings III, V, and VI and 564 SMS. This program began at F. E. Warren (see Table 16.2 for a summary of dates).[28]

## Project Restart

Tests by SAMSO at Malmstrom in 1971 had uncovered a susceptibility to an EMP, which could penetrate the WS-133B facilities at the 564 SMS and the 321 SMW through commercial power lines. Calculations determined that a high-altitude burst of a five-megaton weapon would generate an EMP of 25,000 to 50,000 volts per meter. The susceptible components in the LER included the battery charger and distribution box, while failure of susceptible components in the LSB would result in the missile's guidance-and-control system going into automatic restart or even shutdown. Critical design review was passed in September 1972. Fixes included installing braided sleeves for the power lines, adding filters, and sealing empty conduits.

The first phase of the program began at Grand Forks on 17 October 1973 and was completed with the turnover of LF Lima-22 to SAC control on 4 February 1975. Similar work occurred at 564 SMS, beginning with LF Tango-45 on 3 November 1976. Unfortunately,

TABLE 16.2 Integrated Improvement Program Dates[a]

| WING | DATES |
|---|---|
| 90TH STRATEGIC MISSILE WING, F. E. WARREN AFB | 4 Dec 1972–21 Jan 1975 |
| 91ST STRATEGIC MISSILE WING, MINOT AFB | 13 Sep1974–27 Feb 1976 |
| 564TH STRATEGIC MISSILE SQUADRON, MALMSTROM AFB | 3 Nov 1976–15 Aug 1977 |
| 321ST STRATEGIC MISSILE WING, GRAND FORKS AFB | 10 Oct 1976–23 Mar 1977 |

a) Strategic Air Command Weapon Systems Acquisition, 1964–1979.

the newly emplaced capacitors failed after a short time, causing damage to the electronics and removing LFs from alert. While a new design was being developed, blanks replaced the capacitors and the filter units to keep the facilities on alert.[29]

Facility testing of the fixes were carried out by Boeing using either the simulated electromagnetic ground environment (SIEGE) or the transient omnidirectional radiating unidistant and static simulator (TORUS) in 1974. Both SIEGE and TORUS were used successfully at operational LFs and LCCs.[30]

## Waterstop Modification Ellsworth

Many of the Minuteman LFs had problems with the groundwater created by thawing snow entering the LF and damaging critical equipment. The entry was normally through the rattle space between the launch tube and the LER Level 2 floor. The first attempt at a solution used a polysulfide material called Beta Seal to fill the rattle-space gap. Beta Seal served not only to prevent water intrusion but also cushioned against movement of the launch tube or LER during an attack.

Within two months of the program's completion, 25 of 139 LFs that had been modified had bulging Beta Seals caused by the groundwater pressure. If the seals burst, the subsequent release of water would cause damage to critical equipment, and if the water level was high enough—five inches—it would not only flow through weep holes into the launch tube but also enter the ESA vault and short out the commercial alternating-current power filters, disabling the launch tube sump pump. Even if the seals did not rupture, the bulging material could not serve to cushion the launch tube movement. Fortunately, the weather that year turned out to be the driest in South Dakota in 33 years. Unfortunately, the worst fears were realized when on 11 February 1975, the Beta Seal at LF Lima-11 failed and caused water to cascade onto the missile. The eventual solution was to place hardened foam between the Beta Seal and the EMP mesh, which provided a rigid backstop and prevented the rupture of the seal. This solution was also installed at both Minot and Grand Forks, both of which had groundwater issues in many of their LFs.[31]

## Minuteman III Dust-Hardening

Deployment of dust-hardened Minuteman III missiles began in December 1972 at F. E. Warren during the combined Force Modernization and Integrated Improvement

Modification programs. Dust-hardened missiles were then retrofitted into the rest of the force beginning on 17 September 1973 at Minot, 3 February 1975 at Grand Forks, and 20 January 1975 at Malmstrom. The program was completed on 11 July 1975 with the turnover of 564 SMS at Malmstrom (see Chapter 13).[32]

## Minuteman Integrated Life Extension Program

By mid-1984 Minuteman II missiles had been deployed for nearly 20 years and Minuteman III for nearly 14 years. The first major upgrade program, Force Mod, had been completed in July 1975. Logistical support for both systems was becoming increasingly difficult because several manufacturers had discontinued production of many of the Minuteman-unique items. On 15 December 1985, SAC, the Ballistic Missile Office, and Ogden Air Logistics Center (Ogden ALC) released the Minuteman Long-Range Plan. Within the plan was the Minuteman Integrated Life Extension Program (Rivet MILE). The intent behind Rivet MILE was maintaining the Minuteman II and III missiles and facilities through 2005 by integration of PDM and approved system modifications. Each wing would have a depot facility, preferably on base. A new support structure was developed for each of the wings, which would correct deferred or delayed discrepancies too difficult or complex for the missile wing support teams as well as correct hardness deficiencies and water intrusion problems. Teams from the AFLC conducted the work at all six wings simultaneously.

Rivet MILE was divided into three-year cycles: 1985 to 1988, 1988 to 1991, and 1991 to 1994. Rivet MILE 2010, begun in 1994, was designed to extend the weapon system through 2010. As the need for Minuteman was extended again, the latest Rivet MILE program, which began in 1996, was designed to extend the Minuteman III force to at least 2020. While Rivet MILE was concerned mainly with the improvement of LF and LCC equipment, other programs were folded into it as well to minimize the number of days that LFs and LCCs were off alert.

Rivet MILE began forcewide on 1 June 1985. By 18 November 1985, slightly more than 90 percent of the Cycle 1 tasks had been accomplished in 101 LFs and 16 LCCs.[33]

Modifications for incorporation into the force by Rivet MILE included the Improved Minuteman Physical Security System (IMPSS), the LCC EMP System, and the MESP. A major challenge in this program was scheduling maintenance, Stage II motor changeouts, and guidance system upgrades.[34]

## Minuteman Motor Programmed Depot Maintenance

By 1985 the Minuteman II and III Stage II motors were approaching their 17th year of strategic alert, and a large number of them had degraded liners, also known as the sticky liner problem (see Chapter 8). The Stage II motors had been undergoing PDM activity since the late 1970s. Stage II motors were removed before they reached the 17th year and sent to Ogden ALC, where the propellant was removed, the casings relined, and the stage refueled. In October 1985 the PDM process became part of the Rivet MILE program. By September 1985 356 of the Minuteman II Stage II motors had been refurbished, with only 97 remaining for the PDM process. In October 1985 Ogden ALC began the Minuteman

III Stage II motor PDM process. Since the Minuteman III Stage III had the same problem, Ogden ALC scheduled the PDM process for Stage III to begin in June or July 1986.[35]

## Propulsion Replacement Program

The Propulsion Replacement Program (PRP) was a key part of the effort to extend the service life of Minuteman III to 2020. While the Minuteman solid-propellant motors were much simpler systems than the liquid-propellant engines, unlike a liquid-propellant system, the solid propellants were susceptible to possible aging anomalies. There were two parts to the PRP: replacement of all the booster stages and replacement or repair of the PSRE.[36]

### *Booster PRP*

The booster PRP was necessary to correct problems with the aging propellant, which was hardening and cracking, as well as the already observed internal liner degradation with Stage II. Additionally, the program replaced materials restricted by the Environmental Protection Agency (EPA). Thiokol had monitored aging effects on Stage I and had not established a service life estimate. The Aerojet Stage II motors, which had been remanufactured by Aerojet from 1985 to May 1993, would begin to age out in 2002 for the earliest remanufactured units. The United Technologies Chemical System Division had remanufactured Stage III from 1987 to March 1993, and those manufactured first would end their life cycles in 2004.[37]

The first missile was delivered on 15 April 2001 at Malmstrom, and the program was completed with the installation of the last PRP Minuteman III in LF Echo-08, 341 SMW, on 18 August 2009, completing the program, which cost approximately $2 billion.[38]

### *PSRE Life Extension Program*

There had been 216 flight and 55 static tests of the PSRE since the beginning of the program; the last static firing, in 1985, had successfully tested a 15-year-old unit. While there had only been one in-flight failure, attributed to a manufacturing defect in the arm–disarm switch, potential failure modes like galvanic corrosion and chemical deterioration of the corrosive engine propellants as well as the O-ring seals led to the inclusion of the PSRE in the Minuteman III PRP.[39] In February 2000 TRW was awarded a $107-million, 10-year contract for the program; the contract was for the manufacture of 586 PSRE modification kits.

### *Minuteman III Reentry Vehicle Programs*

#### SINGLE REENTRY VEHICLE

The Single Reentry Vehicle (SRV) Program involved the deployment of new reentry vehicle platforms for Minuteman III. Two options were evaluated: use of mass simulators to replace the removed reentry vehicles on the existing bulkhead, along with necessary software changes, or relocating the SRV into the center of a new bulkhead, again with necessary software changes. The first option did not meet the criteria of START I, which required the destruction of the MIRV bulkhead. START II did not require the destruction of the MIRV bulkhead, but the Air Force decided to follow the START I criteria because not all the former Soviet republics had signed START II. Initially, the SRV program replaced Mark 12 reentry vehicles with the newer Mark 12A reentry vehicle, which had higher performance and additional weapon safety features. Originally, the Mark 12A was to be deployed on all 550 Minuteman III missiles in the early 1980s. Due to the additional weight of the larger Mark 12A warhead and a resulting decrease in range, deployment was limited to 300 Minuteman IIIs. With the pending deactivation of the Peacekeeper force, the Peacekeeper Mark 21 reentry vehicle was also considered for this program (see Chapter 2).[40]

SAFETY ENHANCED REENTRY VEHICLE

By 2005 the Department of Energy was in the process of decertifying the Mark 12 reentry vehicle by 2009. This prompted the DoD to approve the SERV program—the replacement of the Mark 12 reentry vehicle with the Mark 21 reentry vehicles from the Peacekeeper program. The program had originated as part of the National Defense Authorization Act for FY 1993, which stipulated "all warheads that will comprise the post-START II nuclear inventory be equipped with enhanced detonation safety systems, insensitive high explosives and fire-resistant pits." The impetus for the stipulation had come from the 1990 Drell Commission, which reviewed nuclear weapons safety. The Mark 21 warhead contained all recommended features.

The Safety Enhanced Reentry Vehicle/Warhead (SERV/W) Concept Action Group studied the problem and, in a November 1994 memo, recommended a mixed force of 350 Mark 21s and 150 Mark 12As, which would be modified to meet the new requirements. This mixed force would accommodate unforeseen problems with either the warheads or reentry vehicles, along with having the advantage of maintaining both reentry vehicles in the inventory in case the force had to be reconstituted due to treaty issues.[41] The first SERV-Minuteman III was deployed at F. E. Warren in December 2006 and placed on alert on 1 January 2007.[42]

## Minuteman III Guidance Replacement Program

On 29 July 1992, the DoD submitted to Congress the "Minuteman III Life Extension Report," which documented the feasibility and need for extending the Minuteman III program past the year 2010. With Minuteman II already deactivated and Peacekeeper forces in the process of deactivation, Minuteman III would be the remaining land-based component of the strategic triad. In an effort to not repeat the problems of the Minuteman II guidance system as it aged, the DoD proposed the GRP.

The Minuteman III NS20 MGS was continuously monitored during ground alert, and therefore detailed information on instrument performance provided the ability to identify system degradation before it became a problem. Originally planned to replace the entire NS20, in December 1992 the program was subsequently divided into two phases. Phase I would replace the aging electronics to correct anticipated availability, reliability, and maintenance problems based on experience with the Minuteman II system. This included replacement of the D37D computer, P92 amplifier, and MGS controller. The PIGA was the single high–failure rate item, and the report recommended an upgrade to the PIGA to increase field reliability. The gyrocompass assembly (GCA) bearing system would be replaced with an improved design as well as a new optical electrical resolver. These upgrades or replacements would serve to correct the effects of aging in the NS20 MGS. The updated NS20 was renamed the NS50. Phase II provided for the replacement of the inertial measurement unit to improve reliability and increase the accuracy of Minuteman III to that of the Peacekeeper system.[43]

Faced with the proposed reduction in FY 1995 funding, it was doubtful if the Air Force would have the funding for GRP. When the Air Force Space Command took over responsibility for the GRP, a review of the Phase I program was conducted that confirmed the validity of the program and thus the cost. The result was that funding was made available. However, on 25 June 1993, the General Accounting Office (GAO) issued a highly critical report of the Phase I plan, noting that the Minuteman III guidance system's reliability had improved over recent years and the failure rate was still acceptable, with no discernible increase in MTBF. Furthermore, the report was highly critical of the use of Minuteman II failure information to support the need for the GRP Phase I for Minuteman III.[44] An independent assessment was conducted by the Reliability Analysis Center, which effectively countered the GAO's arguments, and the Air Force began the GRP Phase I program (see Chapter 10).[45]

The first NS50 MGS was installed on 3 August 1999 in LF India-09, 341st Missile Wing, Malmstrom AFB. The GRP program at Malmstrom was completed on 4 December 2007 at LF Charlie-10. The 91st Missile Wing, Minot AFB, completed the NS50 installation in January 2008, and the 90th Missile Wing, F. E. Warren AFB, completed the fleetwide upgrade on 18 February 2008.[46]

## *ICBM Security Modernization Program*

This program was comprised of three subprograms: Fast B-Plug, LF Concrete Enhancement, and the Remote Visual Assessment Program.

### FAST B-PLUG

The B-plug is a piston-like security feature at the LFs to prevent unauthorized access to the LER using the PAS (see Chapter 6). The modification allowed the plug to rise 100 inches within 15 seconds of activation, rising to a final height of 148 inches and extending 12 lock bolts within 30 seconds to prevent unauthorized access.[47]

### LF CONCRETE ENHANCEMENT

This program called for extension of the existing concrete apron near the primary access cover to delay unauthorized access to the LF. Work began in May 2004 with a planned completion date of 2009.[48]

### REMOTE VISUAL ASSESSMENT

This program, also known as Prairie Hawk, consisted of the installation of two high-resolution television cameras at each LF to help security forces at the missile alert facility to determine if there were unauthorized personnel or hostile activities in the vicinity of the LFs. The new system prevented unnecessary dispatch of a security team if the outer zone system was triggered by birds or rabbits.[49]

The operational facilities were modified extensively due to two factors. The first was the need to accommodate the missiles as they were deployed, since Minuteman II and III missiles were physically different from the Minuteman I series. The second factor incorporated the first in many ways as changes were made to the LFs and, to a lesser extent, the LCFs to adapt to the new threats of improved performance of Soviet ICBMs (Table 16.3).

TABLE 16.3 Selected Minuteman Modification Completion Dates[a]

| MODIFICATION PROGRAM | WING/SQUADRON | | | | | | |
|---|---|---|---|---|---|---|---|
| | I | II | III | IV | V | VI | 564 SMS |
| FORCE MODERNIZATION WS-133A-M | 1969 | 1973 | 1971 | 1967 | 1975 | --- | --- |
| UPGRADE SILO | 1979 | 1973[b] | 1976 | 1980 | 1975 | 1977 | 1977 |
| HARDNESS MODIFICATION | 1979 | 1973 | 1976 | --- | --- | 1977 | 1977 |
| COMMAND DATA BUFFER | --- | --- | 1976 | --- | 1975 | 1977 | 1977 |
| HYBRID EXPLICIT FLIGHT PROGRAM | --- | --- | 1976 | --- | 1976 | 1977 | 1977 |
| IMPROVED LAUNCH CONTROL SYSTEM | 1979 | --- | --- | 1980 | --- | --- | --- |
| REACT | 1996 | --- | 1996 | --- | 1995 | --- | 1996 |
| SINGLE REENTRY VEHICLE | --- | --- | --- | --- | 2001 | --- | --- |
| GUIDANCE REPLACEMENT PROGRAM | --- | --- | 2008 | --- | 2008 | --- | 2007 |
| SAFETY ENHANCED REENTRY VEHICLE | --- | --- | 2010 | --- | 2010 | --- | 2010 |

a) Adapted from Minuteman Weapon System History and Description 2001, Hill AFB; Strategic Air Command Weapon Systems Acquisition 1964–1979; 90th Space Wing History, 1 January–31 December 1998; "History of the Air Force Space Command, 1 January 2007–31 December 2008"; date of completion of program; b) partial silo upgrade.

★ ★ ★

As discussed in Chapter 2, implementation of strategic arms limitation treaties, coupled with hard-to-support subsystems, led to the deactivation of Minuteman II and a reduction in the force level of Minuteman III. Implementation of these changes is the final chapter in the Minuteman weapon system story.

# 17

# FORCE REDUCTION

## Minuteman II Deactivation and Dismantlement

The first indication of the possible deactivation and dismantlement of the force of 450 Minuteman IIs took place on 19 November 1989, when the *New York Times* reported the Air Force plan. On 13 January 1990, the *Washington Post* reported that Secretary of Defense Dick Cheney had tentatively approved the Air Force proposal to deactivate Minuteman II. The deactivation would be contingent upon successful negotiations underway on a new strategic weapons treaty.[1] On 4 February 1991, Secretary of the Air Force Donald Rice and Air Force Chief of Staff Gen. Merrill McPeak announced that major strategic weapon changes were going to take place due to continued budget restraints and expectations of the new limitations coming from ongoing arms reduction treaty negotiations. It was the Air Force's intention to begin retiring Minuteman II missiles in FY 1992, which meant beginning on 1 October 1991.[2]

At this point only three bases still had Minuteman II missiles: Malmstrom, Whiteman, and Ellsworth. There were three major criteria that SAC used to evaluate which bases would be deactivated and dismantled or converted as a force consolidation move. The first was range. Malmstrom's LFs, on average, were located at 3,900 feet, Whiteman's at 800 feet, and Ellsworth's at 2,600 feet elevation. Therefore, the altitude advantage was significantly in Malmstrom's favor. Perhaps more importantly, Malmstrom was approximately 100 nm from the Canadian border, while Whiteman was 690 nm, and Ellsworth was 300 nm distant. The second criterion was the level of upgrades at each of the three wings. Whiteman and Malmstrom had received all the system upgrades to increase system hardness and retargeting capability, while, for reasons unclear in all my research, Ellsworth had received only some of the modifications. If SAC were to decide to add the modifications to the Ellsworth system and retain it, this would require an additional investment of over $1 billion, in addition to the cost for subsequent conversion of the system to handle Minuteman III. This was also the case, to a lesser extent, at Whiteman for conversion to Minuteman III. Malmstrom already had 50 Minuteman III missiles in the 564 SMS. While the 564 SMS facilities were significantly different from the Minuteman II facilities, trained Minuteman III missile combat crews and maintenance personnel, as well as maintenance facilities and equipment, already existed at Malmstrom. Thus, Ellsworth and Whiteman were chosen for deactivation and dismantlement (code name Rivet Dome II), and Malmstrom would undergo conversion from Minuteman II to III (code name Rivet Add). The decision to deactivate and dismantle Ellsworth before Whiteman was due to a congressional mandate,

Public Law 101–510, which stated, "The Secretary of the Air Force shall provide that the installation which receives the last operational upgrade for the Minuteman II missile system shall be the installation from which the last Minuteman II missile is retired."[3]

## 44th Missile Wing, Ellsworth AFB

On 19 April 1991, the Air Force published a Notice of Intent in the *Federal Register* to prepare an EIS for the deactivation of the Minuteman II system at Ellsworth. A town hall meeting in Rapid City, South Dakota, on 7 May 1991, provided an opportunity for input from the local communities, which generated a list of 19 major concerns. While most were relatively minor, several were of major import: disposition of the hardened intersite cable system, fences, gates, and azimuth markers; traffic issues with the heavy equipment necessary for the deactivation on the rural roads; potential impact on private and public wells from the demolition of the LF headworks; and impact on small electricity cooperatives and their customers from the loss of Air Force revenue from electricity use for the Minuteman II system.[4]

On 27 September 1991, President George H. W. Bush's announcement on national television of his decision to immediately deactivate the Minuteman II force of 450 missiles caught politicians by surprise.[5] START I, which had been signed on 31 July 1991, stipulated that deactivation and dismantlement of missiles to meet treaty requirements would be phased over a seven-year period beginning on 5 December 1994. Senators from the affected states questioned the wisdom of accelerating the deactivation process, which resulted in considerable debate prior to the ratification of the treaty on 1 October 1992.[6]

At 1536Z on 28 September 1991, all Minuteman II LCCs received a series of four EAMs from the secretary of defense via the Primary Alerting System with instructions to take all Minuteman II missiles off strategic alert and lock the SCS at each LF in the safe position.[7] The same day, teams of maintenance, security police, and operations personnel began working around the clock at Ellsworth, Malmstrom, and Whiteman to dissipate the launch codes in the LCCs and reconfigure the SCS, preventing rotation to the armed position.[8]

The final EIS for the deactivation of the Minuteman II force at Ellsworth was released on 1 October 1991. The deactivation process took place in four phases: Phase I consisted of removal of the reentry vehicles, guidance sets, and missile stages. Phase II entailed removal of the classified components, materials that would be hazardous or toxic wastes, and reusable components; the shutting and safing of all systems; and the placement of the LCFs and LFs into caretaker status. Phase III involved closure of the LCF waste disposal facilities, removal or closure-in-place of the underground storage tanks at both locations, and the deactivation and dismantlement of specific LF and LCF equipment. Phase IV involved the final disposal of the LFs and LCFs. Phases I and II were accomplished by the Air Force, Phase III by an outside contractor, and Phase IV was accomplished by the Air Force and the assisting federal agencies.

## Launch Facilities

Approximately four to six LFs per month were processed through Phases I and II. Deactivation began with the 67th Missile Squadron (67 MS), followed by the 66 MS, and ending with the 68 MS.[9]

### PHASE I REMOVAL OF THE MISSILE

The removal of the reentry vehicle and missile guidance system officially began the LF deactivation process. This phase did not involve any new equipment or procedures since the protocols for missile, missile guidance system components, and reentry vehicle transportation from the LFs were already in place as part of routine maintenance. Missile removal had been planned to start on 15 November 1991 but did not actually begin until the removal of the first missile from LF Golf-02 on 3 December 1991. The last missile was removed from LF Lima-07 on 7 April 1994.[10]

### PHASE II SELECTIVE REMOVAL OF FACILITY COMPONENTS

The removal of salvageable items took approximately 10 to 14 days, with classified equipment the first to be returned to base. The ballistic gas generators were salvaged for use at other bases. Fluids were drained from the guidance cooling and hydraulic systems in the LSB. Some of the brine chillers were removed for future use, with the rest being salvaged by the dismantlement contractors. The lead–acid batteries that served as backup power supply for the weapon system were removed for use on base or released to the Defense Reutilization and Marketing Office for resale. A limited number of the LSB diesel–electric generators were removed by the Air Force, with the rest salvaged by the dismantlement contractor. The diesel emergency startup batteries, lube oil tank, and diesel day tank were also removed. All power filters and capacitors containing polychlorinated biphenyls (PCBs) were removed. At the end of Phase II, the launcher closure was returned to the closed position and remained there while the site was in caretaker status awaiting demolition.

### PHASE III SITE DISMANTLEMENT

Before dismantlement activities began, each site underwent extensive sampling for potential contamination. Lead-based paint had been used on the launch tube, and cadmium electroplating had been used on the PAS seals. The sodium chromate solution used in the MGS cooling system and ethylene glycol brine solution used in the ground equipment cooling, as well as the PCBs found in various electronic equipment, were also major concerns.

Shortly after dismantlement began, PCBs were discovered in underground storage tank coatings. In May 1994 the Air Force, the EPA, South Dakota Department of Environment and Natural Resources, and the Army Corps of Engineers worked together to evaluate the extent of nonliquid PCB contamination at all the Minuteman II facilities at both Ellsworth and Whiteman. After extensive sampling at five LFs at Ellsworth, the decision was made to expand analysis of the LSB waterproof coating to include more than 100 LFs. The result was that approximately 20 percent of the reported values exceeded the action level of 50 parts per million (ppm) established for the sites. Since no established action level existed for nonliquid matrix PCBs, the 50-ppm limit was agreed upon by the agencies involved, and on

16 November 1995, a Federal Facilities Compliance Agreement was reached between the Air Force, the EPA, and the state agencies. The conclusion was that a low potential existed for PCBs in coating materials to impact surrounding soils. However, future development would be restricted at LF sites:

> Future real property transfer documents will contain deed restrictions that restrict future owners from performing subsurface development. Specifically, no drilling, excavation, trenching, or digging that exceeds two feet below existing grade shall be allowed without advanced approval by the appropriate EPA Regional Office. Installation of water wells will be prohibited.[11]

Surface-level features also were removed or demolished: the IMPSS antennas, area lighting poles, power poles, concrete markers, pull boxes, antenna bases, transporter-erector landing gear pads and similar concrete features, the launcher closure door and rail slab, and wing walls. The azimuth monuments were removed only at the landowner's request. The hardened intersite cable system remained in the ground. Marker posts and gates were removed at the discretion of the landowner. Receive-and-transmit antenna structures and the UHF antenna were removed and the openings sealed and abandoned in place. Water wells that were operating prior to deactivation were left for use by the owner as long as sampling revealed no contamination above the maximum contaminant level established by the EPA. This phase also included closure of the LCF waste disposal lagoon and removal or closure-in-place of the underground diesel storage tanks.[12]

The treaty stipulated two methods for demolition of the LF: mechanical excavation to a depth of no less than 27.4 feet or explosives to a depth of no less than 20.6 feet. All demolition of the LF headworks at Ellsworth was done with explosives. Fifty-six holes, varying in depth from 5.25 (on the LER roof) to 21.5 feet (LER walls) were drilled using an Ingersoll-Rand ECM 360 drilling rig positioned to drill between the reinforcing steel. The holes were then filled with a total of approximately 500 pounds of dynamite. The launcher closure had 45 holes drilled 2.8 feet in depth, filled with approximately 70 pounds of dynamite. Tests were run at both LF Golf-02 and LF Hotel-10 to finalize the headwork demolition protocol. After these tests a decision was made to simply bury the launcher closure door as it was not thick enough to properly contain the explosive energies required to fracture the door without launching the debris beyond the fence onto private property, which was not allowed by contract. The LSB demolition was done mechanically.[13]

To facilitate demolition, the upper launch tube was cut at the level of the lower LER floor and salvaged. The steel liner of the two LER levels was cut and scored to make removal easier after the blast. After the explosive demolition, the site structures were still recognizable, just broken up. Mechanical excavators were used to separate approximately 100 tons of steel and rebar and dump the concrete rubble down the remaining 53-foot-deep lower launch tube section. The steel was salvaged by the dismantlement contractor.

The treaty did not stipulate any details beyond the depth of the demolition by either means. Originally, a 26-foot-diameter, 2-foot-thick concrete-and-steel cap was to be placed over the launch tube, but the dismantlement contractor submitted a Value Engineering Proposal to replace the cast-in-place concrete with a two-piece precast cap for just the launch tube at the depth required by the treaty. The final work prior to the observation period was to excavate a 75-foot-diameter observation cone for satellite or on-site observation (Figures 17.1 and 17.2).[14]

The treaty timeline for dismantlement, 180 days, began when the launcher closure was left in the open position. The first LF implosion at Ellsworth took place on 4 April 1994, at LF Hotel-10. Demolition of the headworks and grading of the observation cone at LF Hotel-10 was completed on 31 May 1994, and the LF was returned to Ellsworth Civil Engineering on 3 October 1996. LF Oscar-07 was returned on 29 October 1997, marking the end of a nearly six-year process.[15]

## Launch Control Facilities

Deactivation of the LCCs took place after each squadron's LFs were in the final stages of dismantlement. Classified material was removed, as well as the lead–acid weapon system batteries, PCB-containing filters, and other potential contaminants. The dismantlement contractor was permitted to salvage items from the LCC, and then the blast door was welded shut. The vestibule in front of the LCC blast door and the entire elevator shaft were filled with rubble, sand, gravel, and dirt and compacted to within one to two feet of the top of the elevator shaft, then a reinforced concrete cap was placed over the shaft to prevent settlement and deny future access to the abandoned LCC structure. The air ducts from the environmental control system room to the LCC blast valves were filled and sealed with a two-foot cap of reinforced concrete.[16]

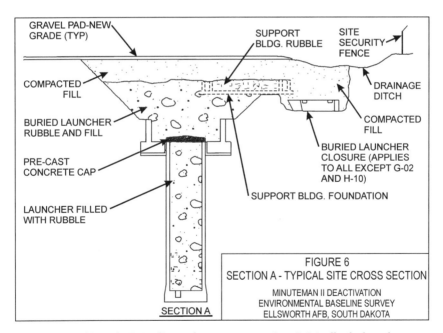

FIGURE 17.1 Typical LF site dismantlement cross-section. Originally, the launcher closure was to be destroyed by explosives, but this turned out to be inefficient after two attempts at LF Golf-02 and Hotel-10, so the launcher closure was buried instead. *Adapted from Golf-Flight Launch Facilities Environmental Baseline Survey, courtesy of Tim Pavek.*

FIGURE 17.2 The lower launch tube ready for concrete cap placement. The LER debris was removed and the reinforcing steel salvaged. The resulting excavation was to the level of the top of the lower launch tube. *Photograph courtesy of Tim Pavek.*

## *351st Missile Wing, Whiteman AFB, Missouri*

On 14 November 1991, the Air Force published a Notice of Intent in the *Federal Register* to prepare an EIS for deactivation of the Minuteman II system at Whiteman. A public hearing was held on 10 December 1991, where citizens voiced their concerns and questioned the Air Force about the process. On 13 January 1992, the JCS approved acceleration of the deactivation schedule, and the draft EIS was published on 17 April 1992. A second public hearing on the draft EIS was held on 19 May 1992.[17] On 1 May 1992, SAC issued Minuteman II Deactivation Plan, 351st Missile Wing.[18] The final EIS was published in August 1992, and community input was provided through meetings on 21 and 22 October 1992. Again, the major concerns at the public meetings were possible damage to water wells from the use of explosives in the demolition phase of the dismantlement as well as contamination due to water intrusion into the rubble-filled launch tubes and subsequent leaching of heavy metals and PCBs into the water table.[19]

Deactivation began on 28 September 1991 and was completed within 48 hours. Dismantlement followed the same four phases that had taken place at Ellsworth.[20] The first missile was removed from LF India-11 on 7 October 1992. The final missile pull was at LF Juliet-03 on 18 May 1995.[21] Dismantlement used the same four phases as Ellsworth.[22] The first demolition took place on 8 December 1993, at LF India-02.[23] In September 1994 analysis of waterproofing asphaltic coatings for the LFs and underground storage tanks revealed the presence of PCBs. Dismantlement was placed on hold pending a solution. The solution was a Federal Facilities Compliance Agreement between the Air Force, the EPA, and the Missouri Department of Natural Resources. The agreement listed restrictions on the property to avoid disturbing hazardous substances.[24] Demolition resumed in June 1995, and dismantlement was completed with the implosion of LF Hotel-11 on 15 December 1997.[25]

## *Conversion to Minuteman III*

On 15 April 1991, the Air Force announced the decision to retire the 150 Minuteman II missiles at Malmstrom and replace them with Minuteman III missiles. Malmstrom had earlier undergone the Force Modernization modifications, so the only modification needed to the LFs was reinforcement of the headworks to support the heavier Minuteman III missile during installation and removal operations. Timing for the conversion depended on progress with the ongoing START negotiations, as well as approval by President Bush and Congress.[26]

On 23 September 1991, SAC issued Program Plan 91-7, detailing plans to deactivate the Minuteman II system at Malmstrom and convert the LFs and LCCs for deployment of Minuteman III. No mention was made of the source of the Minuteman III missiles.[27] On 13 November 1991, the deactivation of Minuteman II began with the removal of the missile at LF Juliet-03, 23 miles northeast of Power, Montana. Deactivation was completed on 10 August 1995 with the removal of the last operational Minuteman II at LF Kilo-11 near Harlowton, Montana.[28]

The deployment of Minuteman III at the 341 MW began on 17 November 1992 when the first Minuteman III, previously stored at Malmstrom, was emplaced in LF Juliet-09, 12th Missile Squadron (12 MS), the first of 30 to be emplaced, prior to the transfer of 120 missiles from an as-yet-unnamed wing.[29] By June 1994 the initial conversion of 30 LFs of the 341st Missile Wing had been completed.[30] Conversion of the remaining 120 LFs began on 21 October 1995 and was completed with the emplacement of the last Minuteman III from Grand Forks in LF Bravo-08 on 15 May 1998. LF Golf-15, 447 SMS, relinquished the last of the 150 Minuteman IIIs from Grand Forks AFB on 4 June 1998.[31]

# Limited Deactivation of Minuteman III

### DEACTIVATION OF 321ST MISSILE GROUP, GRAND FORKS AFB

On 5 October 1995, after an 18-month delay due to congressional mandates and the decision of the Base Realignment and Closure Commission, the first of 120 Minuteman III missiles to be transferred from the 321st Missile Group was removed from LF Alpha-04, 446 MS, and readied for shipment to Malmstrom, with the remaining 30 shipped for temporary storage to Hill AFB and the Camp Navajo Depot, Arizona (see Chapter 2 for details on the deactivation decision).[32]

The available details on the deactivation and dismantlement of the 321st Missile Group's Minuteman III facilities indicate that they were similar to the process for Minuteman II at Ellsworth and Whiteman. Deactivation Phases I and II began with the 446 MS, followed by the 448 MS and 447 MS. Dismantlement was placed on hold as a result of the 18 June 1997 decision that the Grand Forks missile sites might be used for a National Missile Defense system.[33] Final dismantlement began with the first implosion on 6 October 1999 of LF Alpha-04 and was completed on 24 August 2001 with implosion of LF Hotel-22.[34]

### DEACTIVATION OF THE 564TH MISSILE SQUADRON, MALMSTROM AFB

Deactivation again took place in four phases, with the first three nearly identical to those used in the deactivation of Minuteman II at Ellsworth and Whiteman and Minuteman III at Grand Forks.[35] On 12 July 2007, the first of the 50 missiles was removed from LF Sierra-38. The final missile was removed from LF Tango-41 on 28 July 2008.[36] Phase IV involved excavating a pit in which to bury the launcher closure door. The launch tube was then filled with dirt and gravel and sealed with a concrete cap after a 60-day inspection (Figure 17.3.)[37]

★ ★ ★

The decision to deactivate Minuteman II in 1991, well before strategic arms limitation treaty requirements, raised the ire of the affected states' congressional leaders and began a move in Congress to place restrictions on similar decisions in the appropriation bills for the Defense Department. Realistically, however, Minuteman II was becoming hard to support and, with the apparent end of the Cold War, unnecessary. Dismantlement of 10 percent of the Minuteman III force was a cost-saving measure as well as one approach to meeting further strategic arms limitations treaty requirements (Table 17.1).

As of this writing, the Minuteman III force is scheduled to be replaced by a new system, the Ground-Based Strategic Deterrent (GBSD), by approximately 2030. If funded, the new system will maintain the land-based ICBM leg of the strategic triad. It remains to be seen what the fate of Minuteman III will be if GBSD becomes the victim of budget cuts.

FIGURE 17.3 The launcher closure door is pulled from the top of LF Romeo-29 on 25 February 2014 and buried in a nine-foot-deep hole. This was the first of the 50 Minuteman III LFs of the 564 SMS, 341 SMW, Malmstrom AFB, to undergo dismantlement in accordance with the New START treaty requirements. *Official USAF photograph, Senior Airman Katrina Heikkinen.*

TABLE 17.1  Minuteman Deployment and Deactivation Summary[a]

| MODIFICATION | WING/ SQUADRON | | | | | | |
|---|---|---|---|---|---|---|---|
| | I | II | III | IV | V | VI | 564 SMS |
| WS-133A DEPLOYMENT | 1963 | 1963 | 1964 | 1964 | 1965 | ------ | ------ |
| WS-133B DEPLOYMENT | ------ | ------ | ------ | ------ | ------ | 1966 | 1967 |
| MINUTEMAN II OPERATIONAL | 1969 | 1973 | ------ | 1967 | ------ | 1966 | 1967 |
| MINUTEMAN II DEACTIVATION | 1995[b] | 1994[c] | ------ | 1995[c] | ------ | ------ | ------ |
| MINUTEMAN III PARTIAL DEACTIVATION | ----- | ------ | ------ | ------ | ------ | 1998[c] | 2008[c] |
| MINUTEMAN III WARM SILO | 2016 | ------ | 2015 | ------ | 2017 | ------ | ------- |

a) Adapted from Minuteman Weapon System History and Description 2001; date of completion of program; b) last Minuteman III emplaced in launcher; c) last missile removed from launcher.

# EPILOGUE

The April 2010 Nuclear Posture Review Report stated that the Minuteman III Service Life Extension Program (SLEP) would, as mandated by Congress, "maintain a sufficient supply of launch test assets and spares to sustain the deployed force of such missiles through 2030."[1]

As pointed out in a RAND Corporation study of 2014, *The Future of the US Intercontinental Ballistic Missile Force,* one of many factors determining the service's life was the need for missiles for test launches to evaluate longevity, readiness, and reliability of the system. If the operational force was reduced to 420 missiles, the test inventory would be depleted by 2030. If the operational force were reduced to 400 missiles, the test inventory could be extended to 2035 and even further with reduction in number of test launches. The RAND study concluded that sustaining Minuteman III through additional SLEPs was "a relatively inexpensive way to retain ICBM capabilities." In reaching this conclusion, the RAND report compared alternatives such as Minuteman IV, V rail-mobile, and V road-mobile to continued support of modernized Minuteman III.[2]

Taking the RAND report and other internal studies into consideration, the Air Force concluded that by 2030, the 60-year-old Minuteman III force would be at the limits of sustainability. It made more sense to take advantage of current technology rather than try to recreate systems that are 35 to 50 years old. After several years of additional study, on 21 August 2017, the Air Force announced that Boeing ($349 million) and Northrop Grumman ($328 million) were selected over Lockheed Martin to conduct a Technology Maturation and Risk Reduction system design for the GBSD program. The contract covered a three-year period during which the Air Force was to solidify system requirements that were still in flux, encourage technological innovation, and reduce cost and schedule risks through proven designs and approaches to Minuteman III replacement.[3] One central factor was a desire to retain certain facets of the ground system infrastructure, which remain true national treasures pioneered during the system's development.[4]

On 13 December 2018, the Congressional Budget Office issued a report describing an option for reducing the deficit between 2019 and 2028. The report urged the cancellation of development and production of a new missile for the GBSD. Ground equipment improvement would take place, but the ground-based deterrent would end with the retirement of Minuteman III when the SLEP programs were no longer cost-effective. The major argument in support of this option was that the likelihood of a large-scale disabling nuclear strike was now much lower than during the Cold War. Additionally, sufficient warheads would be available as long as the SLBM force remained undetectable. The arguments against this option were that it would contribute to strategic instability as well as lead to nuclear proliferation if the retirement of the ICBM force led allies to develop their own nuclear arsenals.[5]

On 28 March 2019, in testimony before the House Armed Services Subcommittee on the Strategic Forces, Gen. John Hyten, commander, USSTRATCOM, reiterated that the

ICBM leg of the nuclear triad denies the adversary the opportunity to preemptively destroy our nuclear arsenal with a small-scale strike. Furthermore, "Our ICBMs create enormous targeting problems for our adversaries, requiring a massive raid that would be impossible to hide and would guarantee their own demise. With its range, payload, accuracy, and speed the ICBM is critical to our nation's deterrent strategy."[6]

On 16 July 2019, the Air Force released a request for proposals for the GBSD system engineering and manufacturing development phase, which included five production lot options to produce and deploy the weapon system.[7] On 25 July 2019, Boeing announced it was withdrawing from the bidding process because it did not believe it would be competing on a level playing field with Northrop Grumman, which had recently purchased Orbital ATK, a major producer of solid propellants.[8]

The first Minuteman III squadron, the 741st Strategic Missile Squadron (741 SMS), 91 SMW, Minot AFB, North Dakota, achieved full operational status on 29 December 1970.[9] The 49 years of deployment and relevance as a strategic nuclear deterrent force are a tribute to the officers, enlisted, civil servants, and contractors who devoted their skills and careers to continually upgrade, sustain, and develop emerging tactics for the most stabilizing element in our nation's nuclear triad. Just how the Minuteman facilities and the missiles themselves were extended to nearly five times their original service life is testimony to the design ingenuity of the engineers who developed, deployed, and continue to keep the system viable well into the twenty-first century. This in itself is another story that needs to be told.

# APPENDIX A

# CONSTRUCTION, ACCEPTANCE, AND ACTIVATION SUMMARIES

TABLE A.1  341 SMW, Malmstrom AFB, Construction and Acceptance Milestone Dates

| 10TH STRATEGIC MISSILE SQUADRON | FLIGHT | START[a] | COMPLETED[a] | TURNOVER TO SAC[b] |
|---|---|---|---|---|
| | A | 15 Mar 61 | 15 Dec 61 | 24 Oct 62 |
| | B | 1 Apr 61 | 2 Apr 62 | 30 Nov 62 |
| | C | 10 Apr 61 | 16 Apr 62 | 4 Jan 63 |
| | D | 19 Apr 61 | 30 Apr 62 | 9 Feb 63 |
| | E | 24 Apr 61 | 30 Apr 62 | 28 Feb 63 (operational)[c] |
| 12TH STRATEGIC MISSILE SQUADRON | | | | |
| | F | 26 Apr 61 | 1 Jun 62 | 6 Feb 63 |
| | G | 9 May 61 | 21 Jun 62 | 19 Mar 63 |
| | H | 23 May 61 | 3 Jul 62 | 27 Mar 63 |
| | I | 5 Jun 61 | 16 Jul 62 | 29 Apr 63 |
| | J | 25 Jun 61 | 24 Jul 62 | 15 May 63 (operational) |
| 490TH STRATEGIC MISSILE SQUADRON | | | | |
| | K | 8 Nov 61 | 10 Aug 62 | 5 Apr 63 |
| | L | 9 Dec 61 | 23 Aug 62 | 4 May 63 |
| | M | 8 Jan 61 | 31 Aug 62 | 25 Jun 63 |
| | N | 8 Feb 62 | 20 Sep 62 | 28 Jun 63 |
| | O | 8 Feb 62 | 27 Sep 62 | 3 Jul 63 (operational) |

a) "Historical Summary, 1 January–30 June 1963," Corps of Engineers Ballistic Missile Construction Office; b) "History of the Site Activation Task Force, Malmstrom AFB, MT, 1 January–3 July 1963"; c) date squadron became operational as well.

TABLE A.2   44 SMW, Ellsworth AFB, Construction and Acceptance Milestone Dates

| 66TH STRATEGIC MISSILE SQUADRON | FLIGHT | START[a] | COMPLETED[a] | TURNOVER TO SAC[b] |
|---|---|---|---|---|
| | A | 16 Sep 61 | 23 Mar 63 | 9 Sep 63 |
| | B | 21 Aug 61 | 24 Oct 62 | 24 Jun 63 |
| | C | 10 Sep 61 | 14 Feb 63 | 19 Aug 63 |
| | D | 29 Aug 61 | 8 Dec 62 | 30 Jun 63 |
| | E | 5 Sep 61 | 8 Jan 63 | 30 Jul 63 |
| 67TH STRATEGIC MISSILE SQUADRON | | | | |
| | F | 6 Oct 61 | 5 Feb 63 | 29 Aug 63 |
| | G | 2 Mar 62 | 16 Mar 63 | 12 Sep 63 |
| | H | 9 Mar 62 | 1 Apr 63 | 4 Oct 63 |
| | I | 27 Sep 61 | 25 Dec 62 | 23 Jul 63 |
| | J | 16 Oct 61 | 19 Jan 63 | 10 Aug 63 |
| 68TH STRATEGIC MISSILE SQUADRON | | | | |
| | K | 16 Nov 61 | 18 Jan 63 | 5 Sep 63 |
| | L | 28 Oct 61 | 26 Dec 62 | 15 Aug 63 |
| | M | 9 Mar 62 | 23 Feb 63 | 18 Sep 63 |
| | N | 2 Apr 62 | 9 Mar 63 | 11 Oct 63 |
| | O | 9 Apr 62 | 21 Apr 63 | 23 Oct 63 |

a) "Historical Summary, 1 January–30 June 1963," Corps of Engineers Ballistic Missile Construction Office; b) "History of the Site Activation Task Force, Ellsworth AFB, SD, July–October 1963."

TABLE A.3   455 SMW, Minot AFB, Construction and Acceptance Milestone Dates

| 740TH STRATEGIC MISSILE SQUADRON | FLIGHT | START[a] | COMPLETED[a,b] | TURNOVER TO SAC[b] |
|---|---|---|---|---|
| | A | 6 Feb 62 | 5 Feb 63 | 12 Oct 63 |
| | B | 18 Jan 62 | 9 Mar 63 | 6 Nov 63 |
| | C | 28 Feb 62 | 5 Apr 63 | 26 Nov 63 |
| | D | 23 Feb 62 | 7 May 63 | 31 Dec 63 |
| | E | 5 Apr 62 | 28 May 63 | 24 Jan 64 |
| 741ST STRATEGIC MISSILE SQUADRON | | | | |
| | F | 13 Feb 62 | 30 Mar 63 | 19 Nov 63 |
| | G | 14 Mar 62 | 12 Apr 63 | 24 Dec 63 |
| | H | 22 Mar 62 | 15 May 63 | 11 Jan 64 |
| | I | 19 Apr 62 | 7 Jun 63 | 4 Jan 64 |
| | J | 20 Apr 62 | 22 Jun 63 | 27 Feb 64 |
| 742ND STRATEGIC MISSILE SQUADRON | | | | |
| | K | 27 Mar 62 | 27 Apr 63 | 26 Nov 63 |
| | L | 29 Mar 62 | 29 May 63 | 17 Jan 64 |
| | M | 4 Apr 62 | 29 May 63 | 10 Feb 64 |
| | N | 20 Apr 62 | 28 Jun 63 | 4 Mar 64 |
| | O | 3 May 62 | 15 Jul 63 | 19 Mar 64 |

a) "Historical Summary 1 January–30 June 1963," Corps of Engineers Ballistic Missile Construction Office; b) "History of SATAF Detachment #21, 1 January–30 April 1964, Minot AFB, ND"; "History of SATAF Detachment #21, 1 July–31 December 1963, Minot AFB, ND."

### TABLE A.4  351 SMW, Whiteman AFB, Construction and Acceptance Milestone Dates

| 508TH STRATEGIC MISSILE SQUADRON | FLIGHT | START[a] | COMPLETED[a] | TURNOVER TO SAC[b] |
|---|---|---|---|---|
| | A | 29 May 62 | 30 Aug 63 | 21 Apr 64 |
| | B | 2 Apr 62 | 26 Jun 63 | 21 Feb 64 |
| | C | 12 Apr 62 | 13 Jul 63 | 16 Mar 64 |
| | D | 18 Apr 62 | 5 Aug 63 | 9 Apr 64 |
| | E | 24 Apr 62 | 26 Aug 63 | 30 Apr 64 |
| 509TH STRATEGIC MISSILE SQUADRON | | | | |
| | F | 3 May 1962 | 11 Sep 63 | 19 Jun 64 |
| | G | 16 Apr 62 | 27 July 63 | 25 May 64 |
| | H | 24 Apr 62 | 13 Aug 63 | 26 Mar 64 |
| | I | 18 Jun 62 | 26 Sep 63 | 23 Jun 64 |
| | J | 4 Jun 62 | 12 Oct 63 | 12 May 64 |
| 510TH STRATEGIC MISSILE SQUADRON | | | | |
| | K | 24 May 62 | 16 Sep 63 | 6 May 64 |
| | L | 4 Jun 62 | 12 Oct 63 | 12 May 64 |
| | M | 20 Jun 62 | 17 Oct 63 | 6 Jun 64 |
| | N | 30 Jun 62 | 7 Nov 63 | 30 Jun 64 |
| | O | 27 Jun 62 | 14 Sep 63 | 25 May 64 |

a) "History of Construction Activities and Contract Administration Phases Encountered by Whiteman Area Office Corps of Engineers during Construction of Minuteman Strategic Missile Wing IV"; b) Air Force Global Strike Command History Office, Barksdale AFB, LA.

## TABLE A.5  90 SMW, F. E. Warren AFB, Construction and Acceptance Milestone Dates

| 319TH STRATEGIC MISSILE SQUADRON | FLIGHT | START[a] | COMPLETE[a] | TURNOVER TO SAC[b] |
|---|---|---|---|---|
| | A | 25 Oct 62 | 7 Nov 63 | 24 Jul 64 |
| | B | 5 Nov 62 | 27 Nov 63 | 5 Aug 64 |
| | C | 5 Nov 62 | 10 Nov 63 | 21 Aug 64 |
| | D | 7 Nov 62 | 30 Dec 63 | 3 Sep 64 |
| | E | 13 Nov 62 | 9 Jan 64 | 18 Sep 64 |
| 320TH STRATEGIC MISSILE SQUADRON | | | | |
| | F | 16 Nov 62 | 23 Jan 64 | 9 Oct 64 |
| | G | 26 Nov 62 | 7 Feb 64 | 28 Oct 64 |
| | H | 1 Dec 62 | 18 Feb 64 | 13 Nov 64 |
| | I | 8 Dec 62 | 3 Mar 64 | 4 Dec 64 |
| | J | 30 Dec 62 | 16 Mar 64 | 18 Dec 64 |
| 321ST STRATEGIC MISSILE SQUADRON | | | | |
| | K | 19 Nov 62 | 26 Mar 64 | 15 Jan 65 |
| | L | 18 Dec 62 | 8 Apr 64 | 1 Feb 65 |
| | M | 21 Jan 63 | 13 Apr 64 | 16 Feb 65 |
| | N | 21 Jan 63 | 23 Apr 64 | 8 Mar 65 |
| | O | 26 Jan 63 | 30 Apr 64 | 22 Mar 65 |
| 400TH STRATEGIC MISSILE SQUADRON | | | | |
| | P | 7 Jan 63 | 25 May 64 | 6 Apr 65 |
| | Q | 23 Jan 63 | 3 Jun 64 | 26 Apr 65 |
| | R | 18 Jan 63 | 15 Jun 64 | 11 May 65 |
| | S | 10 Jan 63 | 26 Jun 64 | 2 June 65 |
| | T | 15 Jan 63 | 8 Jul 64 | 15 Jun 65 |

a) "Historical Summary, 1 January 1963–30 June 1963," Corps of Engineers Ballistic Missile Construction Office; b) Air Force Global Strike Command History Office.

TABLE A.6   321 SMW, Grand Forks AFB,
Construction and Acceptance Milestone Dates

| 446TH STRATEGIC MISSILE SQUADRON | FLIGHT | START[a] | COMPLETED[a] | TURNOVER TO SAC[b] |
|---|---|---|---|---|
| | A | May 64 | 11 Oct 65 | 22 Jun 66 |
| | B | | 22 Oct 65 | 16 Jul 66 |
| | C | | 30 Oct 65 | 5 Aug 66 |
| | D | | 26 Sep 65 | 3 Jun 66 |
| | E | | 10 Sep 65 | 17 May 66 |
| 447ST STRATEGIC MISSILE SQUADRON | | | | |
| | F | Mar 64 | 25 Aug 65 | 25 Apr 66 |
| | G | | 14 Mar 65 | 31 Oct 65 |
| | H | | 4 May 65 | 16 Dec 65 |
| | I | | 1 Aug 65 | 25 Jan 66 |
| | J | | 13 Aug 65 | 3 Apr 66 |
| 448TH STRATEGIC MISSILE SQUADRON | | | | |
| | K | Apr 64 | 10 Nov 65 | 22 Nov 66 |
| | L | | 12 Nov 65 | 21 Sep 66 |
| | M | | 4 Nov 65 | 1 Sep 66 |
| | N | | 18 Nov 65 | 14 Oct 66 |
| | O | | 24 Nov 65 | 2 Nov 66 |

a) Dates are for substantial completion, "History of Grand Forks Area during Construction of Wing VI Minuteman II ICBM"; b) "Minuteman Historical Summary, 1966–1967."


TABLE A.7   564 SMS, Malmstrom AFB,
Construction and Acceptance Milestone Dates

| 564TH STRATEGIC MISSILE SQUADRON | FLIGHT | START[a] | COMPLETED[a] | TURNOVER TO SAC[b] |
|---|---|---|---|---|
| | P | 8 Mar 65 | 12 Jul 66 | 18 Jan 67 |
| | Q | -- | 5 Aug 66 | 10 Feb 67 |
| | R | -- | 29 Aug 66 | 6 Mar 67 |
| | S | -- | 23 Sep 66 | 29 Mar 67 |
| | T | -- | 16 Oct 66 | 21 Apr 67 |

a) "History of Malmstrom Area during Construction of Collocated Squadron Number 20 Minuteman II ICBM Facilities"; b) "Space and Missile Systems Organization: A Chronology, 1954–1979"; "564th Strategic Missile Squadron Chronology."

TABLE A.8  Activation Summary for the 341 SMW and 564 SMS, Malmstrom AFB

| 341ST STRATEGIC MISSILE WING | |
|---|---|
| DATE | EVENT[a] |
| 15 Jul 1961 | 341 SMW activated. |
| 1 Dec 1961 | 10 SMS, 341 MIMS activated. |
| 13 Nov 1961 | First launch facility completed, accepted by SATAF. |
| 15 Dec 1961 | First flight of Minuteman IA LFs accepted by SATAF. |
| 1 Mar 1962 | 12 SMS activated. |
| 1 May 1962 | 490 SMS activated. |
| 23 Jul 1962 | First production Minuteman arrives at Malmstrom. |
| 27 Jul 1962 | First Minuteman emplaced in LF Alpha-09. |
| 17 Sep 1962 | Construction of 150 LFs and 10 LCCs completed. |
| 24 Oct 1962 | Alpha Flight turned over to SAC. |
| 27 Oct 1962 | Alpha Flight placed on modified strategic alert during Cuban Missile Crisis. |
| 11 Dec 1962 | Minuteman weapon system becomes operational with the acceptance of the second flight, beginning continuous alert. |
| Oct 1963 | 341 SMW is complete and on full alert. |

| 564TH STRATEGIC MISSILE SQUADRON | |
|---|---|
| DATE | EVENT[b] |
| 1 Apr 66 | 564 SMS activated. |
| 26 Oct 66 | Final inspection and acceptance by SATAF. |
| 21 Apr 67 | Declared operational by SAC. |
| 3 May 67 | The last Minuteman II was put on alert in LF Tango-41, and the 564 SMS, achieving full alert status, was declared fully operational on schedule. The deployment of the force of 1,000 Minuteman ICBMs was complete. This brought the strength of the 34l SMW at Malmstrom AFB to 200 Minuteman missiles and 50 Minuteman IIs, collocated with 150 Minuteman I missiles. |
| 5 May 67 | A formal turnover ceremony completed the deployment of the programmed 1,000 Minuteman force. The 564th was both the first (Atlas) and the last (Minuteman) operational ICBM squadron activated in the United States. |

a) "Space and Missile Systems Organization Chronology"; b) "564th Strategic Missile Squadron Chronology"; "Minuteman Historical Summary, 1966–1967."

## TABLE A.9 Activation Summary for the 44 SMW, Ellsworth AFB, and 455 SMW, Minot AFB

### 44TH STRATEGIC MISSILE WING

| DATE | EVENT[a] |
|---|---|
| 1 Jan 62 | 44 SMW activated. |
| 1 Jul 62 | 66 SMS activated. |
| 1 Aug 62 | 67 SMS activated. |
| 1 Sep 62 | 68 SMS activated. |
| 4 Apr 63 | First Minuteman IB operational missile emplaced. |
| 24 Jun 63 | 66 SMS Flight B accepted by SAC. |
| 30 Jul 63 | 66 SMS Flight B on alert. |
| 30 Sep 63 | 66 SMS declared operational. |
| 7 Oct 63 | 150th missile delivered. |
| 23 Oct 63 | 44 SMW accepted three weeks ahead of schedule. |
| 24 Oct 63 | 44 SMW declared fully operational by SAC. |

### 455TH STRATEGIC MISSILE WING

| DATE | EVENT[b] |
|---|---|
| 1 Nov 62 | 455 SMW activated. |
| 1 Nov 62 | 740 SMS activated. |
| 1 Dec 62 | 741 SMS activated. |
| 1 Jan 62 | 742 SMS activated. |
| 12 Jul 63 | Assembly and checkout begins. |
| 9 Sep 63 | First missile emplaced at LF Alpha-02. |
| 12 Oct 63 | 740 SMS Alpha Flight turned over to SAC. |
| 24 Jan 64 | 740 SMS turned over to SAC. |
| 27 Feb 64 | 741 SMS turned over to SAC. |
| 19 Mar 64 | 742 SMS turned over to SAC, 455 SMW accepted five days ahead of schedule. |
| 21 Mar 64 | 455 SMW declared operational by SAC. |

a) "From Snark to Peacekeeper: A Pictorial History of Strategic Air Command Missiles," "History of the Site Activation Task Force, Ellsworth AFB, SD, July 1963–October 1963," Strategic Air Command Missile Chronology, 1939–1973, "Space and Missile Systems Organization: A Chronology, 1954," "Minuteman Historical Summary, 1962–1963." b) "From Snark to Peacekeeper," "455th Strategic Missile Wing History, 1–30 November," *Minuteman Service News, May–June 1973.*

TABLE A.10  Activation Summary for the 351 SMW, Whiteman AFB, and 90 SMW, F. E. Warren AFB

| 351ST STRATEGIC MISSILE WING | |
|---|---|
| DATE | EVENT[a] |
| 1 Feb 63 | 351 SMW activated. |
| 1 May 63 | 508 SMS activated. |
| 1 Jun 63 | 509 SMS activated. |
| 1 Jul 63 | 510 SMS activated. |
| 20 Apr 64 | 508 SMS accepted by SAC. |
| 25 May 64 | 509 SMS accepted by SAC. |
| 30 Jun 64 | 510 SMS accepted by SAC. |
| | Wing IV accepted 22 days ahead of schedule, 351 SMW declared operational by SAC. |

| 90TH STRATEGIC MISSILE WING | |
|---|---|
| DATE | EVENT[b] |
| 1 Jul 63 | 90 SMW activated. |
| 1 Oct 63 | 319 SMS activated. |
| 2 Oct 63 | 319 SMS LF Alpha-6 turned over to SATAF. |
| 8 Jan 64 | 320 SMS activated. |
| 8 Apr 64 | 321 SMS activated. |
| 1 Jul 64 | 400 SMS activated. |
| 15 Jun 65 | 90 SMW accepted 22 days ahead of schedule. |
| 30 Jun 65 | 90 SMW declared operational by SAC. |

a) History of Construction Activities and Contract Administration Phases Encountered by Whiteman Area Office Corps of Engineers during Construction of Minuteman Strategic Missile Wing IV, Vol. I, Space and Missile Systems Organization Chronology; *Minuteman Service News, May–June 1973*. b) "90th Strategic Missile Wing History, September 1963, October–December 1964, April–June 1965," *Minuteman Service News, May–June 1973*.

TABLE A.11  Activation Summary for the 321 SMW, Grand Forks AFB

| DATE | EVENT[a] |
|---|---|
| 1 Nov 64 | 321 SMW activated. |
| 1 Feb 65 | 447 SMS activated. |
| 1 Jul 65 | 448 SMS activated. |
| 5 Aug 65 | First Minuteman II missile arrives at Grand Forks AFB. |
| 7 Aug 65 | First Minuteman II missile emplaced at LF Golf-15. |
| 10 Sep 65 | Last Minuteman II missile emplaced in Flight Golf, LF Golf-20. |
| 15 Sep 65 | 449 SMS activated. |
| 25 Apr 66 | 447 SMS declared fully operational by SAC. |
| 22 Nov 66 | 321 SMW declared operational by SAC. |
| 7 Dec 66 | Wing VI accepted 16 days ahead of schedule. |

a) 321st Strategic Missile Wing History, January–March 1965, 321st Strategic Missile Wing History, July–September 1965; "A History of Strategic Arms Competition, 1945–1972, Vol. 2," A Handbook of Selected US Weapon Systems; *Minuteman Service News, May–June 1973*.

# APPENDIX B

# FLIGHT TEST PROGRAMS

TABLE B.1  Minuteman IA Flight Test Program,
Atlantic Missile Range, 1961–1963[a]

| MISSILE # | LC | LAUNCH DATE | REMARKS |
|---|---|---|---|
| 1961 | | | |
| FTM 401 | 31A | 1 Feb 1961 | Successful launch and flight, 3,999 nm, reentry vehicle impacted 2.4 nm long and 0.4 nm left, near Ascension Island. |
| FTM 402 | 31A | 19 May 1961 | Successful launch, missile broke up at 98 seconds into flight, Stage II nozzle control unit failure. |
| FTM 403 | 31A | 27 Jul 1961 | Successful launch and flight until premature Stage III thrust termination, impact 425 nm short of target. |
| FTM 404 | 32B | 30 Aug 1961 | Unsuccessful launch, Stage I and II separation charges fired prematurely simultaneously with Stage II ignition just as the missile started to emerge from the silo. |
| FTM 405 | 32B | 17 Nov 1961 | Successful launch and flight, reentry vehicle impact 2,993 nm as planned. |
| FTM 406 | 31B | 18 Dec 1961 | Successful launch and flight, reentry vehicle impact at programmed target. |
| 1962 | | | |
| FTM 407 | 31B | 5 Jan 1962 | Successful flight, high trajectory, reentry vehicle impact at programmed midrange target. |
| FTM 408 | 31B | 25 Jan 1962 | Successful flight, high trajectory, reentry vehicle impact at programmed midrange target. |
| FTM 410 | 31B | 15 Feb 1962 | Successful flight, skirt removal test, reentry vehicle impact at programmed target, 3,389 nm. |
| FTM 409 | 31B | 8 Mar 1962 | Successful flight, reentry vehicle impact at programmed target, 3,000 nm. |
| FTM 411 | 31B | 22 Mar 1962 | Successful flight, reentry vehicle impact at programmed target, last of PAM telemetry series. |
| FTM 412 | 32B | 24 Apr 1962 | Successful launch, missile destroyed 52 seconds into flight, failure in aft dome closure, impacted 10 nm off Cape Canaveral. |
| FTM 413 | 32B | 11 May 1962 | Successful flight, reentry vehicle impact at programmed target. |
| FTM 414 | 32B | 18 May 1962 | Successful flight, reentry vehicle impact at programmed target. |
| FTM 415 | 32B | 8 Jun 1962 | Successful flight, reentry vehicle impact at programmed target. |

| MISSILE # | LC | LAUNCH DATE | REMARKS |
|---|---|---|---|
| 1962 (CONTINUED) | | | |
| FTM 416 | 32B | 29 Jun 1962 | Successful flight, reentry vehicle impact at programmed target. |
| FTM 418 | 32B | 9 Aug 1962 | Successful launch, flight terminated after 40 seconds due to premature separation command, launcher was Malmstrom configuration. |
| FTM 421 | 32B | 18 Sep 1962 | Successful flight, minimum range trajectory, reentry vehicle impact at 9A MILS Net (Antigua) target; missile had all operational hardware in Wing I configuration. |
| FTM 420 | 31B | 19 Sep 1962 | Successful flight, minimum-range trajectory, reentry vehicle impact at 9A MILS Net (Antigua) target; missile had all operational hardware in Wing I configuration. |
| FTM 422 | 31B | 17 Oct 1962 | Successful launch, missile destroyed eight seconds into flight, failure in guidance system; missile had all operational hardware in Wing I configuration. |
| FTM 421A | 31B | 19 Nov 1962 | Successful flight, reentry vehicle impact at programmed target, full range, missile had all operational hardware in Wing I configuration. |
| FTM 424 | 32B | 7 Dec 1962 | Successful flight, reentry vehicle impact at programmed target Broad Ocean Area Zone 3, full range; missile had all operational hardware in Wing I configuration, except Mark 11 reentry vehicle flown. |
| FTM 426 | 32B | 20 Dec 1962 | Successful launch, missile malfunction due to Stage II motor casing burn through just prior to Stage II–III separation, no destruct signal sent, reentry vehicle impacted at one-third range, missile had all operational hardware in Wing I configuration except for the Mark 11 reentry vehicle, launched with missile offset in silo for test purposes. |
| 1963 | | | |
| FTM 421B | 32B | 20 Feb 1963 | Successful launch and flight, reentry vehicle impact in target area 3,132 nm, 0.32 nm short and 0.41 nm right; Wing I operational configuration except for Mark 11 reentry vehicle, last Wing I configured missile launch from AMR. |

a) Except where noted, all flights carried the Mark 5 reentry vehicle, and all launch dates are local; "Minuteman Historical Summary, 1962–1963"; "Atlantic Missile Range Index of Missile Launchings: Supplement II FY 62, July 1961–June 1962."

TABLE B.2  Minuteman IB Flight Test Program, Atlantic Missile Range, 1962–1964[a]

| MISSILE # | LC | LAUNCH DATE | REMARKS |
|---|---|---|---|
| 1962 | | | |
| FTM 424 | 32B | 7 Dec 1962 | Successful launch and flight, programmed for impact in the Broad Ocean Area Zone Three, first flight of Wing II Mark 11 reentry vehicle. |
| FTM 423 | 31B | 14 Dec 1962 | Successful launch, first flight to Ascension Island 12 MILS Net, Wing II configuration except for Mark 5 reentry vehicle flown, Wing II launcher configuration. |
| 1963 | | | |
| FTM 423A | 31B | 7 Jan 1963 | Successful launch; midway through Stage III operation, a burn through of the aft closure occurred, resulting in the Mark 5 reentry vehicle impacting 1,990 nm short of the target; launcher was Wing II configuration. |
| FTM 419 | 32B | 23 Jan 1963 | Successful launch and flight, reentry vehicle impact at programmed target, Wing II Stage I and II motors, Wing I Stage III motor, Mark 11 reentry vehicle, Wing I launcher configuration. |
| FTM 425 | 31B | 18 Mar 1963 | Successful launch, normal flight until Stage III motor burn through at 127 seconds into flight, range safety destruct, Wing II interim configuration with cylindrical Stage I skirt. |
| FTM 419A | 32B | 27 Mar 1963 | Successful launch and flight, Wing II configured missile, impacted in target area. |
| FTM 425A | 31B | 10 Apr 1963 | Successful launch and flight, Wing II configured missile, impacted in target area, slightly short and left. |
| FTM 425B | 32B | 18 May 1963 | Successful launch, missile self-destructed 45 seconds into flight due to Stage I aft closure burn through. |
| FTM 428 | 31B | 28 May 1963 | Successful launch and flight, first flight of interim Wing II configuration with Stage III retro and tumble rockets, Stage III impacted 75 nm short of reentry vehicle impact. |
| FTM 427 | 32B | 5 Jun 1963 | Successful launch and flight, interim Wing II configuration, reentry vehicle impacted in Ascension Island Splash Net target area, 4,390 nm, slightly left and long of target. |
| FTM 429 | 31B | 27 Jun 1963 | Successful launch and flight, interim Wing II configuration, reentry vehicle impacted in Broad Ocean Area 4, 3,500 nm, slightly right and long of target. |
| FTM 431 | 32B | 1 Jul 1963 | Successful launch and flight, operational Wing II configuration except for Stage I, which was interim Wing II configuration, reentry vehicle impacted in Ascension Island Splash Net, 4,385 nm, slight short and to the left of target, first Minuteman to use MISTRAM tracking system. |
| FTM 430 | 32B | 16 Jul 1963 | Unsuccessful launch, electrical malfunction Stage I nozzle control unit beginning at 1.4 seconds, at 10 seconds self-destruct of Stage I, command destruct of Stage II and III. |

| MISSILE # | LC | LAUNCH DATE | REMARKS |
|---|---|---|---|
| 1963 (CONTINUED) | | | |
| FTM 432 | 31B | 24 Jul 1963 | Successful launch and flight, Wing II configuration, reentry vehicle impacted close to target, Broad Ocean Area Zone 2, 1,860 nm. |
| FTM 433 | 32B | 5 Aug 1963 | Successful launch, Wing II configuration, Stage III malfunction six seconds into Stage III flight due to propellant–engine case bond failure, successful test of dual target capability during Stage I flight. |
| FTM 435 | 31B | 27 Aug 1963 | Successful launch and flight with reentry vehicle impact in Broad Ocean Area Zone 2, 1,860 nm. |
| FTM 434 | 32B | 7 Nov 1963 | Successful launch; the missile performed an inside loop without breaking up due to an improperly connected Stage I nozzle control unit actuator. |
| FTM 446 | 31B | 13 Nov 1963 | Successful launch and flight, primary objectives met, first of the "Fly Three" series of Wing II missiles modified into intermediate Wing VI WS-133B configuration with Wing VI–type Stage II motor with liquid-injection thrust vector control. |
| FTM 447 | 32B | 18 Dec 1963 | Successful launch and flight, second "Fly 3" series of Wing II missiles modified into intermediate WS-133B configuration with Wing VI–type Stage II motor with liquid-injection thrust vector control; reentry vehicle impacted in the Ascension Island target area, 5,000 nm down range, slightly long and left. |
| 1964 | | | |
| FTM 438 | 32B | 16 Jan 1964 | Successful launch and flight, intermediate Wing IV configuration, first completely successful demonstration of dual target capability at AMR. |
| FTM 448 | 31B | 28 Jan 1964 | Successful launch and flight, last "Fly 3" series of Wing II missiles modified into intermediate WS-133B configuration with Wing VI–type Stage II motor with liquid-injection thrust vector control, impacted in the Ascension Island MILS Net target area. |
| FTM 436 | 32B | 12 Feb 1964 | Successful launch and flight, Wing IV configuration, trajectory shaped to provide near-horizontal missile attitude at thrust termination to provide for decoy impact in Ascension Island MILS Net, reentry vehicle impacted in target area, 5,000 nm, first flight with penetration aids. |
| FTM 437 | 31B | 25 Feb 1964 | Successful launch and flight, Wing IV configuration, trajectory shaped to provide near-horizontal missile attitude at thrust termination to provide for decoy impact in Ascension Island MILS Net, reentry vehicle impacted in target area, 5,000 nm. |

| MISSILE # | LC | LAUNCH DATE | REMARKS |
|---|---|---|---|
| 1964 (CONTINUED) | | | |
| FTM 439 | 32B | 27 Feb 1964 | Successful launch and flight with reentry vehicle impact in target area. |
| FTM 440 | 32B | 13 Mar 1964 | Successful launch and flight until thrust termination, Wing IV configuration, thrust termination port No. 3 failed to release, resulting in a 40-degree-per-second pitch rate, reentry vehicle impacted five nm short of target in Ascension Island MILS Net. |
| FTM 442 | 31B | 20 Mar 1964 | Successful launch and flight, interim Wing IV configuration, followed program trajectory in the reentry vehicle, impacted in the Ascension Islands MILS Net. |
| FTM 441 | 32B | 30 Mar 1964 | Successful launch and flight, interim Wing IV configuration, followed program trajectory in the reentry vehicle, impacted in the Ascension Islands MILS Net. |
| FTM 443 | 31B | 7 Apr 1964 | Successful launch and flight, interim Wing IV configuration, followed program trajectory in the reentry vehicle, impacted in the Ascension Islands MILS Net. |
| FTM 444 | 31B | 24 Apr 1964 | Successful launch and flight, interim Wing IV configuration, followed program trajectory in the reentry vehicle, impacted in the Ascension Islands MILS Net. |
| FTM 445 | 31B | 29 Sep 1964 | Successful launch, but at 165 seconds into flight, range safety commanded destruction of missile, last R&D launch of the Minuteman IB. |

a) Unless otherwise noted, all reentry vehicles were Mark 11; "Minuteman Historical Summary, 1962–1963, 1964–1965."

TABLE B.3  Minuteman II Flight Test Program, Atlantic Missile Range, 1964–1968[a]

| MISSILE # | LC | LAUNCH DATE | REMARKS |
|---|---|---|---|
| 1964 | | | |
| FTM 449 | 32B | 24 Sep 1964 | Successful launch and flight, primary objectives met, Mark 11 reentry vehicle impacted as planned. |
| FTM 450 | 32B | 29 Oct 1964 | Successful launch and flight, primary objectives met, Mark 11 reentry vehicle impacted as planned. |
| FTM 451 | 32B | 15 Dec 1964 | Successful launch and flight, primary objectives met, Mark 11 reentry vehicle impacted as planned in the Ascension MILS Net. |
| FTM 452 | 31B | 18 Dec 1964 | Successful launch and flight, primary objectives met, Mark 11 reentry vehicle impacted as planned, first test of high roll-rate capability. |
| 1965 | | | |
| FTM 453 | 31B | 28 Jan 1965 | Successful launch and flight, primary objectives met, Mark 11 reentry vehicle impacted as planned, second Roll Control System test for Stage III. |
| FTM 455 | 32B | 7 May 1965 | Successful launch and flight, primary objectives met, Mark 11 reentry vehicle impacted in Ascension Broad Ocean Area as planned. |
| FTM 456 | 32B | 25 May 1965 | Successful launch and flight, fully configured Wing VI missile, primary objectives met, Mark 11 reentry vehicle impacted as planned, carried first hardened subsystems except for the reentry vehicle. |
| FTM 457 | 32B | 3 Aug 1965 | Successful launch and flight, fully configured Wing VI missile, primary objectives met, Mark 11 reentry vehicle impacted as planned. |
| FTM 454 | 31B | 23 Aug 1965 | Successful launch and flight, Mark 11 reentry vehicle impacted in target area, a high counterclockwise roll rate for Stage I was successfully demonstrated, and roll perturbations for Stage II were also successfully demonstrated. |
| FTM 458 | 32B | 23 Sep 1965 | Successful launch and flight, primary objectives met, Mark 11 reentry vehicle impacted as planned. |
| FTM 459 | 31B | 1 Oct 1965 | Successful launch and flight, primary objectives met, Mark 11 reentry vehicle impacted as planned. |
| 1966 | | | |
| FTM 461 | 32B | 5 Jan 1966 | Successful launch and flight, primary objectives met, Mark 11 reentry vehicle impacted as planned. |
| FTM 460 | 31B | 10 Feb 1966 | Successful launch and flight, primary objectives met, Mark 11 reentry vehicle impacted as planned. |
| FTM 462 | 31B | 31 Mar 1966 | Successful launch and flight, primary objectives met, Mark 11 reentry vehicle impacted as planned. |
| FTM 463 | 32B | 8 July 1966 | Successful launch and flight, first flight of Mark 12 reentry vehicle at Cape Kennedy, reentry vehicle disintegrated due to spin reversal, minimum reaction time (MRT) launch. |

| MISSILE # | LC | LAUNCH DATE | REMARKS |
|---|---|---|---|
| 1967 | | | |
| FTM 464 | 32B | 17 Jan 1967 | Successful launch and flight; Mark 12 did not separate from Stage III until reentry but impacted close to target, MRT launch. |
| FTM 465 | 32B | 24 Feb 1967 | Successful launch and flight, primary objectives with powered flight met, Mark 12 test flight, no further information on reentry vehicle performance available, MRT launch. |
| FTM 466 | 32B | 17 Aug 1967 | Successful launch and flight, primary objectives with powered flight met, Mark 12 test flight, no further information on reentry vehicle performance available, MRT launch. |
| FTM 467 | 32B | 6 Nov 1967 | Successful launch and flight, primary objectives with powered flight met, Mark 12 test flight, no further information on reentry vehicle performance available, MRT launch. |

a) "Minuteman Historical Summary, 1964–1965, 1966–1967, 1967–1968"; the Atlantic Missile Range was renamed, effective 15 May 1964.

TABLE B.4  Minuteman III Flight Test Program, Air Force Eastern Test Range, 1968–1970[a]

| MISSILE # | LC | LAUNCH DATE | REMARKS |
|---|---|---|---|
| 1968–1969 | | | |
| FTM 201 | 32B | 16 Aug 1968 | Successful launch and flight to target, all objectives met, Block III missile guidance set. |
| FTM 202 | 32B | 24 Oct 1968 | Successful launch, thrust termination exhaust plume damaged the PBV. |
| FTM 203 | 31B | 26 Mar 1969 | Successful launch, premature thrust termination. |
| FTM 204 | 31B | 22 Apr 1969 | Successful launch and flight, all objectives met. |
| FTM 205 | 32B | 27 May 1969 | Successful launch and flight, all objectives met. |
| FTM 206 | 32B | 25 Jun 1969 | Successful launch, guidance system anomalies after thrust termination prevented reentry vehicles from impacting in target area |
| FTM 207 | 31B | 31 Jul 1969 | Successful launch and flight to target. |
| FTM 208 | 31B | 23 Sep 1969 | Successful launch and flight to target. |
| FTM 209 | 32B | 6 Nov 1969 | Successful launch and flight to target. |
| FTM 210 | 32B | 4 Dec 1969 | Successful launch, flight anomaly, no further details. |
| 1970 | | | |
| FTM 212 | 32B | 13 Mar 1970 | Successful launch and flight, no further details available, first Block IV flight. |
| FTM 213 | 32B | 3 Apr 1970 | Successful launch and flight, no further details available, Block IV. |
| FTM 214 | 32B | 29 Apr 1970 | Successful launch and flight, no further details available, Block IV. |
| FTM 215 | 32B | 27 May 1970 | Successful launch and flight, no further details available, Block IV. |
| SPECIAL TEST MISSILE PROGRAM | | | |
| STM 1E | 32B | 16 Sep 1970 | Successful launch and flight to target. |
| STM 3E | 32B | 2 Dec 1970 | Successful launch and flight to target. |
| STM 2E | 32B | 14 Dec 1970 | Successful launch and flight to target. |

a) "Minuteman Historical Summary, 1968–1969, 1970–1971"; All flights up to FTM 212 used the Block III missile guidance set; "Minuteman III Flight Test History," 20th Air Force History Office.

# APPENDIX C

# OPERATIONAL FLIGHT TEST AND EVALUATION PROGRAMS

TABLE C.1  Minuteman IA Follow-on Operational Test Program, Vandenberg AFB, 1965–1966[a]

| DATE OF LAUNCH/ EXERCISE | FTM | ATTEMPTED LAUNCH | DESCRIPTION |
| --- | --- | --- | --- |
| SHUTTLE TRAIN | FTM 677 | 24 August 1965 | Successful, CE 0.512 nm. |
| PILOT ROCK | FTM 509 | 25 August 1965 | Successful launch, Stage III failed to ignite due to failure of the arm–disarm switch. |
| ARCTIC HOLIDAY | FTM 523 | 16 November 1965 | Ground abort, failure of Stage II nozzle control unit. |
| GRAND RIVER | FTM 516 | 14 December 1965 | Successful, CE 1.01 nm. |
| ROCKY POINT | FTM 673 | 7 December 1965 | Ground abort, D17 computer failed to receive launch command. |
| SEA DEVIL | FTM 529 | 24 February 1966 | Successful, first salvo launch, CE 1.8 nm. |
| BROAD ARROW | FTM 629 | 24 February 1966 | Successful, first salvo launch, CE 2.1 nm. |
| TULIP TREE | FTM 569 | 21 March 1966 | Successful, CE 1.61 nm. |
| WHITE BOOK | FTM 649 | 25 March 1966 | Successful, CE 1.42 nm. |
| SAGE GREEN | FTM 710 | 16 May 1966 | Successful, CE 0.757 nm. |
| NIGHT STAND | FTM 622 | 31 May 1966 | Successful, CE 0.791 nm. |
| WHITE ARC | FTM 522 | 16 August 1966 | Data not available. |
| TOWN DOCTOR | FTM 621 | 22 August 1966 | Data not available. |
| RED SPIDER | FTM 571 | 20 September 1966 | Data not available. |
| GROVE HILL | FTM 578 | 3 October 1966 | Data not available. |

a) "History of First Strategic Aerospace Division, Strategic Air Command, January–June 1966," Vol. 1; Vandenberg AFB Launch Summary, 1958–2003.

TABLE C.2  Minuteman IB Demonstration and Shakedown Operation Program, Vandenberg AFB[a]

| DATE OF LAUNCH/ EXERCISE | FTM | ATTEMPTED LAUNCH | DESCRIPTION |
|---|---|---|---|
| ANSWERMAN | 854 | 30 October 1963 | Abort, the second launch console key did not work, not operationally significant, sortie would have launched upon expiration of War Plan Timer but was canceled. |
|  |  | 2 November 1963 | Hang fire, launcher environmental protection system door failed to lower into place, Vandenberg-specific, not considered an operational system failure. |
|  |  | 20 November 1963 | Hang fire, combat test launch instrumentation failure, nonoperational equipment failure. |
| BIG CIRCLE | 842 | 27 November 1963 | Successful flight, Mark 11 Mod 5B impacted in target area. |
| ANSWERMAN | 854 | 13 December 1963 | Successful flight, Mark 11 Mod 5B impacted in target area. |
| CLOCK WATCH | 883 | 30 December 1963 | Hang fire, launcher environmental protection system door failed to lower into place, Vandenberg-specific, not considered an operational system failure. |
|  |  | 11 January 1964 | Initial Launch command was made at 2256:58 Zulu 10 January, the Inhibit Launch command was sent, and the missile guidance set returned to normal strategic alert at 2301:23 Zulu, as expected, after which a Launch command was made at 2306:33 Zulu with the War Plan Timer set to 53.8 minutes, which was allowed to run out as per operational conditions followed by terminal countdown and launch. |
| DOUBLE BARREL | 870 | 16 January 1964 | Successful flight, Mark 11 Mod 5B impacted in target area. |
| ECHO HILL | 808 | 30 January 1964 | Successful flight, Mark 11 Mod 5B impacted in target area, RV failed to arm. |

a) "Demonstration and Shakedown Operation Test Program Final Report, LGM-30B Minuteman Weapon System."

TABLE C.3 Summary of Successful Minuteman IB OT Mixed Marble II Exercises[a]

| MARK 5 MOD 5B DENUCLEARIZED W59 2A WARHEAD | | | | | |
|---|---|---|---|---|---|
| SORTIE | FTM | LAUNCH DATE | BURST HEIGHT (FT) | | CE(NM) |
| | | | PLANNED | ACTUAL | |
| GEORGIA BOY | 973 | 9 July 1964 | 5,317 | 5,561 | 1.633 |
| ORANGE CHUTE | 848 | 6 November 1964 | 5,317 | 5,440 | 0.263 |
| MARK 11 MOD 3 DENUCLEARIZED W56 TYPE 2A WARHEAD | | | | | |
| PRONTO ROSE | 1142 | 8 February 1965 | 6,019 | 6,362 | 0.290 |
| SILVER CLOUD | 1115 | 18 May 1965 | 6,020 | 6,759 | 0.521 |
| STAR DUST | 1069 | 6 July 1965 | 6,020 | 6,261 | 0.833 |

a) "Minuteman B (LGM-30B) Operational Test Program Final Report, First Strategic Aerospace Division History July–December 1965."


TABLE C.4 Minuteman IB Operational Test Program, Mark 5 and Mark 11 Reentry Vehicle Recovery Flights to the Eniwetok Lagoon[a]

| MARK 5 | | | | |
|---|---|---|---|---|
| EXERCISE NAME | FTM | DATE OF LAUNCH | TARGET | DESCRIPTION |
| FIVE POINTS | 1007 | 7 July 1964 | Option 18 | Stage I nozzle control unit failure, range safety destroyed at T+8.5 seconds.[b] |
| OLD FOX | 861 | 13 July 1964 | Option 18 | Successful launch and flight, Mark 5 Mod 5A with modified spin fins and EWO training kit successfully recovered, 0.17 nm right and 0.86 nm long. |
| MARK 11 | | | | |
| ROSY FUTURE | 1057 | 18 December 1964 | Option 18 | Successful launch and flight, Mark 11 Mod 5B, ballast training kit, 0.329 nm right, 0.835 nm short, successfully recovered. |

a) "Minuteman B (LGM-30B) Operational Test Program Final Report, First Strategic Aerospace Division History July–December 1965"; (b) The FIVE POINTS flight test report gives the reentry vehicle as a Mark 11 reentry vehicle, while the Operational Training program summary indicates in the test objectives that FIVE POINTS was used with a Mark 5 reentry vehicle for recovery; "Operational Test Exercise Report: FIVE POINTS, Sortie Number 99-64-MB-13, Minuteman 'B' LGM-30B AFSN 64-272/FTM 1007, 29 February 1964."

TABLE C.5  Minuteman IB Follow-on Operational Test Program Impact Results[a]

| EXERCISE NAME | FTM | DATE OF LAUNCH | RV | DOWN-RANGE (NM) | CROSS-RANGE (NM) | CE (NM) |
|---|---|---|---|---|---|---|
| ANCHOR POLE | 1243 | 27 January 1966 | Mark 11 | 2.383 long | 0.080 left | 2.384 |
| CREEK BED | 1387 | 12 February 1966 | Mark 11 | 0.403 short | 0.388 right | 0.559 |
| BAIT CAN | 1362 | 8 March 1966 | Mark 11A | 1.062 long | 0.142 right | 1.072 |
| CLEAN SLATE | 1211 | 11 March 1966 | Mark 11A | 0.099 short | 0.135 right | 0.168 |
| ARROW FEATHER | 1258 | 4 April 1966 | Mark 11 | 0.151 short | 0.387 right | 0.460 |
| FLY BURNER | 1347 | 4 April 1966 | unavailable | 0.972 long | 0.189 right | 0.990 |
| GAY CROWD | 1312 | 15 April 1966 | Mark 11A | 0.337 long | 0.418 right | 0.537 |
| ECHO CANYON | 1252 | 22 April 1966 | Mark 11A | 0.522 long | 0.198 right | 0.558 |
| LACE STRAP | 1388 | 2 May 1966 | Mark 11A | 0.152 short | 0.157 right | 0.218 |
| DOCK WORKER | 1191 | 16 May 1966 | Mark 11 | 0.456 short | 0.246 left | 0.518 |
| TIGHT DRUM | 1360 | 17 May 1966 | Mark 11A | 1.134 short | 0.667 left | 1.321 |
| GREEN PEA | 1270 | 31 May 1966 | Mark 11A | 0.096 long | 0.229 right | 0.248 |
| FOUR ACES | 1240 | 2 June 1966 | Mark 11A | 2.269 short | 0.526 left | 2.239 |
| EBONY ANGEL | 1261 | 10 June 1966 | unavailable | 0.985 short | 0.298 right | 1.029 |

a) "History of the First Strategic Aerospace Division, Strategic Air Command, January–June 1966"; "Advanced Missile Signature Center Database of ICBM Reentry Vehicles"; "History of the First Strategic Aerospace Division, January–June 1966."

TABLE C.6  Reentry System Launch Program, Minuteman IB Flight List[a]

| PROGRAM | DATES |
|---|---|
| System Technology Reentry Experiments Program (STREP) | 8 October 1982–10 June 1984 |
| Large Ballistic Reentry Vehicle (LBRV) | 7 January 1983 |
| Homing Overlay Experiment (HOE) | 7 February 1983–10 June 1984 |
| ICBM Penetration Aids (IPA) | 18 October 1984–17 March 1986 |
| Technology Demonstration Maneuvering Reentry Vehicle (TDMaRV) | 19 January 1988 |
| Maneuvering System Technology (MaST) | 14 February 1990 |
| Airborne Surveillance Testbed (AST) | 25 October 1992 |
| Target Development Test (TDT) | 15 June 1993 |

a) Vandenberg AFB Launch Summary, 1958–2003.

TABLE C.7 Wing VI Minuteman II Research and Development Flight Tests[a]

| CATEGORY I, II | | | | | |
|---|---|---|---|---|---|
| EXERCISE NAME | FTM | LF | LAUNCH DATE | RV | DESCRIPTION |
| REBEL RANGER | 2001 | 21 | 18 Aug 65 | 11A | Cat. I, successful flight to target. |
| DICE SPOT | 2002 | 22 | 6 Oct 65 | 11 | Cat. I, successful flight to target. |
| LOW TREE | 2009 | 21 | 9 Nov 65 | 11A | Cat. I, successful flight to target, heatshield test flight. |
| PUSH PULL | 2017 | 24 | 16 Dec 65 | 11 | Cat. II, successful flight to target. |
| RESTLESS DRIFTER | 2053 | 26 | 18 Jan 66 | 11A | Cat. II, successful flight to target, ground-burst fusing. |
| CALAMITY JANE | 2053 | 25 | 16 Feb 66 | 11A | Cat. II, successful flight to target, air-burst fusing. |
| GIN BABY III | 2386 | 05 | 17 Nov 67 | 11B | Cat. I, Mark I penetration aid development flight, successful. |
| GIN BABY V | 2435 | 05 | 28 Dec 67 | 11B | Cat. I, Mark I penetration aid development flight, successful. |
| MARK 12 DEVELOPMENT | | | | | |
| FAINT CLICK | 2049 | 08 | 17 Mar 66 | 12 | Mark 12 development flight, successful launch and flight, JLGM-30F, reentry vehicle failure due to overheated nose tip. |
| TATTERED COAT | 2045 | 08 | 26 Jul 66 | 12 | Mark 12 development flight, successful launch and flight, JLGM-30F, reentry vehicle failure due to overheated nose tip, spin reversal. |
| BUSY FELLOW | 2095 | 21 | 11 May 67 | 12 | Mark 12 development flight, successful launch and flight, JLGM-30F, first successful reentry vehicle performance. |
| BUSY JOKER | 2337 | 21 | 15 Jul 67 | 12 | Mark 12 development flight, successful launch and flight, JLGM-30F, reentry vehicle performance unavailable. |
| BUSY LOBBY | 2071 | 08 | 21 Nov 67 | 12 | Mark 12 development flight, successful launch, in-flight anomalies, JLGM-30F, reentry vehicle performance unavailable. |

a) "Minuteman Historical Summary, 1966–1967"; "Chronology of the Ballistic Missile Organization, 1945–1990"; Advanced Missile Signature Center Database of ICBM Reentry Vehicles, February 2001; Minuteman Flight Test Record, Vandenberg AFB History Office.

TABLE C.8  Wing VI Minuteman II Demonstration
and Shakedown Operation Flight Tests[a]

| EXERCISE NAME | FTM | LF | LAUNCH DATE | RV | DESCRIPTION |
|---|---|---|---|---|---|
| STAR BRIGHT | 2061 | 24 | 1 Aug 66 | 11A | Successful flight to target, remote target change achieved, system rejected a Launch command deliberately sent before Launch Enable commanded, reentry vehicle fell short. |
| CAREER GIRL | 2056 | 23 | 26 Aug 66 | 11A | Successful flight to target, reentry vehicle fell short. |
| GOLDEN AGE | 2068 | 22 | 2 Nov 66 | 11A | Range destruct at T+9 seconds. |
| OLYMPIC TRIALS 1 | 2137 | 22 | 21 Apr 67 | 11A | Successful flight, accuracy data not available. |
| OLYMPIC TRIALS 2 | 2122 | 24 | 19 May 67 | 11A | Successful flight, accuracy data not available. |
| OLYMPIC TRIALS 3 | 2370 | 22 | 20 Jul 67 | 11B | Successful flight, accuracy data not available. |
| OLYMPIC TRIALS 4 | 2445 | 24 | 21 Oct 67 | 11B | Successful flight, accuracy data not available. |
| OLYMPIC TRIALS 5 | 2457 | 24 | 23 Dec 67 | 11B | Successful flight, accuracy data not available. |
| OLYMPIC TRIALS 6 | 2437 | 25 | 10 Jan 68 | 11B | Successful flight to impact, improved accuracy. |
| OLYMPIC TRIALS 7 | 2448 | 22 | 25 Jan 68 | 11B | Successful flight to impact, improved accuracy. |
| OLYMPIC TRIALS 8 | 2464 | 24 | 2 Feb 68 | 11B | Successful flight to impact, improved accuracy. |
| OLYMPIC TRIALS 9 | 2522 | 25 | 29 Mar 68 | 11B | Successful flight to impact, improved accuracy. |
| GIANT BLADE 1 | 2503 | 22 | 30 Apr 68 | unk | Launch abort due to Stage II, safe-arm circuit failure. |
| GIANT BLADE 2 | 2556 | 25 | 21 Nov 68 | unk | Successful launch, Stage III in-flight failure. |

a) "Minuteman Historical Summary, 1966–1967"; "Advanced Missile Signature Center Database of ICBM Reentry Vehicles"; Minuteman Flight Test Record, Vandenberg AFB History Office.

TABLE C.9  Wings I, II, IV Minuteman II Force Modernization Flight Tests[a]

| CATEGORY I | | | | | |
|---|---|---|---|---|---|
| EXERCISE NAME | FTM | LF | LAUNCH DATE | RV | DESCRIPTION |
| SUPREME CHIEF | 2022 | 05 | 22 Jan 66 | 11 | Cat. I, successful launch, Stage II control problem during flight. |
| FAINT CLICK | 2049 | 08 | 17 Mar 66 | 12 | Cat. I, successful flight to target, reentry vehicle broke up at reentry. |
| FOX TRAP | 2083 | 08 | 24 Jun 66 | 11A | Cat. I, successful flight to target. |
| TATTERED COAT | 2045 | 08 | 26 Jul 66 | 12 | Cat. I, successful flight to target, reentry vehicle broke up at reentry. |
| BONUS BOY | 2113 | 08 | 11 Jan 67 | 12 | Cat. I, successful flight to target, reentry vehicle broke up at reentry. |
| EMERGENCY ROCKET COMMUNICATION SYSTEM | | | | | |
| WATER TEST | 2036 | 05 | 13 Dec 66 | ERCS | Cat. I, successful flight, first 494L flight. |
| SYCAMORE TREE | 2087 | 05 | 2 Feb 67 | ERCS | Cat. I, successful flight, 494L flight. |
| BUSY MISSILE | 2143 | 08 | 17 Apr 67 | ERCS | Cat. I, successful flight, 494L flight. |
| DEMONSTRATION AND SHAKEDOWN OPERATIONS | | | | | |
| GIANT FIST 1 | 2589 | 04 | 8 Jul 68 | unk | Successful flight to target. |
| GIANT FIST 2A | 2191 | 04 | 24 Oct 68 | unk | Successful flight to target. |
| GIANT FIST 4 | 2537 | 05 | 13 Nov 68 | unk | Successful launch, Stage III anomaly. |
| SSTM 050 | 2504 | 02 | 29 Jan 69 | 11B | Successful flight to target, test of modified Stage III nozzle control unit. |
| GIANT FIST 5 | 2208 | 04 | 2 Feb 69 | 11C | Successful flight to target. |
| SSTM 051 | 2252 | 05 | 7 Mar 69 | 11C | Successful flight to target, test of modified Stage III nozzle control unit. |
| SSTM 052 | 2287 | 25 | 18 Apr 69 | unk | Successful flight to target, test of modified Stage III nozzle control unit. |
| GIANT FIST 3 | 2642 | 04 | 12 Mar 69 | 11C | Successful flight to target. |

a) "Minuteman Historical Summary, 1966–1967"; "Minuteman Historical Summary, 1968–1969"; "Advanced Missile Signature Center Database of ICBM Reentry Vehicles"; "Vandenberg AFB Launch Summary, 1958–2003."

TABLE C.10  Minuteman II Operational Test and Special Test Program Summary[a]

| PHASE I | | # | COMMENTS |
|---|---|---|---|
| | Improved Minuteman | 25 | Includes 10 ALCS flights. |
| | Force Modernization | 13 | Includes 1 ALCS flight. |
| | ERCS | 1 | |
| | Special Tests | | |
| |    Mark 1A penetration aids | 3 | |
| |    SAMSO SSTM 053 | 1 | |
| |    SAMSO SSTM 060, 061 | 2 | |
| Total | | 45 | |

| PHASE II | | # | COMMENTS |
|---|---|---|---|
| | Improved Minuteman | 6 | |
| | Force Modernization | 55 | |
| | ERCS | 8 | Emergency Rocket Communication System |
| | Special Tests | | |
| |    GIANT PATRIOT | 2 | Test of Operational Base Launch Safety System. |
| |    Ogden Special Launch Targets | 1 | |
| |    MSLS | 10 | Multi-Service Launch System |
| Total | | 82 | |

a) McDowell, "JSR Launch Vehicle Database, 2019 Dec 28 Edition."

# APPENDIX D

## AIRBORNE LAUNCH CONTROL CENTER PANELS

These pages from technical order manual T. O. 21M-LGM30F-1-7, Airborne Launch Control System Operation Instructions, give a detailed description of the ALCS launch control panels and illustrate the location of the ALCS ERCS Control Panel.

Permission to use received from Global Strike Command 13 November 2019.

T.O. 21M-LGM30F-1-7

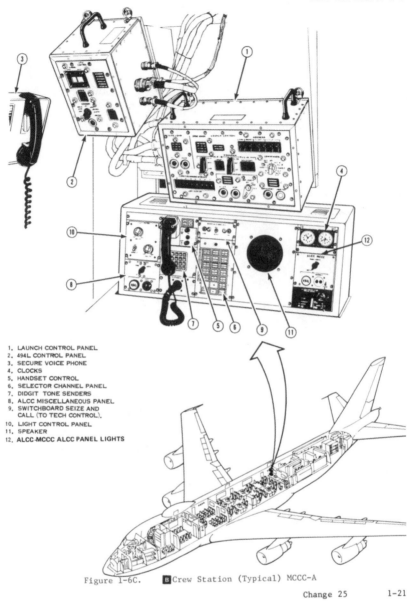

1. LAUNCH CONTROL PANEL
2. 494L CONTROL PANEL
3. SECURE VOICE PHONE
4. CLOCKS
5. HANDSET CONTROL
6. SELECTOR CHANNEL PANEL
7. DIDGIT TONE SENDERS
8. ALCC MISCELLANEOUS PANEL
9. SWITCHBOARD SEIZE AND CALL (TO TECH CONTROL).
10. LIGHT CONTROL PANEL
11. SPEAKER
12. ALCC-MCCC ALCC PANEL LIGHTS

Figure 1-6C.  B Crew Station (Typical) MCCC-A

T.O. 21M-LGM30F-1-7

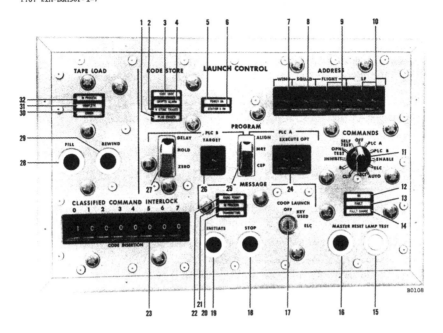

| NO. | CONTROL/INDICATOR | FUNCTION |
|---|---|---|
| | CODE STORE group | |
| 1 | PLUG ERASED warning light (red) | Illuminates when no volatile keying assembly is installed in code processor or when the memory of either VKA installed in code processor has been erased. Warning light will be out whenever both VKAs, with memories intact, are installed in code processor. |
| 2 | V STORE ERASED warning light (red) | Illuminates when code processor has successfully performed an erased test with an all-zero cryptovariable in storage. Code processor automatically clears cryptovariable register and performs an erased test when power is first turned on. Cycling code processor power after turn-on has the same result. An erased test can be commanded manually at code control panel. |
| 3 | CODE USED warning light (red) | The CODE USED warning light is not used however light may illuminate intermittently. |
| 4 | CRYPTO ALARM warning light (red) | Illuminates when: (1) code processor equipment malfunctions, (2) verify test fails, (3) sum check fails, (4) an erase test fails, or (5) crypto alarm override is in effect with both ALCC switch and CRYPTO ALARM OVERRIDE switch on and CLASSIFIED COMMAND CONTROL switch positioned to ENABLE. |

Figure 1-12. Launch Control Panel (Sheet 1 of 4)

1-30  Change 17

T.O. 21M-LGM30F-1-7

| NO. | CONTROL/INDICATOR | FUNCTION |
|---|---|---|
| | Power and interlock logic group | |
| 5 | POWER ON indicator (green) | Illuminates when ALCC power supply is on. |
| 6 | STATION 3 ON caution light (amber) | Illuminates when ALCC switch is ON with CLASSIFIED COMMAND CONTROL switch positioned to ENABLE. |
| 7 | WING selector | Used to insert binary address code into launch control panel register. Positions are OFF and numerals 1 through 7. |
| 8 | SQUAD selector | Used to insert binary address code into launch control panel register. Code is used to determine correct two-frequency tone in message address. Positions are OFF and numerals 1 through 4. |
| 9 | FLIGHT selectors (two) | Used to insert binary address into launch control panel register for one of five flights within wing and squadron. Left-hand selector positions are OFF and letters A through J. Right-hand selector positions are OFF and letters K through T. |
| 10 | LF address selectors | Used to insert binary address code into launch control panel register. Address of individual LF or ALL-CALL address (entire squadron) may be selected. Left-hand (tens) selector positions are OFF, numerals 0 through 5, and ALL. Right-hand selector positions are OFF and numerals 0 through 9. |
| 11 | COMMANDS select switch | |
| | ECT position | Used with exercise program to select launch control panel test. |
| | CLIP position | Used to select CLIP command with operational program. |
| | B position | Not used. |
| | INHIBIT position | Used to select inhibit command with operational program. |
| | OPNL TEST position | Used with both programs to select a test air-to-ground message to all Minuteman and Peacekeeper missiles for check of air-to-ground communication capability. Message is addressed to a squadron and response is monitored at all LCCs in squadron. |
| | SELF-TEST position | Used with both programs to verify airborne launch system operation. A program test routine sequences panel indicators in a timed order and pattern to monitor system responses. |
| | OFF position | Used during equipment turn-on and turn-off and when a command selection is not required. |
| | PLC-A position | Used with operational program to select preparatory launch command-A (PLC-A) message for transmission to LFs. |
| | PLC-B position | Used with operational program to select preparatory launch command-B (PLC-B) message for transmission to a single LF. |
| | ENABLE position | Used with operational program to select an enable command for one or more LFs. |

Figure 1-12. Launch Control Panel (Sheet 2 of 4)

APPENDIX D

T.O. 21M-LGM30F-1-7

| NO. | CONTROL/INDICATOR | FUNCTION |
|---|---|---|
| 11 Cont | ELC position | Used with operational program to select execute launch command (ELC) to launch a previously-enabled missile. This command is applicable to all WS-133 and WS-118 series configurations. Launch may be commanded to a squadron with two simultaneous (within 2.5 sec) key-actuated launch votes. |
| | AUTO position | Used with operational program to select a combined enable-launch message in a single instruction to all WS-133 and WS-118 series configurations. Automatic launch may be commanded for one LF or a squadron with two simultaneous (within 2.5 seconds) key-actuated launch votes. |
| | Equipment status group | |
| 12 | GO indicator (green) | Illuminates when a manually-initiated ALCC test routine is successfully completed. |
| 13 | FAULT warning light (red) | Illuminates when a manual or automatic self-test of launch control equipment has failed. Also illuminates when a fault is encountered during normal operations. Fault location identification, and number may be determined through an octal number readout at the ALCC data processor code control panel. |
| 14 | FAULT CHANGE caution light (amber) | Illuminates for succeeding faults after FAULT warning light has illuminated. Goes out when fault readout has been accomplished. |
| 15 | LAMP TEST pushbutton | When pressed, provides lamp test activation to all launch control panel fixed legend indicators and warning lights except POWER ON. |
| 16 | MASTER RESET pushbutton | Pressing pushbutton initializes DPU logic and restarts DPU program. |
| | MESSAGE group | |
| 17 | COOP LAUNCH key switch | Initiates execute launch or auto command when used in conjunction with cooperative launch switch on launch monitor panel (within 2.5 seconds time interval). |
| 18 | STOP pushbutton | If pressed immediately after initiate, stops message transmission at completion of minimum sequence. If pressed after minimum sequence, stops message repetition. Also provides test routine reset following self-test of launch control group. |
| 19 | INITIATE pushbutton | When pressed, initiates transfer of launch control panel stored binary data settings to ALCC data processor for message assembly. INITIATE pushbutton has no effect for selection of ELC or AUTO commands. |
| 20 | TRANSMITTING indicator (white) | Illuminates during weapon system and 494L message modulation of UHF radio carrier by waveform converter. |
| 21 | IN PROCESS indicator (white) | Illuminates during message processing and goes out after minimum message has been transmitted. |
| 22 | RADIO PERMIT indicator (green) | Illuminates to indicate RADIO switch on launch monitor panel is positioned to ON. |

Figure 1-12. Launch Control Panel (Sheet 3 of 4)

T.O. 21M-LGM30F-1-7

| NO. | CONTROL/INDICATOR | FUNCTION |
|---|---|---|
| 23 | CLASSIFIED COMMAND INTERLOCK/ CODE INSERTION selectors | Used to insert eight octal digits into LAUNCH CONTROL panel register. Each of eight rotary switch selectors can be set to eight settings numbered from 0 through 7. An octal 8-digit word has been combined with each encrypted classified command stored in the DATA PROCESSOR unit memory; prior to decryption, this word is recombined with encrypted command when appropriate settings are selected on interlock switches. Decryption cannot be accomplished until this recombining operation is performed. The eight switches must be set to 10000000 positions for all peacetime operation except manual self-test. |
|   | PROGRAM group |   |
| 24 | PLC-A EXECUTE OPT selector | Allows selection of 100 options (00 through 99) for programming a squadron to a specific execution plan. Selector consists of separate tens and units thumbwheel switches with OFF and numeric 0 through 9 settings. |
| 25 | PLC A/B ALIGN switch | Used for selection of circular error probability (CEP) or minimum reaction time (MRT). With cover down, switch is set to CEP. When cover and switch are in up position, MRT is selected. ALIGN selection is used for both PLC-A and PLC-B commands. For Peacekeeper PLC-A and PLC-B, the ALIGN selection must be CEP. |
| 26 | PLC-B TARGET selector | Used to select a specific target stored in memory of addressed missile. Positions are OFF and 1 through 8. Target selection code is addressed to a single launch facility only. |
| 27 | PLC-B DELAY switch | Used in conjunction with PLC-B command to select either immediate launch or infinite delay. With cover down, switch is set to DELAY-0. When cover and switch are in up position, infinite DELAY-HOLD instruction is commanded. Delay setting is selective to a single launch facility only. |
|   | TAPE LOAD group |   |
| 28 | FILL pushbutton | When pressed, starts advancement of tape and ALCC DATA PROCESSOR read-in of tape data. Tape fill continues until an automatic error halt or complete halt is encountered. |
| 29 | REWIND pushbutton | When pressed, tape is rewound on a supply spool. When rewind is complete, stop is automatic. |
| 30 | ERROR warning light (red) | Illuminates during tape fill in event of discrepancy in either tape parity or tape verify. Fill process is terminated upon error detection and ERROR warning light remains illuminated until rewind is instructed, or MASTER RESET is pressed. |
| 31 | COMPLETE indicator (green) | Illuminates upon completion of tape fill and goes out upon initiation of rewind. Re-illuminates upon completion of rewind and goes out when MASTER RESET pushbutton is pressed. |
| 32 | IN PROCESS indicator (white) | Illuminates intermittently during tape advance (fill) or continuously during rewind. Goes out automatically when tape motion ceases upon completion of fill, rewind, or due to an error halt. |

Figure 1-12. Launch Control Panel (Sheet 4 of 4)

APPENDIX D

T.O. 21M-LGM30F-1-7

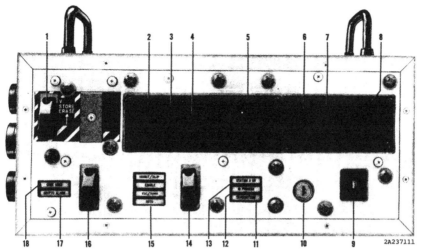

| NO. | CONTROL/INDICATOR | FUNCTION |
|---|---|---|
| | CODES group | |
| 1 | V STORE ERASE switch | Positioning switch up removes power from code processor and volatilizes cryptovariable in code processor storage. Returning switch down restores power to code processor, causing an automatic erased test. Removal of code processor power will cause erasure of VKA memory if VKAs are installed and armed. |
| | ADDRESS group | |
| 2 | WING display | Single-digit readout of wing address numbers 1 through 7 as selected on LAUNCH CONTROL panel. |
| 3 | SQUAD display | Single-digit readout of squadron address numerals 1 through 4 as selected on LAUNCH CONTROL panel. |
| 4 | FLIGHT displays | Two-digit readout of letters A through T inclusive as selected on LAUNCH CONTROL panel. |
| 5 | LF displays | Two-digit readout units and tens for numerals 0 through 59 as selected on LAUNCH CONTROL panel. |
| | PROGRAM group | |
| | PLC-B | |
| 6 | DELAY display | 0 or HOLD readout as selected on LAUNCH CONTROL panel. |
| 7 | TARGET display | Single-digit target numerals 1 through 8 readout as selected on LAUNCH CONTROL panel. |

Figure 1-13. Launch Monitor Panel (Sheet 1 of 2)

1-34  Change 17

AIRBORNE LAUNCH CONTROL CENTER PANELS

| NO. | CONTROL/INDICATOR | FUNCTION |
|---|---|---|
|  | PLC-A |  |
| 8 | EXECUTE OPT display | Two-digit readout of units and tens for numerals 00 through 99 as selected on LAUNCH CONTROL panel. |
| 9 | LAMP TEST switch | Used to check operation of ADDRESS and PROGRAM displays and fixed legend indicators. Switch positions are OFF, digits 0 through 9, and ALL. |
| 10 | COOP LAUNCH key switch | Initiate execute launch command (ELC) and automatic launch command (AUTO) when rotated in conjunction with cooperative launch switch on LAUNCH CONTROL panel (within 2.5 second time interval). |
|  | RADIO group |  |
| 11 | TRANSMITTING indicator (white) | Illuminates when waveform converter is transmitting a weapon system command or 494L message. |
| 12 | IN PROCESS indicator (white) | Illuminates while DPU is processing a manually-initiated test or command. During weapon system command transmission, indicator goes out when minimum sequence has been transmitted. |
| 13 | STATION 3 ON caution light (amber) | Illuminates when ALCC switch is set to ON with CLASSIFIED COMMAND CONTROL switch position to ENABLE. |
| 14 | RADIO switch | Used to control carrier of UHF transmitter selected for ALCC use by interconnecting box TRANSMITTER selector switch. ON position turns transmitter carrier on to enable UHF transmission, OFF position disables carrier. RADIO switch also regulates transfer of command data and voice from waveform converter to UHF transmitter. ON position allows data and voice transmission. OFF position disconnects waveform converter output from transmitter input to prevent carrier modulation. RADIO PERMIT indicator on LAUNCH CONTROL panel illuminates when RADIO switch is ON. Setting RADIO switch to ON interrupts waveform converter test during interruptible self test routine. |
| 15 | INHIBIT/CLIP, ENABLE, ELC/TIMER, and AUTO indicators (white) | Four fixed-legend command indicators are provided for indication of LAUNCH CONTROL panel classified command selection. Setting of COMMANDS selector switch on LAUNCH CONTROL panel enables illumination circuitry. |
| 16 | CRYPTO ALARM OVERRIDE switch | Used for system checkout during cryptovariable loading and for fault isolation. When this switch is set to OVERRIDE with the ALCC switch on and the CLASSIFIED COMMAND CONTROL switch set to ENABLE, code processor internal interlocks are deactivated. Overriding these interlocks makes it possible to use a malfunctioning code processor in an attempt to decrypt classified commands stored in ALCC DATA PROCESSOR memory. Overriding a malfunctioning code processor may not result in correct decryption of an encrypted word. |
| 17 | CRYPTO ALARM warning light (red) | Illuminates when: (1) code processor equipment malfunctions, (2) verify test fails, (3) sum check fails, (4) an erased test fails, or (5) crypto alarm override is in effect with both ALCC switch and CRYPTO ALARM OVERRIDE switch on and CLASSIFIED COMMAND CONTROL SWITCH set to ENABLE. |
| 18 | CODE USED warning light (red) | CODE USED warning light is not used, however light may illuminate intermittently. |

Figure 1-13. Launch Monitor Panel (Sheet 2 of 2)

# NOTES

## Abbreviations

| | |
|---|---|
| AFHRA | Air Force Historical Research Agency. The publicly available online research tool, Air Force History Index, is only current to 2000. Use of the leading zeros in the IRISNUM citation is necessary when using the Air Force History Index. IRISREF is the old reference system and only included where an IRISNUM is not available. Air Force History Index can be searched at www.airforcehistoryindex.org. |
| DDRS | Declassified Documents Reference System, also known as Declassified Documents Online. University of California, Santa Barbara, originated the system, and now Gale runs access through a subscription service, www.gale.com. |
| DDEL | Dwight D. Eisenhower Presidential Library |
| DNSA | Digital National Security Archive, operated by George Washington University, www.nsarchive.gwu.edu. Another version is run by ProQuest, www.proquest.com/databases/dnsa. |
| DTIC | Defense Technology Information Center, https://discover.dtic.mil/ |
| EO 13526 | Executive Order 135256 was issued on 29 December 2009 and established a uniform system for declassified national security information. Redacted copy declassified before Executive Order 135256. |
| IAW | in accordance with |
| NARA | National Archives and Records Administration |
| Scribd | online document resource, www.scribd.com |

## Chapter 1

1. Dik A. Daso, *Hap Arnold and the Evolution of American Airpower* (Smithsonian Institution Press, 2000), 153–154.
2. Jacob Neufeld, *The Development of Ballistic Missiles in the United States Air Force, 1945–1960* (Office of Air Force History, Washington, DC, 1990), 35. George P. Kennedy, *Vengeance Weapon 2: The V-2 Guided Missile* (National Air and Space Museum, Washington, DC, 1984), 30–35. While British intelligence had knowledge of the V-2 as early as 1939, examples were not captured until mid-1944, and launches against Allied targets did not begin until 6 September 1944, with the first success on 8 September 1944 against Paris. This makes Arnold's comments even more prescient.
3. Theodore von Kármán, *Prophecy Fulfilled: "Toward New Horizons" and Its Legacy*, edited by Michael Gorn (Air Force History and Museums Program, 1994), 2, 9.
4. Ibid., 12.
5. *Hearings before the Committee on Military Affairs*, 79th Cong. (1945), 77–78.
6. *Hearings before the Special Committee on Atomic Energy*, 79th Cong. (1945), 179.
7. Carl A. Spaatz, "Airpower in the Atomic Age," *Collier's Weekly*, 8 December 1945, 11–12.
8. *Hearings before the Special Committee on Atomic Energy*, 79th Cong. (1945), 180–181. Kennedy, *Vengeance Weapon 2*, 40, 51. Nearly 3,000 V-2s had been used against large cities or industrial areas

of Europe: an estimated 1,115 V-2s on London and its suburbs, as well as an estimated 1,780 on Paris, Liege, and Antwerp. There were relatively few attacks on tactical targets due to the inaccuracy of the V-2 guidance system.

9. Henry P. Arnold, "Air Power for Peace," *National Geographic,* February 1945, 187–188. Arnold also contributed a chapter to *One World or None*, edited by Dexter Masters and Katharine Way (McGraw Hill Book Company, 1946), 26–32, where he went into more detail on the issues brought up in the *National Geographic* article.
10. Vannevar Bush, *Modern Arms and Free Men* (Simon and Schuster, New York, NY, 1949), 83–87.
11. "The Development of Long-Range Guided Missiles 1945–1959," Declassified Documents Reference System (DDRS), CK2349514815, 10–12. Elliott V. Converse III, *Rearming for the Cold War 1945–1960, History of Acquisition in the Department of Defense Volume I* (Historical Office, Office of the Secretary of Defense, Washington, DC, 2012), 118. K. T. Keller, "Final Report of the Director of Guided Missiles, 17 September 1953" (Office of the Secretary of Defense), Dwight D. Eisenhower Presidential Library (DDEL), K. T. Keller Collection, Box 1, 1947–1958, 3.
12. Robert Cutler, *No Time for Rest* (Little, Brown, 1965, 1966), 296.
13. David L. Snead, *The Gaither Committee, Eisenhower and the Cold War* (Ohio State University Press, 1999), 25.
14. Ibid., 28–30.
15. IBMS was the first acronym for intercontinental ballistic missiles, but it was changed to ICBM to avoid confusion with International Business Machines.
16. Neufeld, *Development of Ballistic Missiles*, 259–260.
17. Ibid., 106.
18. Ibid., 107–108.
19. Snead, *The Gaither Committee*, 35–36.
20. James R. Killian, "Technological Capabilities Panel Report, Vol. I," 38. A copy of the report is located at the Dwight D. Eisenhower Presidential Library; White House Office, Office of the Special Assistant for National Security Affairs: Records, 1952–61 NSC Series, Briefing Notes Subseries Box 16, folders Killian Report—Technological Capabilities Panel (1)-(2).
21. Dwight D. Eisenhower, *Waging Peace: 1956–1961* (Doubleday, Garden City, NY, 1965), 208.
22. *Deterrence and Survival in the Nuclear Age (the Gaither Report of 1957), Joint Committee and Defense Production Congress of the United States*, 94th Cong. (US Government Printing Office, 1976), 17. This report remained classified until 1973.
23. Bush, *Modern Arms and Free Men*, paperback edition (MIT Press, 1959), foreword.
24. Neufeld; Harry G. Stine, *ICBM: The Making of the Weapon That Changed the World*, (Orion Books, 1991). Ernest G. Schwiebert, *A History of the US Air Force Ballistic Missiles*, (Praeger, 1964). Gary B. Conine, *Not for Ourselves Alone: The Evolution and Role of the Titan II Missile in the Cold War* (CreateSpace Independent Publishing Platform, 2015). Eugene M. Emme, *The History of Rocket Technology* (Wayne State University Press, 1964). Christopher J. Gainor, "The United States Air Force and the Emergence of the Intercontinental Ballistic Missile 1945–1954" (Thesis, University of Alberta, Canada, 2011). Max Rosenberg, "The Air Force and the National Guided Missile Program, 1944–1950" (USAF Historical Division Liaison Office, 1964).
25. David K. Stumpf, *Titan II: A History of a Cold War Missile Program* (University of Arkansas Press, Fayetteville, AR, 2000).
26. Francis X. Ruggiero, "Missileers' Heritage" (Air Command and Staff College, Air University, Maxwell AFB, AL, 1981), 90–91.

## Chapter 2

1. Charles H. Eger, "Minuteman Financial History 1955–1963" (Minuteman Special Project Office, undated), National Archives and Records Administration (NARA), Washington, DC, Samuel C. Phillips Collection, Folder 1, Box 11, 5–8.
2. Robert F. Piper, *The Development of the SM-80 Minuteman*, Vol. I and II (Historical Office, Deputy Commander for Aerospace Systems, Air Force Systems Command, April 1962), Air Force Historical Research Agency (AFHRA), IRISNUM 897232, K243–012–05. This is the original source of the full document, which remains classified. A declassified version is available from the Digital National Security Archive (DNSA) as Document NH00024. Vol. II, "Memo, Brig Gen C. M. McCorkle, Asst C/S for Guided Missiles, To: Dir Guided Missiles, OSD, 21 Feb 1958, Subject: ICBM Force Objectives," Document 53, 2. W. M. Holaday, "Air Force Ballistic Missile Objectives," 27 February 1958, memorandum, Memos February 1958," personal collection of Gen. B. A. Schriever, AFHRA, IRISNUM 01040256, Reel 35259. The collection is classified; the memorandum cited is declassified IAW EO 13526. The memorandum is referenced in Piper's history as Document 57; the declassified copy does not contain this memo in the supporting documents.
3. A. S. Low, "Minuteman Program Summary, 27 October 1959," DNSA, NH00677, 1.
4. Discussion at the 439th National Security Council Meeting Held on 1 April 1960, DDRS, CK2349158797, 20.
5. Bernard A. Schriever, "Letter to: Clark A. Millikan, California Institute of Technology, 3 April 1960," personal collection of Gen. B. A. Schriever, AFHRA, IRISNUM 1040303, Reel 35263, Frame 404.
6. Charles Lauritsen, "Lauritsen Committee Report to Lt. Gen. Bernard A. Schriever Concerning the United States Air Force ICBM Program, 31 May 1960," personal collection of Bernard Schriever, AFHRA, IRISNUM 01040303, Reel 35263, 1-5, 1-13. The report includes the working papers explaining the conclusions.
7. Piper, *Development of the SM-80*. Vol. I, 52.
8. TWX, From COMDR AFBMD, To COS USAF and COMDR ARDC, 21 February 1959, personal collection of Bernard A. Schriever, AFHRA, IRISNUM 1040303, Reel 35263, no pagination.
9. J. C. Hopkins, "The Development of the Strategic Air Command, 1946–1981, A Chronological History" (Office of the Historian, Headquarters Strategic Air Command, 1 July 1982), 82, author's collection. Minutes of the 56th Meeting of the Air Force Ballistic Missile Committee, 2 August 1960, AFHRA, IRISNUM 02055699, K243.8631-2030, 1960, 1–6. This document is classified; the extract used is declassified IAW EO 13526.
10. J. Ponturo, "Analytical Support for the Joint Chiefs of Staff: The WSEG Experience, 1948–1976" (Institute for Defense Analysis Study S-507, July 1979), Defense Technology Information Center (DTIC), ADA090946, Appendix C, 1.
11. WSEG Report No. 50: Evaluation of Strategic Offensive Weapon Systems, 27 December 1960 (Weapons Systems Evaluation Group, Office of the Secretary of Defense), DNSA, NH00422, 10. This document is classified; the extract used is declassified IAW EO 13526. The only existing copy of this document was found at the John F. Kennedy Presidential Library by the author in 2000, and it took six years to get it reviewed and partially declassified.
12. Ponturo, "Analytical Support," 176–177.
13. WSEG Report No. 50, Appendix B to Enclosure F, 43.
14. *Military Construction Authorization 1962, Hearings before the Committee on Armed Services,* 87th Cong. (1961) 22, 371, 989 (testimony of Gen. A. M. Minton).
15. Recommended Long Range Nuclear Delivery Forces 1963–1967, 23 September 1961, draft memorandum from Secretary of Defense McNamara to President Kennedy, Foreign Relations of the United States, 1961–1963, Vol. VIII, National Security Policy, Part 2, Document 46, 2.

16. Carl Kaysen, Memorandum for the President, 27 December 1961, DDEL, President Office Files, Air Force, 12/1961, Box 69, 2.
17. "Recommended Long Range Nuclear Delivery Forces, 1963–1967," 3.
18. C. Kaysen, Secretary McNamara's Recommendation for Long-Range Nuclear Forces, 1963–1967, memorandum, 11 October 1961, DNSA, NH00433, 3. *Statement of Secretary of Defense Robert S. McNamara before the Senate Committee on Armed Services,* 4 April 1961, DNSA, NH00426, 11.
19. Recommended FY 1964–FY 1968 Strategic Retaliatory Forces, draft memorandum, Lyndon B. Johnson Library, Reference File, Draft Presidential Memorandum (DOD) 1961–1969, 2.
20. Recommended FY 1965–FY 1969 Strategic Retaliatory Forces, memorandum, Lyndon B. Johnson Library, Reference File, Draft Presidential Memorandum (DOD) 1965–1969, I-2.
21. Ibid., I-4.
22. Ibid., I-4.
23. Ibid., I-12–14, I-25–26, I-30.
24. Ibid., I-4.
25. "SAC Missile Chronology 1939–1988" (Office of the Historian, Headquarters Strategic Air Command, Offutt AFB, NE, 1 May 1990), 45. Desmond Ball, *Politics and Force Levels: Strategic Missile Program of the Kennedy Administration* (University of California Press, 1980), 251. Ball relates that the decision was made on 22 December 1964, during a meeting with President Johnson, LeMay, McNamara, and Kermit Gordon at Johnson's ranch in Texas. McNamara and LeMay got into an argument over the 1,000 McNamara felt were sufficient and Lemay's desire for 1,200. To break the impasse, Johnson asked Gordon, his budget director, what he thought was right, and Gordon said 900 was all that was necessary. Johnson had met with Gordon earlier and knew that 900 was the number he would suggest. McNamara and LeMay disagreed, but Johnson was able to reach an agreement on 1,000 missiles. Recommended FY 1966–FY 1970 Strategic Retaliatory Forces, 3 December 1964, memorandum, Lyndon B. Johnson Presidential Library, Reference File, Draft Presidential Memorandum (DOD) 1965–1969, 11–13, 22, 30, 36.
26. Recommended FY 1966–FY 1970 Strategic Retaliatory Forces, 3 December 1964, 11–13, 22, 30, 36.
27. *Hearings on Military Posture and H. R. 4016, to Authorize Appropriations during Fiscal Year 1966 for Procurement of Aircraft, Missiles, and Naval Vessels, and Research, Development, Test, and Evaluation, for the Armed Forces, and for Other Purposes,* 85th Cong. 211 (1965) (testimony of Deputy Secretary of Defense Cyrus Vance).
28. Recommended FY 1967–FY 1971 Strategic Retaliatory Forces, draft memorandum, Lyndon B. Johnson Library, Reference File, Draft Presidential Memorandum (DOD) 1961–1969, 40.
29. *Senate Armed Services Committee Hearings on Military Authorization and Defense Appropriations for Fiscal Year 1967,* 247–248 (1966) (testimony of Secretary of Defense R. S. McNamara).
30. Recommended FY 1968–FY 1972 Strategic Retaliatory Forces, draft memorandum, 22 September 1966, Lyndon B. Johnson Library, Reference File, Draft Presidential Memorandum (DoD) 1961–1969, 3.
31. "History of Strategic Air Command, Fiscal Year 1970," Historical Study No. 117, Vol. II, Narrative (Office of the Historian, Headquarters Strategic Air Command, 20 April 1971, DNSA, NT00923, 231. This document is classified; the extract used is declassified IAW EO 13526.
32. Ibid., 231. "Strategic Air Command Weapons Systems Acquisition: 1964–1979" (Office of the Historian, Headquarters Strategic Air Command, 28 April 1980), AFHRA, IRISNUM 01114095, K416.04-39, Vol. I, 196. This document is classified; the extract used is declassified IAW EO 13526.
33. "History of Strategic Air Command, Fiscal Year 1969," Historical Study No. 116, Vol. II, Narrative (Office of the Historian, Headquarters Strategic Air Command, March 1970), DNSA, NT00502, 247A. This document is classified; the extract used is declassified IAW EO 13526.
34. "History of Strategic Air Command, Fiscal Year 1970," 231.

35. "Strategic Air Command Weapons Systems Acquisition," 201. No further information is available on Minuteman IIIA.
36. "History of Strategic Air Command, 1 July 1975–31 December 1976," Historical Study No. 161, Vol. II, Narrative (Office of the Historian, Headquarters Strategic Air Command, 15 July 1977), DNSA, NT02287, 465. This document is classified; the extract used is declassified IAW EO 13526.
37. "History of the Strategic Air Command Fiscal Year 1975," Historical Study No. 153, Vol. 1, Narrative (Office of the Historian, Headquarters Strategic Air Command, 10 February 1976), AFHRA, IRISNUM 1010404, TSHOA7618, V.1, Reel 31271, 309. This document is classified; the extract used is declassified IAW EO 13526.
38. "Strategic Air Command Weapons Systems Acquisition," 202.
39. *Fiscal Year 1972 Authorization for Military Procurement*, Senate Committee on Armed Services, 68 (1971) (testimony of Secretary of Defense Melvin Laird). "Strategic Air Command Weapons Systems Acquisition," 199. This reference gives 14 January 1971 as the date the 450 Minuteman II and 550 Minuteman III force level decision was made. "History of the Strategic Air Command, Fiscal Year 1975," Historical Study No. 153, Vol. I, 309.
40. *Senate Resolution 241, Disapproving the Basing Mode for the MX Missile,* 97th Cong. 1 (1981). Ralph W. Holm, "Chronology of the Ballistic Missile Organization: 1945-1990," Vol. I, Narrative (History Office Ballistic Missile Organization, August 1993),164. The author thanks Jonathan McDowell for a copy of this comprehensive document.
41. Department of Defense Authorization Act of 1983, Pub. L. No. 95-252, 97th Cong. 722 (1978). *MX Missile Basing Mode, Communication from the President of the United States,* 97th Cong. 1 (1982).
42. Holm, "Chronology of the Ballistic Missile Organization," 168. "Peacekeeper in Minuteman Silos Draft Environmental Impact Statement" (History Office, 90th Strategic Missile Wing, F. E. Warren AFB, WY, October 1983), cover sheet.
43. *Congressional Record* (Sen.), 23 May 1985, 13544–13545. *Congressional Record* (H. R.) 18 September 1985, 24126.
44. "90th Strategic Missile Wing History: January–March 1986" (History Office, 90th Strategic Missile Wing, 21 June 1986), ix–x. "90th Strategic Missile Wing History: January–December 1988" (History Office, 90th Strategic Missile Wing, 15 October 1988). These documents are classified; the extracts used are declassified IAW EO 13526. The author thanks Mike Byrd, 20th Air Force historian, for these documents.
45. Peter Johnson, "MAFB Cuts? Air Force Making Reductions Nationwide," *Great Falls Tribune*, 6 February 1991, 1; "Mid-1990s for Missile Loss Here? Deactivation of Minuteman IIs Not Expected to Start Immediately," *Great Falls Tribune*, 7 February 1991, 1A; "Way Cleared for Missiles," *Great Falls Tribune*, 6 June 1992, 1B.
46. "Treaty on Further Reduction and Limitation of Strategic Offensive Arms," US–USSR, 31 July 1991, Article II.
47. George H. W. Bush, "Address to the Nation on Reducing United States and Soviet Nuclear Weapons," 27 September 1991, American Presidency Project, University of California, Santa Barbara.
48. Minuteman II Deactivation/Minuteman III Conversion, 341st Missile Wing, Program Plan 91-7, 23 September 1991 (Headquarters, Strategic Air Command, Offutt AFB, NE), author's collection. National Defense Authorization Act for Fiscal Years 1992 and 1993, Pub. L. No. 102-190, 102nd Cong. 1313 (1991).
49. "History of the Air Force Space Command, 1 January 1992–31 December 1993," Chapter II (History Office, Air Force Space Command, CO), On Alert Archive, 119–120. This document is classified; the extract used is declassified IAW EO 13526.
50. Peter Johnson, "Modernizing Mission, First of New Missiles Put in Empty Silo," *Great Falls*

*Tribune*, 18 November 1992, 1C. "History of the Air Force Space Command, 1 January 1992–31 December 1993," 122. This document is classified; the extract used is declassified IAW EO 13526.

51. Amy Woolf, "US Strategic Nuclear Forces: Background, Developments, and Issues," updated 3 April 2007, Congressional Research Service, RL 33640, 8–13.
52. Steve Foss, "Moving Missiles, It's the Beginning of the End for ICBMs at Grand Forks Air Force Base," *Grand Forks Herald*, 5 October 1995, A1.
53. *Hearing before the Committee on Armed Services*, 103rd Cong. 27 (1995) (testimony of John M. Deutsch, Deputy Secretary of Defense).
54. "Defense Base Closure and Realignment Commission Report," 1 July 1995, Chapter 1, 101. The report can be found at www.brac.gov/docbrowse1995.aspx, accessed 12 July 2019.
55. Foss, "Moving Missiles," A1.
56. Melissa Phillips, "F. E. Warren Finishes Treaty Requirement Early," Air Force Space Command News Service, 10 August 2001, 1.
57. Ibid., 2.
58. "Special Briefing on Nuclear Posture Review," Department of Defense Press Briefing, 9 January 2002, news transcript, 5–6.
59. Amy Woolf, "Nuclear Arms Control: The Strategic Offensive Reductions Treaty," 3 January 2007," Congressional Research Service Report for Congress, RL 31448, 1–31.
60. "History of the Air Force Space Command, 1 January–31 December 2006" (History Office Air Force Space Command, Peterson AFB, CO, 2006), On Alert Archive, 115. This document is classified; the extract used is declassified IAW EO 13526.
61. *United States Subcommittee on Strategic Forces, Committee on Armed Services*, 109th Cong. 95, 121–122, 135 (2006) (testimony of Gen. James E. Cartwright). Quadrennial Defense Review Report (Office of the Secretary of Defense, February 6, 2006), 49.
62. John Warner National Defense Authorization Act for Fiscal Year 2007, Pub. L. No. 109-364, 109th Cong. 2115 (2006).
63. Peter Johnson, "Missile Dismantled, Warhead Removed," *Great Falls Tribune*, 14 July 2007, 1A. John Turner, 341st Missile Wing Public Affairs Office, 13 September 2018.
64. "The Long Pole of the Nuclear Umbrella," Senate ICBM Coalition (Washington, DC, 4 November 2009), 5.
65. "Nuclear Posture Review Report" (Department of Defense, 6 April 2010), 23, 26.
66. *Senate Committee on Foreign Relations Report, Treaty with Russia on Measures for Further Reduction and Limitation of Strategic Offensive Arms (the New START Treaty)*, 111th Cong. 12 (2010).
67. *Status of United States Strategic Forces, Hearings before the Subcommittee on Strategic Forces of the Committee of Armed Services* (H. R.), 112th Cong. 78–79 (2011) (testimony of Gen. C. Robert Kehler).
68. Ibid.
69. *Hearing before Committee on Armed Services House of Representatives, Fiscal Year 2015 National Defense Authorization Budget Request from the Department of the Air Force*, 113th Cong. 23 (2014) (testimony of Secretary of the Air Force Deborah Lee James).
70. "Report on Plan to Implement the Nuclear Force Reductions, Limitations, and Verification and Transparency Measures Contained in the New START Treaty Specified in Section 1042 of the National Defense Authorization Act for Fiscal Year 2012," 8 April 2014, www.archive.defense.gov/documents/new-start-implementation-report.pdf, accessed 15 June 2019, 3. "Fact Sheet on US Nuclear Force Structures under New START Treaty," https://archive.defense.gov/documents/FactSheetonUSNuclearForceStructureundertheNewSTARTTreaty.pdf, 1, accessed 6 May 2019.

71. Jenn Rowell, "Defense Department to Remove 50 Minuteman Missiles from Silos at Three Bases, Including Malmstrom," *Great Falls Tribune*, 8 April 2014, 1A.
72. Carl Levin and Howard P. "Buck" McKeon National Defense Authorization Act for Fiscal Year 2015, Pub. L. No. 113-295, 113th Cong., 3651 (2014).
73. Esther Willet, "AF Meets New START Requirements" (90th Missile Wing Public Affairs, 28 June 2017). Email to author from Kenneth Vantiger, Air Force Global Strike Command Treaty Office, 29 October 2019.
74. Email to author from Rex C. Ellis, chief, Treaty Compliance Office, 90th Missile Wing, 24 September 2018.

# Chapter 3

1. Charles H. Eger Jr., "Minuteman Financial History, 1955–63," Systems Program Office, National Archive and Records Administration (NARA), Samuel C. Phillips Collection, Box 11, Ballistic Systems Division, Folder 11, 8–9.
2. "Strategic Air Command Operations in the Cuban Missile Crisis of 1962," Historical Study No. 90, Vol. 1, DNSA CU01379, 66.
3. "SAC Missile Chronology: 1939–1988" (Office of the Historian, Headquarters Strategic Air Command, Offutt AFB, NE, 1 May 1990), 37.
4. Ruggiero, "Missileers' Heritage," 55, 65, 67.
5. Charles Lauritsen, "Lauritsen Committee Report to Lieutenant Gen. Bernard Schriever Concerning the United States Air Force Minuteman Program, 15 June 1961," DNSA, NH00730, 3.
6. Scott Sagan, "SIOP-62: The Nuclear War Plan Briefing to President Kennedy," *International Security* 12, no. 1 (Summer 1987): 35.
7. Desmond Ball and Jeffrey Richelson, eds., *Strategic Nuclear Targeting* (Cornell University Press, Ithaca, NY, 1986), 55.
8. Sagan, "SIOP-62," 22, 29, 44–50.
9. Ibid., 37–38.
10. Ibid., 38–39. Ball and Richelson, *Strategic Nuclear Targeting*, 62–63.
11. Collin S. Gray, "The Future of Land-Based Missile Forces," *Adelphi Papers*, no. 140, (International Institute for Strategic Studies, London, UK, 1978), 32. Chuck Hansen, *The Swords of Armageddon: US Nuclear Weapons Development Since 1945*, Vol. VII (Chukelea Publications, Sunnyvale, CA, October 1995), 358–360. "History of Strategic Air Command Fiscal Year 1970," Historical Study No. 117, Vol. II, Narrative (Office of the Historian, Headquarters Strategic Air Command, 20 April 1997). Redacted copy, DNSA, NT0093, 231.
12. Robert S. McNamara, *Annual Defense Department Report, Fiscal Year 1965* (Washington DC, US Government Printing Office, 1967), 12.
13. Ball and Richelson, *Strategic Nuclear Targeting*, 74–75. Terry Teriff, *The Nixon Administration and the Making of US Nuclear Strategy* (Cornell University Press, Ithaca, NY, 1995), 207–8.
14. "Policy Guidance for the Employment of Nuclear Weapons, 3 April 1974," (Office of the Secretary of Defense, 1974), DNSA, NT01738, Appendix A, A-7. This document was declassified IAW EO 12958, Section 3.6.
15. *Hearings before a Subcommittee of the Committee on Appropriations* (H. R) 95th Cong. (1977), Department of Defense Appropriations for 1978, 152, 193–212.
16. John M. Collins, *US-Soviet Military Balance Concepts and Capabilities: 1960–1980* (McGraw-Hill Publications Company, 1980), 452–53.
17. Stumpf, *Titan II*, 264.

18. William J. Perry, *Annual Report to the President and the Congress: February 1995* (US Government Printing Office, Washington, DC, 1995), 166.
19. William S. Cohen, *Annual Report to the President and the Congress: 2001* (US Government Printing Office, Washington, DC, 2000), 275.
20. "2010 Nuclear Posture Review Report," Department of Defense, 23. https://dod.defense.gov/News/Special-Reports/NPR/, accessed 12 May 2019. John Warner National Defense Authorization Act for Fiscal Year 2007, Pub. L. 109-364, 109th Cong. (2006), 2114.
21. Graham Spinardi, *From Polaris to Trident: The Development of US Fleet Ballistic Missile Technology* (Cambridge University Press, New York, NY, 1994), 58.

## Chapter 4

1. An extended version of this chapter was published as "Minuteman: Solid Propellant Comes of Age," *Quest: The History of Spaceflight Quarterly* 26, no. 3 (2019): 15–53.
2. The proper units for specific impulse are "pounds-seconds/pounds," canceled out to give seconds, which is incorrect but accepted. The author thanks Dr. Leonard Caveny for bringing this to my attention.
3. J. S. Billheimer and F. R. Wagner, "The Morphological Continuum in Solid Propellant Grain Design," IAF Paper P-94, *19th Congress of the International Astronautical Federation*, New York, NY, 13–19 October 1968. Grain terminology in solid-propellant motor nomenclature is a holdover from gun performance and is related to the shape of the cordite or similar propellant grains, or pieces, of propellant. Early in the development of the more modern gun propellants, such as cordite, it was determined that the grain shape and size could be varied to provide tailored burn rates for the various gun types. These included rods, perforated rods, cubes, plates, and a variety of other shapes. This was one of the unique advantages of these new propellants.
4. Frank H. Winter, *The First Golden Age of Rocketry: Congreve and Hale Rockets of the 19th Century* (Smithsonian Institute Press, 1990), 50, 55, 71–72, 197–199, 206. This book is a fascinating description of the early science of solid-propellant rocketry.
5. R. H. Goddard, "A Method for Reaching Extreme Altitudes," *Smithsonian Miscellaneous Collection* 71, no. 2 (1919), 1.
6. Goddard, "A Method for Reaching," 6. This multistage rocket idea was patented by Goddard in 1919: Rocket apparatus, US Patent No. 1,102,653, filed 1 October 1913 and issued 7 July 1919.
7. Milton Lehman, *This High Man: The Life of Robert H. Goddard* (Farrar, Strauss, and Company, New York, NY, 1963), 78.
8. Mike Gruntman, *Blazing the Trail: The Early History of Spacecraft and Rocketry* (American Institute of Aeronautics and Astronautics, July 2004), 178.
9. Goddard, "A Method for Reaching," 1, 32, 34, and note 19, 55–56.
10. Theodor von Kármán, *The Wind and Beyond: Theodor von Kármán, Pioneer in Aviation and Pathfinder in Space* (Little, Brown and Company, 1967), 234–239.
11. F. J. Malina and A. O. Smith, "Flight Analysis of a Sounding Rocket," *Journal of the Aeronautical Sciences* 5, no. 5 (1938): 199–202.
12. H. S. Tsien and F. J. Malina, "Flight Analysis of a Sounding Rocket with Special Reference to Propulsion by Successive Impulses," *Journal of Aeronautical Sciences* 6 (1938): 50–58.
13. Frank Malina, "Origins and First Decade of the Jet Propulsion Laboratory," in *The History of Rocket Technology* by Eugene Morlock Emme and W. M. Bland (Wayne State University Press, 1964), 52.
14. von Kármán, *The Wind and Beyond*, 243.
15. Malina, "Origins and First Decade," 52–53.
16. Ibid.

17. John W. Parsons and Edward S. Forman, "Experiments with Powder Motors for Rocket Propulsion by Successive Impulses," *Astronautics* 9, no. 43 (1939): 4–11.
18. Malina, "Origins and First Decade," 54. T. von Kármán and F. J. Malina, "Characteristics of the Ideal Solid Propellant Rocket Motor," JPL Report No. 1–4, 1940, in *Collected Works of Theodore von Kármán,* Vol. IV, 1940–51 (Butterworths Scientific Publications, 1956), 94–106.
19. F. Malina, "US Army Corps Jet Propulsion Research Project, GALCIT Project No.1, 1939–1946: A Memoir," in *History of Rocketry and Astronautics: Proceedings of the Third through Sixth History Symposia of the International Academy of Astronautics*, 1986, Vol. 2, 163, 169. Page 169 shows the four equations Malina solved.
20. Malina, "Origins and First Decade," 54–56.
21. "JATO Study," *Engineering and Science Monthly* IX, no. 7, 7–8. F. J. Malina and J. W. Parsons, "Results of Flight Tests of the Ercoupe Airplane Auxiliary Jet Propulsion Supplied by Solid Propellant Jet Units," GALCIT Project No. 1, Report No. 9, 2 September 1941 (Guggenheim Aeronautics Laboratory, California Institute of Technology), DTIC, ADA 800514, 1–94.
22. J. I. Shafer, "Solid Rocket Propulsion," Chapter 16 in *Space Technology,* edited by H. S. Seifert (John Wiley and Sons, New York, NY, 1959), 5.
23. C. R. Koppes, *JPL and the American Space Program: A History of the Jet Propulsion Laboratory* (Yale University Press, 1982), 12–13.
24. Malina, "Origins and First Decade,," 59.
25. von Kármán, *Wind and Beyond*, 264.
26. Malina, "Origins and First Decade," 60.
27. Koppes, *JPL and the American Space Program*, 19.
28. F. J. Malina, "America's First Long-Range-Missile and Space Exploration Program: The ORDCIT Project of the Jet Propulsion Laboratory, 1943–1946: A Memoir," in *History of Rocketry and Astronautics: Proceedings of the Third through Sixth History Symposia of the International Academy of Astronautics*, Vol 2, 1986, 346. Koppes, *JPL and the American Space Program*, 20.
29. Gregory P. Kennedy, *The Rockets and Missiles of White Sands Proving Ground 1945–1958* (Schiffer Military History, Atglen, PA, 2009), 14.
30. Malina, "America's First," 353–354. M. T. Cagle, *History of the Sergeant Weapon System*, Historical Monograph Project No. AMC 55M (US Army Missile Command, 1971), 4. This is an excellent history of the Sergeant weapon system and highly recommended.
31. Shafer, "Solid Propellant Propulsion," 16-07.
32. E. S. Sutton, "From Polymers to Propellants to Rockets, A History of Thiokol," Paper 99-2929, presented at *35th American Institute of Aeronautics and Astronautics Joint Propulsion Conference,* Los Angeles, CA, 20–24 June 1999, 5.
33. Shafer, "Solid Rocket Propulsion," 16-2. Prior to the development of case-bonded propellants, a steel cage, called a trap, suspended the propellant charge within the case to prevent it being expelled.
34. T. Carroll, "Historical Origins of the Sergeant Missile Powerplant," in *History of Rocketry and Astronautics* (Univelt, San Diego, CA, 1989), 130–134.
35. "A History of the Development of British Rockets, 1935–1941," Projectile Development Establishment Report PDE 1935–41, UK Ministry of Supply, British National Archives, AVIA 41/207, November 1945, 1–7. A. D. Crow, "The Rocket as a Weapon of War in the British Forces," *Proceedings of the Institute of Mechanical Engineers*, No. 158, 1948, 15–21. H. J. Poole, "Charge Shapes for Plastic U. P. Propellants," Advisory Council Report 799/U. P. Propellant Subcommittee Report 13, UK Ministry of Supply (OSRD Liaison Office Ref. No. W40-9N) British National Archives, DSIR 23/29534, 26 April 1961, 1–9.
36. H. J. Poole, "Extracts from Personal Notes of Dr. H. J. Poole Relative to His Visit to Various Stations in the U.S.A., October 1943," British National Archives, WO 195/5557, 2–3.

37. R. J. Thompson and R. R. Newton, "Design of the High Velocity Rocket Vicar," Allegany Ballistics Laboratory Final Report, Series W, Number 21 (OSRD Ref. No. 5793), December 1945, 7. J. Beek Jr. and H. Siller, "Some Fundamental Properties of Star-Perforated Charges," ABL memorandum to R. E. Gibson, November 1945, reproduced as Appendix B in W. H. Avery and J. Beek Jr., "Propellant Charge Design of Solid Fuel Rockets," Allegany Ballistics Laboratory, Final Report, Series B, Number 4 (OSRD Ref. No. 5890), June 1946.
38. Carroll, "Historical Origins," 135.
39. E. W. Price, C. L. Horine, and C. W. Snyder, "Eaton Canyon: A History of Rocket Motor Research and Development in the Caltech-NRDC-Navy Rocket Program, 1941–1946," *34th AIAA/ASME/SAE/ASEE Joint Propulsion Conference and Exhibit*, Cleveland, OH, 13–15 July 1998, 1–28. Elizabeth Babcock, *Magnificent Mavericks: History of the Navy at China Lake, California, Volume 3* (Naval Historical Center, Washington Navy Yard, November 2008), 30–31, 145.
40. M. Summerfield, J. I. Shafer, H. L. Thackwell Jr., and C. E. Bartley, "The Applicability of Solid Propellants to High Performance Rocket Vehicles," *Astronautics*, October 1962, 50–56. This is a reprint of the original report of the same title, JPL Memorandum 4-17, 1 October 1947.
41. H. L. Thackwell Jr. and J. I. Shafer, "High-Acceleration Flight Tests of a 6-inch Rocket to Test a Polysulfide-Rubber-Base Propellant: Phase 1, Ambient Temperature Firings," JPL Progress Report 4-55, 8 January 1948. J. I. Shafer, H. L. Thackwell Jr., "High-Acceleration Flight Tests of a 6-inch Rocket to Test a Polysulfide-Rubber-Base Propellant: Phase 2, High and Low Temperature Firings," JPL Progress Report 4-76, 15 November 1948. H. L. Thackwell Jr. and J. I. Shafer, "The Applicability of Solid Propellants to Rocket Vehicles of V-2 Size and Performance," JPL Memorandum 4-25, 1 July 1948.
42. Carroll, "Historical Origins," 140–141.
43. Sutton, "From Polymers to Propellants," 11. T. Carroll, transcript of telephone conversation with R. W. Porter, 7 August 1972.
44. Sutton, "From Polymers to Propellants," 11. J. W. Wiggins, "Hermes: Milestone in US Aerospace Progress," *Aerospace Historian* 21, no. 1 (March 1974): 36.
45. Wiggins, "Hermes: Milestone," 37.
46. Cagle, *History of the Sergeant*, 15.
47. Wiggins, "Hermes: Milestone," 37–40.
48. Cagle, *History of the Sergeant*, 20, 23.
49. Ibid., 27, 37.
50. Ibid., 104–107.
51. Lester Weil, Solid propellant compositions and processes for making same, US Patent 2,966,403, filed 6 September 1950 and issued 27 December 1960.
52. H. S. Tsien, "Future Trends in the Design and Development of Solid and Liquid Fuel Rockets," *Towards New Horizons* Vol. VI, Part 4, 91.
53. A. S. Leonard, "Some Possibilities for Rocket Propellants, Part III," *Journal of the American Rocket Society* 2 (1947): 14.
54. Babcock, *Magnificent Mavericks*, 166. It is not clear in the reference how large the wires were and what their distribution was in the propellant: "the burning rate of a propellant could be increased by putting aluminum wires in it."
55. Philip K. Reilly, *The Rocket Scientists: Achievement in Science, Technology, and Industry at Atlantic Research Corporation* (Vantage Press, New York, NY, 1999), 44, 49–50. Rumbel and Henderson had found the use of long, thin wire embedded centrally in the grain made end-burning grains much more efficient, with a substantially increased burning rate. This is a somewhat enthusiastic history of the Atlantic Research Corporation. Having said that, the author placed copies of key grants and papers in the appendices, making research much easier. Keith E. Rumbel, "Poly (Vinyl Chloride)

Plastisol Propellants," Appendix 4 in *Propellants Manufacturing, Hazard, and Testing, Advances in Chemistry Series 88* (American Chemical Society, Washington, DC, 1969), 52.
56. J. D. Hunley, "Launch Vehicle Technology Collection, Solids, 1946–2006" (NASA Headquarters, Washington, DC), #19416, Box 17, 1.
57. A. C. Scurlock, K. E. Rumbel, and M. L. Rice, Solid polyvinyl chloride propellants containing metal, US Patent 3,107,186, filed 6 August 1953 and issued 15 October 1963.
58. J. D. Hunley, "Evolution of Large Solid Propellant Rocketry in the United States," *Quest: The History of Spaceflight Quarterly* 6. no. 1 (1998): 26. G. Spinardi, *From Polaris to Trident: The Development of US Fleet Ballistic Missile Technology* (Cambridge University Press, 1994), 29.
59. Beryllium has been used in satellite solid-propellant motors where the toxicity in the exhaust products is not a concern.
60. Charles B. Henderson, personal communications, May 2019.
61. Charles B. Henderson, personal communications, June 2019. J. D. Hunley, *The History of Solid-Propellant Rocketry: What We Do and Do Not Know* (American Institute of Aeronautics and Astronautics, Inc., 1999), 7.
62. R. D. Launius and D. R. Jenkins, *To Reach the High Frontier, A History of US Launch Vehicles* (University of Kentucky Press, 2002), 242. A. C. Scurlock and K. E. Rumbel, Matrix propellant formulations containing aluminum, US Patent 3,407,100, filed 28 October 1960 and issued 22 October 1968, Figure 1. J. F. Sparks and M. P. Friedlander III, "Fifty Years of Solid Propellant Technical Achievements at Atlantic Research Corporation," Paper A99-31568 presented at *35th AIAA Joint Propulsion Conference*, Los Angeles, CA, 20–24 June 1999, 8. Robert Gordon, *Aerojet: The Creative Company* (Aerojet History Group, 1995), Chapter IV, 83, 86. Figure IV-41 gives a graphic analysis of delivered specific impulse versus time, from which an estimate of a 14 percent increase can be calculated. In K. Klager, "Historical Breakthroughs and Their Effects on Solid Rocket Performance," *AIAA Meeting*, Monterey, CA, 14 July 1989. Figure 17 lists a 15 percent increase, with the previous figure being the same graphic as Figure IV-41. Combining data from Endnote 59, Figure 1, and Endnote 61, there is direct evidence for the 15 percent value. E. W. Price, "Comments on Role of Aluminum in Suppressing Instability in Solid Propellant Rocket Motors," *AIAA Journal* (May 1971): 987–990.
63. E. S. Sutton, "From Polysulfides to CTPB Binders: A Major Transition in Solid Propellant Binder Chemistry," paper presented at *AIAA/SAE/ASME 20th Joint Propulsion Conference*, Cincinnati, OH, 11–13 June 1984, 7.

# Chapter 5

1. Memorandum for the Secretary of the Army, Secretary of the Navy, Secretary of the Air Force, the Chairman of the Joint Chiefs of Staff, Assistant Secretary of Defense (R&D), Assistant Secretary of Defense (Comptroller), Subject: Intercontinental Ballistic Missile (ICBM) and Intermediate Range Ballistic Missile (IRBM) Programs, 8 November 1955, DNSA, NH00546, 4.
2. Memorandum from the Director of the Policy Planning Staff (Bowie) to the Acting Secretary of State, 7 September 1955, Foreign Relations of the United States, 1955–1957, Volume XXIX, 110. Approximately 180 programs had this same designation, causing the ICBM program to be seriously delayed.
3. J. M. Grimwood and F. Strowd, "History of the Jupiter Missile Program" (US Army Ordnance Missile Command, 1962), 23–35. Memorandum for the Chairman, Joint Army–Navy Ballistic Missiles Committee, Subj: IRBM #2 Program, Deputy Secretary of Defense, 20 December 1955, National Archives and Research Administration (NARA), 1.

4. C. Walker, *Atlas: The Ultimate Weapon* (Apogee Books, Ontario, Canada, 2005), 17–18.
5. R. A. Fuhrman, "Fleet Ballistic Missile System: Polaris to Trident," Paper 78-355, presented at *American Institute of Astronautics and Aeronautics 14th Annual Meeting*, February 1978, 2.
6. Gregory P. Kennedy, *The Rockets and Missiles of White Sands Proving Ground* (Schiffer Military History, Atglen, PA, 2009), 46.
7. Ibid., 74, 80, 85, 162. M. Rosen, *The Viking Rocket Story* (Harper Brothers, New York, NY, 1955). The entire book is worth reading—a highly detailed history of the program.
8. Grimwood and Strowd, "History of the Jupiter," 32. E. J. Kelly, Report of Silverstein Committee Meeting at ABMA, 23 July 1956, AFHRA, IRISNUM 01040241, K168.717-93, Reel 35258, 1005–1018.
9. The Scientific Advisory Committee's First Report on Ballistic Missiles to the Secretary of Defense, 11 February 1956, Office of the Secretary of Defense, DNSA NH00553, 2–3.
10. Ibid., 2.
11. William H. Avery, "The Talos Booster Rockets," *John Hopkins APL Technical Digest* 3, no. 2 (April–June 1982): 135–137.
12. Babcock, *Magnificent Mavericks*, 315–316.
13. Monthly Report on Progress of ICBM and IRBM Programs (Report No. 4), 31 March 1956, DDEL, Ann Whitman Series, Folder: Guided Missiles (6), Box 16, 3.
14. J. P. McManus, "A History of the FBM System" (Lockheed Space and Missile Company, 1989), A-2. Author's collection.
15. Babcock, *Magnificent Mavericks*, 331–334.
16. G. Weir, *An Ocean in Common* (Texas A&M University Press, 2001), 277.
17. See Chapter 4 for discussion of the evolution of solid-propellant performance. The dates do not coincide exactly with the improvement revealed in 1958, but the timeline suggests the Navy was aware of theoretical improvement in specific impulse with added aluminum.
18. H. York, *Race to Oblivion: A Participant's View of the Arms Race* (Simon and Schuster: New York, NY, 1970), 88–89. Given that the accuracy requirement was 0.1 percent of the range and, using this calculation, an accuracy of 1.5 nm for the FBM would have required a much smaller warhead for the same blast damage (author's calculation), York says, perhaps tongue in cheek, that a megaton, a million tons of TNT equivalent, was just a nice, round number.
19. Spinardi, *From Polaris to Trident*, 29–31. Thomas Reed, *At the Abyss: An Insider's Story of the Cold War* (Ballantine Books, New York, NY, 2005), 117.
20. Grimwood and Strowd, "History of the Jupiter," 34.
21. J. Davis, Memorandum for RDZP, Subject: Missile Nomenclature, 3 October 1955, AFHRA, IRISNUM 01040231, 168.717-83, Reel 35258, 324. Piper, *Development of the SM-80*, Vol. I, 7, footnote.
22. Hall had several years of experience with the problems associated with solid-propellant development, having worked to improve solid JATO devices both in cost and performance in the early 1950s. The first models of the new B-47 bomber required JATO augmentation when fully fueled and armed. Existing JATO units were underpowered and expensive, so Hall and his group at Wright Air Development Center's Non-Rotating Engine Branch had undertaken a study to find a less expensive and higher performance propellant combination. Hall had hands-on experience with current aspects of solid propellants, albeit in small rockets. Edward N. Hall, "Oral History Interview of Colonel Edward N. Hall, 11 July 1989," AFHRA, IRISNUM 01105443, K239.0512-1820 C.1, Reel 44388, 14.
23. Piper, *Development of the SM-80*, Vol. II, Memo, Lt. Col. E Hall, Ch, Prop Group, to Col. C. Terhune, D/Cmdr, Technical Ops, 16 June 1958 Subj: Solid Propellant Rockets for TBM, DNSA NH0024 Document 3, 1–2.
24. Ibid., Vol. II, Letter: Brig. Gen. Schriever to Prof. D. F. Gunder 15 June 1955, Document 2, 1.

25. Ibid., Vol. II, Memorandum For: Lt. Colonel Hall, Subject: Solid Propellant Rockets for TBM, 24 June 1955, Document 4, 1.
26. Ibid., Vol. I, xiii.
27. "The Scientific Advisory Committee's First Report on Ballistic Missiles to the Secretary of Defense, 11 February 1956, Office of the Secretary of Defense," DNSA NH0053, 2–3.
28. Piper, *Development of the SM-80*, Vol. II, From: Cmdr Western Development Division, To: Cmdr HQS ARDC, Attn: Lt. Colonel P. G. Ayres, 23 Feb 1956, Document 9, 1–6; Vol. I, 8–9.
29. Ibid., Vol. II, TWX, WDTLP-2-6-E, Col. C. H. Terhune, D/Cmdr, Technical Ops, to Cmdr, ARDC, 28 February 1956, Document 9, 1–6.
30. Ibid., Vol. II, Letter: Maj. Gen. B. A. Schriever, Cmdr, WDD, To: Cmdr ARDC, 31 May 1956, Subject: Solid Propellant Research Program, Document 14, 1.
31. Robert Gordon, ed., *Aerojet: The Creative Company* (Steward F. Cooper Company, 1995), Section IV, 15.
32. William True, personal communication, 31 May 2016. Gordon, *Aerojet: The Creative Company*, Section IV, 15–16.
33. As shown in the previous chapter, the first experiments demonstrating the efficacy of aluminum powder in solid propellants to improve specific impulse had taken place at ARC and had been shared with other industry representatives at a conference.
34. Minutes of the Fourth Meeting of the OSD Ballistic Missile Committee, 15 February 1956, 13 March 1956, Office of the Secretary of Defense, DNSA NH00557, 1. Edward N. Hall, *The Art of Destructive Management* (E. Hall, Vanguard Press, New York, NY, 1984). Hall's interviews and his book contain colorful, highly personal opinions and often conflict with official documents and other histories. Nonetheless, they are worth reading if for no other reason than gaining an appreciation of his character at the time.
35. Minutes of the Fifth Meeting OSD Ballistic Missiles Committee, 20 March 1956, DNSA NH 00560, 1–4.
36. Piper, *Development of the SM-80*, Vol. II, Letter, Maj. Gen. B. A. Schriever, Cmdr, WDD to Cmdr, ARDC, 31 May 1956, Document 14, 1.
37. Hall, *The Art of Destructive Management*, 52, 59, 60.
38. Cagle, *History of the Sergeant Weapon System*, Appendix III, 1.
39. D. Dyer and D. Sicilia, *The Labors of a Modern Hercules* (Harvard Business Press: Boston, MA, 1990), 258.
40. K. P Werrell, *The Evolution of the Cruise Missile* (Air University Press: Maxwell AFB, AL, 1985), 108–112. Matador and its variant, Mace, were deployed in Europe from 1955 to 1962 and 1959 to 1969, respectively; in Korea from 1959 to 1962 (Matador), and Okinawa, 1961 to 1969 (Mace), and in Taiwan (ROC) from 1959 to 1962 (Matador). Piper, *Development of the SM-80*, Vol. II, Letter: Lt. Gen. W. H. Tunner, Cmdr, USAFE, To: Dir of Requirements Hq USAF, 18 July 1956, Subject: Qualitative Operational Requirement for a Tactical Ballistic Missile, Document 17, 1–4.
41. Ibid., 2.
42. 15th Monthly Report on Progress of ICBM and IRBM Programs, 28 February 1957, Department of Defense, DDRS CK2349145607, 21.
43. Piper, *Development of the SM-80*, Vol. II, Letter: Lt. Gen. T. S. Power, Cmdr, ARDC, To Cmdr, WDD, 1 April 1957, Subject: Solid Propellant IRBM; Letter: Lt. Gen. D. L. Putt, D/CS Dev., To Cmdr, ARDC 20 March 1957, Document 18, 1–3.
44. Ibid.
45. 40 KS-100,000; 40 seconds firing time; K stands for rubberized polymeric propellant, S stands for solid propellant, and 100,000 designated thrust in pounds of force.
46. 16th Monthly Report on Progress of ICBM and IRBM Programs, 31 March 1957, Department of Defense, DDRS, CK2349149174, 21–22.

47. Piper, *Development of the SM-80*, Vol. II, Letter: Maj. Gen. J. F. Ferguson, Dir of Requirements, D/CS Dev, To Cmdr, ARDC, 23 April 1957, Subject: Tactical Surface to Surface Missiles, Document 19, 1–2.
48. 17th Monthly Report on Progress of ICBM and IRBM Programs, 30 April 1957, Department of Defense, DDRS, CK2349685638, 18–19.
49. 19th Monthly Report on Progress of ICBM and IRBM Programs, 30 June 1957, Department of Defense, DDRS, CK2349685667, 19.
50. Piper, *Development of the SM-80*, Vol. II, Memo, Col. P. Blasingame, Dir, WS-107A-2, To: Col. Terhune, D/Cmdr, Technical Ops, 17 Jul 1957, Subject: ICBM Solid Propellant Program, Document 21, 1–2.
51. Piper, *Development of the SM-80*, Vol. I, 14. See also Vol. II, GOR No. 161, 18 July 1957, Subject: General Operational Requirement Short Range Ballistic Missile Weapon System, Document 22, 1–4.
52. 20th Monthly Report on Progress of ICBM and IRBM Programs, 31 July 1957, Department of Defense, DDRS, CK2349685702, 15.
53. R. F. Bacher, "Report of a Panel Which Met to Study the Future Developments in Ballistic Missiles," 8 August 1957, AFHRA, unaccessioned collection, Ballistic Missile Organization, Box F4, unclassified, 4, 10.
54. Piper, *Development of the SM-80*, Vol. I, 15.
55. 21st Monthly Report on Progress of ICBM and IRBM Programs, 21 August 1957, Department of Defense, DDRS CK2349145648, 19.
56. Piper, *Development of the SM-80*, Vol. II, Letter: Maj. Gen. J. W. Sessums, V/Cmdr, ARDC, To: Maj. Gen. B. A. Schriever, Cmdr. AFBMD, 16 Aug 1957, no subject, Document 26, 1.
57. Ibid., Vol. II, Interoffice letter, Col. B. P. Blasingame, Dir, WS-107A-2, to Col C H Terhune, D/Cmdr, Technical Ops, 19 September 1957, Subject: New Ballistic Missile System Office, Document 27, 1–2.
58. Ibid., Vol. II, Letter: Maj. Gen. B. A. Schriever, Cmdr, AFBMD, To: Maj. Gen. J. W. Sessums, V/Cmdr, ARDC, 20 September 1957, no subject, Document 28, 1–3.
59. Ibid., Vol. II, Letter: Lt. Gen. S. E. Anderson, Cmdr, ARDC, to Gen. T. D. White, C/S USAF, 4 October 1957, no subject, Document 30, 1–2.
60. D. L. Snead, *The Gaither Committee, Eisenhower and the Cold War* (Ohio State University Press: Columbus, 1999), 44. "Deterrence and Survival in the Nuclear Age," 7 November 1957, DNSA, NH01329, 17.
61. Piper, *Development of the SM-80*, Vol. II, Memo, Col. E. N. Hall, Dir, Solid Ballistic Weapons, For the Record, 8 November 1957, Subject: Study to be Performed by Lockheed on the Conduct of the Development and Production Program for a Weapon System to Satisfy AF GOR No. 161, Document 33, 1–2. Reed, *At the Abyss*, 89. Hall's selection of the name reflected his desire to keep a low profile while developing the program. It was reminiscent of the Q-ships used in World War I and II as "unarmed" decoys to lure U-boats to the surface to sink the apparently unarmed ship by gunfire, only to find that the ship had uncovered deck guns with which to sink the submarine. The analogy was that under cover of the unassuming name of System Q, Hall could lie in wait and, at the right moment, overwhelm the Ramo-Wooldridge engineers and win the day for "his" System Q.
62. D. Dyer, *TRW: Pioneering Technology and Innovation since 1900* (Harvard Business School Press: Boston, MA, 1998), 198.
63. Piper, *Development of the SM-80*, Vol. II, Ltr. Col. E. N. Hall, Dir Solid Ballistic Weapons, To: Mr. A. F. Donovan and Mr. H. Lawrence (STL), 22 November 1957, Subject: Silo Configuration for Solid Rocket, Document 34, 1.
64. R. Neal, *Ace in the Hole: The Story of the Minuteman Missile* (Doubleday and Company, Inc.: Garden City, NY, 1962), 86–87. Hall, *The Art of Destructive Management*, 63.
65. "Proceedings of the Special Projects Office Steering Task Group—Task II, First Meeting,

25–26 July 1957," DNSA, NH00581, 49. See also endnote 67. Dyer and Sicilia, *Labors of a Modern Hercules*, 318–319.
66. Piper, *Development of the SM-80*, Vol. II, Memo, Col. E. N. Hall, Asst for Solid Ballistic Weapons, to Col C H Terhune, D/Cmdr, Weapon Systems, 27 December 1957, Subject: R/W Final Report on Advanced Guidance Systems Study, Phase I, Document 37, 1–2.
67. Piper, *Development of the SM-80*, Vol. II, Memo, Lt. Col. F. K. Bagby, Solid Ballistic Weapons Dir, For the Record, Low Cost "Silo" Launcher, Document 38, 1.
68. Ibid., Vol. II, Intercontinental Ballistic Missile Scientific Advisory Committee to the Secretary of the Air Force, Meeting 16 and 17 December 1957, Document 35, 3–4.
69. 25th Monthly Report on Progress of ICBM and IRBM Programs, 31 December 1957, DNSA, NH0063, 19–20. *House Supplemental Defense Appropriations for 1958*, 85th Cong. (1958), 193–196.
70. Piper, *Development of the SM-80*, Vol. II, Memo, Col. E. N. Hall, Asst for Solid Ballistic Weapons, To: Industrial Facilities, BMO, 30 December 1957, Subject: Industrial Facilities Program for Weapon System "Q," Document 40, 1.
71. 36th Meeting of the Office of the Secretary of Defense Ballistic Missile Committee, 7 January 1958, DNSA, NH00604, 4.
72. Memorandum of Conference with the President, 6 February 1958, DNSA, MS00253, 2.
73. Neil Sheehan, *A Fiery Peace in a Cold War* (Random House: New York, NY, 2009), 412. Prior to the briefing, apparently Schriever and Terhune thought a catchier name was necessary. Sentry, Sentinel, or Minuteman had been suggested by Hall and others.
74. Charles H. Eger Jr., "Minuteman Financial History, 1955–63," Systems Program Office, NARA, Samuel C. Phillips Collection, Box 11, Ballistic Systems Division, Folder 11, 8–9.
75. Ibid., iii–iv.
76. Ibid., 6–13.
77. Memorandum for the Record, Subject: Secretary of Defense Review of Air Force Ballistic Missile Proposals, 10 February 1958, AFHRA, IRISNUM 01040356, Reel 035259, 1–2. See also Piper, *Development of the SM-80*, Vol. II, Memo, Col. R. E. Soper, Cmdr Ballistic Div., AFCGM, For the Record, 10 February 1958, Subject: Secretary of Defense Review of Air Force Ballistic Missile Proposals, Document 50, 1–2. Sheehan, *A Fiery Peace*, 414–416. The actual detailed briefing materials for Hall's series of presentations remain classified. Several versions of the initial higher headquarters' briefings on Minuteman exist. The gist was that he briefed Terhune on his program plan, and then Terhune brought Hall into Schriever's office. Schriever then called Lt. Gen. Putt, Air Force deputy chief of staff, and told him they wanted to come to Washington to brief him on Minuteman later that month, January 1958. Schriever also asked Putt to set up additional briefings, including the secretary of the Air Force, James Douglas. The briefing was delayed until 6 February 1958, with the first one being the Air Force Council with Gen. Curtis LeMay, vice chief of staff, as its chairman. After a long and detailed presentation, LeMay agreed to the Minuteman concept. Next was Secretary Douglas, also successful. The last was Secretary of Defense Neil McElroy. LeMay and Douglas joined Hall, Terhune, and Schriever in the meeting, where LeMay was vigorously supportive of the concept. McElroy agreed, and then the real work began.
78. Piper, *Development of the SM-80*, Vol. II, Memo, Mr. R. J. Douglas, SAF, to Mr. N. H. McElroy, S/Def, 10 February 1958, Subject: Air Force Ballistic Missile Objectives, Document 49, 1–2.
79. Memorandum for Dr. Killian, Subject: Technical Progress and Actions Required in the Long-Range Ballistic Missile Program, 13 February 1958, DNSA, NH00611, 2.
80. Memorandum by the Chief of Staff of the Air Force for the Joint Chiefs of Staff on Evaluation of the "Minuteman" Ballistic Missile Program, 18 February 1958, DNSA, NH00613, 1–2.
81. Minutes of the 38th Meeting of the OSD Ballistic Missiles Committee Held 19 February 1958, DNSA, NH00612, 1.

82. Piper, *Development of the SM-80*, Vol. II, Memo, Brig. Gen. C. M. McCorkle, Asst C/S for Guided Missiles, To: Dir Guided Missiles, OSD, 21 February 1958, Subject: ICBM Force Objectives, Document 53, 2.
83. Ibid., Vol. II, Memo, Col. E. N. Hall, Asst for Weapon System 'Q,' to Col. C. H. Terhune, D/Cmdr, Weapons Systems, 24 February 1958, Comments on AFCGM Memo for Director of Guided Missiles, OSD, dated 21 February 1958 (58-01227), Document 55, 1.
84. Ibid., Vol. II, Letter: Maj. Gen. B. A. Schriever, Cmdr, AFBMD, to Cmdr, ARDC, 24 February 1958, Subject: Ballistic Missile Program Recommendations, Document 54, 2–4. Eger, "Minuteman Financial History," 25.
85. Ibid., Vol. II, Memorandum, W. M. Holaday, Dir/GM, DOD, To SAF/J H Douglas, 27 February 1958, Subject Air Force Ballistic Missile Objectives, Document 57, 1. Chronology of Significant Events and Decisions Relating to the US Missile and Earth Satellite Development Programs, Supplement I, October 1957 through October 1958, DNSA, NH0010, 41.
86. From: HED USAF, WASHDC, to: COMDR AFMD (ARDC) Inglewood, CA, 18 February 1958, AFHRA, IRISNUM 01040356, Reel 0035259, 1.
87. Piper, *Development of the SM-80*, Vol. II, TWX from: Cmdr ARDC Andrews AFB (Lt. Col. Wortham) to Com AFBMD, Inglewood, CA (Col Terhune), 4 March 1958, Document 63, 1; TWX from HEDUSAF, to: Cmdr AFDBMD, Info: Cmdr ARDC, Cmdr AMC, Cmdr SAC, Cmdr AFMTC, 18 June 1958, Document 81, 1.
88. Memorandum for Dr. J. R. Killian Jr., from: The Ballistic Missiles Panel, Subject: Whither Ballistic Missile Systems, 4 March 1958, DNSA, NH00616, 7–9.
89. Letter: J. R. Killian to President Eisenhower, 8 March 1958, DDEL, Whitman File, Administration Series, Box 17, Folder: Guided Missiles, 1–5.
90. Memorandum of Conference with the President, 10 March 1958, DDRS CK2349256057, 1.
91. Ibid., 1–2.
92. Piper, *Development of the SM-80*, Vol. II, Solid Propellant (Minuteman), Summary of Presentation on 14 March 1958 to the Scientific Advisory Committee, Document 56, 1–7. This projected deployment date was essentially met to the month.
93. Piper, *Development of the SM-80*, Vol. II, Memorandum for the Chief of Staff, from: Maj. Gen. McCorkle, Asst C/S for Guided Missiles, Subject: Minuteman Funds for Contracts and Industrial Facilities, Document 73, 1.
94. Neal, *Ace in the Hole*, 96–97. York, *Race to Oblivion*, 97–98.
95. Eger, "Minuteman Financial History," 26.
96. Piper, *Development of the SM-80*, Vol. II, Proposed Preliminary Operational Concept for Minuteman, Document 76, 1–10.
97. Interoffice Correspondence, to: Maj. Gen. B. A. Schriever, from: Louis G. Dunn, Subject: STL Recommendations Concerning Management of Minuteman Program, 28 March 1958, AFHRA, IRISNUM 01040315, Reel 0035264, Frames 372–375, 1–5.
98. Piper, *Development of the SM-80*, Vol. II, Memorandum for Colonel Terhune, Subject: STL Role in Minuteman Program, 9 April 1958, Document 77, 1.
99. Ibid., Vol. I, 32. Dyer, *TRW*, 206–208.
100. Letter: Dr. G. B. Kistiakowsky, Scientific Advisor to the President, from: Dr. William McEwan, No Subject, 28 May 1958, DDEL, White House Office, Office of the Special Assistant for Science and Technology, Box 5, Folder: Solid Propellants, May 1958, 1–2.
101. Memorandum for: Dr. J. R. Killian, from: Ballistic Missile Panel, Subject: Status of Ballistic Missile Programs, 17 June 1958, DDEL, White House Office, Office of Staff Secretary, Subject Series, Department of Defense, Subseries, Box 6: Military Doctrine and Organization, Missiles and Satellites, Vol. II (3) 16 March–30 June 1958, 10–13. Somewhat intriguing to the author is the fact that, while drafted and apparently sent to Dr. Bode on that date, the same memo, at least as far as Minuteman is concerned, was sent somewhere on 18 July 1958 and revised on 11 August 1958, as indicated in this file.

102. Ibid.
103. Ibid.
104. Piper, *Development of the SM-80*, Vol. II, Proposed Preliminary Operational Concept for Minuteman, Document 76, 1–10.
105. Ibid., Vol. II, Memorandum for the Chairman, OSD Ballistic Missile Committee, from: J. H. Douglas, SAF, Subject Minuteman Development Plan and FY 1959 Program, 11 August 1958, Document 87, 1–2; Vol. I, 36–38. "Four Companies Assigned Work on Solid-Fuel Minuteman," *New York Times*, 20 July 1958, 21. Eger, "Minuteman Financial History," 32.
106. Letter from Dr. Clark B. Millikan, Dir/GAL, to: W. H. Holaday. Dir/GM SOD, 24 July 1958, AFHRA, IRISNUM 01040260, 168.7171-112, Reel 035259, 1–2.
107. Eger, "Minuteman Financial History," 33.
108. Letter from H. W. Bode, Dir/Research, Bell Laboratories, to: G. B. Kistiakowsky, Dept of Chemistry, Harvard University, 1 August 1958, DNSA, NH00631, 1–3.
109. Piper, *Development of the SM-80*, Vol. II, Memorandum for the Chairman, Joint Chiefs of Staff, from W. H. Holaday, Dir/GM SOD, Subject Minuteman Program, 7 August 1958, Document 86, 1–2.
110. Eger, "Minuteman Financial History," 36.
111. Piper, *Development of the SM-80*, Vol. II, General Operational Requirement for a Quick Reaction Intercontinental Ballistic Missile System, Document 85.
112. Ibid., Vol. II., Letter, Maj. Gen. B. A. Schriever, Cmdr AFBMD, to: C/S, USAF, Subject Ballistic Missile Programming Information, 29 August 1958, Document 88, 1–6.
113. Eger, "Minuteman Financial History," 36–37.
114. "US to Hold Back Extra Arms Funds and Trim Forces," *New York Times*, 12 September 1958, 1.
115. Piper, *Development of the SM-80*, Vol. II, Memorandum, W. M. Holaday, Dir/GM, DOD, to SAF, Subject: Minuteman Program, 17 September 1958, Document 91.
116. Ibid., Vol. II, Memorandum, Maj. Gen. B. L. Funk, Dep Dir of Ballistic Missile Proc and Prod (Cmdr AMC-BMC, Inglewood) to Maj. Gen. B. A. Schriever, Cmdr, AFBMD, Subject Minuteman Facilities Funding, Document 90, 1.
117. "National Implications of Atomic Parity," Naval Warfare Analysis Group Study No. 5, 20 November 1957, DNSA, NH00094 3.
118. W. Gunston, *The Illustrated Encyclopedia of the World's Rockets and Missiles* (Salamander Books, Ltd.: London, UK, 1979), 40–41. Maj. Gen. J. B. Medaris and A. Gordon, *Countdown to Decision* (G. P. Putnam's Sons: New York, NY, 1960), 302.
119. V. Adm. J. H. Sides, "Presentation by the Director, Weapons Systems Evaluation Group to the National Security Council on the Subject of Offensive and Defensive Weapons," 13 October 1958, DNSA, Nuclear History 1955–1968, NH00411, 17–18.
120. Memorandum for Mr. Gordon Gray, from: J. R. Killian, Subject Possible Further Studies Relevant to the NSC Action 1994 on the Evaluation of Offensive and Defensive Weapons Systems, 29 October 1958, DNSA, NH01345, 2.
121. Piper, *Development of the SM-80*, Vol. II, Letter, Lt. Gen. S. E. Anderson, Cmdr/ARDC, from: Brig. Gen. O. J. Ritland, VCmdr/AFBMD, 5 November 1958, Document 92, 1–2; Memorandum for Col. Terhune, from Col. Otto Glasser, Director, Weapon System 133A, 7 November 1958, Document 93, 1.
122. Eger, "Minuteman Financial History," 40–42.
123. Piper, *Development of the SM-80*, Vol. I, 46.
124. Ibid., Vol. II, Subject WS 133A Priority, to Commander ARDC, from Col. John R. V. Dickinson, Asst Dir/Research and Development, DOD, 22 December 1958, Document 98, 1.
125. Personal papers, Maj. Gen. B. A. Schriever, AFHRA, IRISNUM 01040266, 168.7171-118, Reel 35259, 1.
126. Eger, "Minuteman Financial History," 43.
127. Piper, *Development of the SM-80*, Vol. I, 50–59.

128. Minutes of the 48th Meeting OSD Ballistic Missiles Committee, 13 May 1959, DNSA, NH 00663, 2–3.
129. Eger, "Minuteman Financial History," 47.
130. Piper, *Development of the SM-80*, Vol. I, 53.
131. Report by the J5 to the Joint Chiefs of Staff on Minuteman Program, 21 August 1959, DDRS, CK 2349456775, 1968–1969.
132. Eger, "Minuteman Financial History," 53.

# Chapter 6

1. "Minuteman Historical Summary, 1958–1959" (The Boeing Company, Seattle, WA, 1974), D2-26485, Sheet II, 1.
2. Ibid., Sheet II, 4.
3. Thomas White, "SAC and the Ballistic Missile," in *The United States Air Force Report on the Ballistic Missile*, edited by Kenneth Gantz (Doubleday and Company, Garden City, NY, 1958), 186.
4. "Defense: The Second Generation," *Time*, 10 March 1958, 16–17.
5. "Minuteman Historical Summary, 1958–1959," Sheet II, 6.
6. Ibid., Sheet II, xii, 5–17.
7. Ibid., Sheet XI, 5.
8. Ibid., Sheet XI, 6.
9. Ibid., Sheet XI, 7–8.
10. Ibid., Sheet XI, 10–11.
11. Ibid., Sheet XI, 13–14.
12. Piper, *Development of the SM-80*, 57–58.
13. Ibid., 83; Vol. II, Letter, Lt. Gen. R. C. Wilson, DCS/D, USAF, to Gen. C. E. LeMay, Vice C/S, USAF, 22 December 1959, Subject: Vice Chief of Staff Meeting of 7 December 1959, Supporting Document 128, 1.
14. Ibid., Vol. II, Ltr, Lt. Gen. B. A. Schriever, Cmdr, ARDC, to Gen. T. D. White, C/S USAF, 21 March 1960, Subj: Minuteman Industrial Facilities, Document 135, 1.
15. "Minuteman Historical Summary, 1958–1959," Sheet XI, 16. "History of Hill Air Force Base" (History Office, Ogden Air Logistics Center, Hill AFB, UT), 182. "Minuteman Program Summary, 27 October 1959" (Air Force Directorate of Research and Development, Headquarters, United States Air Force), DNSA, NH00677, 1–2.
16. *Statement of the Secretary of Defense Robert S. McNamara before the Senate Committee on Armed Services* (1961), DNSA, NH00426, 12. "Strategic Air Command Weapon Systems Acquisition 1964–1979," Vol. I, AFHRA, IRISNUM 01114095, K416.04-39 V.1, Reel 45868, 184.
17. "Minuteman Historical Summary, 1960–1961" (The Boeing Company, Seattle, WA, 1974), D2-26485, Sheet XI, 12–17.
18. Memorandum, Dir., Joint Staff, to the JCS. JCS 2277/1, 17 October 1957, DNSA, NH00590, 2–6.
19. Frederick J. Shaw, ed., *Locating Air Force Base Sites, History's Legacy* (Air Force History Museums Program, United States Air Force, Washington, DC, 2004), 65.
20. Piper, *Development of the SM-80*, Vol. I, 116.
21. *Civil Defense Part III: Relation to Missile Programs, Hearings before a Subcommittee of the Committee on Government Operations* (H. R.) 86th Cong. (1961), 137–138, 150.
22. Clyde Littlefield, "The Site Program, 1961, Vol. III, History of Deputy Commander Aerospace Systems" (DCAS History Office, 1962), AFHRA, IRISNUM 897237, K243.012-8, Vol. 1, Reel 23733, Supporting Document 94, 1–3.

23. Chuck Walker, *Atlas: The Ultimate Weapon* (Apogee Space Books, 1971), 281. Stumpf, *Titan II: A History*, 105.
24. "History of the 341st Strategic Missile Wing, 1–31 December 1961," Vol. 1, AFHRA, IRISNUM 456129, KWG341Hi, Reel 14470, 23. "Strategic Air Command Missile Chronology 1939–1973" (Office of the Historian, Headquarters Strategic Air Command, Offutt AFB, NE, 2 September 1975), AFHRA, IRISNUM 01011276, K416.052-10, Reel 31400, no pagination, reference by date.
25. Piper, *Development of the SM-80*, Vol. II, GOR-171, 6 August 1958, Subj: For Quick Reaction ICBM System, Document 85, 1–4. Eger, "Minuteman Financial History," III. Bernard Nalty, "United States Air Force Ballistic Missile Programs: 1962–1964" (Office of Air Force History, April 1966). Redacted copy, DDRS, CK2349654709, 17.
26. Neal, *Ace in the Hole*, 68.
27. Charles Lauritsen, "Lauritsen Committee Report on Minuteman," AFHRA, IRISNUM 01040303, 168.7171-155, Reel 35263, 9–10.
28. Theodore Runyon, Minutes of the 56th Meeting of the Air Force Ballistic Missile Committee, 2 August 1960, AFHRA, K243.8631-2030, IRISNUM 02055699, 3–4. This document is classified; the extract used is declassified IAW EO 13526.
29. Littlefield, "The Site Program," 68–69. TWX, MCG 031, AMC to BMC, 5 January 1961, Supporting Document 17, 1.
30. Ibid., TWX, MCG 031, 1.
31. Ibid., Ltr, Col. S. G. Phillips, Dep Cmdr for Minuteman, BSD, to Maj. Gen. T. P. Gerrity, Cmdr, BSD, 28 April 1961, Subj: Minuteman Sites; Atch to Ltr, Status of Minuteman Site Exploration Program, Supporting Document 50, 2-3, 69.
32. Piper, *Development of the SM-80*, Vol. I., 70.
33. Littlefield, "The Site Program," 71. "Minuteman Bases Ring Cheyenne," *New York Times*, 10 December 1962, 10.
34. "Minuteman Deployment Study" (Headquarters Strategic Air Command, Offutt AFB, NE, 1962), Muir S. Fairchild Research Information Center, Maxwell AFB, AL, M-U 39147-21, 1–72.
35. Richard Coulter, "Ltr to Brig. Gen. S. Phillips, 15 July 1962, Subj: Siting," Samuel C. Phillips Papers, NARA, Box 9, Folder 7.
36. Holm, "Chronology of the Ballistic," 87, 90. The author thanks Jonathan McDowell for a copy of this document. History of Grand Forks Area during Construction of Wing VI Minuteman II ICBM, Corps of Engineers Ballistic Missile Construction Office, Grand Forks AFB, ND, 1 May 1966 (Army Corps of Engineers History Office, Fort Belvoir, MD), Military Files XVIII-10-1, Box 10, Bd 1, Appendix I-2, 71.
37. Ruggiero, "Missileers' Heritage," 65.
38. W. R. Large Jr., "Ballistic Missile Hardening Study, 10 July 1958" (Air Force Materiel Center History Office, Wright-Patterson AFB, OH), Titan Missile Museum National Historic Landmark (TMM), Sahuarita, AZ, David K. Stumpf Collection, 29,35.
39. "Missile Site Separation, October 1959" (Advanced Planning Office, Deputy Commander Ballistic Missiles, Air Force Ballistic Missile Division, Air Materiel Command, Wright-Patterson AFB, OH), TMM, David K. Stumpf Collection, ii.
40. Ibid., 54.
41. Ibid., 3.
42. Nalty, "United States Air Force Ballistic Missile Programs 1962–1964," 3. Littlefield, "The Site Program," Ltr, Maj. Gen. R. M. Montgomery, Asst Vice Chief of Staff, USAF, to SAC, AFLC, AFSC, D, and BSD, 16 November 1961, Subj: ICBM Site Selection Criteria, Supporting Document 125, 2–3. History of Construction Activities and Contract Administration Phases Encountered by the Whiteman Area Office Corps of Engineers, CEBMCO, during Construction of Minuteman

Strategic Missile Wing IV, Whiteman AFB, MO (Corps of Engineers History Office, Beltsville, MD), Military Files XVIII, Box 112, Bd 3, III-11.

43. S. Glasstone and P. J. Dolan, eds., *The Effects of Nuclear Weapons*, Third Edition (Department of Defense and Energy Research and Development Administration, 1977), 253–256. "Missile Site Separation," 6.
44. Glasstone and Dolan, *Effects of Nuclear Weapons*, 233–234.
45. Design Analysis: Weapon System 107 A-2, Titan II Operational Base Facilities, Phase II Construction, Volume II, 5 January 1961, TMM, David K. Stumpf collection, 2.
46. Nalty, "United States Air Force Ballistic Missile Programs 1962–1964," 3. Holm, "Chronology of the Ballistic," 76. Weapons Systems Evaluation Report No. 50: Evaluation of Strategic Offensive Weapons Systems, 27 December 1960 (Weapons Systems Evaluation Group, Office of the Department of Defense Research and Evaluation), DNSA, NH00422, Appendix B to Enclosure F, Estimated Cost of Minuteman Weapon System, 41.
47. Piper, *Development of the SM-80*, Vol. II, TWX, WDF-29-8-43, Hq AFBMD, to Hq ARDC, 29 August 1960," Supporting Document 145, 2.
48. Ibid., Ltr, Col. J. L. McCoy, Deputy Commander, Ballistic Missiles, to Lt. Gen. O. J. Ritland, Commander, Air Force Ballistic Missile Division, Air Research and Development Command, 5 May 1960, Sub: Increased Hardness for Minuteman Silo Launcher, Supporting Document 141, 1; "Minuteman Chronology 1955–1961" (Historian's Office, Headquarters, Air Force Ballistic Missile Division, April 1961), Air Force Historical Support Division, Joint Base Anacostia–Bolling, Washington, DC, 84.
49. Weapon System Operation Instructions, Wing II, USAF Series LGM30B Missile, Technical Operation Manual, T. O. 21M-LGM30B-1-3, 1 April 1969, Association of Air Force Missileers (AAFM) Collection, Section I-7.
50. Littlefield, "The Site Program," TWX, BSQ-10-9-2h, Maj. Gen. T. P. Gerrity, Cmdr, BSD, to Lt. Gen. B. A. Schriever, Cmdr AFSC, 10 September 1961, Supporting Document 106, 1.
51. Ibid.
52. Ibid., 80–81.
53. Eger, "Minuteman Financial History," 7.
54. Ibid., 49–50.
55. "Minuteman Historical Summary, 1960–1961," Sheet I, 11–12.
56. "Effect of Depth Increase, Launch Tube, Wing V, WS-133A, 21 May 1963" (The Boeing Company, Seattle, WA, 1963), DTIC 407476, 7–8.
57. WS-133A Technical Facilities, First Operational Deployment Area, Malmstrom AFB, MT (Parsons–Wenzel Architects & Engineers, Los Angeles, CA, and Great Falls, MT, 10 October 1960), AAFM Collection, Sheets S-19, S-20, S-21, S-21B, S-32, S-32.1, S-33, S-38.
58. "Suspension System Comparison," *Minuteman Service News*, March–April 1963, 12–13. The author thanks Monte Watts for the digital collection of the *Minuteman Service News*.
59. Ibid. "Study of Shock Isolation for Hardened Structures, 17 June 1966" (Office of the Chief of Engineers, Washington, DC), DTIC, AD0639303, 568–571.
60. "Minuteman III Is Coming," *Minuteman Service News*, May–June 1968, 3.
61. "What's Force Mod?" *Minuteman Service News*, November–December 1965, 11–12.
62. "MM III Suspension Cable Links," *Minuteman Service News*, March–April 1970, 11–12.
63. "Wing V Missile Support System," *Minuteman Service News*, May–June 1973, 6–8.
64. Technical Order 21M-LGM30F-2-22, Organizational Maintenance Instructions, Missile Umbilicals and Missile Suspension System (VAFB and Wing II), USAF Series LGM30F, 30 November 1992, AAFM Collection, 1-1.
65. Minuteman Illustrated Technical Requirements, D2-31384-1 (The Boeing Company, Seattle, WA), Sheet 173.
66. Robert C. Anderson, "Minuteman to MX, the ICBM Revolution," *Quest* 3 (Autumn 1979): 42.

67. "History of Minuteman Construction Wing II Ellsworth Area Engineer Office US Army Corps of Engineers, 1 August 1961–31 August 1963" (Army Corps of Engineers History Office, Fort Belvoir, MD), Military Files XVIII, Box 9 BD 3, 154.
68. Luther Stenvik, *The Agile Giant: A History of the Minuteman Production Board* (The Boeing Company, Seattle, WA, 1966), 53. "Summary Report: Minuteman Flexibility and Safety Study Group, 27 September 1961" (Department of Defense), AFHRA, IRISNUM 2054414, K243.8631-745, Section 4, 7. This document is classified; the extract used is declassified IAW EO 13526.
69. WS-133B Technical Facilities, Collocated Squadron No. 20, Part II of III Architectural–Structural, Malmstrom AFB, MT (Ralph M. Parsons Architects–Engineers, Los Angeles, CA, 4 January 1965), AAFM Collection, Sheet S-23, S-25, M-8. "Minuteman II Collimator Alignment," *Minuteman Service News*, January–February 1965, 5. "What's New for Minuteman II," *Minuteman Service News*, July–August 1964, 3–4. "Study of Shock Isolation," 581.
70. WS-133A Technical Facilities, First Operational Deployment Area, Part IV of IV, Malmstrom AFB, MT, Sheet S-23, S-25, M-8.
71. "Debriefing Exercise, 12 July 1963," Site Activation Task Force Detachment 18, Malmstrom AFB, MT, AFHRA, IRISNUM 919736, K243.0121-18, Reel 26199, 62.
72. Technical Order 21M-LGM30G-1-10, Weapon System Operation Instructions (VAFB, Wing I Squadron Four, Wing VI), USAF Series LGM30G Missile, 1 March 1978, January 1975, AAFM Collection, 1-9.
73. WS-133A Technical Facilities, First Operational Deployment Area, Part IV of IV, Malmstrom AFB, MT, AAFM Collection, Sheet A-13. Technical Order 21M-LGM30B-1-3, Weapon System Operation Instructions, Wing II, USAF Series LGM30B Missile, 1 April 1969, AAFM Collection, 1-8.
74. WS-133A Technical Facilities, First Operational Deployment Area, Minot AFB, ND, AAFM Collection, Sheets A-8, S-57.
75. Ibid., Sheet S-59.
76. Nalty, "United States Air Force Ballistic Missile Programs 1962–1964," 17.
77. Technical Order 21M-LGM30F-1-6, Weapon System Operation Instructions, VAFB; Wing I Squadron Four; Wing VI, USAF Series LGM30F and LGM30G Missile, 29 January 1975, AAFM Collection, 1-9. WS-133B Technical Facilities, Collocated Squadron No. 20, Part II of III Architectural–Structural, Malmstrom AFB, MT (Ralph M. Parsons Architects–Engineers, Los Angeles, CA, 4 January 1965), AAFM Collection, Sheet S-90 to S-100.
78. Technical Order 21M-LGM30B-1-3, 1-8.
79. WS-133A Technical Facilities, First Operational Deployment Area, Minot AFB, ND (Parsons–Stavern Architects–Engineers, Los Angeles, CA, and Rapid City, SD, 15 November 1961), AAFM Collection, Sheet A-1.
80. WS-133A Technical Facilities, First Operational Deployment Area, Malmstrom AFB, MT, Sheet S-2, S-4, S-9.
81. "Study of Shock Isolation," 591, 598.
82. Technical Order 21M-LGM30B-1-3, 1-9 to 1-10.
83. WS-133A Technical Facilities, First Operational Deployment Area, Malmstrom AFB, MT, Sheet S-2, S-4, S-9, S-11.
84. Ibid, Sheet S-11. Technical Order 21M-LGM30F-1-6, Section 1, 80.
85. WS-133A Technical Facilities, First Operational Deployment Area, Minot AFB, ND, Sheets S-52, S-53, S-54, S-55, S-56, M-6, M-18, M-21.
86. Mitch Cannon, personal communications, December 2017.
87. Technical Order 21M-LGM30F-1-6, Section 1, 37–42.
88. WS-133B Technical Facilities, Collocated Squadron No. 20, Part II of III Architectural–Structural, Malmstrom AFB, MT, AAFM Collection, Sheets S-32, S-44. Technical Order 21M-LGM30F-1-6, Section 1, 37.

## Chapter 7

1. Harry E. Goldsworthy, Lt. Gen., USAF (ret.), "ICBM Site Activation," *Aerospace Historian* 29, no.3 (1982): 154.
2. Ibid., 155.
3. Ibid., 156–158.
4. "Air Force Intercontinental Ballistic Missile Construction Program," *Hearings before the Subcommittee of the Committee on Appropriations* (H. R.) 87th Cong. (1961), 220.
5. "Air Force Intercontinental Ballistic Missile Construction Program," Report by the Committee on Appropriations (H. R.), 3 March 1961, 3.
6. Goldsworthy, "ICBM Site Activation," 154–161.
7. Piper, *Development of the SM-80*, Vol. I, 78.
8. Littlefield, "The Site Program," 71. "Minuteman Historical Summary, 1960–1961," D2-26485, Sheet XV, 5.
9. Littlefield, "The Site Program," Minuteman Construction, 6 January 1961, Supporting Document 22, 1.
10. Littlefield, "The Site Program," 71–73.
11. "History of Minuteman Construction, Wing II, Ellsworth Area Engineer Office, US Army Corps of Engineers, 1 August 1961–31 August 1963" (Corps of Engineers History Office, Beltsville, MD), Military Files XVIII, Box 9, Bd 3, 38–178. Unless otherwise noted, the descriptions of construction are from this report.
12. Ibid., 170.
13. Ibid., 44–46.
14. Ibid., 47–48.
15. Ibid., 50–51.
16. Ibid., 53–56.
17. Ibid., 56.
18. Ibid., 57–60.
19. Ibid., 61–70.
20. Ibid., 81–82.
21. Ibid., 82.
22. Ibid., 45.
23. Ibid., 62–65.
24. Ibid., 65.
25. Ibid., 89.
26. Ibid., 89.
27. Ibid., 145–146.
28. WS-133A Minuteman Missile Facilities, Vol. I, Minot AFB, Minot, ND (US Army Corps of Engineers, HQ, Office of History, Fort Belvoir, MD), Military Files XVIII, Box 11, Bd 2, 114–115.
29. "Warren Area Minuteman History" (US Army Corps of Engineers, HQ, Office of History), Military Files XVIII, Box 12, Bd 1, 80–81. History of Grand Forks Area during Construction of Wing VI Minuteman II ICBM (US Army Corps of Engineers, HQ, Office of History), Military Files XVIII, Box 10, Bd 1, 18. History of Malmstrom Area during Construction of Collocated Squadron Number 20 Minuteman II ICBM Facilities, 31 December 1966, (US Army Corps of Engineers, HQ, Office of History), Military Files XVIII, Box 10, Bd 5, 4–3.
30. History of Grand Forks, 27–28.
31. Minuteman Technical Facilities, Malmstrom AFB, Great Falls, MT (US Army Corps of Engineers, HQ, Office of History), Military Files XVIII, Box 10, Bd 3, Part 7, 1–6.
32. "Minuteman Historical Summary, 1958–1959," Sheet VIII, 1–7.

33. "Minuteman Historical Summary, 1960–1961" (The Boeing Company, Seattle, WA, 1974), D2-26485, Sheet XIII, 1–4.
34. Ibid., 4–7.
35. Minuteman Technical Facilities, Malmstrom AFB, Great Falls, MT, Part 7, 1. "Strategic Air Command Missile Chronology, 1939 to 1973" (Office of the Historian, Headquarters, Strategic Air Command, September 1975), no page numbers, referenced by date. The author thanks Greg Ogletree for a copy of this document.
36. T. J. Hayes III, "Engineers Dig in for Defense," *Army Information Digest,* November 1962, 25–27. Minuteman Technical Facilities, Malmstrom AFB, Part 7, 2.
37. 341st Strategic Missile Wing and 341st Combat Support Group, 1–30 November 1961, Malmstrom AFB, AFHRA, IRISREF N0427, K-WG-341, Vol. I, 19.
38. 341st Strategic Missile Wing and 341st Combat Support Group, 1–31 December 1961, AFHRA, IRISREF N0427, K-WG-341 Vol. I, 3.
39. 341st Strategic Missile Wing and 341st Combat Support Group, 1–31 January 1962, AFHRA, IRISREF N0427 K-WG-341 Vol. I, 69–76.
40. Minuteman Technical Facilities, Malmstrom AFB, Military Files XVIII, Box 10, Bd 5, Part 3, 3.
41. "Final Report: Project High Climber Missile Site Transportation Service Test, 15 January 1962," Headquarters, 341st Strategic Missile Wing, Malmstrom AFB, MT, AFHRA, IRISREF N0427, 15.
42. The De Havilland Beaver L-20 was redesignated U-6A in 1962. The Helio Super Courier L-28 was redesignated U-10 in 1962.
43. "Final Report: Project High Climber," 56.
44. 341st Strategic Missile Wing and 341st Combat Support Group, 1–31 March 1962, Malmstrom AFB, AFHRA, IRISREF N0427, 4, 25–30.
45. 10th Strategic Missile Squadron, 341st Strategic Missile Wing and 341st Combat Support Group, 1–30 April 1962, Malmstrom AFB, AFHRA, IRISREF N0427, 15.
46. Ibid., Operations Plan (Test Hop) 404-2 341st Strategic Missile Wing Missile Combat Crew 24-Hour Shift Test, Document 6, 1–2, 8.
47. 341st Strategic Missile Wing and 341st Combat Support Group, 1–30 April 1962, 4, 25–30.
48. 341st Strategic Missile Wing and 341st Combat Support Group, 1–31 May 1962, Malmstrom AFB, AFHRA, IRISREF N0428, 25–30. "From Snark to Peacekeeper: A Pictorial History of Strategic Air Command Missiles" (Office of the Historian, Headquarters Strategic Air Command, Offutt AFB, NE, 1 May 1990), 29.
49. 341st Strategic Missile Wing and 341st Combat Support Group, 1–31 May 1962, Operations, 9.
50. Ibid., Operations Plan (Test Hop) 405-62, 341st Strategic Missile Wing Missile Combat Crew 48-Hour Shift Test, Document 1, 2.
51. 341st Strategic Missile Wing and 341st Combat Support Group, 1 June–31 July 1962, AFHRA, IRISNUM 456135 Reel 14471, 40–43.
52. Ibid., 28.
53. Ibid., 21–24.
54. 341st Strategic Missile Wing and 341st Combat Support Group, 1–31 August 1962, 11–13.
55. Ibid., SM-80 Minuteman Program Programming Plan 2-61, Progress Report No. 15, Report of August 1962, AFHRA, IRISREF N0428, 9,15.
56. Ibid., 33.
57. 341st Strategic Missile Wing and 341st Combat Support Group, 1–30 September 1962, AFHRA, IRISREF N0428, 27.
58. Maintenance Plan: Demonstration and Acceptance, Inventory and Turnover Minuteman Weapon System Number 2-62, 341st Strategic Missile Wing and 341st Combat Support Group, 1–31 May 1962, Malmstrom AFB, AFHRA, IRISREF N0428, Annex C: Tab A.

59. 341st Strategic Missile Wing and 341st Combat Support Group, 1–30 November 1962, AFHRA, IRISREF N0428, 13; "Speech by Lt. Gen. T. P. Gerrity," DCS/Systems and Logistics, Headquarters, United States Air Force, 11 December 1962, 341st Strategic Missile Wing and 341st Combat Support Group, 1–30 December 1962, AFHRA, IRISREF N0428, Document 3, 7.
60. 341st Strategic Missile Wing and 341st Combat Support Group, 1–30 December 1962, 11. See Chapter 14 for further description of Emergency Combat Capability.
61. "Minuteman Historical Summary, 1962–1963" (The Boeing Company, Seattle, WA, 1975), D2-26485, Sheet XI, 47.
62. 341st Strategic Missile Wing and 341st Combat Support Group History, 1–31 October 1963, AFHRA, IRISNUM 00456152, K-WG-341-HI, 9. "Strategic Air Command Missile Chronology, 1939 to 1973," no pagination, reference by date. "History of the Site Activation Task Force, Malmstrom AFB, MT, 1 January–3 July 1963," AFHRA, IRISNUM 919730, K243.0121-17, 75.
63. Ibid., 82, 85.
64. "History of the Strategic Air Command, Fiscal Year 1985" (History Office, Air Force Space Command), On Alert reference collection, Frame 599.
65. "Minuteman Historical Summary, 1958–1959," Sheet IX, 17.

# Chapter 8

1. Cagle, *History of the Sergeant Weapon System*, 278–279. Author's collection. R. Smelt, "Lockheed X-17 Rocket Test Vehicle and Its Applications," *American Rocket Society Journal* 29, no. 8 (1959): 566. J. P. McManus, "A History of the FBM System" (Lockheed Missiles and Space Company, Inc., 1989), A-13. The author thanks Robert Wertheim, R. Adm., USN (ret.), for a copy of this document. L. H. Caveny, R. L. Geisler, R. A. Ellis, and T. L. Moore, "Solid Rocket Enabling Technologies and Milestones in the United States," *Journal of Propulsion and Power* 19, no. 6 (2003): 1038–1066. This excellent article served as a resource for the subsequent details of this chapter.
2. Piper, "Minuteman Chronology," 14.
3. "Propulsion System Development for Solid Propellant ICBM, Vol. V, Part 1: Summary," Thiokol Chemical Corporation, author's collection, 5–6. This document is classified; the extract used is declassified IAW EO 13526.
4. Karl Klager, "The Interaction of the Efflux of Solid Propellants with Nozzle Materials," *Propellants and Explosives*, Vol. 2 (John Wiley & Sons, 1977), 55–63.
5. "Proceedings of the Special Projects Office Steering Task Group, Task II, First Meeting, 25–26 July 1957," DNSA NH00581, 49–50.
6. "History of the Arnold Engineering Development Center, Vol. 1, Narrative, 1 July–31 December 1959," AFHRA, IRISNUM 00476712, K215.16 Vol. I, Reel 14847, 68.
7. Ibid., 46.
8. McManus, "History of the FBM," A-11, A-14, A-16.
9. Cagle, "History of the Sergeant," 278.
10. McManus, "History of the FBM," A-5.
11. "History of the Arnold Engineering Development Center, 1 July–31 December 1959," 68–76.
12. Piper, "Minuteman Chronology," no page numbers, referenced by date, July 1958.
13. E. S. Sutton, "From Polysulfides to CTPB Binders—A Major Transition in Solid Propellant Binder Chemistry," AIAA 84-1236, *AIAA/SAE/ASME 20th Joint Propulsion Conference*, Cincinnati, OH, 11–13 June 1984, 2–3.
14. Specific impulse varies directly with the square root of the ratio of flame temperature to the average molecular weight of the product gases. The addition of aluminum not only increases the flame temperature but also decreases the average molecular weight of the product gases.

15. Sutton, "From Polysulfides," 6–8. C. Boyars and K. Klager, *Propellants: Manufacture, Hazards, and Testing, Advances in Chemistry Series 88* (American Chemical Society, 1969), 124.
16. J. Thirkill, personal interview, 20 April 2017. Thirkill was Thiokol's Minuteman Stage I program manager from 1958 to 1962.
17. Yield strength is a measurement of a physical property of a metal, and the temperature range given is particular to the test.
18. "Propulsion System Development," 6–10. C. W. Bert and W. S. Hyler, "Structures and Materials for Solid Propellant Motor Cases," in *Advances in Space Science and Technology*, Vol. 8, edited by F. I. Ordway III (Academic Press, Cambridge, MA, 1966), 147.
19. Transcript of the film *Forging Marks Missile Progress*, Ladish Company, courtesy of Periscope Film LLC, 1959. I. Stone, "Minuteman ICBM Solid Motor Stages Enter Production Stage," *AWST*, 27 August 1962, 56.
20. "Notch Sensitivity-Barrier to Solid Rockets," *Missiles and Rockets*, 8 June 1959, 23–25. Personal communication with Jill Peri and Chris Peri, material scientists, April 2017.
21. The de Laval nozzle design accelerates the subsonic exhaust to supersonic speeds in a purely axial direction, i.e., along the axis of the motor.
22. Refrasil is a trademark for a woven glass fabric that is a highly heat-resistant insulation.
23. Stone, "Minuteman ICBM," 56. McManus, "History of the FBM," A-12.
24. "History of the Arnold Engineering Development Center, 1 July–31 December 1959," 73. "History of the Arnold Engineering Development Center, Vol. 1, July–December 1960," AFHRA, IRISNUM 476717, K215.16, Reel 14848, Vol. I, Narrative, II-14. This document is classified; the extract used is declassified IAW EO 13526.
25. Piper, "Minuteman Chronology," 44–76; Vol. II, Flash Report, Thiokol Stage I Engine Firing, 25 July 1960, Document 163, 1.
26. Ibid., Vol. I, 173. Stone, "Minuteman ICBM," 62.
27. Elmer Graesser, telephone interview, April 2018. Graesser was a Hercules Powder Company engineer who worked on the Minuteman I and II Stage III design and manufacture. For maximum efficiency the exhaust gases need to exit the nozzle as close as possible along the axis of the rocket motor (nonaxial thrust is wasted energy). Ludwig Prandtl, a German scientist, was the discoverer of the boundary layer effect and instrumental in the development of the study of the fluid–gas dynamic equations that enabled nozzle design to be most efficient, i.e., axial, a further refinement of the de Laval design. Theodore von Kármán was one of his students. See also George P. Sutton and Oscar Biblarz, *Rocket Propulsion Elements*, Seventh Edition (Wiley-Interscience Publications, New York, NY, 2001), 79.
28. Piper, "Minuteman Chronology," referenced by date, July 1958.
29. Both the Air Force and the Navy benefitted from Aerojet working on both programs. While Col. Hall maintained it was a one-way relationship, with the Navy greatly benefitting from the Air Force's work, the final report from Aerojet on the Large Rocket Feasibility Program clearly indicates otherwise. "Study of Large Solid-Propellant Rocket Motors, Volume I of III, Final Report: April 1956–March 1958," Section I, 1–3. This document is classified; the extract used is declassified IAW EO 13526. Author's collection. The author thanks William True for the months-long process to get this document partially declassified.
30. Robert L. Duerksen and Joseph Cohen, Solid propellant with polyurethane binder, US Patent 3,793,099, filed 31 May 1960 and issued 19 February 1974, 1.
31. I. Stone, "Aerojet Second Stage Must Withstand Heaviest Stress," *AWST*, 3 September 1962, 71. J. S. Billheimer and F. R. Wagner, "The Morphological Continuum in Solid Propellant Grain Design," IAF Paper P-94, *19th Congress of the International Astronautical Federation*, 13–19 October 1968, 378–381.
32. "Minuteman Weapon System History and Description, August 1990" (Hill AFB, UT, 1990), 35.

33. H. L. Podell and R. J. Kotfila, "Design and Fabrication of Titanium Rocket Chambers," *American Rocket Society Solid Propellant Rocket Conference Proceedings*, Salt Lake City, UT, 1–3 February 1961, 1–17.
34. Weapon System Operation Instructions (Wing II) USAF Series LGM30B Missile, T. O. 21M-LGM30B-1-3 (formerly 21M-LGM30A-1), Association of Air Force Missileers Collection, Section 1-3.
35. Stone, "Aerojet Second Stage," 71.
36. Christian M. Frey and Earl D. Shank, Rocket motor, US Patent 3,212,257; filed 25 March 1959 and issued 19 October 1965, 1. The listed objects of the invention were to increase the amount of propellant in the motor without increasing the diameter or overall length, provide for a reduction in weight of the overall nozzle assembly, provide for the utilization of nozzle materials that had not been found satisfactory due to poor tensile strength, provide for the utilization of nozzle materials that had been found unsatisfactory due to thermal shock.
37. A. E. Wetherbee Jr., Directional control means for a supersonic vehicle, US Patent 2,943,821, filed 5 December 1950 and issued 5 July 1960, 1–2.
38. G. F. Hausmann, Directional control means for rockets or the like, US Patent 3,143,856, filed 30 July 1963 and issued 11 August 1964, 1.
39. G. R. Richards and J. W. Powell, "Titan 3 and Titan 4 Space Launch Vehicles," *Journal of the British Interplanetary Society,* 46 (1993): 129.
40. Minuteman Flight Test Program, P-FT-3.6-00, August 1989 (ICBM Systems Program Management Division, Ogden Air Logistics Center, Hill AFB, UT), 31. Author's collection.
41. Flight Control Group Student Study Guide (Chanute Technical Training Center, Chanute AFB, IL, 6 December 1966), 2. The author thanks Bob Coambes for a copy of this document.
42. Piper, "Minuteman Chronology," 44–76.
43. R. A. Fuhrman, "The Fleet Ballistic Missile System: Polaris to Trident," *Journal of Spacecraft and Rockets* 15, no. 5 (1978): 278.
44. Neal, *Ace in the Hole*, 135–136.
45. Piper, "Minuteman Chronology," 7.
46. George P. Kennedy, *Rockets and Missiles of White Sands Proving Ground 1945–58* (Schiffer Publishing, Atglen, PA, 2009), 106, 107, 130, 143. D. Dyer and D. B. Sicilia, *Labors of a Modern Hercules* (Harvard Business School Press, Cambridge, MA, 1990), 228–245, 257. F. A. Dean, "The Unified Talos," *John Hopkins APL Technical Digest* 3. no. 2 (1982): 124.
47. Dyer and Sicilia, *Labors of a Modern*, 320. R. E. Young, "History and Potential of Filament Winding," presented at *13th Annual Technical and Management Conference, Reinforced Plastics Division, Society of the Plastics Industry*, Chicago, IL, 4 February 1958, Section 15-C, 1–6.
48. Graesser, telephone interview, September 2015.
49. Piper, "Minuteman Chronology," 90–104. "History of the Arnold Engineering Development Center, July–December 1960," II-14.
50. I. Stone, "Hercules Stage 3 Uses Glass Fiber Case," *AWST,* 10 September 1962, 162.
51. Graesser, telephone interview.
52. "Third Stage Motor-LGM-30G," *Minuteman Service News*, July–August 1971, 5–7.
53. J. D. Mockenhaupt, "Performance Characteristics and Analysis of Liquid Injection Thrust Vector Control," TM-16-SRO, 15 March 1965 (Aerojet General, Sacramento, CA), DTIC, AD856135, 3.
54. "Third Stage Motor-LGM-30G," 7. Bernard C. Nalty, "United States Air Force Ballistic Missile Programs, 1967–1968" (Office of Air Force History, 1969), DDRS, CK2349655277, 54–55. Redacted copy.
55. F. Culick, M. V. Heitor, and J. H. Whitelaw, eds., *Unsteady Combustion* (Kluwer Academic Publishers, Dordrecht, The Netherlands, 1996), 184–185. R. L. Geisler, "A Global View of the Use

of Aluminum Fuel in Solid Rocket Motors," Paper AIAA 2002-3748, presented at *38th AIAA/ASME/SAE/ASEE Joint Propulsion Conference*, Indianapolis, IN, 7–10 July 2002, 1.

56. "History of Strategic Air Command, 1 July 1975–31 December 1976," Historical Study No. 161, Vol. II, Narrative (Office of the Historian, Headquarters Strategic Air Command, 15 July 1977), DNSA, NT02287, 415–416. Redacted copy.

57. Ibid., 417–418.

58. Daniel Ruchonnet, "MIRV: A Brief History of Minuteman and Multiple Reentry Vehicles" (Lawrence Livermore Laboratory, February 1976), DNSA, NT02157, 23–25. Ted Greenwood, *Making the MIRV: A Study of Defense Decision-Making* (Ballenger Publishing Company, Cambridge, MA, 1975), 159. This book is a comprehensive study of the politics involved in the development of the MIRV concept.

59. "Proceedings of the Special Projects Office Steering Task Group: Task II-Monitor and Sponsor the Fleet Ballistic Missile Development Program, 14th meeting, 3 September 1959," DNAS NH00676, 2–3. Greenwood, *Making the MIRV*, 42–43.

60. Jonathan McDowell, JSR Launch Vehicle Database, www.planet4589.org/space/lvdb/thor, retrieved August 2019. Robert A. McDonald and Sharon K. Moreno, "Grab and Poppy: America's Early ELINT Satellites" (History Office, National Reconnaissance Office, Chantilly, VA, 2005), 1, 18.

61. Herbert F. York, "The Origins of MIRV," in *The Dynamics of the Arms Race*, edited by David Carlton and Carlo Schaerf (Halstead Press, New York, NY, 1975), 23–35.

62. B. T. Feld, T. Greenwood, G. W. Rathjens, and S. Weinberg, eds., *Impact of New Technologies on the Arms Race* (MIT Press, Cambridge, MA, 1970), 44. While seemingly insignificant, a vertical velocity error of three feet/second upward at warhead release has two effects: the shape of the trajectory is modified, and the time to impact is increased by 0.65 second. If the target is on the equator, this equals a 1,000-foot miss to the west; if not on the equator, the miss is less by a factor of the cosine of the latitude.

63. John B. Peller, "Events Leading up to Minuteman III," *North American Aviation Retirees Newsletter*, September 1995, 13–14.

64. Holm, "Chronology of the Ballistic," 106. The author thanks Jonathan McDowell for a copy of this document.

65. "Propulsion System Rocket Engine: Minuteman III," *Minuteman Service News*, September–October 1970, 3–6.

66. Robert W. Hovey, "Cork Thermal Protection Design Data for Aerospace Vehicle Ascent Flight," *Journal of Spaceflight* 2, no. 3 (1965): 300.

67. Ibid.

68. "Ablative Overcoat for Minuteman," *Minuteman Service News*, August 1963, 7. "The Appearance of Minuteman," *Minuteman Service News*, March–April 1969, 3–4.

69. Minuteman WS-133B and WS-133A-M Illustrated Figure A List (The Boeing Company, Seattle, WA), Sheets 225–231.

70. Michael S. Krogen, email communication, 11 February 2019. Krogen is a senior specialist, Program Management Defense Advanced Programs, Aerojet-Rocketdyne, Huntsville, AL.

71. "Interstage Ordnance," *Minuteman Service News*, July–August, 1965, 3. 1 STRAD LGM-30B OT Exercise Report, "Sweet Talk," 2 July 1965, "History of the 1st Strategic Aerospace Division, Strategic Air Command, July–December 1965," Vol. III, Supporting Document Nine, 17–18; A Compendium of Structural Joints for Assembly, Field and Flight Separation on Missiles (The Boeing Company, Seattle, WA, July 1968), D2-125911-1, DTIC, AD840998, Sheets 34–37.

72. "Minuteman Skirt Change," *Minuteman Service News*, September 1963, 2. "Minuteman Historical Summary, 1964–1965" (The Boeing Company, Seattle, WA, 1974), D2-26485, Sheet VII, 2.

73. Minuteman WS-133B and WS-133A-M Illustrated Figure A List, Sheet 238.

## Chapter 9

1. This chapter was published in extended form as "Reentry Vehicle Development Leading to the Minuteman Avco Mark 5 and 11," *Air Power History*, Fall 2017, 13–36.
2. T. C Lin, "Development of US Air Force Intercontinental Ballistic Missile Weapon Systems," *Journal of Spacecraft and Rockets* 40, no. 4 (2003): 503–506.
3. D. D. Baals and W. R. Corliss, *Wind Tunnels of NASA* (National Aeronautics and Space Administration, Scientific and Technical Information Office, 1981), SP-440, 55. J. R. Hanson, *Engineer in Charge: A History of the Langley Aeronautical Laboratory, 1917–1958* (National Aeronautics and Space Administration, Scientific and Technical Information Office, 1986), SP-4305, 343.
4. C. Wagner, "The Skin Temperature of Missiles Entering the Atmosphere at Hypersonic Speed" (US Department of the Army, Ordnance Research and Development Division), Technical Report No. 60, October 1949, 1.
5. H. J. Allen and A. J. Eggers Jr., "A Study of the Motion and Aerodynamic Heating of Missiles Entering the Earth's Atmosphere at High Supersonic Speeds," National Advisory Committee on Aeronautics, 1953, RM A53D28, 17, 25–27. These authors published several updated versions of this seminal paper. It is not clear to the author what was updated; the conclusions in each of the sequential papers appear to remain the same. E. P. Hartman, *Adventures in Research, A History of Ames Research Center 1940–1965* (National Aeronautics and Space Administration, Scientific and Technical Information Office, 1970), SP-4302, 216–218.
6. H. J. Allen, "Hypersonic Flight and the Re-Entry Problem," *Journal of the Aeronautical Sciences* 25, no. 4, 1958, 222, Fig. 9.
7. Theodore von Kármán, "Aerodynamic Heating: The Temperature Barrier in Aeronautics," in *Proceedings of the Symposium on High Temperature—A Tool for the Future,* Berkeley, CA, June 1956 (Menlo Park, CA: Stanford Research Institute, 1956), 140–142. *CRC Handbook of Chemistry and Physics, 52nd edition, 1971–72* (The Chemical Rubber Company, Cleveland, OH, 1971), D-142.
8. A. J. Eggers, C. F. Hansen, and B. E. Cunningham, "Stagnation-Point Heat Transfer to Blunt Shapes in Hypersonic Flight, Including Effects of Yaw," National Advisory Committee for Aeronautics, 1958, Technical Note 4229, 1–2.
9. Von Kármán, "Aerodynamic Heating," 142.
10. D. L. Schmidt, "Ablative Plastics for Reentry Thermal Environments" (United States Air Force, Air Research and Development Command, Wright Air Development Division, 1961), Report 60-862, 4–5.
11. Stumpf, *Titan II*, 13–36.
12. P. Juneau, "Composite Materials, Ablative," *Encyclopedia of Chemical Technology*, Third Edition, Vol. 1 (Wiley Blackwell, 1978), 10–26.
13. E. Stuhlinger, "Army Activities in Space," *IRE Transactions in Military Electronics* MIL, no. 2 (April–July 1960): 65.
14. *Organization and Management of Missile Programs, Eleventh Report by the Committee on Government Operations: Hearings before a Subcommittee of the Committee on Government Operations* (H. R.) 86th Cong. (1959), 108.
15. E. J. Redman and L. Pasuk, "Pressure Distribution on an ABMA Jupiter Nose Cone (13.3 Degrees Semi-Vertex Angle) at Nominal Mach Numbers 5, 6, 7, and 8," NAVORD Report 4486, U. S. Naval Ordnance Laboratory Aeroballistic Research Report 381, 27 September 1957, DTIC, AD0158516, 6. H. C. Dubose and R. C. Bauer, "Transonic Dynamic Stability Tests of a Half-Scale Model of the Jupiter W-14 Re-Entry Configuration," Arnold Engineering Development Center TN-58-1, February 1958, 3.

16. J. M. Grimwood and F. Strowd, "History of the Jupiter Missile System" (US Army Ordnance Missile Command, 1962), DTIC, AD778200, Appendix 8, 8–1. "Explorers in Orbit," Army Ballistic Missile Agency, 10 November 1958, National Archives and Record Administration (NARA), Record Group 156, Stack Area 290, Row 902, Compartment 9, Shelf 1, Box 10, 38–42, 112–119. W. R. Lucas and J. E. Kingsbury, A Brief Review of the ABMA Ablation Materials Program, DSM-TM-7-60, 12 May 1960, Army Ballistic Missile Agency, Redstone Arsenal, AL (Redstone Scientific Information Center, Redstone Arsenal, AL), 6.
17. M. C. Cleary, "Army Ballistic Missile Programs at Cape Canaveral, 1953–1988" (45th Space Wing History Office, Patrick AFB, FL, 2006), 52. Grimwood and Strowd, "History of the Jupiter," Appendix 8, 8-1.
18. Ibid., Appendix 9, 9-2, 9-4.
19. W. von Braun, Letter to D. L. Schmidt, Technical Manager for Ablative Materials, Air Force Materials Laboratory, Wright-Patterson AFB, OH, 14 September 1969, NARA, SE, RG 255, 01-0002, Box 24, 2. Grimwood and Strowd, "History of the Jupiter," 65. Cutting off the top of a right circular cone generates a frustum. J. C. Brassell, "Jupiter: Development Aspects and Deployment" (Mobile Air Material Area, Brookley AFB, AL, Historical Office, Office of Information, 1962), AFHRA, IRISNUM 474833, K205.0504-2, Vol 1, Reel 14643, 30.
20. C. Hansen, *The Swords of Armageddon: US Nuclear Weapons Development Since 1945*, Vol. VII (Chukelea Publications, Sunnyvale, CA, 1995), 323. M. Yaffee, "Two Approaches Used in First Production Nose Cones," *AWST*, 12 May 1958, 61.
21. J. Neufeld, "The Development of Ballistic Missiles," Appendix 1: The Tea Pot Committee Report, 259–260. *Organization and Management of Missile Programs: Hearings before a Subcommittee of the Committee on Government Operations* (H. R.) (1959) 383.
22. Monthly Report on Progress of ICBM and IRBM Programs, Report 21, 31 August 1957, Department of Defense, Office of the Director of Defense Research and Engineering, DDRS, CK2349145648, 2.
23. W. E. Green, "The Development of the SM-68 Titan" (History Office, Deputy Commander for Aerospace Systems, Air Force Systems Command, August 1962), Vol. II. Supporting Documents, Letter, Brig Gen O J Ritland, V Cmdr, AFBMD, to Cmdr, ARDC Subject: Reorientation of Nose Cone Program, 22 August 1958, Document 57, 1–2.
24. A. V. Levy, "Evaluation of Reinforced Plastic Material in High Speed Guided Missile and Power Plant Application," *Plastics World* 14 (March 1956): 10–11.
25. I. J. Gruntfest and L. H. Schenker, "Behavior of Materials at Very High Temperatures," *Industrial and Engineering Chemistry* 50 (October 1958): 75A-76A. G. W. Sutton, "Ablation of Reinforced Plastics in Supersonic Flow," *Journal of Aero/Space Sciences* (May 1960): 378. Refrasil is approximately 90 percent silicone oxide, made by leaching out the lower flux oxides of E glass fibers found in fiberglass.
26. M. C. Adams and E. Scala, "The Interaction of High Temperature Air with Materials During Reentry," *Proceedings of an International Symposium on High Temperature Technology*, Asilomar Conference Grounds, CA, 6–9 October 1959 (McGraw-Hill, New York, NY, 1960), 54–60. M. Yaffee, "Ablation Wins Missile Performance Gain," *Aviation Week*, 18 July 1960, 57. Flight Summary Report Series D Atlas Missiles (General Dynamics/Astronautics, San Diego, CA, 21 June 1961), DTIC, AD0833337, 8-36.
27. Study on the Future of AFBMD/MCO/SACMIKE/STL (R-W) Co., AFHRA, IRISNUM 01040315, 168.7171-167, Reel 35264, 74–79.
28. J. W. Powell, "Thor-Able and Atlas Able," *Journal of the British Interplanetary Society* 37 (1984): 219–220, 222. "Thor Able Launched in Nose Cone Test," *AWST*, 28 July 1958, 27. D. L. Schmidt, "Ablative Plastics for Thermal Protection" (Wright Air Development Division, US Air Force,

WADD Report 60-862, August 1968), 58. M. Morton, "Progress in Reentry Vehicle Development" (General Electric Missile and Space Vehicle Department, Philadelphia, PA, 2 January 1961), 11. The author thanks Don Schmidt for copies of these papers.

29. W. M. Arms, *Thor, the Workhorse of Space—A Narrative History* (McDonnell Douglas Astronautics Company, Huntington Beach, CA, 31 July 1972), 4-1. The author thanks Joel Powell for a copy of this report.

30. Smithsonian news release, no title, 14 May 1959. "Recoverable Data Capsule Designed for Thor and Atlas Test Missiles," *AWST*, 24 November 1958, 71–72.

31. The author thanks Craig Brunetti for the Avco RVX-1-5 measurements.

32. Arms, *Thor, the Workhorse*, 4-4, 4-6. M. Morton, "Thor-Able and Atlas Reentry Recovery Programs" (General Electric Defense Electronics Division, Missile and Space Vehicle Department, Philadelphia, PA, no date), 1–7. The author is indebted to Don Schmidt for this document. Schmidt, "Ablative Plastics for Reentry," 58–59.

33. "Flight Summary Report, Series D Atlas Missiles," 2-2.

34. W. T. Barry, personal interview and correspondence with author, February 1997. Barry was a materials scientist consultant at the General Electric Space Sciences Laboratory in Philadelphia during the development of the Titan II Mark 6 reentry vehicle. The Series 100 plastic ablation process was patented by Barry (US Patent 3,177,175) in 1965.

35. While not explicitly described as such, the available pictures of the recovered vehicle appear to show material similar to the RVX-2A reentry vehicle, which did have these materials.

36. Morton, "Thor-Able and Atlas," 22–23.

37. "Flight Summary Report, Series D Atlas Missiles," 8-38.

38. Ibid., 8–5 to 8–11.

39. Ibid., 8–2.

40. Ruggiero, "Missileers' Heritage," 34–69.

41. Progress of ICBM and IRBM Programs, April, May, June 1960 (Office of the Director of Defense Research and Engineering, Department of Defense), DDRS, CK2349126998, 17.

42. Green, "Development of the SM-68," 25.

43. "Flight Test Summary, Series D Atlas Missiles," 8-32 to 8-37.

44. Progress of ICBM and IRBM Programs, April, May, June 1960, 17.

45. Ruggiero, "Missileers' Heritage," 34–69. Neufeld, "Development of Ballistic Missiles," 234, 276.

46. McDowell, JSR Launch Vehicle Database. McDowell's database is considered the gold standard of civilian missile launch records.

47. Memos, May 1958, personal collection of Gen. B. A. Schriever, AFHRA, IRISNUM 01040258, K168.7171-110, Reel 35259, 1–2.

48. Piper, *Development of the SM-80*, Vol. II, Supporting Documents, Proposed Preliminary Operational Concept for Minuteman, 8 April 1958, Document 76, 6.

49. Ibid., Letter, Col H. L. Evans, Asst D/Cmdr, Space Sys, to Col C. H. Terhune D/Cmdr, Ballistic Missiles, 28 May 1958, Subject: Minuteman Warheads and Reentry Vehicles, Document 80, 1–2.

50. Personal interview with Secretary of the Air Force Thomas Reed, May 2015. The companies were listed in his personal diary entry for Monday, 23 June 1958.

51. "Space and Missile Systems Organization: A Chronology, 1954–1979," US Air Force, Space Division (Chief of Staff History Office, October 1979), 57. The author thanks Jonathan McDowell for a copy of this document.

52. Piper, *Development of the SM-80*, Vol. II, Supporting Documents, Cancellation of the Light Minuteman Reentry Vehicle, 9 December 1958, Document 94, 1.

53. Phil Fote, personal interview, May 2017. Piper, *Development of the SM-80*, Vol. I, 190–195.

54. McDowell, JSR Launch Vehicle Database.

55. Operational Test Report: ELM BRANCH, Minuteman IA, LGM-30A, FTM 631, 1st Strategic

Aerospace Division, AFHRA, IRISNUM 00918812, Reel 26122, 2. J. C. Hopkins, "The Development of the Strategic Air Command 1946–1981: A Chronological History" (Office of the Historian, Headquarters Strategic Air Command, 1 July 1982), DTIC ADA120491, 112, 157.

56. Piper, *Development of the SM-80*, Vol. I, 190–196.
57. Fote, personal interview. M. L. Yaffee, "Mark 11A Hardened against Air Bursts," *AWST*, 24 August 1964, 50–60.
58. Fote, personal interview. Yaffee, "Mark 11A Hardened," 50–59.
59. Fote, personal interview.
60. McDowell, JSR Launch Vehicle Database.
61. Operational Test Report: NICKED BLADE, Minuteman Missile IB, LGM-30B, FTM 1101, 1st Strategic Aerospace Division, AFHRA, IRISNUM 00918790, Reel 26121, 4. Fote, personal interview.
62. "History of Strategic Air Command, Fiscal Year 1970," Historical Study 117, Vol. II (Office of the Historian, Headquarters, Strategic Air Command, 20 April 1971), DNSA, NT00923, 231. "A History of Strategic Arms Competition 1945–1972," June 1976 (United States Air Force Supporting Studies), DNSA, SE00542, 362. Hopkins, "Development of the Strategic," 191.

# Chapter 10

1. Arms, *Thor, the Workhorse*, Section 33, 4. The AC Spark Plug Division of General Motors built the inertial guidance system for Thor. This document lists the flight as 9 December 1957 and elsewhere as 19 December 1957. Local newspapers report a successful launch on 19 December. The author thanks Joel Powell for a copy of this report.
2. John W. Bullard, "History of the Redstone Missile System" (History Office Army Missile Command, 15 October 1965), 100, 164. This document is available from the DTIC, ADA434109. Richard H. Parvin, *Inertial Navigation* (Van Nostrand Company, Princeton, NJ, 1962), 100. Dr. Walter Haeussermann recalls the flight on 22 September 1955 as completely successful, while Bullard's "History of the Redstone Missile System," 164, indicates that the flight had a control malfunction due to overheating in the tail section. Grimwood and Strowd, "History of the Jupiter," 65, Appendix 9, 9-2. Available from DTIC, AD778200. The Jupiter inertial guidance design was based on the successful Redstone inertial guidance set and presented relatively few problems during the development of the program.
3. "Flight Summary Report, Series D Atlas Missiles," Section 2, 27.
4. Piper, *Development of the SM-80*, Vol. I, 183.
5. Ibid., Proposed Preliminary Operational Concept for Minuteman, Vol. II, Supporting Documents, Document 76, 6, 7.
6. Robert Burnett, National Air and Space Museum Oral History Project, 10 January 1990, 9.
7. "The Second Generation," *Time*, 10 March 1958, 9.
8. Susan Boltinghouse, email correspondence, 15 September 2018.
9. Robert Knox, personal interview, April 2018. The N5B Inertial Navigation System was a technical development program funded by the Air Force and was not used in an operational system.
10. Harold Engebretson, "The Minuteman I Guidance and Control System Proposal," *North American Aviation Retirees Bulletin*, Winter 2008, 10–12. Marshall McMurran, *Achieving Accuracy: A Legacy of Computers and Missiles* (Xlibris Corporation, Bloomington, IN, 2008), 276–277.
11. Piper, *Development of the SM-80*, Vol. I, 36. The other companies were Nortronics Division of Northrop Aircraft Inc., Minneapolis-Honeywell Regulator Company, Kearfott Company, Litton Industries, Emerson Electric Manufacturing Company, AC Spark Plug Division of General Motors Corporation, American Bosch Arma Corporation, and Sperry Rand Corporation.

12. J. S. Ausman and M. Wildman, "How to Design Hydrodynamic Gas Bearings," *Production Engineering* 28, no. 25 (1957): 103–106. Knox, personal interview.
13. Donal B. Duncan and Joseph V. Boltinghouse, Free rotor gyroscope, US Patent 3,251,233, filed 21 February 1957 and issued 17 May 1966. This appears to be the first patent for the free-rotor gyroscope at Autonetics. An improved free-rotor gyroscope was patented by John M. Slater and Joseph C. Boltinghouse: Free rotor gyroscope motor and torquer drives, US Patent 3,025,708, filed 19 December 1958 and issued 20 March 1962.
14. John M. Slater, *Inertial Guidance Sensors* (Reinhold Publishing Corporation, 1964), 35–37, 98–99.
15. There are a number of excellent books on the history of inertial guidance use in ballistic missiles; for example, Donald MacKenzie's *Inventing Accuracy* (MIT Press, Cambridge, MA, 1993) and Marshall McMurran's *Achieving Accuracy*.
16. Piper, *Development of the SM-80*, Vol. I, 61.
17. J. M. Wuerth, "The Impact of Guidance Technology on Automated Navigation," *Navigation: Journal of the Institute of Navigation* 14, no. 3 (1967): 330.
18. Richard Stranix, "Minuteman Reliability: Guide for Future Component Manufacturing," *Electronic Industries*, December 1960, 89–104. This is a comprehensive description of the steps Autonetics required of its suppliers.
19. Wuerth, "Impact of Guidance Technology," 330. "The Industry Week," *Missiles and Rockets*, 18 June 1962, 37.
20. John M. Slater, Doyle E. Wilcox, Darwin L. Freebairn, and Walter L. Pondrom Jr., Inertial velocity meter, US Patent 3,077,782, filed 2 April 1956 and issued 19 February 1963.
21. Philip J. Klass, "Minuteman Guidance and Control Part 2: Minuteman Guidance Built for Long Life," *AWST*, 5 November 1962, 75.
22. J. M. Wuerth, "The Evolution of Minuteman Guidance and Control," *Navigation: Journal of the Institute of Navigation* 23, no. 1 (1976): 69.
23. Robert Nease and Dan Hendrickson, "A Brief History of Minuteman Guidance and Control" (Rockwell Defense Electronics, Autonetics Division, March 1995), Section 1-7, 13.
24. Ibid., Section 1-8.
25. C. F. Adams, S. W. Cogan, and E. W. Nealon, "Beryllium in Inertial Navigation," *Proceedings of the Beryllium Conference*, Vol. I, 23–25 March 1970, National Materials Advisory Board Publication 272, Division of Engineering, National Research Council (National Academy of Sciences–National Academy of Engineering, Washington, DC, July 1970), 249–279. The discussion of the use of beryllium in the G6B4 gyroscope and the stable platform is adapted from this paper.
26. Ibid., 290.
27. The term "scrunched" refers to the orientation of the meters, which are roughly centered on the direction of the missile trajectory; i.e., the input axes line up in a conical orientation pointed toward the boost trajectory. Personal communication, Robert Knox, 9 August 2019.
28. Luther L. Stenvick, *The Agile Giant: A History of the Minuteman Production Board* (The Boeing Company, Seattle, WA, 1966), 73, 80. Piper, *Development of the SM-80*, Vol. I, 185–189.
29. Knox, personal communication. McMurran, *Achieving Accuracy*, 275–276.
30. WS-133A/B Weapon System Briefing FETM-M-12B-SA-28, Autonetics Division, North American Aviation, no date, 22. The author thanks Jay Bogess for this document.
31. L. Parker Temple III, *Implosion: Lessons from National Security, High Reliability Spacecraft, Electronics, and the Forces Which Changed Them* (IEEE Press, John Wiley & Sons, 2013), 58.
32. Charles H. Beck, "Investigation of Minuteman D17B Computer Reutilization" (Tulane University School of Engineering, January 1971), 2, 24–25. The number of resistors is often misquoted as 509. Available from DTIC AD722476. Piper, *Development of the SM-80*, Vol. I, 188. "Raising the Longevity Record," *Minuteman Service News*, October–December 1972, 1.
33. Nease and Hendrickson, "Brief History of Minuteman," Section 1-10. "Minuteman Weapon System

Familiarization: Function of Guidance Control Systems" (Chanute Technical Training Center, Chanute AFB, IL, April 1966), 16–17. The author thanks Bob Coambes for a copy of this document.

34. Knox, personal communication, 16 June 2018.
35. Klass, "Minuteman Guidance," 75.
36. Knox, personal communication, 16 June 2018.
37. "Minuteman Historical Summary, 1962–1963" (The Boeing Company, Seattle, WA, 1975), D2-26485, Sheet I-19. Except where otherwise indicated, the material in this section comes from Nease and Henderson, "Brief History of Minuteman," Section 2, 2–24.
38. Robert Knox, personal communication, 21 August 2019. The NS17 missile guidance set did not have true gyrocompass capability, and this quote from "A Brief History of Minuteman Guidance" has been corrected. The gyrocompass assembly, or GCA capability, was found in the NS20.
39. Adams et al., "Beryllium in Inertial," 255–265.
40. Robert Knox, personal communications, 2 June 2018.
41. R. H. Parvin, *Inertial Navigation*, 95–96. R. E. Hopkins, Fritz K. Mueller, and Walter Haeussermann, "The Pendulous Integrating Gyroscope Accelerometer (PIGA) from the V2 to Trident D5, the Instrument of Choice," Paper AIAA 2001-4288, presented at *Guidance, Navigation, and Control Conference and Exhibit*, Montréal, Canada, 6–9 August 2001, 5.
42. Michele Sapuppo, telephone interview, 11 May 2018. Hopkins et al., "Pendulous Integrating Gyroscope," 6.
43. John Hepfer, transcript of interview by Donald MacKenzie, 29 July 1985, 13. The author thanks Donald MacKenzie for a copy of the interview.
44. Frank DeArmond and Ralph Lombardo, "Minuteman II Aerospace Ground Equipment Including Self-Test Features and Status Display Techniques," in *Automation and Electronic Test Equipment, Vol. 3,* edited by David M. Goodman (New York University Press, April 1967), 102. The author thanks Jay Bogess for a copy of this document.
45. Knox, personal communication, 16 June 2018.
46. Bernard C. Nalty, "USAF Ballistic Missile Programs 1967–1968" (USAF Historical Division Liaison Office, March 1967), Air Force Historical Support Division, Joint Base Anacostia–Bolling, Washington DC, 34. This document is classified; the extract used is declassified IAW EO 13526.
47. "Minuteman Is Top Semiconductor User," *AWST*, 26 July 1965, 83.
48. Rex Pay, "Circuit Weights Slashed in Advanced Minuteman," *Missiles and Rockets*, 2 March 1964, 35.
49. Arnold J. Borofsky, "Component Quality Assurance Programs for Micro Miniature Electronic Components for Minuteman II," *Third Annual Symposium on the Physics of Failure in Electronics, Institute of Electronic and Electrical Engineers*, 29 September–1 October 1964, 1–14.
50. Richard Platzek, "Microelectronics and Minuteman," Technical Papers, *WESCON 63: Frontiers in Electronics,* San Francisco, CA, 20–23 August 1963, 2.
51. C. W. Scott and P. H. Eisenberg, "Failure Analysis of Minuteman Integrated Circuit Failures, Rome Air Development Center (RADC) Technical Report 669-457, May 1970," 1. Available from DTIC as AD870725. This report gives a highly detailed discussion of the types of failures and detection of these failures.
52. Nease, personal interview, April 2018. Nease recalled that the purple plague was the least of their worries during the period of reliability problems.
53. "Minuteman Historical Summary, 1966–1967" (The Boeing Company, Seattle, WA, 1975), D2-26485, Sheet XVI-4, XVI-18.
54. Ibid., Sheet XVI-21.
55. "History of the Strategic Air Command 1967," Historical Study No. 106, Vol. 1, Narrative (History and Research Division Headquarters, Strategic Air Command, March 1968). Redacted copy, 309.
56. "Minuteman Historical Summary, 1966–1967," Sheet XVI-42.
57. Nalty, "USAF Ballistic Missile Programs: 1967–1968," 35.

58. "Minuteman Historical Summary, 1964–1965" D2-26485, Sheet I-25.
59. Nalty, "USAF Ballistic Missile Programs: 1967–1968," 36.
60. Ibid., 43.
61. Holm, "Chronology of the Ballistic," 106. Author's collection. The author thanks Jonathan McDowell for a copy of this document.
62. Frederick Shaw Jr., "Minuteman, Strategic Air Command Weapons Systems Acquisition: 1964–1979" (Office of the Historian Headquarters Strategic Air Command, 18 April 1980), AFHRA, IRIS 01114095, K416.04–39 Vol. 1, Reel 45868, 194.
63. Nease and Hendrickson, "Brief History of Minuteman," Section 1-1.
64. "History of the Strategic Air Command, FY 1970," Historical Study No. 117, Vol. II, Narrative (Office of the Historian, Headquarters, Strategic Air Command, 20 April 1971). Redacted copy, DNSA, NT00923, 274–275. Nalty, "USAF Ballistic Missile Programs, 1967–1968," 36.
65. PBPS/PSRE is used for consistency.
66. Wuerth, "Evolution of Minuteman Guidance," 64–75.
67. Nease and Hendrickson, "Brief History of Minuteman," Section 3-5.
68. Ibid., Section 3-1 to 3-33.
69. Shaw, "Minuteman, Strategic Air," 263.
70. MacKenzie, *Inventing Accuracy*, 319–320.
71. Nease and Hendrickson, "Brief History of Minuteman," Section 3-38.
72. Ibid., Section 3-40.
73. Ibid., Section 3-41-3-44.
74. Ibid., Section 3-46-3-47.
75. Minuteman III Guidance Replacement Program Reference, no date, Autonetics North American Rockwell, 1–37. Author's collection.

# Chapter 11

1. Richard K. Burkard, "Geodesy for the Layman" (US Air Force Aeronautical Chart and Information Center, St. Louis, MO, 1968), 35–40.
2. Ibid., 53–62. D. J. Warner, "Political Geodesy: The Army, Air Force and World Geodetic System of 1960," *Annals of Science* 59 (2002): 384–386.
3. B. Louis Decker, personal communications, March 2016.
4. Placement of the Station A point depended on the location of the sight tube, which changed at Wing VI and 564 SMS LFs.
5. John Allen, personal communications, September, December 2015.
6. M. M. Bennett and P. W. Davis, "Minuteman Gravity Modeling," *American Institute of Aeronautics and Astronautics* (1976), 1. Gordon Barnes, Lt. Col., USAF (ret.), personal interview and communications, June 2015 to August 2019. Decker, personal communications, June 2015 and March 2016.
7. Burkard, "Geodesy for the Layman," 52. Bennett and Davis, "Minuteman Gravity Modeling," 400–404.
8. Burkard, "Geodesy for the Layman," 28–52. B. Szabo, "The Significance of the World Geodetic Datum to Long-Range Navigation and Guidance Systems," in *Symposium: Size and Shape of the Earth* (Ohio State University, Columbus, OH, November 1956), 99–102. Decker, personal communications, March 2016.
9. Garrett Moore, CWO, USA (ret.), personal interview, June 2015.
10. Andy Doll, SMSgt., USAF (ret.), personal communications, January 2016. An autocollimator generates a narrow beam of collimated, or parallel, light. An autocollimator works by projecting an image onto a target mirror and measuring the deflection of the returned image against a scale,

either visually or by means of an electronic detector. A visual autocollimator can measure angles as small as 0.5 arc second, while an electronic autocollimator can be up to 100 times more accurate. The autocollimator used with Minuteman was considered an electronic autocollimator.

11. Ibid., December 2015, January 2016, August, September 2019. The theodolite telescope has two internal alignment aids. A reticle like a rifle scope has an internal cross device that is illuminated by an external light source called a gauss. The gauss image is projected out the front of the theodolite and reflected back into the theodolite from the target transfer mirror. Using horizontal and vertical drive knobs on the theodolite, the operator aligns the telescope reticle to the gauss to perfectly align the theodolite to the target transfer mirror.
12. Ibid., James W. Barnard, personal communication, October 2018.
13. John Mills, TSgt., USAF (ret.), personal communication, January 2016. The original ALS was developed in 1969 to 1970 and implemented at the wings in 1973 and 74.
14. Nease and Hendrickson, "Brief History of Minuteman," Section 2-6.
15. Mills, personal communication.
16. Rik Gervais, Col., USAF (ret.), personal communications, December 2015, January 2016.
17. D. A. Day, J. M. Logsdon, and B. Latrell, eds., *Eye in the Sky: The Story of the Corona Spy Satellites* (Smithsonian, Washington, DC, 1998), 201–208. Deborah J. Warner, "Political Geodesy: The Army, the Air Force, and the World Geodetic System of 1960," *Annals of Science* 59 (2002): 386.
18. B. Louis Decker, "World Geodetic System 1984," *Fourth International Geodetic Symposium on Satellite Positioning*, University of Texas, Austin, 29 April–2 May 1986, 1–3.

# Chapter 12

1. The term "battleship" refers to a rugged version of a missile built for repeated use in developing systems such as handling, transportation, and launcher equipment.
2. C. N. Hill, *A Vertical Empire: The History of the British Rocketry Programme* (Imperial College Press, London, 2012), 125–137. Barrie Ricketson and E. T. B. Smith, "Work Supporting the Development of an Underground Launching System for Blue Streak," Rocket Propulsion Establishment, British National Archives, AVIA 68/23, October 1958, 7.
3. David K. Stumpf, *Titan II*, 26, 44–47, 177. Detailed Design Specifications for Model SM-68B Missile, including Addendum for XSM-68B (The Martin Company, Denver, CO, 1961), Stumpf Collection, Titan Missile Museum Archive, Sahuarita, AZ, 21.
4. "Minuteman Historical Summary, 1958–1959" (The Boeing Company, Seattle, WA, 1974), D2-26485, Sheet IX, 1, 5, 12. Robert G. Bowlin, "One-Third Scale Test Silo Launcher Development Project WS-133A (Minuteman)," "History of the Air Force Flight Test Center, 1 July 1959–31 December 1959, Vol. 4," AFHRA IRISNUM 489378, K286.69-35 V.4, 1–41.
5. "Minuteman Historical Summary, 1958–1959," Sheet IX, 2, 13.
6. Robert Knox, personal communication, 16 June 2018.
7. "Minuteman Historical Summary, 1958–1959," Sheet IX, 17, 24.
8. "Minuteman Historical Summary, 1960–1961," Sheet I, 11,12. STMs 106, 108, and 110 were used for nonflight tests.
9. Ibid., Sheet X, 5.
10. Air Force Space and Missile Museum, Cape Canaveral Air Force Station, http://afspacemuseum.org/ccafs/CX3132/, retrieved June 2018.
11. Piper, "Minuteman Chronology, December 1955–April 1961," 96. *Rocketry in the 1950s*, NASA Historical Report No. 36, 28 October 1971 (US Government Printing Office, 1971), 78.
12. Piper, *Development of the SM-80*, Vol. I, 201–202. "Minuteman Historical Summary, 1960–1961," Sheet II-15. Marven R. Whipple, Atlantic Missile Range (Eastern Test Range) Index of Missile

Launchings: July 1961–June 1962, AFETR Historical Publication Series 64–132 (Historical Division, Office of Information, Patrick AFB, FL), 20. The author thanks Jonathan McDowell for a copy of this document.

13. "Minuteman Historical Summary 1960-1961," Sheet II, 16–17. This document has a typographical error for the launch time. It gives 11:30 p.m., which should be 1:30 p.m. "AF Minuteman Explodes during Silo Launching," *Bakersfield Californian*, 30 August 1961, 5. This article gives the correct launch time of 1:30 p.m. EDT.
14. Ibid., Sheet II, 20. "Chronology of the Ballistic Missile Organization," 95.
15. All launch information for the Atlantic Missile Range for 1962 to 1963 comes from either "Minuteman Historical Summary, 1962–1963" (The Boeing Company, Seattle, WA, 1975), D2–26485, Sheet VI-1-53, or Whipple, "Atlantic Missile Range," 1–16.
16. Jack Hilden, Capt., USAF (ret.), personal interview, February 2016.
17. Alan MacDonald, personal interview, September 2016.
18. "Minuteman Historical Summary, 1962–1963," Sheet VI, 33.
19. G. J. Tatnall and K. W. Foulke, "Joint Air Force–Navy Supersonic Rain Evaluations of Materials," Technical Report AFML-TR-67-164 (Air Force Materials Laboratory, Air Force Systems Command, Wright-Patterson AFB, OH, 1967), 120.
20. Mark Cleary, *The 6555th Missile and Space Launches through 1970*, 45th Space Wing History Office (Progressive Management Publications, 1991), Sheet 8–3.
21. "Minuteman Historical Summary, 1962–1963," Sheet VI, 47–49.
22. Ibid., Sheet VI, 51. "Minuteman Historical Summary, 1964–1965," Sheet VII, 1–2. "Minuteman Skirt Change," *Minuteman Service News*, no. 8, September 1963, 2.
23. "Minuteman Skirt Change," 4.
24. "Minuteman Historical Summary, 1962–1963," Sheet I, 19.
25. "Minuteman Historical Summary, 1964–1965," Sheet VII, 6.
26. Cleary, *6555th Missile*, Chapter 3, Sheet 8, 4.
27. "Minuteman Historical Summary, 1964–1965," Sheet VII, 12–21. Whipple, "Atlantic Missile Range," 13–15. All subsequent Patrick AFB flight test information for Minuteman II comes from these sources unless otherwise noted.
28. Whipple, "Atlantic Missile Range," 16.
29. "Chronology of the Ballistic Missile Organization: 1945–1990," 365.
30. "Minuteman Historical Summary, 1964–1965," Sheet VII, 1–22; "Minuteman Historical Summary, 1966–1967" (The Boeing Company, Seattle, WA, 1975), D2–26485, Sheet VIII-1-10; "Minuteman Historical Summary, 1968–1969" (The Boeing Company, Seattle, WA, 1976), D2–26485, Sheet I, 1–3.
31. "Minuteman Historical Summary, 1970–1971" (The Boeing Company, Seattle, WA, 1976), D2–26485 Sheet X, 2, 8.
32. "Minuteman Historical Summary, 1960-1961," Sheet IV, 28–29.
33. Robert S. McNamara, Memorandum for the President, Subject: Recommended Fiscal Year 1966–1970 Programs for Strategic Offensive Forces, Continental Air and Missile Defense Forces and Civil Defense, DNSA, NH0045, 30, 46–50.
34. Bernard C. Nalty, "USAF Ballistic Missile Programs, 1965" (Historical Division Liaison Office, March 1967), DDRS, CK2349654796, 43–45.
35. "Strategic Air Command Missile Chronology: 1939–1973" (Office of the Historian, Headquarters Strategic Air Command, 2 September 1975). Author's collection; no pages are given, and information is located by date. The author thanks Greg Ogletree for a copy of this document.
36. "Space and Missile Systems Organization: A Chronology, 1954–1979" (History Office, Chief of Staff, Space Division, October 1979), DNSA MS00303, 166.
37. "Minuteman Historical Summary, 1968–1969," Sheet I, 7, 13.
38. Nease and Hendrickson, "Brief History of Minuteman," Section 3, 10.

39. "History of the Strategic Air Command, FY 1969, Historical Study No. 116, Vol. II, Narrative (Office of the Historian, Headquarters, Strategic Air Command, 2 May 1974). Redacted copy, DNSA, NT00502, 279–280.
40. "Minuteman Historical Summary, 1970–1971," Sheet V, 33.

## Chapter 13

1. "Minuteman Historical Summary, 1962–1963," Sheet X, 1, 2.
2. "SAC Missile Chronology," 29. "Minuteman Historical Summary,1960–1961," Sheet XII, 16. "Minuteman Historical Summary, 1962–1963," Sheet X, 4.
3. "Support Equipment Goes Underground—LCF," *Minuteman Service News*, April–March 1964, 3. Author's collection. "Minuteman Historical Summary, 1962–1963," Sheet X, 6–8.
4. "Support Equipment Goes Underground," 16.
5. The launch facility nomenclature at Vandenberg changed from the 394 AXX to the LF-XX designation in January 1964.
6. "History of the 1st Strategic Aerospace Division, Strategic Air Command, 1 July–31 December 1962," Vol. II, Supporting Documents, AFHRA, IRISNUM 00425434, K-DIV-1-HI V.2, Reel 11110, Supporting Document 110, no title. This document is classified; the extract used is declassified IAW EO 13526.
7. "Minuteman Historical Summary, 1962–1963," Sheet X, 22–23.
8. "History of the 1st Strategic Aerospace Division, Strategic Air Command, 1 July–31 December 1962," Vol. I, Narrative, 159–161. This document is classified; the extract used is declassified IAW EO 13526. Mitch Cannon, CMSgt., USAF, (ret.), personal interview, August 2017.
9. "History of the 1st Strategic Aerospace Division, Strategic Air Command, 1 July–31 December 1962," Vol. I, 161–163. Cannon, personal interview.
10. "Space and Missile Systems Organization: A Chronology, 1954–1979" (Space Division, Chief of Staff, History Office), DNSA, MS00303, 115.
11. "Minuteman Historical Summary, 1962–1963," Sheet X, 30–31.
12. Ibid., 36–37.
13. Ibid., 36. "History of the 1st Strategic Aerospace Division, Strategic Air Command, 1 July–31 December 1962," Vol. II, Supporting Document 112, no title. This document is classified; the extract used is declassified IAW EO 13526.
14. "Minuteman Historical Summary, 1962–1963," Sheet X, 38.
15. Testing/Evaluation of Systems, Subsystems and Equipments, Air Force Regulation No. 80–14 (Department of the Air Force, Washington, DC, 1963), Muir S. Fairchild Research Information Center, Maxwell AFB, AL, 13. Minuteman Historical Summary, 1962–1963, Sheet X, 41.
16. "Demonstration and Shakedown Operation Test Program Final Report," LGM30B Minuteman Weapon System, no date (1st Strategic Aerospace Division, Strategic Air Command, Vandenberg AFB, CA), AFHRA, IRISNUM 00425439, K-DIV-1-HI V.2, 62.
17. Holm, "Chronology of the Ballistic," 371.
18. Testing/Evaluation of Systems, 13.
19. "Minuteman Historical Summary, 1962–1963," Sheet X, 49. "SAC Missile Chronology 1939–1988," 41. "History of the 341st Strategic Missile Wing," Malmstrom AFB, MT, October 1963, AFHRA, IRISNUM 00456152, Reel 14473, Supporting Document 22, no page given. This document is classified; the extract used is declassified IAW EO 13526.
20. Operational Test Exercise Report: Drag Chute, Sortie Number 99-64-M-3, Minuteman "A," LGM-30A AFSN 63-124/FTM 664, 31 October 1963, AFHRA, IRISNUM 00918797, K416.85-33, Reel 26122, 12–13.
21. Operational Test Exercise Report: Box Seat, Sortie Number 99–64-M-9, Minuteman "A," LGM-30A

AFSN 63-115/FTM 636, 29 February 1964, AFHRA, IRISNUM 00918776, K416.85-12, Reel 26121, 19–20. Operational Test Exercise Report: Brass Ring, Sortie Number 99-64-M-10, Minuteman "A," LGM-30A AFSN 63-073/FTM 581, 29 February 1964, AFHRA, IRISNUM 00918813, K416.85-49, Reel 26122, 2, 19–20.

22. Test/Evaluation of System, 3.
23. Film of the launch is extraordinary to watch as the missile climbs toward the capsule. Science Photo Library, Clip K006/0021, www.sciencephoto.com, accessed 20 September 2017.
24. "History of the 1st Strategic Aerospace Division, Strategic Air Command, 1 January–30 June 1966," Vol. I, Narrative, AFHRA, IRISNUM 00425453, K-DIV-1-HI, Reel 11114, 61, 62. This document is classified; the extract used is declassified IAW EO 13526. LGM30A Exercise Report Rocky Point, "History of the 1st Strategic Aerospace Division, Strategic Air Command, 1 July–31 December 1965," Vol. III, Supporting Documents, AHFRA IRISNUM 00425448, K-DIV-1-HI, Reel 11114, Supporting Document 28, 13.
25. "History of the 1st Strategic Aerospace Division, Strategic Air Command, 1 January–30 June 1966," Vol. I, Narrative, 63–65. This document is classified; the extract used is declassified IAW EO 13526.
26. Estimate for distance to land made with Google Earth. "History of the 1st Strategic Aerospace Division: January–June 1966, Vol. II, Document 37, 8; This document is classified; the extract used is declassified IAW EO 13526.
27. "Minuteman Historical Summary, 1962–1963," Sheet X, 68–69.
28. Ibid., Sheet X, 48.
29. Demonstration and Shakedown Operation Test Program Final Report, "History of the 1st Strategic Aerospace Division, 1 January–1 June 1964," 1, 57.
30. Ibid., 3.
31. Ibid., 64.
32. Ibid., 60, 61, 74, TWIX ECHO HILL Reentry Vehicle, 1–2.
33. Ibid., 62.
34. "History of the 1st Strategic Aerospace Division, Strategic Air Command, January–June 1966," Vol. I, Narrative, 1. This document is classified; the extract used is declassified IAW EO 13526.
35. Ibid., 2–3.
36. "History of the 1st Strategic Aerospace Division, Strategic Air Command, January–June 1966," Vol. I, 48–51, 87–88. LGM30B OT Final Report, "History of the 1st Strategic Aerospace Division, Strategic Air Command, July–December 1965," Vol. III, Document 8, 36–37. This document is classified; the extract used is declassified IAW EO 13526.
37. LGM30B OT Final Report, 3.
38. John M. Street, "Eniwetok Atoll, Marshall Islands" (Pacific Missile Range, Point Mugu, CA, 30 August 1960), 22. Author's collection.
39. Phil Fote, personal interview, May 2017. Fote was a design engineer for the Mark 5 and Mark 11 series.
40. Operational Test Exercise Report: Sweet Talk, LGM-30B, AFSN 64-063/FTM 799, 2 July 1965, AFHRA, IRISNUM 425448, K-Div-1-HI, Reel 11114, Document Nine, 2.
41. LGM30B OT Final Report, Document Eight, 3.
42. Ibid., 32.
43. "History of the 1st Strategic Aerospace Division, Strategic Air Command, January–June 1966," Vol. I, 77. This document is classified; the extract used is declassified IAW EO 13526.
44. Ibid., 36–46.
45. Cannon, personal interview.
46. "History of the 1st Strategic Aerospace Division, Strategic Air Command, January–June 1966," Vol. I, 72. This document is classified; the extract used is declassified IAW EO 13526.
47. Ibid., 75–76.

48. Ibid., 71.
49. McDowell, JSR Launch Vehicle Database. The author thanks Jonathan McDowell for the use of his comprehensive flight test records and Mitch Cannon for converting them into an Excel format. "History of the Strategic Air Command, FY 1970," Historical Study No. 117, Vol. II, Narrative (Office of the Historian, Headquarters Strategic Air Command, 20 April 1971), DNSA, NT00923, 251. Redacted copy.
50. Nalty, "USAF Ballistic Missile Programs, 1967–1968," 32–34.
51. "History of the Strategic Air Command, FY 1969, Historical Study No. 116, Vol. II, Narrative (Office of the Historian, Headquarters, Strategic Air Command, March 1970), DNSA, NT00502, 255–258. This document is classified; the extract used is declassified IAW EO 13526.
52. "History of the Strategic Air Command, FY 1970," Historical Study No. 117, Vol. II, 249–250.
53. Quick Look Report on the Attempted Launch of GLOWING SAND, "History of the First Strategic Aerospace Division, Strategic Air Command, January–June 1968," Vol. III, Supporting Documents (History Office, Vandenberg AFB), AFHRA IRISNUM 00425473, K-DIV-1-HI, Reel 11117, Document 52, 14–15. This document is classified; the extract used is declassified IAW EO 13526.
54. Minuteman I Reentry Systems Launch Program Handbook for Payload Designers, May, 1974, Revision C, April 1979 (Boeing Aerospace Archives, Seattle, WA), D2-26240-1, 1–1.
55. Dan Melzer, personal interview, April 2016. Melzer was a TRW guidance engineer.
56. Jeffrey Geiger, "Vandenburg AFB Launch Summary: 1958–2003" (Office of History, 30th Space Wing, February 2003), 47–63.
57. Holm, "Chronology of the Ballistic," 91, 94, 97. Bernard Nalty, "United States Air Force Ballistic Missile Programs: 1964–1966" (Office of Air Force History, March 1967). Redacted copy, DDRS CK 2349654796, 6–27. Geiger, "Vandenberg AFB," 60–76.
58. McDowell, JSR Launch Vehicle Database.
59. "Minuteman Historical Summary, 1962–1963," Sheet X, 19; Sheet XVI, 1. "Minuteman Historical Summary, 1964–1965," Sheet X, 50; Sheet XIII, 25–27, 32.
60. "History of the 1st Strategic Aerospace Division, January–June 1966," Vol. II, Narrative, 118. This document is classified; the extract used is declassified IAW EO 13526.
61. Nalty, "United States Air Force Ballistic Missile Programs,1967–1968," 46–47.
62. "Minuteman Historical Summary, 1966–1967," Sheet IX, 1–34. "Minuteman Flight Test Summary" (History Office, Vandenberg AFB, CA), 1–11.
63. "Minuteman Historical Summary, 1968–1969," Sheet II, 25.
64. "Minuteman Historical Summary, 1966–1967," Sheet IX, 16.
65. Ibid., Sheet II, 27–41.
66. Neufeld, "Development of Ballistic Missiles," 14–16. McDowell, JSR Launch Vehicle Database.
67. Neufeld, "Development of Ballistic Missiles," 16–17. McDowell, JSR Launch Vehicle Database. Geiger, "Vandenberg AFB," 26–82.
68. "History of the Strategic Air Command, FY 1969," Historical Study No. 116, Vol. II, Narrative (Office of the Historian, Headquarters, Strategic Air Command, 2 May 1974). Redacted copy, DNSA, NT00502, 286.
69. "History of the 1st Strategic Aerospace Division: July 1969–June 1970," Vol. III: Supporting Documents 77–150 (History Office, Strategic Air Command, Vandenberg AFB, CA), AFHRA IRISNUM 00425483, K-DIV-1-HI, V. III, Exhibit 77–80. This document is classified; the extract used is declassified IAW EO 13526.
70. "History of the Strategic Air Command, FY 1974," Historical Study No. 142, Vol. 1 Narrative (Office of the Historian, Headquarters, Strategic Air Command, 28 January 1974). Redacted copy, DNSA, NT01975, 295.
71. McDowell, JSR Launch Vehicle Database.
72. Ibid.

73. Ibid.
74. "History of the Strategic Air Command, FY 1969," Historical Study No. 116, Vol. II, Narrative (Office of the Historian, Headquarters, Strategic Air Command, 2 May 1974). Redacted copy, DNSA, NT00502, 272. Neufeld, "Development of Ballistic Missiles," 5. Nalty, "USAF Ballistic Missile Programs, 1962–1964," 50–51.
75. McDowell, JSR Launch Vehicle Database.
76. "History of the Strategic Air Command, 1 July 1975–31 December 1976," Historical Study No. 161, Vol. II, Narrative (Office of the Historian, Headquarters Strategic Air Command, 15 July 1977). Redacted copy, DNSA, NT02287, 417, 452.
77. I. Longstaff, S. Elms, K. Goussak, and C. Christoff, "The Litton LN-100 Advanced Technology Medium Accuracy Lightweight Inertial Navigation System," Paper 4416, presented at *Guidance, Navigation and Control Conference*, 10–12 August 1992, no page numbers given. McDowell, JSR Launch Vehicle Database.
78. "Minuteman Historical Summary, 1968–1969," Sheet II, 42, 45.
79. Ibid., Sheet II, 72, 75.
80. "Minuteman Historical Summary, 1970–1971," Sheet II, 29, 32, 35.
81. Ibid., Sheet II, 46, 58.
82. Ibid., Sheet II, 64, 71, 73, 78, 80, 102.
83. "History of the Strategic Air Command, FY 1973," Historical Study No. 124, Vol. 1, Narrative (Office of the Historian, Headquarters, Strategic Air Command, 2 May 1974). Redacted copy, DNSA, NT01755, 345.
84. "Minuteman Dual Launches" (History Office, Vandenberg AFB), 2.
85. "History of the Strategic Air Command, FY 1973," 346.
86. "History of the Strategic Air Command, FY 1970," 256.
87. Lorna Arnold and Katherine Pyne, *Britain and the H-bomb*, (Palgrave, New York, NY, 2001), 96.
88. "History of the Strategic Air Command, FY 1970," 259–260.
89. "History of the Strategic Air Command, FY 1969," 286, 288.
90. Space and Missile Test Center Capability Summary Handbook (Headquarters, Space and Missile Test Center, Vandenberg AFB, November 1974), Section 4, 3; Section 6, 1–3.
91. Environmental Assessment Missile Impacts, Illeginni Island, Kwajalein Missile Range, Kwajalein Atoll, December 1977, DTIC, ADA340912, 2–3. "History of the 1st Strategic Aerospace Division, 1 January–31 December 1979," Vol. IV, Supporting Documents, AFHRA IRISNUM 01039851, K-DIV-1-HI. Exhibits 60–84, no page numbers. This document is classified; the extract used is declassified IAW EO 13526.
92. McDowell, JSR Launch Vehicle Database.
93. Ibid.
94. "Minuteman Flight Test History, 1963–2015" (History Office, Vandenberg AFB, CA), 10–34. McDowell's JSR Launch Vehicle Database updates this list to 2019.
95. Ibid., 31.
96. Director Operational Test and Evaluation FY 2000 Annual Report (Department of Defense, February 2001), Section V, 131–134, hereafter referred to as DOT&E. DOT&E FY 2002 Annual Report (Department of Defense, 2003), 279–280.
97. "History of Strategic Air Command: 1 July 1975–31 December 1976," 448.
98. DOT&E FY 2000, Section V, 133.
99. "History of the Strategic Air Command, FY 1974," 292.
100. "Strategic Air Command Weapon Systems Acquisition: 1964–1979" (Office of the Historian, Headquarters, Strategic Air Command, 28 April 1980), AFHRA IRISNUM 01114095, K41 6.04–39 V. 1, 199.
101. Ibid., 321.

102. "History of the Strategic Air Command, FY 1973," 307.
103. "Minuteman Partial Flight Test History" (20th Air Force History Office, F. E. Warren AFB), Section 14, 36–37. This document is classified; the extract used is declassified IAW EO 13526.
104. "History of the Strategic Air Command, FY 1975," 323–324, 327.
105. McDowell, JSR Launch Vehicle Database. Robert Nease, personal communication, 17 August 2018. "Strategic Air Command Weapons Systems Acquisition: 1964–1979," Vol. I, 262.
106. "Space and Missile Systems Organization: A Chronology, 1954–1979" (History Office, Chief of Staff, Space Division, October 1979), DNSA, MS00303, 241, 242, 246. Nease and Hendrickson, "Brief History of Minuteman," Section 3, 38. Author's collection. "Minuteman III Partial Flight Test History" (History Office, 20th Air Force, F. E. Warren AFB, CA), Section 14, 36–37. This document is classified; the extract used is declassified IAW EO 13526.
107. "History of the Strategic Air Command, 1 July 1975–31 December 1976," 448.
108. Nalty, "USAF Ballistic Missile Programs, 1967–1968," 46.
109. "History of the Strategic Air Command, FY 1974," Historical Study No. 142, Vol. 1, Narrative (Office of the Historian, Headquarters, Strategic Air Command, 28 January 1974). Redacted copy, DNSA, NT01975, 266–272.
110. Ibid.
111. James I. Metcalf, Arnold A. Barnes Jr., and Michael J. Kraus, "Final Report of PVM-4 and PVM-3 Weather Documentation, 19 February 1975" (Air Force Cambridge Research Laboratories, Hanscom AFB, MA), DTIC, ADB004427, Section 3, 38.
112. James I. Metcalf, Arnold A. Barnes Jr., and Michael J. Kraus, "Final Report of STM-8W Weather Documentation, 11 April 1975" (Air Force Cambridge Research Laboratories, Hanscom AFB, MA), DTIC, ADB006666, Section 3, 34.
113. James I. Metcalf, Arnold A. Barnes Jr., and Michael J. Kraus, "Final Report of PVM-5 Weather Documentation, 28 May 1975" (Air Force Cambridge Research Laboratories, Hanscom AFB, MA), DTIC, ADB006667, Section 7, 46.
114. James I. Metcalf, Michael J. Kraus, and Arnold A. Barnes Jr., "Final Report of OT-45, PVM-8 and RVTO Weather Documentation, 23 July 1975" (Air Force Cambridge Research Laboratories, Hanscom AFB, MA), DTIC, ADB011352, Section 3, 55.
115. James I. Metcalf, Michael J. Kraus, and Arnold A. Barnes Jr., "Final Report of PVM-6 and PVM-7 Weather Documentation, 11 September 1975" (Air Force Cambridge Research Laboratories, Hanscom AFB, MA), DTIC, ADB011353, Section 3, 58.
116. "Minuteman III Partial Flight Test History," Section 14, 37. This document is classified; the extract used is declassified IAW EO 13526.
117. Amy F. Woolf, "US Strategic Forces: Background, Developments, and Issues" (Congressional Research Service Report, 8 September 2006), CRS-9.
118. McDowell, JSR Launch Vehicle Database. Amy F. Woolf, "US Strategic Forces: Background, Developments, and Issues" (Congressional Research Service Report, 3 April 2007), CRS-13. "History of the 30th Space Wing, 1 January-31–December 2005" (History Office, Vandenberg AFB, CA), 96–99. This document is classified; the extract used is declassified IAW EO 13526.

## Chapter 14

1. "SAC Missile Chronology," 34, 233.
2. "History of the 44th Strategic Missile Wing, January–March 1965," AFHRA, IRISNUM 00450524, K-WG-44-HI, Reel 13743, Exhibit 11, 27–30. This document is classified; this extract was declassified IAW EO 13526.
3. Ibid., 2–4. "1st Strategic Aerospace Division History: January–June 1965," Vol. 1, AHFRA,

IRISNUM 00425447, K-DIV-1-HI, Reel 11114, 94–96. This document is classified; this extract was declassified IAW EO 13526.

4. "Minuteman Gets an 'A' in Newell Test Shot," *Rapid City Daily Journal*, March 1, 1965, 1–2.
5. Allan Martens, personal interview, 10 May 2017.
6. "Minuteman Gets an 'A,'" 1–2.
7. "History of the 44th Strategic Missile Wing, January–March 1965," 29.
8. "1st Strategic Aerospace Division History, January–June 1966," Vol. 1, AHFRA, K-DIV-1-HI, IRISNUM 00425453, Reel 11114, 119–124. This document is classified; this extract was declassified IAW EO 13526.
9. Bernard Nalty, "United States Air Force Ballistic Missile Programs, 1967–1968" (Office of Air Force History, September 1969), DDRS CK2349655277, 28–29. This document is redacted.
10. "Minuteman Historical Summary, 1967–1968" (The Boeing Company, Seattle, WA, 1974), D2–26485, Sheet XIII, 48.
11. "History of the Strategic Air Command, FY 1969," Historical Study No. 116, Vol. II, Narrative (Office of the Historian, Headquarters Strategic Air Command, March 1970), DNSA, 262–263. This document is redacted.
12. Nalty, "USAF Ballistic Missile Programs, 1967–1968," 30.
13. Ibid., 30. "History of the Strategic Air Command, FY 1969," 262–263.
14. Ibid., 272.
15. Ibid., 274–278.
16. "History of the 390th Strategic Missile Wing, April 1963," AFHRA, IRISNUM 00457361, Reel 10286, K-WG-390-HI-Vol. I, 25. This document is classified; this extract was declassified IAW EO 13526. "History of the Strategic Air Command, FY 1970," Historical Study No. 117, Vol. II, Narrative (Office of the Historian, Headquarters, Strategic Air Command, 20 April 1971), DNSA, NT00923, 261. This document is redacted.
17. AFHRA IRISNUM 01059543, microfilm 38412 abstract. The op plan number was in the title of the abstract, which is unclassified, but the document itself is classified.
18. "1st Strategic Aerospace Division History, January–June 1966," Vol. 1, 125.
19. "History of the Strategic Air Command, FY 1970," 261.
20. Ibid., 127. Shaw, "Minuteman, Strategic Air," Vol. 1, Reel 45868, 224. This document is classified; this extract was declassified IAW EO 13526.
21. "History of the Strategic Air Command, FY 1969," 278, 225–233.
22. Ibid., 273–278.
23. "History of the Strategic Air Command, FY 1970," 265–266.
24. "History of the Strategic Air Command, FY 1971" (Office of the Historian, Headquarters, Strategic Air Command, 13 January 1992), DNSA, NT02290, 301–302.
25. W. W. Hickman, "Giant Patriot Destruct," *Combat Crew*, December 1972, 26–27.
26. "History of the Strategic Air Command, FY 1973," Historical Study No. 124, Vol. 1, Narrative (Office of the Historian, Headquarters, Strategic Air Command, 2 May 1974), DNSA, NT01755, 353–355.
27. "Minuteman II Operational Base Launch Planned," press release, 28 December 1973, Office of the Assistant Secretary of Defense for Public Affairs. "History of Strategic Air Command FY 1974," Historical Study No. 142, Vol. X, Chapter V, AFHRA, IRISNUM 1005879, K416. 01–142, Reel 0030688, Exhibit 71. Holm, "Chronology of the Ballistic," Vol. I, 126. Author's collection. J. G. Sestak, "Giant Patriot," *Combat Crew*, May 1974, 20–24.
28. "History of the Strategic Air Command, FY 1974," Historical Study No. 142, Vol. 1, Narrative (Office of the Historian, Headquarters, Strategic Air Command, 28 January 1974), DNSA, NT01975, Exhibit 71, Slides 9–17.

29. There is no evidence these launches took place, as the flights during this period are labeled operational test launches, which describe missiles that are not altered in this fashion.
30. "Minuteman II Operational Test Program," *Commanders Digest*, 31 January 1974, 2, 6–8. "GIANT PATRIOT Minuteman Operational Base Launch Program," 13 August 1974, briefing slides. The author thanks Greg Ogletree for a copy of this document. "History of the Strategic Air Command, FY 1974," 302.
31. *Congressional Record*, (S.) 6 June 1974, 18060–18062.
32. Ibid.
33. A. Ernest Fitzgerald, *The High Priests of Waste* (W. W. Norton and Company, Inc., New York, NY, 1972), 120–124.
34. "History of the Strategic Air Command, FY 1974," 308.
35. "History of the 321st Strategic Missile Wing and 804th Combat Support Group," Grand Forks AFB, ND, July–September 1970, AFHRA IRISNUM 00455859, K-WG-321-HI, Vol. I, Reel 14441, 68–71. This document is classified; the extract was declassified IAW EO 13526. "History of the First Strategic Aerospace Division, July 1972–June 1973," Vol. I, Narrative (History Office, Vandenberg AFB, CA, 29 April 1974), AFHRA, IRISNUM 00903210, K-DIV-1-HI V. 1, Reel 24563, 219. This document is classified; the extract was declassified IAW EO 13526.
36. "History of the 321st Strategic Missile Wing and 804th Combat Support Group, Grand Forks AFB, ND, July–September 1970," 321st Strategic Missile Wing Giant Profit Exercise Summary Report, 1 November 1970, Exhibit 50, 3. This document is classified; the extract was declassified IAW EO 13526.
37. Ibid., 9–28.
38. Ibid., 29.
39. Ibid., 34–35. Mitch Cannon, personal communication, 1 July 2018. "History of the Strategic Air Command, FY 1973," 357.
40. "History of the 321st," 41–42.
41. Ibid., 5.
42. "History of the Strategic Air Command, FY 1973," 357.
43. Shaw, "Minuteman, Strategic Air," 238.
44. Ibid., 239–240.
45. "History of the Strategic Air Command, FY 1973," 357–361.
46. Cannon, personal communication. Much later in the program, a special test wafer was inserted between the reentry vehicle and the guidance-and-control section, completely isolating the operational reentry vehicle. This change reduced the cost and security issues involved with moving an operational reentry vehicle to and from the LF.
47. "History of the Strategic Air Command, FY 1974," 309–315.
48. Ibid., 309–317.
49. Ibid., 317.
50. "History of the Strategic Air Command, 1 July 1975–31 December 1976," Historical Study No. 161, Vol. 1, Narrative" (Office of the Historian, Headquarters Strategic Air Command, 15 July 1977), DNSA, NT02287, 458. Redacted copy.
51. Cannon, personal communication.

# Chapter 15

1. Piper, *Development of the SM-80*, Supporting Documents, "Proposed Preliminary Operational Concept for Minuteman, 8 April 1958," Document 76, 6.

2. G. B. Kistiakowsky, Memorandum for the President, Subject: Problems Involved in the Minuteman Program, 12 February 1960, DNSA, NH00689, 1–2. "Principal Technical Problems of Minuteman, 12 February 1960," DDEL, White House Office, Office of the Special Assistant for Science and Technology (James R. Killian and George B. Kistiakowsky) Records, 1957–1961, Box 12, Folder Missiles January–June 1960, 1–2.
3. J. Ponturo, "Analytical Support for the Joint Chiefs of Staff: The WSEG Experience, 1948–1976" (Institute for Defense Analysis, July 1979), DTIC, ADA090946, 176–177.
4. WSEG Report 50: Evaluation of Strategic Offensive Weapons Systems, 27 December 1959, DNSA, NH00422, 7. Many details of the report's findings on command-and-control issues for Minuteman remain classified. This document is classified; the extract used is declassified IAW EO 13526.
5. Piper, *Development of the SM-80*, Vol. I, 69.
6. G. A. Reed, "US Defense Policy, US Air Force Doctrine, and Strategic Nuclear Weapon Systems, 1958–1964: The Case of the Minuteman ICBM," 1986, ProQuest Dissertations and Theses, Thesis 8720847, 196.
7. WSEG Report 50, Enclosure F, Appendix B, DNSA, NH00422, 46. "Minuteman Historical Summary, 1958–1959" (Boeing Aircraft Corporation, 1974), Boeing Archives, D2–26485, Sheet XI, 8. The decision for 10 LFs per LCC had been made by May 1959.
8. Ponturo, "Analytical Support," 176.
9. Lauritsen Committee Report to Lt. Gen. Bernard Schriever Concerning the United States Air Force Minuteman Program, 15 June 1961, DNSA, NH00730, 2. The distance measurements were calculated by the author using Google Earth.
10. Ibid., 4.
11. J. H. Rubel, Memorandum for the Assistant Secretary of the Air Force (Research and Development), Subject: Minuteman, 2 November 1960, AFHRA, IRISNUM 01040303, 168.7171–155, Reel 35263, Frames 406–407.
12. J. H. Rubel, *Doomsday Delayed: USAF Strategic Weapons Doctrine and SIOP-62, 1959–1962* (Hamilton Books, 2008), 5–16. "History of Strategic Air Command 1961," SAC Historical Study No. 89, January 1962, Scribd, 95–96. Redacted copy.
13. Lauritsen Committee Report, AFSC Comments on Lauritsen Committee Findings, 2.
14. Piper, *Development of the SM-80*, Vol. I, 70.
15. Rubel, *Doomsday Delayed*, 5–16. Piper, *Development of the SM-80*, Vol. I, 71–73. Reed, "US Defense Policy," 197–198.
16. Summary Report: Minuteman Flexibility and Safety Study Group, 27 September 1961 (Department of Defense), AFHRA IRISNUM 2054414, K243.8631–745, Section 4, 7. This document is classified; the extract used is declassified IAW EO 13526. Piper, *Development of the SM-80*, Vol. I, 71–72.
17. Reed, "US Defense Policy," 199.
18. Unless otherwise noted, the remainder of the launch command section comes from written correspondence with Mitch Cannon, CMSgt., USAF (ret.), May 2019.
19. Today's weapon system safety rules force LCC code dissipation for single LCC control of a squadron. This is most likely to occur when the squadron is split for either modification or code change. Mitch Cannon, personal communication, May 2019.
20. Proposed Safety Rules for the HSM-80A (Minuteman)/MK 5 RV Weapon System, 18 October 1962, NARA, CCS 4615 (2 October 1962), JCS 1962, Annex A: 5–6.
21. Modernized Minuteman Command-and-Control Systems, WS-133A-M, 5 October 1967 (Minuteman Systems Engineering, Aero-Space Group, Missile and Information Systems Division, Boeing, Seattle, WA), 21. Author's collection.
22. Monte Watts, personal communications, September 2019. Watts is a staff systems engineer, Minuteman III Command-and-Control and Training Systems, Northrup Grumman Technology

Services. Unless otherwise noted, the material in the REACT discussion comes from this communication.

23. "SAC Missile Chronology," 92. "History of the Strategic Air Command, 1990" (Office of the Historian, Headquarters Strategic Air Command, Offutt AFB, NE), Air Force Space Command On Alert Archive, 416–417. This document is classified; the extract used is declassified IAW EO 13526.

24. Robert J. Boyd, "Emergency Rocket Communications System (ERCS) (Program 279/494L), Strategic Air Command Weapon Systems Acquisition, 1964–1979" (Headquarters Strategic Air Command History Office, 28 April 1980), AFHRA, IRISNUM 01114095, K416.04-39, V. I, Reel 45868, 364.

25. Ibid., 365.

26. J. D. Hunley, *The Development of Propulsion Technology for US Space-Launch Vehicles: 1926–1991* (Texas A&M University Press, College Station, TX, 2007), 72–73.

27. BEAN STALK launches No. 1–6, 6595th Aerospace Test Wing. The author thanks Jonathan McDowell for a copy of this document.

28. Boyd, "Emergency Rocket Communications," 366, 369. Minuteman ERCS, http://www.siloworld.net/ICBM/MM/ERCS/ercs.htm, retrieved 15 June 2018. "Alert Operations and the Strategic Air Command, 1957–1991" (Office of the Historian, Headquarters Strategic Air Command, Offutt AFB, NE, 7 December 1991), 27–28.

29. Boyd, "Emergency Rocket Communications," 367–368.

30. Cannon, personal communication, 15 March 2019. Cannon had 16 years' experience with the ERCS program at the 351st Strategic Missile Wing, Whiteman AFB, and Headquarters SAC and 44 years of experience with Minuteman. He was a graduate of the Bendix ERCS course at Ann Arbor, Michigan. Unless otherwise noted, the ERCS discussion is based on Cannon's experience and information from the ERCS Orientation Handbook, 30 September 1976, author's collection.

31. Boyd, "Emergency Rocket Communications," 368–369. "Vandenberg AFB Launch Summary, 1958–2003" (30th Space Wing History Office, Vandenberg AFB, CA, 2003), 7–25. "Minuteman Historical Summary, 1966–1967," Sheet IX, 13.

32. "Vandenburg AFB Launch Summary," 25.

33. "History of the Strategic Air Command, 1 July 1975–31 December 1976," Historical Study No. 161, Vol. 1, Narrative (Office of the Historian, Headquarters Strategic Air Command, 15 July 1977), Air Force Historical Support Division, Joint Base Anacostia–Bolling, Washington, DC, 109. ERCS Operational Test Flight GM-9, FTM 2225, July 1990, Bendix Aerospace Systems Division, Ann Arbor, MI, DTIC, ADC022794.

34. Cannon, personal communications.

35. *Testimony of Secretary of Defense Robert McNamara*, Senate Committee on Foreign Relations, 88th Cong., (1963), 98–102.

36. Greg Ogletree, Maj., USAF (ret.), "A History of PACCS, ACCS and ALCS Part 1," *Strategic Air Command Airborne Command Control Association Newsletter* 1, no. 2 (October 1995): 2.

37. "Strategic Air Command Weapon Systems Acquisition," Vol. I, 411.

38. Ibid., 412.

39. Seymour L. Zeiberg, "MX: The Full Perspective," *Defense* (American Forces Information Service, Arlington, VA, 1980), 7. This publication was hard to find, but it was published by the Defense Department as a monthly journal, Gov Doc: D 2.15/3, GPO Item No: 0312B; LC: UA23.A1; Dewey: 355/.00. Ernest R. May, John D. Steinbruner, and Thomas W. Wolf, "History of the Strategic Arms Competition 1945–1972, Vol. 3, A Handbook of Selected Soviet Weapon and Space Systems (United States Air Force, June 1976), 162–167. The author thanks Hans Kristiansen for providing me with these pages. Deployment of the SS-9 Mod 1 and Mod 2 began in early 1966. "History of the Strategic Arms," 500.

40. "Space and Missile Systems Organization: A Chronology, 1954–1979" (Space Division, Chief of Staff History Office, 1979), 165.
41. "Strategic Air Command Weapon Systems Acquisition," 413.
42. Ibid., 413–414. "494L and ALCS Launches at Vandenberg AFB" (History Office, Vandenberg AFB, CA), 1.
43. Ibid. "Strategic Air Command Weapon Systems Acquisition," 196.
44. "History of Strategic Air Command, January–June 1968," Historical Study No. 112, Vol. I, Narrative (History and Research Division, Headquarters Strategic Air Command, February 1969), AFHRA, IRISNUM 1120106, K416.01–112, V. I, Reel 40267, 41. This document is classified; the material used is declassified IAW EO 13526.
45. "History of Strategic Air Command, January–June 1968," 39–41.
46. Robert Parker, Maj. Gen., USAF (ret.), personal interview, 12 October 2018. Greg Ogletree, Maj., USAF (ret.), "The 4th Airborne Command and Control Squadron," *SAC Airborne Command Control Association Newsletter* 6, no. 3 (November 2000): 2.
47. Ronald Gray, Brig. Gen., USAF (ret.), personal interview, 12 October 2018. Gary L. Curtin, Maj. Gen., USAF (ret.), personal interview, 12 October 2018.
48. ALCS Weapon System Study Guide, 1 October 1994 (Airborne Launch Control System Combat Crew Training School, Headquarters Air Force Space Command, Offutt AFB, NE), 46. The author thanks Greg Ogletree for a copy of this document.
49. ALCS Weapon System Study Guide, 88–91.
50. Ibid., 91.
51. Ibid., 89.
52. "History Report Chronologies" (75th Air Base Wing, Office of History, 1965–2000, Hill AFB, UT), no page given, citation by date.
53. "HETF Aids ALCS Launches," *Minuteman Service News*, July–August 1971, 8–9. ALCS Weapon System Study Guide, 90.
54. Ibid., 88.
55. "Minuteman Historical Summary, 1970–1971" (Boeing Company, Seattle, WA, 1976), Sheet II, 73. "494L and ALCS," 1–3.
56. "The 4th Airborne Command and Control Squadron," 2.
57. Greg Ogletree, Maj., USAF (ret.), "The 2nd Airborne Command and Control Squadron," *SAC Airborne Command Control Association Newsletter* XVII, no. 2 (June 2011): 4.
58. Scott D. Sagan, *The Limits of Safety: Organizations, Accidents, and Nuclear Weapons* (Princeton University Press, NJ, 1993), 81–91.
59. Robert S. Norris, "The Cuban Missile Crisis: A Nuclear Order of Battle, October/November 1962," a presentation at the Woodrow Wilson Center, Washington, DC, 24 October 2012. Raymond L. Garthoff, *Reflections on the Cuban Missile Crisis* (The Brookings Institution, Washington, DC, 1989), 20, 209.
60. "Radio and Television Address to the American People on the Soviet Arms Build-up in Cuba," 22 October 1962 (John F. Kennedy Library, Boston, MA), JFKWHA-142-001. Proclamation 3504: "Interdiction of the Delivery of Offensive Weapons to Cuba," 23 October 1962 (John F. Kennedy Library, Boston, MA), JFKPOF-041–019, 1.
61. "Strategic Air Command Operations," Vol. 1, 25, 35. This document is classified, the material used is declassified IAW EO 13526.
62. 341st Strategic Missile Wing Operations Plan 402-63, Emergency Combat Capability, 341st Strategic Missile Wing and 341st Combat Support Group History, 1–31 December 1962, AFHRA, IRISNUM 00456138, K-WG-341-HI, V. I, Reel 14471, Document 4: 5. "Strategic Air Command Operations," Vol. 4, Chapter 3, Tab 8. This document is classified; the material used is declassified IAW EO 13526.

63. Ibid., 1. 341st Strategic Missile Wing and 341st Combat Support Group History, 1–31 December 1962, Document 4: 5, 6. Proposed Safety Rules for the HSM-80A (Minuteman)/MK 5 RV Weapon System, Appendix A, Operational Concept of the HSM-80A (Minuteman)/MK 5 RV/MK 59 MOD 0 Weapon System, 5–6, NARA, CCS 4615 (2 October 1962), JCS 1962, Annex A.
64. 341st Strategic Missile Wing and 341st Combat Support Group History, Cuban Crisis Annex, 1–31 October 1962, Malmstrom AFB, AFHRA, IRISNUM 00456139, K-WG-341-HI, V.2, Reel 14471, 1. 341st Strategic Missile Wing and 341st Combat Support Group History, 1–30 September 1962, Malmstrom AFB, AFHRA, IRISNUM 00456137, K-WG-341-HI, Reel N0428, 24. "Strategic Air Command Operations," Vol. 4, Chapter 3, Tab 9.
65. "Strategic Air Command Operations," Vol. 1, vii, 25, 35.
66. 341st Strategic Missile Wing and 341st Combat Support Group History, Cuban Crisis Annex, 1–31 October 1962, 4, 6. "History of the 3901st Strategic Missile Evaluation Squadron, Strategic Air Command, November to January 1962–63," AFHRA, IRISNUM 00419382, K-SQ-EVAL-3901-HI, Reel 12474, 3–5.
67. "Strategic Air Command Operations," Vol. 4, Chapter 3, Tab 8, 2.
68. 341st Strategic Missile Wing and 341st Combat Support Group History, 1–31 December 1962, Document 4: 5, 6; Annex A, 2.
69. Proposed Safety Rules for the HSM-80A (Minuteman)/MK 5 RV Weapon System, 18 October 1962, Annex A: 5, 6.
70. Ibid., Appendix A: 4.
71. Ibid., Appendix B: 12–13.
72. Samuel Goodwin, Col., USAF (ret.), telephone interview, 29 March 2018. Goodwin was a missile combat crew commander at Malmstrom during the crisis. 341st Strategic Missile Wing and 341st Combat Support Group History, Cuban Crisis Annex, 1–30 November 1962, 3. 341st Strategic Missile Wing and 341st Combat Support Group History, Cuban Crisis Annex, 1–31 October 1962, 3.
73. "Strategic Air Command Operations," Vol. 4, Chapter 3, Tab 25.
74. Ibid., Tab 26.
75. Ibid., 5.
76. Ibid., 3. "Strategic Air Command Operations," Vol. 1, 2.
77. 341st Strategic Missile Wing and 341st Combat Support Group History, Cuban Crisis Annex, 1–30 November 1962, 3. Other accounts give 20 November as the date that SAC returned to DEFCON 3. Dates used here are from official Air Force sources.
78. "Minuteman Historical Summary, 1962–1963," Sheet X, 14, 19–20.
79. Ibid., Sheet X, 22–23.
80. Ed Gill, Col., USAF (ret.), telephone interview, 26 June 2018. Gill was the targeting officer and recalls bringing only four missiles to alert. Gill, personal communication, 23 June 2018. Gill firmly denies reports of the targets being changed to Cuba. Robert J. Serling, *Legend and Legacy: The Story of Boeing and Its People* (St. Martin's Press, 1992), 174.
81. "History of the 6595th Aerospace Test Wing, 22 October–20 November 1962" (History Office, Vandenberg AFB, CA), Titan Missile Museum Archive, David K. Stumpf Collection, S-8 80100, 1–11. The 6595th Aerospace Test Wing historian's LF nomenclature is incorrect and has been corrected in the text.

## Chapter 16

1. "History of Strategic Air Command, 1 January–31 December 1985," Historical Study No. 214, Vol. II, Narrative (History Office, Headquarters Strategic Air Command), Air Force Space

Command On Alert Archive, 547. This document is classified; the extract used is declassified IAW EO 13526.

2. William R. Large Jr., "Ballistic Missile Hardening Study," 10 July 1958 (Office of the Assistant Commander-In-Chief, Strategic Air Command), Titan Missile Museum Archive, David K. Stumpf Collection, Box S4, Folder 8004, 53–56. Edward Cohen and A. Bottenhofer, "Test of German Underground Personnel Shelters, Operation Plumbbob," Project 30.7 (Amman and Whitney, New York, NY, July 1970), DTIC, AD691407, 20–21, 53–75, 193–194.

3. Howard W. Wampler, Gerald G. Leigh, Myron E. Furbee, and Frank E. Seusy, "A Status and Capability Report on Nuclear Black Air Blast Simulation Using HEST," *Proceedings of the Nuclear Blast and Shock Simulation Symposium*, 28–30 November 1978, Vol. I (Defense Nuclear Agency, Washington, DC, December 1978), DTIC, ADA073766, 32. The original name for the test series was Gas Bag Hardness Test. The name was later changed to HEST and will be used in the rest of the discussion. "Designing Facilities to Resist Nuclear Weapons Effects: Hardness Verification," TM 5-858-6, August 1984 (Headquarters Department of the Army), Army Publishing Directorate, Chapter 4, 3; Appendix B, 1.

4. Nalty, "USAF Ballistic Missile Programs, 1964–1966," 28–29.

5. "Minuteman Historical Summary, 1964–1965," Sheet XIX, 14. "Minuteman Historical Summary, 1966–1967," Sheet X, 2, 3, 18, 28.

6. Ibid., Sheet X, 8, 17–18, 38. Nalty, "USAF Ballistic Missile Programs, 1967–1968," 31.

7. Nalty, "USAF Ballistic Missile Programs, 1967–1968," 31–32. "Minuteman Historical Summary, 1966–1967," Sheet X, 32, 39.

8. "Minuteman Historical Summary, 1966–1967," Sheet X, 39–40, Sheet XVI, 40. "LER Repositioning Technique Demonstrated," *Minutemen Service News*, January–February 1968, 10. While not explicitly stated in the article, the timing of the article in early 1968 and the completion of the repair in September 1967 implied to the author that this was the discussion of the repair process. Efforts to confirm this have been unsuccessful.

9. Nalty, "USAF Ballistic Missile Programs, 1967–1968," 31–32.

10. Ibid., 33. "History of the Strategic Air Command, FY 1969," Historical Study No. 116, (Office of the Historian, Headquarters Strategic Air Command, March 1970), DNSA, NT00502, 298–299.

11. Holm, "Chronology of the Ballistic," 96, 104.

12. "Minuteman III Is Coming," *Minuteman Service News*, May–June 1968, 3.

13. "What's Force Mod," *Minuteman Service News*, November–December 1965, 11–12.

14. Holm, "Chronology of the Ballistic," 109.

15. "Periscope for Minuteman III, Wing III," *Minuteman Service News*, Issue 449, March–April 1970, 10.

16. "History of the Strategic Air Command, FY 1969," Historical Study No. 116, 335. This document is classified; the extract used is declassified IAW EO 13526.

17. "Space and Missile Systems Organization: A Chronology, 1954–1979" (Office of History, Headquarters, Space Division, October 1979), DNSA, MS00303, 240. "History of the Strategic Air Command, 1 July 1975–31 December 1976," Historical Study No. 161, Vol. II, Narrative (Office of the Historian, Headquarters Strategic Air Command, 15 July 1977), DNSA, NT02287, 466. Redacted copy.

18. Neufeld, "USAF Ballistic Missile Programs," 25–32.

19. "LF Upgrades Wing VI and Squad XX, no date" (TRW-Deployment Engineering), 3. Author's collection.

20. Ibid., 25

21. Ibid., 23–61.

22. Ibid., 108–132.

23. Ibid., 64–91.

24. "Strategic Air Command Weapon Systems Acquisition, 1964–1979" (Office of the Historian,

Headquarters Strategic Air Command, 28 April 1980), AFHRA, IRISNUM 01114095, Reel 45868, K416.04–39, Vol. 1, 246.

25. Ibid., 244–249.
26. "Space and Missile Systems," 241, 242, 246.
27. "History of the Strategic Air Command, 1 July 1975–31 December 1976," 409. Redacted copy.
28. Ibid., 395.
29. "History of the Strategic Air Command, FY 1975," Historical Study No. 153, Vol. I, Narrative (Office of the Historian, Headquarters Strategic Air Command, 10 February 1976), AFHRA, IRISNUM 01010404, Reel 31271, TS-HOA-76–18, 319, 323. This document is classified; the extract used is declassified IAW EO 13526.
30. "TORUS," *Minuteman Service News*, September–October 1974, 8–9; "Siege," *Minuteman Service News*, September–October 1975, 10–12.
31. "History of the Strategic Air Command, FY 1975," 324–326.
32. Ibid., 323–324, 327.
33. "History of the Strategic Air Command, 1 January–31 December 1985," Historical Study No. 214, Vol. II, Narrative (History Office, Headquarters Strategic Air Command), Air Force Space Command On Alert Archive, 347–348. This document is classified; the extract used is declassified IAW EO 13526.
34. "History of the Strategic Air Command, 1980" (History Office, Headquarters Strategic Air Command), Air Force Space Command On Alert Archive, 331–332. This document is classified; the extract used is declassified IAW EO 13526.
35. "History of the Strategic Air Command, 1 January–31 December 1985," History Study No. 214, Vol. II, Narrative (History Office, Headquarters Strategic Air Command, Offutt AFB, NE), Air Force Space Command On Alert Archive, 544–549. This document is classified; the extract used is declassified IAW EO 13526.
36. "Minuteman III Life Extension Program: A Report to Congress," 29 July 1992, 1. Author's collection.
37. "History of the Air Force Space Command, 1 January 1992–31 December 1993" (History Office, Air Force Base Command, Peterson AFB, CO), Air Force Space Command On Alert Archive, 146–147. This document is classified; the extract used is declassified IAW EO 13526.
38. Dillon White, "Propulsion Replacement Program Complete," *Great Falls Tribune*, 23 August 2009, 1.
39. "Minuteman III Life Extension Program," 11.
40. "Minuteman III Life Extension Program," 8. Department of Defense Appropriations for 1980, *Hearings before a Subcommittee of the Committee on Appropriations* (H. R.) 96th Cong. (1979), 499.
41. "History of the Air Force Space Command, 1 January 1994–31 December 2003" (History Office, Air Force Space Command, Peterson AFB, CO), Air Force Space Command On Alert Archive, 153–155. This document is classified; the extract used is declassified IAW EO 13526. "History of the Air Force Space Command, 1 January 2004–31 December 2005" (History Office, Air Force Space Command, Peterson AFB, CO), Air Force Space Command On Alert Archive, 64. This document is classified; the extract used is declassified IAW EO 13526.
42. "History of the Air Force Space Command, 1 January 2007–31 December 2008" (History Office, Air Force Space Command, Peterson AFB, CO), Air Force Space Command On Alert Archive, 199. This document is classified; the extract used is declassified IAW EO 13526. "History of the 90th Space Wing, 1 January 2005–31 December 2007," 47.
43. "Minuteman III Life Extension Program," 9.
44. "ICBM Modernization: Minuteman III Guidance Replacement Program Has Not Been Adequately Justified," 25 June 1993 (General Accounting Office, Washington, DC), 3.
45. "An Independent Assessment of the Need for the Minuteman III Guidance Replacement Program," Reliability Analysis Center, August 1993, iii–iv. This document was referenced on page 136 of a

declassified extract of the "History of the Air Force Space Command, 1 January 1992–31 December 1993."

46. "History of the Air Force Space Command, 1 January 2007–31 December 2008" (History Office, Air Force Space Command, Peterson AFB, CO), Air Force Space Command On Alert Archive, 198. This document is classified; the extract used is declassified IAW EO 13526.

47. "History of the Air Force Space Command, 1 January 2004–31 December 2005" (History Office, Air Force Space Command, Peterson AFB, CO), Air Force Space Command On Alert Archive, 67. This document is classified; the extract used is declassified IAW EO 13526.

48. Ibid., 67–68.

49. Ibid., 68.

## Chapter 17

1. Stephen Engelberg, "Air Force Offers to Close 15 Bases and Scrap Missiles," *New York Times*, 19 November 1989, 1. Peter Johnson, "Missile Cuts Get Skeptical Response," *Great Falls Tribune*, 16 January 1990, 1A. R. Jeffrey Smith and Molly Moore, "Air Force Calls for Scrapping Minuteman II, Cheney Orders Freeze on Hiring Civilians," *Washington Post*, 13 January 1990, A1.

2. Peter Johnson, "MAFB Cuts? Air Force Making Reductions Nationwide," *Great Falls Tribune*, 6 February 1991, 1A; "Mid-1990s for Missile Loss Here? Deactivation of Minuteman IIs Not Expected to Start Immediately," *Great Falls Tribune*, 7 February 1991, 1A.

3. Pub. L. 101-510, National Defense Authorization Act for FY 1991, 101st Cong., 1775. Final Environmental Impact Statement, Deactivation of the Minuteman II Missile Wing at Ellsworth AFB, SD, October 1991 (Department of the Air Force, Headquarters Strategic Air Command, Offutt AFB, NE), Executive Summary, 23–27.

4. *Federal Register*, 19 April 1991, Vol. 56, no. 76, 16077. Final Environmental Impact Statement; Deactivation of the Minuteman II Missile Wing at Ellsworth AFB, SD, October 1991, Section I, 6.

5. President George H. Bush, "Address to the Nation on Reducing United States and Soviet Nuclear Weapons, 27 September 1991," American Presidency Project, University of California, Santa Barbara.

6. *Treaty between the United States of America and the Union of Socialist Soviet Republics on Further Reduction and Limitation of Strategic Offensive Arms*, 31 July 1991 (Bureau of Arms Control, Verification and Compliance, October 2001 edition), Article II.

7. 44th Commemorative Committee, *Aggressor Beware, A Brief History of the 44th Missile Wing 1952–1994* (Grelind Printing, Rapid City, SD, 1994), 43–44.

8. Ibid., 44.

9. Deactivation of the Minuteman II Missile System: Final Deactivation Environmental Plan, Ellsworth AFB, SD, September 1993 (Headquarters Strategic Air Command, Offutt AFB, NE), Executive Summary, 1.

10. Minuteman ICBM National Historic Landmark Nomination, no date given, Section 8, 47. Missile Dismantlement Status, 44th Missile Wing, 4 November 1997, 1–3. The author thanks Tim Pavek, Ellsworth AFB Civil Engineering Office, for this document.

11. Hotel-Flight Launch Facilities Environmental Baseline Survey: Minuteman II Deactivation Site Disposals, 20 July 2000, Ellsworth AFB, SD, 19–21. The author thanks Tim Pavek, Ellsworth AFB Civil Engineering Office, for this document.

12. Deactivation of the Minuteman II Missile System, Executive Summary, 3.

13. Launch Facility Golf-02 Deactivation Checklist, 46. The author thanks Tim Pavek, Ellsworth AFB Civil Engineering Office, for this document. Scott Gustafson, Demtech engineer in charge of demolition, personal communication, 23 October 2018.

14. Golf-Flight Launch Facilities Environmental Baseline Survey, Minuteman II Deactivation Site Proposals, Ellsworth AFB, SD, 19 April 2000, 6. The author thanks Tim Pavek, Ellsworth AFB Civil Engineering Office, for this document. Gustafson, personal communication.
15. Missile Dismantlement Status, 3.
16. Deactivation of the Minuteman II Missile System: Final Deactivation Environmental Plan, Ellsworth AFB, SD, September 1993 (Headquarters Strategic Air Command, Offutt AFB, NE), Section 2, 19–20.
17. *Federal Register*, 14 November 1991, Vol. 56, no. 220, 57–79.
18. Minuteman II Deactivation, 351st Missile Wing Program Plan 92-12, 1 May 1992, (Headquarters Strategic Air Command, Offutt AFB, NE), 2.
19. "Wing Deactivation to Be Discussed at Public Meeting," *Sedalia Democrat*, 18 October 1992, 1. Final Environmental Impact Statement: Deactivation of the Minuteman II Missile Wing at Whiteman AFB, MO (Headquarters, Air Combat Command, Langley AFB, VA), Executive Summary, 5.
20. Minuteman II Deactivation and 351st Missile Wing Deactivation, Program Plan 92-12, (Headquarters Strategic Air Command, Offutt AFB, NE), 5.
21. Richard E. Rice, Sgt., US Air Force, "Closeout History of the 351st Strategic Missile Wing," 1 January–31 July 1995, Vol. I of VII, 1 August 1995, Appendix U, 81–85. Author's collection.
22. Minuteman II Deactivation, 351st Missile Wing Inactivation, 5.
23. John Reidy, "First Silo Demolition Proves Something Less Than a Blast," *Sedalia Democrat*, 9 December 1992, 1.
24. "Minuteman II Missile Sites Fact Sheet, 24 March 2011," Missouri Department of Natural Resources, 1–2.
25. Wally Kennedy, "Last Minuteman II Silo in Missouri Imploded," *Joplin Globe*, 16 December 1997, 1A. Associated Press, "Whiteman Suspends Silo Dismantlement," *Sedalia Democrat*, 28 September 1994, 4. Samantha Beeman, "Missile Silo Teardown Expected to Resume," *Sedalia Democrat*, 28 April 1995, 1.
26. Peter Johnson, "Air Force Wants Missile Upgrade, More Refueling Tankers for Base Here," *Great Falls Tribune*, 16 April 1991, 1A.
27. Program Plan 91-7, Minuteman II Deactivation/Minuteman III Conversion, 341st Missile Wing, 23 September 1991 (Headquarters Strategic Air Command, Offutt AFB, NE).
28. Peter Johnson, "Out of the Ground, into History: Malmstrom Crews Take First of 150 Minuteman out of Silo for Dismantling," *Great Falls Tribune*, 14 November 1991, 1A. Wayne Arnst, "Base Retires Last of Nuclear Class," *Great Falls Tribune*, 11 August 1995, 1A.
29. Peter Johnson, "Modernizing Mission, First of New Missiles Put in Empty Silo," *Great Falls Tribune*, 18 November 1992, 1C; "Malmstrom Ready to Add New Missiles," *Great Falls Tribune*, 9 October 1995, 2B.
30. "History of the 341st Missile Wing, July 1994–December 1994," Vol. I of II, Malmstrom AFB, 16. This document is classified; the extract used is declassified IAW EO 13526.
31. "Malmstrom Starts Installing Missiles," *Grand Forks Herald*, 21 October 1995, A6. Associated Press, "Last of 200 Minuteman III Silos Filled," *Kalispell Daily Inter-Lake*, 18 May 1998, A7.
32. The 321st Missile Wing was redesignated a group on 1 July 1994.
33. Draft Environmental Impact Statement, Minuteman III Missile System Dismantlement, Grand Forks AFB, ND, May 1998, Executive Summary, Section 1, 2–3. Author's collection.
34. Steve Foss, "Moving Missiles, It's the Beginning of the End for ICBM's at Grand Forks Air Force Base," *Grand Forks Herald*, 5 October 1995, A1. Sue Ellyn Scaletta, "Last One Out," *Grand Forks Herald*, 4 June 1998, A1. "Last Minuteman III Silo Destroyed," *Space and Missile Times*, 31 August 2001, 3.
35. Environmental Assessment Minuteman III Deactivation, Malmstrom AFB, MT, May 2007, DTIC A496029, Section 2, 6. Designation changed to Missile Squadron on 1 September 1991.

36. Peter Johnson, "Missile Dismantled, Warhead Removed," *Great Falls Tribune*, 14 July 2007, 1A. John Turner, personal communication, 341st Missile Wing Public Affairs Office, 13 September 2018.
37. Katrina Heikkinen, "New START: 564 MS Silos Have Been Eliminated," 4 March 2014, Public Affairs, 341st Missile Wing, Malmstrom AFB. John Turner, "Demolition of Final 'Deuce' Squadron Missile Launcher Is a New Start Milestone," 6 August 2014, Public Affairs, 341st Missile Wing, Malmstrom AFB.

## Epilogue

1. Department of Defense, *2010 Nuclear Posture Review Report*, 23.
2. Lauren Caston, Robert S. Leonard, Christopher A. Mouton, Chad J. R. Ohlandt, S. Craig Moore, Raymond E. Conley, and Glenn Buchan, *The Future of the US Intercontinental Ballistic Missile Force* (RAND Corporation, Santa Monica, CA, 2014), xviii, 107–109.
3. "Boeing Awarded Design Work for New Intercontinental Ballistic Missile," Boeing news release, 21 August 2017. "Boeing, Northrup Move Forward on Next-Gen ICBM Program; Lockheed Out," 21 August 2017, www.defensenews.com/space, retrieved: July 2019.
4. Wilson Brissett, "Replacing Minuteman," *Air Force Magazine,* February 2018, 31.
5. Congressional Budget Office, *Cancel Development and Production of the New Missile in the Ground-Based Strategic Deterrent Program*, 13 December 2018, 1–4.
6. *Fiscal Year 2020 Priorities for Department of Defense Nuclear Forces, House Armed Services Subcommittee on Strategic Forces,* 116th Cong., 28 March 2019 (testimony of Gen. John E. Hyten, Commander, United States Strategic Command), transcript, 4.
7. Leah Bryant, "Air Force Releases Request for Proposals for New ICBM System," Air Force Nuclear Weapons Center Public Affairs, 16 July 2019.
8. Aaron Gregg, "Boeing Dropped Out of Massive Pentagon Nuclear Missile Program, Citing Unfair Competition," *Washington Post,* 25 July 2019, digital edition. Steve Trimble, "Competing Proposals Emerged as GBSD Faces Sole-Source Award Decision," *Aviation Week and Space Technology*, 19 August–1 September 2019, 24.
9. Strategic Air Command Missile Chronology: 1939–1973 (Office of the Historian, Headquarters Strategic Air Command, 2 September 1975), no page numbers given.

# BIBLIOGRAPHY

Listed here are only writings that have been used in the making of this book. They are divided into General, Boeing Company, and US government categories and further subdivided into topic sections. Within the sections, entries are generally listed alphabetically; exceptions are noted below. In the case where there is no author for an article or report, sources are listed alphabetically by title at the beginning of the section followed by the alphabetized author list.

Boeing Company Minuteman Historical Summaries are listed chronologically. *Minuteman Service News* articles are listed alphabetically by title, as are the reports.

Wing histories and reports from the histories are listed chronologically within the particular wing section. As-built drawings are listed alphabetically by wing location. Technical orders are listed alphabetically according to missile variant. It must be made clear that the Air Force Global Strike Command classification review included reference to the technical orders listed as well as the as-built drawings and as such have been approved for use in this book.

Since the sources of the documents are given in the endnotes, only general source information is given in the bibliography, with the three exceptions being journal, magazine, and newspaper articles.

## General

### Books and Chapters

BOOKS

*Aggressor Beware, A Brief History of the 44th Missile Wing 1952–1994* (Grelind Printing, Rapid City, SD, 1994).

*CRC Handbook of Chemistry and Physics,* 52nd Edition, 1971–72 (The Chemical Rubber Company, Cleveland, OH, 1971).

*History of Hill Air Force Base* (History Office, Ogden Air Logistics Center, Hill AFB, UT, 1988).

Arnold, Lorna, and Katherine Pyne. *Britain and the H-bomb* (Palgrave, New York, NY, 2001).

Baals, D. D., and W. R. Corliss. *Wind Tunnels of NASA,* SP-440.

Babcock, Elizabeth. *Magnificent Mavericks: History of the Navy at China Lake, California, Volume 3* (Naval Historical Center, Washington Navy Yard, November 2008).

Ball, Desmond. *Politics and Force Levels: Strategic Missile Program of the Kennedy Administration* (University of California Press, 1980).

Ball, Desmond, and Jeffrey Richelson, eds. *Strategic Nuclear Targeting* (Cornell University Press, Ithaca, NY, 1986).

Boyars, C., and K. Klager. *Propellants: Manufacture, Hazards, and Testing, Advances in Chemistry Series 88* (American Chemical Society, 1969).

Burkard, Richard K. *Geodesy for the Layman* (US Air Force Aeronautical Chart and Information Center, St. Louis, MO, 1968).

Bush, Vannevar. *Modern Arms and Free Men* (Simon and Schuster, New York, NY, 1949).

Collins, John M. *US–Soviet Military Balance Concepts and Capabilities: 1960–1980* (McGraw-Hill Publications Company, 1980).

Conine, Gary B. *Not for Ourselves Alone: The Evolution and Role of the Titan II Missile in the Cold War* (CreateSpace Independent Publishing Platform, 2015).

Culick, F., M. V. Heitor, and J. H. Whitelaw, eds. *Unsteady Combustion* (Kluwer Academic Publishers, Dordrecht, The Netherlands, 1996).

Cutler, Robert. *No Time for Rest* (Little, Brown, 1965, 1966).

Daso, Dik A. *Hap Arnold and the Evolution of American Airpower* (Smithsonian Institution Press, 2000).

Day, D.A., J. M. Logsdon, and B. Latrell, eds. *Eye in the Sky: The Story of the Corona Spy Satellites* (Washington, DC, 1998).

Dyer, D. *TRW: Pioneering Technology and Innovation since 1900* (Harvard Business School Press, Boston, MA, 1998).

Dyer, D., and D. B. Sicilia. *Labors of a Modern Hercules* (Harvard Business School Press, Cambridge, MA, 1990).

Eisenhower, Dwight D. *Waging Peace, 1956–1961* (Doubleday, Garden City, NY, 1965).

Emme, Eugene. *The History of Rocket Technology* (Wayne State University Press, 1964).

Fitzgerald, A. Ernest. *The High Priests of Waste* (W. W. Norton and Company Inc., New York, NY, 1972).

Gantz, Kenneth, ed. *United States Air Force Report on the Ballistic Missile* (Doubleday and Company, Garden City, NY, 1958).

Garthoff, Raymond L. *Reflections on the Cuban Missile Crisis* (The Brookings Institution, Washington, DC, 1989).

Glasstone, S., and P. J. Dolan, eds. *The Effects of Nuclear Weapons,* Third Edition (Department of Defense and Energy Research and Development Administration, 1977).

Gordon, Robert. *Aerojet: The Creative Company* (Aerojet History Group, 1995).

Greenwood, Ted. *Making the MIRV: A Study of Defense Decision-Making* (Ballinger Publishing Company, Cambridge, MA, 1975).

Gruntman, Mike. *Blazing the Trail: The Early History of Spacecraft and Rocketry* (American Institute of Aeronautics and Astronautics, July 2004).

Gunston, W. *The Illustrated Encyclopedia of the World's Rockets and Missiles* (Salamander Books, Ltd., London, UK, 1979).

Hall, Edward N. *The Art of Destructive Management* (E. Hall, Vanguard Press, New York, NY, 1984).

Hansen, Chuck. *The Swords of Armageddon: US Nuclear Weapons Development Since 1945,* Volume VII (Chukelea Publications, Sunnyvale, CA, October 1995).

Hill, C. N. *A Vertical Empire: The History of the British Rocketry Programme* (Imperial College Press, London, UK, 2012).

Hunley, J. D. *The Development of Propulsion Technology for US Space-Launch Vehicles: 1926–1991* (Texas A&M University Press, College Station, TX, 2007).

Kennedy, G. P. *Rockets and Missiles of White Sands Proving Ground 1945–58* (Schiffer Publishing, Atglen, PA, 2009).

Kennedy, George P. *Vengeance Weapon 2: The V-2 Guided Missile* (National Air and Space Museum, Washington, DC, 1984).

Koppes, Clayton R. *JPL and the American Space Program: A History of the Jet Propulsion Laboratory* (Yale University Press, New Haven, CT, 1982).

Launius, R. D., and D. R. Jenkins. *To Reach the High Frontier, A History of US Launch Vehicles* (University of Kentucky Press, 2002).

Ley, Willy. *Rockets, Missiles, and Men in Space* (Viking Press, Inc., New York, NY, 1968).

MacKenzie, Donald. *Inventing Accuracy* (MIT Press, Cambridge, MA 1993).

Masters, Dexter, and Katharine Way, eds. *One World or None* (McGraw-Hill Book Company, 1946).

McDonald, Robert A., and Sharon K. Moreno. *Grab and Poppy: America's Early ELINT Satellites* (History Office, National Reconnaissance Office, Chantilly, VA, 2005).

McMurran, Marshall. *Achieving Accuracy: A Legacy of Computers and Missiles* (Xlibris Corporation, Bloomington, IN, 2008).

Medaris, J. B., and A. Gordon. *Countdown to Decision* (G. P. Putnam's Sons: New York, NY, 1960).

Neal, R. *Ace in the Hole: The Story of the Minuteman Missile* (Doubleday and Company Inc., Garden City, NY, 1962).

Nease, Robert, and Dan Hendrickson. *A Brief History of Minuteman Guidance and Control* (Rockwell Defense Electronics Autonetics Division, March 1995).

Neufeld, Jacob. *The Development of Ballistic Missiles in the United States Air Force, 1945–1960* (Office of Air Force History, Washington, DC, 1990).
Parvin, R. H. *Inertial Navigation* (D. Van Nostrand Company, Inc., Princeton, NJ, 1962).
Reed, Thomas. *At the Abyss: An Insider's Story of the Cold War* (Ballantine Books, New York, NY, 2005).
Reilly, Philip K. *The Rocket Scientists: Achievement in Science, Technology, and Industry at Atlantic Research Corporation* (Vantage Press, New York, NY, 1999).
Rosen, M. *The Viking Rocket Story* (Harper Brothers, New York, NY, 1955).
Rosenberg, Max. *The Air Force and the National Guided Missile Program, 1944–1950* (USAF Historical Division Liaison Office, 1964).
Rubel, J. H. *Doomsday Delayed: USAF Strategic Weapons Doctrine and SIOP-62, 1959–1962* (Hamilton Books, 2008).
Sagan, Scott D. *The Limits of Safety: Organizations, Accidents, and Nuclear Weapons* (Princeton University Press, NJ, 1993).
Schwiebert, Ernest G. *A History of the US Air Force Ballistic Missiles* (Praeger, 1964, 1965).
Seifert, H. S., ed. *Space Technology* (John Wiley and Sons, New York, NY, 1959).
Serling, Robert J. *Legend and Legacy: The Story of Boeing and Its People* (St. Martin's Press, 1992).
Shaw, Frederick J., ed. *Locating Air Force Base Sites, History's Legacy* (Air Force History Museums Program, United States Air Force, Washington, DC, 2004).
Sheehan, Neil. *A Fiery Peace in a Cold War* (Random House, New York, NY, 2009).
Slater, John M. *Inertial Guidance Sensors* (Reinhold Publishing Corporation, 1964).
Snead, David L. *The Gaither Committee, Eisenhower and the Cold War* (Ohio State University Press, 1999).
Spinardi, Graham. *From Polaris to Trident: The Development of US Fleet Ballistic Missile Technology* (Cambridge University Press, New York, NY, 1994).
Stenvik, Luther. *The Agile Giant: A History of the Minuteman Production Board* (The Boeing Company, Seattle, WA, 1966).
Stine, Harry G. *ICBM: The Making of the Weapon That Changed the World* (Orion Books, 1991).
Stumpf, David K. *Titan II: A History of a Cold War Missile Program* (University of Arkansas Press, Fayetteville, 2000).
Temple III, L. Parker. *Implosion: Lessons from National Security, High Reliability Spacecraft, Electronics, and the Forces Which Changed Them* (IEEE Press, John Wiley & Sons, 2013).
Teriff, Terry. *The Nixon Administration and the Making of US Nuclear Strategy* (Cornell University Press, Ithaca, NY, 1995).
von Kármán, Theodore. *The Wind and Beyond: Theodore von Kármán, Pioneer in Aviation and Pathfinder in Space* (Little, Brown & Company, 1967).
von Kármán, Theodore. *Prophecy Fulfilled: "Toward New Horizons" and Its Legacy*, edited by Michael Gorn (Air Force History and Museums Program, 1994).
Walker, C. *Atlas: The Ultimate Weapon* (Apogee Books, Ontario, Canada, 2005).
Weir, G. *An Ocean in Common* (Texas A&M University Press, 2001).
Werrell, K. P. *The Evolution of the Cruise Missile* (Air University Press, Maxwell AFB, AL, 1985).
Winter, Frank H. *The First Golden Age of Rocketry: Congreve and Hale Rockets of the 19th Century* (Smithsonian Institute Press, 1990).
York, H. *Race to Oblivion: A Participant's View of the Arms Race* (Simon and Schuster, New York, NY, 1970).

CHAPTERS

Bert, C. W., and W. S. Hyler. "Structures and Materials for Solid Propellant Motor Cases." In *Advances in Space Science and Technology*, edited by F. I. Ordway III, Vol. 8 (Academic Press, Cambridge, MA, 1966).
Carroll, T. "Historical Origins of the Sergeant Missile Powerplant." In *History of Rocketry and Astronautics* (Univelt, San Diego, CA, 1989).
DeArmond, Frank, and Ralph Lombardo. "Minuteman II Aerospace Ground Equipment Including

Self-Test Features and Status Display Techniques." In *Automation and Electronic Test Equipment*, Vol. 3, edited by David M Goodman (New York University Press, April 1967).

Gray, Collin S. "The Future of Land-Based Missile Forces," *Adelphi Papers* No. 140, (International Institute for Strategic Studies, London, UK, 1978), 1–44.

Juneau, P. "Composite Materials, Ablative," *Encyclopedia of Chemical Technology*, Vol. 1, Third edition, (1978) 10–26.

Klager, Karl. "The Interaction of the Efflux of Solid Propellants with Nozzle Materials," *Propellants and Explosives*, Vol. 2 (John Wiley & Sons, 1977), 55–63.

Rumbel, Keith E. "Poly (Vinyl Chloride) Plastisol Propellants." Appendix 4, in *Propellants Manufacturing, Hazard, and Testing, Advances in Chemistry Series 88* (American Chemical Society, Washington, DC, 1969).

Tsien, H. S. "Future Trends in the Design and Development of Solid and Liquid Fuel Rockets." In *Towards New Horizons*, Vol. VI, 1945, Part 4 (Headquarters Air Material Command, Wright Field, Dayton, OH, 1946).

von Kármán, T., and F. J. Malina. "Characteristics of the Ideal Solid Propellant Rocket Motor," JPL Report No. 1–4, 1940. In *Collected Works of Theodore von Kármán*, Vol. IV 1940–51, 1956 (Butterworths Scientific Publications, 1956).

York, Herbert F. "The Origins of MIRV." In *The Dynamics of the Arms Race*, edited by David Carlton and Carlo Schaerf (Halstead Press, New York, NY, 1975).

## *Conferences/Symposium Proceedings*

Adams, C. F., S. W. Cogan, and E. W. Nealon. "Beryllium in Inertial Navigation," in *Proceedings of the Beryllium Conference* (Vol. I), 23–25 March 1970 (National Academy of Sciences–National Academy of Engineering, Washington, DC, July 1970), 249–279.

Adams, M. C., and E. Scala. "The Interaction of High Temperature Air with Materials During Reentry," in *Proceedings of an International Symposium on High Temperature Technology*, Pacific Grove, CA, 6–9 October 1959, 54–60.

Billheimer, J. S., and F. R. Wagner. "The Morphological Continuum in Solid Propellant Grain Design," IAF Paper P-94, in *19th Congress of the International Astronautical Federation*, 13–19 October 1968, 378–397.

Borofsky, Arnold J. "Component Quality Assurance Programs for Micro Miniature Electronic Components for Minuteman II," paper presented at *Third Annual Symposium on the Physics of Failure in Electronics*, Institute of Electronic and Electrical Engineers, 29 September–1 October 1964, 1–14.

Decker, B. Louis. "World Geodetic System 1984," paper presented at *Fourth International Geodetic Symposium on Satellite Positioning*, University of Texas, Austin, 29 April–2 May 1986, 1–3.

Hopkins, R. E., Fritz K. Mueller, and Walter Haeussermann. "The Pendulous Integrating Gyroscope Accelerometer (PIGA) from the V2 to Trident D5, the Instrument of Choice," AIAA 2001-4288, paper presented at *Guidance, Navigation, and Control Conference and Exhibit*, Montréal Canada, 6–9 August 2001, 5.

Hunley, J. D. "The History of Solid-propellant Rocketry: What We Do and Do Not Know," paper presented at *35th AIAA, ASME, SAE, ASEE Joint Propulsion Conference and Exhibit*, Los Angeles, CA., 20–24 June 1999, 7.

Longstaff, I., S. Elms, K. Goussak, and C. Christoff. "The Litton LN-100 Advanced Technology Medium Accuracy Lightweight Inertial Navigation System," Paper 4416 presented at *Guidance, Navigation and Control Conference*, 10–12 August 1992, no page numbers given.

Podell, H. L., and R. J. Kotfila. "Design and Fabrication of Titanium Rocket Chambers," paper presented at *American Rocket Society Solid Propellant Rocket Conference*, Salt Lake City, UT, 1–3 February 1961, 1–17.

Price, E. W., C. L. Horine, and C. W. Snyder. "Eaton Canyon: A History of Rocket Motor Research and Development in the Caltech-NRDC-Navy Rocket Program, 1941–1946," paper presented at *34th*

*AIAA/ASME/SAE/ASEE Joint Propulsion Conference and Exhibit*, Cleveland, OH, 13–15 July 1998, 1–28.

Sparks, J. F., and M. P. Friedlander III. "Fifty Years of Solid Propellant Technical Achievements at Atlantic Research Corporation," Paper A99-31568, presented at *35th AIAA Joint Propulsion Conference*, Los Angeles, CA, 20–24 June 1999, 8.

Sutton, E. S. "From Polysulfides to CTPB Binders—A Major Transition in Solid Propellant Binder Chemistry," Paper AIAA 84-1236, presented at AIAA/SAE/ASME 20th Joint Propulsion Conference, Cincinnati, OH, 11–13 June 1984, 2–3.

Sutton, E. S. "From Polymers to Propellants to Rockets, A History of Thiokol," Paper 99–2929, presented at *35th American Institute of Aeronautics and Astronauts, Joint Propulsion Conference*, Los Angeles, CA, 20–24 June 1999, 5.

Szabo, B. "The Significance of the World Geodetic Datum to Long-Range Navigation and Guidance Systems," in *Symposium: Size and Shape of the Earth* (Ohio State University, Columbus, OH, November 1956), 99–102.

von Kármán, T. "Aerodynamic Heating—The Temperature Barrier in Aeronautics," in *Proceedings of the Symposium on High Temperature—A Tool for the Future* (Stanford Research Institute, Menlo Park, CA, 1956), 140–142.

Wampler, Howard W., Gerald G. Leigh, Myron E. Furbee, and Frank E. Seusy. "A Status and Capability Report on Nuclear Air Blast Simulation Using HEST," in *Proceedings of the Nuclear Blast and Shock Simulation Symposium* (Defense Nuclear Agency, Washington, DC, December 1978), Vol. 1, 32.

Young, R. E. "History and Potential of Filament Winding," paper presented at *13th Annual Technical and Management Conference, Reinforced Plastics Division, Society of the Plastics Industry*, Chicago, IL, 4 February 1958, Section 15-C, 1–6.

## *In-House Publications*

"Newest Member of the Hercules Family," *Hercules Chemist* 35, 1959.

Slater, J. M. "20 Years of Inertial Navigation at North American Aviation" (North American Aviation, Downey, CA, 1966)

## *Interviews and Email Correspondence*

INTERVIEWS

Barnes, Gordon, Lt. Col., USAF (ret.). Personal interview, 12 June 2015.
Cannon, Mitch, CMSgt., USAF (ret.). Personal interview, 6 August 2015.
Curtin, Gary L., Maj. Gen., USAF (ret.). Personal interview, 12 October 2018.
Decker, B. Louis. Personal interview, 12 June 2015.
Doll, Andy, SMSgt., USAF (ret.). Personal interview, 11 April 2016.
Fote, Phil. Personal interview, 6 May 2016 and 15 June 2017.
Gill, Ed, Col., USAF (ret.). Telephone interview, 26 June 2018.
Goodwin, Samuel, Col., USAF (ret.). Telephone interview, 29 March 2018.
Graesser, Elmer. Telephone interview, 15 April 2017.
Gray, Ronald, Brig. Gen., USAF (ret.). Personal interview, 12 October 2018.
Gustafson, Scott. Telephone interview, 23 October 2018.
Henderson, Charles B. Telephone interview, 10 May 2019.
Hilden, Jack, Capt., USAF (ret.). Personal interview, 5 February 2016.
Knox, Robert. Personal interview, 10 April 2018.
MacDonald, Alan. Personal interview, 15 September 2016.
Martens, Allan. Personal interview, 10 May 2017.
Mills, John, TSgt., USAF (ret). Personal interview, 20 September 2019.
Moore, Garrett, CWO, USA (ret.). Personal interview, 15 June 2015.
Nease, Robert. Personal interview, 9 April 2018.

Parker, Robert, Maj. Gen., USAF (ret.). Personal interview, 12 October 2018.
Reed, Thomas, Secretary of the Air Force. Personal interview, 15 May 2015.
Sapuppo, Michele. Telephone interview, 12 May 2018.
Tange, Leigh, CMSgt., USAF (ret.). Personal interview, 10 May 2016.
Thirkill, J. Personal interview, 20 April 2017.
True, William. Personal interview, 31 May 2016.
Watts, Monte. Personal interview, August 2015.

EMAIL CORRESPONDENCE

Allen, John. September, December 2015.
Barnard, James W. October 2018, August 2019.
Barnes, Gordon, Lt. Col., USAF (ret.). July, August 2015, May 2016.
Boltinghouse, Susan. September 2018.
Cannon, Mitch, CMSgt., USAF (ret.). June 2015, November 2019.
Doll, Andy, SMSgt., USAF (ret.). April 2016–October 2019.
Ellis, Rex C., Chief, Treaty Compliance Office, 90th Missile Wing. 24 September 2018.
Krogen, Michael S. 11 February 2019.
Gervais, Rik, Col., USAF (ret.). December 2015, January 2016.
Gill, Ed, Col., USAF (ret.). 30 April 2018.
Goodwin, Samuel, Col., USAF (ret.). 29 April 2018.
Graesser, Elmer. April 2017–December 2018.
Henderson, Charles B. May 2019.
Knox, Robert. June 2018, 21 August 2019.
Mills, John, TSgt., USAF (ret.). January 2016–August 2019.
Moore, Garrett, CWO, USAF (ret.). June 2015–August 2019.
Peri, Jill, and Chris Peri. June 2017.
Tange, Leigh, CMSgt., USAF (ret.). May 2016–October 2019.
True, William. May 2016–April 2019.
Watts, Monte. June 15–November 2019.

## *Journal Articles*

Allen, H. J. "Hypersonic Flight and the Re-Entry Problem," *Journal of the Aeronautical Sciences* 25, no. 4: 223.
Anderson, Robert C. "Minuteman to MX, the ICBM Revolution," *Quest* 3 (Autumn 1979): 42.
Ausman, J. S., and M. Wildman. "How to Design Hydrodynamic Gas Bearings," *Production Engineering* 28, no. 25: 103–106.
Bennett, M. M., and P. W. Davis. "Minuteman Gravity Modeling," Paper 76–1960, *American Institute of Aeronautics and Astronautics*, 1976, 1.
Caveny, L. H., R. L. Geisler, R. A. Ellis, and T. L. Moore. "Solid Rocket Enabling Technologies and Milestones in the United States," *Journal of Propulsion and Power* 19, no. 6 (2003): 1038–1066.
Crow, A. D. "The Rocket As a Weapon of War in the British Forces," *Proceedings of the Institute of Mechanical Engineers*, no. 158 (1948): 15–20.
Furhman, R. A. "The Fleet Ballistic Missile System: Polaris to Trident," *Journal of Spacecraft and Rockets* 15, no. 5 (1978): 278.
Goldsworthy, Harry E., Lt. Gen., USAF (ret.). "ICBM Site Activation," *Aerospace Historian* 29, no. 3: 154.
Gruntfest, I. J., and L. H. Schenker. "Behavior of Materials at Very High Temperatures," *Industrial and Engineering Chemistry* 50 (1958): 75A-76A
Hayes III, T. J. "Engineers Dig in for Defense," *Army Information Digest*, November 1962, 25–27.
Hickman, W. W. "Giant Patriot," *Combat Crew*, December 1972, 26–27.
Hovey, Robert W. "Cork Thermal Protection Design Data for Aerospace Vehicle Ascent Flight," *Journal of Spaceflight* 2, no. 3 (1965): 300.

Hunley, J. D. "Evolution of Large Solid Propellant Rocketry in the United States," *Quest: The History of Spaceflight Quarterly* 6. no. 1, 26.
Leonard, A. S. "Some Possibilities for Rocket Propellants, Part III," *Journal of the American Rocket Society* no. 2 (1947): 14.
Lin, T. C. "Development of US Air Force Intercontinental Ballistic Missile Weapon Systems," *Journal of Spacecraft and Rockets* 40, no. 4 (2003): 491–509.
Malina, F. J. "US Army Corps Jet Propulsion Research Project, GALCIT Project No.1, 1939–1946: A Memoir," in *History of Rocketry and Astronautics: Proceedings of the Third through Sixth History Symposia of the International Academy of Astronautics*, 1986, Vol. 2, 163, 169.
Malina, F. J. "America's First Long-Range-Missile and Space Exploration Program: The GALCIT Ordnance Project of the Jet Propulsion Laboratory, 1943–1946: A Memoir," in *History of Rocketry and Astronautics: Proceedings of the Third through Sixth History Symposia of the International Academy of Astronautics*, 1986, Vol. 2, 346.
Parsons, John W., and Edward S. Forman. "Experiments with Powder Motors for Rocket Propulsion by Successive Impulses," *Astronautics* 9, no. 43 (1939): 4–11.
Powell, J. W. "Thor-Able and Atlas Able," *Journal of the British Interplanetary Society* 37 (1984): 219–225.
Price, E. W. "Comments on Role of Aluminum in Suppressing Instability in Solid Propellant Rocket Motors," *AIAA Journal*, May 1971, 987–990.
Richards, G. R., and J. W. Powell. "Titan 3 and Titan 4 Space Launch Vehicles," *Journal of the British Interplanetary Society* 46 (1993): 122–144.
Sagan, Scott. "SIOP-62: The Nuclear War Plan Briefing to President Kennedy," *International Security* 12, no. 1 (1987): 35.
Sestak, J. G. "Giant Patriot," *Combat Crew*, 19 May 1974, 20–24.
Smelt, R. "Lockheed X-17 Rocket Test Vehicle and Its Applications," *American Rocket Society Journal* 29, no. 8 (1959): 566.
Stuhlinger, E. "Army Activities in Space," *IRE Transactions in Military Electronics* MIL, no. 2 (1960): 65.
Stumpf, David K. "Reentry Vehicle Development Leading to the Minuteman Mark 5 and 11," *Air Power History*, Fall 2017, 13–36.
Stumpf, David K. "Minuteman: Solid Propellant Comes of Age," *QUEST: A History of Spaceflight Quarterly* 26, no. 3 (2019): 13–53.
Summerfield, M., J. I. Shafer, H. L. Thackwell Jr., and C. E. Bartley. "The Applicability of Solid Propellants to High Performance Rocket Vehicles," *Astronautics*, October 1962, 50–56.
Sutton, G. W. "Ablation of Reinforced Plastics in Supersonic Flow," *Journal of Aero/Space Sciences*, May 1960, 378.
Tsien, H. S., and F. J. Malina. "Flight Analysis of a Sounding Rocket with Special Reference to Propulsion by Successive Impulses," *Journal of Aeronautical Sciences* 6 (1938): 50–58.
Warner, D. J. "Political Geodesy: The Army, Air Force and World Geodetic System of 1960," *Annals of Science* 59 (2002): 384–386.
Wiggins, J. W. "Hermes: Milestone in US Aerospace Progress," *Aerospace Historian* 21, no. 1 (1974): 34–40.
Wuerth, J. M. "The Impact of Guidance Technology on Automated Navigation," *Navigation: Journal of the Institute of Navigation* 14, no. 3 (1967): 328–339.
Wuerth, J. M. "The Evolution of Minuteman Guidance and Control," *Navigation: Journal of the Institute of Navigation* 23, no. 1 (1976): 64–75.
Zeiberg, Seymour L. "MX: The Full Perspective," *Defense* (American Forces Information Service, Arlington, VA, 1980), 1–16.

## Launch Databases

574th Flight Test Wing, Vandenberg AFB, "494L and ALCS Launches at Vandenberg AFB."
Cleary, Mark. 45th Space Wing History Office, "The 6555 ATW Missile and Space Launches through 1970."
Geiger, Jeffrey. Vandenberg AFB, "Vandenberg Air Force Base Launch Summary: 1958–2003."
Geiger, Jeffrey. Vandenberg AFB, "Minuteman Flight Test History 1963–2015."

McDowell, Jonathan. "JSR Launch Vehicle Database, 2019 Dec. 28 Edition," http://www.planet4589.org/space/lvdb/index.html., last accessed November 2019.

Whipple, Marven R. "Atlantic Missile Range (Eastern Test Range) Index of Missile Launchings: July 1963-June 1964."

## Magazine Articles

"Minuteman is Top Semiconductor User," *Aviation Week and Space Technology*, 26 July 1965, 83.

"Minuteman II Operational Test Program," *Commanders Digest*, 31 January 1974, 2, 6–8.

"Notch Sensitivity-Barrier to Solid Rockets," *Missiles and Rockets*, 8 June 1959, 23–25.

"Recoverable Data Capsule Designed for Thor and Atlas Test Missiles," *Aviation Week*, 24 November 1958, 71–72.

"Thor Able Launched in Nose Cone Test," *Aviation Week*, 28 July 1958, 27.

Klass, Philip J. "Minuteman Guidance and Control Part 2: Minuteman Guidance Built for Long Life," *Aviation Week and Space Technology*, 5 November 1962, 75.

Levy, A. V. "Evaluation of Reinforced Plastic Material in High Speed Guided Missile and Power Plant Application," *Plastics World*, March 1956, 10–15.

Pay, Rex. "Circuit Weights Slashed in Advanced Minuteman," *Missiles and Rockets*, 2 March 1964, 35.

Stone, I. "Minuteman ICBM Solid Motor Stages Enter Production Stage," *Aviation Week and Space Technology*, 27 August 1962, 56.

Stone, I. "Aerojet Second Stage Must Withstand Heaviest Stress," *Aviation Week and Space Technology*, 3 September 1962, 7.

Stone, I. "Hercules Stage 3 Uses Glass Fiber Case," *Aviation Week and Space Technology*, 10 September 1962, 162.

Stranix, Richard. "Minuteman Reliability: Guide for Future Component Manufacturing," Electronic Industries, December 1960, 89–104.

Yaffee, M. L. "Two Approaches Used in First Production Nose Cones," *Aviation Week*, 12 May 1958, 6.

Yaffee, M. L. "Ablation Wins Missile Performance Gain," *Aviation Week*, 18 July 1960, 57.

Yaffee, M. L. "Mark 11A Hardened against Air Bursts," *Aviation Week and Space Technology*, 24 August 1964, 50–60.

## Newsletters

Engebretson, Harold. "The Minuteman I Guidance and Control System Proposal," *North American Aviation Retirees Bulletin,* Winter 2008.

Ogletree, Greg, Maj., USAF (ret.). "A History of PACCS, ACCS and ALCS Part 1," *Strategic Air Command Airborne Command Control Association Newsletter* 1, no. 2, October 1995.

Ogletree, Greg, Maj., USAF (ret.). "The 4th Airborne Command and Control Squadron," *Strategic Air Command Airborne Command Control Association Newsletter* 6, no. 3, November 2000.

Ogletree, Greg, Maj., USAF (ret.). "The 2nd Airborne Command and Control Squadron," *Strategic Air Command Airborne Command Control Association Newsletter* XVII, no. 2, June 2011.

Peller, John B. "Events Leading up to Minuteman III," *North American Aviation Retirees Newsletter,* September 1995.

## Newspaper Articles

"Last Minuteman III Silo Destroyed," *Space and Missile Times*, 31 August 2001, 3.

"Last of 200 Minuteman III Silos Filled," *Kalispell Daily Inter-Lake,* 18 May 1998,, A7.

"Malmstrom Starts Installing Missiles," *Grand Forks Herald*, 21 October 1995, A6.

"Minuteman Bases Ring Cheyenne," *New York Times*, 10 December 1962, 10.

"Minuteman Gets an 'A' in Newell Test Shot," *Rapid City Daily Journal*, 1 March 1965, 1–2.

"US to Hold Back Extra Arms Funds and Trim Forces," *New York Times*, 12 September 1958, 1.

"Whiteman Suspends Silo Dismantlement," *Sedalia Democrat*, 28 September 1994, 4.

"Wing Deactivation to Be Discussed at Public Meeting," *Sedalia Democrat*, 18 October 1992, 1.
Arnst, Wayne. "Base Retires Last of Nuclear Class," *Great Falls Tribune*, 11 August 1995, 1A.
Beeman, Samantha. "Missile Silo Teardown Expected to Resume," *Sedalia Democrat*, 28 April 1995, 1.
Engelberg, Stephen, "Air Force Offers to Close 15 Bases and Scrap Missiles," *New York Times*, 19 November 1989, 1.
Foss, Steve. "Moving Missiles, It's the Beginning of the End for ICBMs at Grand Forks Air Force Base," *Grand Forks Herald*, 5 October 1995, A1.
Johnson, Peter. "Missile Cuts Get Skeptical Response," *Great Falls Tribune*, 16 January 1990, 1A.
Johnson, Peter. "MAFB Cuts? Air Force Making Reductions Nationwide," *Great Falls Tribune*, 6 February 1991, 1A.
Johnson, Peter. "Mid-1990s for Missile Loss Here? Deactivation of Minuteman IIs Not Expected to Start Immediately," *Great Falls Tribune*, 7 February 1991, 1A.
Johnson, Peter. "Air Force Wants Missile Upgrade, More Refueling Tankers for Base Here," *Great Falls Tribune*, 16 April 1991, 1A.
Johnson, Peter. "Out of the Ground, into History: Malmstrom Crews Take First of 150 Minuteman out of Silo for Dismantling," *Great Falls Tribune*, 14 November 1991, 1A.
Johnson, Peter. "Way Cleared for Missiles," *Great Falls Tribune*, 6 June 1992, 1B.
Johnson, Peter. "Modernizing Mission, First of New Missiles Put in Empty Silo," *Great Falls Tribune*, 18 November 1992, 1C.
Johnson, Peter. "Malmstrom Ready to Add New Missiles," *Great Falls Tribune*, 9 October 1995, 2B.
Johnson, Peter. "Missile Dismantled, Warhead Removed," *Great Falls Tribune*, 14 July 2007, 1A.
Kennedy, Wally. "Last Minuteman II Silo in Missouri Imploded," *Joplin Globe*, 16 December 1997, 1A.
Reidy, John. "First Silo Demolition Proves Something Less Than a Blast," *Sedalia Democrat*, 9 December 1992, 1.
Rowell, Jenn. "Defense Department to Remove 50 Minuteman Missiles from Silos at Three Bases, Including Malmstrom," *Great Falls Tribune*, 8 April 2014, 1A.
Scaletta, Sue Ellyn. "A Last One Out," *Grand Forks Herald*, 4 June 1998, A1.
Smith, R. Jeffrey, and Molly Moore. "Air Force Calls for Scrapping Minuteman II, Cheney Orders Freeze on Hiring Civilians," *Washington Post*, 13 January 1990, A1.
White, Dillon. "Propulsion Replacement Program Complete," *Great Falls Tribune*, 23 August 2009, 1A.

## *Research Laboratories*

### JET PROPULSION LABORATORY/
### GUGGENHEIM AERONAUTICAL LABORATORY

Malina, F. J., and J. W. Parsons. "Results of Flight Tests of the Ercoupe Airplane Auxiliary Jet Propulsion Supplied by Solid Propellant Jet Units," GALCIT Project No. 1, No. 1, Report No. 9, 2 September 1941.
Shafer, J. I., and H. L. Thackwell Jr. "High-Acceleration Flight Tests of a 6-inch Rocket to Test a Polysulfide-Rubber-Base Propellant: Phase 2, High and Low Temperature Firings," JPL Progress Report 4-76, 15 November 1948.
Thackwell Jr., H. L., and J. I. Shafer. "High-Acceleration Flight Tests of a 6-inch Rocket to Test a Polysulfide-Rubber-Base Propellant: Phase 1, Ambient Temperature Firings," JPL Progress Report 4-55, 8 January 1948.
Thackwell Jr., H. L., and J. I. Shafer. "The Applicability of Solid Propellants to Rocket Vehicles of V-2 Size and Performance," JPL Memorandum 4-25, 1 July 1948.

### ALLEGANY BALLISTICS LABORATORY

Beek Jr., J., and H. Siller. "Some Fundamental Properties of Star-Perforated Charges," November 1945, reproduced as Appendix B in W. H. Avery and J. Beek Jr., "Propellant Charge Design of Solid Fuel Rockets," Allegany Ballistics Laboratory, Final Report, Series B, Number 4 (OSRD Ref. No. 5890), June 1946.

Thompson, R. J., and R. R. Newton. "Design of the High Velocity Rocket Vicar," Allegany Ballistics Laboratory Final Report, Series W, Number 21 (OSRD Ref. No. 5793), December 1945.

## Speeches/Presentations

Bush, George H. "Address to the Nation on Reducing United States and Soviet Nuclear Weapons, 27 September 1991." American Presidency Project, University of California, Santa Barbara.
Kennedy, John F. "Radio and Television Address to the American People on the Soviet Arms Build-up in Cuba, 22 October 1962." American Presidency Project, University of California, Santa Barbara.
Norris, Robert S. "The Cuban Missile Crisis: A Nuclear Order of Battle, October–November 1962." Presentation at the Woodrow Wilson Center, Washington, DC, 24 October 2012.

## Theses and Academic Research Reports

### THESES

Gainor, Christopher J. "The United States Air Force and the Emergence of the Intercontinental Ballistic Missile 1945–1954" (Thesis, University of Alberta, Canada, 2011).
Reed, G. A. "US Defense Policy, US Air Force Doctrine and Strategic Nuclear Weapon Systems, 1958–1964: The Case of the Minuteman ICBM" (PhD Diss., Duke University, 1986).

### RESEARCH REPORTS

Ruggiero, F. "Missileers' Heritage," Student Research Report, Air Command and Staff College, 2065–81.
Simpson, Charles G. "The Impact of Reduced Flight Test Rates on the Strategic Missile Leg of the Triad," Research Report 478, March 1978 (Air War College, Air University, Maxwell AFB, AL, 1978).

# Boeing Company

## Minuteman Historical Summaries

"Minuteman Historical Summary, 1958–1959" (The Boeing Company, Seattle, WA, 1974).
"Minuteman Historical Summary, 1960–1961" (The Boeing Company, Seattle, WA, 1974).
"Minuteman Historical Summary,1962–1963" (The Boeing Company, Seattle, WA, 1975).
"Minuteman Historical Summary, 1964–1965" (The Boeing Company, Seattle, WA, 1975).
"Minuteman Historical Summary, 1966–1967" (The Boeing Company, Seattle, WA, 1975).
"Minuteman Historical Summary, 1968–1969" (The Boeing Company, Seattle, WA, 1976).
"Minuteman Historical Summary, 1970–1971" (The Boeing Company, Seattle, WA, 1976).

## Minuteman Service News

"Ablative Overcoat for Minuteman," *Minuteman Service News*, August 1963, 7.
"HETF Aids ALCS Launches," *Minuteman Service News*, July–August 1971, 8–9.
"Interstage Ordnance," *Minuteman Service News*, July–August 1965, 3.
"Minuteman II Collimator Alignment," *Minuteman Service News*, January–February 1965, 5.
"Minuteman III Is Coming," *Minuteman Service News*, May–June 1968, 3.
"Minuteman Skirt Change," *Minuteman Service News*, September 1963, 2.
"MM III Suspension Cable Links," *Minuteman Service News*, March–April 1970, 11–12.
"Periscope for Minuteman III, Wing III," *Minuteman Service News*, Issue 449, March–April 1970, 10.
"Propulsion System Rocket Engine: Minuteman III," *Minuteman Service News*, September–October 1970, 3–6.
"Raising the Longevity Record," *Minuteman Service News*, October–December 1972, 1.
"SIEGE," *Minuteman Service News*, September–October 1975, 10–12.

"Support Equipment Goes Underground-LCF," *Minuteman Service News*, April–March 1964, 3.
"Suspension System Comparison," *Minuteman Service News*, March–April 1963, 12–13.
"The Appearance of Minuteman," *Minuteman Service News*, March–April 1969, 3–4.
"Third Stage Motor-LGM-30G," *Minuteman Service News*, July–August 1971, 5–7.
"TORUS," *Minuteman Service News*, September–October 1974, 8–9.
"What's Force Mod," *Minuteman Service News*, November–December 1965, 11–12.
"What's New for Minuteman II," *Minuteman Service News*, July–August 1964, 3–4.
"Wing V Missile Support System," *Minuteman Service News*, May–June 1973, 6–8.

## *Reports*

"A Compendium of Structural Joints for Assembly, Field, and Flight Separation on Missiles, 1968."
Development Engineering Inspection, 1960.
"Effect of Depth Increase, Launch Tube, Wing V, WS-133A, 21 May 1963."
"Minuteman As a Space Launch Vehicle, 25 July 1990."
"Minuteman Burner II Integration Study, 1971."
Minuteman I Reentry Systems Launch Program Handbook for Payload Designers, May 1974.
Minuteman Illustrated Technical Requirements, no date.
Minuteman WS-133B and WS-133A-M Illustrated Figure A List, no date.
Modernized Minuteman Command-and-Control Systems, WS-133A-M, 5 October 1967.
RSLP Minuteman I Replacement Study, May 1978.
WS-133A-M Modernized Minuteman Command-and-Control Systems, 1 August 1966.
WS-133A-M Upgrade Wings III and V Minuteman Command-and-Control Systems, October 1972.

## British National Archive

"A History of the Development of British Rockets, 1935–1941," Projectile Development Establishment Report PDE 1935–41, UK Ministry of Supply, British National Archives, AVIA 41/207, November 1945.
Poole, H. J. "Charge Shapes for Plastic U. P. Propellants," Advisory Council Report 799/U. P. Propellant Subcommittee Report 13, UK Ministry of Supply, OSRD Liaison Office Ref. No. W40–9N British National Archives, DSIR 23/29534, 26 April 1961.
Poole, H. J. "Extracts from Personal Notes of Dr. H. J. Poole Relative to His Visit to Various Stations in the USA, October 1943," British National Archives, WO 195/5557, 1944, 2–3.
Ricketson, Barrie, and E. T. B. Smith. "Work Supporting the Development of An Underground Launching System for Blue Streak," *Rocket Propulsion Establishment*, British National Archives, AVIA 68/23.

## US Government

### *Congressional*

CONGRESSIONAL RECORD
S. Doc., 6 June 1974, 18060–18062.
S. Doc., 23 May 1985, 13544–13545.
H. R. Doc., 18 September 1985, 24126.

CONGRESSIONAL RESEARCH SERVICE
Woolf, Amy F. "US Strategic Forces: Background, Developments, and Issues," Congressional Research Service Report, 8 September 2006.

Woolf, Amy. "Nuclear Arms Control: The Strategic Offensive Reductions Treaty," Congressional Research Service Report, 3 January 2007.
Woolf, Amy F. "US Strategic Forces: Background, Developments, and Issues," Congressional Research Service Report, 3 April 2007.
Woolf, Amy. "US Strategic Nuclear Forces: Background, Developments, and Issues," Congressional Research Service, updated 3 April 2007.

HOUSE OF REPRESENTATIVES

*Supplemental Defense Appropriations for 1958*, 85th Cong., 13 January 1958.
*Organization and Management of Missile Programs: Hearings before a House Subcommittee of the Committee on Government Operations*, 86th Cong. (1959).
*Organization and Management of Missile Programs, 11th Report by the House Committee on Government Operations: Hearings before a Subcommittee of the Committee on Government Operations*, 86th Cong. (1959).
*Air Force Intercontinental Ballistic Missile Construction Program, Hearings before the House Subcommittee of the Committee on Appropriations*, 87th Cong. (1961).
*Air Force Intercontinental Ballistic Missile Construction Program, Report by the House Committee on Appropriations*, 87th Cong., 3 March 1961.
*Civil Defense Part III: Relation to Missile Programs, Hearings before a House Subcommittee of the Committee on Government Operations*, 86th Cong. (1961).
*Hearings before the House Subcommittee on Appropriations, Department of Defense Appropriations for 1962*, 87th Cong. (1961) (testimony of Robert S. McNamara).
*House Hearings on Military Posture and H. R. 4016, to Authorize Appropriations during Fiscal Year 1966 for Procurement of Aircraft, Missiles, and Naval Vessels, and Research, Development, Test, and Evaluation, for the Armed Forces, and for Other Purposes*, 85th Cong. (1965) (testimony of Deputy Secretary of Defense Cyrus Vance).
*Department of Defense Appropriations for 1978, Hearings before a House Subcommittee of the Committee on Appropriations*, 95th Cong. (1977) (testimony of John B. Walsh).
*Department of Defense Appropriations for 1980, Hearings before a House Subcommittee of the Committee on Appropriations*, 96th Cong., Part 2 (1979).
*The Status of United States Strategic Forces, Hearings before the House Subcommittee on Strategic Forces of the Committee of Armed Services*, 112th Cong. (2011) (testimony of Gen. C. Robert Kehler).
*Hearing before the House Committee on Armed Services, Fiscal Year 2015 National Defense Authorization Budget Request from the Department of the Air Force*, 113th Cong., 14 March 2014 (testimony of Secretary of the Air Force Deborah Lee James).

SENATE

*Hearings before the Senate Committee on Military Affairs*, 79th Cong. (1945).
*Hearings before the Senate Special Committee on Atomic Energy*, 79th Cong., Part 1 (1945).
*Hearings before the Senate Committee on Armed Services, Military Construction Authorization 1962*, 87th Cong. (1961) (testimony of Gen. A. M. Minton).
*Statement of Secretary of Defense Robert S. McNamara before the Senate Committee on Armed Services*, 4 April 1961.
*Nuclear Test Ban Treaty, Hearings before the Senate Committee on Foreign Relations*, 88th Cong. (1963) (testimony of Secretary of Defense Robert S. McNamara).
*Senate Armed Services Committee Hearings on Military Authorization and Defense Appropriations for Fiscal Year 1967*, 89th Cong. (1966) (testimony of Secretary of Defense Robert S. McNamara,).
*Senate Committee on Armed Services, Fiscal Year 1972 Authorization for Military Procurement*, 92nd Cong., 15 March 1971 (testimony of Secretary of Defense Melvin Laird).
Senate Resolution 241, *Disapproving the Basing Mode for the MX Missile*, 97th Cong., 2 November 1981.
*Hearing before the Senate Committee on Armed Services*, 103rd Cong., 22 September 1995 (testimony of Deputy Secretary of Defense John M. Deutsch).

Senate Subcommittee on Strategic Forces, Committee on Armed Services, 109th Cong., 29 March 2006 (testimony of Gen. James E. Cartwright).

Senate Committee on Foreign Relations Report, *Treaty with Russia on Measures for Further Reduction and Limitation of Strategic Offensive Arms (the New START Treaty)*, 111th Cong., 1 October 2010.

OTHER

*Air Force Statistical Digest*, Defense Information Technology Center, by date.

*Deterrence and Survival in the Nuclear Age (the Gaither Report of 1957), Joint Committee and Defense Production Congress of the United States*, 94th Cong., 1 January 1976.

*MX Missile Basing Mode, Communication from the President of the United States*, 97th Cong., 29 November 1982.

Public Law 97-252, *Department of Defense Authorization Act, 1983*, 97th Cong., 8 September 1982.

Public Law 101-510, *National Defense Authorization Act for Fiscal Year 1991*, 101st Cong., 5 November 1990.

Public Law 102-190, *National Defense Authorization Act for Fiscal Years 1992 and 1993*, 102nd Cong., 5 December 1991.

*Minuteman III Life Extension Program: A Report to Congress*, 29 July 1992.

Public Law 109-364, *John Warner National Defense Authorization Act for Fiscal Year 2007*, 109th Cong., 17 October 2006.

*Report on Plan to Implement the Nuclear Force Reductions, Limitations, and Verification and Transparency Measures Contained in the New START Treaty Specified in Section 1042 of the National Defense Authorization Act for Fiscal Year 2012*, 8 April 2014.

Public Law 113-295, *Carl Levin and Howard P. "Buck" McKeon National Defense Authorization Act for Fiscal Year 2015*, 113th Cong., 19 December 2014.

## NASA/NACA Reports

"Rocketry in the 1950s." NASA Historical Report No. 36, 28 October 1971.

Allen, H. J., and A. J. Eggers Jr. "A Study of the Motion and Aerodynamic Heating of Missiles Entering the Earth's Atmosphere at High Supersonic Speeds," NACA 1953, RM A53D28.

Eggers Jr., A. J., C. F. Hansen, and B. E. Cunningham. "Stagnation-Point Heat Transfer to Blunt Shapes in Hypersonic Flight, Including Effects of Yaw," NACA 1958, Technical Note 4229.

Hanson, J. R. *Engineer in Charge: A History of the Langley Aeronautical Laboratory, 1917–1958*, SP-4305.

Hartman, E. P. *Adventures in Research, A History of Ames Research Center, 1940–1965*, NASA Center History Series, SP-4302.

## United States Patents

Duerksen, Robert L., and Joseph Cohen. Solid propellant with polyurethane binder, US Patent 3,793,099, filed 31 May 1960 and issued 19 February 1974.

Duncan, Donal B., and Joseph V. Boltinghouse. Free rotor gyroscope, US Patent 3,251,233, filed 21 February 1957 and issued 5 May 1966.

Frey, Christian M., and Earl D. Shank. Rocket motor, US Patent 3,212,257, filed 25 March 1959 and issued 19 October 1965.

Goddard. Robert H. Rocket apparatus, US Patent 1,102,653, filed 1 October 1913 and issued 7 July 1919.

Hausmann, G. F. Directional control means for rockets or the like, US Patent 3,143,856, filed 30 July 1963 and issued 11 August 1964.

Scurlock, A. C., K. E. Rumbel, and M. L. Rice. Solid polyvinyl chloride propellants containing metal, US Patent 3,107,186, filed 6 August 1953 and issued 15 October 1963.

Scurlock, A. C., and K. E. Rumbel. Matrix propellant formulations containing aluminum, US Patent 3,407,100, filed 28 October 1960 and issued 22 October 1968.

Slater, John M., and Joseph C. Boltinghouse. Free rotor gyroscope motor and torquer drives, US Patent 3,025,708, filed 19 December 1958 and issued 20 March 1962.

Slater, John M., Doyle E. Wilcox, Darwin L. Freebairn, and Walter L. Pondrom Jr. Inertial velocity meter, US Patent 3,077,782, filed 2 April 1956 and issued 19 February 1963.

Weil, Lester. Solid propellant compositions and processes for making same, US Patent 2,966,403, filed 6 September 1950 and issued 27 December 1960.

Wetherbee Jr., A. E. Directional control means for a supersonic vehicle, US Patent 2,943,821, filed 5 December 1950 and issued 5 July 1960.

# Department of Defense

## *Air Force*

CHRONOLOGIES

"Ballistic Missile Organization Chronology: 1945–1990," Vol. I, Narrative.

"Chronology of Significant Events and Decisions Relating to the US Missile and Earth Satellite Development Programs: Supplement III," 1 November 1959–31 October 1960.

"Minuteman Chronology: December 1955–April 1961."

"Space and Missile Systems Organization: A Chronology, 1954–1979."

"Strategic Air Command Missile Chronology: 1939–1973."

"Strategic Air Command Missile Chronology: 1939–1988."

FLIGHT TEST REPORTS

ERCS Operational Test Flight GM-9, FTM 2225, July 1990, Bendix Aerospace Systems Division, Ann Arbor, MI.

LGM-30B OT Exercise Report, SWEET TALK, 2 July 1965, "History of the 1st Strategic Aerospace Division, Strategic Air Command, July–December 1965," Vol. III, Supporting Document Nine.

Operational Test Report: ELM BRANCH, Minuteman IA, LGM-30A, FTM 631, 29 June 1964, 1st Strategic Aerospace Division.

Operational Test Report: NICKED BLADE, Minuteman Missile IB, LGM-30B, FTM 1101, 9 December 1964, 1st Strategic Aerospace Division.

Demonstration and Shakedown Operation Test Program Final Report, LGM-30B Minuteman Weapon System.

Operational Test Exercise Report: DRAG CHUTE, Sortie Number 99–64-M-3, Minuteman "A," LGM-30A AFSN 63-124/FTM 664, 31 October 1963.

Operational Test Exercise Report: BOX SEAT, Sortie Number 99–64-M-9, Minuteman "A," LGM-30A AFSN 63-115/FTM 636, 29 February 1964.

Operational Test Exercise Report: BRASS RING, Sortie Number 99–64-M-10, Minuteman "A," LGM-30A AFSN 63-073/FTM 581, 29 February 1964.

1st STRAD LGM-30A Exercise Report ROCKY POINT, "History of the 1st Strategic Aerospace Division, Strategic Air Command, 1 July-31–December 1965, Vol. III," Supporting Documents.

Demonstration and Shakedown Operation Test Program Final Report, "History of 1st Strategic Aerospace Division, 1 January–1 June 1964."

LGM-30B Operational Test Final Report, "History of the 1st Strategic Aerospace Division, Strategic Air Command, July–December 1965, Vol. III," Supporting Documents.

# HISTORIES

## Arnold Engineering Development Center

"History of the Arnold Engineering Development Center, 1 July–31 December 1959."
"History of the Arnold Engineering Development Center, Vol. 1, July–December 1960."

## Deputy Commander for Aerospace Systems

Green, W. E. "The Development of the SM-68 Titan," August 1962.
Littlefield, Clyde. "The Site Program, 1961," Vol. III, September 1962.
Piper, Robert F. *The Development of the SM-80 Minuteman*, Vol. I and II, April 1962.

## Hill AFB

History of Construction Activities and Contract Administration Phases Encountered by Whiteman Area Office Corps of Engineers during Construction of Minuteman Strategic Missile Wing IV, no date.
"History of the Site Activation Task Force, Malmstrom AFB, MT, 1 January–3 July 1963."
"History of the Site Activation Task Force, Ellsworth AFB, SD, July–October 1963."
"History of SATAF Detachment #21, 1 January–30 April 1964, Minot AFB, ND."
"History of SATAF Detachment #21, 1 July–31 December 1963, Minot AFB, ND."
The Minuteman Weapon System History and Description, July 2001, Hill AFB, UT.

# SPACE COMMAND

"History of the Air Force Space Command, 1 January 1992–31 December 1993."
"History of the Air Force Space Command, 1 January 1994–31 December 2003."
"History of the Air Force Space Command, 1 January 2004–31 December 2005."
"History of the Air Force Space Command, 1 January 200731 December 2008."

# STRATEGIC AIR COMMAND

"History of Strategic Air Command 1961," SAC Historical Study No. 89, January 1962.
"Strategic Air Command Operations in the Cuban Crisis of 1962," Historical Study No. 90, Vol. 1.
"History of Strategic Air Command, 1990 Strategic Air Command Operations in the Cuban Crisis of 1962," Historical Study No. 90, Vol. 4.
"History of Strategic Air Command 1967," Historical Study No. 106.
"History of Strategic Air Command, January–June 1968," Historical Study No. 112.
"History of Strategic Air Command, FY 1969," Historical Study No. 116.
"History of Strategic Air Command, FY 1970," Historical Study No. 117.
"History of Strategic Air Command, FY 1971."
"History of Strategic Air Command, FY 1973," Historical Study No. 124.
"History of Strategic Air Command, FY 1974," Historical Study No. 142.
"History of Strategic Air Command, FY 1975," Historical Study No. 153.
"History of Strategic Air Command, 1 July–31 December 1976," Historical Study No. 161.
"History of Strategic Air Command, 1980," Historical Study No. 184.
"History of Strategic Air Command, 1 January–31 December 1985," Historical Study No. 214.

# 1ST STRATEGIC AEROSPACE DIVISION

"History of the 1st Strategic Aerospace Division, Strategic Air Command, 1 July–31 December 1962," Vol. I, Narrative.
"History of the 1st Strategic Aerospace Division, Strategic Air Command, 1 July–31 December 1962," Vol. II, Supporting Documents.

"History of the 1st Strategic Aerospace Division, Strategic Air Command, January–June 1964," Vol. II, Supporting Documents.
"History of the 1st Strategic Aerospace Division, Strategic Air Command, January–June 1965," Vol. 1.
"History of the 1st Strategic Aerospace Division, Strategic Air Command, 1 January–30 June 1966," Vol. I, Narrative.
"History of the 1st Strategic Aerospace Division, Strategic Air Command, July 1969–June 1970," Vol. III, Supporting Documents.
"History of the 1st Strategic Aerospace Division, Strategic Air Command, 1 January–31 December 1979," Vol. IV, Supporting Documents.

STUDY GUIDES

Minuteman Weapon System Familiarization: Function of Guidance Control Systems" (Chanute Technical Training Center, Chanute AFB, IL, April 1966).

WING HISTORIES AND REPORTS

*30th Space Wing*

"History of the 30th Space Wing, 1 January–31 December 2005."

*44th Strategic Missile Wing*

"History of the 44th Strategic Missile Wing, January–March 1965."

*90th Strategic Missile Wing (Space Wing)*

Peacekeeper in Minuteman Silos Draft Environmental Impact Statement, October 1983.
"90th Strategic Missile Wing History: January–March 1986."
"90th Strategic Missile Wing History: January–December 1988."
"History of the 90th Space Wing, 1 January 2005–31 December 2007."

*321st Strategic Missile Wing*

"History of the 321st Strategic Missile Wing and 804th Combat Support Group, Grand Forks AFB, ND, July–September 1970."

*341st Strategic Missile Wing*

341st Strategic Missile Wing and 341st Combat Support Group, 1–30 November 1961, Malmstrom AFB.
341st Strategic Missile Wing and 341st Combat Support Group, 1–31 December 1961.
341st Strategic Missile Wing and 341st Combat Support Group, 1–31 January 1962.
Final Report: Project High Climber Missile Site Transportation Service Test, 15 January 1962, Headquarters, 341st Strategic Missile Wing.
341st Strategic Missile Wing and 341st Combat Support Group, 1–31 March 1962.
10th Strategic Missile Squadron, 341st Strategic Missile Wing, and 341st Combat Support Group, 1–30 April 1962.
Missile Combat Crew 24-Hour Shift Test Operations Plan (Test Hop) 404B 62, 341st Strategic Missile Wing.
341st Strategic Missile Wing and 341st Combat Support Group, 1–30 April 1962.
341st Strategic Missile Wing and 341st Combat Support Group, 1–31 May 1962.
341st Strategic Missile Wing and 341st Combat Support Group, 1 June–31 July 1962.
341st Strategic Missile Wing and 341st Combat Support Group, 1–31 August 1962.
SM-80 Minuteman Program Programming Plan 2- 61, Progress Report No. 15, Report of August 1962.
341st Strategic Missile Wing and 341st Combat Support Group, 1–30 September 1962.
341st Strategic Missile Wing and 341st Combat Support Group History, Cuban Crisis Annex, 1–31 October1962.

341st Strategic Missile Wing and 341st Combat Support Group History, Cuban Crisis Annex, 1–30 November 1962.
341st Strategic Missile Wing and 341st Combat Support Group, 1–30 December 1962.
341st Strategic Missile Wing Operations Plan 402-63, Emergency Combat Capability, 341st Strategic Missile Wing and 341st Combat Support Group History, 1–31 December 1962.
341st Strategic Missile Wing and 341st Combat Support Group History, 1–31 October 1963.
Minuteman II Deactivation/Minuteman III Conversion, 341st Missile Wing, Program Plan 91-7, 23 September 1991.
"History of the 341st Missile Wing: July–December 1994," Vol. I.
Environmental Assessment Minuteman III Deactivation, Malmstrom AFB, MT, May 2007.

*351st Strategic Missile Wing*

Minuteman II Deactivation, 351st Missile Wing Program Plan 92-12, Whiteman AFB, 1 May 1992.
"Closeout History of the 351st Strategic Missile Wing, 1 January–31 July 1995," Volume I of VII, 1 August 1995.

*390th Strategic Missile Wing*

"History of the 390th Strategic Missile Wing, April 1963."

## USAF BALLISTIC MISSILE PROGRAM

Nalty, Bernard C. "USAF Ballistic Missile Programs, 1962–1964," April 1966.
Nalty, Bernard C. "USAF Ballistic Missile Programs, 1964–1966," March 1967.
Nalty, Bernard C. "USAF Ballistic Missile Programs 1966–1968," September 1969.
Neufeld, Jacob. "USAF Ballistic Missile Programs 1969–1970," June 1971.
Rosenberg, Max. "USAF Ballistic Missiles 1958–1959," July 1960.

## DEACTIVATION

*44th Strategic Missile Wing*

Program Plan 91-7, Minuteman II Deactivation/Minuteman III Conversion, 341st Missile Wing, 23 September 1991.
Final Environmental Impact Statement, Deactivation of the Minuteman II Missile Wing at Ellsworth AFB, SD, October 1991.
Deactivation of the Minuteman II Missile System: Final Deactivation Environmental Plan, Ellsworth AFB, SD, September 1993.
Missile Dismantlement Status, 44th Missile Wing, 4 November 1997.
Hotel-Flight Launch Facilities Environmental Baseline Survey: Minuteman II Deactivation Site Disposals, 20 July 2000.
Golf-Flight Launch Facilities Environmental Baseline Survey, Minuteman II Deactivation Site Proposals, Ellsworth AFB, SD, 19 April 2000.

*321st Strategic Missile Wing*

Draft Environmental Impact Statement: Minuteman III Missile System Dismantlement, Grand Forks AFB, ND, May 1998.

*351st Strategic Missile Wing*

Minuteman II Deactivation and 351st Missile Wing Deactivation, Program Plan 92-12, May 1992.
Final Environmental Impact Statement: Deactivation of the Minuteman II Missile Wing at Whiteman AFB, MO, August 1992.

*564th Strategic Missile Squadron*
Environmental Assessment Minuteman III Deactivation, Malmstrom AFB, MT, May 2007.

AS-BUILT DRAWINGS

WS-133A Technical Facilities, First Operational Deployment Area, Minot AFB, ND, Association of Air Force Missileers Collection, Sheets A-8, S-57.

WS-133A Technical Facilities, First Operational Deployment Area, Malmstrom AFB, MT (Parsons-Wenzel Architects–Engineers, Los Angeles, CA, and Great Falls, MT, 10 October 1960), Association of Air Force Missileers Collection, Sheets S-19, S-20, S-21, S-21B, S-32, S-32.1, S-33, S-38.

WS-133B Technical Facilities, Collocated Squadron No. 20, Part II of III Architectural B Structural, Malmstrom AFB, MT (Ralph M. Parsons Architects–Engineers, Los Angeles, CA, 4 January 1965), Association of Air Force Missileers Collection, Sheet S-23, S-25, M-8.

REPORTS/STUDIES

"Alert Operations and the Strategic Air Command, 1957–1991" (Office of the Historian, Headquarters Strategic Air Command, Offutt AFB, NE, 7 December 1991).

Detailed Design Specifications for Model SM-68B Missile, Including Addendum for XSM-68B, The Martin Company, Denver, CO, 1961.

Environmental Assessment, Missile Impacts, Illeginni Island, Kwajalein Missile Range, Kwajalein Atoll, December 1977.

Flight Summary Report, Series D Atlas Missiles (General Dynamics/Astronautics, San Diego, CA, 21 June 1961).

"ICBM Modernization: Minuteman III Guidance Replacement Program Has Not Been Adequately Justified" (General Accounting Office, Washington, DC, 25 June 1993).

Space and Missile Test Center Capability Summary Handbook, Headquarters, Space and Missile Test Center, Vandenberg AFB, CA, November 1974.

WS-133A/B Weapon System Briefing FETM-M-12B-SA-28, Autonetics Division, North American Aviation.

Arms, W. M. *Thor, the Workhorse of Space: A Narrative History* (McDonnell Douglas Astronautics Company, Huntington Beach, CA, 31 July 1972).

Bowlin, Robert G. One-Third Scale Task Silo Launcher Development Project WS 133A (Minuteman), "History of the Air Force Flight Test Center, 1 July–31 December 1959," Vol. 4.

Brassell, J. C. "Jupiter: Development Aspects and Deployment," Mobile Air Material Area, Brookley AFB (Historical Office, Office of Information, 1962).

Cleary, M. C. "Army Ballistic Missile Programs at Cape Canaveral, 1953–1988" (45th Space Wing History Office, Patrick AFB, FL, 2006).

Cohen, Edward, and A. Bottenhofer. "Test of German Underground Personnel Shelters, Operation Plumbbob," Project 30.7 (Ammann and Whitney, New York, NY, 1957).

Dubose, H. C., and R. C Bauer. "Transonic Dynamic Stability Tests of a Half-Scale Model of the Jupiter W-14 Re-Entry Configuration" (US Air Force, Arnold Engineering Development Center), AEDC TN-58-1, February 1958.

Large Jr., William R. "Ballistic Missile Hardening Study (Headquarters Strategic Air Command, Offutt AFB, NE, 10 July 1958)."

McManus, J. P. "A History of the FBM System" (Lockheed Missiles and Space Company, Inc., 1989).

Metcalf, James I., Arnold A. Barnes Jr., and Michael J. Kraus. "Final Report of PVM-4 and PVM-3 Weather Documentation, 19 February 1975."

Metcalf, James I., Arnold A. Barnes Jr., and Michael J. Kraus. "Final Report of STM-8W Weather Documentation, 11 April 1975."

Metcalf, James I., Arnold A. Barnes Jr., and Michael J. Kraus. "Final Report of PVM-5 Weather Documentation, 28 May 1975."

Metcalf, James I., Michael J. Kraus, and Arnold A. Barnes Jr. TM-16-SRO, 15 March 1965.

Metcalf, James I., Michael J. Kraus, and Arnold A. Barnes Jr. "Final Report of OT-45, PVM-8 and RVTO Weather Documentation, 23 July 1975."

Mockenhaupt, J. D. "Performance Characteristics and Analysis of Liquid Injection Thrust Vector Control," TM-16-SRO (Aerojet General, Sacramento, CA, 15 March 1965).

Morton, M. "Progress in Reentry Vehicle Development" (General Electric Missile and Space Vehicle Department, Philadelphia, PA, 2 January 1961).

Morton, M. "Progress in Reentry Vehicle Development, Final Report of PVM-6 and PVM-7 Weather Documentation, 11 September 1975."

Redman, E. J., and L. Pasuk. "Pressure Distribution on an ABMA Jupiter Nose Cone (13.3 Degrees Semi-Vertex Angle) at Nominal Mach Numbers 5, 6, 7, and 8," NAVORD Report 4486, US Naval Ordnance Laboratory Aeroballistic Research Report 381, 27 September 1957.

Ruchonnet, Daniel. "MIRV: A Brief History of Minuteman and Multiple Reentry Vehicles" (Lawrence Livermore Laboratory, February 1976).

Schmidt, D. L. "Ablative Plastics for Reentry Thermal Environments," Report 60–862 (United States Air Force, Air Research and Development Command, Wright Air Development Division,1961).

Schmidt, D. L. "Ablative Plastics for Thermal Protection," WADD Report 60–862 (US Air Force, Wright Air Development Division, August 1968).

Scott, C. W., and P. H. Eisenberg. Failure Analysis of Minuteman Integrated Circuit Failures, Rome Air Development Center (RADC) Technical Report 669–457, May 1970.

Shaw Jr., Frederick. Minuteman, Strategic Air Command Weapons Systems Acquisition: 1964–1979 (Office of the Historian, Headquarters Strategic Air Command, 18 April 1980).

TECHNICAL ORDERS

Technical Order 21M-LGM30B-1-3, Weapon System Operation Instructions (Wing II), USAF Series LGM30B Missile, 1 April 1969.

Technical Order 21M-LGM30F-1-6, Weapon System Operation Instructions (VAFB, Wing I Squadron Four; Wing VI), USAF Series LGM30F and LGM30G Missile, 29 January 1975.

Technical Order 21M-LGM30F-2-22, Organizational Maintenance Instructions, Missile Umbilicals and Missile Suspension System (VAFB and Wing II), USAF Series LGM30F, 30 November 1992.

Technical Order 21M-LGM30G-1-10, Weapon System Operation Instructions, (VAFB, Wing I Squadron Four, Wing VI), USAF Series LGM30G Missile, 1 March 1978.

STUDY GUIDES

Minuteman Weapon System Familiarization: Function of Guidance and Control, Student Study Guide (Chanute Technical Training Center, Chanute AFB, IL, 12 April 1966).

Minuteman Weapon System Familiarization: Flight Control Group, Student Study Guide, (Chanute Technical Training Center, Chanute AFB, IL, 12 April 1966).

# Army

## *Construction Engineering Research Laboratory*

Nowlan, Patrick, and Roy McCullough. Cold War Properties Evaluation Phase II: Inventory and Evaluation of Minuteman, M-X Peacekeeper, and Space Tracking Facilities at Vandenberg AFB, CA, June 1997.

## *Historic American Engineering Record*

Ensore, Susan I., Julie L. Webster, Angela M. Fike, and Martin J. Stupich. Level II Documentation of Launch Complex 31/32, Cape Canaveral Air Force Station, Florida, December 2008.

## Histories

**CORPS OF ENGINEERS BALLISTIC MISSILE CONSTRUCTION OFFICE**

"Historical Summary 1 January–30 June 1963," Corps of Engineers Ballistic Missile Construction Office.

**MALMSTROM AFB**

History of Malmstrom Area during Construction of Collocated Squadron Number 20 Minuteman II ICBM Facilities, 31 December 1966.
Minuteman Technical Facilities, Malmstrom AFB, Great Falls, MT.

**ELLSWORTH AFB**

"Historical Summary 1 January–30 June 1963," Corps of Engineers Ballistic Missile Construction Office, Ellsworth AFB.
"History of Minuteman Construction Wing II," Ellsworth Area Engineer Office, US Army Corps of Engineers, 1 August 1961–31 August 1963.

**MINOT AFB**

WS-133A Minuteman Missile Facilities, Vol I, Minot AFB, Minot, ND.
WS-133A Minuteman Missile Facilities, Vol II, Minot AFB, Minot, ND.

**WHITEMAN AFB**

History of Construction Activities and Contract Administration Phases Encountered by the Whiteman Area Office Corps of Engineers, CEBMCO, during Construction of Minuteman Strategic Missile Wing IV, Whiteman AFB, MO.
History of Construction Activities and Contract Administration Phases Encountered by Whiteman Area Office Corps of Engineers during Construction of Minuteman Strategic Missile Wing IV.

**F. E. WARREN AFB**

Warren Area Minuteman History, US Army Corps of Engineers Ballistic Missile Construction Office.

**GRAND FORKS AFB**

History of Grand Forks Area during Construction of Wing VI Minuteman II ICBM.

## Reports/Studies

"Designing Facilities to Resist Nuclear Weapons Effects: Hardness Verification, TM 5-858-6, August 1984."
"Explorers in Orbit," Army Ballistic Missile Agency, 10 November 1958.
Ordnance Engineering Design Handbook, Ballistic Missile Series, Trajectories, Ordnance Corps Pamphlet ORDP 20-284, no date. DTIC AD389298
"Study of Shock Isolation for Hardened Structures, 17 June 1966."
Wagner, C. "The Skin Temperature of Missiles Entering the Atmosphere at Hypersonic Speed," Technical Report No. 60, October 1949.

*Weapon System Histories*

Bullard, John W. "History of the Redstone Missile System" (History Office Army Missile Command, 15 October 1965).

Cagle, M.T. *History of the Sergeant Weapon System*, Historical Monograph Project No. AMC 55M (US Army Missile Command, 1971), 4.

*Yearbook*

"25 Years of Minuteman Deterrence, Malmstrom's Missiles of October Celebration 1962-1987" (History Office, 341st Strategic Missile Wing, Malmstrom AFB, MT, 1987).

## Office of the Secretary of Defense

*Reports/Studies*

"Defense Base Closure and Realignment Commission Report, 1 July 1995."
"Director Operational Test and Evaluation FY 2000 Annual Report."
"Director Operational Test and Evaluation FY 2002 Annual Report."
"Nuclear Posture Review Report, April 2010" (Department of Defense, 6 April 2010).
"Policy Guidance for the Employment of Nuclear Weapons, 3 April 1974" (Office of the Secretary of Defense).
"Quadrennial Defense Review Report" (Office of the Secretary of Defense, 6 February 2006).
Weapon System Evaluation Group Report Number 50: Evaluation of Strategic Offensive Weapon Systems, 27 December 1960.
Boyd, Robert J. "Emergency Rocket Communications System (ERCS)," (Program 279/494L), Strategic Air Command Weapon Systems Acquisition: 1964–1979.
Cohen, William S., *Annual Report to the President and the Congress: 2001* (US Government Printing Office, Washington, DC, 1995).
Converse III, Elliott V. *Rearming for the Cold War 1945–1960, History of Acquisition in the Department of Defense*, Vol. I.
Eger, Jr., Charles H. "Minuteman Financial History, 1955–63," Systems Program Office.
Fletcher, James. "Summary Report: Minuteman Flexibility and Safety Study Group, 27 September 1961" (Department of Defense).
Keller, K. T. "Final Report of the Director of Guided Missiles, 17 September 1953" (Office of the Secretary of Defense).
Killian, James R. "Technological Capabilities Panel Report, Vol. I."
Lauritsen, Charles. "Lauritsen Committee Report to Lt. Gen. Bernard A. Schriever Concerning the United States Air Force ICBM Program, 31 May 1960."
Lauritsen, Charles. "Lauritsen Committee Report to Lt. Gen. Bernard Schriever Concerning the United States Air Force Minuteman Program, 15 June 1961."
McNamara, Robert S. *Annual Defense Department Report, Fiscal Year 1965* (US Government Printing Office, Washington, DC, 1967).
Perry, William J., *Annual Report to the President and the Congress: February 1995* (US Government Printing Office, Washington, DC, 1995).
Ponturo, J. "Analytical Support for the Joint Chiefs of Staff: The WSEG Experience, 1948–1976," Institute for Defense Analysis Study S-507, July 1979.

# INDEX

ablative heat dissipation, 186
ablative materials/coatings, 177–78, 178*t*, 190–96
ablative reentry vehicle systems, 188–96
Able (monkey), 189
Able RTV program, 192
Able-Star booster, 175
accelerometers: angular accelerometers, 215; pendulous integrating gyro-accelerometer (PIGA), 221, 222*f*, 223, 226, 230, 231, 396–97; vibrating string accelerometers (VSAs), 213; VM4 velocity meter, 206, 208, 213
accidental launch, 277, 337, 339, 375
accuracy: *see* targeting accuracy
Accuracy, Reliability, Supportability Improvement Program (ARSIP), 227–28
acrylic materials, 254
acrylonitrile additives, 160
AC Spark Plug, 223
Adelman, Barney, 66
Advanced Ballistic Reentry System (ABRES) Program, 293
Advanced Inertial Reference Sphere (AIRS), 305, 306
advanced manned strategic aircraft (AMSA), 23
advanced reentry test vehicle (ARTV), 192
Aerojet Engineering Corporation, 3, 44, 52–53
Aerojet-General Corporation: engine thrust research, 249; funding obstacles, 78; Minuteman motor designations, 164*t*; Minuteman production facilities, 87; missile assembly-and-recycle facilities, 86; motor casing manufacturing processes, 66, 157; motor fabrication contracts, 74, 84, 172; motor testing, 62–63, 168; nozzle design and development, 231; offset nozzle simulators, 162*f*; rocket propulsion research, 165; silo development research, 249; solid propellant research and development, 52–53, 60, 78; SR73-AJ-1, 172–73; Stage II motor development and testing, 164*t*, 165–68, 169*t*, 395; Stage III motor development and testing, 164*t*, 168, 170, 172, 173*t*, 174
Aerophysics Allison, 197
Aerospace Guidance and Metrology Center Depot, Newark Air Force Station, Ohio, 226
Aerospace Test Wing (6595 ATW), 154, 377
Aerowrap technique, 166
AFGHAN RUG, 278
Airborne Astro Graphic Camera System, 303
Airborne Command Control Squadron (4 ACCS), 366
Airborne Launch Control Center (ALCC): command-and-control systems, 369; deactivation, 371; deployment plans, 366; equipment modifications and upgrades, 358; Follow-on Operational Test Program (FOT), 371; Launch Control Panel/Console, 367*f*, 368*f*; mobile transporters/launchers, 267; operational flight and evaluation programs, 369–70, 371; simulated launches, 332, 333, 334; technical manuals, 437–44; UHF radio systems, 358, 360, 366
Airborne Launch Control System (ALCS): EC-135 aircraft, 365–66, 365*f*, 371; flight test programs, 365–66, 369–70, 371, 436*t*; Follow-on Operational Test Program (FOT), 371; historical perspective, 340, 363–65; Launch Control Panel/Console, 367*f*; Launch Monitor Panel, 368*f*; Minuteman II, 264, 299, 357, 365–66; Minuteman III, 301
airborne power, 216, 217*t*
Airborne Surveillance Testbed (AST), 432*t*
air-conditioning systems: blast effects, 382, 384–85; HEST (high-explosive simulation technique) testing, 384; launch control equipment building (LCEB), 116, 120, 122–23; launch control support building (LCSB), 112; launcher equipment room (LER), 108; launch tubes, 132; LCCs (launch control centers), 115–16, 118, 151
Aircraft Laboratory Project MX-121, 44
aircraft rocket (AR), 47
aircraft windshield deicing, 43
AIR CRUSADE, 273, 274, 277, 376–77
Air Force Ballistic Missile Committee (AFBMC), 18, 71, 73, 79–80, 87, 89–90
Air Force Ballistic Missile Division (AFBMD): command and control systems, 337; construction operations, 127, 128; contractor responsibilities, 83, 197; flight test programs, 192; funding impacts, 78; ICBM program review, 17; Minuteman reentry vehicles, 196–98; missile assembly-and-recycle facilities, 87; research and development programs, 71, 74–75, 80; site separation and dispersal recommendations, 91, 92; system development programs, 65; *see also* Ballistic Systems Division (BSD); Western Development Division (WDD)

519

Air Force Flight Test Center (AFFTC), 147, 250–51, 255, 257
Air Force Logistics Command (AFLC), 128, 292, 302
Air Force Missile Test Center (AFMTC), 255
Air Force Systems Command (AFSC), 128, 267
Air Force (USAF): *see* US Air Force (USAF)
airframe components and production: Bumblebee anti-aircraft missile booster, 57; centerline determination, 242; comparison chart, 180*f*; digital angular accelerometers, 206, 215; flight test programs, 198, 254, 265*f*, 319, 322; insulation materials, 254; interstages, 178–79, 180*f*; Minuteman general characteristics, 69*t*; modernization and upgrade programs, 25, 84, 229; modified operational missiles (MOMS), 329; Polaris missile program, 159*f*; production statistics, 179*t*; reentry vehicles, 198, 201; Sergeant missiles, 57, 61–62, 157; Silo Test Missile (STM), 254; skirts, 179, 180*f*; solid propellant missile systems, 60–61, 66, 250; V-2 long-range guided ballistic missile, 55
Air Materiel Command (AMC), 44, 45, 87, 94, 126, 128
*Air Quarterly Review*, 83
Air Research and Development Command (ARDC), 10, 18, 60, 62, 63, 65, 71, 126, 128, 190
Alaska, 91
alignment blocks, 206
alignment mechanisms, 215, 231, 233, 235, 301, 386
Allegany Ballistics Laboratory (ABL), 42, 47, 57, 62, 63, 66, 74, 168, 170
Allen, H. Julian, 183, 186
Allied-Schaffer Construction Company, 295
all-inertial guidance systems, 205; *see also* guidance-and-control systems; inertial guidance systems
All-Weather Impact Location System (AWILS), 286, 290
Alpha Draco sounding rocket program, 255
Altus Air Force Base (Texas), 12*f*
aluminum additives, powdered, 51–53, 53*f*, 63, 160, 165, 173–74
AMERICAN BEAUTY, 274, 277, 377
American Bosch ARMA Corporation, 213
American Samoa, 324
Ames Aeronautical Laboratory (California), 183, 186, 190
ammonium perchlorate oxidizer, 48, 52, 53, 53*f*, 63, 160
Anaheim Hills, California, 246
ANCHOR POLE, 289, 290, 432*t*
Anderson Air Force Base (Guam), 91
Anderson, Samuel E, 65

Andrus, Cecil, 326
angular accelerometers, 215
ANSWERMAN, 430*t*
Anti-Ballistic Missile (ABM) defense system, 174–75
Anti-Ballistic Missile (ABM) Treaty, 28
Antigua, 422*t*
A-plug, 111
Arcite polyvinyl chloride plastisol composite propellant, 51–52
ARCTIC HOLIDAY, 281, 429*t*
Arde Inc., 162
arms limitation negotiations, 24, 26–29, 36
Army Air Corps, 43, 45
Army Ballistic Missile Agency (ABMA), 188
Army Ballistic Missile Defense Program, 294
Army Corps of Engineers: construction operations, 125–28; Minuteman wing locations, 238; PCB contamination, 401
Army–Navy Joint Ballistic Missiles Committee, 55, 59
Army Special Test Program HK (homing kill), 294
Arnold Air Force Base (Tennessee), 158, 188
Arnold Engineering Development Center (AEDC), 158, 161, 188
Arnold, Henry "Hap," 5–7, 42–43
ARROW FEATHER, 432*t*
asbestos-reinforced phenolic resins, 188
Ascension Island, 194, 257, 263, 265, 421–26*t*
asphalt–potassium perchlorate castable propellant, 44, 46
assemble and test (A&T): base location considerations and criteria, 87–91; contractor responsibilities, 84–87; launch control equipment building (LCEB), 95, 113*f*, 115*f*, 118, 120, 120*f*, 121*f*, 122, 122–23, 123*f*; launch control facility (LCF), 95, 112, 113*f*, 114*f*, 115*f*, 403; launch control support building (LCSB), 95, 112, 113*f*, 115–16; launcher closure, 105–8, 105*f*, 107*f*, 109*f*; launcher equipment room (LER), 102–4, 104*f*, 105*f*; launcher support building (LSB), 95, 96, 97*f*, 108–11, 110*f*, 111*f*; launch facility security system, 111; launch tubes, 96, 98–99, 98*f*, 105*f*; missile assembly-and-recycle facilities, 85–87, 85*f*; operational facility design and criteria, 94–96; prospective deployment locations, 84, 85–87, 85*f*; site separation and dispersal recommendations, 91–92, 93*t*; survivability criteria, 95, 95*t*; *see also* LCCs (launch control centers)
astrogeodetic surveys, 238–39
astronomic azimuth, 239
Atlantic Missile Range (AMR): flight test programs, 205, 421–27*t*; full-range operational base launches, 321; launch failures, 277–78;

Minuteman development and testing programs, 150, 255, 257, 259–64; Stage I motor development and testing, 162; Stage III motor development and testing, 168, 172
Atlantic Research Corporation (ARC), 51, 52, 61
Atlas missile program: Air Force reentry vehicle designations, 184t; Atlas D, 125, 126, 183, 195, 198, 202; Atlas E, 126, 198; Atlas F, 13f, 66, 125, 198, 321; base location considerations and criteria, 88; ceramics-based ablative materials, 191; construction operations, 125; design considerations and criteria, 92; effectiveness, 33, 71; first-generation heatsink reentry vehicles, 183, 190; flight test programs, 191, 194–96, 198, 202, 205; force objectives, 70t; full-range operational base launches, 321; funding estimates, 70t; launch control facility (LCF), 112; launch facility criteria, 89; missile assembly-and-recycle facilities, 85; missile suspension systems, 99; operational characteristics, 74; program management, 73; rocket propulsion research, 60, 62; silo launch facilities, 66; site separation and dispersal recommendations, 93t; strategic alerts, 33; task force study and recommendations, 8, 9–11; Truman administration, 8; see also reentry vehicles
Atlas Scientific Advisory Committee, 10
Atomic Energy Commission (AEC), 57, 103, 197, 198, 323
attack scenarios, 92; design considerations and criteria, 92–94; estimated Soviet threat, 93t; hardening studies and specifications, 91–92, 94, 95t
attitude control: Mark 11 reentry vehicle, 201; Stage I motor development and testing, 161; Stage II motor development and testing, 166–68, 169f; Stage III motor development and testing, 169f, 172
autocollimators: coarse positioning procedures, 244; fine alignment procedures, 244; flight test programs, 301; Force Modernization plans, 385, 386; functional role, 478n10; GI-T1-B azimuth gyroscope, 218–221; Improved Launch Control System (ILCS), 227; launch azimuth and alignment determinations, 239–47, 242f, 243f, 377; missile centerline determination, 244; NS10 series guidance-and-control system, 215, 218–19; NS17 guidance-and-control system, 228f; NS20 guidance-and-control system, 231, 233; operational flight and evaluation programs, 284; transfer mirror, 244
AUTO command, 369, 370
Autonetics: see North American Aviation, Autonetics Division
Avcoat coatings, 177–78, 178t, 195, 200f, 263

Avcoite, 191, 193, 196, 201
Avco Manufacturing Corporation: ablative coatings, 177–78, 178t, 191, 195, 200t, 263; Air Force reentry vehicle designations, 184t; contractor responsibilities, 74, 84; external insulation, 177; first-generation heatsink reentry vehicles, 187f, 190; Mark 4 reentry vehicle tests, 195–96, 197f, 199f; Mark 5 reentry vehicle, 196–198; Mark 11 reentry vehicle tests, 198; Minuteman reentry vehicles, 196–98, 199f; RVX-1 reentry vehicle test, 193, 193f, 194f
axial engines, 176
axial nozzles, 158, 159f, 469n27
Ayres, L. F., 60
azimuth gyroscopes, 219–20, 231, 235
azimuth laying set (ALS), 245–47


B-17 bombers, 5, 43
B-47 bombers, 23, 456n22
B-52 bombers, 23, 36, 83
Bacher Committee/Report, 64, 65
Bacher, Robert F., 64
Bagby, F. K., 66–67
BAIT CAN, 432t
Baker Island, 303
Baker (monkey), 189
ballistic coefficient ($\beta$), 186, 190
Ballistic Missile Office, 394
Ballistic Missile Organization, 25
Ballistic Missile Scientific Advisory Committee: see Scientific Advisory Committee on Ballistic Missiles
Ballistic Systems Division (BSD): guidance-and-control systems, 207; hardness design studies, 380; Minuteman II flight test programs, 295, 319; organizational change, 128; safety evaluations, 375–76
ballistic trajectory, 183
Bankston, L. T., 166–67
BARE FOOT: see GIANT FOOT
Barry, W. T., 474n34
Bartley, Charles, 46, 47, 48
base location considerations and criteria, 87–91, 338–39
Base Realignment and Closure Commission, 27, 406
Battelle Memorial Institute, 191
Beale Air Force Base (California): full-range operational base launches, 321; geographic location, 12f
BEAN STALK, 356
Bell Aerospace, 176, 230
Bendix, 166, 223

beryllium: first-generation heatsink reentry vehicles, 187*f*; gyroscope housings, 209, 211, 213; interstage design, 267; platform structural design, 220–21; spacers, 201; stable platform design, 220–21; as thermal protection material, 188, 191; toxicity, 52, 455*n*59
Beta Seal, 393
BIG BARK, 289
BIG CIRCLE, 282, 283–84, 430*t*
Big Stoop flight test vehicle, 57, 58*f*
binders, 47, 53
bipropellant systems, 165–66, 167*f*, 168, 170
bird sanctuaries, 303
Birnie Island, 302, 303, 304*f*
Bishop, Robert, 30
black powder propellant, 41
Black, Sivalls, and Bryson, 172
Blasingame, Benjamin Paul, 63, 64–65
blast doors, 141–42, 141*f*, 142*f*, 147, 403
blast tubes, 166
blast valves: construction material requirements, 144*t*; deactivation and dismantlement, 403; launch control equipment building (LCEB), 120, 120*f*, 122, 123*f*; launcher support building (LSB), 111; LCCs (launch control centers), 114*f*, 116, 122, 140, 142, 380, 403; modernization efforts, 384
Blue Book computer system, 213–14
Blue Scout Junior Program 279L (MER-6A), 356–57, 357*f*
Blue Streak IRBM program, 249
Bode, Hendrik W., 8, 17, 75, 460*n*101
body bending load, 60
Boeing Aircraft Company: ALCC flight test program, 370; bomber production, 83; command-and-control systems, 356; contractor responsibilities, 3, 78, 83–87, 147, 216, 228–29; demonstration and acceptance procedures, 154; development and testing programs, 150–51, 154, 249, 257, 260, 263, 370; EC-135 aircraft, 365; external insulation, 177; Ground-Based Strategic Deterrent (GBSD) program, 409, 410; Growth Minuteman (Minuteman IIB), 267; guidance alignment survey monuments, 238; guidance-and-control systems, 305, 377; HEST (high-explosive simulation technique) testing, 380, 383; in-silo launch environment tests, 249, 250–52, 250*f*; interstages, 178; launch facility and launch control center design changes, 339–40; modernization and upgrade programs, 295; modification programs, 24; operational flight and evaluation programs, 271–72, 288; Project Restart, 393; simulated combat launch capability (SCLC), 332–33; test target recommendations, 272

Boeing Development Center, 249
Boltinghouse, Joseph, 206, 211, 476*n*13
BONUS BOY, 435*t*
Booher, Bob, 214
booster assets, 28
booster replacement and repair programs, 395
boron oxide, 52
Bosch ARMA Corporation, 213
Bothwell, Frank E., 57
Boushey, Homer A., Jr., 44
Bowser, Duane, 243*f*
bow shock waves, 184*f*
BOX SEAT, 279–80
B-plug, 103, 106*f*, 111, 397
BRASS RING, 279
Brezhnev, Leonid I., 24
BROAD ARROW, 281, 429*t*
Broad Ocean Scoring Systems, 298–99
Broad Ocean Area target zone, 422–24*t*, 423*t*, 426*t*
Brown, Harold, 229, 320
Brush Beryllium Company, 211
bubble levels, 206
BULL MARKET, 289
Bumblebee anti-aircraft missile booster, 57, 170
Buna-N (butyl rubber compound), 161
Buna-S (butyl rubber compound), 46
Burke, Arleigh, 34, 57
Burnett, J. Robert, 205
Bush, George H. W., 26, 400, 405
Bush (George H. W.) administration, 26
Bush, George W., 28, 312, 371
Bush (George W.) administration, 27–29
Bush, Vannevar, 5, 6–7, 11, 43
Business Leader launch program, 324
BUSY FELLOW, 296–97, 366, 433*t*
BUSY JOKER, 433*t*
BUSY LOBBY, 433*t*
BUSY MISSILE, 366, 435*t*
BUSY MUMMY, 366

C-47 aircraft, 152
CALAMITY JANE, 433*t*
California Institute of Technology (Caltech), 6, 42–45
Camp Navajo Depot (Arizona), 27, 406
Cannon, Mitch, 489*n*30
Canton Island, 302, 303, 304*f*, 326, 327*f*
Cape Canaveral, 177, 192, 198, 251*f*, 421*t*
Cape Kennedy, 267, 281, 283, 426*t*
Cape Town, South Africa, 194
carbon dioxide, 52
carbon monoxide, 52
carboxyl-terminated polybutadiene propellants, 166, 172

CAREER GIRL, 297, 434*t*
Carswell Air Force Base (Texas), 91
Cartwright, James E., 28
Casmalia Express, 293, 295*f*
CASTLE, 9
Castle Rock, South Dakota, 316
Category I (engineering test program), 257, 271, 272–74, 277–78, 282, 283, 296, 435*t*
Category II (flight test and launch programs), 271, 278, 282, 283, 296, 301, 376
Category III (operational system test and evaluation programs), 271, 278–79, 296
C-band Airborne Tracking Safety System, 323
CEDAR LAKE, 279
centerline determination, 242–43, 244
ceramic materials, 188, 191
chaff clouds, 300, 303
chaff dispensers, 37*t*, 201, 293, 299–300, 302–4
Charyk, Joseph V., 127
Cheney, Dick, 399
Cheyenne, Wyoming, 86; *see also* F. E. Warren Air Force Base (Wyoming)
China, as missile target, 33, 34
Christmas Island, 56
Christmas Island hydrogen bomb tests, 302
Church, Frank, 326
circular error probable (CEP): estimated Soviet threat, 93*t*; Expanded Execution Plans (EEP) program, 235; Guidance Upgrade Program (GUP), 234; impact analysis, 290–92, 291*f*; military requirements, 9; Minuteman general characteristics, 69*t*; Minuteman II flight test programs, 299; Minuteman III flight test programs, 268, 299; missile system specifications, 63, 76, 205; Polaris missile program, 67; Preparatory Launch Command (PLC) capability, 348; reentry vehicles, 298; solid propellant missile systems, 62, 65; target location accuracy, 33, 36, 37*t*, 219, 247–48, 261*f*, 272, 279, 281, 284, 288–89
civil defense policies, 10–11
Clark University, 41
CLEAN SLATE, 289, 432*t*
Clements, William P., Jr., 328
Clifford, Clark, 320
Clinton administration, 27
CLOCK WATCH, 430*t*
cloud sampling patterns, 310–11, 311*f*
cluster-rocket launch techniques, 61
Cohen, Joseph, 165
Cohen, William, 36
Cold/Heat Soak program, 292–93
combat readiness training, 152–54
Combat Support Group Civil Engineering Squadron (804), 331

Combat Targeting Team (CTT), 240, 240*f*, 243*f*, 246
Combat Test Launch Instrumentation, 284
Combat Training Launch Instrumentation, 287, 322
combustion instability, 50, 51, 53
combustion reaction products, 52–53
command-and-control systems: Airborne Launch Control System (ALCS), 363–66, 365*f*, 367*f*, 369–70; automation programs, 353–54; basic process, 340–41; Command Console, 350*f*; Communication Console, 349*f*; developmental evolution, 340–43; Force Modernization plans, 386, 390; Headquarters Air Force, 390; historical perspective, 337–40; launch command procedures, 341–46; Launch Control Panel/Console, 343, 344*f*, 345*f*, 348*f*, 355*f*, 359*f*, 367*f*; Launch Monitor Panel, 368*f*; Minuteman IA, 343–46, 344*f*, 345*f*; Minuteman IB, 346–47; Minuteman II, 347–48, 352–54, 356; Preparatory Launch Command (PLC) capability, 347–48, 354; Rapid Execution and Combat Targeting (REACT) Program, 353–54, 355*f*, 356, 398*t*; readiness verification, 341; remote targeting, 352–53, 390–91; rogue launches, 337; Secure Enable command, 353; Status Console, 351*f*; strategic targeting plans, 34, 35, 36; system configurations, 343–48; target and launch time selections, 341, 343, 346–47; technical manuals, 437–44
Command Data Buffer (CDB) program: command-and-control systems, 352–54; flight test programs, 247; guidance-and-control systems, 104, 232–34, 246; Hybrid Explicit software, 309; launch control system, 227; launcher closure, 107; missile suspension systems, 99, 100*f*, 102; modification completion dates, 398*t*; remote targeting, 390–92
Common/PACER LINK satellite communication system, 371
Component Quality Assurance Program (CQAP), 225
Comprehensive Test Ban Treaty, 285, 363
concrete enhancement program, 397
Congreve, William, 41, 43, 46
conical internal-burning grain design, 41, 43, 46–47
Conolly, Richard, 8
Conolon 505, 190
Conrad, Kent, 29
Conrad, Pete, 281
construction operations: backfill operations, 134–38, 141, 145, 151–53; blast doors, 141–42, 141*f*, 142*f*, 147; construction costs, 150*t*; construction material requirements, 144*t*; excavation

methods, 145–46, 148f, 149f; final bids and contracts, 128–30; inspections and turnover, 147; interior structural and mechanical work, 138, 141–42, 147; issues and concerns, 125–28, 144–47; launch control facility (LCF), 139–42, 140f, 141f, 142f, 143f; launch control support building (LCSB), 143–44; launch facility, 130, 132, 133f, 134–38, 135f, 136f, 137f, 138f; Seattle Test Program (STP) III, 150–51; start and completion dates, 131t; timeline, 126t; weather impacts, 144–45
controlled response criteria, 95, 95t
Convair, 9, 10
Cooke Air Force Base (California), 190; *see also* Vandenberg Air Force Base (California)
cooling systems: ablation processes, 188; categories, 186; ground equipment, 401; guidance-and-control systems, 388, 401; gyroscopes, 221; launch control equipment building (LCEB), 120, 122–23; LCCs (launch control centers), 115–16, 118, 142, 151; payload operations, 361
Cooper, Gordon, 281
copper heat sink, 188
core-and-slotted-tube modified end burners, 170, 171f
cork materials, 177–78, 178t, 254, 263, 267, 307, 307f
Cornell University, 59, 188
Cornhusker Ordnance Plant (Nebraska), 87
Corona imagery, 247
Corporal guided missile, 48
Corps of Engineers: *see* Army Corps of Engineers
Corps of Engineers Ballistic Missile Construction Office (CEBMCO), 94, 126–27, 128, 129–30, 131t, 147
CREEK BED, 432t
cruise missiles, 62
Cuban Missile Crisis: Minuteman command-and-control systems, 340; Minuteman demonstration and acceptance procedures, 154; Minuteman program, 371–77; strategic alerts, 33, 277
Curate, 47
Cushman, Henry R., 316
CYH 77 propellant, 170
cylindrical grain propellants, 41, 43, 166, 172

D17B computer system, 213–14, 224f
D37 series computer system, 220, 223–24, 228f, 230, 231–32, 233f, 234, 266, 396
Davis-Monthan Air Force Base (Arizona): full-range operational base launches, 321; geographic location, 12f

DC 2106, 190
DEFCON declarations, 372–75; *see also* Cuban Missile Crisis
Defense Appropriations Bill (1975), 326
Defense Extraordinary Priority (DX) rating, 80
Defense Meteorological Satellite Program, 310
deflection of the vertical, 238–39
Dekker, A. O., 52
de Laval nozzle design, 42, 161, 166, 469n21
delay launch mode, 374
Del E. Webb Corporation, 129–30
Delta Guidance sequence, 234
De-MIRVing Treaty, 26, 27, 28, 30
demonstration and acceptance procedures, 154
Demonstration and Shakedown Operation (DASO) program: Force Modernization flight tests, 435t; Force Modernization Program, 295–98; Minuteman IA, 278–79; Minuteman IB, 282–84, 287, 430t; Minuteman II, 434t; Minuteman III, 301; wing facilities, 296, 297
Department of Defense: base location considerations and criteria, 88–89; Minuteman planned force levels, 18–20, 20t, 21t, 22t; warhead development, 198
Department of Defense Authorization Act (1986), 25
deployment position radar system, 303
depressed trajectories, 60, 175
Desimone property, Seattle, Washington, 150
deterrence policies, 78
DIAL RIGHT, 279
DICE SPOT, 433t
digital flight control systems, 206; *see also* guidance-and-control systems
DMCCC (deputy missile combat crew commander): ALCC missile combat crew, 366, 368f, 370; command-and-control systems, 343, 345, 345f, 347, 348, 349f, 351f, 352–53; Launch Enable System (LES), 374; Long Life I, 316, 318f
DOCK BELL, 286, 287
DOCK WORKER, 432t
DOUBLE BARREL, 430t
double-base propellants, 168, 170
Douglas/Goodyear, 197
Douglas, James, 68, 459n77
downrange bias, 234–35
DRAG CHUTE, 279
Draper Instrument Laboratory, 213, 221, 223
Drell Commission/Report, 396
Dubridge, Lee, 10
Duerksen, Robert L., 165
Duncan, Donal, 206
Dunn, Louis G., 8, 48, 59, 73

dust-hardening programs, 178*t*, 180*f*, 307–8, 307*f*, 308*f*, 312, 393–94
Dust Modification Performance program, 305, 307–8, 307*f*, 308*f*
Dyess Air Force Base (Texas), 12*f*

E-6B Mercury aircraft, 371
Eastern Europe strategic targeting plans, 36
Eastern Test Range (ETR), 225, 268–69, 292
Eaton Canyon, Pasadena, California, 47
EBONY ANGEL, 289, 432*t*
EC-135 aircraft, 365–66, 365*f*, 371
ECHO CANYON, 432*t*
ECHO HILL, 282, 284, 430*t*
economic targets, 34, 88
Edwards Air Force Base (California): launcher closure testing, 150; launch tube evaluation facilities, 94, 96; motor testing, 155; silo development and testing programs, 250–52, 250*f*, 251*f*, 252*f*, 253*f*, 254–55, 254*f*, 257; testing facilities construction, 79–80, 147
Edwards, Jack, 36
Eggers, A. J., Jr., 183, 186
Eisenhower administration: force level development and evolution, 80; Minuteman program force level development and evolution, 17–18, 72, 74–75, 78–80, 337; retaliation policies, 337; strategic missile systems, 8–11, 34, 55, 62, 67
Eisenhower, Dwight D., 55, 62, 67, 72
ejecta zone, 94, 94*t*
electrical surge arrester (ESA), 151
electromagnetic pulse (EMP), 388, 392–93
Elkton, Maryland, 159
Ellsworth Air Force Base (South Dakota): activation summary, 418*t*; ALCS-equipped aircraft, 366; base location considerations and criteria, 90, 339; command-and-control systems, 390; construction and acceptance milestone dates, 412*t*; construction operations, 125, 126*t*, 130, 131*t*, 142*f*, 145, 147; cutbacks and consolidations, 26; deactivation and dismantlement, 399, 400–403, 403*f*, 404*f*; excavation methods, 145; Force Modernization plans, 386, 387*t*; geographic location, 12*f*; groundwater intrusion problems, 393; launch azimuth and alignment determinations, 246–47; launch codes, 400; launcher equipment room (LER), 103; LCCs (launch control centers), 116*f*, 117*f*; Minuteman geographical statistics, 131*t*; Minuteman inventory, 25, 296, 399; missile range estimates, 39*f*; missile suspension systems, 389; operational base launch (OBL) programs, 315; operational flight and evaluation programs, 285; operational wing locations and field layouts, 89, 89*f*, 90; PCB contamination, 401; reentry vehicles, 201; short-range operational base launches, 316, 317*f*, 319; simulated electronic launch Minuteman (SELM), 334; site activation process, 152; *see also* Wing II
Elmendorf Air Force Base (Alaska), 91
Emergency Action Message (EAM), 340, 353, 354, 369
Emergency combat capability (ECC): DEFCON declarations, 372–74, 377; launch command procedures, 333, 347; launch control panels, 329, 346; Minuteman IA alert status, 24, 33; modified operational missiles (MOMS), 329, 333
Emergency Rocket Communications System (ERCS): battery activation, 361; Blue Scout Junior Program 279L (MER-6A), 356–57, 357*f*; communications payloads, 360–61; Data Transfer Unit (DTU), 358–59, 359*f*; deactivation, 363; equipment modifications and upgrades, 358; flight organization and nomenclature, 129*t*; Force Modernization flight tests, 435*t*; ground operations, 358–60; historical perspective, 340, 356; launch control facility (LCF), 358–60, 359*f*, 362*f*; Minuteman payload summary, 37*t*; Minuteman Program 494L, 357–63; operational flight and evaluation programs, 298, 299, 436*t*; operational test launches, 361, 362*f*, 363, 364*f*, 365; payload operations, 361, 362*f*
emergency war order (EWO) procedures, 153
ENABLE configuration and command: ALCC flight test program, 369, 370; basic process, 341–42; DEFCON declarations, 372–74; Minuteman IA, 343–44; Minuteman II, 352, 353; Secure Enable command, 353
end-burning propellants, 43, 46, 170, 454*n*55
Enderbury Island, 302, 303, 304*f*
Engineering Test Program (Category I): *see* Category I (Engineering Test Program)
Eniwetok Atoll: geographic characteristics, 276*f*; launch azimuth and alignment determinations, 96; as operational test target, 272, 274*t*, 278–83, 286, 290, 297, 299, 302
Environmental Protection Agency (EPA), 395, 401–2, 405
Enzi, Michael, 29
epoxy-polyamide resins, 177
Ercoupe NC286655, 44, 45*f*
ERCS (Emergency Rocket Communications System): *see* Emergency Rocket Communications System (ERCS)
estimated Soviet fatalities and industrial destruction, 22–23, 22*t*
ethylcellulose, 47

EXECUTE LAUNCH command: ALCC flight test program, 369, 370; basic process, 342; DEFCON declarations, 372–74; Minuteman IA, 343, 344–45; Minuteman II, 352
Expanded Execution Plans (EEP) program, 235
experimental solid-propellant vehicle (XSPV), 47, 48
Explicit Guidance scheme, 234, 309
exposure time, 92
external insulation, 177–78

FAINT CLICK, 296, 433*t*, 435*t*
Fairchild Air Force Base (Washington), 12*f*
fallout limits, 34
Fast B-Plug, 397
Federal Facilities Compliance Agreement, 402, 405
Ferguson, James, 63
F. E. Warren Air Force Base (Wyoming): activation summary, 419*t*; ALCS-equipped aircraft, 366; base location considerations and criteria, 91; command-and-control systems, 355*f*, 390, 392; construction and acceptance milestone dates, 415*t*; construction operations, 125, 126*t*, 131*t*; dust-hardening programs, 308, 393–94; excavation methods, 146, 149*f*; flight organization and nomenclature, 128; Force Modernization plans, 386, 387*t*; Geodetic Survey Squadron (1381 GSS), 238; geographic location, 12*f*; Guidance Replacement Program (GRP), 397; HEST (high-explosive simulation technique) testing, 380, 382, 382*f*; Integrated Improvement Program, 392, 393*t*; launch facility status, 30, 31*f*; Minuteman geographical statistics, 131*t*; Minuteman inventory, 24, 25, 27, 29; Minuteman modification program, 312; missile suspension systems, 388; operational base launch (OBL) programs, 322; operational wing locations and field layouts, 89*f*; Peacekeeper missile program, 25–26; reentry vehicle development and testing program, 190; SERV-Minuteman III deployment, 396
fiberglass-reinforced plastics, 188, 189, 190–91
FINE SHOW, 278
first-strike capability, 19
Fitzgerald, A. Ernest, 328
FIVE POINTS, 286, 287, 431*t*
five-point star-grain propellant design, 48–50, 50*f*
Flax, Alexander, 228
Fleet Ballistic Missile Program, 175
Fletcher Committee/Report, 103, 151–52, 277, 339–40, 374
Fletcher, James C., 339

flight control systems, 206, 215–16, 220; *see also* guidance-and-control systems
Flight Development Evaluation (FDE) program, 305
flight organization and nomenclature, 128, 129*t*
flight test and launch programs (Category II): *see* Category II (flight test and launch programs)
Flight Test Missile (FTM): launch failures, 277–78, 281, 282, 290–93, 295*f*, 305; Mark 12 reentry vehicle, 296–97; Minuteman IA, 198, 257, 258*f*, 259–62, 259*f*, 273, 274, 277–82, 421–22*t*; Minuteman IB, 262–64, 282–90, 292–93, 366, 423–25*t*; Minuteman II, 264–67, 265*f*, 295–300, 365–66, 369–71, 376–77, 426–27*t*; Minuteman III, 268, 300–306, 308–11, 313*f*, 428*t*; reentry vehicles, 203*f*
FLY BURNER, 432*t*
Follow-on Operational Test Program (FOT): ALCC flight test program, 371; Minuteman IA, 280–82, 289; Minuteman IB, 289–91, 291*f*, 432*t*
Folsom, California, 87
Forbes Air Force Base (Kansas), 12*f*
force level development and evolution: Bush (George H. W.) administration, 26; Bush (George W.) administration, 27–29; Clinton administration, 27; deployment plans, 17–18; Eisenhower administration, 17–18, 79, 80; estimated Soviet fatalities and industrial destruction, 22–23, 22*t*; Ford administration, 24–25; Johnson administration, 20–24, 21*t*, 22*t*; Kennedy administration, 18–20, 20*t*, 21*t*; Nixon administration, 24; Obama administration, 29–31; objectives and funding estimates, 70*t*; planned force levels, 19*t*; Reagan administration, 25–26
Force Modernization Program: air-conditioning systems, 123; command-and-control systems, 342, 347, 352, 390; Command Data Buffer (CDB) program, 99, 232–34, 352–54, 390–92; dust-hardened missiles, 308; flight test programs, 297–98, 319, 435*t*; force level recommendations, 23, 24–25; guidance-and-control systems, 226, 227, 370; Hill Engineering Test Facilities (HETF), 370; Integrated Improvement Program, 390; missile fleet modernization plans, 385–86; missile suspension systems, 101, 102, 387, 390*f*; modification completion dates, 398*t*; operational base launch (OBL) programs, 323; operational flight and evaluation programs, 436*t*; physical facilities, 377; Preparatory Launch Command (PLC) capability, 347–48, 354; purpose, 101; remote targeting capability,

390–91; wing facilities, 295–98, 370, 385–96, 387t; WS-133A-M Weapon System, 227, 296, 297–98, 385–86; *see also* Minuteman II; modernization and upgrade programs

force reduction: Minuteman II deactivation and dismantlement, 399–405; Minuteman III conversion program, 405; Minuteman III limited deactivation and dismantlement, 406, 407f, 407t

Ford administration, 24–25
Ford Aeronutronic, 197
Ford Aerospace, 353
Ford, Gerald R., 24
Forman, Edward S., 42, 43
Fort Collins, Colorado, 239
FOUR ACES, 290, 432t
four-point star-grain propellant design, 165
FOX TRAP, 435t
Francis E. Warren Air Force Base (Wyoming): *see* F. E. Warren Air Force Base (Wyoming)
free-rotor gas-bearing gyroscopes, 206–7, 208, 209, 210f, 211, 212f, 213, 476n13
Freon, 167, 169f, 172, 230, 264
Fuller, John, 60–61
Fuller-Webb Construction, 3, 150t
full first-strike capability, 19
full-range operational base launches, 321–26, 328
fungicide-impregnated cork materials, 307, 307f
Funk, Benjamin I., 78

G6B4 free-rotor gas-bearing gyroscope, 209, 210f, 211, 212f, 213; *see also* free-rotor gas-bearing gyroscopes
Gagan Island, 312
Gaither Committee/Report, 10–11, 65
Gaither, Horace Rowan, Jr., 10
GALCIT propellants, 44, 46
Gardner, Trevor, 8, 17
Gary, Indiana, 132
Gates, Robert, 337
Gates, Thomas, 34
GAY CROWD, 432t
Gemini V spacecraft, 281
General Accounting Office (GAO) report, 397
General American Transportation Corporation, 379
General Electric: ablative materials research, 191; Able RTV program, 192–94; Air Force reentry vehicle designations, 184t; electronic component reliability, 208; first-generation heatsink reentry vehicles, 190; Mark 2 heatsink reentry vehicle, 187f, 190; Mark 3 reentry vehicle, 190, 191, 194, 195, 197f; Mark 12 reentry system heat shield manufacturing procedures, 266, 309–10; Project Hermes, 48; RVX-2 reentry vehicle, 194–95

General Motors, Allison Division, 162
General Operational Requirement: *see* GOR-161, GOR-171
General Telephone Electronics/General Dynamics, 353, 356
Geodetic and Geophysical (G&G) Error Budget, 248
geodetic and gravimetric surveys, 237–38, 240f, 241f, 246
Geodetic Survey Squadron (1 GSS), 331
Geodetic Survey Squadron (1381 GSS), 237–38, 246
George A. Fuller Company, 129
GEORGIA BOY, 285, 286, 431t
Georgia Institute of Technology, 191
Gerrity, Thomas P., 126–27
Gervais, Rik, 247
GIANT BALL, 369, 370
GIANT BLADE, 297, 434t
Giant Boost program, 320, 332
GIANT FIST, 297–98, 435t
GIANT FOOT, 321, 324
GIANT MOON, 174, 363
GIANT PACE SELM test, 334–35, 334f, 369, 370
GIANT PATRIOT, 299, 322, 324, 325, 327f, 436t
GIANT PROFIT, 329–32
Giant Roar program, 320
Gilpatric, Roswell L., 338, 357
gimballed engines, 176
gimballed nozzles, 158, 161, 164f, 172
gimballed platforms, 220–21, 222f, 228f, 231, 233
GIN BABY, 433t
GI-T1-B azimuth gyroscope, 219, 221, 227, 233, 265
Glasgow Air Force Base (Montana), 89, 91
glass filament–wound motor cases, 62, 157, 168, 171f, 172
GLASS WAND, 282, 288
Glenn L. Martin Company, 55
glide trajectory, 183
GLORY TRIP, 298–99, 302–6, 310–11, 313f, 369
GLOWING SAND, 292–93
Goddard, Robert H., 41–42, 43
GOLD DUKE, 282
GOLDEN AGE, 297, 434t
Goldsworthy, Harry E., 151
Goodpaster, Andrew, 67
Goodyear Aircraft Corporation, 189, 197
GOR-161, 63, 64–65
GOR-171, 76, 77f, 79
Gorbachev, Mikhail, 26

Gordon, Kermit, 448*n*25
GRAB satellite: *see* Solar Radiation/Galactic Radiation and Background (GRAB) satellite
graduated deterrence, 78
grain propellant designs: basic concepts, 452*n*3; burning rates, 454*n*55; double-base propellants, 170; historical research, 43, 46, 74; research and development programs, 46–50, 50*f*, 160; Stage II motor development and testing, 165–66, 167*f*; Stage III motor development and testing, 170, 171*f*; *see also* star-grain propellant design
Grand Central Rocket Company, 60
Grand Forks Air Force Base (North Dakota): activation summary, 420*t*; ALCS-equipped aircraft, 366; base location considerations and criteria, 91; construction and acceptance milestone dates, 416*t*; construction operations, 126*t*, 131*t*, 140*f*, 144*t*; dust-hardening programs, 308, 394; electromagnetic pulse (EMP) susceptibility, 392; excavation and foundation operations, 146–47; Force Modernization plans, 387*t*; geographic location, 12*f*; groundwater intrusion problems, 393; HEST (high-explosive simulation technique) testing, 382–85, 384*f*; Integrated Improvement Program, 393*t*; Minuteman geographical statistics, 131*t*; Minuteman inventory, 24, 25, 27, 28, 296; missile conversion program, 405; missile deactivation and dismantlement, 406; missile deployment plans, 225–26, 245; missile suspension systems, 386, 388; modernization and upgrade programs, 347; modified operational missiles (MOMS), 329; operational wing locations and field layouts, 89*f*, 90; Project Restart, 392; reliability improvement program, 225; short-range operational base launches, 319, 320; weather impacts, 146–47; *see also* Wing VI
Grand Island, Nebraska, 87
GRAND RIVER, 281, 429*t*
GRAND TOUR, 278
Grant, Al, 221
graphite materials, 52, 63, 157–58, 161, 166, 172
gravimetric surveys and data, 238–39
gravitational trajectories, 328
Gray, Gordon, 79
Great Britain, 249, 302–3
Great Falls, Montana, 128
GREEN PEA, 432*t*
Green River, Utah, 293
Ground-Based Interceptor Program, 300
Ground-Based Strategic Deterrent (GBSD), 406, 409, 410
ground maintenance response (GMR), 341

ground test missiles (GTM), 251–52
groundwater intrusion problems, 393
GROVE HILL, 429*t*
Growth Minuteman (Minuteman IIB), 267
Guam Island, 91, 274*t*, 363
Guggenheim Aeronautical Laboratory, California Institute of Technology (GALCIT, Caltech), 42, 43, 45
guidance alignment survey monuments, 238, 241*f*
guidance-and-control systems: accuracy investigations, 305; alignment mechanisms, 104; all-inertial guidance systems, 205; bias errors, 305–6; contractor responsibilities, 84; DEFCON declarations, 377; design and development, 205–35; electromagnetic pulse (EMP) susceptibility, 392; estimated costs, 75, 79; flight test programs, 255; Force Modernization plans, 226, 227, 370; Guidance Replacement Program (GRP), 235, 305, 396–97; HEST (high-explosive simulation technique) testing, 382; historical research, 57; launch azimuth and alignment determinations, 239–48, 240*f*, 241*f*, 242*f*, 243*f*; launch failures, 257, 259, 277–78, 281; NS10 series guidance system, 205–16, 239; NS17 guidance-and-control system, 104, 216, 218–21, 223–28, 228*f*; NS50 guidance-and-control system, 305, 397; operational performance requirements, 76; quality control issues, 259; short-range operational base launches, 316; System Q, 66; umbilical retraction system, 102; Upgrade Silo modification program, 388; *see also* NS20 guidance-and-control system; targeting accuracy
Guidance Improvement Program (GIP), 234
Guidance Replacement Program (GRP), 235, 305, 396–97, 398*t*
Guidance System Evaluation program, 305, 308–9
Guidance Upgrade Program (GUP), 234–35
Guided Missile Group (6555 GMG), 257, 263–64
Gunder, Dwight, 59, 60
gyroscopes: azimuth gyroscopes, 219–20, 231, 235; azimuth laying set (ALS), 245–47; free-rotor gas-bearing gyroscopes, 206–7, 208, 209, 210*f*, 211, 212*f*, 213, 476*n*13; GI-T1-B azimuth gyroscope, 219, 221, 227, 233, 265; guidance-and-control systems, 206, 208, 209, 210*f*, 211, 212*f*, 213, 218–20; gyrocompass assembly (GCA), 397; pendulous integrating gyro-accelerometer (PIGA), 213, 219, 220, 221, 222*f*, 223, 226, 230, 231; portable gyrocompass assembly (PGA), 245; self-alignment technique calibration (SATCAL), 231; silo development and testing programs, 252

HAC/RMPE Software Support Facility (HSSF), 356
Hale, William, 41, 43, 46
Hall, Edward N.: launch facility criteria, 96; Minuteman program briefings, 3, 59–60, 67–69, 155, 248, 459*n*77; Minuteman program concept, 3, 17, 84, 85, 96, 123, 337; Minuteman program support facilities, 123–24; program name selection, 71, 458*n*61; silo launch test facilities, 65–66; solid propellant missile systems, 42, 59–62, 65–70, 250, 456*n*22; *see also* System Q
Hampton, Gerald E., 316, 317
Hampton, Virginia, 183
hardened intersite cable system (HICS), 108, 135, 154, 155, 329, 400, 402
hardening programs and specifications: communications systems, 140, 329, 343; computer systems, 223, 266; dust-hardening programs, 178*t*, 180*f*, 307–8, 307*f*, 308*f*, 312, 393–94; electromagnetic pulse (EMP) susceptibility, 388, 392; flight test programs, 426*t*; guidance-and-control systems, 266, 298; hardened-and-dispersed squadrons, 18–19, 79, 83, 127; hardened mobility, 79; launch control equipment building (LCEB), 114*f*, 118, 122; launchers, 205; launcher support building (LSB), 109, 111*f*; LCCs (launch control centers), 140, 271, 273, 379–80; LFs (launch facilities), 25, 88, 92, 271, 273, 278, 376; missile readiness, 76; missile suspension systems, 95; modification completion dates, 398*t*; reentry vehicles, 201, 298; silo launch test facilities, 67–68, 72, 79, 92; siting and facility design, 91–92, 94, 95*t*; solid propellant missile systems, 83; Soviet targets, 175; Soviet threat estimates, 91–92, 93*t*; targeting strategies, 66, 216; Upgrade Silo modification program, 388; wing facilities, 103, 108–9, 111*f*, 114*f*, 118, 271, 273, 384, 385
Hardness Verification Review Panel, 380
Hard Rock Silo (HRS) Program, 386–87; *see also* Upgrade Silo modification program
Harlowton, Montana, 405
Harned, Ralph, 168
Hastings, Nebraska, 86, 87
Hatfield, Mark, 326
Hausman, George F., 166
Hawaii, 91, 363
H. C. Smith Construction Company, 383
heading sensitivity testing (HST) program, 246, 247
heatsink cooling, 186
heatsink reentry vehicles, 183, 186, 187*f*, 190
heavy bombers, 29–30, 33, 43, 302

helicopter transportation, 152
Hemsley, R. T., 339
Henderson, Charles B., 52, 454*n*55
Hercules Powder Company: contractor responsibilities, 3, 60, 62, 74, 84; Minuteman motor designations, 164*t*; Minuteman production facilities, 87; motor case fabrication, 66, 157, 172; rocket propulsion research, 168, 170; solid propellant development, 63, 173, 174; Stage III motor development and testing, 164*t*, 168, 171*f*, 172, 173*t*, 174
Hermes A-1/A-2 missiles, 48–49
HEST (high-explosive simulation technique), 379–80, 381*f*, 382–85, 382*f*, 383*f*, 384*f*, 387
HEY DAY, 282
high-altitude testing, 55, 158–59
high-energy double-base propellants, 168, 170
Higher Authority Communication/Rapid Message Processing Element (HAC/RMPE), 353–54, 356
*The High Priests of Waste* (Fitzgerald), 328
Hill Air Force Base (Utah): ALCC flight test program, 369, 370–71; Minuteman III production, 24, 26; Minuteman III storage, 27, 406; missile assembly-and-recycle facilities, 87, 89, 153, 216, 226, 301; missile repairs, 319, 320; modification programs, 301, 316, 329
Hill Engineering Test Facilities (HETF), 370–71
Hnyde (unsymmetrical dimethylhydrazine), 189
Holaday, William, 17, 70, 74, 75, 78
Holifield, Chet, 10, 88
Homing Overlay Experiment (HOE) Program, 300, 432*t*
Honest John surface-to-surface missile, 57, 170
Honeywell, 223, 230, 232, 268, 305, 308–9
Hound Dog cruise missile, 213
House Armed Services Committee: defense appropriations, 23, 30, 328; Subcommittee on the Strategic Forces, 409–10
House Committee on Government Operations, 88
House Subcommittee on Military Construction, 127
House Subcommittee on Military Operations, 10, 88
House Subcommittee on Strategic Forces, 30
Howland Island, 303
Hull Island, 302, 303, 304*f*
Humfeld, Harold E., 319
Hunsaker, Jerome, 43
Hybrid Explicit Flight Program (HEFP), 234, 309, 392, 398*t*
Hybrid Explicit software/Hybrid Explicit guidance, 232, 234, 305, 309
hydrogen bomb tests, 302

hydrospin motor casings, 66
Hyland, Lawrence A., 8, 17
Hypalon fungicide, 177
hypergolic propellants, 176
Hyten, John, 409–10

ICBM Operations Software Sustainment Program (IOSSP), 356
ICBM Penetration Aids (IPA), 432*t*
Illeginni Island, 303
Improved Digital Computer Unit, 232, 308–9
Improved Launch Control System (ILCS), 227, 246, 352–54, 356, 392, 398*t*
Improved Minuteman (Minuteman II), 18, 20, 216, 264, 340, 436*t*; *see also* Minuteman II
Improved Minuteman Physical Security System (IMPSS), 394
industrial targets, 22*t*, 34, 88
inertial guidance systems: accuracy, 248; alignment mechanisms, 227, 244, 245, 247–48; contractor responsibilities, 206; development issues, 337; Explicit Guidance scheme, 234; flight test programs, 205; gyroscopes, 208, 213; high-reliability program, 208; intercontinental ballistic missiles (ICBMs), 237; Jupiter missile program, 475*n*2; modification programs, 300; self-contained systems, 9, 63; solid propellant missile systems, 57, 71, 235; stable platform design, 220–21; System Q, 66; targeting accuracy, 237; task force recommendations, 9; Thor-Able II program, 192; *see also* guidance-and-control systems; NS10 series guidance-and-control system; NS17 guidance-and-control system; NS20 guidance-and-control system
INFINITE TIME DELAY, 319
in-flight safety systems, 299
INHIBIT LAUNCH command: ALCC flight test program, 369, 370; basic process, 342–43; functional role, 277; launch modes, 274; Minuteman IA, 346; Minuteman II, 352
Initial Defense Communications Satellite Program, 175
initial yield, 76
integrated circuits, 223, 225, 227, 230–32, 301
Integrated Demonstration Flight (IDF)-1, 305
Integrated Improvement Program, 390, 392, 393*t*
intercontinental ballistic missiles (ICBMs): arms limitation negotiations, 24, 26–29, 36; base location considerations and criteria, 87–91, 338–39; delivery capabilities, 36; deterrence policies, 409–10; estimated Soviet threat, 93*t*, 363–64; flight times, 33; force objectives, 70*t*; funding estimates, 70–73, 70*t*, 75, 78; historical perspective, 8–10, 33; operational performance requirements, 76, 77*f*, 78; range estimates, 39*f*; rocket propulsion research, 41–53, 53*f*, 55–57, 59–81, 83; security modernization program, 397–98; strategic alerts, 33; strategic targeting plans, 33–36; targeting accuracy, 237–38; task force study and recommendations, 8–11; thermonuclear warheads, 57; *see also* Minuteman program; Polaris missile program
intermediate-range ballistic missiles (IRBMs), 8, 33, 55, 70, 79
internal grain propellant designs: historical research, 43, 46, 74; research and development programs, 46–50, 50*f*, 160; Stage II motor development and testing, 165–66, 167*f*; Stage III motor development and testing, 170; *see also* star-grain propellant design
interstages, 178–79, 180*f*, 252*f*, 263, 308*f*, 325, 327*f*
Inyokern, California, 49*f*

James, Deborah Lee, 30
Jason sounding rocket program, 255
Jennings, Al, 214
jet-assisted takeoff (JATO) research, 5, 43–46, 45*f*, 456*n*22
jetevators, 64, 67, 158, 159*f*
Jet Propulsion Laboratory (JPL), 45–49, 51
jet vanes, 158, 188
Johnson administration: force level development and evolution, 20–24, 21*t*, 22*t*; strategic missile systems, 35
Johnson, Lyndon B., 35, 448*n*25
Johnston Island, 292
Joiner, W. H., 44–45
Joint Army–Navy–Air Force Solid-Propellant Group, 52, 74
Joint Chiefs of Staff (JCS): Cuban Missile Crisis, 372–73; DEFCON declarations, 372–73; force and production level report, 80, 337; missile deployment plans, 18, 19, 22, 75, 338; strategic targeting plans, 34
Joint Strategic Target Planning Staff (JSTPS), 34
JPL-121 (polysulfide-based propellant), 47–48
Jupiter missile program: ceramics-based ablative materials, 191; flight test programs, 205; initial operational capability (IOC), 11; Jupiter-C program, 189, 190; Jupiter-S rocket, 57, 58*f*, 59, 61; logistics support, 79; retaliatory effectiveness, 71; rocket propulsion research, 55, 57, 58*f*, 60; second generation ablative reentry vehicle system, 183, 188–90

Kaiser Corporation, 94
Kansas Ordnance Plant, 87
Kantrowitz, Arthur, 188
Kauai, Hawaii, 91
Kaysen, Carl, 19
Kehler, C. Robert, 29–30
Keller, Kaufman T., 7–8
Kelly bars, 146
Kennan, George, 8
Kennedy administration: force level development and evolution, 18–20, 20*t*, 21*t*; strategic missile systems, 34–35, 338
Kennedy, John F., 18, 19, 34–35, 372
Kiewit and Sons' Company: *see* Peter Kiewit and Sons' Company
Killian Committee/Report, 10, 55, 72
Killian, James R., 10, 67, 72, 79
Kincaid, John F., 170
King, Claude, 214
Kirtland Air Force Base (New Mexico), 380, 381*f*
Kissinger, Henry, 35, 325
Kistiakowsky, George B., 8, 34, 67, 72, 74, 337
Knox, Robert, 214, 221
Kravitsky, Stephen, 307*f*
Kwajalein Atoll: cloud sampling patterns, 310, 311*f*; Emergency Rocket Communications System (ERCS), 363; geographic characteristics, 275*f*; launch azimuth and alignment determinations, 96; as operational test target, 272, 274*t*, 291, 293, 297, 299, 302, 303, 305; Safety-Enhanced Reentry Vehicle (SERV) program, 312; signal coverage, 363; weather conditions, 310, 311
Kwajalein Missile Range, 303

LACE STRAP, 432*t*
Ladish Forging Company, 160
Laird, Melvin, 24, 25, 324
Langley Aeronautical Laboratory (Virginia), 183, 186, 190
"lard tub," 50
Large Ballistic Reentry Vehicle (LBRV), 432*t*
Large Solid Rocket Feasibility Program, 61, 66, 157, 159, 165, 168
Larson Air Force Base (Washington): geographic location, 12*f*; operational wing locations and field layouts, 89
latitude and longitude measurements, 238–39
launch azimuth and alignment determinations, 239–48, 240*f*, 241*f*, 242*f*, 243*f*
launch codes, 316, 319, 342, 370, 373, 374, 376, 400
launch control equipment building (LCEB): air-conditioning systems, 116, 120, 122–23; blast protection, 122; construction material requirements, 144*t*; construction operations, 142, 145*f*; cut-away drawing, 123*f*; floor plan, 121*f*; Force Modernization plans, 386; hardening programs, 95; HEST (high-explosive simulation technique) testing, 382, 383*f*; launch control facility (LCF), 112; modernization and upgrade programs, 386; operational facility design and criteria, 95, 96; siting and facility design, 95, 113*f*, 115*f*, 118, 120, 120*f*, 121*f*, 122–23, 123*f*; wing facilities, 112, 113*f*, 114*f*, 115*f*, 118, 120, 120*f*, 122–23
launch control facility (LCF): blast doors, 141–42, 141*f*, 142*f*; components, 112; construction material requirements, 144*t*; construction operations, 151, 153, 155, 295; deactivation and dismantlement, 400, 402, 403; DEFCON declarations, 372, 375; Emergency Rocket Communications System (ERCS), 358, 359*f*, 362*f*; estimated travel times, 152; excavation and foundation operations, 139–41, 140*f*, 141*f*, 142*f*, 143*f*; flight organization and nomenclature, 129*t*; geographic locations, 273*f*; Inhibit command, 346; interior structural and mechanical work, 141–42; modernization and upgrade programs, 319–20; modification programs, 398; operational facility design and criteria, 94; operational readiness training (ORT), 152, 153; security systems, 103; site separation and dispersal recommendations, 92; siting and facility design, 95, 112, 113*f*, 114*f*, 115*f*, 403; standby missile combat crews, 372; typical site plan, 114*f*, 115*f*; wing facilities, 112, 113*f*, 128
Launch Control Panel/Console: Airborne Launch Control System (ALCS), 367*f*; DEFCON declarations, 373–74; Emergency Rocket Communications System (ERCS), 359*f*; Inhibit Launch switch, 374; Minuteman IA, 343–46, 344*f*, 345*f*; Minuteman II, 348*f*, 349*f*, 350*f*, 351*f*, 355*f*; technical manuals, 437–44
launch control support building (LCSB): air-conditioning systems, 115–16; construction material requirements, 144*t*; construction operations, 143–44; Force Modernization plans, 386; hardening programs, 95; siting and facility design, 95, 112, 113*f*, 115–16; wing facilities, 112, 113*f*
Launch Enable Logic Unit (LELU), 340
Launch Enable System (LES), 374–75, 376
launcher closure: construction material requirements, 144*t*; construction operations, 135–36, 137*f*; debris bin system, 387–88, 389*f*;

DEFCON declarations, 373–74, 376; door revetment, 109*f*; door tracks, 107*f*; flight test programs, 277; hardness and overpressure improvements, 388; HEST (high-explosive simulation technique) testing, 385; missile deactivation and dismantlement, 401, 403, 403*f*, 406, 407*f*; modified operational missiles (MOMS), 329–31, 335*f*; siting and facility design, 105–8, 105*f*, 107*f*, 109*f*; testing programs, 150; Upgrade Silo modification program, 387–88, 389*f*

launcher equipment room (LER): azimuth laying set (ALS) testing, 246; backfill operations, 135–38; construction material requirements, 144*t*; construction operations, 130, 132, 134, 135, 135*f*, 136, 136*f*; deactivation and dismantlement, 404*f*; development and testing programs, 150–51; electromagnetic pulse (EMP) susceptibility, 392; excavation and foundation operations, 134, 135*f*, 136*f*, 147; groundwater intrusion problems, 393; HEST (high-explosive simulation technique) testing, 382–83; interior structural and mechanical work, 138–39; missile suspension systems, 105*f*; security modernization program, 397; shock-isolated floors, 389–90, 391*f*; simulated electronic launch Minuteman (SELM), 333; siting and facility design, 94, 98–99, 100, 102–4, 104*f*, 105*f*; underground layout, 98*f*; Upgrade Silo modification program, 387, 389–90, 391*f*; upper level layout, 104*f*

launchers, 96, 97*f*, 98*f*, 130, 132, 144*t*

launcher support building (LSB): construction operations, 130, 132, 134, 135; deactivation and dismantlement, 401, 402; electromagnetic pulse (EMP) susceptibility, 392; excavation and foundation operations, 134, 147; facility characteristics, 97*f*; hardening programs, 95; interior structural and mechanical work, 139; operational facility design and criteria, 95, 96; siting and facility design, 95, 96, 97*f*, 108–11, 110*f*, 111*f*

launch facility radio test (LFRT), 369–70

launch facility security system, 111

Launch Monitor Panel, 368*f*

launch tubes: deactivation and dismantlement, 402, 404*f*, 406; excavation and foundation operations, 130, 132, 133*f*, 134, 146, 149*f*; Force Modernization plans, 385; groundwater intrusion problems, 393; interior structural and mechanical work, 138–39; launch azimuth and alignment determinations, 377; operational facility design and criteria, 96, 98–99; siting and facility design, 96, 98–99, 98*f*, 105*f*; underground layout, 98*f*; Upgrade Silo modification program, 388; *see also* guidance-and-control systems; missile suspension systems

Lauritsen, Charles C., 8, 17

Lauritsen Committee/Report, 17–18, 90, 338–39

Lawrence, E., 52

LCCs (launch control centers): ALCC flight test program, 370; attack scenarios, 92; base location considerations and criteria, 90, 91; blast doors, 141–42, 141*f*, 142*f*, 147; blast valves, 114*f*, 116, 122, 140, 142, 144*t*, 380, 403; command and control systems, 337–38; command-and-control systems, 392; Command Console, 350*f*; Communication Console, 349*f*; construction material requirements, 144*t*; deactivation and dismantlement, 400, 403, 405; DEFCON declarations, 372–75, 377; design considerations and criteria, 92–94; development and testing programs, 150–51; Emergency Rocket Communications System (ERCS), 358–60, 362*f*; estimated Soviet threat, 93*t*; excavation and foundation operations, 139–41, 140*f*, 141*f*, 142*f*, 143*f*; flight organization and nomenclature, 128, 129*t*; floor plan, 116*f*; Force Modernization plans, 385–86, 390; hardening studies and specifications, 91–92, 95*t*, 271; hardness design parameters, 379–80, 382; interior structural and mechanical work, 141–42, 147; launch control facility (LCF), 113*f*, 114*f*, 115*f*; Launch Control Panel/Console, 344*f*, 345*f*, 348*f*, 355*f*, 373–74; launch design implementation changes, 340; long-range plans, 394; Minuteman operational requirements, 84, 86; missile conversion program, 405; modernization and upgrade programs, 347; modified operational missiles (MOMS), 329, 332; operational facility design and criteria, 94–96, 112, 115–18, 116*f*, 117*f*, 119*f*; operational flight and evaluation programs, 271–74; Rapid Execution and Combat Targeting (REACT) Program, 353–54, 355*f*, 398*t*; remote targeting capability, 390–91; security systems, 111; short-range operational base launches, 316–17, 318*f*, 319–20; simulated electronic launch Minuteman (SELM), 333; site separation and dispersal recommendations, 91–92, 93*t*; Status Console, 351*f*; survivability criteria, 95, 95*t*; typical site plan, 114*f*, 115*f*

LeMay, Curtis, 22, 70, 88, 90, 125–26, 448*n*25, 459*n*77

Lemnitzer, Lyman L., 35

Leonard, Arthur S., 51

LFs (launch facilities): ALCC flight test program, 369–70; arms limitation negotiations, 24, 26–29; attack scenarios, 92; azimuth laying set (ALS) testing, 246; base location considerations and criteria, 90, 91; Command Console, 350*f*; Communication Console, 349*f*; concrete enhancement program, 397; deactivation and dismantlement, 400, 401–3, 403*f*, 404*f*, 405; DEFCON declarations, 372–75, 377; deployment area criteria, 89; design considerations and criteria, 92–94; Emergency Rocket Communications System (ERCS), 358–60, 359*f*, 362*f*; estimated costs, 75; estimated Soviet threat, 93*t*; facility characteristics, 96, 97*f*; flight organization and nomenclature, 128, 129*t*; Force Modernization plans, 385–86, 390; geodetic and gravimetric surveys, 237–39; groundwater intrusion problems, 393; guidance alignment survey monuments, 238, 239, 241*f*, 242*f*; hardening studies and specifications, 91–92, 94, 95*t*, 271; hardness design parameters, 379–80, 382–83, 386–88; launch azimuth and alignment determinations, 239–45, 240*f*, 241*f*, 243*f*; launch control facility (LCF), 95, 112, 113*f*, 114*f*, 115*f*, 403; Launch Control Panel/Console, 344*f*, 345*f*, 348*f*, 355*f*; launch control support building (LCSB), 95, 112, 113*f*, 115–16; launch design implementation changes, 340; launcher closure, 105–8, 105*f*, 107*f*, 109*f*, 135–36, 137*f*; long-range plans, 394; Minuteman development and testing programs, 255, 256*f*, 257, 258*f*, 259–69, 259*f*, 261*f*; missile conversion program, 405; modified operational missiles (MOMS), 329–31; operational capabilities, 72, 73; operational facility design and criteria, 94–96; operational flight and evaluation programs, 271–74; operational status, 26–30, 31*f*; Peacekeeper missile program, 25–26, 30; programmed depot maintenance (PDM), 30–31; Project Restart, 392–93; remote targeting capability, 390–91; remote visual assessment programs, 397–98; security systems, 111, 397; short-range operational base launches, 316–17, 319–20; silo launch facilities, 23, 65–67; simulated electronic launch Minuteman (SELM), 333; site separation and dispersal recommendations, 91–92, 93*t*; siting and facility design, 86; Status Console, 351*f*; survivability criteria, 95, 95*t*; targeting accuracy, 237–38; Upgrade Silo modification program, 386–88; *see also* construction operations; launch control equipment building (LCEB); launcher equipment room (LER); launcher support building (LSB); launch tubes

limited-attack scenarios, 35

Limited Treaty Banning Nuclear Weapon Tests in the Atmosphere, Outer Space, and Under Water, 379

*The Limits of Safety* (Sagan), 371

Lincoln Air Force Base (Nebraska), 12*f*
liquid-injection thrust vector control (LITVC), 166–67, 169*f*, 172, 263, 264, 301
liquid oxygen, 43
liquid propellants: historical research, 41–44, 51, 55–56, 58*f*; Jupiter missile program, 55, 58*f*; operational challenges, 55–56, 71; propulsion system rocket engines (PSREs), 176; Titan II missile, 11; *see also* Polaris missile program
lithium oxide, 52
Little Rock Air Force Base (Arkansas), 12*f*
Litton Systems, 300
Livermore Radiation Laboratory, 57
Lockheed Martin, 353, 356, 409
Lockheed Missiles and Space Company, 56, 65
lofting, 99
Long Life I, 316–17, 317*f*, 318*f*, 319
Long Life II, 319–20, 332
long-range guided missiles: arms limitation negotiations, 24, 26–29, 36; base location considerations and criteria, 87–91, 338–39; Bush (George H. W.) administration, 26; Bush (George W.) administration, 27–29; Clinton administration, 27; cutbacks, consolidations, and deactivation, 26–30; delivery capabilities, 36; Eisenhower administration, 8–11, 17–18, 34, 55, 62, 67, 72, 337; estimated Soviet threat, 93*t*, 363–64; force objectives, 70*t*; Ford administration, 24–25; funding estimates, 70–73, 70*t*, 75, 78; historical perspective, 5–11; Johnson administration, 20–24, 35; Kennedy administration, 18–20, 20*t*, 21*t*, 34–35; Nixon administration, 24, 35–36; Obama administration, 29–31; operational performance requirements, 76, 77*f*, 78; propulsion systems, 45; range estimates, 39*f*; Reagan administration, 25–26; rocket propulsion research, 41–53, 53*f*, 55–57, 59–81, 83; security modernization program, 397–98; strategic alerts, 33; strategic targeting plans, 33–36; thermonuclear warheads, 57; Truman administration, 7–8; *see also* Minuteman program; Polaris missile program
LONG SHOT, 288
LOOKING GLASS, 366, 371
Loral Command and Control Systems, 353
Los Alamos National Laboratory, 59
Lowry Air Force Base (Colorado): base location considerations and criteria, 91; geographic location, 12*f*
LOW TREE, 433*t*
LP-2 polymer, 46

M55/M55A1 (Thiokol), 159–62, 164*t*, 165*t*
M56/M56A1 (Aerojet), 164*t*, 165–68, 169*t*

M57/M57A1 (Hercules), 164*t*, 173, 173*t*
Mace tactical cruise missile, 457*n*40
Magna, Utah, 87
magnesium additives, powdered, 51, 52, 53*f*
magnesium honeycomb reinforced ceramics, 191
magnetic logic digital differential analyzer (DDA) computers, 206
Mahoney, William C., 247
Maine, 91
Malina, Frank J., 42, 43, 44, 45
Malmstrom Air Force Base (Montana): activation summary, 417*t*; ALCS-equipped aircraft, 366; alert status, 155; base location considerations and criteria, 89–90, 91, 338; construction and acceptance milestone dates, 411*t*, 416*t*; construction operations, 125, 126*t*, 131*t*; Cuban Missile Crisis, 371, 372–76; cutbacks, consolidations, and deactivation, 26, 27, 28–29; deactivation and dismantlement, 399, 405, 407*f*; Demonstration and Shakedown Operation (DASO) program, 278–79; dust-hardening programs, 308, 394; electromagnetic pulse (EMP) susceptibility, 392; Emergency Rocket Communications System (ERCS), 363; excavation and foundation operations, 146; facility characteristics, 97*f*; Force Modernization plans, 386, 387*t*, 405; geographic location, 12*f*; Guidance Replacement Program (GRP), 397; Integrated Improvement Program, 393*t*; launch azimuth and alignment determinations, 246; launch codes, 400; launch control facility (LCF), 113*f*; launcher equipment room (LER), 103; launch facility status, 30, 31*f*; Minuteman geographical statistics, 131*t*; Minuteman inventory, 24, 25, 27, 28, 29, 33, 296, 326, 399, 405; missile conversion program, 405; missile deployment plans, 17, 18; missile range estimates, 39*f*; missile suspension systems, 99, 388; modernization and upgrade programs, 347; operational base launch (OBL) programs, 315, 321, 323–24, 326, 327*f*, 328; operational readiness training (ORT), 373; operational wing locations and field layouts, 89, 89*f*; Propulsion Replacement Program (PRP), 395; simulated electronic launch Minuteman (SELM), 334; site activation process, 151–54; strategic alerts, 33; weather impacts, 144–45; *see also* Strategic Missile Squadron (564 SMS); Strategic Missile Wing (341 SMW); Wing I
Maneuvering System Technology (MaST), 432*t*
manganin, 208
manned aircraft, 6, 7, 23
manned silo/launch control complexes, 3
Mansfield, Mike, 326, 328

MAPLE GROVE, 287, 288
March Air Force Base (California), 366
March Field, Riverside, California, 44
Mark 1 reentry vehicle, 183, 184*t*, 186, 299–300, 436*t*
Mark 2 reentry vehicle, 183, 184*t*, 186, 187*f*, 190, 192
Mark 3 reentry vehicle, 184*t*, 190, 191, 194, 195, 197*f*
Mark 4 reentry vehicle, 184*t*, 190, 191, 194, 195–96, 197*f*, 199*f*, 293
Mark 5 reentry vehicle: Air Force designations, 184*t*; characteristics, 203*f*; comparison chart, 180*f*; DEFCON declarations, 377; deployment history, 203*t*; flight test programs, 196–98, 199*f*, 200*f*, 273, 278, 279–81, 285, 286, 287, 423*t*; full-range operational base launches, 321; impact analyses, 289; Minuteman planned force levels, 23, 24; Mixed Marble II test program, 431*t*; Operational Testing (OT) program, 431*t*; payload summary, 37*t*; range, accuracy, and reliability estimates, 37*t*; reentry effects, 200*f*
Mark 6 reentry vehicle, 184*t*, 194–95
Mark 7 reentry vehicle, 184*t*
Mark 8 reentry vehicle, 184*t*
Mark 9 reentry vehicle, 184*t*
Mark 10 reentry vehicle, 184*t*
Mark 11 reentry vehicle: Air Force designations, 184*t*; attitude control, 201; characteristics, 203*f*; circular error probable (CEP) estimates, 229, 290–92; comparison chart, 180*f*; Demonstration and Shakedown Operation (DASO) program, 430*t*; deployment history, 203*t*; Emergency Rocket Communications System (ERCS), 358, 359, 360; fabrication processes, 201; flight test programs, 198, 201–2, 262, 282–87, 290, 296, 297, 422*t*, 423*t*, 426*t*; Follow-on Operational Test Program (FOT), 432*t*; full-range operational base launches, 321, 324; impact analyses, 289; improvement plans, 229; launch failures, 297; Minuteman planned force levels, 23, 24; Mixed Marble II test program, 431*t*; operational test flights, 290–92; Operational Testing (OT) program, 431*t*; payload summary, 37*t*; penetration aids, 300; range, accuracy, and reliability estimates, 37*t*; short-range operational base launches, 316
Mark 12 reentry vehicle: Air Force designations, 184*t*; circular error probable (CEP) estimates, 229; deployment history, 203*t*; Dust Modification Performance program, 308*f*; flight test programs, 266, 268, 296–97, 303–4, 426–27*t*; heat shield evaluation tests,

309–10; improvement plans, 228–29; multiple independently targetable reentry vehicles (MIRVs), 175–76, 229, 268; Natural Hazards Program, 310; payload summary, 37t; range, accuracy, and reliability estimates, 37t; Reentry Vehicle Performance/Nose Tip Redesign program, 305, 309–11; research and development programs, 433t; Safety-Enhanced Reentry Vehicle (SERV) Program, 305, 312, 396; Single Reentry Vehicle (SRV) Program, 396; weapon system testing, 296–97; weather impacts, 310–11, 311f; *see also* Minuteman III

Mark 13 reentry vehicle, 184t
Mark 14 reentry vehicle, 184t
Mark 15 reentry vehicle, 184t
Mark 16 reentry vehicle, 184t
Mark 17 reentry vehicle, 184t
Mark 18 reentry vehicle, 184t
Mark 19 reentry vehicle, 184t
Mark 20 reentry vehicle, 184t
Mark 21 reentry vehicle: Air Force designations, 184t; guidance-and-control systems, 27; payload summary, 37t; Safety-Enhanced Reentry Vehicle (SERV) program, 312; Single Reentry Vehicle (SRV) Program, 396
Mark 500 Evader, 293–94
Mark 56 reentry vehicle, 285
Mark, J. Carson, 59
Marquardt Aircraft Company, 190, 191
Marshall Islands, 272; *see also* Eniwetok Atoll; Kwajalein Atoll
Martens, Allan R., 316–17, 318f
Martin Marietta, 175
Massachusetts Institute of Technology (MIT), 5, 43
Matador tactical cruise missile, 457n40
Maui, Hawaii, 91, 303
McCall, Tom, 326
MCCC (missile combat crew commander): ALCC missile combat crew, 366, 367f, 370; command-and-control systems, 344f, 345–46, 348f, 350f; Long Life I, 316, 318f
McConnell Air Force Base (Kansas), 12f
McConnell, John P., 384
McCorkle, Charles M., 69–70, 72
McCormack, James, 8
McDonnell Aircraft Corporation, 197
McElroy, Neil, 3, 67, 68, 459n77
McEwan, William S., 51, 74
McMahon, Brien, 7
McMillan, Frank M., 46
McMurran, Marshall, 214
McNamara, Robert S.: Airborne Launch Control System (ALCS), 365; force level recommendations, 18–24, 338, 339, 448n25; Force Modernization Program, 385, 386; hardness design studies, 380; Mark 12 reentry system approval, 229; Minuteman II improvements, 267–68, 340; Minuteman III approval, 176, 229, 268; Minuteman inventory recommendations, 91; Minuteman production facilities, 87; operational base launch (OBL) programs, 315, 321; Senate committee testimony, 363–64; strategic targeting plans, 35

McPeak, Merrill, 26, 399
mean time between failure (MTBF), 207, 226, 228
Medford, Oregon, 325
Medvedev, Dmitry, 29
melamine: *see* fiberglass-reinforced plastics
metal additives, powdered, 51–53, 53f, 63, 160, 165, 173–74
metallized propellants: *see* powdered metal additives
metal motor cases, 157
Metcalf, Lee, 326
Midcourse Optical Station (Hawaii), 303
Midway Island chain, 272, 274t, 292
military targets, 34, 38f, 88
Miller, Fred S., 44
Millikan, Clark B., 8, 17, 74–75
Millikan, Robert, 43
Mills, John, 246
Mills, Mark M., 44
MILSTAR communication satellite, 363
Minimum Reaction Time (MRT) target change method, 219, 220, 266, 426–27t
Minot Air Force Base (North Dakota): activation summary, 418t; ALCS-equipped aircraft, 366, 369; base location considerations and criteria, 90; construction and acceptance milestone dates, 413t; construction operations, 126t, 131t; dust-hardening programs, 308, 394; excavation methods, 145–46; Force Modernization plans, 386, 387f; geographic location, 12f; groundwater intrusion problems, 393; Guidance Replacement Program (GRP), 397; Integrated Improvement Program, 393t; launch facility status, 30, 31f; Minuteman geographical statistics, 131t; Minuteman inventory, 25, 27, 29; missile installation, 232; missile range estimates, 39f; missile suspension systems, 388; modified operational missiles (MOMS), 332; operational flight and evaluation programs, 285; operational status, 410; operational wing locations and field layouts, 89f, 90; weather impacts, 145; *see also* Wing III
Minton, A. M., 129
Minuteman Bench Test Program, 333
Minuteman Command Control System (MICCS), 390

Minuteman Configuration Control Board, 128
Minuteman Extended Survivable Power (MESP) modification, 104, 394
Minuteman Flexibility and Safety Group, 339
Minuteman I: airframe components and production, 179, 179t; astronomic azimuth, 239; autocollimators, 241–42, 245; computer systems, 213–14, 224f; correlation chart, 16t; external insulation, 177–78; flight test programs, 255, 256f, 257, 258f, 259–64, 259f; force level recommendations, 20, 21t, 22, 22t, 23, 24; high-altitude testing, 159; interstages, 179, 180f; launch azimuth and alignment determinations, 239–45, 242f; LCCs (launch control centers), 271; missile rotation, 244; motor designations, 164t; motor specifications, 165t; NS10 series guidance system, 205–16; operational test targets, 272; projected service life, 155; propellant processing, 158; research and development programs, 257, 259–64; site design considerations and criteria, 93; Stage I motor development and testing, 157–62, 162f, 163f, 164f, 164t; Stage II motor development and testing, 164t, 165–68, 166; Stage III motor development and testing, 164t, 171f; Stage I/Stage II motor casings, 92, 157; survivability criteria, 95t; targeting accuracy, 239–45; technological challenges, 157–59

Minuteman IA: Air Force reentry vehicle designations, 184t; airframe components and production, 179t, 180f; base location considerations and criteria, 89–90, 338; command-and-control systems, 343–46, 344f, 345f; Demonstration and Shakedown Operation (DASO) program, 278–79; external insulation, 177–78; flight sequence, 280t; flight test programs, 198, 257, 258f, 259–62, 259f, 261f, 421–22t; Follow-on Operational Test Program (FOT), 280–82, 289; force level recommendations, 23, 31f; Force Modernization plans, 385; full-range operational base launches, 321; guidance-and-control systems, 209, 226; launch facility designations, 272t; launch modes, 374; LCCs (launch control centers), 271, 343–46; missile comparison diagram, 58f; missile suspension systems, 101; modification programs, 24, 293; nozzle control units, 215–16; operational flight and evaluation programs, 272–74, 277–82, 280t, 429t; Operational Testing (OT) program, 279–80; payload summary, 37t; projected service life, 155, 379; range, accuracy, and reliability estimates, 37t, 39f; reentry vehicles, 198, 203f; site activation process, 153, 154, 155; skirts, 179, 180f; Stage I motor development and testing, 162, 164f, 262; Stage II motor development and testing, 166, 262; Stage III motor development and testing, 172, 262; strategic alerts, 33; targeting activation procedures, 244–45; test targets, 272; warhead packages, 35; *see also* Mark 5 reentry vehicle

Minuteman IB: activation summary, 418t; Advanced Ballistic Reentry System (ABRES) Program, 293; Air Force reentry vehicle designations, 184t; airframe components and production, 179t, 180f; Army Special Test Program HK (homing kill), 294; circular error probable (CEP) estimates, 229; Cold/Heat Soak program, 292–93; command-and-control systems, 346–47; Demonstration and Shakedown Operation (DASO) program, 282–84, 287, 430t; external insulation, 177–78; flight test programs, 202, 262–64, 423–25t; Follow-on Operational Test Program (FOT), 289–91, 291f, 432t; force level recommendations, 20, 23, 25, 31f; Force Modernization plans, 385; full-range operational base launches, 321; guidance-and-control systems, 209, 226, 263; Hill Engineering Test Facilities (HETF), 370; impact analyses, 288–89, 290; in-flight problem areas, 288; launch facility designations, 272t; LCCs (launch control centers), 271; Minuteman Reentry System Launch Program (RSLP), 293, 294f; missile suspension systems, 101; Mixed Marble II test program, 285–86, 431t; modification programs, 24; nozzle control units, 215–16; operational flight and evaluation programs, 282–94, 366; Operational Testing (OT) program, 285, 431t; payload summary, 37t; projected service life, 155, 379; range, accuracy, and reliability estimates, 37t, 39f; Reentry Body Supplemental Flight Test Program, 293–94; Reentry System Launch Program, 432t; reentry vehicles, 201, 202, 203f; Safeguard System Target Test Program, 293; short-range operational base launches, 315–17, 319; skirts, 179, 180f, 263; Stage I motor development and testing, 162, 164f, 262–64, 282; Stage II motor development and testing, 166, 262–64; Stage III motor development and testing, 172, 262–64, 282; strategic alerts, 33; targeting activation procedures, 244–45; warhead packages, 35

Minuteman II: Accuracy, Reliability, Supportability Improvement Program (ARSIP), 227–28; activation summary, 420t; Air Force reentry vehicle designations, 184t; airframe components and production, 179t, 180f; autocollimators, 242; circular error probable (CEP) estimates, 229;

command-and-control systems, 347–48, 348f, 349f, 350f, 351f, 352–54, 356; Command Console, 350f; Communication Console, 349f; computer systems, 223–24, 224f, 228f; correlation chart, 16t; cutbacks, consolidations, and deactivation, 26; deactivation and dismantlement, 371, 396, 399–400, 407t; Demonstration and Shakedown Operation (DASO) program, 434t; electronic component reliability, 227–28; Emergency Rocket Communications System (ERCS), 298, 299, 357–63, 436t; estimated Soviet fatalities and industrial destruction, 22–23, 22t; flight reliability problems, 173–74; flight test programs, 426–27t; force level recommendations, 18, 20–23, 21t, 22t, 24, 25, 26, 31f; Force Modernization flight tests, 435t; Force Modernization plans, 385–86, 387t, 390, 436t; full-range operational base launches, 321–26, 328; guidance-and-control systems, 104, 216, 218–21, 223–28, 228f, 246; HEST (high-explosive simulation technique) testing, 382; Hill Engineering Test Facilities (HETF), 370; Improved Launch Control System (ILCS), 227, 246, 392; improvement plans, 227–29; launch azimuth and alignment determinations, 242, 242f, 245–47; launch facility designations, 272t; launch failures, 319–20; LCCs (launch control centers), 271, 347, 352, 390; long-range plans, 394; missile rotation, 243–44; missile suspension systems, 99, 101; mobile range safety system (MRSS), 324–26; modernization and upgrade programs, 295, 298, 347; modified operational missiles (MOMS), 328–32; motor designations, 164t; motor programmed depot maintenance (PDM) process, 394–95; motor specifications, 165t, 169t; Multi-Service Launch System (MSLS) program, 300; nozzle control units, 224–25; NS17 guidance system, 216, 218–21, 223–28, 245, 296; Operational Base Launch Safety System (OBLSS), 322–25, 323f; operational flight and evaluation programs, 295–300, 365–66, 436t; operational test targets, 272; payload summary, 37t; penetration aids, 299–300; post-boost control systems (PBCSs), 229; projected service life, 155, 379; range, accuracy, and reliability estimates, 37t, 39f; Rapid Execution and Combat Targeting (REACT) Program, 353–54, 355f, 356, 398t; reliability improvement program, 225; research and development programs, 264–67, 433t; short-range operational base launches, 319–20; simulated electronic launch Minuteman (SELM), 325, 332–35, 334f; software status and authentication system, 227; special flight programs, 299–300; special test missiles (STMs), 298, 436t; Stage II motor development and testing, 164t, 166, 167f, 169f; Stage III motor development and testing, 164t, 171f, 172; Status Console, 351f; survivability criteria, 95t

Minuteman III: Advanced Inertial Reference Sphere (AIRS), 305, 306; Air Force reentry vehicle designations, 184t; airframe components and production, 178, 179t, 180f; arms limitation negotiations, 23, 26–29; autocollimators, 242; azimuth alignment, 235; circular error probable (CEP) estimates, 229; command data buffer (CDB) program, 232–33; computer systems, 223–24, 224f, 228f, 230, 231–32, 233f, 234, 396; correlation chart, 16t; cost effectiveness, 23; cutbacks, consolidations, and deactivation, 26–30; delivery capabilities, 36; Demonstration and Shakedown Operation (DASO) program, 301; deployment improvements, 232–35; downrange bias, 234–35; dust-hardening programs, 393–94, 398t; Dust Modification Performance program, 305, 307–8, 307f, 308f; estimated service life, 409; Expanded Execution Plans (EEP) program, 235; Flight Development Evaluation (FDE) program, 305; flight reliability problems, 174; flight test programs, 428t; force level recommendations, 23, 24, 25, 26, 28–29, 31f; Force Modernization plans, 385–86, 387t, 390, 409; guidance-and-control systems, 222f, 228–35, 228f, 233f, 246, 301, 305; Guidance Improvement Program (GIP), 234; Guidance Replacement Program (GRP), 235, 305, 396–97, 398t; Guidance System Evaluation program, 305, 308–9; Guidance Upgrade Program (GUP), 234–35; Hill Engineering Test Facilities (HETF), 370; Hybrid Explicit Flight Program (HEFP), 392; Hybrid Explicit software/Hybrid Explicit guidance, 234; impact measurement system, 304f; interstages, 178, 180f; launch azimuth and alignment determinations, 242, 247; launch facility designations, 272t; launch tubes, 96; LCCs (launch control centers), 390; Life Extension Program, 36; limited deactivation and dismantlement, 406, 407f, 407t; long-range plans, 394; missile conversion program, 405; missile rotation, 244; missile suspension systems, 99, 101; motor designations, 164t; motor programmed depot maintenance (PDM) process, 394–95; motor specifications, 165t; Natural Hazards Program, 310; nozzle control units, 231; operational flight and evaluation programs, 296, 300–312; Operational Testing (OT) program, 326, 327f; operational test targets, 272; origins, 229;

INDEX 537

payload summary, 37*t*; Peacekeeper missile program, 25–26; penetration aids, 303; post-boost control systems (PBCSs), 174–76, 177*f*, 180*f*, 229, 267; production closure, 24; Product Verification Missile (PVM) program, 305, 306, 308, 309–10, 311*f*; projected service life, 155, 379; Propellant Replacement Program (PRP), 305, 306; Propulsion Replacement Program (PRP), 395; propulsion system rocket engines (PSREs), 176, 177*f*, 230, 231, 309, 395; range, accuracy, and reliability estimates, 37*t*, 39*f*; Reentry Vehicle Performance/Nose Tip Redesign program, 305, 309–11; research and development programs, 265*f*, 267–69; Safety-Enhanced Reentry Vehicle (SERV) Program, 305, 312, 396, 398*t*; Service Life Extension Program (SLEP), 409; Single Reentry Vehicle (SRV) Program, 395–96, 398*t*; Special Test Missile (STM) program, 305, 306, 307–10; Stage II motor development and testing, 164*t*, 166, 169*f*; Stage III motor development and testing, 164*t*, 168, 169*f*, 170, 172–76, 229, 230; targeting parameters, 234, 298–99, 302; weather impacts, 310–11, 311*f*; *see also* NS20 guidance-and-control system

Minuteman Integrated Life Extension Program (Rivet MILE): *see* Rivet MILE

Minuteman program: ablative coatings, 177–78, 178*t*; activation summaries, 417–20*t*; Airborne Launch Control System (ALCS), 363–66, 365*f*, 369–70; Air Force reentry vehicle designations, 184*t*; airframe components and production, 178–79, 179*t*, 180*f*; alert status, 155; arms limitation negotiations, 24, 26–29, 36; attack scenarios, 92; base location considerations and criteria, 87–91, 338–39; Bench Test Program, 333; Bush (George H. W.) administration, 26; Bush (George W.) administration, 27–29; circular error probable (CEP) estimates, 229; Clinton administration, 27; construction operations, 125–47, 126*t*, 131*t*, 144*t*, 150*t*; contractor responsibilities, 84–87, 125–28, 147, 150–51; correlation chart, 16*t*; cost estimates, 69*t*, 75, 86; Cuban Missile Crisis, 371–77; cutbacks, consolidations, and deactivation, 26–30; deactivation and dismantlement, 399–406, 407*f*, 407*t*; Defense Extraordinary Priority (DX) rating, 80; demonstration and acceptance procedures, 154; design considerations and criteria, 92–94; development and testing programs, 150–51, 154, 255, 257, 259–69, 261*f*; effectiveness, 79; Eisenhower administration, 72, 74–75, 78–80, 337; estimated Soviet fatalities and industrial destruction, 22–23, 22*t*; exposure time estimations, 92; external insula-

tion, 177–78; flight organization and nomenclature, 128, 129*t*; flight reliability problems, 173–74; force level development and evolution, 17–31, 69–72, 70*t*, 78–80; Ford administration, 24–25; funding estimates and obstacles, 24, 70–73, 70*t*, 75, 78; Geodetic and Geophysical (G&G) Error Budget, 248; geographical statistics, 131*t*; geographic distribution, 12*f*; government briefings, 459*n*77; hardware research and development programs, 249–52; high-altitude testing, 159; Hill Engineering Test Facilities (HETF), 370–71; historical perspective, 3, 11, 33–36, 59–81, 83; Kennedy administration, 18–20, 20*t*, 21*t*; launch azimuth and alignment determinations, 239–48; launch control facility (LCF), 95, 112, 113*f*, 114*f*, 115*f*, 403; launch control support building (LCSB), 95, 112, 113*f*, 115–16; launcher closure, 105–8, 105*f*, 107*f*, 109*f*, 135–36, 137*f*; launch facility criteria, 89; launch facility designations, 272*t*; launch facility security system, 111; Life Extension Program, 36; long-range plans, 394; missile assembly-and-recycle facilities, 85–87, 85*f*; missile suspension systems, 94, 95, 96, 99–102, 100*f*, 105*f*; modification completion dates, 398*t*; modification programs, 293, 294*f*; motor designations, 164*t*; motor programmed depot maintenance (PDM) process, 394–95; motors and airframe, 157–81; motor specifications, 165*t*, 169*t*, 173*t*; multiple independently targetable reentry vehicles (MIRVs), 175–76; Nixon administration, 24; Obama administration, 29–31; operational characteristics, 67–70, 69*t*, 72–73, 74, 79, 81, 84; operational facility design and criteria, 94–96; operational flight and evaluation programs, 69–73, 272–74, 274*t*, 277–82; operational performance requirements, 76, 77*f*, 78; operational readiness training (ORT), 152–54, 373; payload summary, 37*t*; penetration aids, 299–300, 303; planned force levels, 19*t*; program management considerations, 73; projected service life, 155, 379; prospective deployment locations, 84, 85–87, 85*f*; rail-mobile squadrons, 17, 18–19; range, accuracy, and reliability estimates, 37*t*, 39*f*; Reagan administration, 25–26; research and development plans, 74–76, 78–80, 249–52; simulated combat launch capability (SCLC), 332–33; site activation, 147, 150–54; site separation and dispersal recommendations, 91–92, 93*t*, 128, 152; siting and facility design, 83–124; solid propellants, 11; Stage I motor development and testing, 157–62, 162*f*, 163*f*, 164*f*, 164*t*, 260–64; Stage II motor development and testing, 157, 164*t*, 165–68, 262–64; Stage III motor development and testing, 157, 164*t*, 168, 170, 171*f*, 173*t*, 174–77; strategic alerts, 33, 375, 376, 377; survivability criteria, 95, 95*t*; technological challenges, 157–59; thrust termination, 158; *see also* command-and-control systems; Force Modernization Program; guidance-and-control systems; hardening programs and specifications; launch control equipment building (LCEB); launcher equipment room (LER); launcher support building (LSB); launch tubes; LCCs (launch control centers); LFs (launch facilities); modernization and upgrade programs; NS10 series guidance-and-control system; NS17 guidance-and-control system; Polaris missile program; reentry vehicles; targeting accuracy

Minuteman Propulsion Long-Range Service Life Analysis Program, 174

Minuteman Reentry System Launch Program (RSLP), 293, 294*f*

missile assembly-and-recycle facilities, 85–87, 85*f*, 89, 153, 216, 226, 301

missile centerline determination, 242–43, 244

Missile Defense Agency targets, 28

Missile Guidance Set (MGS): dust-hardening programs, 308*f*; electronics system, 224–25, 227, 396; environmental controls, 215; flight control systems, 215–16, 220; flight test programs, 179, 309, 328; improved computer units, 268; inertial guidance systems, 220; missile deactivation and dismantlement, 401; modification programs, 230, 235, 305, 396–97; post-boost control systems (PBCSs), 176, 180*f*; reliability program, 225–26, 227, 232; *see also* NS20 guidance-and-control system

Missile Impact Locator System (MILS) Net, 263, 265, 422–25*t*

Missile Impact Sound Fixing and Ranging System (MISS) hydrophones, 283

Missile Maintenance Squadron (341 MIMS), 151, 154, 417*t*

Missile Maintenance Team (MMT), 243

Missile Performance Measurement System, 306

Missile Procedures Trainer (MPT), 353

missile suspension systems: flight test programs, 260, 262, 266, 277, 306; functional role, 99–102; guidance-and-control systems, 215; hardening programs, 95; launcher equipment room (LER), 105*f*; launch tubes, 132, 277; Stage I skirt, 179; Upgrade Silo modification program, 387, 388–89, 390*f*; wing facilities, 94–96, 99–101, 100*f*, 376, 386, 388–89

Missile Training Squadron (394 MTS), 273

missile transporter-erector pads and pylons, 135–38, 137*f*, 147

Missouri Department of Natural Resources, 405

Mixed Marble II test program, 285–86, 431*t*

mobile range safety system (MRSS), 324–26

modernization and upgrade programs: Command Data Buffer (CDB) program, 390–92, 398*t*; groundwater intrusion problems, 393; Guidance Replacement Program (GRP), 235, 305, 396–97, 398*t*; HEST (high-explosive simulation technique), 379–80, 381*f*, 382–85, 382*f*, 383*f*, 384*f*, 387; Hybrid Explicit Flight Program (HEFP), 234, 309, 392, 398*t*; Integrated Improvement Program, 390, 392, 393*t*; Minuteman III dust-hardening programs, 393–94, 398*t*; modification completion dates, 398*t*; Project Restart, 392–93; Propulsion Replacement Program (PRP), 395; remote visual assessment programs, 397–98; Rivet MILE, 30, 390*f*, 394; Safety-Enhanced Reentry Vehicle (SERV) Program, 305, 312, 396, 398*t*; security modernization program, 397–98; Single Reentry Vehicle (SRV) Program, 395–96, 398*t*; Upgrade Silo modification program, 174, 386–87, 389*f*, 398*t*; *see also* Force Modernization Program

modified operational missiles (MOMS), 324, 328–32, 335*f*

Moffett Field, California, 183

Moneyhon, Dale B., 316, 317, 318*f*

monocoque tanks, 55

monomethylhydrazine, 176

Morrison-Hardeman-Perini-Leavell Construction, 150*t*

Morrison-Knudsen Company Inc., 146, 150*t*

Moscow Treaty, 28

motor casings: development and testing programs, 51, 66–67, 71, 74, 147; dynamic pressures, 61; high-altitude testing, 159; lining materials, 44; manufacturing processes, 66, 68*t*; Minuteman I, 92; motor testing, 161–62; operational base launch (OBL) programs, 325, 327*f*; operational challenges, 74; silo development and testing programs, 251; solid propellants, 43, 52, 59; Stage I motor development and testing, 160, 327*f*; Stage II motor development and testing, 166, 169*t*; Stage III motor development and testing, 168, 170, 171*f*, 172, 173*t*; Stage I/Stage II motor casings, 92, 157; technological challenges, 61–62, 61–64, 157

motor development and testing: *see* internal grain propellant designs; rocket propulsion research; solid propellants; Stage I development; Stage II development; Stage III development

motor programmed depot maintenance (PDM) process, 394–95

Mountain Home Air Force Base (Idaho), 12*f*

multiple independently targetable reentry vehicles (MIRVs), 24, 26–29, 175–76, 228, 229
Multi-Service Launch System (MSLS) program, 300
mutually assured destruction policy, 35
MX missile program, 8, 25

National Academy of Sciences Committee on Army Air Corps Research, 43
National Academy of Sciences Committee on Undersea Warfare, 57
National Advisory Committee for Aeronautics (NACA), 5, 183
National Air and Space Museum, 189
National Bureau of Standards Station WWV (Colorado), 239
National Defense Authorization Act, 26, 28–29, 30, 396
National Defense Research Committee, 5
National Engineering Science Company, 94
National Research and Defense Council (NRDC), 47
National Security Council (NSC), 8, 19, 55
national security policies, 8, 19, 29
Natural Hazards Program, 310
Navaho cruise missile program, 8, 255
Naval Ammunition Depot (Nebraska), 87
Naval Ordnance Laboratory, 188
Naval Ordnance Test Station (NOTS), China Lake, California, 47, 51, 56–57, 74, 166–67
Naval Powder Factory, 74
Naval Research Laboratory, 55
Naval Warfare Analysis Group, 78
Navy Bureau of Ordnance, 52
Navy Special Projects Office, 56
Nebraska Ordnance Plant, 87
neoprene, 46
Nevada Test Site (Nevada National Security Site), 379
Newark Air Force Station (Ohio), 226
Newell, South Dakota, 316
New START Treaty, 29, 30
Newton, R. R., 47
Nez Perce Indian Reservation (Idaho), 326
NICKED BLADE, 286, 287, 288
NIGHT STAND, 429*t*
NIGHT TICKET, 289
Nike missile program: booster rockets, 170; guidance-and-control systems, 206; Nike-X missile, 288; Nike-Zeus missile, 197; sounding rockets, 62; Truman administration, 7
nitrocellulose, 47, 63
nitrogen gas, 52
nitrogen tetroxide, 176

nitroglycerin, 47, 63, 170
Nitze, Paul H., 323
Nixon administration: force level development and evolution, 24; strategic missile systems, 35–36
Nixon, Richard M., 324
normal launch mode, 374
North American Aviation, Autonetics Division: azimuth laying set (ALS) testing, 246, 247; contractor responsibilities, 3; guidance-and-control systems, 18, 74, 84, 206–7, 211, 213–16, 221, 223, 225–26, 229–30, 232; propulsion system rocket engines (PSREs), 176; quality control issues, 259; supersonic bomber contract, 83; targeting accuracy, 175
North American Datum 1927 (NAD 27), 237, 238
North American Rockwell Corporation, 176
Northrop Grumman, 356, 409, 410
North Star observations, 239
Nose Cone Division, Space Systems, 196–97
nozzles: axial nozzles, 158, 159*f*, 469*n*27; control units, 215–16, 224–25, 231, 257, 435*t*; Dust Modification Performance program, 308*f*; flight test programs, 421*t*; gimballed nozzles, 158, 161, 164*f*, 172; graphite materials, 52, 63, 157–58, 161, 166, 172; high-altitude testing, 159; historical research, 42, 43; modernization and upgrade programs, 298; motor testing, 161–62, 162*f*, 164*f*; offset nozzles, 161, 162*f*; operational challenges, 74; operational flight and evaluation programs, 306; Polaris missile program, 67, 158, 159*f*; Silo Test Missile (STM), 252, 254; Stage I development and testing, 157, 161–62, 162*f*, 163*f*, 164*f*, 261–62, 288; Stage II motor development and testing, 166–68, 169*f*, 169*t*, 257, 263; Stage III motor development and testing, 170, 171*f*, 172, 173*t*, 298; submerged nozzles, 166–67; technological challenges, 157–58
NS10 series guidance-and-control system: airborne power, 216, 217*t*; alignment mechanisms, 215; angular accelerometers, 215; astronomic azimuth, 239; computer systems, 213–14, 224*f*; developmental research, 205–7; electromechanical equipment, 208–9; electronic components, 207–8, 211*f*; environmental controls, 215; flight control systems, 215–16; free-rotor gas-bearing gyroscopes, 208, 209, 210*f*, 211, 212*f*, 213; major component descriptions, 209, 211, 211*f*, 212*f*, 213; nozzle control units, 215–16; operational flight and evaluation programs, 282; operational guidelines, 217*t*; reliability improvement program, 207–8; stable platform design, 209, 210*f*,

211f; targeting accuracy, 239; velocity meters, 208, 210f, 213

NS17 guidance-and-control system: accuracy issues, 296; Accuracy, Reliability, Supportability Improvement Program (ARSIP), 227–28; computer systems, 220, 223–24, 224f, 228f, 233f; electronic component reliability, 227–28; equipment changes, 220–21, 223–25; GI-T1-B azimuth gyroscope, 219, 221, 227; Improved Launch Control System (ILCS), 227, 246, 392; improvements and upgrades, 227; launch azimuth and alignment determinations, 246–47; new concepts and mechanizations, 218–20; nozzle control units, 224–25; pendulous integrating gyro-accelerometer (PIGA), 221, 222f, 223, 226; pre-existing concepts and mechanizations, 216, 218; reliability improvement program, 225–27; software status and authentication system, 227; stable platform design, 220–21, 222f, 228f

NS20 guidance-and-control system: alignment mechanisms, 231, 233, 235; autocollimator system, 231, 233; command data buffer (CDB) program, 232–33; computer systems, 230, 231–32, 233f, 234, 396; deployment improvements, 232–35; development and testing programs, 230–32; downrange bias, 234–35; Expanded Execution Plans (EEP) program, 235; flight test programs, 268, 301; Guidance Improvement Program (GIP), 234; Guidance Replacement Program (GRP), 235, 305, 396–97; Guidance System Evaluation program, 305, 308–9; Guidance Upgrade Program (GUP), 234–35; Hybrid Explicit software/Hybrid Explicit guidance, 234; nozzle control units, 231; pendulous integrating gyro-accelerometer (PIGA), 231; post boost propulsion system (PBPS), 230, 231, 232; reliability improvement program, 232; stable platform design, 222f, 231, 233, 233f; system modifications and upgrades, 230; targeting parameters, 234

NS50 guidance-and-control system, 305, 397

nuclear policy reviews, 27

Nuclear Posture Review Report, 27, 28, 29, 36, 409

Nuclear Test Ban Treaty, 315, 321

nuclear warheads: *see* thermonuclear warheads; warheads

nuclear weapons: estimated Soviet threat, 92–93, 363; Mixed Marble II test program, 285–86; operational base launch (OBL) programs, 321; security concerns, 103; solid propellant missile systems, 59; strategic targeting plans, 33–36, 57; testing programs, 379–80; *see also* Cuban Missile Crisis; warheads

Nuclear Weapons Safety Group, 103
Nuclear Weapons Systems Safety Group, 328
Nunn–Warner Amendment (1985), 25

Obama administration, 29–31
Obama, Barack, 29
OBLSS (Operational Base Launch Safety System): *see* Operational Base Launch Safety System
OCEAN VIEW II, 321
Odd Squad: *see* Strategic Missile Squadron (564 SMS)
Oeno Island, 96, 274t, 298–99
offensive weapons systems, 18, 19t, 377, 378
Office of Defense Mobilization, Science Advisory Committee, 10
Office of Scientific Research and Development, 47
Office of Secretary of Defense Ballistic Missile Committee (OSD-BMC), 17, 61, 62, 63–64, 67, 68, 72, 79–80
offset nozzles, 161, 162f
Offutt Air Force Base (Nebraska): ALCS-equipped aircraft, 366; Blue Scout Junior Program 279L (MER-6A), 356; geographic location, 12f
Ogden Air Logistics Center, 174, 300, 394–95
Ogden Air Materiel Area Depot, 87, 333
Ogden Special Launch, 300
OLD FOX, 286, 287, 301, 431t
OLYMPIC TRIALS, 290–92, 297, 366, 434t
ON TARGET, 321
operational base launch (OBL) programs: full-range operational base launches, 321–26, 328; historical perspective, 315; major objectives, 323, 325; Operational Base Launch Safety System (OBLSS), 322–25, 323f; operational base missile test programs, 328; range safety systems, 321–24; short-range operational base launches, 315–17, 317f, 318f, 319–20
Operational Base Launch Safety System (OBLSS), 322–25, 323f
operational base missile test programs: government funding issues, 326, 328; modified operational missiles (MOMS), 324, 328–32, 335f; operational base launch (OBL) programs, 315–17, 319–26, 328; simulated electronic launch Minuteman (SELM), 325, 332–35, 334f, 335f, 369, 370
operational flight and evaluation programs: Advanced Ballistic Reentry System (ABRES) Program, 293; Army Special Test Program HK (homing kill), 294; Boeing Aircraft Company, 271–72; Category I (Engineering Test Program), 257, 271, 272–74, 277–78, 282, 283, 296, 435t; Category II (flight test and

launch programs), 271, 278, 282, 283, 296, 301, 376; Category III (operational system test and evaluation programs), 271, 278–79, 296; Cold/Heat Soak program, 292–93; Demonstration and Shakedown Operation (DASO) program, 278–79, 282–84, 287, 295–98, 301, 430*t*; Emergency Rocket Communications System (ERCS), 298, 299, 436*t*; Flight Development Evaluation (FDE) program, 305; Follow-on Operational Test Program (FOT), 280–82, 289–91, 291*f*, 432*t*; hardening studies and specifications, 271; impact analyses, 288–89, 290; impact measurement system, 304*f*; in-flight problem areas, 288; in-flight safety systems, 299; launch failures, 277–78, 281, 282, 290–93, 295*f*, 305; Minuteman IA, 272–74, 277–82, 280*t*, 429*t*; Minuteman IB, 282–94; Minuteman II, 295–300, 436*t*; Minuteman III, 296, 300–312; Minuteman Reentry System Launch Program (RSLP), 293, 294*f*; Multi-Service Launch System (MSLS) program, 300; Natural Hazards Program, 310; Operational Testing (OT) program, 279–80, 285, 431*t*; Reentry Body Supplemental Flight Test Program, 293–94; Reentry System Launch Program, 432*t*; Safeguard System Target Test Program, 293; special flight programs, 299–300; special test missiles (STMs), 298, 436*t*; test targets, 272–74, 274*t*, 298–99; weapon system testing, 296–98; *see also* Vandenberg Air Force Base (California)

Operational Readiness Training and Combat Training Launch (ORT/CTL) program, 271

operational readiness training (ORT), 152–54, 373

Operational Suitability Test Facility, 14*f*

operational system test and evaluation programs (Category III), 271

Operational Testing (OT) program: Minuteman IA, 279–80; Minuteman IB, 285, 431*t*

Operation Plumbbob, 379

Operation Pushover, 56

Operation Sandy, 56

Operations Ground Program (OGP), 356

optional attack plan scenarios, 35–36

ORANGE CHUTE, 285, 286, 431*t*

Orbital ATK, 410

Ordnance/California Institute of Technology 1 (ORDCIT-1), 45

Orlando Air Force Base (Florida), 238

overpressure: blast protection, 118, 122; exposure time, 92, 377; hardening studies, 94, 108, 109, 388; high-explosive simulation technique (HEST) test, 379–80, 382; Upgrade Silo modification program, 386; weapon effects, 92–93, 94*t*, 116

oxidizers: ammonium perchlorate, 48, 53, 53*f*, 63, 160; content percentage, 50; corrosive properties, 61; ignition preparations, 176; LP-2 polymer, 46; plastisol process, 51; powdered metal additives, 52, 63; storability, 43; toxicity, 61

P92 amplifier, 220, 225, 228*f*, 230, 233*f*, 396

Pacer Kite program, 290–92

Pacific Missile Range (PMR), 150, 272–74, 274*t*; *see also* Vandenberg Air Force Base (California)

PAINTED WARRIOR, 286, 287, 288

paranitrophenol fungicide, 177

Parry Island, 286

Parsons, John W., 42, 43, 44

Patrick Air Force Base (Florida), 84, 150, 255, 256*f*, 257, 259–69, 259*f*

PAVE PEPPER, 305, 309

Peacekeeper missile program, 25–26, 30, 36, 184*t*, 312, 353, 370

Pease Air Force Base (New Hampshire), 90

pendulous integrating gyro-accelerometer (PIGA), 213, 219, 220, 221, 222*f*, 223, 226, 230, 231, 396–97

penetration aids, 299–300, 436*t*; *see also* Minuteman III

perfluorohexane, 167

periscope alignment system, 231, 301, 386

Perry, William J., 27, 36

Personnel Access System (PAS) shaft, 103, 106*f*, 136*f*, 137, 138*f*

perturbation self-alignment technique (PSAT), 227

Peter Kiewit and Sons' Company, 95, 130, 150*t*

phenolic resin–asbestos ablative material, 188, 189

phenolic resin plastics, 188, 189, 190–91, 194–95, 198, 201

Phillips Petroleum Company, 60

Phillips, Samuel C., 257, 274, 339

Phoenix Islands, 96, 274*t*, 302–3, 304*f*, 326

Photomapping Group (1370), 238

pilotless aircraft, 6, 7

PILOT ROCK, 281, 289, 429*t*

Plant 77 (Hill Air Force Base): Minuteman III production, 24; missile assembly-and-recycle facilities, 87, 226; modification programs, 301, 316, 319, 329

plastic zone, 93–94, 94*t*

plastisol process, 51–52

Plattsburgh Air Force Base (New York): Atlas

F rocket, 13*f*; full-range operational base launches, 321; geographic location, 12*f*
Point Arguello, California, 356, 357*f*
Polaris missile program: airframe design, 66; Explicit Guidance scheme, 234; first-generation heatsink reentry vehicles, 183; force level recommendations, 23, 79; historical perspective, 55–56, 59–81; motor casings, 157, 168, 172; multiple independently targetable reentry vehicles (MIRVs), 175; nozzles, 67, 158, 159*f*; operational characteristics, 67–68; Polaris A-1 missile, 36, 58*f*, 158, 159*f*, 183; propellant processing, 158; retaliatory effectiveness, 71; rocket propulsion research, 62–64, 68*t*; thrust termination, 158
polybutadiene acrylic acid–acrylonitrile polymer (PBAN), 160
polybutadiene acrylic acid (PBAA) polymer propellants, 160
polychlorinated biphenyls (PCBs), 401–2, 403
polypropylene, 254
polysulfide-based propellants, 46–48, 53, 160
polyurethane propellants, 53, 64, 165
polyvinyl chloride propellants, 51–52, 53*f*
Poole, Harold J., 47
portable gyrocompass assembly (PGA), 245
Porter, Richard W., 48
Poseidon missiles: arms limitation negotiations, 23; delivery capabilities, 36; Explicit Guidance scheme, 234; reentry vehicles, 228
Post-Attack Command-and-Control System (PACCS), 364, 365*f*, 366
post-boost control systems (PBCSs), 174–76, 177*f*, 180*f*, 229, 267
post boost propulsion system (PBPS), 230, 231, 232, 301
potassium perchlorate oxidizer, 48
powdered metal additives, 51–53, 53*f*, 63, 160, 165, 173–74
Power, Montana, 405
Power, Thomas S., 10, 62, 373
P-plug, 220
Prague, Czech Republic, 29
Prairie Hawk program, 397–98
Prandtl, Ludwig, 469*n*27
Precisely Guided RTV flights, 192
Precision Deployment System (PDS), 175
Preparatory Launch Command (PLC) capability, 347–48, 354, 369, 370
Price, Edward W., 47
Primacord explosives, 379–80, 382, 383*f*
Primary Alerting System, 373
Prince Edward Islands, 194
Princeton University, 41

Private A (XF10S1000-A) solid-propellant missile, 45–46
Product Verification Missile (PVM) program, 305, 306, 308, 309–10, 311*f*
programmed depot maintenance (PDM), 30–31
programmed trajectories, 268, 317*f*, 324
Project 117L/Corona imagery, 247
Project Confirm, 154
Project Gas Bag, 379
Project Giant Boost, 320
Project Hermes, 48
Project High Climber, 152
Project Long Life, 315–17, 317*f*, 318*f*, 319
Project Mercury, 57
Project Nobska, 57, 59, 62
Project Restart, 392–93
Project Shoe Lace, 247–48
Project Solarium, 8
Project Test Hop, 152–53, 154
Project Viking, 55–56
PRONTO ROSE, 285, 286, 431*t*
propellant processing challenges, 158
Propellant Replacement Program (PRP), 305, 306
propellants: *see* liquid propellants; rocket propulsion research; solid propellants
Propulsion Replacement Program (PRP), 395
propulsion system rocket engines (PSREs), 28, 176, 177*f*, 230, 231, 309, 395
Public Law 97-252, 25
Puckett, Alan E., 8
PURPLE LIGHT, 285
purple plague, 225, 477*n*52
PUSH PULL, 433*t*
Putin, Vladimir, 28
Putt, Donald L., 62, 64, 459*n*77
pyrolytic/charring ablation method, 183


Q Guidance scheme, 234
Q-ships, 458*n*61
QUAntized INtegrating Torquer (QUAINT) computer, 206
QUICK JUMP, 280
QUICK NOTE, 286, 287


RaD-58B phenolic resin, 198
RaD 58E phenolic resin, 195
RaD-60 phenolic resin, 201
radar, 43
radar systems, 303, 304*f*, 310–11, 311*f*
radiant cooling, 186
radioactive fallout, 88
radio-inertial guidance systems, 66

Ralph M. Parsons Company, 94
ramjet propulsion, 45, 190
Ramo-Wooldridge Corporation: contractor responsibilities, 10, 73, 83; operational deployment dates, 72; solid propellant missile systems, 59–60, 64–66; *see also* Thompson-Ramo-Wooldridge (TRW)
RAND Corporation, 339, 409
Rapid City, South Dakota, 132, 400
Rapid Execution and Combat Targeting (REACT) Program, 353–54, 355*f*, 356, 398*t*
Raynesford, Montana, 151
Reaction Motors, 55
Reagan administration, 25–26
Reagan, Ronald, 25
rebar, 134, 135, 136*f*, 137, 137*f*, 140, 145*f*, 402
REBEL RANGER, 433*t*
red fuming nitric acid, 43
RED SPIDER, 429*t*
Redstone Arsenal, Alabama, 48, 51, 52, 74, 160
Redstone missile, 188, 189, 205
Reentry Body Identification Group, 174
Reentry Body Supplemental Flight Test Program, 293–94
Reentry Management System V phased-array radar, 303
Reentry System Launch Program, 432*t*
Reentry Vehicle Performance/Nose Tip Redesign program, 309–11
reentry vehicles: ablative material research, 190–96; Able RTV program, 192; aerodynamic heating and shock waves, 183, 184*f*, 185–86; Air Force designations, 184*t*; ballistic coefficient ($\beta$), 186, 190; circular error probable (CEP) estimates, 229, 290–92; comparison chart, 180*f*; contractor responsibilities, 74, 84; cooling categories, 186; deactivation and dismantlement, 401; DEFCON declarations, 377; Demonstration and Shakedown Operation (DASO) program, 430*t*; deployment history, 203*t*; design and development, 196–98, 199*f*, 200*f*, 201–2; Dust Modification Performance program, 308*f*; Emergency Rocket Communications System (ERCS), 358, 359, 360; first-generation heatsink reentry vehicles, 186, 187*f*; flight test programs, 191–96, 205, 255, 266, 268, 273, 278–87, 290, 296–97, 303–4, 421–28*t*; Follow-on Operational Test Program (FOT), 432*t*; Force Modernization flight tests, 435*t*; full-range operational base launches, 321, 324; historical research, 57, 183, 185–86; impact analyses, 289; launch failures, 297; Minuteman general characteristics, 69*t*; Minuteman II research and development programs, 433*t*; Mixed Marble II test program, 431*t*; modeling errors, 292; multiple independently targetable reentry vehicles (MIRVs), 175–76, 228, 229, 268; nose shape, 183, 184*f*, 185, 186; operational test flights, 290–94; Operational Testing (OT) program, 431*t*; propulsion system rocket engines (PSREs), 176; pyrolytic/charring ablation method, 183; Reentry Body Supplemental Flight Test Program, 293–94; Reentry Vehicle Performance/Nose Tip Redesign program, 305, 309–11; research and development programs, 196–98, 199*f*, 200*f*, 201–2, 203*t*; RVX-1 reentry vehicle test, 192–94, 193*f*, 194*f*; RVX-2 reentry vehicle series tests, 194–95; RVX-3 reentry vehicle tests, 195–96; RVX-4 reentry vehicle tests, 195–96; Safety-Enhanced Reentry Vehicle (SERV) Program, 305, 312, 396, 398*t*; second generation ablative reentry vehicle system, 188–96; short-range operational base launches, 316; silo development and testing programs, 252*f*; Single Reentry Vehicle (SRV) Program, 395–96, 398*t*; Thor-Able 0 program, 192; trajectory characteristics, 183, 185–86, 185*f*; weapon system testing, 296–97; weather impacts, 310–11, 311*f*; wind tunnel tests, 188; *see also* specific reentry vehicle
Reese Air Force Base (Texas), 91
reference mirror alignment verification (RMAV), 240, 240*f*, 241–42, 241*f*, 243*f*, 245–46
reference mirror azimuth determination (RMAD), 245–47
Refrasil–phenolic materials, 188, 191, 193, 195, 196
Remote Data Change Target (RDCT) function, 354
Remote Retargeting system, 390
remote visual assessment programs, 397–98
Republic Aviation, 197
research and development programs: Air Force Ballistic Missile Division (AFBMD), 71, 74–75, 80; grain propellant designs, 46–50, 50*f*, 160; launch failures, 257, 259; Minuteman I, 257, 259–64; Minuteman II, 264–67, 265*f*, 433*t*; Minuteman III, 267–69; reentry vehicles, 196–98, 199*f*, 200*f*, 201–2, 203*t*, 433*t*; silo development research, 249–52, 250*f*, 251*f*, 252*f*, 253*f*, 254–55, 254*f*, 256*f*, 257, 258*f*, 259–69
resin-based ablative materials, 188
RESTLESS DRIFTER, 433*t*
retaliation policies, 34–35, 337
retrorocket spacers, 201
Rice, Donald, 26, 399
Rice, M. L., 52
Richardson, Al, 49*f*
ripple fire option, 274, 277, 279, 281, 339, 343, 374

Ritland, O. J., 190
Rivet Add, 26, 399
Rivet Dome II, 26, 399
Rivet MILE, 30, 390f, 394
road and bridge improvement costs, 90
rocket-assisted takeoff (RATO), 42, 43
Rocket Propulsion Establishment (Great Britain), 249
rocket propulsion research, 41–53, 53f, 55–57, 59–81, 68t, 165, 168
rocketry history, 41–43
Rocket System Launch Program, 28
Rocky Hill, New Jersey, 172
ROCKY POINT, 281, 429t
rogue launches, 277, 337
roll control: see attitude control
rolled motor casings, 66, 157
Rosholdt, E., 52
ROSY FUTURE, 286, 287, 431t
Roy, Montana, 128
Rubel, R. H., 339
Rubens, Idaho, 326
Rumbel, Keith E., 51–52, 454n55
Runit Island, 286
rupture zone, 93, 94t
RV-10-A program, 50, 51
RVX-1 reentry vehicle test, 192–94, 193f, 194f
RVX-2A reentry vehicle tests, 195
RVX-2 reentry vehicle series tests, 194–95
RVX-3 reentry vehicle tests, 195–96
RVX-4 reentry vehicle tests, 195–96

Safeguard System Target Test Program, 293
Safety Control Switch (SCS), 331, 332, 374–75, 400
Safety-Enhanced Reentry Vehicle (SERV) Program, 305, 312, 396, 398t
Safety Enhanced Reentry Vehicle/Warhead (SERV/W) Concept Action Group, 396
safing tone, 343–44
Sagan, Scott, 371
SAGE GREEN, 429t
salvo fire option, 274, 277, 281, 313f, 339, 343, 374
Sammons, Ronald, 243f
SAMSO: see Space and Missile Systems Organization (SAMSO)
Sapuppo, Michele, 223
satellite deployments, 175
Schaefer, Herbert S., 316, 317
Schilling Air Force Base (Kansas): base location considerations and criteria, 91; geographic location, 12f
Schlesinger, James R., 24, 25, 325, 326, 328, 386

Schonka, Joe M., 319
Schriever, Bernard A.: Atlas missile program, 10; estimated program funding, 70, 78; ICBM program review, 17–18; Minuteman program briefings, 67, 337–38, 459n77; Minuteman program management considerations, 70–71, 73, 74; research and development plans, 76, 78, 79; solid propellant missile systems, 59, 61, 64–65; Strategic Missiles Evaluation Group (Teapot Committee), 8
Schultz, R. D., 52
Science Advisory Committee, Office of Defense Mobilization, 10, 65
Scientific Advisory Committee on Ballistic Missiles, 56, 60, 68, 71, 74
Scott, Austin, 30
Scurlock, A. C., 52
SEA DEVIL, 281, 429t
Seamans, Robert C., Jr., 323–24
Seattle Test Program (STP) III, 150–51, 370
Seattle, Washington, 147; see also Boeing Aircraft Company
second generation ablative reentry vehicle system, 188–96
Secure Enable command, 353
Security Control Center, 112, 118
security systems, 96, 97f, 103, 111, 397–98
self-alignment technique calibration (SATCAL), 231
Self-Contained Range Safety Abort System (SCRSAS), 322, 324
semi-monocoque construction, 178, 179
Senate Armed Services Committee: defense appropriations, 23, 328; strategic missile system reductions, 28
Senate Committee on Foreign Relations, 363–64
Sensitive Command Network (SCN), 340
Sentinel Lace, 247
Sergeant missiles: airframe components and production, 57, 61–62, 157; booster rockets, 160; solid propellants, 189; sounding rockets, 48; tactical ballistic missile (TBM) system, 51, 157, 158
Service Life Extension Program (SLEP), 409
Sessums, John W., Jr., 64–65
Settlemire, Lawrence, 46
Shafer, John I., 46, 47, 48
Shell Development Company, 46
Shelter-Based Minuteman, 267
Sheppard Air Force Base (Texas), 91
Sheppard Committee/Report, 127–28
Sheppard, Harry R., 127
Ships Inertial Navigation System, 206
shock absorbers: see missile suspension systems
shock-isolated floors, 389–90, 391f

shock-isolation systems: *see* missile suspension systems
shock waves: aerodynamic heating, 183, 185; blast effects, 109; bow shock waves, 184*f*; fluid injection experiments, 166–67; HEST (high-explosive simulation technique) testing, 380, 381*f*, 384
short-range operational base launches, 315–17, 317*f*, 318*f*, 319–20
SHORT ROUND, 320
Shuey, Henry M., 170
SHUTTLE TRAIN, 281, 429*t*
Sides, John H., 79
silo launch test facilities: Air Force Flight Test Center (AFFTC), 255, 256*f*, 257; attack scenarios, 92; design considerations and criteria, 92–94; design recommendations, 65–67; Edwards Air Force Base (California), 250–52, 250*f*, 251*f*, 252*f*, 253*f*, 254–55, 254*f*, 257; facility characteristics, 96; force level recommendations, 23; operational facility design and criteria, 94–96; Patrick Air Force Base (Florida), 255, 256*f*, 257, 259–69, 259*f*; prospective deployment locations, 85–87; research and development programs, 249–52, 250*f*, 251*f*, 252*f*, 253*f*, 254–55, 254*f*, 256*f*, 257, 258*f*, 259–69; site separation and dispersal recommendations, 93*t*
Silo Test Missile (STM), 252, 252*f*, 254–55, 254*f*
SILVER CLOUD, 285, 286, 431*t*
Simons, James A., 319
simulated combat launch capability (SCLC), 332–33
simulated electromagnetic ground environment (SIEGE), 393
simulated electronic launch Minuteman (SELM), 325, 332–35, 334*f*, 335*f*, 369, 370
simulated missile launch programs: modified operational missiles (MOMS), 324, 328–32, 335*f*; simulated electronic launch Minuteman (SELM), 325, 332–35, 334*f*, 335*f*
Single Integrated Operational Plan (SIOP), 34–35
single-reentry vehicle platforms, 27
Single Reentry Vehicle (SRV) Program, 395–96, 398*t*
single-rocket launch techniques, 61
site activation, 147, 150–54
Site Activation Task Force (SATAF), 126, 127, 147, 151–54, 372, 375
siting and facility design: attack scenarios, 92; base location considerations and criteria, 87–91, 338–39; contractor responsibilities, 83–87; design considerations and criteria, 92–94; estimated Soviet threat, 93*t*, 363–64; hardening programs and specifications, 91–92,

94, 95*t*; launch control equipment building (LCEB), 95, 113*f*, 115*f*, 118, 120, 120*f*, 121*f*, 122–23, 123*f*; launch control facility (LCF), 95, 112, 113*f*, 114*f*, 115*f*, 403; launch control support building (LCSB), 95, 112, 113*f*, 115–16; launcher closure, 105–8, 105*f*, 107*f*, 109*f*; launcher equipment room (LER), 102–4, 104*f*, 105*f*; launcher support building (LSB), 95, 96, 97*f*, 108–11, 110*f*, 111*f*; launch facility security system, 111; launch tubes, 96, 98–99, 98*f*, 105*f*; missile assembly-and-recycle facilities, 85–87, 85*f*; operational facility design and criteria, 94–96; prospective deployment locations, 84, 85–87, 85*f*; site separation and dispersal recommendations, 91–92, 93*t*, 128; survivability criteria, 95, 95*t*; *see also* construction operations; LCCs (launch control centers)
six-point star-grain propellant design, 160, 163*f*, 164
skip-entry trajectory, 183
Skirt Removal Test Program, 260
skirts, 179, 180*f*, 263
Sky bolt missile program, 184*t*
Slater, John M., 211, 476*n*13
Smith, Apollo M. O., 42
Smith, Levering, 56–57
SMOKY RIVER, 287, 288
SM (strategic missile)-80: *see* Minuteman program
SNAP ROLL, 285
Snark missile program, 8, 170
Software Status Authentication System (SSAS) 1, 227
Solar Radiation/Galactic Radiation and Background (GRAB) satellite, 175
Solid Propellant Information Agency, 74
solid propellants: advantages, 249–50; Aerojet-General Corporation, 165; bipropellant systems, 165–66, 167*f*, 168, 170; Cold/Heat Soak program, 292–93; flight test programs, 302; high-altitude testing, 159; historical research, 41–53, 45*f*, 53*f*, 55–57, 58*f*, 59–81, 83, 456*n*22; motor testing, 161–62; operational advantages, 56–57, 59–63; Propellant Replacement Program (PRP), 305, 306; Propulsion Replacement Program (PRP), 395; response capabilities, 11; status report, 68*t*; technological challenges, 158; temperature impacts, 284; Thiokol Chemical Corporation, 159–60
sound fixing and ranging (SOFAR) bomb device, 192–93, 279, 283
sounding rockets, 41–42, 43, 47–48, 55, 62, 192, 255
South Dakota: *see* Ellsworth Air Force Base (South Dakota)

South Dakota Department of Environment and Natural Resources, 401
Soviet-bloc countries, 34–35
Soviet R-36 (8K67), 365
Soviet Union: Cuban Missile Crisis, 33, 371–72; estimated fatalities and industrial destruction, 22–23, 22*t*; estimated threat to U.S. sites, 93*t*, 363–64; missile base locations, 38*f*; missile deployment estimates, 17; as missile target, 19, 33, 34–36
Spaatz, Carl, 6–7
Space and Missile Systems Organization (SAMSO): electromagnetic pulse (EMP) susceptibility, 392; flight reliability testing, 173–74; Force Modernization plans, 386; Geodetic and Geophysical (G&G) Error Budget, 248; limited-range launches, 320; remote targeting, 390; simulated combat launch capability (SCLC), 332–33; special test missiles (SSTMs), 298, 300, 306, 308, 310–11, 436*t*
Space Technology Laboratories (STL): advanced reentry test vehicle (ARTV), 192; contractor responsibilities, 3, 73, 83, 375; flight test programs, 192, 295; guidance-and-control systems, 205, 213–14; ICBM program review, 17; modification programs, 263; nozzle design and development, 162; research and development programs, 75; silo launch test facilities, 65–66; solid propellant missile systems, 75
Sparrow missile program, 7
Spartan launch complex, 303
Special Test Missile (STM) program: flight test programs, 268–69, 298, 300, 306–11, 428*t*; Force Modernization flight tests, 435*t*; operational flight and evaluation programs, 298, 436*t*; *see also* Space and Missile Systems Organization (SAMSO)
Special Weapons Laboratory, 379
specific impulse: high-altitude testing, 159; historical research, 41, 50–53, 53*f*, 63–65; jetevators, 158; Minuteman general characteristics, 69*t*; motor testing, 161–62, 162*f*; nozzle design and development, 159, 162*f*, 172; Polaris missile program, 66; polybutadiene acrylic acid (PBAA) polymer propellants, 160; solid propellant development, 63, 68*t*, 71, 74; variability, 468*n*14
SPEED KING, 286, 288
Spiralloy, 170, 171*f*, 172
spiral wound motor casings, 66, 170, 171*f*, 172
splash detection radar system, 303, 304*f*
Sprint launch complex, 303
Sputnik launches, 11, 67, 174
Squadron 20: *see* Strategic Missile Squadron (564 SMS)
squadrons: *see* flight organization and nomenclature
SR19-AJ-1, 164*t*, 165–68, 169*t*
SR73-AJ-1: manufacturing production, 172, 173; motor designations, 164*t*; motor specifications, 173*t*; propellants, 172
SS-4 medium-range ballistic missiles, 371–72
SS-9: *see* Soviet R-36 (8K67)
stable platforms, 209, 210*f*, 211*f*, 220–21, 222*f*, 228*f*, 231, 233
Stage I development: attitude control, 161; Dust Modification Performance program, 308*f*; flight test programs, 421*t*, 423–24*t*; Growth Minuteman (Minuteman IIB), 267; in-flight problem areas, 288; interstages, 178–79, 180*f*, 263, 327*f*; motor casings, 157, 160, 327*f*; motor designations, 164*t*; motor specifications, 165*t*; motor testing, 157–62, 162*f*, 163*f*, 164*f*, 164*t*, 165*t*, 254, 261–64, 394–95; nozzles, 161–62, 162*f*, 215, 225, 261–62, 288; operational base launch (OBL) programs, 315–17, 319–20, 326, 327*f*; operational flight and evaluation programs, 282, 289, 292–93; premature ignition, 257, 259*f*; propellant processing, 158; service life estimates, 395; short-range operational base launches, 315–17, 319–20; silo development and testing programs, 252, 260–67; skirts, 179, 180*f*, 263
Stage II development: accelerometers, 215; attitude control, 166–68, 169*f*; Dust Modification Performance program, 308*f*; flight test programs, 421*t*, 423–24*t*; Growth Minuteman (Minuteman IIB), 267; in-flight problem areas, 288; interstages, 178–79, 180*f*, 252*f*, 263; long-range plans, 394; motor casings, 157, 166; motor programmed depot maintenance (PDM) process, 394–95; motor specifications, 169*t*; motor testing, 157, 165–68, 254, 394–95; nozzles, 166–68, 169*f*, 215, 225, 257, 263; operational base launch (OBL) programs, 325; operational flight and evaluation programs, 289, 292, 306; premature ignition, 257, 259*f*; service life estimates, 395; short-range operational base launches, 316; silo development and testing programs, 251, 252*f*, 262–67
Stage III development: accelerometers, 215; airframe components and production, 178–79; attitude control, 169*f*, 172; Dust Modification Performance program, 308*f*; Emergency Rocket Communications System (ERCS), 361; flight reliability problems, 173–74; flight test programs, 268, 421*t*, 423–24*t*; Force Modernization flight tests, 435*t*; Growth Minuteman (Minuteman IIB), 267; interstages, 178–79, 180*f*, 252*f*, 263; motor casings, 157,

168, 170, 171*f*, 172; motor specifications, 173*t*; motor testing, 170, 172, 230, 254, 394–95; nozzles, 170, 171*f*, 172, 215, 225, 298; operational flight and evaluation programs, 282, 287–88, 289, 292, 301, 306; post-boost control systems (PBCSs), 174–76, 177*f*, 180*f*; premature ignition, 257, 259*f*; propellants, 168, 170, 171*f*, 302; service life estimates, 395; short-range operational base launches, 316; silo development and testing programs, 251, 252*f*, 262–69; thrust termination, 158

Stage I/Stage II motor casings, 92, 157

Standard Oil Company of Indiana, 60

Standby Missile Combat Crew, 373–74

STAR BRIGHT, 297, 434*t*

STAR DUST, 285, 286, 431*t*

star-grain propellant design: bipropellant systems, 165, 167*f*; burn patterns, 50*f*; development and testing programs, 47–50; five-point configuration, 48–50, 50*f*; four-point configuration, 165; six-point configuration, 64, 160, 163*f*; Stage I motor development and testing, 317*f*; Stage II motor development and testing, 165–66, 167*f*; Stage III motor development and testing, 170

START I/II Treaties, 26, 27, 28, 30, 36, 312, 395–96, 400

STATE PARK, 282

static firing tests, 47–48, 50, 67, 161, 315, 395

Status Console, 351*f*

Staver, R. B., 45

Stead Air Force Base (Nevada), 91

steel motor cases, 92, 147, 157, 161

"step" rockets, 42

stop-launch capability, 339

storable liquid propellants, 11

Strategic Aerospace Division (STRAD), 290

Strategic Air Command (SAC): Airborne Launch Control System (ALCS), 363–65, 365*f*; alert status, 155; azimuth laying set (ALS) testing, 246, 247; base deactivation criteria, 399, 405; base location recommendations, 88, 90; Cuban Missile Crisis, 372–77; cutbacks, consolidations, and deactivation, 26; deactivation, 371; DEFCON declarations, 372–73; demonstration and acceptance procedures, 154; Demonstration and Shakedown Operation (DASO) program, 282–83; Emergency Rocket Communications System (ERCS), 356; force level recommendations, 448*n*25; full-range operational base launches, 321, 324, 325–26; historical perspective, 11; Minuteman inventory, 24; Minuteman Long-Range Plan, 394; missile assembly-and-recycle facilities, 86, 87; modified operational missiles (MOMS), 329–31; operational flight and evaluation program objectives, 285–88, 292, 302–3; Post-Attack Command-and-Control System (PACCS), 364; Preparatory Launch Command A (PLC-A), 329–31; range safety systems, 322–23; short-range operational base launches, 316, 320; simulated combat launch capability (SCLC), 332–33; strategic targeting plans, 34

Strategic Arms Limitation Talks (SALT I/II), 24, 26, 312, 328

Strategic Missile Evaluation Squadron (3901 SMES), 247, 373

Strategic Missile Integration Complex, 370; *see also* Hill Engineering Test Facilities (HETF)

Strategic Missiles Evaluation Group (Teapot Committee), 8–9

Strategic Missile Squadron (10 SMS), 3, 33, 153, 411*t*, 417*t*

Strategic Missile Squadron (12 SMS), 152, 153, 405, 411*t*, 417*t*

Strategic Missile Squadron (66 SMS), 401, 412*t*, 418*t*

Strategic Missile Squadron (67 SMS), 366, 401, 412*t*, 418*t*

Strategic Missile Squadron (68 SMS), 316, 401, 412*t*, 418*t*

Strategic Missile Squadron (310 SMS), 371

Strategic Missile Squadron (319 SMS), 25, 415*t*, 419*t*

Strategic Missile Squadron (320 SMS), 415*t*, 419*t*

Strategic Missile Squadron (321 SMS), 415*t*, 419*t*

Strategic Missile Squadron (400 SMS), 25, 415*t*, 419*t*

Strategic Missile Squadron (446 SMS), 27, 406, 416*t*

Strategic Missile Squadron (447 SMS), 27, 319, 329, 405, 406, 416*t*, 420*t*

Strategic Missile Squadron (448 SMS), 406, 416*t*, 420*t*

Strategic Missile Squadron (449 SMS), 420*t*

Strategic Missile Squadron (490 SMS), 153, 155, 411*t*, 417*t*

Strategic Missile Squadron (491 SMS), 91

Strategic Missile Squadron (508 SMS), 414*t*, 419*t*

Strategic Missile Squadron (509 SMS), 414*t*, 419*t*

Strategic Missile Squadron (510 SMS), 357–60, 414*t*, 419*t*

Strategic Missile Squadron (556 SMS), 321

Strategic Missile Squadron (564 SMS): activation summary, 417*t*; command-and-control systems, 347, 351*f*, 392; Command Console, 350*f*; construction and acceptance milestone dates, 416*t*; construction costs, 150*t*; construction operations, 126*t*, 131*t*; dust-hardening programs, 394; electromagnetic pulse (EMP) susceptibility, 392; excavation and foundation operations, 146; facility characteristics, 96, 97*f*; flight organization and nomenclature, 128; Force Modernization plans, 385, 386, 387*t*; geographical statistics, 131*t*; Hill Engineering Test Facilities (HETF), 370; Hybrid Explicit Flight Program (HEFP), 392; Integrated Improvement Program, 392, 393*t*; launch control equipment building (LCEB), 122–23; launch control facility (LCF), 112, 113*f*, 115*f*, 271; launch control support building (LCSB), 144; launcher closure, 107; launcher equipment room (LER), 103–4; launcher support building (LSB), 109–11, 111*f*; launch tubes, 96, 98; LCCs (launch control centers), 115*f*, 117–18, 271; Minuteman deactivation and dismantlement summary, 407*t*; Minuteman inventory, 24, 25, 28, 29, 91, 296, 326, 399; missile deactivation and dismantlement, 30, 406, 407*f*; missile suspension systems, 99, 100*f*, 101, 102, 388; modernization and upgrade programs, 298, 347; modification completion dates, 398*t*; operational base launch (OBL) programs, 323, 326, 327*f*; Project Restart, 392–93; Status Console, 351*f*; survivability criteria, 95*t*; typical site plan, 115*f*

Strategic Missile Squadron (740 SMS), 413*t*, 418*t*

Strategic Missile Squadron (741 SMS), 366, 410, 413*t*, 418*t*

Strategic Missile Squadron (742 SMS), 413*t*, 418*t*

strategic missile systems: arms limitation negotiations, 24, 26–29, 36; base location considerations and criteria, 87–91, 338–39; Bush (George H. W.) administration, 26; Bush (George W.) administration, 27–29; Clinton administration, 27; cutbacks, consolidations, and deactivation, 26–30; delivery capabilities, 36; Eisenhower administration, 8–11, 17–18, 34, 55, 62, 67, 72, 337; estimated Soviet threat, 93*t*, 363–64; force objectives, 70*t*; Ford administration, 24–25; funding estimates, 70–73, 70*t*, 75, 78; historical perspective, 5–11; Johnson administration, 20–24, 35; Kennedy administration, 18–20, 20*t*, 21*t*, 34–35, 338; Nixon administration, 24, 35–36; Obama administration, 29–31; operational performance requirements, 76, 77*f*, 78; range estimates, 39*f*; Reagan administration, 25–26; rocket propulsion research, 41–53, 53*f*, 55–57, 59–81, 83; security modernization program, 397–98; strategic alerts, 33; targeting plans, 33–36; thermonuclear warheads, 57; Truman administration, 7–8; *see also* Minuteman program; Polaris missile program

Strategic Missile Wing (10 SMW), 113*f*

Strategic Missile Wing (44 SMW): activation summary, 418*t*; construction and acceptance milestone dates, 412*t*; construction operations, 125, 126*t*, 130, 131*t*; deactivation and dismantlement, 400–403; geographical statistics, 131*t*; guidance-and-control systems, 247; short-range operational base launches, 316; simulated electronic launch Minuteman (SELM), 334; *see also* Wing II

Strategic Missile Wing (90 SMW): activation summary, 419*t*; construction and acceptance milestone dates, 415*t*; construction operations, 131*t*; geographical statistics, 131*t*; Guidance Replacement Program (GRP), 397; HEST (high-explosive simulation technique) testing, 380, 382; Integrated Improvement Program, 393*t*

Strategic Missile Wing (91 SMW): construction operations, 126*t*; geographical statistics, 131*t*; Guidance Replacement Program (GRP), 397; heading sensitivity testing (HST) program, 246; Integrated Improvement Program, 393*t*; operational status, 410

Strategic Missile Wing (320 SMW), 24

Strategic Missile Wing (321 SMW): activation summary, 420*t*; construction and acceptance milestone dates, 416*t*; construction operations, 126*t*, 131*t*; electromagnetic pulse (EMP) susceptibility, 392; geographical statistics, 131*t*; Integrated Improvement Program, 393*t*; Minuteman inventory, 27, 28; missile deactivation and dismantlement, 406; missile deployment plans, 245; modernization and upgrade programs, 347; modified operational missiles (MOMS), 329; short-range operational base launches, 319, 320

Strategic Missile Wing (341 SMW): activation summary, 417*t*; alert status, 155; combat readiness declaration, 3; construction and acceptance milestone dates, 411*t*; construction operations, 125, 126*t*, 131*t*; DEFCON declarations, 371, 372–73, 376; geographical statistics, 131*t*; Guidance Replacement Program (GRP), 397; launch control facility (LCF), 113*f*; Launch Control Panel/Console, 344*f*, 345*f*; Minuteman inventory, 24, 26, 27, 91; missile conversion program, 405; missile deactivation and dismantlement, 407*f*; operational base launch (OBL) programs, 315, 321, 326; operational readiness training (ORT), 373; Propulsion Replacement Program (PRP), 395; simulated electronic launch Minuteman (SELM), 334; site activation process, 151–54; strategic alerts, 33; *see also* Wing I

Strategic Missile Wing (351 SMW): activation summary, 419*t*; construction and acceptance milestone dates, 414t; construction operations, 126t, 131t; deactivation and dismantlement, 404–5; geographical statistics, 131t; short-range operational base launches, 319–20; simulated electronic launch Minuteman (SELM), 334

Strategic Missile Wing (390 SMW), 321

Strategic Missile Wing (455 SMW): activation summary, 418t; construction and acceptance milestone dates, 413t; construction operations, 126t, 131t

Strategic Missile Wing (490 SMW), 24

Strategic Offensive Reductions Treaty (SORT), 28

strategic targeting plans, 22–23, 33–36

Straza Industries, 166

strip-wound motor casings, 67

strontium perchlorate, 172, 230

submarine-launched ballistic missiles (SLBMs): arms limitation negotiations, 24, 26, 28, 29–30; ballistic missile programs, 56–57; delivery capabilities, 36; guidance-and-control systems, 206; liquid propellants, 55; Polaris missile program, 59, 62, 67

submerged nozzles, 166–67

the Suicide Club, 42

Summerfield, Martin, 47, 48

Support Information Network (SIN) telephone, 108

SUPREME CHIEF, 435t

surface overpressure, 93, 94, 94t

SURF SPRAY, 287, 288

survivability criteria, 95, 95t

SWEET TALK, 286, 287–88

SYCAMORE TREE, 435t

Sydney Island, 302, 304f

Sylvania, 332–33

System Q, 59, 65, 66, 67, 458n61; *see also* Minuteman program

systems engineering and technical development (SETD) contractors, 73

System Technology Reentry Experiments Program (STREP), 432t

tactical ballistic missiles (TBMs), 59, 62, 79

Talos supersonic surface-to-air and surface-to-surface missile program, 57, 58f, 170

tape-wound Refrasil, 195, 196, 201

Target and Alignment Team, 240, 243f, 246, 247

Target Development Test (TDT), 432t

targeting accuracy: activation procedures, 244–45; geodetic and gravimetric surveys, 237–39, 240f, 241f, 246; guidance alignment survey monuments, 238, 241f; launch azimuth and alignment determinations, 239–48, 240f, 241f, 242f, 243f; location accuracy, 247–48, 298–99, 302; missile centerline determination, 242–43, 244; missile rotation, 243–44; photographic imagery, 247; remote targeting, 352–53, 390–91; target selection, 341

targeting plans, strategic, 22–23, 33–36

TATTERED COAT, 433t, 435t

Tau Island, 324

Teapot Committee: *see* Strategic Missiles Evaluation Group (Teapot Committee)

Technological Capabilities Panel: *see* Killian Committee/Report

technological challenges: high-altitude testing, 158–59; Minuteman I, 157–59; motor casings, 61–64, 157; nozzles, 157–58; propellant processing, 158; thrust termination, 158; thrust vector control, 158

Technology Demonstration Maneuvering Reentry Vehicle (TDMaRV), 432t

Teller, Edward, 57, 59

Terhune, Charles, 59, 60, 63, 66, 70, 73, 459n77

Terrier missile program, 7, 170

tethered launches, 252, 252f, 254f

Texas, 84, 86, 91

Texas Instruments, 225

Thackwell, Henry L., 46, 47, 48–49

theodolites, 239, 239–44, 240f, 245, 247, 479n11

thermonuclear warheads: full-range operational base launches, 321; historical research, 11, 57, 59; safety concerns, 75

Thiokol Chemical Corporation: contractor responsibilities, 3, 74; deployment schedules, 18; funding obstacles, 78; Minuteman motor designations, 164t; Minuteman production facilities, 87; missile assembly-and-recycle facilities, 86; motor case fabrication, 157; motor fabrication contracts, 84; nozzle development and fabrication, 157, 161–62, 162f, 163f, 255, 260; polysulfide-based propellants, 46, 48–49, 51, 53, 60; rocket propulsion research, 159–60; Sergeant development program, 51; Stage I motor development and testing, 157–62, 162f, 163f, 164f, 164t, 261–62, 395; Stage III motor development and testing, 172, 173, 173t, 174

Thompson-Ramo-Wooldridge (TRW), 288, 292, 395

Thompson, R. J., 47

Thor missile program: Able RTV program, 192; Air Force reentry vehicle designations, 184t; Explicit Guidance scheme, 234; first-generation heatsink reentry vehicles, 183, 190; flight test programs, 191, 205; initial operational capability (IOC), 11; military responsibilities, 55, 61; post-boost control systems (PBCSs), 175; program management, 73; retaliatory effectiveness, 71; rocket propulsion research, 60, 62, 63; RVX-1 reentry vehicle test, 192–94, 193*f*; Thor-Able 0 program, 192; Thor-Able II program, 192

three-axis platforms, 206, 221

three-axis VM4 velocity meter, 206, 208, 213

thrust: *see* rocket propulsion research

thrust termination: core-and-slotted-tube modified end burners, 170, 171*f*; Emergency Rocket Communications System (ERCS), 361; flight test programs, 257, 268, 278, 301, 421*t*, 425*t*, 428*t*; high-altitude testing, 159; operational challenges, 74; solid propellant missile systems, 61, 63–64, 68*t*; Stage III motor development and testing, 170, 172; technological challenges, 158

thrust vector control: gimballed nozzles, 158, 161, 164*f*, 172; jetevators, 64, 158, 159*f*; liquid-injection thrust vector control (LITVC), 166–67, 169*f*, 172, 263, 264, 301; operational flight and evaluation programs, 306; solid propellant missile systems, 68*t*; submerged nozzles, 166; technological challenges, 158

Thunderbird (flight test vehicle), 48, 49*f*

THUNDER VALLEY, 289

TIGHT DRUM, 432*t*

*Time* magazine, 83

Tinker Air Force Base (Oklahoma), 91

Titan I missile program: Air Force reentry vehicle designations, 184*t*; base location considerations and criteria, 88; ceramics-based ablative materials, 191; construction operations, 125; design considerations and criteria, 92; effectiveness, 33, 71; flight test programs, 191, 194, 196; force objectives, 70*t*; full-range operational base launches, 321; funding estimates, 70*t*; geographic distribution, 12*f*; historical perspective, 9, 11; launch control facility (LCF), 112; launch facility criteria, 89; missile assembly-and-recycle facilities, 85; missile suspension systems, 99; operational characteristics, 74; operational facility design and criteria, 94; rocket propulsion research, 61, 63; silo launch facilities, 66; site separation and dispersal recommendations, 93*t*; strategic alerts, 33; Titan I missile, 14*f*

Titan II missile program: Air Force reentry vehicle designations, 184*t*; base location considerations and criteria, 88; blast doors, 147; construction operations, 125; deactivation, 36; design considerations and criteria, 92; effectiveness, 33; external insulation, 177; flight test programs, 194–95; full-range operational base launches, 321; geographic distribution, 12*f*; guidance-and-control systems, 205; historical perspective, 11; launch control facility (LCF), 112; launch environment and sequence, 249; launch facility criteria, 89; missile assembly-and-recycle facilities, 85; missile suspension systems, 99; operational facility design and criteria, 94, 96; penetration aids, 299; site separation and dispersal recommendations, 93*t*; storable liquid propellants, 11; strategic alerts, 33; Titan II missile, 15*f*; umbilical retraction system, 15*f*, 102

Titan missile program: Air Force reentry vehicle designations, 184*t*; base location considerations and criteria, 88; ceramics-based ablative materials, 191; construction operations, 125; design considerations and criteria, 92; flight test programs, 191, 194, 196; force level recommendations, 79; launch control facility (LCF), 112; launch facility criteria, 89; missile assembly-and-recycle facilities, 85; missile suspension systems, 99; operational facility design and criteria, 94; penetration aids, 299; post-boost control systems (PBCSs), 175; program management, 73; site separation and dispersal recommendations, 93*t*

titanium alloys, 166, 307, 307*f*, 308*f*

TM-61 Matador cruise missile, 62

TOP RAIL, 288–89

torpedoes, 59

TOWN DOCTOR, 429*t*

tracking radar systems, 303

trajectories: accuracy prediction systems, 229; aerodynamic heating, 183, 261; Blue Scout Junior Program 279L (MER-6A), 356; characteristics, 185*f*, 297–99; depressed trajectories, 60, 175; ERCS payloads, 358, 361, 362*f*; exposure time, 92; flight control systems, 215; flight test programs, 260, 263, 268, 284, 292, 316, 327*f*, 421–25*t*; gravitational trajectories, 328; guidance-and-control systems, 205, 213–14, 216, 217*t*, 230, 305, 361; Hybrid Explicit software/Hybrid Explicit guidance, 234; launch failures, 293; multiple independently targetable reentry vehicles (MIRVs), 230; operational base launch (OBL) programs, 328; programmed trajectories, 268, 317*f*, 324; radar cross-sections, 201; reentry vehicles, 9, 175, 189, 195, 268; targeting accuracy, 237, 238; Titan I missiles, 61; UHF radio systems, 356

transfer mirror, 244

transient omnidirectional radiating unidistant and static simulator (TORUS), 393

Transit IIA satellite, 175

transpiration and film cooling, 186

Transtage, 175

INDEX    551

Trident missiles: arms limitation negotiations, 24, 29; delivery capabilities, 36; operational test flights, 293–94
TRIM CHIEF, 279
Truman administration, 7–8
Tsien, Hsue-Shen, 42, 44, 45, 51
TULIP TREE, 429*t*
tungsten inserts, 310–11
Tunner, William, 62
two-launch-vote protocol, 273, 274, 277, 369, 377

U-2 reconnaissance aircraft, 371–72
U-boats, 458*n*61
UHF radio systems, 356, 358, 360, 366, 369–70, 371; *see also* Emergency Rocket Communications System (ERCS)
umbilical retraction system: DEFCON declarations, 376; disconnect mechanisms, 102, 260, 297, 333, 342; environmental controls, 215; flight test programs, 216, 260, 264, 277, 297, 317, 320, 329–32, 333, 385; Force Modernization plans, 385; guidance-and-control systems, 99, 102, 211*f*, 215, 320; HEST (high-explosive simulation technique) testing, 385; launch tubes, 99, 330; missile suspension systems, 389; modernization programs, 386, 389; Stage I skirt, 179; Titan II missile program, 15*f*, 102
undersea warfare, 57
unfired ceramic material, 188
United Aircraft Corporation, 166
*The United States Air Force Report on the Ballistic Missile: Its Technology, Logistics and the Strategy*, 83
United States Army Air Forces, 5, 7, 43, 45
United Technologies Corporation, 173, 173*t*, 395
unmanned aircraft, 6, 7
unmanned launch facilities, 3
unreinforced phenolic resin plastics, 194–95
Upgrade Silo modification program: hardness and overpressure improvements, 388; historical perspective, 386–88; launcher closure debris system bins, 387, 388, 389*f*; launcher equipment room (LER), 387, 389–90, 391*f*; missile suspension systems, 102, 387, 388–89, 390*f*; modification completion dates, 398*t*; Stage III motor development and testing, 174
urban targets, 22–23, 34, 88
US Air Force Aeronautical Chart and Information Center (ACIC), 237, 247–48
US Air Force (USAF): ballistic missile programs, 67; base location considerations and criteria, 88–89, 91, 338–39; Geodetic Survey Squadron (1381 GSS), 237–38, 246; Minuteman planned force levels, 18–20, 19*t*, 20*t*, 21*t*, 22*t*, 69–73; Minuteman wing locations, 238; multiple independently targetable reentry vehicles (MIRVs), 175–76; operational flight and evaluation program objectives, 285–88, 292; PCB contamination, 401–2; Peacekeeper missile program, 25–26; Photomapping Group (1370), 238; Polaris missile program, 67; reentry vehicle designations, 184*t*; rocket propulsion research, 55–56, 59–64, 72–76, 78–80, 83; Scientific Advisory Board, 323; second generation ablative reentry vehicle system, 189–93; Special Weapons Laboratory, 379
US Army: ablative reentry vehicle systems, 188–90; operational flight and evaluation program objectives, 288, 294; program support, 78, 79, 80; rocket propulsion research, 55–57
US Army Map Service, 237
US Coast and Geodetic Survey, 238
US Navy: operational flight and evaluation programs, 293–94; Polaris missile program, 67, 81; program support, 78, 80; rocket propulsion research, 55–57, 59–64, 72, 74
USS *George Washington*, 36
USS *Midway*, 56
USS *Norton Sound*, 56
US Strategic Command (USSTRATCOM), 28, 29, 30, 356, 371
Utah, 87; *see also* Hill Air Force Base (Utah); Ogden Air Logistics Center
Utah Launch Complex, Green River, Utah, 293

V-1 long-range guided ballistic missile, 44
V-2 long-range guided ballistic missile, 6–7, 48, 55–56, 221, 237, 445*n*2
Vance, Cyrus, 23
Vandenberg Air Force Base (California): ALCC flight test program, 369, 370; ALCS-equipped aircraft, 366; DEFCON declarations, 377; Demonstration and Shakedown Operation (DASO) program, 430*t*; Emergency Rocket Communications System (ERCS), 363; facility characteristics, 96; full-range operational base launches, 321, 322, 324–25, 326; geographic location, 12*f*; heading sensitivity testing (HST) program, 246, 247; launch azimuth and alignment determinations, 246, 247, 298–99; launcher closure, 108, 109*f*; launch facilities, 272*t*, 273*f*; launch failures, 277–78, 281, 282, 295*f*; Minuteman development and testing programs, 150, 154; Minuteman inventory, 376; Minuteman Reentry System Launch Program (RSLP), 293, 294*f*; modified opera-

tional missiles (MOMS), 328; Multi-Service Launch System (MSLS) program, 300; operational flight and evaluation programs, 271–74, 277–90, 293, 295–312, 313*f*, 376–77, 429*t*; range safety systems, 322–24; reentry vehicle development and testing program, 296; short-range operational base launches, 320; strategic alerts, 33, 377; test and evaluation launch programs, 28, 30, 269; Titan I missile, 14*f*; Titan II missile, 15*f*; weapon system testing, 296; *see also* Cuban Missile Crisis; Pacific Missile Range (PMR); Western Test Range (WTR)
Vela satellites, 175
velocity meters, 206, 208, 210*f*, 213, 221
VELVET TOUCH, 278
Vening–Meinesz formula, 239
Vernier Velocity Unit, 175
VERsatile Digital ANalyzer (VERDAN) computer, 213–14
vertical velocity error, 471*n*62
vibrating string accelerometers (VSAs), 213
Vibrin 135, 190
Vicar high-speed rocket, 47
VIOLET RAY, 285, 288
Virginia, 91
VM4 velocity meter, 206, 208, 213, 221
voice reporting signal assembly (VRSA), 341
von Kármán, Theodore, 6, 42–43, 44, 45, 185
von Neumann, John, 8, 18


W49 nuclear weapon, 103
W-56 warhead, 35
W-59 warhead, 35
W-62 warhead, 312
W-87 warhead, 27
Wagner, Carl, 183
Wahoo, Nebraska, 87
Wake Island, 274*t*
Walker Air Force Base (New Mexico), 12*f*
Walsh, John B., 35–36
WAR AXE, 279
warheads: aerodynamic heating, 181, 186; contractor responsibilities, 73; cost estimates, 86; deployment criteria, 75; design and development programs, 198, 264; emergency combat capability (ECC), 346; ERCS payloads, 358; flight test programs, 189–90, 195–98, 246, 281, 284–87; force level recommendations, 17, 29; hardened mobility, 79; historical research, 11, 57, 59, 183; Minuteman general characteristics, 69*t*; Mixed Marble II test program, 431*t*; multiple independently targetable reentry vehicles (MIRVs), 24, 26–29, 175–76, 228, 229; operational base launch (OBL) programs, 315, 321, 324; payload summary, 37*t*; performance specifications, 9, 66, 67, 75, 228; safety systems, 396; Sergeant missiles, 51; site design considerations and criteria, 92–94; solid propellant missile systems, 45, 57, 59, 63, 65, 67; Soviet missiles, 364, 365, 372; START I/II Treaties, 26, 27–29, 312; strategic missile program recommendations, 10; surface overpressure, 93, 94*t*; targeting strategies, 35, 93, 409; thermonuclear warheads, 11, 57, 59, 75, 321; Titan missile program, 33, 286; War Reserve warheads, 346, 373; yield strength, 57, 59, 63, 80, 93; *see also* reentry vehicles
Warm Silo Agreement, 30
War Plan switch, 343, 374
Warren Air Force Base (Wyoming): *see* F. E. Warren Air Force Base (Wyoming)
*Washington Post*, 399
water, 52
WATER TEST, 363, 435*t*
Weapons System Evaluation Group (WSEG), 18, 19*t*, 337–38
Weapons System Integration and Vibration Laboratory, 249
Weapon System Integration Laboratory, 147, 150
weapon systems: *see* intercontinental ballistic missiles (ICBMs); Minuteman program; solid propellants; specific missile system; WS-133A-M Weapon System; WS-133B Weapon System
weather conditions, 310–11, 311*f*
Weil, Les, 51
welded motor casings, 66, 157
WELL DONE, 279
Welling, Allan, 126–27
Westcott, Buckinghamshire, Great Britain, 249
Western Development Division (WDD), 10, 42, 59–63, 65; *see also* Air Force Ballistic Missile Division (AFBMD)
Western Test Range (WTR), 225, 269, 274*t*, 292; *see also* Pacific Missile Range (PMR); Vandenberg Air Force Base (California)
Westinghouse, 225
Wetherbee, A. E., Jr., 166
WHITE ARC, 429*t*
WHITE BOOK, 429*t*
WHITE GLOVE, 287, 288
Whiteman Air Force Base (Missouri): activation summary, 419*t*; ALCS-equipped aircraft, 366; base location considerations and criteria, 90; construction and acceptance milestone dates, 414*t*; construction operations, 126*t*, 131*t*; deactivation and dismantlement, 399, 404–5; Emergency Rocket Communications System (ERCS), 363; Force Modernization plans,

386, 387t; geographic location, 12f; launch azimuth and alignment determinations, 246; launch codes, 400; launch control equipment building (LCEB), 120f, 121f; Launch Control Panel/Console, 348f; Minuteman geographical statistics, 131t; Minuteman inventory, 24, 296, 357–58, 399; missile assembly-and-recycle facilities, 87; missile suspension systems, 388; operational flight and evaluation programs, 285; operational wing locations and field layouts, 89f, 90; PCB contamination, 401; short-range operational base launches, 319–20; simulated electronic launch Minuteman (SELM), 334; *see also* Wing IV

White Oak, Maryland, 188

White Sands Proving Grounds, New Mexico, 48, 56, 324

White, Thomas D., 10, 65, 69, 70, 83

Wichita Falls, Texas, 86

Wichita, Kansas, 83

Wiesner, Jerome B., 8, 17–18

Wilson, Charlie, 55, 59

Winchester, Idaho, 326

windshield deicing, 43

wind tunnel tests, 184f, 188, 198, 252f

Wing I: activation summary, 417t; ALCS-equipped aircraft, 366; alert status, 155; construction costs, 150t; construction operations, 125, 126t; deployment plans, 17–18; facility characteristics, 97f; flight organization and nomenclature, 128, 129t; flight test programs, 422t; Force Modernization flight tests, 435t; Force Modernization plans, 385, 386, 387t; geographic location and field layouts, 89, 89f; Improved Launch Control System (ILCS), 227, 398t; launch control equipment building (LCEB), 118; launch control facility (LCF), 113f, 114f, 271; launch control support building (LCSB), 112, 115–16, 143; launch design implementation changes, 340; launcher closure, 107; launcher equipment room (LER), 102–3; launcher support building (LSB), 108, 110f; launch tubes, 98–99; LCCs (launch control centers), 112, 114f, 115–16, 271; Minuteman deactivation and dismantlement summary, 407t; Minuteman inventory, 20, 21–22, 296; missile deployment plans, 226; missile suspension systems, 99–101, 100f; modernization and upgrade programs, 295, 296; modification completion dates, 398t; operational facility design and criteria, 94; operational flight and evaluation programs, 279–80; Seattle Test Program (STP) III, 150–51; site activation process, 151–52; site separation and dispersal recommendations, 92, 128; software status and authentication system, 227; survivability criteria, 95, 95t; typical site plan, 114f

Wing II: activation summary, 418t; ALCS-equipped aircraft, 366; construction and acceptance milestone dates, 412t; construction costs, 150t; construction operations, 125, 126t, 130; cutbacks and consolidations, 26; facility characteristics, 97f; flight organization and nomenclature, 128, 129t; flight test programs, 263, 423–24t; Force Modernization flight tests, 435t; Force Modernization plans, 385, 386, 387t; geographic location and field layouts, 89f; launch control equipment building (LCEB), 118; launch control facility (LCF), 113f, 114f, 271; launch control support building (LCSB), 112, 115–16, 143; launcher closure debris system bins, 388; launcher equipment room (LER), 102–3; launcher support building (LSB), 108, 110f; launch tubes, 98–99; LCCs (launch control centers), 112, 114f, 115–16, 271; LFs (launch facilities), 130; Minuteman deactivation and dismantlement summary, 407t; Minuteman inventory, 20, 21–22, 25, 296; missile suspension systems, 99, 100f, 101; modernization and upgrade programs, 296, 298; modification completion dates, 398t; operational facility design and criteria, 94; operational flight and evaluation programs, 285; site separation and dispersal recommendations, 92; survivability criteria, 95, 95t; typical site plan, 114f

Wing III: ALCS-equipped aircraft, 366; construction costs, 150t; facility characteristics, 97f; flight organization and nomenclature, 128, 129t; Force Modernization plans, 385, 386, 387t; geographic location and field layouts, 89f; Integrated Improvement Program, 392; launch control equipment building (LCEB), 118, 120, 122, 142, 145f; launch control facility (LCF), 112, 113f, 114f, 271; launch control support building (LCSB), 112, 143; launcher closure, 107; launcher equipment room (LER), 102–3; launcher support building (LSB), 108–9; launch tubes, 98–99; LCCs (launch control centers), 114f, 116, 271; Minuteman deactivation and dismantlement summary, 407t; Minuteman inventory, 20, 21–22, 25; missile suspension systems, 99, 100f, 101, 102; modernization and upgrade programs, 296, 298; modification completion dates, 398t; operational flight and evaluation programs, 285, 309; site separation and dispersal recommendations, 92; software status and authentication system, 227; survivability criteria, 95, 95t; typical site plan, 114f

Wing IV: activation summary, 419t; ALCS-equipped aircraft, 366; construction costs, 150t; facility characteristics, 97f; flight organization and nomenclature, 128, 129t; flight test programs, 425t; Force Modernization flight tests, 435t; Force Modernization plans, 385, 386, 387t; geographic location and field layouts, 89f; Improved Launch Control System (ILCS), 227, 398t; launch control equipment building (LCEB), 118, 120, 122, 142, 145f; launch control facility (LCF), 112, 113f, 114f, 271; launch control support building (LCSB), 112, 143; launcher closure, 107; launcher equipment room (LER), 102–3; launcher support building (LSB), 108–9; launch tubes, 98–99; LCCs (launch control centers), 114f, 116, 271; Minuteman deactivation and dismantlement summary, 407t; Minuteman inventory, 296; missile deployment plans, 226; missile suspension systems, 99, 100f, 101, 102; modernization and upgrade programs, 296, 298; modification completion dates, 398t; operational flight and evaluation programs, 285; site separation and dispersal recommendations, 92; software status and authentication system, 227; survivability criteria, 95t; typical site plan, 114f

Wing V: ALCS-equipped aircraft, 366; command data buffer (CDB) program, 232–33; computer systems, 214; construction costs, 150t; dust-hardened Minuteman III missiles, 308; excavation methods, 148f; facility characteristics, 97f; flight organization and nomenclature, 128, 129t; Force Modernization plans, 385, 386, 387t; geographic location and field layouts, 89f; Hill Engineering Test Facilities (HETF), 370; Integrated Improvement Program, 392; launch control equipment building (LCEB), 118, 120, 122, 142, 145f; launch control facility (LCF), 112, 113f, 114f, 271; launch control support building (LCSB), 112, 143; launcher closure, 107, 388; launcher equipment room (LER), 102–3; launcher support building (LSB), 108–9; launch tubes, 96, 98–99; LCCs (launch control centers), 114f, 116, 271; Minuteman deactivation and dismantlement summary, 407t; Minuteman inventory, 20, 21–22, 24, 25; missile suspension systems, 99, 100f, 101, 102; modernization and upgrade programs, 295, 296, 298; modification completion dates, 398t; site separation and dispersal recommendations, 92; survivability criteria, 95t; typical site plan, 114f

Wing VI: activation summary, 420t; ALCS-equipped aircraft, 366; blast doors, 141f; construction costs, 150t; construction material requirements, 144t; Demonstration and Shakedown Operation (DASO) program, 297, 434t; facility characteristics, 97f; flight organization and nomenclature, 128, 129t; flight test programs, 264–66, 297, 424t, 426t; Force Modernization plans, 385, 386, 387t; geographic location and field layouts, 89f; Integrated Improvement Program, 392; launch control equipment building (LCEB), 122–23; launch control facility (LCF), 112, 113f, 115f, 271, 295; launch control support building (LCSB), 144; launcher closure, 107; launcher equipment room (LER), 103–4; launcher support building (LSB), 109–11, 111f; launch tubes, 96, 98–99; LCCs (launch control centers), 115f, 117–18, 271; Minuteman deactivation and dismantlement summary, 407t; Minuteman II research and development programs, 433t; Minuteman inventory, 24, 25, 27, 296; missile deployment plans, 226; missile suspension systems, 99, 100f, 101, 102; modernization and upgrade programs, 298, 347; modification completion dates, 398t; site separation and dispersal recommendations, 92; survivability criteria, 95t; typical site plan, 115f; weapon system testing, 296

wing facilities: facility characteristics, 96, 97f; flight organization and nomenclature, 128, 129t; Force Modernization plans, 295–98, 370, 385–86, 387t; geographical statistics, 131t; guidance alignment survey monuments, 238; launch control equipment building (LCEB), 112, 113f, 114f, 115f, 118, 120, 120f, 122–23; Minuteman deactivation and dismantlement summary, 407t; Minuteman program correlation chart, 16t; missile deactivation and dismantlement, 401; missile deployment plans, 17–18, 21–22; missile suspension systems, 94–96, 99–101, 100f, 376, 386, 388–89; modernization and upgrade programs, 295, 296, 298, 347; *see also* specific wing

WINTER BREW, 286, 287

WIRE NET, 321, 322

Woods Hole Oceanographic Institute, Woods Hole, Massachusetts, 57

Worcester Polytechnic Institute, 41

World Geodetic System 1960 (WGS 60), 237, 238

Wright Air Development Center, Dayton, Ohio, 61, 456n22

Wright Field, Dayton, Ohio, 44

Wright-Patterson Air Force Base (Ohio), 190, 206

WS-133A-M Weapon System: Communication Console, 349f; flight test programs, 297–98, 320; Force Modernization Program, 227, 296,

297–98, 385–86; funding estimates, 79–80; historical perspective, 71; internal communication systems, 28; launch control centers, 273f; Launch Control Panel/Console, 348f; Minuteman deactivation and dismantlement summary, 407t; modification completion dates, 398t; operational base launch (OBL) programs, 323, 324, 326; operational facility construction contracts, 150t; operational ground programs, 235; program correlation chart, 16t; Rapid Execution and Combat Targeting (REACT) Program, 353; typical site plan, 115f

WS-133B Weapon System: electromagnetic pulse (EMP) susceptibility, 392; flight test programs, 263, 296–97, 320, 424t; guidance-and-control systems, 216; internal communication systems, 28; launch control centers, 119f, 273f; launch control equipment building (LCEB), 123f; Minuteman deactivation and dismantlement summary, 407t; modification programs, 227, 263, 385; operational base launch (OBL) programs, 323, 324, 326; operational facility construction contracts, 150t; operational flight and evaluation programs, 295–97; operational ground programs, 235; program correlation chart, 16t; Rapid Execution and Combat Targeting (REACT) Program, 353; Status Console, 351f; typical site plan, 115f; *see also* Minuteman II

WSCE system, 353, 354, 356

Wyoming, 86; *see also* F. E. Warren Air Force Base (Wyoming)

X-17 sounding rocket, 192
XLR10-RM-2 rocket engine, 55
XW-56X1 warhead, 198
XW-59 warhead, 198

Yeltsin, Boris, 26, 312
yield strength, 469n17
Yokota Air Base, Japan, 363
York, Herbert, 67, 72, 339
Young Development Laboratories, 170
Young, Richard, 62, 170

Zuckert, Eugene, 88, 90, 91, 267, 339

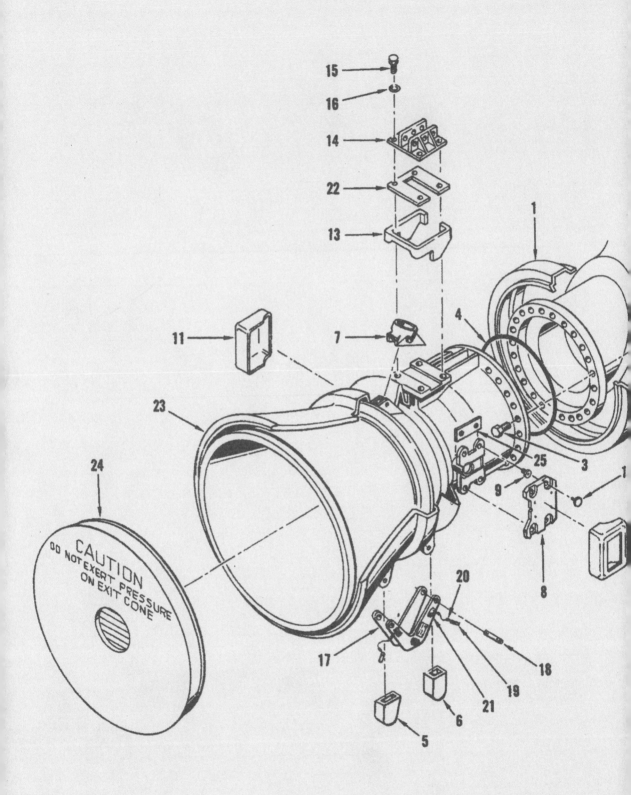